EVOLUTION OF VIRULENCE IN EUKARYOTIC MICROBES

EVOLUTION OF VIRULENCE IN EUKARYOTIC MICROBES

Edited by

L. DAVID SIBLEY
Washington University School of Medicine
St. Louis, MO

BARBARA J. HOWLETT
The University of Melbourne
Australia

JOSEPH HEITMAN
Duke University Medical Center
Durham, NC

WILEY-BLACKWELL
A JOHN WILEY & SONS, INC., PUBLICATION

Front Cover: Top left: Sexual fruiting body of *Pyrenophora tritici-repentis* the cause of tan spot of wheat (image credit Kasia Rybak). Top right: Immunofluorescence image of intracellular tachyzoites of *Toxoplasma gondii*, a common opportunistic pathogen of humans (image credit Jennifer Gordon). Bottom left: Zygospore of *Mucor circinelloides*, a dimorphic fungal plant pathogen (image credit Soo Chan Lee): used with permission PLoS Pathog 7(6): e1002086. Bottom right: Intracellular trophozoite of *Plasmodium falciparum*, the cause of malaria, within red blood cell (image credit Wandy Beatty).

Back Cover: Top: Electron micrograph of blood stream trypanosomes of *Trypanosoma brucei* (image credit Wandy Beatty). Bottom: Yeast and hyphal forms of *Candida tropicalis* (image credit Ying-Lien Chen)

Published by John Wiley & Sons, Inc., Hoboken, New Jersey
Published simultaneously in Canada

Library of Congress Cataloging-in-Publication Data:

Sibley, L. David.
Evolution of virulence in eukaryotic microbes / editors, L. David Sibley, Barbara J. Howlett, Joseph Heitman.
p. cm.
Includes index.
ISBN 978-1-118-03818-5 (hardback)
1. Protista. 2. Eukaryotic cells. 3. Microbial genomics. 4. Virulence (Microbiology)–Genetic aspects. I. Howlett, Barbara J. II. Heitman, Joseph. III. Title.
QR74.5.S53 2012
579–dc23

2011051095

Printed in Singapore.

10 9 8 7 6 5 4 3 2 1

CONTENTS

PREFACE

This volume assembles for the first time a collection of chapters on the diversity of eukaryotic microorganisms and the influence of evolutionary forces on the origins and emergence of their virulence attributes. In selecting the topics for this volume, we have highlighted examples from three important, divergent groups of eukaryotic microorganisms that cause disease in animals and plants. These include oomycetes, the cause of serious blights in plants, such as the Irish potato famine, protozoan parasites of humans, including *Plasmodium*, the causative agent of malaria, and fungi that cause diseases in animals and plants. Although phylogenetically diverse in the eukaryotic tree of life, each of these groups has adopted pathogenic lifestyles on plant or animal hosts that often result in serious diseases. Traditionally, studies on these three groups of organisms have been pursued by independent scientific communities. Hence, our goal for this volume is to serve as a bridge that links findings from these disparate communities, resulting in a synthesis of knowledge that voices common themes in evolution and pathogenesis.

The overarching theme of this volume is how pathogenic determinants of eukaryotic microorganisms have evolved in concert with their hosts, often overcoming innate and adaptive immune mechanisms. Host defenses are not prominently featured here as there are many volumes covering these aspects of the host–microbe interaction. Rather, our focus is on pathogenic determinants and evolutionary processes that have shaped microbial pathogen genomes and the resulting complex biology they orchestrate. The initial chapters cover general, broad principles of evolution, phylogenetics, and genetic exchange. In many cases, our appreciation of the virulence traits of pathogens has been driven by genetic approaches, either classical or molecular. Hence, one of the subthemes for the book is how the development of genetic tools has fostered the identification and functional analyses of virulence determinants. An equally important subtheme is how pathogens exchange genetic material in nature, including via classical or modified meiotic processes, horizontal gene transfer, and sexual cycles including those that are cryptic or even unisexual. Since genetic exchange provides the means to shuffle genomes and to acquire new determinants, these processes play a central role in the evolution of virulence. The combined treatment of different inheritance systems highlights common mechanisms of evolutionary adaptations that led to the emergence of the capacity to cause disease in diverse hosts.

In comparison to viral and bacterial pathogens, eukaryotic microorganisms are characterized by large genome sizes, complex biological life cycles, and shared biological features with their hosts. These properties combine to create challenges in studying eukaryotic pathogens, both in the laboratory and in animal and plant models. However, these challenges are counterbalanced by recent advances in genetics, comparative genomics, phylogenetics, and evolutionary analyses that have accelerated progress in defining the molecular basis of virulence in eukaryotic pathogens. In the contributions contained herein, these advances illuminate how coevolution between hosts and pathogens has shaped their interactions and the resulting outcome in terms of pathogenesis. Common themes between different pathogenic microbes are illustrated, providing a broad framework for formulating future studies.

We are grateful to all of the authors for their excellent series of contributed chapters, and we hope that the resulting volume will enrich discussions and even drive novel research on eukaryotic pathogens. As with all such publication ventures, this synthesis across disciplines is an experiment. We welcome you as readers to communicate to us ways in which we may have succeeded and also ways in which the focus of our efforts might be sharpened in possible future editions.

L. David Sibley
Barbara J. Howlett
Joseph Heitman

ACKNOWLEDGMENTS

We thank Eileen Wojciechowski for outstanding administrative assistance and Cecelia Shertz for editing prowess. We are greatly indebted in particular to Janet Whealen, without whom this volume would not have been possible. More than anyone, Jan was responsible for keeping editors and authors on track and for reminding us of both important formatting aspects and timelines. We thank Dr. Karen Chambers, Editor, Life Science, Wiley-Blackwell, for the initial concept this volume and for helping us navigate the complexity of a large publishing house. We are also grateful to Anna Ehlers and other members of the Wiley-Blackwell editorial team for assistance with compiling the final volume. Finally, we acknowledge each of our families for their forbearance during this project.

CONTRIBUTORS

JAMES W. AJIOKA, PHD, Department of Pathology, University of Cambridge, Cambridge, UK

FRANCISCO J. AYALA, PHD, Department of Ecology and Evolutionary Biology, University of California, Irvine, CA

GUUS BAKKEREN, PHD, Pacific Agri-Food Research Centre, Agriculture and Agri-Food Canada, Summerland, BC, Canada

J. DAVID BARRY, PHD, Wellcome Trust Centre for Molecular Parasitology, Institute of Infection, Immunity and Inflammation, College of Medical, Veterinary and Life Sciences, University of Glasgow, Glasgow, UK

RICHARD J. BENNETT, PHD, Department of Molecular Microbiology and Immunology, Brown University, Providence, RI

JIM L. BEYNON, PHD, School of Life Sciences, The University of Warwick, Warwick, UK

PAUL R. J. BIRCH, PHD, Division of Plant Sciences, University of Dundee at James Hutton Institute, Invergowrie Dundee, UK

JOHN C. BOOTHROYD, PHD, Department of Microbiology and Immunology, Stanford University School of Medicine, Stanford, CA

GERALDINE BUTLER, PHD, UCD School of Biomolecular & Biomedical Science, Conway Institute, University College Dublin, Belfield, Ireland

RICHARD CARTER, PHD, School of Biological Sciences, University of Edinburgh, Edinburgh, UK

ARTURO CASADEVALL, MD, PHD, Department of Microbiology and Immunology, Albert Einstein College of Medicine, Bronx, NY

HSIAO-HAN CHANG, Department of Organismic and Evolutionary Biology, Harvard University, Cambridge, MA

MARY E. COATES, PHD, School of Life Sciences, University of Warwick, Warwick, UK

LEAH E. COWEN, PHD, Department of Molecular Genetics, University of Toronto, Toronto, ON, Canada

RICHARD CULLETON, PHD, Malaria Unit, Institute of Tropical Medicine, Nagasaki University, Nagasaki, Japan

RACHEL F. DANIELS, Department of Immunology and Infectious Diseases, Harvard School of Public Health, Boston, MA

KIRK W. DEITSCH, PHD, Department of Microbiology and Immunology, Weill Medical College of Cornell University, New York, NY

MANOJ T. DURAISINGH, PHD, Harvard School of Public Health, Boston, MA

JEFFREY D. DVORIN, MD, PHD, Children's Hospital Boston, Boston, MA

Ananias A. Escalante, PhD, Center for Evolutionary Medicine and Informatics, The Biodesign Institute, Arizona State University, Tempe, AZ

Rhys A. Farrer, BSc, MSc, Department of Infectious Disease Epidemiology, Imperial College, London, UK

Michael T. Ferdig, PhD, Eck Institute for Global Health, Department of Biological Sciences, University of Notre Dame, Notre Dame, IN

Matthew C. Fisher, PhD, Department of Infectious Disease Epidemiology, Imperial College, London, UK

Monica A. Garcia-Solache, MD, PhD, Department of Microbiology and Immunology, Albert Einstein College of Medicine, Bronx, NY

Neil A. R. Gow, PhD, School of Medical Sciences, Institute of Medical Sciences, University of Aberdeen, Aberdeen, UK

Daniel L. Hartl, PhD, Department of Organismic and Evolutionary Biology, Harvard University, Cambridge, MA

Joseph Heitman, MD, PhD, Departments of Molecular Genetics and Microbiology, Pharmacology & Cancer Biology & Medicine, Duke University Medical Center, Durham, NC

Jessica A. Hill, MSc, Department of Molecular Genetics, University of Toronto, Toronto, ON, Canada

Samantha J. Hoot, PhD, Seattle Biomedical Research Institute, Seattle, WA

Barbara J. Howlett, PhD, School of Botany, The University of Melbourne, Melbourne, VIC, Australia

Christina M. Hull, PhD, Department of Biomolecular Chemistry and Department of Medical Microbiology and Immunology, University of Wisconsin, School of Medicine and Public Health, Madison, WI

Christopher D. Huston, MD, Division of Infectious Diseases, University of Vermont College of Medicine, Burlington, VT

Alexander Idnurm, PhD, Division of Cell Biology and Biophysics, School of Biological Sciences, University of Missouri-Kansas City, Kansas City, MO

Andrew P. Jackson, DPhil, Pathogen Genomics, Wellcome Trust Sanger Institute, Hinxton, UK

Satomi Kato, PhD, Center for Tropical and Emerging Global Diseases, University of Georgia, Athens, GA

Jessica C. Kissinger, PhD, Department of Genetics, Institute of Bioinformatics, and Center for Tropical and Emerging Global Diseases, University of Georgia, Athens, GA

H. Corby Kistler, PhD, USDA ARS Cereal Disease Laboratory, University of Minnesota, St. Paul, MN

Martin Kolisko, PhD, Centre for Comparative Genomics and Evolutionary Bioinformatics, Department of Biology, Dalhousie University, Halifax, NS, Canada

Emilia K. Kruzel, PhD, Department of Biomolecular Chemistry, University of Wisconsin, School of Medicine and Public Health, Madison, WI

Jean-Paul Latge, PhD, Institut Pasteur, Unité des *Aspergillus*, Paris, France

Wenjun Li, PhD, Department of Molecular Genetics and Microbiology, Duke University Medical Center, Durham, NC

John M. Logsdon, Jr., PhD, Department of Biology, University of Iowa, Iowa City, IA

Michael Lorenz, PhD, Department of Microbiology and Molecular Genetics, University of Texas-Houston, Houston, Texas

LI-JUN MA, PHD, College of Nature Sciences, University of Massachusetts, Amherst, MA

ANNETTE MACLEOD, PHD, University of Glasgow, College of Medical, Veterinary and Life Sciences, Glasgow, UK

LIAM J. MORRISON, PHD, University of Glasgow, College of Medical, Veterinary and Life Sciences, Glasgow, UK

LAETITIA MUSZKIETA, PHD, Institut Pasteur, Unité des *Aspergillus*, Paris, France

DANIEL E. NEAFSEY, PHD, Broad Institute, Cambridge, MA

KIRSTEN NIELSEN, PHD, Department of Microbiology, University of Minnesota, Minneapolis, MN

RICHARD P. OLIVER, PHD, Australian Centre for Necrotrophic Fungal Pathogens, Curtin University, Perth, WA, Australia

PETER R. PREISER, PHD, Division of Molecular Genetics and Cell Biology, School of Biological Sciences, Nanyang Technological University, Singapore

MARTIJN REP, PHD, Molecular Plant Pathology, Swammerdam Institute for Life Sciences, University of Amsterdam, Amsterdam, The Netherlands

JEAN BEAGLE RISTAINO, PHD, Department of Plant Pathology, North Carolina State University, Raleigh, NC

ANDREW J. ROGER, PHD, Centre for Comparative Genomics and Evolutionary Bioinformatics, Department of Biochemistry and Molecular Biology, Dalhousie University, Halifax, NS, Canada

BENJAMIN M. ROSENTHAL, SD, Animal Parasitic Disease Laboratory, Agriculture Research Service, USDA, Beltsville, MD

PARDIS C. SABETI, DPHIL, MD, FAS Center for Systems Biology, Harvard University, Cambridge, MA

ELIZABETH SAVELKOUL, Department of Biology, University of Iowa, Iowa City, IA

STEPHEN F. SCHAFFNER, PHD, Broad Institute, Cambridge, MA

L. DAVID SIBLEY, PHD, Department of Molecular Microbiology, Washington University School of Medicine, St. Louis, MO

ALASTAIR G. B. SIMPSON, PHD, Centre for Comparative Genomics and Evolutionary Bioinformatics, Department of Biology, Dalhousie University, Halifax, NS, Canada

UPINDER SINGH, MD, Division of Infectious Diseases and Geographic Medicine, Departments of Internal Medicine and Microbiology and Immunology, Stanford University School of Medicine, Stanford, CA

JOSEPH SMITH, PHD, Seattle Biomedical Research Institute, Seattle, WA

JASON E. STAJICH, PHD, Department of Plant Pathology and Microbiology, University of California, Riverside, CA

WILLIAM J. STEINBACH, MD, Department of Pediatrics and Department of Molecular Genetics and Microbiology, Duke University, Durham, NC

XIN-ZHUAN SU, PHD, Malaria Functional Genomics Section, Laboratory of Malaria and Vector Research, National Institute of Allergy and Infectious Disease, National Institutes of Health, Rockville, MD

ANDY TAIT, PHD, University of Glasgow, College of Medical, Veterinary and Life Sciences, Glasgow, UK

JOHN C. TAN, PHD, Genomics Core Facility, Eck Institute for Global Health, University of Notre Dame, Notre Dame, IN

SARAH K. VOLKMAN, SCD, Department of Immunology and Infectious Disease, Harvard School of Public Health, Boston, MA

THEODORE C. WHITE, PHD, School of Biological Sciences, University of Missouri-Kansas City, Kansas City, MO

DYANN F. WIRTH, PHD, Department of Immunology and Infectious Disease, Harvard School of Public Health, Boston, MA

JOHN C. WOOTTON, PHD, NCBI, NLM, NIH, Computational Biology Branch, National Center for Biotechnology Information, National Library of Medicine, National Institutes of Health, Bethesda, MD

JIANPING XU, PHD, Department of Biology, McMaster University, Hamilton, ON, Canada

PART I

GENERAL OVERVIEWS

CHAPTER 1

POPULATION GENETICS AND PARASITE DIVERSITY

HSIAO-HAN CHANG, RACHEL F. DANIELS, and DANIEL L. HARTL

MUTATION

Mutation encompasses a wide range of changes from substitution of single nucleotides to relocation of entire segments of chromosomes. Single-nucleotide polymorphisms (SNPs) include transitions and transversions that exchange bases within or across purine and pyrimidine classes. Some nucleotide substitutions can affect the codon for an amino acid without changing the amino acid (a synonymous SNP). When the mutant codon does code for a different amino acid, the SNP is nonsynonymous. Some mutations insert or delete one or a small number of nucleotide bases and shift the reading frame of the translational machinery, causing a usually nonfunctional product to be translated from the RNA transcript.

On a larger genetic scale than single-nucleotide changes, copy number variations (CNVs) are mutations caused by mobile genetic elements like insertion sequences, transposons, or retroelements. Genes can also be duplicated through chromosomal mutations such as transpositions and translocations; like the other types of CNVs, these mutations place genes out of their normal genetic context and away from their normal regulatory elements. The new copies may also be less prone to correction by processes such as gene conversion, leaving them free to accumulate further mutations. A special class of copy number polymorphisms consists of minisatellites and microsatellites with variable numbers of tandem repeats of relatively short stretches of DNA.

Mutation can have significant consequences on populations. For example, the effects of nonsynonymous SNPs and CNVs have been associated with reduced drug sensitivity in *Plasmodium falciparum*, the apicomplexan parasite that causes most malaria deaths. The implications of these and other types of mutations in populations of *P. falciparum* and other eukaryotic microbes will be discussed in greater detail in later chapters.

At the beginning of population genetics, the only way to observe mutations was through direct polymorphic evidence: Mendel's vaunted peas, Kettlewell's melanic moths, and Galton's amazing catalog of continuous traits all owed their phenotypic variation to different versions of genes where mutations had caused the organisms

Evolution of Virulence in Eukaryotic Microbes, First Edition. Edited by L. David Sibley, Barbara J. Howlett, and Joseph Heitman.

to appear different from each other. Increasingly sophisticated techniques such as allozyme (also known as allelic isozyme) and restriction fragment length polymorphism (RFLP) analyses allowed more direct studies of protein and DNA differences between individuals within populations and among species. These early experiments formed the basis for many current theories in population genetics and remain informative to the present day. The recent emergence of cost-effective technologies such as DNA and protein arrays has changed the scale and accelerated the discovery of diversity from single genes or proteins to genomes and proteomes in massively multiplexed studies involving larger numbers of individuals within and among populations. Advances in sequencing technologies have further enhanced the deluge of new data available to researchers. With lowered costs and increased coverage, entire genomes can be sequenced for a complete picture of polymorphic sites across populations and time. Further technological advances like hybrid selection and single-molecule amplification mean that pathogens and other microorganisms—including eukaryotic microbes—that cannot be separated from host material or cultured in sufficient quantities to allow easy sequencing under previous technologies can now be sequenced directly from patients or the environment. These technologies do not rely on *in vitro* culturing systems and offer researchers glimpses of the population genetics of pathogens whose mechanisms of virulence may be intractable for study in model organisms.

GENETIC DIVERSITY AND RANDOM GENETIC DRIFT

Knowing the extent of genetic diversity allows us to make comparisons among populations in space or time. Studies of genetic differences among populations and the mechanisms that bring them about can reveal evolutionary history and can predict future evolutionary events. Genetic diversity is measured in several standard ways, so comparisons can be made within and among populations. Heterozygosity, originally developed to quantify allozyme differences, calculates the proportion of loci that are heterozygous. This flexible and straightforward metric allows testing between variable locations in a single diploid organism, among individuals in a population, and between species. With access to genetic sequences, other measures of genetic diversity are more commonly used. One measure, nucleotide diversity (typically symbolized as π), is the average proportion of nucleotides that differ between any randomly sampled pair of sequences. A second measure of DNA sequence variation is the number of segregating sites (usually symbolized as S) (Kimura and Crow, 1964).

While cataloging the magnitude of genetic differences offers ways to compare populations, it does little to inform evolutionary history or relationships among populations. In order to recognize and understand the forces that change the frequency of alleles in a population, it is convenient to compare an actual population with an idealized model population in which only well-defined evolutionary forces are at work. In the simplest models, the ideal population is infinitely large, mating is random, and generations do not overlap. In the absence of mutation, migration, and selection, the allele frequencies remain constant generation after generation, and with only two alleles at a locus, the genotype frequencies in a diploid organism are given by the familiar Hardy–Weinberg principle as p^2, $2pq$, and q^2, where p and q are allele frequencies and $p + q = 1$.

The next level of complexity—still in an idealized model—is to allow mutation to take place and to assume that the population is finite in size. The finite size allows random fluctuations in allele frequency to take place from generation to generation. The mutations are assumed to have negligible effects on survival and reproduction, which is called selective neutrality. Under these assumptions, the idealized model is known as the neutral model and often serves as a null hypothesis for comparison with observed data. Tajima's D is one test statistic that we can use to test differences between the observed and idealized situations. Tajima's D (Tajima, 1989) can be easily calculated from sequences of selected loci of many individuals in a population through measures of sequence variation π and S/a, where

$$a = \sum_{i=1}^{n-1} \frac{1}{i}$$

and n is sample size. Rejection of the neutral model, where the observed value is significantly different from the expected value of Tajima's D (which is 0), can identify genes in which forces such as natural selection may be important; however, departures from neutrality can also result from the departures from any other assumptions made in the idealized model such as population growth.

In a finite population, random genetic drift changes allele frequencies over time through stochastic sampling variation that can change the relative representation of an allele in the next generation. Gametes from each parent that successfully combine to form offspring are a subset of the total possible from each parent; which particular set of alleles combines with those from another gamete is entirely random. Also random is the chance that an individual carrying a particular allele will survive and successfully reproduce. A formal model for the effects of random genetic drift was proposed independently by Wright and Fisher (Fisher, 1930; Wright, 1931) and hence is known as the Wright–Fisher model. The key simplifying assumptions are a finite and constant population size (N), nonoverlapping generations, and equal likelihood of reproduction of each individual. In its simplest form, the model also assumes no new mutation and deals only with those that already exist in the population. With these assumptions, the probability that the copy number of an allele will be k in a diploid population in the next generation is

$$\frac{(2N)!}{k!(2N-k)!} p^k q^{2N-k},$$

where p is the frequency of the allele in the current population and $q = 1 - p$. The formula here could be further used to predict the influence of genetic drift and population dynamics, such as the expected frequency of mutations in a population or the expected time for an allele to become fixed or lost by genetic drift.

Once a mutation enters the population, its possible fates are limited to either survival or extinction. With no selection, its fate is determined by purely random fluctuations in allele frequency (random genetic drift). A new mutation in a diploid population of size N has an initial allele frequency of $1/(2N)$. The odds are greatly against the allele establishing itself in the population. For a large population, the probability that the allele will be lost in the next generation is 0.368. The odds against surviving to the second or fifth generation after introduction are 0.532 and 0.732, respectively. On average, an allele that is destined to be lost will do so in approximately $2\ln(2N)$ generations. Although the odds are strongly against any new neutral mutation remaining in the population for very long, there are so many new mutations that some of them do become established, and a few even drift to fixation. In the neutral model, the probability that an allele eventually becomes fixed is equal to its frequency in the population. Fixation of a new neutral allele takes a long time, however. A mutation at initial frequency $1/(2N)$ that is destined to be fixed requires an average of $4N$ generations for this to occur (Kimura and Ohta, 1969). Through the accumulation of mutant alleles and their random changes in a frequency due to genetic drift, an equilibrium is eventually established with predictable values of π and S and other population parameters, as well as a predictable probability distribution of allele frequencies (called the allele frequency spectrum). It is against these predictions that actual data are compared to reveal genes that deviate from the assumptions of the model. For example, deviations caused by selection for biological attributes or drug resistance would be of interest.

THE NEUTRAL THEORY

Kimura formalized the Wright–Fisher model in the neutral theory, which proposes that the primary cause of evolution in populations is through the mechanism of genetic drift on neutral alleles (Kimura, 1968; King and Jukes, 1969). Kimura developed this theory based on thinking at the time that the human genome might contain at least a million genes. With the degree of polymorphism of protein-coding genes then known, most segregating mutations must be selectively neutral and must confer no apparent advantage to individuals with different alleles; otherwise, the genetic load from segregating mutations would be so large that the population could not survive. Kimura also posited that even if non-neutral mutations emerged, the vast majority would be deleterious and quickly lost from the population through negative selection. One strong example used to defend this theory was the degeneracy of the codons mapped to amino acids. Kimura hypothesized that most gene products that are found in the possible genetic landscape are at some form of an optimal state and that most changes are simply variations on a theme that has been successful for many generations. In this theory, the rate of loss of neutral alleles is balanced by the mutation rate (μ) of these same neutral alleles and variation is maintained in the population. We now know that in eukaryotes with large genomes, the vast majority of the DNA does not code for proteins, and hence, the neutral theory has been extended to apply to most noncoding sequences. Because of its simplicity and ability to make specific predictions about such things as the allele frequency spectrum, the neutral theory has served as a critical null hypothesis in molecular population genetics and genomics.

MUTATION AND SELECTION

The obvious alternative to the neutral theory is natural selection. First proposed by Darwin and subsequently modified to reflect updated scientific knowledge, this theory acts on individuals to change the characteristics of an entire population. Under natural selection, in contrast to the neutral theory, selection on allelic variation nonrandomly changes allele frequency owing to differences in the relative fitness of the individuals. Fitness is measured by survivorship and fecundity—the ability to survive to reproduce, the number of successful matings, and the number of offspring produced per mating. Positive selection increases the likelihood of an allele passing to the next generation through increased fitness, while negative selection is deleterious to these chances. There are three general types of selection that work to change allele frequencies in distinctive ways. First, directional selection favors one extreme variant over the other genotypes and pushes the allele frequency toward one homozygote. If the new mutation is favored, the directional selection is positive; if the original type is favored over the mutation, negative or purifying selection occurs. Next, balancing selection chooses an intermediate solution over extremes. A well-known example of balancing selection is the fitness advantage of having one copy of the sickle-cell mutation to help protect against severe illness caused by *P. falciparum* malaria. Homozygous nonmutant genotypes are more likely to suffer severe infections, but homozygous mutant genotypes have severe anemia. The heterozygous genotype, therefore, has an advantage over both homozygous genotypes. The third general type of selection is diversifying selection, which emphasizes maximum allelic diversity. Many genes in the immune system and pathogen virulence factors that interact with the immune system are thought to be under diversifying selection.

Even when a beneficial mutation appears in a population, random genetic drift plays an important role in determining its fate, especially in the early generations. For a new mutation with a selective advantage of 1%, for example, the likelihood that the mutation will be lost in the first, second, or fifth generation is only slightly smaller than under a neutral model: The theoretical probabilities of loss when N is large are 0.364, 0.526, and 0.725, respectively. When the selective advantage s is small, the likelihood of ultimate fixation in the population is only twice the selective advantage.

Selection can indirectly affect neighboring genes. When strong positive selection changes the frequency of an allele, the advantageous form sweeps through the population. If this selective sweep happens in a short period of time, neighboring alleles are carried along as genetic hitchhikers before genetic recombination can break up the allelic associations. One measure to test for the presence of selective sweeps is linkage disequilibrium, which is the nonrandom association of alleles in a chromosome. Under the neutral model, genes are in equilibrium when no particular combination of alleles (a haplotype) is more prevalent than would be expected by chance. Under selective pressure, including selective sweeps, linkage disequilibrium occurs when the frequency of certain haplotypes rises above random expectation.

Proponents of Darwin's theory of natural selection at first saw the neutral theory as being in direct opposition to the tenets of selection, and for about a decade, disagreements between the two camps appeared as polemics supporting either selection or neutrality as a universal mechanism of evolution. These grand theoretical battles are mostly in the past, and modern researchers are more concerned with assessing the relative importance of selection and neutrality in various actual situations. In this instance, progress in sequencing technology and other aspects of genomics were important to help the field of evolutionary genetics escape the impasse. We should note here an important modification of the neutral theory known as the nearly neutral theory, which recognizes that many mutations may indeed have selective effects of their own but stresses that these effects are often small in relation to the population size (Ohta, 1992).

EFFECTIVE POPULATION SIZE

When a population geneticist refers to population size, it is usually the effective population size that is intended. The effective size of a population, sometimes informally called the "breeding size," corresponds to the size of the ideal population discussed earlier that would have the same magnitude of random genetic drift as an actual population (Wright, 1931). The effective size of a population is usually smaller than the actual size and is sometimes much smaller. Symbolized as N_e, the effective population size is an important characteristic of a population that determines the effects of genetic drift interacting with selection and other processes. More formally, the effective size of a population is defined as the size of an ideal Wright–Fisher population that has the same properties with respect to genetic drift as the real population does. The effective size is often more relevant to evolutionary processes such as adaptation, including the evolution of drug resistance, than the actual population size.

The detailed consequences of random genetic drift can be measured in several distinct ways, and depending on the way random genetic drift is measured, three kinds of effective population sizes have been defined: variance effective size, eigenvalue effective size, and inbreeding effective size. Variance effective size is the size of an ideal population that would produce the same variance of allele frequency from one generation to the next as the actual population; eigenvalue effective size considers the rate of loss of heterozygosity across generations; and inbreeding effective size focuses on the probability that alleles are identical by descent.

One complication is that the three types of effective size can differ in certain demographic models. Sjödin et al. (2005) introduced the coalescent effective size and demonstrated that, if it exists, the coalescent effective size is the most general because it allows all aspects of random genetic drift to be described by Kingman's coalescent process (Sjödin et al., 2005). The coalescent effective size is defined as the size of an ideal population that has the same properties of the coalescence process as an actual population. Coalescence is the convergence of the ancestral lineages of two alleles to a common ancestral allele. In an ideal population that follows Kingman's coalescent process, the time for two lineages to coalesce follows an exponential distribution with rate $1/2N$, and hence in an actual population, the coalescent effective size can be estimated as half the reciprocal of the rate of coalescence. Coalescent effective size does not always exist: It exists only when the factors that affect N_e take place on different timescales from coalescence events.

Factors Affecting Effective Population Size

Even though an actual population may be very large, its effective population size could be small. Various factors affect effective population size. First, variation in offspring number reduces N_e, and the effective population size equals approximately the actual population size divided by variance of offspring number. Second, inbreeding decreases effective size, and the inbreeding effective size is equal to the actual population size divided by $(1 + F_I)$, where F_I is the inbreeding coefficient. Third, N_e also depends on the ratio of females to males. If a population consists of N_m males and N_f females, the effective population size is

$$\frac{4N_m N_f}{N_m + N_f}.$$

That is, the effective population size decreases as the sex ratio becomes more skewed. Moreover, effective population sizes for sex chromosomes and autosomes are different. Effective sizes for X chromosome and Y chromosome are

$$\frac{9N_m N_f}{4N_m + 2N_f} \quad \text{and} \quad N_f/2,$$

respectively. In addition, if the population size changes through time, the effective population size is approximately the harmonic mean of the actual sizes. Finally, in the case of subdivided populations, the effective size of the ensemble of populations could be (but not always, depending on migration situation) greater than the actual size because the probability of coalescence between individuals from different subpopulations is smaller than that in an ideal population. (For more details, see Hartl and Clark, 2007.)

The Strength of Selection Relative to Random Genetic Drift

Effective population size is a key parameter that determines the rate of evolution.

The relative importance of selection and random genetic drift depends on the absolute value of the product of effective population size N_e and selection coefficient s. If $|N_e s| \gg 1$, selection dominates over genetic drift and beneficial mutations are more likely to become fixed in the population, while deleterious mutations are efficiently eliminated. If $|N_e s| \ll 1$, selection is swamped and the frequency of non-neutral mutations fluctuates like neutral mutations. If $|N_e s| \approx 1$, selection and genetic drift both play important roles in determining the evolutionary fate of a new mutant allele. These principles not only imply that the efficacy of selection increases with N_e but they also imply that more genes are affected by selection in populations with a larger effective population size. Mutations with the same selection coefficient therefore can have different fates in populations with different effective sizes. Factors that reduce effective size, such as large variance in offspring number, inbreeding, and skewed sex ratio, reduce the efficacy of selection.

How Do We Estimate Effective Size from Sequence Data?

Effective population size can be directly estimated by genetic diversity if the mutation rate of the organism is known. Under neutrality, genetic diversity is proportional to the product of the effective population size and the mutation rate ($\pi = 4N_e\mu$ and $\pi = 2N_e\mu$ for diploid and haploid systems, respectively.) Alternatively, if temporal polymorphism data are available, variance effective population sizes can be estimated by variance in allele frequency changes. If there are large changes in population size, these two methods could give very different estimates. The first N_e represents the effective population size over longer time periods, whereas the second N_e is more related to current population size. If we want to predict the effectiveness of selection in the near future, the current N_e is more suitable, but if inference of past evolutionary processes is the goal, then the long-term N_e is appropriate.

As an example, we may consider the effective population size of *P. falciparum*. Based on the DNA sequence diversity of nuclear and mitochondrial genes, N_e has been estimated to be on the order of 10^5 (Joy et al., 2003; Mu et al., 2005), an order of magnitude larger than the effective population size of humans. It would be interesting to compare this estimate with that of N_e based on changes in allele frequency across generations in contemporary populations as the latter could be considerably smaller. If so, this would mean that selection for drug resistance or virulence may become less effective as the effective population size decreases. The finding would also have implications for increased contributions of random genetic drift in changing allele frequencies in parasite populations.

VARIATION IN MUTATION RATES

In addition to effective population size, mutation rate is an important parameter in evolutionary processes. Mutation rates are important because they influence the speed of adaptation to new environments or selection pressures, the magnitude of mutation load due to deleterious mutations, and other evolutionary dynamics, such as coevolution between host immune systems and pathogenic strains. For neutral mutations, Kimura (1968) showed that the evolutionary rate of neutral substitutions equals the rate of neutral mutations. This important principle is surprisingly simple to derive: The rate of nucleotide substitution is equal to the product of the probability of fixation and the mutation rate. For newly arisen neutral mutations in a diploid population, the fixation rate is $1/(2N_e)$. The total mutation rate equals $2N_e$ times the neutral mutation rate per individual, say, U.

Therefore, the neutral substitution rate is expected to be $1/(2N_e) \times 2N_e \times U = U$.

Many studies suggest that, rather than being uniform across the genome of an organism, mutation rates vary across the genome. Wolfe et al. (1989) found that the rate of synonymous substitution, for which it is widely assumed that $|N_e s| \ll 1$, is not uniform among genes in the mammalian genome. Studies of mammalian repetitive sequence showed a similar variation. Furthermore, Gaffney and Keightley (2005) compared substitutions in repetitive elements in mice and suggested that mutation rates vary on a megabase length scale in the murid genome. On the other hand, Amos (2010) observed small clusters of SNPs in the human genome.

Several confounding factors might affect inference of variation in mutation rate based on sequence data. First, natural selection shapes local nucleotide variation. Genes under positive selection tend to have lower intraspecies polymorphism and higher interspecies divergence, whereas genes under purifying selection are more likely to have lower polymorphism and lower divergence than neutral loci. Second, recombination breaks down associations between linked alleles and makes the history of different regions in the genome different. Polymorphism would vary among different segments that have different times to the most recent common ancestor because of recombination.

To avoid the confounding effects of sequence-based methodology and to achieve more accurate estimates of mutation rate, Lang and Murray (2008) improved the method of estimating mutation rates based on the fluctuation test in yeast and applied this experimental method to the estimation of mutation rates of two genes. The mutation rates were estimated on a per-nucleotide scale and were shown to differ significantly, supporting the view that the mutation rate is not uniform over the yeast genome.

Other researchers have investigated factors that may affect mutation rates across the genome. First, mutation rate is associated with DNA replication timing in humans and is higher in later-replicating regions of the human genome (Stamatoyannopoulos et al., 2009). Second, Amos (2010) suggested that mutations tend to occur near existing polymorphic sites. This author conducted computer simulations of SNPs and compared the simulated pattern with actual SNPs of human chromosome 1 and found that a nonindependence model predicted the frequency and density of the SNP clusters better than the random model. This result suggests that mutations are more likely to occur in regions that are already polymorphic. This mechanism is attractive since it seems to be favored by selection: Polymorphic regions are usually able to tolerate or even favor mutations, and fewer mutations in nonpolymorphic regions can reduce the potential deleterious mutations. This idea will be more convincing if the proposed mechanism can be supported by more direct experimental evidence. If more mutations take place near sites with existing mutations, it also implies that mutations are more likely to happen in genes under balancing and diversifying selection and further suggests that the speed of evolution of antigenic genes could be increased by both mutation and selection.

Variation of mutation rate across the genome reduces effectiveness to identify putatively functional regions. Researchers have been trying to reduce this effect: In coding regions, synonymous changes are usually used as a control for mutation rates by assuming that synonymous changes are effectively neutral. If the numbers of synonymous changes and nonsynonymous changes are both high or are both low, it is likely that they are affected by higher or lower mutation rates in the region. If there are many nonsynonymous changes but few synonymous changes within a gene, the large number of nonsynonymous changes cannot be explained by mutation rates and suggests that positive selection may be acting on this gene. However, these methods are problematic. For example, synonymous sites are not always nearly neutral, and besides non-neutral synonymous sites within coding regions, increasing amounts of noncoding DNA have been found to be functional, making it important to annotate these functional regions in noncoding DNA. To improve the efficiency of identification of putatively functional regions, further work is needed to estimate mutation rates on a fine scale across the genomes.

WHAT CAN WE LEARN FROM POLYMORPHISM AND DIVERGENCE?

Advances in sequencing technology and other aspects of genomics have enhanced the power to study basic evolutionary processes and their relationships to diversity. Sequence data can be used to infer the demographic history of a population, the population structure, and which genes are under selection. It can be used to trace the evolution of antigenic genes in order to understand the evolution of virulence and drug resistance in microbes. In this section, we briefly introduce some methods that are useful in the study of evolution of microbes and we describe what kinds of questions can be answered. Details and many examples of the different methods are discussed in the following chapters.

Linkage disequilibrium would appear if individuals in the population tended to mate nonrandomly due to geographic distance or had mating preference with particular phenotypes. Natural selection can also lead to linkage disequilibrium by favoring specific combinations of alleles at different loci. Therefore, it is important to examine the possibility of population structure before identifying genes under selection. The difference between the effects of natural selection and population structure is that natural selection acts only on some genes, whereas population structure affects the pattern in the whole genome. There are two kinds of approaches for detecting population substructure. One is based on clustering methods, such as those implemented in software packages *Structure* (Pritchard et al., 2000), FRAPPE (Tang et al., 2005), SABER (Tang et al., 2006), and ADMIXTURE (Alexander et al., 2009); the other approach is principal component analysis (PCA), implemented in the SmartPCA application (Patterson et al., 2006).

When population size increases, more new mutations occur, increasing the number of low frequency alleles in the population. Conversely, a reduction in the population size results in a deficit of rare alleles. Since changes of population size affect the relative amount of alleles with different frequencies, an allele frequency spectrum (distribution of allele frequency) can be used to examine whether population size has changed in the recent past. There are some software packages available such as BEAST and *∂a∂i* that can estimate demographic parameters. The method implemented in BEAST is based on Bayesian analysis, whereas *∂a∂i* applies a diffusion approximation to the allele frequency spectrum for demographic inference.

Selection forces can also affect the allele frequency spectrum, but, similar to population structure, any change in population size affects the whole genome, while selection forces only influence a limited number of genes. Any pattern pervasive across the genome is likely caused by demographic changes.

As previously discussed, effective population size can be estimated from genetic diversity if the mutation rate is known or from allele frequency changes if temporal allele frequency data are available. Moment methods, maximum likelihood methods, and Bayesian approaches based on the temporal allele frequency data are all available, and the software package NeEstimator includes many of these.

Identifying Genes Putatively under Selection

Identifying genes under selection has always been a major interest in population genetics. Identifying genes under selection is important in understanding interactions between pathogens and their hosts, as genes related to host–parasite interactions are expected to be subject to natural selection. Inference of selection may also be useful for choosing targets of new therapeutics. To identify genes under selection, investigators typically compare divergence between species, polymorphism within species, or both.

Genes under positive selection are expected to have higher divergence among species and genes under negative selection to be less diverged. In order to control for different mutation rates, the ratio of the number of nonsynonymous substitutions per nonsynonymous site to the number of synonymous substitutions per synonymous site (d_N/d_S) is used as an indicator of selection. The software package PAML can be used to calculate d_N/d_S and to test for statistical significance. $d_N/d_S > 1$ indicates positive selection; $d_N/d_S < 1$ represents purifying selection; and $d_N/d_S = 1$ suggests that the gene is not under selection.

Three approaches are commonly used to detect selection based on data for intraspecies polymorphism. One set of methods is based on the allele frequency spectrum, one set on linkage disequilibrium, and another set on allele frequency differentiation. How different selective forces change the allele frequency spectrum is shown in Figure 1.1. Under negative selection, deleterious mutations are removed from the population and therefore the fraction of low frequency alleles is higher. Under positive selection, mutations are favored and there is an increase in fraction of high

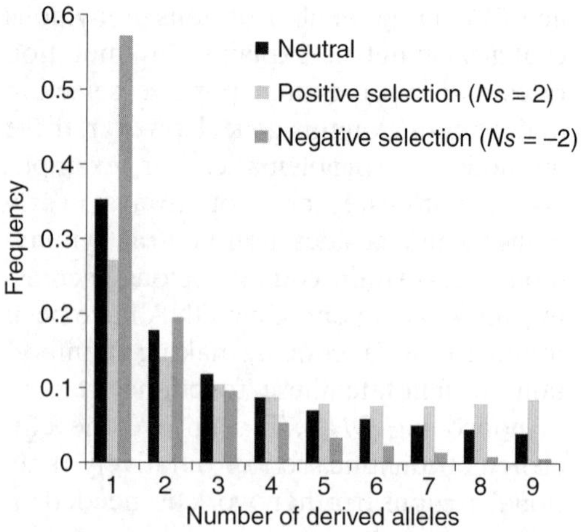

Figure 1.1 Allele frequency spectra under neutrality, positive selection, and negative selection. Equation 11 from Sawyer and Hartl (1992) was used to generate the spectra under positive and negative selections. The example here has sample size equal to 10.

frequency alleles. Because demographic history can alter the null distribution under the neutral hypothesis, if we detect demographic changes in the past, it is better to estimate the null distribution by coalescent simulation rather than to assume a constant population size under the neutral hypothesis. Coalescent simulation of simple demographic models can be done in a program denoted *ms* (Hudson, 2002). Tests for selection based on the allele frequency spectrum include Tajima's D (Tajima, 1989), Fu and Li's D (Fu and Li, 1993), and Fay and Wu's H (Fay and Wu, 2000).

If positive selection is recent, recombination does not have enough time to break down linkage disequilibrium near the selected position, so linkage disequilibrium is higher and haplotype length longer around the selected loci. Methods based on haplotype length and linkage disequilibrium include integrated haplotype score (iHS) (Voight et al., 2006) and cross population extended haplotype homozygosity (XP-EHH) tests (Sabeti et al., 2007). An estimation of recombination rates is needed for applying this kind of method. Recombination rates can be estimated by experimental crosses or from linkage disequilibrium. Programs such as LDhat and PHASE can be used for estimating recombination rates.

Selection changes allele frequency differentiation among populations depending on the type of selection. Regions with local adaptations are more differentiated among populations, whereas regions under uniform balancing selection are less diverged. Allele frequency differentiation among populations is usually summarized by the statistic $F_{ST} = (H_T - H_S)/H_T$, where H_S is the average heterozygosity in subpopulations and H_T is the average heterozygosity in the whole population assuming Hardy–Weinberg equilibrium (Weir and Cockerham, 1984). Since population differentiation could also affect F_{ST}, a null distribution under a proper model of population subdivision should be generated in order to detect loci under selection by conducting F_{ST}-based method. In haploid organisms, H_S and H_T are "virtual" heterozygosities calculated from the observed allele frequencies under the assumption of random mating.

Recently, Grossman et al. (2010) developed a method called composite of multiple signals (CMS) to incorporate different kinds of information, including linkage disequilibrium, allele frequency spectrum, and allele frequency differentiation among populations, to increase the power of detecting recent positive selection. This method successfully increases the resolution of localizing regions under recent positive selection by up to 100-fold.

The widely used McDonald–Kreitman method is based on the number of nonsynonymous and synonymous mutations within and between species and the assumption that synonymous changes are neutral (McDonald and Kreitman, 1991). This test is more robust to the demography since synonymous and nonsynonymous mutations are similarly affected. Under a neutral prediction, the ratio of the number of nonsynonymous polymorphisms to the number of synonymous polymorphisms should be the same as the ratio of the number of nonsynonymous fixations to the number of synonymous fixations. Positive selection is expected to lead to an increased ratio of nonsynonymous-to-synonymous fixations but to have a smaller effect on nonsynonymous-to-synonymous polymorphism; negative selection is expected to result in a reduced ratio of nonsynonymous-to-synonymous fixations but to have a relatively smaller effect on nonsynonymous-to-synonymous polymorphism.

CONCLUSIONS

With the introduction of next-generation sequencing technologies, obtaining sequence data is no longer a limiting factor for large-scale population-level research. Furthermore, third-generation sequencing methods based on semiconductors are emerging. Although these new technologies offer huge advantages and massive amounts of new data, they also bring new tasks. With scale-up of the quantity of data comes a similar scale-up in the potential errors that can be introduced, such as sequencing errors, alignment errors, and missing data. Sequence reads generated by next-generation sequencing methods are short, and thus the difficulty of alignment is challenging. This can be partially solved by using paired-end sequencing, but the alignment for repetitive or highly polymorphic regions is still difficult, and correcting for biases in estimating nucleotide diversity still needs to be solved. Most sequencing errors are visible as singleton alleles observed once per sample and skew the allele frequency spectrum toward singletons. Since some of the methods for inferring demographic history and detecting selection are based on allele frequency spectrum, they are sensitive to sequencing error. Missing data are also a big issue that remains to be solved. Simply ignoring missing data can bias the population genetic inference because often data are not missing at random, but imputation can introduce different biases. When using next-generation sequencing data to study population genetic questions, researchers need to be aware of the biases that sequencing error and missing data introduce. One way to reduce the effect of sequencing error is to incorporate quality scores in the analyses. Rather than using all the sequence information that passes the quality score threshold to make inferences, we can include a quality score that is related to the error probabilities in a statistical model. Lynch (2008, 2009) developed a method to incorporate a quality score to estimate nucleotide diversity and the allele frequency spectrum. Further work to develop tools to account for sequencing errors and missing data remains to be done.

ACKNOWLEDGMENT

This work was supported by National Institutes of Health grants GM079536 and AI075080.

GLOSSARY

Allele Alternative versions of a gene at a specific location (locus) on a chromosome

Balancing Selection Any type of natural selection in which multiple alleles are maintained in the population

CNV Copy number variation—type of genetic variation in which individuals differ in the number of copies of a region of the genome

Coalescence An event that two genetic lineages traced backward in time merge to a single ancestor

Demography Population processes such as change in population size, divergence, or migration that can result in variation within and among populations

Drift (Random Genetic Drift) A change in allele frequency due entirely to stochastic events in a finite population

Eigenvalue A scalar value arising in factor analysis that measures the amount of variation in the variables explained by that factor

Fixation A change in allele frequency that culminates in one allele persisting in a population and all others being lost

Heterozygosity A measure of genetic variation defined as the proportion of heterozygous genotypes at a locus in a population

Linkage Disequilibrium The nonrandom association of alleles from multiple loci, not necessarily on the same chromosome

Negative Selection A type of natural selection in which an allele is deleterious, often leading to a reduction in its allele frequency

Nonsynonymous Mutation A mutation in a coding sequence that changes the amino acid

Positive Selection A type of natural selection in which an allele increases in frequency in a population because individuals carrying that allele are more fit than those carrying alternative alleles

Principal Component Analysis A method of factor analysis that involves data transformation in order to find the components that best explain the variance in the data and reveals its internal structure in a space of fewer dimensions

Selection Another term for Darwin's natural selection—differences in fitness of individuals of different genotypes often resulting in changes in allele frequency

SNP Single-nucleotide polymorphism—type of genetic variation where multiple nucleotides are possible at a given position in the genome

Synonymous Mutation A mutation in a coding sequence that does not alter the amino acid

REFERENCES

Alexander DH, Novembre J, Lange K. 2009. Fast model-based estimation of ancestry in unrelated individuals. Genome Res 19: 1655–1664.

Amos W. 2010. Even small SNP clusters are non-randomly distributed: Is this evidence of mutational non-independence? Proc Biol Sci 277: 1443–1449.

Fay JC, Wu CI. 2000. Hitchhiking under positive Darwinian selection. Genetics 155: 1405–1413.

Fisher RA. 1930. *The Genetical Theory of Natural Selection*. Clarendon Press, Oxford.

Fu YX, Li WH. 1993. Statistical tests of neutrality of mutations. Genetics 133: 693–709.

Gaffney DJ, keightley PD. 2005. The scale of mutational variation in the murid genome. Genome Res 15: 1086–1094.

Grossman SR, Shylakhter I, Karlsson EK, Byrne EH, Morales S, Frieden G, Hostetter E, Angelino E, Garber M, Zuk O, Lander ES, Schaffner SF, Sabeti PC. 2010. A composite of multiple signals distinguishes causal variants in regions of positive selection. Science 327: 883–886.

Hartl DL, Clark AG. 2007. *Principles of Population Genetics*. Sinauer Associates, Sunderland.

Hudson RR. 2002. Generating samples under a Wright-Fisher neutral model. Bioinformatics 18: 337–338.

Joy DA, Feng X, Mu J, Furuya T, Chotivanich K, Krettli AU, Ho M, Wang A, White NJ, Suh E, Beerli P, Su XZ. 2003. Early origin and recent expansion of *Plasmodium falciparum*. Science 300: 318–321.

Kimura M. 1968. Evolutionary rate at the molecular level. Nature 217: 624–626.

Kimura M, Crow JF. 1964. The number of alleles that can be maintained in a finite population. Genetics 49: 725–738.

Kimura M, Ohta T. 1969. The average number of generations until fixation of a mutant gene in a finite population. Genetics 61: 763–771.

King JL, Jukes TH. 1969. Non-Darwinian evolution. Science 164: 788–798.

Lang GI, Murray AW. 2008. Estimating the per-base-pair mutation rate in the yeast *Saccharomyces cerevisiae*. Genetics 178: 67–82.

Lynch M. 2008. Estimation of nucleotide diversity, disequilibrium coefficients, and mutation rates from high-coverage genome-sequencing projects. Mol Biol Evol 25: 2409–2419.

Lynch M. 2009. Estimation of allele frequencies from high-coverage genome-sequencing projects. Genetics 182: 295–301.

Mcdonald JH, Kreitman M. 1991. Adaptive protein evolution at the *Adh* locus in *Drosophila*. Nature 351: 652–654.

Mu J, Awadalla P, Duan J, Mcgee KM, Joy DA, Mcvean GA, Su XZ. 2005. Recombination hotspots and population structure in *Plasmodium falciparum*. PLoS Biol 3: e335.

Ohta T. 1992. The nearly neutral theory of molecular evolution. Annu Rev Ecol Syst 23: 263–286.

Patterson N, Price AL, Reich D. 2006. Population structure and eigenanalysis. PLoS Genet 2: e190.

Pritchard JK, Stephens M, Donnelly P. 2000. Inference of population structure using multilocus genotype data. Genetics 155: 945–959.

Sabeti PC, Varilly P, Fry B, Lohmueller J, Hostetter E, Cotsapas C, Xie X, Byrne EH, Mccarroll SA, Gaudet R, Schaffner SF, Lander ES, Frazer KA, Ballinger DG, Cox DR, Hinds DA, Stuve LL, Gibbs RA, Belmont JW, Boudreau A, Hardenbol P, Leal SM, Pasternak S, Wheeler DA, Willis TD, Yu F, Yang H, Zeng C, Gao Y, Hu H, Hu W, Li C, Lin W, Liu S, Pan H, Tang X, Wang J, Wang W, Yu J, Zhang B, Zhang Q, Zhao H, Zhou J, Gabriel SB, Barry R, Blumenstiel B, Camargo A, Defelice M, Faggart M, Goyette M, Gupta S, Moore J, Nguyen H, Onofrio RC, Parkin M, Roy J, Stahl E, Winchester E, Ziaugra L, Altshuler D, Shen Y, Yao Z, Huang W, Chu X, He Y, Jin L, Liu Y, Sun W, Wang H, Wang Y, Xiong X, Xu L, Waye MM, Tsui SK,

Xue H, Wong JT, Galver LM, Fan JB, Gunderson K, Murray SS, Oliphant AR, Chee MS, Montpetit A, Chagnon F, Ferretti V, Leboeuf M, Olivier JF, Phillips MS, Roumy S, Sallee C, Verner A, Hudson TJ, Kwok PY, Cai D, Koboldt DC, Miller RD, Pawlikowska L, Taillon-Miller P, Xiao M, Tsui LC, et al. 2007. Genome-wide detection and characterization of positive selection in human populations. Nature 449: 913–918.

Sawyer SA, Hartl DL. 1992. Population genetics of polymorphism and divergence. Genetics 132: 1161–1176.

Sjödin P, Kaj I, Krone S, Lascoux M, Nordborg M. 2005. On the meaning and existence of an effective population size. Genetics 169: 1061–1070.

Stamatoyannopoulos JA, Adzhubei I, Thurman RE, Kryukov GV, Mirkin SM, Sunyaev SR. 2009. Human mutation rate associated with DNA replication timing. Nat Genet 41: 393–395.

Tajima F. 1989. Statistical method for testing the neutral mutation hypothesis by DNA polymorphism. Genetics 123: 585–595.

Tang H, Peng J, Wang P, Risch NJ. 2005. Estimation of individual admixture: Analytical and study design considerations. Genet Epidemiol 28: 289–301.

Tang H, Coram M, Wang P, Zhu X, Risch N. 2006. Reconstructing genetic ancestry blocks in admixed individuals. Am J Hum Genet 79: 1–12.

Voight BF, Kudaravalli S, Wen X, Pritchard JK. 2006. A map of recent positive selection in the human genome. PLoS Biol 4: e72.

Weir BS, Cockerham CC. 1984. Estimating F-statistics for the analysis of population structure. Evolution 38: 1358–1370.

Wolfe KH, Sharp PM, Li WH. 1989. Mutation rates differ among regions of the mammalian genome. Nature 337: 283–285.

Wright S. 1931. Evolution in mendelian populations. Genetics 16: 97–159.

CHAPTER 2

EVOLUTION OF MEIOSIS, RECOMBINATION, AND SEXUAL REPRODUCTION IN EUKARYOTIC MICROBES

WENJUN LI,* ELIZABETH SAVELKOUL,* JOSEPH HEITMAN, and JOHN M. LOGSDON, JR.

DEFINING SEX

"Sex," in a cellular sense, is surprisingly difficult to define. Is it the act of individuals mating? Is it the occurrence of meiosis? Is it the generation of novel genotypes? These are all components of sexual reproduction, but using any one of them as the sole defining requirement presents problematic exceptions. For example, mating—conjugation of cells or hyphae between opposite-sex individuals—occurs in *Candida albicans* and produces progeny with novel genotypes, but canonical meiosis has not been documented in this species (Butler et al., 2009; Forche et al., 2008). Alternatively, if meiosis is taken as the defining component of sexual reproduction, this definition will include self-fertilization by automixis—a single cell undergoing meiosis with fusion of its progeny gametes to produce genetically novel, albeit (eventually) homozygous individuals (Archetti, 2004a). Finally, if the generation of novel genotypes were the sole criterion for sexual reproduction, everything from bacterial plasmid exchange to mitotic recombination and mutation itself would have to be included (Cavalier-Smith, 2002). Therefore, for the purposes of this chapter, we will here define sexual reproduction as a two-step process: (i) reduction of ploidy by a parental cell producing progeny cells, gametes, or nuclei; and (ii) restoration of ploidy by the fusion of two cells in syngamy. If this definition seems vague, it is because both steps—ploidy reduction and ploidy restoration—can occur by numerous mechanisms.

The first step of sexual reproduction—ploidy reduction—most frequently occurs when a parental diploid cell undergoes meiosis to produce haploid progeny cells (e.g., gametes). The mechanism of ploidy reduction can range from highly regulated (i.e., meiosis, a single DNA replication round followed by two cell divisions) to largely stochastic (e.g., chromosome loss

*These authors contributed equally to this work.

Evolution of Virulence in Eukaryotic Microbes, First Edition. Edited by L. David Sibley, Barbara J. Howlett, and Joseph Heitman.

with transient aneuploid states in *C. albicans*; Bennett and Johnson, 2003; Butler et al., 2009; Forche et al., 2008). The specific ploidy at each state can also vary across taxa; for example, *C. albicans* alternates between diploid (2N) and tetraploid (4N) states (Forche et al., 2008). Further cell cycle regulation variation is seen in several diverse protist species, which replicate their DNA up to 16N or even 128N before undergoing multiple consecutive meiotic divisions (Archetti, 2004a).

The second step of sexual reproduction—ploidy restoration by syngamy—also has its own set of variations. Taxa can be self-incompatible (as determined by two or more sex-determining loci) or self-compatible (no loci encoding specific mating types, as in *Lodderomyces elongisporus*), or capable of same-sex mating, as are *C. albicans* and *Cryptococcus neoformans* (Alby et al., 2009; Lee et al., 2010; Lin et al., 2005). The fusion of cells and fusion of nuclei also need not always be immediately consecutive: Fungal hyphae can fuse into heterokaryons that contain genetically distinct haploid nuclei, of which only a subset may fuse and generate diploid nuclei (Baptista et al., 2003; Hoffmann et al., 2001). Finally, the time between ploidy reduction and ploidy restoration can also be short or long since taxa vary as to which ploidy state is dominant in their life history (Otto and Mable, 1998).

Sexual reproduction is pervasive throughout eukaryotes, including animals, plants, fungi, and protists. Sexual development of fungi has been extensively studied; indeed, our knowledge of the sexual reproduction of eukaryotic microbes is largely based on model fungi such as *Saccharomyces cerevisiae*. In this chapter, we capitalize upon the well-studied sexual reproduction of fungi to describe the organization and evolution of the *MAT* locus (which determines sexual identity), the genes necessary to undergo meiosis, the biological approaches to identify sexual cycles, and how sexual reproduction affects the evolution of virulence in eukaryotic microbes.

MECHANISMS OF SEX

Mating-Type (*MAT*) Locus and Sexual Identity

One of the crucial determinants of whether sexual reproduction can occur is whether the cells about to undergo syngamy are of compatible mating types. In many—but not all—cases, cells with a common sexual identity or mating type are unable to fuse. In contrast to the sex chromosomes that control sexual development in animals (and some plants), the sexual identity of many eukaryotic microbes is determined by a relatively small genomic region termed the mating-type (*MAT*) locus; this locus contains genes that control mating and cell type identity (Fraser and Heitman, 2005). In most sexual eukaryotic microbes where it has been identified, the *MAT* locus accounts for a relatively small portion of the genome, ranging from <1 to >100 kb and containing 1 to >20 genes. Essential elements of the *MAT* locus include transcription factors with DNA binding motifs that are key regulators of gene expression inducing sexual development. In addition to these transcription factors, the *MAT*-associated regions of basidiomycete fungi also encode pheromone and pheromone receptor (P/R) genes that play key roles in sexual development.

While transcription factor genes are always essential components of the *MAT* locus, the structure of the *MAT* locus is highly diverse because of gene acquisition, gene loss, divergence, conversion, and rearrangement. Even within closely related species, the organization of the *MAT* locus can be highly diverse as demonstrated by comparative analyses of the *MAT* locus of the *Cryptococcus* pathogenic species (Byrnes et al., 2011; Fraser et al., 2004;

Lengeler et al., 2002), some of which will be discussed later.

Transcription Factor Genes Encoded by the *MAT* Locus

The transcription factors of the *MAT* locus regulate a diverse set of target genes involved in sexual developmental processes including cell cycle arrest, cell fusion, nuclear fusion, and meiosis. Notably, the expression of the P/R genes, which initiate sexual reproduction, is controlled by such *MAT*-encoded transcription factors. Three classes of transcription factor genes, homeodomain (HD), alpha domain, and high mobility group (HMG) genes, are commonly encoded by the *MAT* locus. The common feature of these transcription factors is a DNA binding motif that binds to the promoter region of target genes and regulates gene expression. In many ascomycete fungi, one *MAT* locus allele, or idiomorph, harbors an alpha-domain gene and the other *MAT* locus allele contains an HMG domain gene (Raudaskoski and Kothe, 2010). In basidiomycetes and in some ascomycetes (such as *S. cerevisiae* and *C. albicans*) HD genes (HD1 and HD2) determine sexual and cell-type identity.

Recent identification and characterization of the HMG genes encoded in the *MAT* loci of basal zygomycete fungi indicate that the HMG domain gene is the ancestral mode of fungal sexual identity (Gryganskyi et al., 2010; Idnurm et al., 2008; Lee et al., 2008; Li et al., 2011). Interestingly, the transcription factors in both *MAT* locus alleles of the zygomycetes are HMG genes. This suggests that the sexual identity of fungi was ancestrally determined by the HMG domain genes and that the alpha-domain and HD genes were later recruited into the *MAT* locus to replace one or both HMG genes (Casselton, 2008; Dyer, 2008). The later origin of alpha-domain and HD genes relative to the HMG genes may be explained by gene duplication. On the basis of sequence comparisons, phylogenetic analyses, and *in silico* predictions of secondary and tertiary structures of the HMG and alpha-domain genes in ascomycetes, Martin et al. (2010) concluded that the alpha-domain gene may have arisen from an HMG gene ancestor.

Several hypotheses on the origin and evolution of the fungal *MAT* locus genes have been proposed (Casselton, 2008; Dyer, 2008; Lee et al., 2010). In one model, the HMG gene is the ancestral form of the *MAT* locus of fungi. Subsequently, one HMG gene diverged into an alpha-domain gene in ascomycetes, and the HD genes were incorporated into the *MAT* locus and replaced the HMG genes to govern the sexual identity of basidiomycetes. In a second *MAT* evolutionary model, both the HD and HMG genes may have been ancestral with one or the other, or both in some species (e.g., *C. albicans*), emerging as key cell identity factors. However, there are still several gaps in the hypothesized evolutionary history of the *MAT* locus. With the rapidly increasing number of genomic sequences of eukaryotic microbes, further identification and characterization of more *MAT* loci in fungi can fill these gaps and present a clearer map of the plasticity and evolutionary history of the *MAT* locus.

Pheromone/Pheromone Receptor (P/R) Genes

Pheromones are small peptides secreted by individuals of one mating type that trigger mating with individuals of another mating type by acting on the pheromone receptors on the opposite type cell. Interaction of pheromone and pheromone receptors (transmembrane G protein-coupled receptors [GPCRs] on recipient cells) initiates signal transduction and activates transcription factors, such as Ste12 in *S. cerevisiae*. These transcription factors regulate the expression of many genes involved in cell cycle arrest, cell fusion, nuclear fusion, and meiosis, which are key processes of sexual reproduction. Collectively, these interacting pheromones and receptors are encoded by the P/R genes. In heterothallic fungi, P/R genes are important because the interaction of their products is the first step to initiate sexual development by transducing the mating signal.

Two classes of pheromone, **a** and α, have been identified in ascomycete fungi (Jones and Bennett, 2011). The **a** pheromone has a CAAX motif at the C-terminus and is a lipid-modified peptide secreted by a specialized transporter (Ste6), whereas the α pheromone is a simple peptide secreted by the standard protein secretion pathway. In basidiomycetes, all pheromones are of the lipid-modified form (Brown and Casselton, 2001). It is hypothesized that the two classes of pheromones in ascomycetes may have evolutionary advantages over the exclusively lipid-modified peptide pheromones of basidiomycetes. For example, the simple peptide pheromones are more soluble and may mediate longer-distance communication with mating partners, compared to the relatively insoluble lipophilic lipid-modified pheromones, which may signal only over shorter physical distances. However, a recent study found that pheromone asymmetry may not be necessary because either two lipid-modified pheromones or two simple peptide pheromones can support sexual development of *S. cerevisiae*, albeit with reduced mating efficiency (Goncalves-Sa and Murray, 2011).

The number, genomic location, and organization of pheromone genes vary in different species. In basidiomycetes, the P/R genes are encoded by the *MAT* locus and determine cell type identity (Stanton et al., 2010). In ascomycetes, both the **a** and α pheromone genes and their receptors are present in each individual's genome and are not encoded by the *MAT* locus; thus, sexual identity of ascomycetes is determined only by the transcription factor genes at the *MAT* locus. The **a** and α pheromone genes can also vary in copy number between even closely related species. For example, the genome of *S. cerevisiae* encodes two **a** and two α pheromone genes, while its close relative *Saccharomyces pastorianus* has three **a** pheromone genes and four α pheromone genes. In the basidiomycete *C. neoformans* (a human pathogen), the serotype A strain has three α pheromone genes, whereas the serotype D strain has four α pheromone genes (Fraser et al., 2004; Loftus et al., 2005). At least six pheromone genes were identified in another basidiomycete, *Coprinopsis cinerea* (O'Shea et al., 1998). The **a** pheromone gene copy number is typically lower compared with multiple copies of the α pheromone gene; however, the reason and significance are unclear (Martin et al., 2011). Pheromone genes appear to be more diverse than housekeeping genes, exhibiting evidence for both positive diversifying selection and relaxed selective constraints (Martin et al., 2011). However, it is unclear if pheromone gene diversity alone can account for reproductive isolation between different species, even given their variation among species.

Bipolar and Tetrapolar Mating Systems

While the gene content and organization of the *MAT* locus varies considerably among fungal species, mating types are determined by two predominant *MAT* paradigms: bipolar and tetrapolar. Bipolar species have one biallelic *MAT* locus that specifies two mating types, and their union produces progeny of just two mating types. Tetrapolar species have two mating-type loci, both of which must differ for productive mating; their union (A1B1 × A2B2) yields progeny of four mating types (A1B1, A2B2, A1B2, and A2B1), hence the name tetrapolar. In some bipolar mating systems, the HD and P/R loci are linked, which gives rise to two mating types. In tetrapolar mating systems, the HD and P/R loci are unlinked and can even lie on different chromosomes. If both the HD and P/R loci are biallelic, the tetrapolar mating system has four mating types. However, most tetrapolar mating systems, such as those in mushrooms, have several thousands of mating types due to the multiallelic nature of both the HD and P/R encoding *MAT* loci.

Ascomycetes have bipolar mating systems: Their sexual identity is determined solely by transcription factors encoded by the *MAT* locus since both the **a** and α P/R genes are present in each individual's genome. Given that tetrapolar mating systems are exclusively found in basidiomycetes, it is likely that the tetrapolar system emerged after the divergence of these two dikaryotic fungal phyla. Indeed, both bipolar and tetrapolar mating systems are present in basidiomycetes, which allows comparative inferences to be made about the origins of and the transitions between each mating system.

An exemplary system for studying the diversity and evolution of both bipolar and tetrapolar mating systems comes from an extensive study of the human fungal pathogen *C. neoformans* and related species (Bakkeren and Kronstad, 1993; Metin et al., 2010; Xu et al., 2007). *C. neoformans* is one of the most common fungal pathogens, causing opportunistic infections such as pneumonia and meningitis in HIV/AIDS patients. *C. neoformans*, together with its sibling species *Cryptococcus gattii*, comprise the pathogenic *Cryptococcus* species complex (Findley et al., 2009). These pathogenic species are clustered in the *Filobasidiella* clade, which also includes the saprobic species *Tsuchiyaea wingfieldii*, *Cryptococcus amylolentus*, and *Filobasidiella depauperata*. The pathogenic species complex has a bipolar mating system: The HD and P/R loci are linked and many other genes (in total >20) are also present in the *MAT* locus (>100 kb) (Fraser et al., 2004). However, the saprobic species *T. wingfieldii* and *C. amylolentus* (and also *Cryptococcus heveanensis*) have a tetrapolar mating system, and the HD and P/R loci are on different chromosomes and are therefore unlinked (Findley et al., 2012; Metin et al., 2010). These comparative genomic studies provide evidence that the bipolar mating system in the pathogenic species complex has evolved from a tetrapolar mating system in the last common ancestor shared with the closely related sibling saprobic species by translocation, inversions, gene conversion, and gene capture (Findley et al., 2012; Fraser et al., 2004; Metin et al., 2010). Given that the mating type of *C. neoformans* is associated with virulence (Kwon-Chung et al., 1992), it is interesting to consider whether evolution of the bipolar mating system arose concomitant with the emergence of virulence.

An additional example of coextant bipolar and tetrapolar basidiomycete mating systems is in the *Ustilago* genus, which includes a large number of smut fungi causing plant infections (see also Chapter 23, "Sex and Virulence in Basidiomycete Pathogens"). *Ustilago maydis* has a tetrapolar mating system; however, *Ustilago hordei* has a bipolar mating system in which the HD and P/R loci are linked but lie 400–500 kb apart (Bakkeren and Kronstad, 1993). The closely related genus, *Malassezia*, has an apparent bipolar mating system based on genomic sequence inspection of *Malassezia restricta* and *Malassezia globosa* (Xu et al., 2007).

When considering the evolution of bipolar and tetrapolar mating systems of *Cryptococcus* and *Ustilago* species, it is likely that the ancestral mating system of basidiomycetes was a tetrapolar mating system, and extant basidiomycete bipolar mating systems have arisen multiple times independently. However, given that mating systems of both ascomycetes and zygomycetes are bipolar, the original ancestor of basidiomycetes was most likely also bipolar;

thus, tetrapolar species derived from bipolar systems, yet more recent conversions back to bipolarity have subsequently occurred.

One key question is whether any extant bipolar basidiomycete species mirrors the hypothetical bipolar ancestor. Coelho et al. (2010) recently identified a pseudo-bipolar mating system in the red yeast *Sporidiobolus salmonicolor*, an early diverged basidiomycete. The HD and P/R loci of the *MAT* locus are ~1.3 Mb apart from each other in *S. salmonicolor*; however, occasional disruptions of the genetic cohesion of the bipolar *MAT* locus via recombination give rise to new mating types in a process that parallels that of tetrapolar systems (Coelho et al., 2010). This novel mating system in a basal basidiomycete adds additional support to a bipolar or pseudo-bipolar basidiomycete ancestor. Further characterization of the *MAT* locus of other basidiomycete fungi, especially the early diverged Pucciniomycotina lineages, will be important to advance our understanding of mating system evolution.

Homothallic and Heterothallic Mating Systems

The focus on the mechanism of how cells maintain a distinct sexual identity may give the impression that sexual reproduction always occurs between individuals of different mating types, implying obligate outcrossing (reproduction between two genetically distinct individuals). However, fungi are classified as heterothallic or homothallic based on whether they are self-sterile or self-fertile. Species that are heterothallic are self-sterile and mating occurs between partners of different mating types. Species that are homothallic are self-fertile and sexual reproduction can occur during solo monoculture. Mechanistically, homothallism can be achieved by the presence of two *MAT* loci in a single genome (linked or unlinked in the genome), mating-type switching as exemplified by *S. cerevisiae* and *Schizosaccharomyces pombe*, or same-sex mating (unisexual reproduction) as in *C. neoformans* (Lin et al., 2005) and in *C. albicans* (Alby et al., 2009).

Both heterothallism and homothallism are common throughout the fungal kingdom. Basidiomycetes are dominated by heterothallic mating systems. In ascomycetes, both heterothallic and homothallic species sometimes can be found in the same genus. Whether the ancestral mode of fungal sexual reproduction was homothallic or heterothallic is not known. One parsimonious model is that sex was originally homothallic, from which heterothallism then emerged, with other more recent transitions back to homothallism, similar to the bipolar → tetrapolar → bipolar transitions discussed above. Other studies have used comparative analysis data to hypothesize that the ancestral mating system in ascomycetes was heterothallic (Fraser et al., 2007; Inderbitzin et al., 2005; Rydholm et al., 2007; Yun et al., 1999). In at least one case, recent transitions from heterothallism to a derived homothallic state are indicated. Comparison of the sequence and organization of the *MAT* locus among closely related species within the *Cochliobolus* genus suggests that the mechanism for the heterothallism → homothallism transition was an unequal crossover event in a heterothallic *MAT* progenitor that resulted in the fusion of the opposite *MAT* locus alleles (Yun et al., 1999).

Meiosis Overview

Hypothetically, syngamy might occur between cells of any ploidy level as long as they were of compatible mating types. In reality, though, most organisms show a regular alternation between haploid and diploid states. A key mechanism by which ploidy is reduced, thus allowing syngamy to reoccur at a later time, is meiosis. Meiosis consists of a single round of DNA replication, one cell division that separates homologous chromosomes (usually associated with a very high level of interhomolog recombination), and a second cell division that separates sister chromatids. Like most traits in evolution, the molecular requirements for meiosis are generally conserved across eukaryotes but also exhibit frequent derived variations. The following "consensus" mechanism is largely based on a core set of meiosis genes conserved across eukaryotes (Malik et al., 2008; Ramesh et al., 2005; Villeneuve and Hillers, 2001).

Meiosis is initiated by transduction of external and/or internal chemical signals; these signals can originate from various sources, including local crowding (e.g., monogonont rotifers; Stelzer and Snell, 2003), a deficit of essential nutrients (e.g., *S. pombe*; Yamamoto, 1996), or an indication that the organism is a diploid (e.g., the **a**1/α2 complex encoded by the opposite mating-type loci in diploid *S. cerevisiae*; Vershon and Pierce, 2000). The signal transduction eventually leads to the activation or removal of repression of a "master" transcriptional regulator (removal of Rme1 inhibition and activation of Ime1 in *S. cerevisiae*, Ste11 in *S. pombe*; Vershon and Pierce, 2000; Yamamoto, 1996) that initiates transcription of numerous other meiosis genes (Yamamoto, 1996). The details of the meiosis initiation process and its components are not typically conserved.

By contrast, the processes and genes involved in meiotic recombination are generally well-conserved across eukaryotes (Malik et al., 2008; Schurko and Logsdon, 2008). In prophase I, the replicated sister chromatids are bound by cohesin complexes consisting of Smc1, Smc3, and either Rad21/Scc1 (which also functions in mitosis) or Rec8 (which is meiosis specific) (Page and Hawley, 2004). The synaptonemal complex (composed of Hop1 and other proteins) also forms between homologous chromosomes at this time and aids in synapsis (Page and Hawley, 2004). Importantly, the topoisomerase Spo11 creates numerous double-stranded DNA breaks that all must be repaired via intersister or interhomolog recombination before meiosis can proceed (Keeney et al., 1997). Intersister repair (as is the norm during mitosis) proceeds by Rad50- and Mre11-dependent strand resection, coating of single-stranded DNA with replication protein A (RPA), and displacement of RPA by Rad51 with the assistance of several accessory factors (Rad55, Rad57, Rad52) (Gasior et al., 1998; Malik et al., 2008). Intersister repair/recombination does occur during meiosis, but in *S. cerevisiae*, it is partially suppressed by Mek1 (Wu et al., 2010). Conversely, interhomolog repair/recombination occurs at low frequencies during mitosis and is promoted during meiosis by proteins that aid in strand invasion (Dmc1, Hop2, Mnd1) and stabilization/processing of crossovers (Msh4, Msh5, Mer3) between homologous chromosomes (Malik et al., 2008; Nakagawa and Kolodner, 2002; Nishant et al., 2010; Petukhova et al., 2005).

The processes following prophase I vary in their degree of conservation across taxa. Progression out of prophase I can be governed by the pachytene checkpoint, which causes cell cycle arrest if unrepaired double-stranded DNA breaks persist and involves such proteins as Pch2 and Mek1 in *S. cerevisiae* (Wu et al., 2010; Zanders et al., 2011). Lack of inhibition by the pachytene checkpoint triggers additional transcription factors (e.g., Ndt80 in *S. cerevisiae*; Hepworth et al., 1998) that allow progression into metaphase I, although the specific proteins are not highly conserved. In anaphase I, homologous chromosomes are separated by the enzyme separase cleaving most of the cohesin complexes; premature sister chromatid separation is prevented by shugoshin/Sgo1 blocking cleavage of Rec8-containing cohesin complexes at the centromeres (Ishiguro et al., 2010; Katis et al., 2010; Watanabe, 2005). The final unique feature of meiosis compared to mitosis is the transition from meiosis I to meiosis II without a second round of DNA replication; the molecular mechanisms of this stage are still being characterized, but modifications to cell cycle regulation of protein expression (e.g., Kimata et al., 2011) are likely necessary to essentially begin a second M phase immediately and to skip the S phase.

DETECTING SEXUAL REPRODUCTION

Eukaryotic microbes can reproduce both sexually and asexually (facultative sexuality), with further implications to be discussed later. Sexual reproduction generates genetic diversity by recombination of homologous chromosomes during meiosis. This may give rise to novel genotypes that are better adapted to changing environmental niches. However, sex is costly in terms of both time and resources, and well-adapted genotypes may be broken apart by meiotic recombination (Sun and Heitman, 2011). While sexual reproduction is ubiquitous across eukaryotes as a whole, many eukaryotic microbes—many of them well-established pathogens—have been observed to reproduce only asexually. This outcome has two possible explanations: (i) The given species is exclusively asexual or (ii) the given species reproduces primarily asexually but occasionally sexually. Distinguishing these possibilities has historically been difficult. Direct observation of mating or sexual cycles is often challenging because appropriate mating conditions elude definition, especially for species that lack closely related sexual species as a guide. However, with rapid progress in genomics and genetics, an increasing number of eukaryotic microbes have been found to be cryptically sexual. These advances are based on two general types of approaches to detect evidence of the sexual cycle of eukaryotic microbes: direct mating assays and indirect approaches that assess the potential capability for sexual reproduction or its consequences. The indirect approaches discussed here are identification of genes crucial to sexual reproduction (e.g., meiosis genes, *MAT* genes) and population genetics studies to detect genetic recombination.

Meiosis Detection Toolkit

As mentioned above, meiosis is a key hallmark of sexual reproduction, and identification of conserved meiotic genes in the genome is an alternative approach to look for evidence of cryptic sexual cycles in eukaryotic microorganisms. If a gene is essential for a given biological process, the retention of that gene would be consistent with the retention of that biological process. Therefore, the presence of multiple genes known to function in meiosis in the genome of an organism can imply the presence of meiosis at some point in the species' life history (Schurko and Logsdon, 2008).

A commonly used form of ortholog identification is searching genome sequence data. However, in organisms for which genomic sequence is not available, amplification of putative meiotic genes using degenerate primers can be designed on the basis of sequence alignments among closely related species as another way to search for evidence of a sexual cycle (Fig. 2.1).

Many genes necessary for meiosis also function in mitosis (e.g., Smc1, Smc3, Rad21/Scc1, Rad50, Mre11, Rad51; Malik et al., 2008); therefore, their presence in a genome gives no indication as to whether that organism is capable of meiosis (Schurko and Logsdon, 2008). However, other genes encode proteins that function exclusively in meiosis in nearly all organisms for which functional or expression analysis has been conducted (e.g., Dmc1, Hop2, Mnd1, Msh4, Msh5, Pch2, Mek1, Hop1, Mer3, Rec8; Malik et al., 2008; Rockmill and Roeder, 1991; San-Segundo and Roeder, 1999; Schurko and Logsdon, 2008). Eight meiosis-specific genes in particular (Spo11, Hop1, Hop2, Mnd1, Rec8, Dmc1,

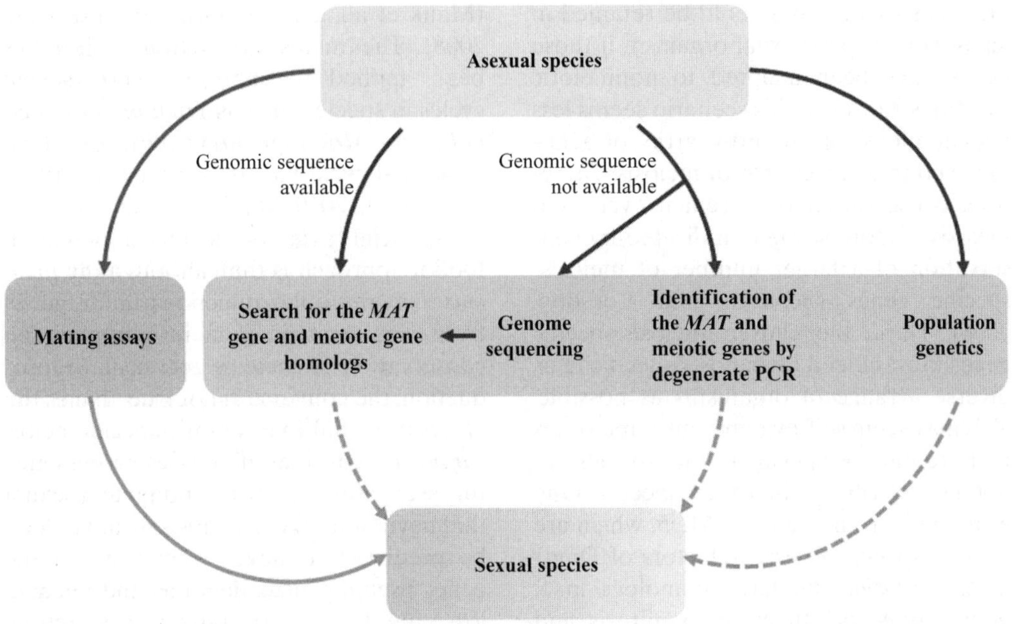

Figure 2.1 Identifying sexual cycles of microbial eukaryotes. Mating assays can provide direct evidence (solid green line) of sexual reproduction based on the observation of sexual structures and the generation of sexual spores. However, mating assays are often challenging because appropriate mating conditions can elude definition, especially for species lacking closely related known sexual species as a guide. Thus, indirect approaches (broken green lines), such as searching for *MAT* locus genes and meiotic gene homologs in the genome, as well as population genetic studies, are practical methods to provide evidence of sexual development in species with currently unknown sexual cycles. The identification of genes that may be required for cell type identity, mating, and meiosis can also support the existence of an extant sexual cycle. Population genetic studies can be informative—a recombinant population may be the result of sexual reproduction, indicating that a given species may reproduce sexually in nature. See color insert.

Msh4, Msh5) are widely sought to predict the capacity for a sexual cycle because these genes are generally conserved in presence and sequence across eukaryotes (Schurko and Logsdon, 2008). The presence of these "meiosis-specific" genes in an organism's genome can often be strong evidence for the presence of meiosis and canonical sexual reproduction at some point in the species' life history (Schurko and Logsdon, 2008).

To make the inference of sexual reproduction based on the presence of meiosis-specific genes, several criteria must be met. First, the orthology of the putative meiosis-specific gene must be validated by phylogenetic analysis. Many of the meiosis-specific genes have paralogs that function in mitosis alone or in both mitosis and meiosis (e.g., Dmc1 and Rad51, Rec8 and Rad21/Scc1, Msh1-6; Malik et al., 2008; Schurko and Logsdon, 2008). These mitotic paralogs are useful because they can be used as outgroup sequences in phylogenetic analysis and allow more rigorous identification of orthology (Schurko and Logsdon, 2008). However, mitotic paralogs also complicate bioinformatic surveys because phylogenetic analysis is implicitly required to reliably conclude the conserved presence of a meiosis-specific gene (implying meiosis) versus a gene with mitotic functions (allowing no conclusion about meiosis). Second, multiple meiosis-specific genes from functionally diverse stages of meiosis must be present. Conceivably, a few individual meiosis-specific genes could be retained in an asexual or parasexual organism if those genes have been adapted to nonmeiotic functions; however, this scenario seems less parsimonious for an entire array of genes involved in diverse steps of meiosis. Therefore, conservation of a sexual cycle and meiosis is more strongly indicated by conservation of a larger number of meiosis-specific genes (Schurko and Logsdon, 2008). Third, the chosen meiosis-specific genes must indeed be meiosis-specific in as diverse a range of organisms as possible. This may seem self-evident, but some genes that are meiosis-specific in one species are not necessarily so in other species. One example is from Sae3 and Mei5, which are meiosis-specific accessory factors of Dmc1 in *S. cerevisiae* but have homologs in *S. pombe* that also function in mitosis and mating-type switching (Haruta et al., 2008). A second example is Ndt80. *S. cerevisiae* has a single copy of this gene, which is a meiosis-specific transcription factor (Hepworth et al., 1998). By contrast, *Neurospora crassa* has Ndt80 and two additional paralogs, none of which are necessary for the completion of meiosis, though they are still implicated in sexual development (Hutchison and Glass, 2010). Fortunately, most of the "meiosis identification toolkit" genes proposed by Schurko and Logsdon (2008)—Spo11, Rec8, Msh4, Msh5, Dmc1, Hop2, Mnd1, Hop1—have not yet been found to have such dramatic changes in function. Where isolated acquisitions of additional functions have occurred (e.g., Msh5 in *Mus* immunoglobin class switching; Sekine et al., 2007), they have not been associated with the loss of the (ancestral) meiotic functions (see De Vries et al., 1999 for *Mus* Msh5). When these considerations are taken into account, the presence of multiple functionally diverse meiosis-specific gene orthologs in a species lends strong support to the hypothesis that meiosis—or a parasexual process very similar to meiosis (parameiosis)—is present (Malik et al., 2008; Schurko and Logsdon, 2008). The meiosis detection toolkit has been applied to support cryptic sexual cycles in species such as *Trichomonas vaginalis* and *Monosiga brevicollis*, for which no sexual cycle has as yet been described (Carr et al., 2010; Malik et al., 2008).

A crucial caveat of the meiosis detection toolkit approach is that, although the *presence* of multiple meiosis-specific genes from diverse steps of meiosis implies the conservation of meiosis and sexual reproduction, the converse is not true; that is, the *absence* of multiple meiosis-specific genes *cannot* be considered conclusive evidence for asexuality. A "true" obligate asexual that never undergoes meiosis would indeed be predicted to have its meiosis-specific genes pseudogenize, degrade, and become lost due to loss of purifying selection. However, meiosis-specific genes have been lost multiple times in known sexual and meiotic organisms, among them *Schizosaccharomyces*, *Neurospora*, *Drosophila*, *Caenorhabditis*, and *Candida lusitaniae* (Butler et al., 2009; Malik et al., 2008; Reedy et al., 2009; Schurko and Logsdon, 2008). Some meiosis-specific genes are lost more often than others: Dmc1, Hop2, Mnd1, Msh4, and Msh5 have been independently lost in several lineages, while no strong evidence for loss of Spo11 has been obtained in any organism yet examined (Malik et al., 2008). This suggests that some genes may be more "indispensable" for meiosis than others. For example, Dmc1 loss-of-function mutants in *S. cerevisiae* (which have greatly reduced sporulation) can be rescued by overexpression of Rad51 (Tsubouchi and Roeder, 2003), which presents a possible mechanism by which some lineages have lost this gene without losing meiosis. By contrast, the function of Spo11—formation of double-stranded DNA breaks (Keeney et al., 1997; Malik et al., 2008)—is not readily provided by other enzymes during meiosis, so its loss may be more difficult (or nearly impossible) to tolerate without also losing meiosis. Alternatively, the nearly pan-eukaryote conservation of some meiosis-specific genes (e.g., Spo11) could imply that they have additional, nonmeiotic functions yet to be determined (Malik et al., 2008). A related caveat is that the hypothesis that true obligate asexuals will have lost all their meiosis-specific genes has yet to be tested for two reasons. First, most putative obligately asexual species (Schurko et al., 2009) do not yet have sequenced genomes to allow comparative genomic analysis of meiosis-specific genes. Second, ascertaining that an organism is exclusively asexual versus facultatively sexual or parasexual can be very challenging (Schurko et al., 2009). The first problem will likely be overcome in time by increasingly cost-effective genome sequencing; the second is more fundamental and difficult to conclusively solve.

Another limitation to the meiosis detection toolkit approach of Schurko and Logsdon (2008) is that its comparative genomics approach (identifying orthologs of meiosis-specific genes conserved in other organisms) is most effective at identifying the presence of conserved meiosis genes. Putative cases of meiosis gene loss can be identified (e.g., if meiosis-specific paralogs are undetectable but non-meiosis-specific paralogs are). However, one must always consider alternative, methodological explanations for failure to detect a given gene: "absence of evidence" is not "evidence of absence." The toolkit approach would be least effective at identifying cases of novel, lineage-specific meiosis gene gain—by definition, such genes would not have orthologs in better-studied model species and could not be identified by queries from a known meiotic gene ortholog sequence. Identification of meiosis genes unique to particular lineages will more likely happen by utilizing transcriptional expression patterns (meiotic upregulation or downregulation), improved prediction of conserved interaction domains with more widely conserved proteins, and meiotic mutant phenotype screens with subsequent molecular characterization. For example, monitoring the expression of meiotic genes (and/or *MAT* genes) under potential mating conditions compared to vegetative growth conditions could detect signs of sexual development.

MAT Locus Identification

The same principles of the meiosis detection toolkit can be applied to identify orthologs of the *MAT* locus by whole-genome sequencing (or polymerase chain reaction [PCR] amplification) and comparative genomics, as exemplified by the identification of the *MAT* locus in the basal zygomycetous fungi such as *Phycomyces blakesleeanus* and Microsporidia (Idnurm et al., 2008; Lee et al., 2008; Fig. 2.1). Little was known at the genetic level about how sex determination is controlled in zygomycetes, even though sexual development has been widely studied in two other fungal phyla (Basidiomycota and Ascomycota). Searching for the sex genes in the genomic sequence of one isolate of *P. blakesleeanus* led to the identification of a homolog of the HMG domain gene in the *MAT/sex* locus that is related to those employed in sex determination in some ascomycete species. The second *Phycomyces MAT* allele was amplified and sequenced from isolates of opposite mating type. Comparison analyses demonstrated that the sexual identity of *P. blakesleeanus* is controlled by a single HMG domain gene with two idiomorphs (Idnurm et al., 2008).

PCR amplification of the opposite *MAT* locus alleles by primers in the highly conserved regions flanking known *MAT* locus alleles and sequencing by primer walking are also commonly applied to identify the second *MAT* locus idiomorph in ascomycetes. *Microsporum gypseum* is a geophilic dermatophyte species. While *M. gypseum* was long known to undergo sexual reproduction, the molecular nature of the *MAT* locus was unknown. Access to the genome sequence for dermatophytes has significantly advanced the characterization of their *MAT* locus. Using comparative genomics approaches, one *MAT* locus allele was identified in the *M. gypseum* genome and was found to exhibit a distinct gene order compared to closely related dimorphic fungi (Li et al., 2010). Amplification using primers to the flanking genes (*APN2*, *SLA2*, and *COX13*) and primer walking identified the opposite *MAT* locus allele. Compared with the related dimorphic fungi, the *MAT* locus of the dermatophytes has a reduced size (~3.5 kb) and altered gene order, which sheds light on the evolution of the *MAT* locus in these two closely aligned groups of important human fungal pathogens (Li et al., 2010).

Other Indirect Approaches

During meiosis, the parental genomes are shuffled to generate recombinant progeny. Sexual species have a higher rate of genetic recombination, which can be detected by population genetics studies employing appropriate genetic markers. Sexual species are typically more genetically diverse than asexual species. In addition, sexual populations have an undifferentiated "brushlike" phylogenetic organization compared to asexual populations, which form highly structured and well-resolved phylogenetic trees (Campbell and Carter, 2006). For example, population genetic studies provide evidence that *Coccidioides immitis*, a dimorphic human fungal pathogen, is sexual even though this has not yet been observed in the lab or in nature (Burt et al., 1996). The closely aligned nonpathogenic species *Uncinocarpus reesii* does have a defined sexual cycle, which may serve as a roadmap for *Coccidioides* (Bowman et al., 1996). Population genetic studies of another human fungal pathogen, *C. neoformans*, suggest **a**–α sexual reproduction occurs in Botswana, where both **a** and α mating types are present (Litvintseva et al., 2003, 2006). In addition, the results of population genetic studies support that same-sex mating occurs in both *C. neoformans* and *C. gattii* (Bui et al., 2008; Hiremath et al., 2008; Lin et al., 2007, 2009; Saul et al., 2008).

Case Studies of Sexual Cycle Identification

With the increasing number of genomic sequences and rapid progress in DNA sequencing technologies, many eukaryotic genome sequences are now available. Searches for both mating-type genes and meiosis-specific genes are a powerful approach to identify the capacity for cryptic sex. *Aspergillus fumigatus* had been regarded as an asexual fungal species until the identification of the *MAT* locus and meiotic genes in the genome suggested sex may be extant (Galagan et al., 2005; Poggeler, 2002). Subsequently, mating assays conducted in the dark on oatmeal agar for 6 months led to the discovery of an extant sexual cycle of *A. fumigatus* that produces viable, fertile, recombinant progeny (O'Gorman et al., 2009). Despite these successes, many other pathogenic fungi are only known to undergo clonal, asexual reproduction. For example, for *C. immitis* and *Candida glabrata*, there is evidence for genetic recombination and gene flow based on population genetic studies, and genomic evidence for sexual machinery, although no sexual cycle has as yet been observed in nature or in the laboratory (Burt et al., 1996; Dodgson et al., 2005; Lott et al., 2010; Wong et al., 2003). *C. albicans* has been observed to mate and to undergo a parasexual cycle, including chromosome loss and Spo11-dependent recombination, but no true meiosis has as yet been reported (Forche et al., 2008). Population genetic evidence does provide hints of recombination in the *C. albicans* population (Graser et al., 1996).

An illustrative example comes from studies of *Paracoccidioides brasiliensis*, a dimorphic fungal species causing disease in Latin America. Since no known sexual cycle has been observed for *P. brasiliensis*, it has long been regarded as asexual. However, population genetics studies identified three discrete "species," S1, PS2, and PS3, in *P. brasiliensis* and provide evidence of recombination in the S1 lineage (Matute et al., 2006). Recent genomic sequencing of three *P. brasiliensis* isolates at the Broad Institute and identification of the *MAT* locus and meiotic genes suggest that this fungus may undergo sexual reproduction (Li et al., 2010; Torres et al., 2010). Both mating types are present in the *P. brasiliensis* population, consistent with a sexual nature (Torres et al., 2010). Even though no mature sexual structures have been observed, the increased expression of *MAT* locus genes and observation of premature sexual structures during coculture of isolates of opposite mating types strongly suggest an extant sexual cycle of *P. brasiliensis* remains to be discovered (Torres et al., 2010).

Given that the *MAT* locus alleles are the major genetic difference between isolates of opposite mating types, comparative genome hybridization (CGH) is an efficient way to identify the *MAT* locus, as demonstrated in recent studies for the social amoeba *Dictyostelium discoideum* (Bloomfield et al., 2010). Unlike most eukaryotes that have two mating types, *D. discoideum* has three mating types and some isolates are even self-fertile. The unusual mating types of *D. discoideum* are intriguing, but the molecular nature of the *MAT* locus that controls sexual reproduction had been previously unknown. The *MAT* locus that defines one mating type (type I) and is absent in another (type II) was pinpointed by CGH assays between *D. discoideum* isolates of different mating types. Two small proteins (107 or 208 amino acids) are encoded by genes in the *MAT* locus and control the sexual identity of *D. discoideum*. Type I mating type expresses one of these small proteins, and type III the other, allowing them to mate with each other. Type II is the interesting third mating type, and it expresses allelic variants of both the type I and type III sex determinants. An individual with the third mating type is fertile with both type I and type III but not with itself or other type II individuals. This study provides a novel model for the evolution of sex in eukaryotes: A third mating type can emerge by conscripting the key sex determinants of both sexes followed by mutations that prevent self-fertility but retain cross-fertility.

EVOLUTIONARY IMPACTS OF SEXUAL REPRODUCTION

Evolution cannot happen unless a population has heritable genetic variation. Although mutations are the ultimate source of variation in both genome sequence and genome structure, a profoundly important additional source of genetic variation is the reassortment of extant alleles from genetically different individuals into novel combinations. This "shuffling" of alleles can free beneficial alleles from an otherwise deleterious genetic background (Peck, 1994) and, separately, can produce novel phenotypes not observed in previous genotypes. Such allelic reassortment can occur by numerous mechanisms, but sexual reproduction is arguably the most widespread method. As a result, whether an organism undergoes sexual reproduction and, if so, at what frequency are important factors in understanding its evolution.

Frequency of Sexual Reproduction

Sexual and asexual reproduction each exhibit advantages and disadvantages. Sexual reproduction generates progeny with a wider range of genotypes (and phenotypes); this is advantageous if one's own fitness is low in the current environment due to genetic drift or fluctuations in selective pressures (Hadany and Comeron, 2008; Hadany and Otto, 2007; Otto and Gerstein, 2006). Yet, sex would appear to be disadvantageous if one's fitness is already high in a little-changing environment (Otto and Gerstein, 2006). Furthermore, outcrossing results in only half of one's genetic material being transmitted, while self-fertilization after meiosis can increase homozygosity of deleterious recessive alleles (Archetti, 2004b). Asexual reproduction avoids these costs as well as the time and energy expenditures of finding a mate (Otto and Gerstein, 2006). However, asexual reproduction can generate novel genetic variation only by mutations, mitotic interhomolog recombination, and (if applicable) parasexual events. This reduction of variation is predicted to make adaptation a challenge for populations in a rapidly changing biotic or abiotic environment (Lively, 2010). In addition, asexual populations also are predicted to accumulate deleterious mutations (Muller, 1964). Given that each strategy has costs and benefits, it should not be terribly surprising that eukaryotes exhibit a wide continuum of sexual reproductive modes and frequencies.

Eukaryotes can be obligately sexual, obligately asexual, and anything in between (Table 2.1). Because both sexual and asexual reproduction can be beneficial under different circumstances, it is not unexpected that many eukaryotes are *facultatively sexual*, capable of both sexual and asexual reproduction (Dacks and Roger, 1999). On an individual level, one could consider the "optimal" frequency of sexual and asexual reproduction to be a function of how much reproductive fitness would be obtained by reproducing by each strategy under the conditions at a given time. Depending on the conditions, that optimum may come from reproduction that is an equal proportion of sexual and asexual, primarily asexual and only rarely sexual, or primarily sexual and only rarely asexual. The wide distribution of at least occasional sexual reproduction across eukaryotes (Dacks and Roger, 1999) predicts that the "optimal" fitness function of most species, on average, would be maximized by having some nonzero amount of sexual reproduction. In other words, facultative sex apparently provides some fitness advantage over being obligately sexual or asexual in most cases.

Despite the expected advantages of retaining the flexibility of both sexual and asexual reproduction, *obligately asexual* species are thought to exist. Many are recently derived lineages that are predicted to succumb to the expected long-term disadvantages of obligate asexuality (Neiman et al., 2009). However, a few are "ancient asexuals" that appear to have persisted for several million years without sexual reproduction (Neiman et al., 2009). One explanation is that these are actually facultatively sexual species that have extremely rare, cryptic sexual reproduction yet to be discovered (Schurko et al., 2009). A second explanation is that parasexual mechanisms provide sufficient genotypic novelty for selection to act upon; a possible example could be the large amount of horizontal gene transfer contributing to bdelloid rotifer genomes (Gladyshev et al., 2008). A third consideration is that many arguments for the disadvantages of asexual reproduction assume that selection will be inefficient on both deleterious and beneficial mutations—resulting in deleterious mutation accumulation and slow adaptation, respectively (Barraclough et al., 2007). However, increased population size is associated with increased effectiveness of selection. The exact relationship between population size and asexuality is not clear—large population sizes may confer tolerance to obligate asexuality as a strategy and/or the rapid reproduction of obligate asexuals

TABLE 2.1 The Continuum of Sexual and Asexual Reproductive Strategies in Eukaryotes*

Strategy	Definition	Alternating Ploidy?	Meiosis?	Outcrossing?	Examples	Advantageous When . . .	Disadvantageous When . . .
1. Exclusively sexual	All reproduction requires meiosis and syngamy; parasexual processes absent	Yes	Yes	Varies	All mammals, most vertebrates (Engelstadter, 2008)	Biotic environment changes rapidly, costs are negligible (Engelstadter, 2008)	Environment is stable; extant high fitness genotypes disrupted by recombination; costs of sex are pronounced (Barton and Charlesworth, 1998; Otto and Gerstein, 2006)
2. Facultatively sexual	Sexual and asexual reproduction both occur in species' life history; strategy can vary cyclically or in response to environmental cues	Yes	Yes	Varies	Most plants, fungi, protists (Dacks and Roger, 1999)	Conditions favoring each reproductive mode occur	Costs of sex are pronounced and competing against exclusive asexuals
2A. "Mostly sexual"	Primary strategy is sexual reproduction, only occasionally asexual	Yes	Yes	Varies	Komodo dragon (Watts et al., 2006), *Nauphoeta* cockroach (Corley et al., 1999)	Sex is generally favored but mate access situationally limited (Watts et al., 2006)	Developmental constraints produce offspring with very low fitness (and sex was an available alternative) (Corley et al., 1999)

(*Continued*)

TABLE 2.1 (*Continued*)

Strategy	Definition	Alternating Ploidy?	Meiosis?	Outcrossing?	Examples	Advantageous When . . .	Disadvantageous When . . .
2B. "Mostly asexual"	Primary strategy is asexual reproduction, only occasionally sexual	Yes	Yes	Varies	Many protists (Dacks and Roger, 1999)	Environment has long periods of stasis and occasional changes	Costs of sex are pronounced and competing against exclusive asexuals
3. Parasexual	No canonical sexual reproduction but have additional sources of ploidy alternation and/or genetic recombination not present in exclusive asexuals. Can be obligate or facultative	Varies	Varies	Varies	*Candida albicans* (Forche et al., 2008)	Some amount of aneuploidy circumstantially advantageous (Pavelka et al., 2010; Rancati et al., 2008; Selmecki et al., 2006, 2008; Sionov et al., 2010)	Aneuploidy highly deleterious
4. Exclusively asexual	All reproduction is by mitosis, budding, vegetative growth, and so on. Novel genotypes arise solely through mitotic recombination and mutations.	No	No	No	Some *Daphnia*, snail, aphid lineages (Neiman et al., 2009); bdelloid rotifers, oribatid mites (Schurko et al., 2009)	Efficient selection and/or little-changing environment (Barraclough et al., 2007)	Selection inefficient on beneficial or deleterious mutations, environment changes rapidly (Barton and Charlesworth, 1998; Engelstadter, 2008; Lively, 2010; Peck, 1994)

*Major reproduction strategies shown by eukaryotes (with a noncomprehensive list of exemplar taxa) are reviewed.

may result in large population sizes. Nevertheless, as long as sufficient amounts of genetic diversity are generated (e.g., mitotic recombination, mutation, parasexuality) and exposed to efficient selection via haploid life stages and/or homozygosis of recessive alleles, obligate asexuals may not always be doomed to rapid extinction.

Obligately sexual species also exist (e.g., humans), but the explanations for their existence vary. Although sexual reproduction can indeed confer higher reproductive fitness than asexual reproduction, it is difficult to envision conditions in which retaining the capability for at least occasional asexual reproduction would be disadvantageous (Hadany and Otto, 2007). Consistent with this expectation, obligate sexuals have a rather limited phylogenetic distribution that consists primarily of multicellular animals (D'Souza and Michiels, 2010). This could suggest that obligate sexuality is optimal under only very specific conditions, for example, under sexual selection (Hadany and Otto, 2007). Another consideration is that the effects of genetic drift could contribute to many individuals having suboptimal fitness in their current environment (Otto and Gerstein, 2006); if this effect is frequent and pervasive, fitness-dependent recombination (Hadany and Otto, 2007) may have contributed to high frequencies of sex in these species. However, this would not necessarily require a population to transition from "frequently sexual" to "obligately sexual." Complete absence of asexual reproduction may therefore be linked to biological constraints in some lineages—required fusion of anisogamous gametes to induce development and/or inviability of progeny that do not inherit genomic imprinting from each of two different parental sexes are two such possible contributing factors (Engelstadter, 2008).

Sex and Genetic Variation

Genetic variation, or different arrangements of alleles and generation of novel genotypes, is generated by numerous stages of the typical sexual reproduction process. Some genotypic rearrangement is obtained directly by the independent assortment of homologous chromosomes during meiosis or by stochastic chromosome loss in some parasexual cycles (Forche et al., 2008). Therefore, even an organism that reproduced exclusively by self-fertilization (fusion of gametes produced by the same individual) would produce genetically novel offspring because the gametes would have different combinations of the diploid parent's chromosomes. For example, a diploid parent may have four chromosomes in two copies each (A1, A2, A3, A4; B1, B2, B3, B4). One meiosis may generate (A1, B2, B3, B4) and (B1, A2, A3, A4) gametes, while an independent meiosis may produce (A1, A2, B3, B4) and (B1, B2, A3, A4) gametes. Fusion of gametes generated by these independent but concurrent meiotic divisions could generate progeny homozygous for entire chromosomes (e.g., a diploid with chromosomes A1B1 B2B2 B3A3 B4A4 formed by fusion of the first and last gametes described in the previous sentence). Parasexual chromosome loss also generates additional variation in the form of the aneuploid intermediate stages (Bennett and Johnson, 2003; Forche et al., 2008; Ni et al., 2011; Reedy et al., 2009). Aneuploidy is usually assumed to be deleterious, but several fungal species show higher fitness under certain stress conditions if aneuploid (Pavelka et al., 2010; Rancati et al., 2008; Selmecki et al., 2006, 2008; Sionov et al., 2010). Finally, alternating ploidy can be directly advantageous if haploids and diploids have differential benefits in certain nutrient environments, so genetic variation may not be the sole benefit (Otto and Mable, 1998).

Another source of genetic variation generated by sexual reproduction is its frequent—but not always obligate—association with outcrossing, the mixing of chromosomes from two genetically distinct individuals. Outcrossing is one of the major costs of sex because each individual will have only half the genetic representation in the offspring that an asexual, clonally reproducing individual would (Maynard Smith, 1978). However, outcrossing can also be a beneficial "gamble"—if an individual is poorly adapted to its environment, it could obtain greater reproductive fitness by producing recombinant progeny with another individual (Hadany and Otto, 2007). In addition to being supported by models (Hadany and Otto, 2007), fitness-dependent recombination has been observed in *Aspergillus nidulans* (Schoustra et al., 2010).

New genotypes can be generated by simply outcrossing and alternating between haploid and diploid stages. However, all known sexual species incorporate some nonzero amount of recombination between homologous chromosomes in at least one sex during meiosis or the equivalent parasexual chromosome reduction cycle (Forche et al., 2008; Rauwolf et al., 2011). Genetic recombination can occur between either sister chromatids (which are genetically identical) or homologous chromosomes (which can have allelic differences), but only recombination between homologous chromosomes produces novel rearrangements of alleles. At least one interhomolog crossover per chromosome is mandatory for successful completion of meiosis in nearly all species (Jones and Franklin, 2006; Martini et al., 2006), since the physical tension of the crossover on the spindle facilitates homolog disjunction and ensures segregation of one homolog per nucleus. This minimum crossover requirement guarantees a very high rate of interhomolog recombination in sexual species compared to obligate asexuals. Some species experience only this minimum number of crossovers per meiosis (e.g., *Caenorhabditis elegans*; Barnes et al., 1995), but others have substantially more homologous recombination per meiosis than is minimally necessary for viable haploid offspring (e.g., *S. cerevisiae*) (Nishant et al., 2010). This may suggest that some lineages may have experienced selection for an increased amount of interhomolog recombination, implying that recombination can be important for reasons beyond the mechanistic constraints of meiosis. Interhomolog recombination can also occur outside of meiosis, such as during mitotic cell divisions, but at a much reduced frequency (500–1000× rarer) than in meiosis (Pontecorvo, 1956). Despite this low rate, mitotic interhomolog recombination has been implicated in contributing to adaptive fixation of beneficial alleles in both theoretical (Mandegar and Otto, 2007) and empirical studies (Schoustra et al., 2007).

Sex and Virulence in Eukaryotic Microbes

Pathogens could be subject to the same benefits of sexual reproduction as other species, such as removal of deleterious mutations from the genome and the generation of population diversity (Sun and Heitman, 2011). Similarly, meiotic recombination during sexual reproduction is considered a major driving force for diversification in natural populations of eukaryotic microbes (Bachtrog, 2003; Hsueh and Heitman, 2008). The generation of genetic variation may produce individuals with altered virulence or increased fitness in new ecological niches (Bachtrog, 2003; Hsueh and Heitman, 2008), which is particularly crucial for pathogenic species. Given that sex is costly in terms of both time and resources, however, a central conundrum is why so many eukaryotic microbes still undergo sexual reproduction. Under appropriate conditions, pathogens would be expected to benefit from asexual reproduction just as do many nonpathogenic species (Table 2.1). Nevertheless, evidence so far is suggestive of a strong significance for sexual reproduction in the virulence of many pathogens.

Sexual reproduction is pervasive and is directly or indirectly linked to virulence in many fungal pathogens. First, sexual spores are hypothesized to be important infectious propagules in some pathogenic fungi, such as the human fungal pathogen *C. neoformans* (Velagapudi et al., 2009), and possibly also for the aquatic chytrid fungus *Batrachochytrium dendrobatidis*, the cause of global amphibian decline and extinction (Morgan et al., 2007). Second, mating type is linked to virulence in *C. neoformans* (Kwon-Chung et al., 1992). Third, sexual development produces offspring in which the parental genetic material has been recombined and can generate new phenotypes affecting virulence. Several exemplar cases are here discussed in more detail.

Case Studies of Sex Affecting Virulence

The potential impact of sexual reproduction has been widely studied in both fungi and other eukaryotic parasites. *C. gattii* is a sibling species of *C. neoformans*, and while both species can infect both immunocompromised and immunocompetent hosts, *C. neoformans* causes most opportunistic infections in HIV/AIDS patients. Four molecular types of *C. gattii* have been described: VGI–VGIV. *C. gattii* VGI and VGII predominantly cause infections in immunocompetent individuals, whereas the VGIII and VGIV molecular types are responsible for most *C. gattii* infections in HIV/AIDS patients. *C. gattii* VGII molecular type is causing an outbreak that began on Vancouver Island in 1999 and has since expanded to the Pacific Northwest region in both Canada and the United States (Byrnes et al., 2009, 2010; Kidd et al., 2004). The major outbreak genotype, VGIIa, is highly clonal, but multilocus sequence typing (MLST) and the identification of a diploid isolate of the same mating types provide evidence that the outbreak genotype may have been generated via same-sex mating to cause the outbreak and its expansion (Fraser et al., 2005). Recently, studies revealed that the *C. gattii* VGIII molecular type is another common agent causing cryptococcosis in HIV/AIDS patients in Southern California (Byrnes et al., 2011). The VGIII population includes two distinct sublineages, VGIIIa and VGIIIb, which differ in virulence in animal models, although both can cause infections in HIV/AIDS patients. Many VGIII isolates are fertile and recombination has been detected in the VGIII global population (Byrnes et al., 2011). These findings suggest sexual reproduction may be an important force maintaining genetic diversity and virulence of the *C. gattii* VGIII population that is causing disease in HIV/AIDS patients.

Toxoplasma gondii is a protozoan parasite infecting warm-blooded vertebrates. *T. gondii* reproduces both sexually and asexually and has a clonal population structure composed principally of three lineages. MLST suggests that the currently predominant genotypes of *T. gondii* originated from sexual recombination between two distinct ancestral isolates (Grigg et al., 2001). Genomic single-nucleotide polymorphism (SNP) typing provides further evidence that a single cross can lead to the emergence and dominance of a novel clonal genotype that may increase prevalence and is linked to outbreaks (Boyle et al., 2006). These studies highlight that genetic exchange via sexual reproduction among extant parasite lineages produces new genotypes that emerge to expand the parasite's host range and cause outbreaks, and in some cases, this involves selfing of highly clonal isolates (Grigg et al., 2001; Heitman, 2006, 2010; Su et al., 2003; Wendte et al., 2010).

Giardia is a single-celled protozoan parasite that colonizes the small intestines of several vertebrates—including humans—to cause giardiasis. No evidence for sexual reproduction of *Giardia* had been found, and it was thought to have lost the ability to undergo sexual reproduction (Adam, 2001). However, population genetic studies and identification of meiotic genes in the genome first suggested that *Giardia* might reproduce sexually (Cooper et al., 2007; Ramesh et al., 2005). The findings of the potential for a sexual cycle prompted the identification of an unusual sexual cycle involving nuclear fusion and genetic exchange between the paired nuclei of this dikaryotic organism, termed diplomixis (Poxleitner et al., 2008).

The parasite protozoan *Trypanosoma cruzi* causes about 20 million cases of Chagas' disease in the Americas. While the *T. cruzi* population structure seems to be highly clonal, *in vivo* experiments revealed extensive genetic exchange or recombination occurred when two biological clones of *T. cruzi* were passaged together through their entire life cycle in mammalian cells (Gaunt et al., 2003). Sexual reproduction of *T. cruzi* has been hypothesized to be similar to the parasexual cycle of *C. albicans*. After cell fusion of diploid cells, a tetraploid intermediate may then undergo chromosome loss to generate aneuploid or diploid progeny (Gaunt et al., 2003). Recent population studies using microsatellite markers also support that sexual reproduction has significantly shaped the population structure of *T. cruzi* and may be associated with outbreaks (Higo et al., 2000; Llewellyn et al., 2009; Ocana-Mayorga et al., 2010).

Collectively, these examples highlight the impact of sexual reproduction on microbial pathogenesis and suggest its impact may be broad and pervasive. If so, many additional examples and paradigms remain to be discovered.

ACKNOWLEDGMENTS

This work was partially supported by National Science Foundation (NSF) grant #1011101 (E.S. and J.M.L.) and National Institutes of Health (NIH) grants AR37, AI39115, and R01 AI50113 (J.H.). Any opinions, findings, and conclusions or recommendations are those of the authors and do not necessarily reflect the views of NSF or NIH.

REFERENCES

Adam RD. 2001. Biology of *Giardia lamblia*. Clin Microbiol Rev 14: 447–475.

Alby K, Schaefer D, Bennett RJ. 2009. Homothallic and heterothallic mating in the opportunistic pathogen *Candida albicans*. Nature 460: 890–893.

Archetti M. 2004a. Loss of complementation and the logic of two-step meiosis. J Evol Biol 17: 1098–1105.

Archetti M. 2004b. Recombination and loss of complementation: A more than two-fold cost for parthenogenesis. J Evol Biol 17: 1084–1097.

Bachtrog D. 2003. Adaptation shapes patterns of genome evolution on sexual and asexual chromosomes in *Drosophila*. Nat Genet 34: 215–219.

Bakkeren G, Kronstad JW. 1993. Conservation of the *b* mating-type gene complex among bipolar and tetrapolar smut fungi. Plant Cell 5: 123–136.

Baptista F, Machado MF, Castro-Prado MA. 2003. Alternative reproduction pathway in *Aspergillus nidulans*. Folia Microbiol (Praha) 48: 597–604.

Barnes TM, Kohara Y, Coulson A, Hekimi S. 1995. Meiotic recombination, noncoding DNA and genomic organization in *Caenorhabditis elegans*. Genetics 141: 159–179.

Barraclough TG, Fontaneto D, Ricci C, Herniou EA. 2007. Evidence for inefficient selection against deleterious mutations in cytochrome oxidase I of asexual bdelloid rotifers. Mol Biol Evol 24: 1952–1962.

Barton NH, Charlesworth B. 1998. Why sex and recombination? Science 281: 1986–1990.

Bennett RJ, Johnson AD. 2003. Completion of a parasexual cycle in *Candida albicans* by induced chromosome loss in tetraploid strains. EMBO J 22: 2505–2515.

Bloomfield G, Skelton J, Ivens A, Tanaka Y, Kay RR. 2010. Sex determination in the social amoeba *Dictyostelium discoideum*. Science 330: 1533–1536.

Bowman BH, White TJ, Taylor JW. 1996. Human pathogeneic fungi and their close nonpathogenic relatives. Mol Phylogenet Evol 6: 89–96.

Boyle JP, Rajasekar B, Saeij JP, Ajioka JW, Berriman M, Paulsen I, Roos DS, Sibley LD, White MW, Boothroyd JC. 2006. Just one cross appears capable of dramatically altering the population biology of a eukaryotic pathogen like *Toxoplasma gondii*. Proc Natl Acad Sci U S A 103: 10514–10519.

Brown AJ, Casselton LA. 2001. Mating in mushrooms: Increasing the chances but prolonging the affair. Trends Genet 17: 393–400.

Bui T, Lin X, Malik R, Heitman J, Carter D. 2008. Isolates of *Cryptococcus neoformans* from infected animals reveal genetic exchange in unisexual, *alpha* mating type populations. Eukaryot Cell 7: 1771–1780.

Burt A, Carter DA, Koenig GL, White TJ, Taylor JW. 1996. Molecular markers reveal cryptic sex in the human pathogen *Coccidioides immitis*. Proc Natl Acad Sci U S A 93: 770–773.

Butler G, Rasmussen MD, Lin MF, Santos MA, Sakthikumar S, Munro CA, Rheinbay E, Grabherr M, Forche A, Reedy JL, Agrafioti I, Arnaud MB, Bates S, Brown AJ, Brunke S, Costanzo MC, Fitzpatrick DA, De Groot PW, Harris D, Hoyer LL, Hube B, Klis FM, Kodira C, Lennard N, Logue ME, Martin R, Neiman AM, Nikolaou E, Quail MA, Quinn J, Santos MC, Schmitzberger FF, Sherlock G, Shah P, Silverstein KA, Skrzypek MS, Soll D, Staggs R, Stansfield I, Stumpf MP, Sudbery PE, Srikantha T, Zeng Q, Berman J, Berriman M, Heitman J, Gow NA, Lorenz MC, Birren BW, Kellis M, Cuomo CA. 2009. Evolution of pathogenicity and sexual reproduction in eight *Candida* genomes. Nature 459: 657–662.

Byrnes EJ, 3rd, Bildfell RJ, Frank SA, Mitchell TG, Marr KA, Heitman J. 2009. Molecular evidence that the range of the Vancouver Island outbreak of *Cryptococcus gattii* infection has expanded into the Pacific Northwest in the United States. J Infect Dis 199: 1081–1086.

Byrnes EJ, 3rd, Li W, Lewit Y, Ma H, Voelz K, Ren P, Carter DA, Chaturvedi V, Bildfell RJ, May RC, Heitman J. 2010. Emergence and pathogenicity of highly virulent *Cryptococcus gattii* genotypes in the northwest United States. PLoS Pathog 6: e1000850.

Byrnes EJ, 3rd, Li W, Ren P, Lewit Y, Voelz K, Fraser JA, Dietrich FS, May RC, Chatuverdi S, Chatuverdi V, Heitman J. 2011. A diverse population of *Cryptococcus gattii* molecular type VGIII in southern Californian HIV/AIDS patients. PLoS Pathog 7: e1002205.

Campbell LT, Carter DA. 2006. Looking for sex in the fungal pathogens *Cryptococcus neoformans* and *Cryptococcus gattii*. FEMS Yeast Res 6: 588–598.

Carr M, Leadbeater BS, Baldauf SL. 2010. Conserved meiotic genes point to sex in the choanoflagellates. J Eukaryot Microbiol 57: 56–62.

Casselton LA. 2008. Fungal sex genes-searching for the ancestors. Bioessays 30: 711–714.

Cavalier-Smith T. 2002. Origins of the machinery of recombination and sex. Heredity 88: 125–141.

Coelho MA, Sampaio JP, Goncalves P. 2010. A deviation from the bipolar-tetrapolar mating paradigm in an early diverged basidiomycete. PLoS Genet 6: e1001052.

Cooper MA, Adam RD, Worobey M, Sterling CR. 2007. Population genetics provides evidence for recombination in *Giardia*. Curr Biol 17: 1984–1988.

Corley LS, Blankenship JR, Moore AJ, Moore PJ. 1999. Developmental constraints on the mode of reproduction in the facultatively parthenogenetic cockroach *Nauphoeta cinerea*. Evol Dev 1: 90–99.

Dacks J, Roger AJ. 1999. The first sexual lineage and the relevance of facultative sex. J Mol Evol 48: 779–783.

De Vries SS, Baart EB, Dekker M, Siezen A, De Rooij DG, De Boer P, Te Riele H. 1999. Mouse MutS-like protein Msh5 is required for proper chromosome synapsis in male and female meiosis. Genes Dev 13: 523–531.

Dodgson AR, Pujol C, Pfaller MA, Denning DW, Soll DR. 2005. Evidence for recombination in *Candida glabrata*. Fungal Genet Biol 42: 233–243.

D'Souza TG, Michiels NK. 2010. The costs and benefits of occasional sex: Theoretical predictions and a case study. J Hered 101(Suppl. 1): S34–S41.

Dyer PS. 2008. Evolutionary biology: Genomic clues to original sex in fungi. Curr Biol 18: R207–R209.

Engelstadter J. 2008. Constraints on the evolution of asexual reproduction. Bioessays 30: 1138–1150.

Findley K, Rodriguez-Carres M, Metin B, Kroiss J, Fonseca A, Vilgalys R, Heitman J. 2009. Phylogeny and phenotypic characterization of pathogenic *Cryptococcus* species and closely related saprobic taxa in the *Tremellales*. Eukaryot Cell 8: 353–361.

Findley K, Sun S, Fraser JA, Hsueh YP, Averette AF, Li W, Dietrich FS, Heitman J. 2012. Discovery of a modified tetrapolar sexual cycle in *Cryptococcus amylolentus* and the evolution of *MAT* in the *Cryptococcus* species complex. PLoS Genet 8: e1002528.

Forche A, Alby K, Schaefer D, Johnson AD, Berman J, Bennett RJ. 2008. The parasexual cycle in *Candida albicans* provides an alternative pathway to meiosis for the formation of recombinant strains. PLoS Biol 6: e110.

Fraser JA, Heitman J. 2005. Chromosomal sex-determining regions in animals, plants and fungi. Curr Opin Genet Dev 15: 645–651.

Fraser JA, Diezmann S, Subaran RL, Allen A, Lengeler KB, Dietrich FS, Heitman J. 2004. Convergent evolution of chromosomal sex-determining regions in the animal and fungal kingdoms. PLoS Biol 2: e384.

Fraser JA, Giles SS, Wenink EC, Geunes-Boyer SG, Wright JR, Diezmann S, Allen A, Stajich JE, Dietrich FS, Perfect JR, Heitman J. 2005. Same-sex mating and the origin of the Vancouver Island *Cryptococcus gattii* outbreak. Nature 437: 1360–1364.

Fraser JA, Stajich JE, Tarcha EJ, Cole GT, Inglis DO, Sil A, Heitman J. 2007. Evolution of the mating type locus: Insights gained from the dimorphic primary fungal pathogens *Histoplasma capsulatum*, *Coccidioides immitis*, and *Coccidioides posadasii*. Eukaryot Cell 6: 622–629.

Galagan JE, Calvo SE, Cuomo C, Ma LJ, Wortman JR, Batzoglou S, Lee SI, Basturkmen M, Spevak CC, Clutterbuck J, Kapitonov V, Jurka J, Scazzocchio C, Farman M, Butler J, Purcell S, Harris S, Braus GH, Draht O, Busch S, D'Enfert C, Bouchier C, Goldman GH, Bell-Pedersen D, Griffiths-Jones S, Doonan JH, Yu J, Vienken K, Pain A, Freitag M, Selker EU, Archer DB, Penalva MA, Oakley BR, Momany M, Tanaka T, Kumagai T, Asai K, Machida M, Nierman WC, Denning DW, Caddick M, Hynes M, Paoletti M, Fischer R, Miller B, Dyer P, Sachs MS, Osmani SA, Birren BW. 2005. Sequencing of *Aspergillus nidulans* and comparative analysis with *A. fumigatus* and *A. oryzae*. Nature 438: 1105–1115.

Gasior SL, Wong AK, Kora Y, Shinohara A, Bishop DK. 1998. Rad52 associates with RPA and functions with rad55 and rad57 to assemble meiotic recombination complexes. Genes Dev 12: 2208–2221.

Gaunt MW, Yeo M, Frame IA, Stothard JR, Carrasco HJ, Taylor MC, Mena SS, Veazey P, Miles GA, Acosta N, De Arias AR, Miles MA. 2003. Mechanism of genetic exchange in American *trypanosomes*. Nature 421: 936–939.

Gladyshev EA, Meselson M, Arkhipova IR. 2008. Massive horizontal gene transfer in bdelloid rotifers. Science 320: 1210–1213.

Goncalves-Sa J, Murray A. 2011. Asymmetry in sexual pheromones is not required for ascomycete mating. Curr Biol 21: 1337–1346.

Graser Y, Volovsek M, Arrington J, Schonian G, Presber W, Mitchell TG, Vilgalys R. 1996. Molecular markers reveal that population structure of the human pathogen *Candida albicans* exhibits both clonality and recombination. Proc Natl Acad Sci U S A 93: 12473–12477.

Grigg ME, Bonnefoy S, Hehl AB, Suzuki Y, Boothroyd JC. 2001. Success and virulence in *Toxoplasma* as the result of sexual recombination between two distinct ancestries. Science 294: 161–165.

Gryganskyi AP, Lee SC, Litvintseva AP, Smith ME, Bonito G, Porter TM, Anishchenko IM, Heitman J, Vilgalys R. 2010. Structure, function, and phylogeny of the mating locus in the *Rhizopus oryzae* complex. PLoS One 5: e15273.

Hadany L, Comeron JM. 2008. Why are sex and recombination so common? Ann N Y Acad Sci 1133: 26–43.

Hadany L, Otto SP. 2007. The evolution of condition-dependent sex in the face of high costs. Genetics 176: 1713–1727.

Haruta N, Akamatsu Y, Tsutsui Y, Kurokawa Y, Murayama Y, Arcangioli B, Iwasaki H. 2008. Fission yeast Swi5 protein, a novel DNA recombination mediator. DNA Repair (Amst) 7: 1–9.

Heitman J. 2006. Sexual reproduction and the evolution of microbial pathogens. Curr Biol 16: R711–R725.

Heitman J. 2010. Evolution of eukaryotic microbial pathogens via covert sexual reproduction. Cell Host Microbe 8: 86–99.

Hepworth SR, Friesen H, Segall J. 1998. NDT80 and the meiotic recombination checkpoint regulate expression of middle sporulation-specific genes in *Saccharomyces cerevisiae*. Mol Cell Biol 18: 5750–5761.

Higo H, Yanagi T, Matta V, Agatsuma T, Cruz-Reyes A, Uyema N, Monroy C, Kanbara H, Tada I. 2000. Genetic structure of *Trypanosoma cruzi* in American continents: Special emphasis on sexual reproduction in Central America. Parasitology 121(Pt 4): 403–408.

Hiremath SS, Chowdhary A, Kowshik T, Randhawa HS, Sun S, Xu J. 2008. Long-distance dispersal and recombination in environmental populations of *Cryptococcus neoformans* var. *grubii* from India. Microbiology 154: 1513–1524.

Hoffmann B, Eckert SE, Krappmann S, Braus GH. 2001. Sexual diploids of *Aspergillus nidulans* do not form by random fusion of nuclei in the heterokaryon. Genetics 157: 141–147.

Hsueh YP, Heitman J. 2008. Orchestration of sexual reproduction and virulence by the fungal mating-type locus. Curr Opin Microbiol 11: 517–524.

Hutchison EA, Glass NL. 2010. Meiotic regulators Ndt80 and Ime2 have different roles in *Saccharomyces* and *Neurospora*. Genetics 185: 1271–1282.

Idnurm A, Walton FJ, Floyd A, Heitman J. 2008. Identification of the *sex* genes in an early diverged fungus. Nature 451: 193–196.

Inderbitzin P, Harkness J, Turgeon BG, Berbee ML. 2005. Lateral transfer of mating system in *Stemphylium*. Proc Natl Acad Sci U S A 102: 11390–11395.

Ishiguro T, Tanaka K, Sakuno T, Watanabe Y. 2010. Shugoshin-PP2A counteracts casein-kinase-1-dependent cleavage of Rec8 by separase. Nat Cell Biol 12: 500–506.

Jones GH, Franklin FC. 2006. Meiotic crossing-over: Obligation and interference. Cell 126: 246–248.

Jones SK, Jr, Bennett RJ. 2011. Fungal mating pheromones: Choreographing the dating game. Fungal Genet Biol 48: 668–676.

Katis VL, Lipp JJ, Imre R, Bogdanova A, Okaz E, Habermann B, Mechtler K, Nasmyth K, Zachariae W. 2010. Rec8 phosphorylation by casein kinase 1 and Cdc7-Dbf4 kinase regulates cohesin cleavage by separase during meiosis. Dev Cell 18: 397–409.

Keeney S, Giroux CN, Kleckner N. 1997. Meiosis-specific DNA double-strand breaks are catalyzed by Spo11, a member of a widely conserved protein family. Cell 88: 375–384.

Kidd SE, Hagen F, Tscharke RL, Huynh M, Bartlett KH, Fyfe M, Macdougall L, Boekhout T, Kwon-Chung KJ, Meyer W. 2004. A rare genotype of *Cryptococcus gattii* caused the cryptococcosis outbreak on Vancouver Island (British Columbia, Canada). Proc Natl Acad Sci U S A 101: 17258–17263.

Kimata Y, Kitamura K, Fenner N, Yamano H. 2011. Mes1 controls the meiosis I to meiosis II transition by distinctly regulating the anaphase-promoting complex/cyclosome coactivators Fzr1/Mfr1 and Slp1 in fission yeast. Mol Biol Cell 22: 1486–1494.

Kwon-Chung KJ, Edman JC, Wickes BL. 1992. Genetic association of mating types and virulence in *Cryptococcus neoformans*. Infect Immun 60: 602–605.

Lee SC, Corradi N, Byrnes EJ, 3rd, Torres-Martinez S, Dietrich FS, Keeling PJ, Heitman J. 2008. Microsporidia evolved from ancestral sexual fungi. Curr Biol 18: 1675–1679.

Lee SC, Ni M, Li W, Shertz C, Heitman J. 2010. The evolution of sex: A perspective from the fungal kingdom. Microbiol Mol Biol Rev 74: 298–340.

Lengeler KB, Fox DS, Fraser JA, Allen A, Forrester K, Dietrich FS, Heitman J. 2002. Mating-type locus of *Cryptococcus neoformans*: A step in the evolution of sex chromosomes. Eukaryot Cell 1: 704–718.

Li CH, Cervantes M, Springer DJ, Boekhout T, Ruiz-Vazquez RM, Torres-Martinez SR, Heitman J, Lee SC. 2011. Sporangiospore size dimorphism is linked to virulence of *Mucor circinelloides*. PLoS Pathog 7: e1002086.

Li W, Metin B, White TC, Heitman J. 2010. Organization and evolutionary trajectory of the mating type (*MAT*) locus in dermatophyte and dimorphic fungal pathogens. Eukaryot Cell 9: 46–58.

Lin X, Hull CM, Heitman J. 2005. Sexual reproduction between partners of the same mating type in *Cryptococcus neoformans*. Nature 434: 1017–1021.

Lin X, Litvintseva AP, Nielsen K, Patel S, Floyd A, Mitchell TG, Heitman J. 2007. *alpha* AD *alpha* hybrids of *Cryptococcus neoformans*: Evidence of same-sex mating in nature and hybrid fitness. PLoS Genet 3: 1975–1990.

Lin X, Patel S, Litvintseva AP, Floyd A, Mitchell TG, Heitman J. 2009. Diploids in the *Cryptococcus neoformans* serotype A population homozygous for the *alpha* mating type originate via unisexual mating. PLoS Pathog 5: e1000283.

Litvintseva AP, Marra RE, Nielsen K, Heitman J, Vilgalys R, Mitchell TG. 2003. Evidence of sexual recombination among *Cryptococcus neoformans* serotype A isolates in sub-Saharan Africa. Eukaryot Cell 2: 1162–1168.

Litvintseva AP, Thakur R, Vilgalys R, Mitchell TG. 2006. Multilocus sequence typing reveals three genetic subpopulations of *Cryptococcus neoformans* var. *grubii* (serotype A), including a unique population in Botswana. Genetics 172: 2223–2238.

Lively CM. 2010. A review of Red Queen models for the persistence of obligate sexual reproduction. J Hered 101(Suppl. 1): S13–S20.

Llewellyn MS, Lewis MD, Acosta N, Yeo M, Carrasco HJ, Segovia M, Vargas J, Torrico F, Miles MA, Gaunt MW. 2009. *Trypanosoma cruzi* IIc: Phylogenetic and phylogeographic insights from sequence and microsatellite analysis and potential impact on emergent Chagas disease. PLoS Negl Trop Dis 3: e510.

Loftus BJ, Fung E, Roncaglia P, Rowley D, Amedeo P, Bruno D, Vamathevan J, Miranda M, Anderson IJ, Fraser JA, Allen JE, Bosdet IE, Brent MR, Chiu R, Doering TL, Donlin MJ, D'Souza CA, Fox DS, Grinberg V, Fu J, Fukushima M, Haas BJ, Huang JC, Janbon G, Jones SJ, Koo HL, Krzywinski MI, Kwon-Chung JK, Lengeler KB, Maiti R, Marra MA, Marra RE, Mathewson CA, Mitchell TG, Pertea M, Riggs FR, Salzberg SL, Schein JE, Shvartsbeyn A, Shin H, Shumway M, Specht CA, Suh BB, Tenney A, Utterback TR, Wickes BL, Wortman JR, Wye NH, Kronstad JW, Lodge JK, Heitman J, Davis RW, Fraser CM, Hyman RW. 2005. The genome of the basidiomycetous yeast and human pathogen *Cryptococcus neoformans*. Science 307: 1321–1324.

Lott TJ, Frade JP, Lockhart SR. 2010. Multilocus sequence type analysis reveals both clonality and recombination in populations of *Candida glabrata* bloodstream isolates from U.S. surveillance studies. Eukaryot Cell 9: 619–625.

Malik SB, Pightling AW, Stefaniak LM, Schurko AM, Logsdon JM, Jr. 2008. An expanded inventory of conserved meiotic genes provides evidence for sex in *Trichomonas vaginalis*. PLoS One 3: e2879.

Mandegar MA, Otto SP. 2007. Mitotic recombination counteracts the benefits of genetic segregation. Proc Biol Sci 274: 1301–1307.

Martin T, Lu SW, van Tilbeurgh H, Ripoll DR, Dixelius C, Turgeon BG, Debuchy R. 2010. Tracing the origin of the fungal α1 domain places its ancestor in the HMG-box superfamily: Implication for fungal mating-type evolution. PLoS One 5: e15199.

Martin SH, Wingfield BD, Wingfield MJ, Steenkamp ET. 2011. Causes and consequences of variability in peptide mating pheromones of ascomycete fungi. Mol Biol Evol 28: 1987–2003.

Martini E, Diaz RL, Hunter N, Keeney S. 2006. Crossover homeostasis in yeast meiosis. Cell 126: 285–295.

Matute DR, Mcewen JG, Puccia R, Montes BA, San-Blas G, Bagagli E, Rauscher JT, Restrepo A, Morais F, Nino-Vega G, Taylor JW. 2006. Cryptic speciation and recombination in the fungus *Paracoccidioides brasiliensis* as revealed by gene genealogies. Mol Biol Evol 23: 65–73.

Maynard Smith J. 1978. *The Evolution of Sex*. Cambridge University Press, Cambridge, England; New York.

Metin B, Findley K, Heitman J. 2010. The mating type locus (*MAT*) and sexual reproduction of *Cryptococcus heveanensis*: Insights into the evolution of sex and sex-determining chromosomal regions in fungi. PLoS Genet 6: e1000961.

Morgan JA, Vredenburg VT, Rachowicz LJ, Knapp RA, Stice MJ, Tunstall T, Bingham RE, Parker JM, Longcore JE, Moritz C, Briggs CJ, Taylor JW. 2007. Population genetics of the frog-killing fungus *Batrachochytrium dendrobatidis*. Proc Natl Acad Sci U S A 104: 13845–13850.

Muller HJ. 1964. The relation of recombination to mutational advance. Mutat Res 106: 2–9.

Nakagawa T, Kolodner RD. 2002. The MER3 DNA helicase catalyzes the unwinding of Holliday junctions. J Biol Chem 277: 28019–28024.

Neiman M, Meirmans S, Meirmans PG. 2009. What can asexual lineage age tell us about the maintenance of sex? Ann N Y Acad Sci 1168: 185–200.

Ni M, Feretzaki M, Sun S, Wang X, Heitman J. 2011. Sex in fungi. Annu Rev Genet 45: 405–430.

Nishant KT, Chen C, Shinohara M, Shinohara A, Alani E. 2010. Genetic analysis of baker's yeast Msh4-Msh5 reveals a threshold crossover level for meiotic viability. PLoS Genet 6: e1001083.

Ocana-Mayorga S, Llewellyn MS, Costales JA, Miles MA, Grijalva MJ. 2010. Sex, subdivision, and domestic dispersal of *Trypanosoma cruzi* lineage I in southern Ecuador. PLoS Negl Trop Dis 4: e915.

O'Gorman CM, Fuller HT, Dyer PS. 2009. Discovery of a sexual cycle in the opportunistic fungal pathogen *Aspergillus fumigatus*. Nature 457: 471–474.

O'Shea SF, Chaure PT, Halsall JR, Olesnicky NS, Leibbrandt A, Connerton IF, Casselton LA. 1998. A large pheromone and receptor gene complex determines multiple B mating type specificities in *Coprinus cinereus*. Genetics 148: 1081–1090.

Otto SP, Gerstein AC. 2006. Why have sex? The population genetics of sex and recombination. Biochem Soc Trans 34: 519–522.

Otto SP, Mable BK. 1998. The evolution of life cycles with haploid and diploid phases. Bioessays 20: 453–462.

Page SL, Hawley RS. 2004. The genetics and molecular biology of the synaptonemal complex. Annu Rev Cell Dev Biol 20: 525–558.

Pavelka N, Rancati G, Zhu J, Bradford WD, Saraf A, Florens L, Sanderson BW, Hattem GL, Li R. 2010. Aneuploidy confers quantitative proteome changes and phenotypic variation in budding yeast. Nature 468: 321–325.

Peck JR. 1994. A ruby in the rubbish: Beneficial mutations, deleterious mutations and the evolution of sex. Genetics 137: 597–606.

Petukhova GV, Pezza RJ, Vanevski F, Ploquin M, Masson JY, Camerini-Otero RD. 2005. The Hop2 and Mnd1 proteins act in concert with Rad51 and Dmc1 in meiotic recombination. Nat Struct Mol Biol 12: 449–453.

Poggeler S. 2002. Genomic evidence for mating abilities in the asexual pathogen *Aspergillus fumigatus*. Curr Genet 42: 153–160.

Pontecorvo G. 1956. The parasexual cycle in fungi. Annu Rev Microbiol 10: 393–400.

Poxleitner MK, Carpenter ML, Mancuso JJ, Wang CJ, Dawson SC, Cande WZ. 2008. Evidence for karyogamy and exchange of genetic material in the binucleate intestinal parasite *Giardia intestinalis*. Science 319: 1530–1533.

Ramesh MA, Malik SB, Logsdon JM, Jr. 2005. A phylogenomic inventory of meiotic genes; evidence for sex in *Giardia* and an early eukaryotic origin of meiosis. Curr Biol 15: 185–191.

Rancati G, Pavelka N, Fleharty B, Noll A, Trimble R, Walton K, Perera A, Staehling-Hampton K, Seidel CW, Li R. 2008. Aneuploidy underlies rapid adaptive evolution of yeast cells deprived of a conserved cytokinesis motor. Cell 135: 879–893.

Raudaskoski M, Kothe E. 2010. Basidiomycete mating type genes and pheromone signaling. Eukaryot Cell 9: 847–859.

Rauwolf U, Greiner S, Mracek J, Rauwolf M, Golczyk H, Mohler V, Herrmann RG, Meurer J. 2011. Uncoupling of sexual reproduction from homologous recombination in homozygous *Oenothera* species. Heredity 107: 87–94.

Reedy JL, Floyd AM, Heitman J. 2009. Mechanistic plasticity of sexual reproduction and meiosis in the *Candida* pathogenic species complex. Curr Biol 19: 891–899.

Rockmill B, Roeder GS. 1991. A meiosis-specific protein kinase homolog required for chromosome synapsis and recombination. Genes Dev 5: 2392–2404.

Rydholm C, Dyer PS, Lutzoni F. 2007. DNA sequence characterization and molecular evolution of *MAT1* and *MAT2* mating-type loci of the self-compatible ascomycete mold *Neosartorya fischeri*. Eukaryot Cell 6: 868–874.

San-Segundo PA, Roeder GS. 1999. Pch2 links chromatin silencing to meiotic checkpoint control. Cell 97: 313–324.

Saul N, Krockenberger M, Carter D. 2008. Evidence of recombination in mixed-mating-type and alpha-only populations of *Cryptococcus gattii* sourced from single eucalyptus tree hollows. Eukaryot Cell 7: 727–734.

Schoustra S, Rundle HD, Dali R, Kassen R. 2010. Fitness-associated sexual reproduction in a filamentous fungus. Curr Biol 20: 1350–1355.

Schoustra SE, Debets AJ, Slakhorst M, Hoekstra RF. 2007. Mitotic recombination accelerates adaptation in the fungus *Aspergillus nidulans*. PLoS Genet 3: e68.

Schurko AM, Logsdon JM, Jr. 2008. Using a meiosis detection toolkit to investigate ancient asexual "scandals" and the evolution of sex. Bioessays 30: 579–589.

Schurko AM, Neiman M, Logsdon JM, Jr. 2009. Signs of sex: What we know and how we know it. Trends Ecol Evol 24: 208–217.

Sekine H, Ferreira RC, Pan-Hammarstrom Q, Graham RR, Ziemba B, De Vries SS, Liu J, Hippen K, Koeuth T, Ortmann W, Iwahori A, Elliott MK, Offer S, Skon C, Du L, Novitzke J, Lee AT, Zhao N, Tompkins JD, Altshuler D, Gregersen PK, Cunningham-Rundles C, Harris RS, Her C, Nelson DL, Hammarstrom L, Gilkeson GS, Behrens TW. 2007. Role for Msh5 in the regulation of Ig class switch recombination. Proc Natl Acad Sci U S A 104: 7193–7198.

Selmecki A, Forche A, Berman J. 2006. Aneuploidy and isochromosome formation in drug-resistant *Candida albicans*. Science 313: 367–370.

Selmecki A, Gerami-Nejad M, Paulson C, Forche A, Berman J. 2008. An isochromosome confers drug resistance in vivo by amplification of two genes, *ERG11* and *TAC1*. Mol Microbiol 68: 624–641.

Sionov E, Lee H, Chang YC, Kwon-Chung KJ. 2010. *Cryptococcus neoformans* overcomes stress of azole drugs by formation of disomy in specific multiple chromosomes. PLoS Pathog 6: e1000848.

Stanton BC, Giles SS, Staudt MW, Kruzel EK, Hull CM. 2010. Allelic exchange of pheromones and their receptors reprograms sexual identity in *Cryptococcus neoformans*. PLoS Genet 6: e1000860.

Stelzer CP, Snell TW. 2003. Induction of sexual reproduction in *Brachionus plicatilis* (Monogononta, Rotifera) by a density-dependent chemical cue. Limnol Oceanogr 48: 939–943.

Su C, Evans D, Cole RH, Kissinger JC, Ajioka JW, Sibley LD. 2003. Recent expansion of *Toxoplasma* through enhanced oral transmission. Science 299: 414–416.

Sun S, Heitman J. 2011. Is sex necessary? BMC Biol 9: 56.

Torres I, Garcia AM, Hernandez O, Gonzalez A, Mcewen JG, Restrepo A, Arango M. 2010. Presence and expression of the mating type locus in *Paracoccidioides brasiliensis* isolates. Fungal Genet Biol 47: 373–380.

Tsubouchi H, Roeder GS. 2003. The importance of genetic recombination for fidelity of chromosome pairing in meiosis. Dev Cell 5: 915–925.

Velagapudi R, Hsueh YP, Geunes-Boyer S, Wright JR, Heitman J. 2009. Spores as infectious propagules of *Cryptococcus neoformans*. Infect Immun 77: 4345–4355.

Vershon AK, Pierce M. 2000. Transcriptional regulation of meiosis in yeast. Curr Opin Cell Biol 12: 334–339.

Villeneuve AM, Hillers KJ. 2001. Whence meiosis? Cell 106: 647–650.

Watanabe Y. 2005. Shugoshin: Guardian spirit at the centromere. Curr Opin Cell Biol 17: 590–595.

Watts PC, Buley KR, Sanderson S, Boardman W, Ciofi C, Gibson R. 2006. Parthenogenesis in *Komodo dragons*. Nature 444: 1021–1022.

Wendte JM, Miller MA, Lambourn DM, Magargal SL, Jessup DA, Grigg ME. 2010. Self-mating in the definitive host potentiates clonal outbreaks of the apicomplexan parasites *Sarcocystis neurona* and *Toxoplasma gondii*. PLoS Genet 6: e1001261.

Wong S, Fares MA, Zimmermann W, Butler G, Wolfe KH. 2003. Evidence from comparative genomics for a complete sexual cycle in the "asexual" pathogenic yeast *Candida glabrata*. Genome Biol 4: R10.

Wu HY, Ho HC, Burgess SM. 2010. Mek1 kinase governs outcomes of meiotic recombination and the checkpoint response. Curr Biol 20: 1707–1716.

Xu J, Saunders CW, Hu P, Grant RA, Boekhout T, Kuramae EE, Kronstad JW, Deangelis YM, Reeder NL, Johnstone KR, Leland M, Fieno AM, Begley WM, Sun Y, Lacey MP, Chaudhary T, Keough T, Chu L, Sears R, Yuan B, Dawson TL, Jr. 2007. Dandruff-associated *Malassezia* genomes reveal convergent and divergent virulence traits shared with plant and human fungal pathogens. Proc Natl Acad Sci U S A 104: 18730–18735.

Yamamoto M. 1996. The molecular control mechanisms of meiosis in fission yeast. Trends Biochem Sci 21: 18–22.

Yun SH, Berbee ML, Yoder OC, Turgeon BG. 1999. Evolution of the fungal self-fertile reproductive life style from self-sterile ancestors. Proc Natl Acad Sci U S A 96: 5592–5597.

Zanders S, Brown MS, Chen C, Alani E. 2011. Pch2 modulates chromatid partner choice during meiotic double-strand break repair in *Saccharomyces cerevisiae*. Genetics 188: 511–521.

CHAPTER 3

PHYLOGENOMIC ANALYSIS

ANDREW J. ROGER, MARTIN KOLISKO, and ALASTAIR G. B. SIMPSON

INTRODUCTION

Phylogenetic methods for the analysis of molecular sequences were first developed in the late 1960s and in the early 1970s. The initial appeal of these approaches came, in part, from their provision of an objective means for comparing the fundamental "digital" molecular sequences of organisms to determine their relationships. Since then, molecular phylogenetic approaches have greatly expanded in popularity and have evolved both in terms of the statistical sophistication of the methods employed and the scale of data to which they can be applied.

In this chapter, we outline the "state of the art" for molecular phylogenetic analyses and show how these methods are being extended to analyses of genomic and transcriptomic data that are now available from across the full breadth of the tree of life. Our goal is not to provide an exhaustive account of the history, theory, and details of molecular phylogenetic methods; interested readers are urged to consult excellent recent books by Felsenstein (2004) and Yang (2006), which provide more comprehensive treatments. Instead, we will provide an overview of the key methods required for performing rigorous phylogenetic and phylogenomic analyses with an emphasis on potential pitfalls of the analyses as well as important considerations for interpreting the results.

A number of distinct kinds of genome-scale comparisons have been described in the literature as "phylogenomics." These include (i) analyses of gene or gene family presence, absence, or dynamics in many genomes in a phylogenetic context; (ii) the analysis of gene content or gene order similarity (i.e., synteny) to determine relationships among the genomes (and therefore the organisms); (iii) phylogenetic analyses of every gene in a genome (along with homologs of each gathered from other species) to discover patterns of relationships in different gene families, thereby leading to inferences of lateral gene transfer (LGT), endosymbiotic gene transfer (EGT), gene duplications, and gene loss events; (iv) gene-by-gene analyses of related genomes aimed at detecting genes under positive selection as virulence factors or in adaptation to specific niches; and finally, (v) the phylogenetic analysis of a data set composed of a large number of genes to infer organismal relationships.

All of these types of analyses can be useful in addressing questions about the relationships between organisms and the

Evolution of Virulence in Eukaryotic Microbes, First Edition. Edited by L. David Sibley, Barbara J. Howlett, and Joseph Heitman.

origins of their metabolisms, cellular properties, life histories, pathogenicity, or morphological traits, and most of them rely explicitly or implicitly on phylogenetic inference methods. In this chapter, we will focus mainly on phylogenetic and phylogenomic methods that involve comparisons of nucleotide/amino acid sequences of genes/proteins to infer their interrelationships. Readers interested in methods for comparing gene content, order, and frequency or genome structure data are advised to consult recent papers on those subjects (e.g., Sawa et al., 2003; Spencer and Sangaralingam, 2009).

In recent years, a number of phylogenomic analysis "pipeline" software tools such as PhyloGenie (Frickey and Lupas, 2004), Scafos (Roure et al., 2007), AMPHORA (Wu and Eisen, 2008), Orthoselect (Schreiber et al., 2009), and iPhy (Jones et al., 2011) have been developed that automate some or all of the steps of phylogenomic analyses. Although these tools have made such analyses more accessible to nonexperts, users should understand the principles and uncertainties underlying the various steps and potential pitfalls that can be encountered. In the following sections, we describe the various steps of phylogenetic and phylogenomic analyses (schematically depicted in Fig. 3.1) and, along the way, we highlight some of the most common pitfalls of these analyses and how to address them. We finish with a discussion of the kind of biological information that can be gleaned from these analyses.

ASSEMBLING DATA SETS FOR ANALYSES

General Considerations

One of the most crucial steps in phylogenetic analysis is the selection of data—taxa or individual gene sequences—to include in the analysis. It is also a step that requires considerable judgment to be exercised by the investigator.

First, and most obviously, it is crucial that appropriate data be selected to address the scientific question posed. For example, if one seeks to determine if two parasites, A and B, are closely related and stem from a common parasitic ancestor, it is important to include representatives of any free-living group that might conceivably be most closely related to them. An example of this is seen in Figure 3.2: The predominantly parasitic lineages of parabasalids (e.g., *Trichomonas*) and diplomonads (e.g., *Giardia* and *Spironucleus*) appear as sister taxa on this phylogeny, a pattern that is consistent with descent from a common parasitic ancestor. Missing from this phylogeny, however, are the several lineages of recently described free-living protozoa (*Carpediemonas*-like organisms ["CLOs"]) that are specifically related to diplomonads in phylogenies based on small numbers of genes (Kolisko et al., 2010; Simpson et al., 2006). The phylogenetic position of CLOs, therefore, calls into question the common parasitic ancestry of parabasalids and diplomonads. Second, many questions require a "rooted" tree—the usual way to infer the root of a phylogenetic tree is to include outgroups in the analysis (outgroups are taxa or sequences known a priori to be less closely related to the taxa/sequences of interest, or ingroup, than all the taxa/sequences of interest are to each other). The most appropriate outgroups would generally be the organisms most closely related to the ingroup, although there are exceptions to consider (see below).

How Many Taxa or Sequences Should Be Included?

Improved taxon sampling is generally associated with greater phylogenetic accuracy (Heath et al., 2008). While it is inadvisable to include a single representative of each group of interest, going to the opposite

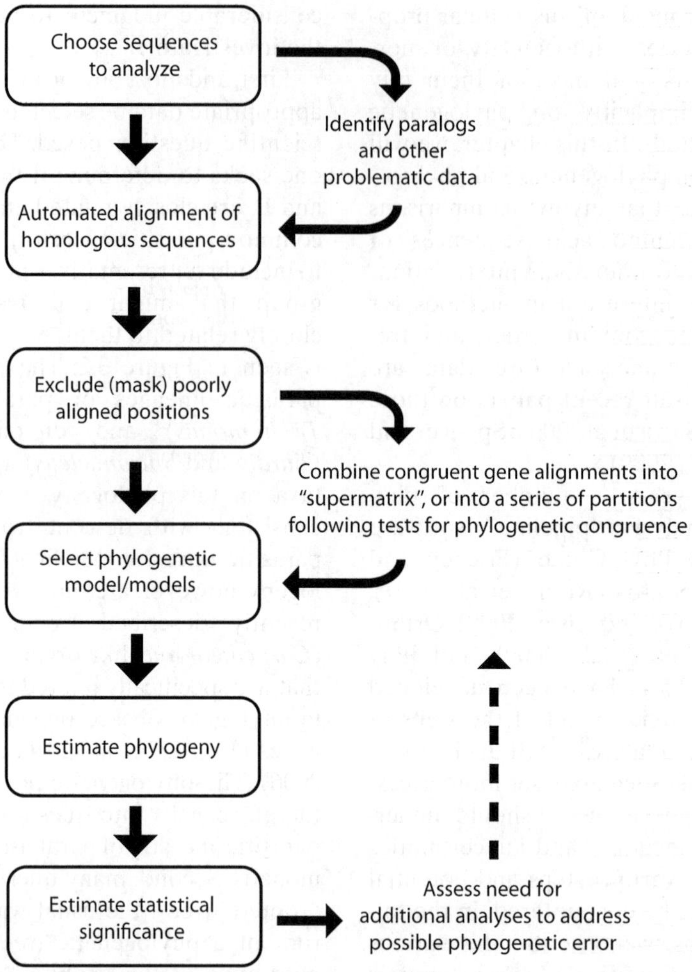

Figure 3.1 A flowchart describing the steps of phylogenetic and phylogenomic analyses discussed in the text.

extreme and including every available sequence or species may also be counterproductive. New algorithmic tools are available that will simultaneously analyze thousands to tens of thousands of sequences (Guindon et al., 2010; Price et al., 2010); however, these data set sizes are typically only achievable at the expense of in-depth searching of tree space for the best supported topology and of assessment of statistical confidence in the results. A compromise is generally necessary, and at present, multigene analyses that address broadscale questions about organismal relationships typically include 50–100 total species, occasionally more (see Fig. 3.2 for an example).

Ideally, one should choose taxa/sequences to represent the groups of interest broadly, and evenly, at a level appropriate for the scientific question to be addressed. Sequences/taxa that are evolving extremely rapidly or have unusual amino acid compositions may foster systematic error (see below), and their inclusion may actually reduce phylogenetic accuracy (Poe, 2003). Such taxa/sequences would normally be excluded, and their

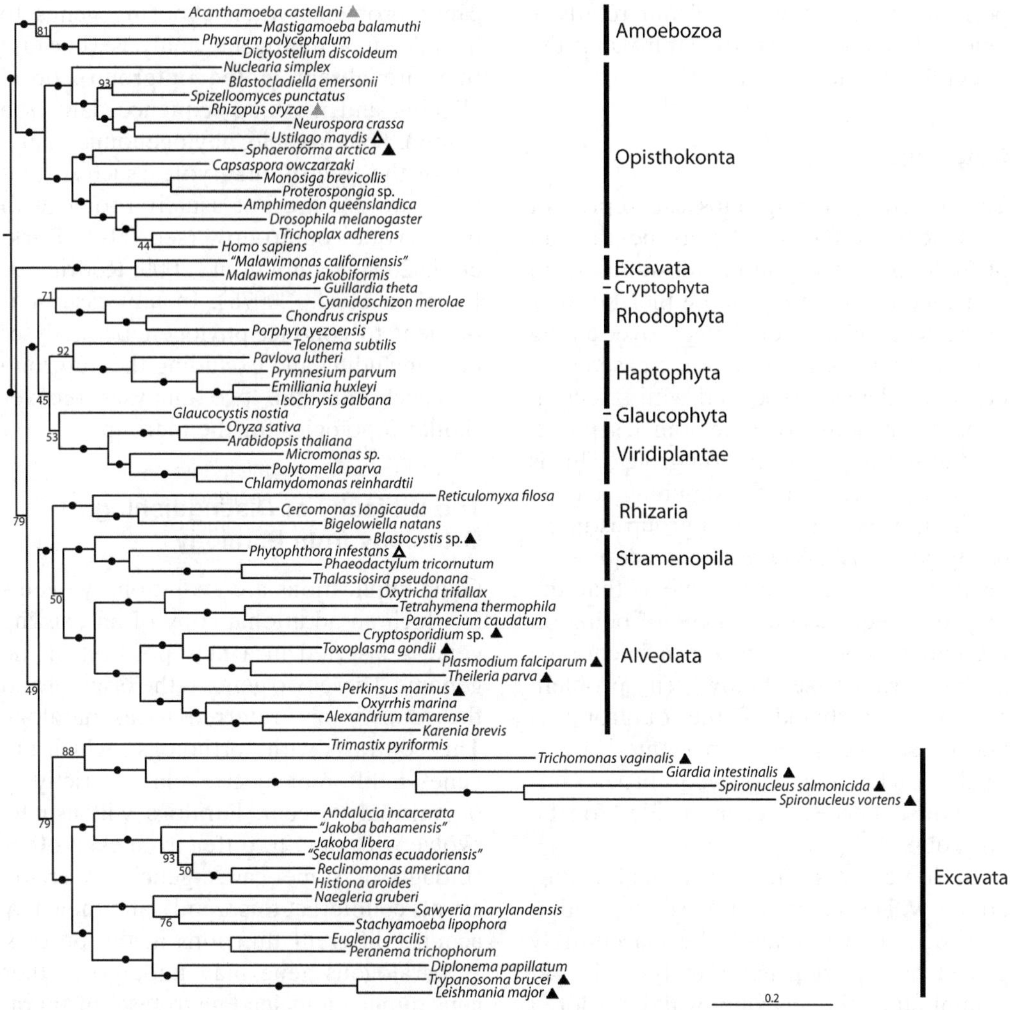

Figure 3.2 A phylogenetic tree of the supergroups of eukaryotes based on a supermatrix of 161 proteins (43,516 amino acid positions) from 72 eukaryotic taxa. All 161 protein alignments were produced using FSA, manually masked, and concatenated into a single-partition supermatrix. The tree was constructed using maximum likelihood implemented in RAxML using the LG + F + Γ substitution model, based on 10 random addition maximum parsimony starting trees. Bootstrap support values (BP) were based on 100 bootstrap replicates in RAxML. Black dots represent bootstrap support (BP) >95%. Black triangles denote animal parasites; gray triangles indicate opportunistic parasites; and empty triangles show plant parasites. Note that the two large groups with longest branches—Excavata and a clade consisting of Rhizaria, Stramenopila, and Alveolata—branch together with low statistical support (BP = 49%). This is an example of a weak LBA artifact grouping the Excavata with the Rhizaria, Stramenopila, and Alveolata clade as argued by Hampl et al. (2009).

positions taken by less aberrant relatives, unless they are critical for addressing the scientific question at hand.

Outgroups

The selection of outgroups can represent a particular difficulty. It may be that all possible outgroups differ substantially in sequence from the ingroup either because of the accumulation of changes over a long period of time or because of increased evolutionary change associated with selection for new functions along the branch separating the ingroup from the outgroup. This is particularly true when examining a gene family in which the studied group is interesting *precisely because* it has acquired a distinct function. This can mean that the outgroup sequences, by virtue of being on the end of a long branch, can encourage analysis artifact (see below). The problem can be exacerbated if the outgroup is poorly sampled (i.e., represented by a single, or a few, sequences), so a broad sampling of sequences is recommended for the outgroup.

In some cases, the only available outgroups will be so different in sequence that they will almost certainly be inaccurately placed by a phylogenetic analysis (i.e., at a minimum, the ingroup will be rooted in an incorrect position). An example of this is seen in attempts to estimate the phylogenetic tree of eukaryotes using multiple nucleus-encoded genes. Homologous sequences from prokaryotes (the logical outgroups) tend to be extremely different from eukaryote sequences. In a phylogenetic analysis, one or the other of the more rapidly evolving eukaryotic lineages tends to group with the prokaryotic outgroup due to the systematic artifact known as long-branch attraction (LBA) (Philippe et al., 2000). In such situations, it is probably counterproductive to include the outgroups in the first place, especially when adding the outgroup to the data set is likely to cause problems when choosing genes to include in a multigene analysis (because they are absent in the outgroup), or in aligning and masking sequence data (see below). For example, phylogenomic analyses of the tree of eukaryotes, such as that in Figure 3.2, are not usually rooted with prokaryotic outgroups (see also Burki et al., 2008; Hampl et al., 2009; Rodriguez-Ezpeleta et al., 2007a). In any case, it is prudent to conduct phylogenetic analyses both including and excluding the outgroup to check that the two analyses recover similar topologies for the ingroup.

The Difficulty of Distinguishing Orthology from Paralogy

Gene duplications are evolutionary events in which an additional copy of an existing gene is inserted in a new position in the genome. These two genes—the original and the "copy"—are referred to as paralogs. This contrasts with "orthologs," which are genes in different species related strictly by organismal descent. Paralogs will usually evolve with time to differ from each other (though in some cases gene conversion events counteract this trend), and they may acquire different functions in the process. A paralogous gene may undergo further gene duplications, leading to cases of "paralogs within paralogs." Conversely, one or other paralog may be lost in a lineage that ancestrally had both forms.

These complex histories of genes may be unraveled using phylogenetic approaches: In fact, discerning relationships within gene families that result from multiple gene duplications is an important rationale for estimating phylogenies for individual genes (along with discerning events of gene transfer—see below). However, because gene duplication represents an instance where the evolutionary histories of organisms and their genes will not coincide, it is important not to mix these different historical signals inadvertently if one is interested only in organismal phylogeny. Thus, the potential for paralogous genes to be present in genomes needs to be taken into account when inferring organismal phylogeny. For example, if there are two or more related genes in a genome, it is not obvious which should be used to represent the species.

Gene duplications may be very recent events within the context of the organisms under study, very ancient, or anything between. Very recent duplications that occurred in the lineage of just one included species (i.e., after the divergence of that species from all others included in the analysis) are not usually problematic. Any one paralog could be used to represent the species—often whichever is the least divergent is chosen.

Cases in which gene duplications occurred in the ancestors of more than one considered species are the bigger problem. To obtain an accurate reflection of organismal history, it is important to select the sequences that will represent the organisms from just one paralog family (such that orthologs are compared). This requires, in turn, that the paralog families are accurately identified. In principle, the presence of multiple similar genes in the species from a particular taxonomic group will serve to easily identify these older paralog groups, and genes could be assigned to paralog groups by similarity. In practice, however, examples of paralogy can be easily overlooked if only one of the two paralogs is known from most species either because of incomplete data from their genomes or because the paralogs have actually been lost. These are common situations. In order to detect such events of paralogy (as well as ancient paralogs, gene transfers, etc., see below), it is common to conduct preliminary phylogenetic analyses of the individual genes that will be included in a phylogenomic analysis. "Ancient" gene duplications that occurred prior to the last common ancestor of all organisms under study yield what are effectively separate genes, and often both genes are included (separately) in multigene analyses.

Elongation factors in eukaryotes are an example of the difficulties with distinguishing paralogs and orthologs. While most eukaryotes have the canonical eukaryotic translation elongation factor 1α (EF-1α), a minority, representing several unrelated groups, have a distantly related elongation factor-like (EFL) protein that may have originated in an ancient gene duplication, possibly prior to, or fairly soon after, the diversification of living eukaryotes (Gile et al., 2009; Kamikawa et al., 2010). Until it was realized that EFL was a paralog of EF-1α, EFL sequences were sometimes included in phylogenetic analyses as if they were true EF-1α sequences (i.e., orthologs). Again, these cases can often be diagnosed by gene phylogenies that include all relevant related genes.

Automated Ortholog Selection

There are semi- or fully automated approaches for selecting orthologs. These most frequently use pairwise sequence similarity scores from basic local alignment search tool (BLAST) searches. For example, the "reciprocal-best-hit" (RBH) approach takes a query gene/protein A from a given genome (or gene set) X and uses it as a query to search another genome (or gene set) Y. If a best-matching protein is returned (call this A'), then a reciprocal BLAST search is done using A' as the query to search genome X. If A is also recovered as the "best hit" in this case, then A and A' are operationally defined as orthologs. However, as paralogy and loss patterns can be quite complex, this approach will not always work. More sophisticated methods such as OrthoMCL (Ostlund et al., 2011), InParanoid (Li et al., 2003), BranchClust (Poptsova and Gogarten, 2007), or OrthoInspector (Linard et al., 2011) have been developed to help identify paralogs (or split sets of genes/proteins into paralogous families), but none is likely to work perfectly (see Trachana et al., 2011 for a review). When using these methods, it is prudent to also perform manual inspection of individual gene/protein trees where possible.

Does Gene Function Matter?

The functions of genes used for phylogenetic studies vary a lot depending on the phylogenetic depths being investigated as well as the specific questions being addressed. If the phylogenetic relatedness of the organisms is the primary focus, then so-called housekeeping genes are usually considered, that is, genes that are constitutively expressed and whose encoded proteins function in basic cellular processes and are essential for normal growth and survival. The rationale for this practice is several-fold. These genes are thought to be less likely to be transferred between species because they sometimes function within larger complexes, so they should better track organismal phylogeny (Leigh et al., 2011b; Wellner et al., 2007). Furthermore, it is expected that housekeeping genes experience more or less the same selective forces in all of the organisms under study and they are often well conserved on the sequence level and are therefore relatively easy to align. By contrast, genes with extremely patchy distributions across the organisms of interest or genes that may be involved in pathogenicity (i.e., virulence factors) or other niche-specific functions may be more likely to be transferred between organisms (Andersson et al., 2007; Mitreva et al., 2009; Richards et al., 2006) and can be subject to wildly different selective forces in different organisms, not to mention large lineage-specific gene family expansions (Sun et al., 2010) with the potential for paralog confusion. These genes are often of great interest but should be analyzed with caution, keeping in mind that their phylogenetic histories may be quite complex and difficult to reconstruct using standard phylogenetic methods that generally assume homogeneity in the evolutionary process across the phylogeny (see the section "Model Realism and LBA").

MULTIPLE ALIGNMENT CONSTRUCTION AND EDITING

Once a set of homologous sequences has been collected, the next step is the determination of which sites in each of the sequences are homologous to those of the other sequences using a multiple alignment method. First, however, a decision must be made regarding nucleotide versus amino acid sequence-based analysis.

Nucleotide versus Amino Acid Alignments

Although protein-coding sequences can either be analyzed at the nucleotide or the amino acid level, the optimal choice remains a matter of debate (discussed in Holder et al., 2008) and depends somewhat on the phylogenetic question addressed. For very recent phylogenetic divergences (e.g., between or within closely related species), amino acid sequences may not vary sufficiently between taxa to provide phylogenetic resolution. In that case, synonymous codon changes in the nucleotide sequences will likely provide a richer source of variation. At the deeper end of the phylogenetic spectrum, synonymous codon changes may have been so frequent as to have "saturated"—the deep phylogenetic signal at these sites being overwritten by more recent substitutions. The danger of analyzing coding regions with saturated synonymous changes is that systematic biases in the kind of substitutions occurring at synonymous sites may manifest (e.g., because of mutational pressure toward high or low G + C content), and these biases may change over the tree such that distantly related organisms end up sharing similar biased patterns (Inagaki et al., 2004). These kinds of nonphylogenetic "signals" in the data may lead to artifacts in phylogenetic inference (discussed in more detail below).

Regardless of whether the ultimate phylogenetic analysis is at the nucleotide or amino acid level, it is generally preferable to first align the amino acid sequences to be compared since most multiple alignment programs do not consider the codon structure of nucleotide sequence data. Once an amino acid alignment is obtained, if a nucleotide-level alignment is desired, then the coding sequences can be back-aligned to correspond to the amino acid-level alignment using software tools, such as RevTrans (Wernersson and Pedersen, 2003).

Multiple Alignment Algorithms and Masking Tools

Multiple alignment algorithms that estimate globally optimal alignments are computationally infeasible for large data sets. Therefore, for many years, heuristic progressive alignment methods were used that aligned progressively more distantly related individual sequences to one or more groups of aligned sequences using a guide tree inferred from a pairwise alignment-based dissimilarity matrix between sequences. The most popular of these is ClustalW (Larkin et al., 2007), a program that employs biochemically reasonable and adjustable schemes for sequence similarity weighting and determining insertion/deletion gap penalties. More recently, a new generation of more sophisticated statistical modeling or machine learning approaches to multiple alignment have been developed including software tools such T-Coffee (Notredame et al., 2000), MUSCLE (Edgar, 2004), MAFFT (Katoh et al., 2005), FSA (Bradley et al., 2009), PROBCONS (Do et al., 2004), Probalign (Roshan and Livesay, 2006), and HMMER3 (Eddy, 2009). Simulation and benchmark performance studies (Bradley et al., 2009; Thompson et al., 2011) indicate that all of these new methods perform consistently better than simpler methods like ClustalW, with comparable performance to each other, although with greatly varying runtimes.

Rather than focusing on slight differences in the performance of these tools, effort is more fruitfully directed at determining the regions of alignments that are well supported (thus recovered by most or all of the best alignment methods) versus regions that are intrinsically difficult to align with confidence. We recommend the removal (masking) of the latter poorly aligned regions prior to phylogenetic analysis so as to avoid the introduction of downstream estimation biases from misaligned sites. For analyses of individual RNA gene or protein gene alignments, this masking step is often done by manual visual inspection by the researcher, a practice that is sometimes criticized as data manipulation. However, manual masking should not pose a problem as long as it is performed prior to tree estimation and care is taken to avoid the influence of the phylogenetic preconceptions of the investigator (e.g., by being conservative in the definition of a well-aligned site). A more objective alternative is to make use of alignment quality scores for individual cells in the multiple alignment that are output by newer multiple alignment programs (e.g., FSA [Bradley et al., 2009]; HMMER3 [Eddy, 2009]) to automate the construction of masks based on quality thresholds of varying stringency. A variety of alignment postprocessing tools are also available that evaluate alignment quality and are useful aids to masking, such as GBLOCKS (Castresana, 2000), heads or tails (HoT) (Landan and Graur, 2007), GUIDANCE (Penn et al., 2010), BMGE (Criscuolo and Gribaldo, 2010), and MANUEL (Blouin et al.,

2009). Rather than treating the alignment as fixed, we recommend that researchers investigate the degree to which alignments vary over the parameter settings of the chosen multiple alignment software and evaluate whether downstream phylogenetic analyses are strongly affected by different masking strategies.

METHODS OF PHYLOGENETIC ESTIMATION

Dealing with Multiple Gene Data

Before discussing the specifics of phylogenetic methods, we will briefly address how the various types of phylogenomic strategies differ at this analysis step. If the goal is to reconstruct the phylogeny of a large number of genes/proteins or families whose precise histories may differ because of gene duplications, LGT and/or EGT, or allelic sorting effects, then phylogenetic analysis of each individual alignment can be conducted. The resulting collection of trees can then be parsed to quantify the frequencies of phylogenetic patterns of interest. To faciliate these analyses, several tools such as PhyloPattern (Gouret et al., 2009) and PhyloSort (Moustafa and Bhattacharya, 2008) can be used to search for specific phylogenetic patterns in large collections of trees.

If, instead, the aim is to infer organismal phylogeny from a large set of orthologous genes, then there are two general options for analysis: (i) the alignments can be concatenated into a large supermatrix from which a tree (or network) can be estimated or (ii) individual phylogenetic analyses of each alignment can be conducted and the results combined by various methods into a supertree or supernetwork. There is some debate as to the relative merits of supermatrix versus supertree methods (see Rannala and Yang, 2008). Generally, while supermatrix methods tend to be more computationally intensive (thus slower) than supertree methods, the latter lack a rigorous statistical approach to propagating the uncertainty of estimated individual gene trees to the final supertree (Rannala and Yang, 2008). For the latter reason, and because of their relative popularity, we focus here on supermatrix analyses. Note that although the estimation of phylogenetic networks (i.e., phylogenetic graphs with reticulations) is becoming more popular, the details of these methods are covered in other recent reviews (e.g., Huson and Scornavacca, 2011); here we focus on the inference of strictly bifurcating phylogenies.

Phylogenetic Inference Paradigms

Widely used molecular phylogenetic methods broadly fall into four categories: (i) maximum parsimony (MP) approaches, (ii) distance matrix methods, (iii) maximum likelihood (ML) methods, and (iv) Bayesian analysis. An in-depth description of precisely how these methods work is beyond the scope of this chapter and readers are urged to consult excellent books by Felsenstein (2004) and Yang (2006) for comprehensive treatments. Instead, we offer a brief account of the various methods, a summary of their strengths and weaknesses, and some useful software implementations.

MP

The MP method seeks to find the tree topology that minimizes the numbers of nucleotide or amino acid substitutions (totaled over all aligned sites) that are needed to explain the observed sequences in the alignment. MP analyses often accord each kind of substitution equal weight in the calculations, although "weighted" parsimony analyses are possible that assign different costs to different kinds of changes. MP methods are implemented in numerous software tools including the PHYLIP package (Felsenstein, 2005), PAUP*

(Swofford, 2002), and MEGA (Tamura et al., 2007). Although the MP algorithm has been used to estimate phylogenies for more than four decades and some researchers defend its use on philosophical grounds (see Felsenstein, 2004), it has undesirable statistical properties. The most famous of these is the LBA form of systematic error (Felsenstein, 1978) whereby two sequences on the ends of "long" branches (by virtue of a high rate of evolution or ancient divergence) will tend to group together artificially. This problem stems from MP's underestimation of the true numbers of substitutions that have occurred. If multiple changes along unrelated lineages give rise to the same nucleotide or amino acid state frequently enough, MP will mistakenly treat these shared derived states as evidence of common ancestry. In practice, MP methods are expected to perform well if substitutions are rare occurrences on each branch in the tree. For many phylogenetic (or phylogenomic) problems, the rates of sequence evolution are large and/or vary greatly and/or taxonomic sampling is uneven, and this condition is not met.

Distance Matrix Methods

Distance methods first estimate the numbers of substitutions that have occurred between all pairs of sequences and then fit a phylogenetic tree to this distance matrix. Tree fitting approaches include methods such as the unweighted pair group method with arithmetic mean (UPGMA) and the neighbor joining (NJ) method (or more sophisticated versions like BIONJ and WEIGHBOR; see Desper and Gascuel, 2006) that progressively cluster taxa based on the distances and then output single phylogenetic trees. Other methods such as least squares or minimum evolution (ME) seek the best-fitting tree by evaluating candidate topologies through a search of tree space (Felsenstein, 2004). Distance matrix methods are relatively fast, allowing the analysis of large data sets with tens of thousands or more sequences (Desper and Gascuel, 2006). However, their accuracy depends heavily on: (i) the correct estimation of distances between sequence pairs, (ii) appropriate assumptions about the evolutionary process and errors in the distances, and (iii) the thoroughness of "tree space" exploration. ML distance estimation using a substitution model that closely approximates the evolutionary process (discussed below) should be accurate if the number of sites in the analysis is large enough (Susko et al., 2004). Of the tree fitting algorithms, UPGMA is generally the least useful as it makes the unwarranted assumption that sequences have evolved according to a strict molecular clock (i.e., substitution rates are fixed over all lineages in the tree). All other methods perform similarly well, with the balanced minimum evolution (BME) method probably the best (Desper and Gascuel, 2006). Distance tree estimation is implemented in PHYLIP, MEGA, and PAUP* with especially fast implementations in tools such as FastME (Desper and Gascuel, 2006). Despite their speed, distance matrix methods rely on information just from pairs of sequences and therefore are expected to be less efficient at extracting phylogenetic signal than the full ML or Bayesian approaches described below. This prediction seems to be borne out by simulation studies (Holder et al., 2008). Consequently, they are typically used only if ML or Bayesian approaches are computationally infeasible for the size of the data sets under examination.

ML

ML and Bayesian methods are typically the methods "of choice" for phylogenetic and phylogenomic analysis and, while their probabilistic approaches are very similar in some respects, they hail from quite different statistical paradigms (see discussion in Felsenstein, 2004; Yang, 2006).

ML seeks the value of the parameter of interest (in this case the tree, T) that maximizes the probability of observing the sequence data (D) in the alignment given a model (M) of sequence evolution (i.e., the T that maximizes $L_T = P(\mathrm{D}|T, M)$). Most commonly, the model of evolution operating over the tree is assumed to be a reversible continuous-time Markov process where the relative rates of nucleotide or amino acid interchange are specified by parameters in a rate matrix **Q**. At the core of the likelihood calculation is the probability that a nucleotide (or amino acid) state i changes to j over a given interval of evolutionary distance corresponding to a branch in the phylogenetic tree of length t; this probability is given by $P_{ij} = [\exp\mathbf{Q}t]_{ij}$. In the ML paradigm, for a given tree topology, the probability of a given site pattern in the alignment is calculated over the entire tree averaged over all possible unknown nucleotide or amino acid states at internal nodes. Typically, different alignment positions are treated as independent observations so that the total probability of the data is the product of the site probabilities (or equivalently, the log-likelihood is the sum of the logarithms of those site probabilities). In order to perform this calculation, however, the values of unknown parameters such as the branch lengths of the tree or parameters of the **Q** matrix must be determined. These are optimized to maximize the likelihood for each tree; the ML estimate of the tree is the one with the globally highest likelihood score. For data sets of more than a dozen or so sequences, there are far too many possible tree topologies to exhaustively explore tree space for the ML solution, and heuristic tree-searching methods are almost always used.

A variety of ML software tools are available that differ in their speed, the extent to which tree space is searched, and the models available for analysis. While PHYLIP, PAUP*, and MEGA have ML implementations, the fastest methods currently in development include GARLI (Zwickl, 2006), PhyML (Guindon et al., 2010), RAxML (Stamatakis, 2006), and FastTree (Price et al., 2010). Note that these programs often allow for tree searching from multiple distinct starting trees and allow choices of tree rearrangement algorithms to explore tree space more or less exhaustively. Since it can be difficult to find a globally optimal solution, it is recommended that, computation time permitting, tree space should be searched as exhaustively as possible from multiple starting points.

Bayesian Analysis

The Bayesian approach also takes into account the likelihood of the tree calculated in exactly the same way as described above. However, it differs from ML in two key respects. First, the optimality criterion is the posterior probability of the tree (T) given the data (D) and the model (M), which is determined using Bayes formula: $P(T|\mathrm{D}, M) = P(T)P(\mathrm{D}|T, M)/P(\mathrm{D})$, where $P(T)$ is the prior probability of the tree and associated parameters, $P(\mathrm{D}|T, M)$ is the likelihood, and $P(\mathrm{D})$ is the probability of the data over all trees and parameters. In Bayesian phylogenetics, prior probability distributions on parameters (e.g., $P(T)$ above) are usually chosen to be diffuse (roughly "noninformative") to avoid strongly affecting the posterior probabilities of trees. In principle, in the absence of prior information, the posterior probability is supposed to mostly be influenced by the fit of the model and tree to the data through the likelihood function. Unfortunately, diffuse priors are sometimes unexpectedly more informative than intended (see Yang, 2006).

The second key difference between ML and Bayesian methods is that, rather than optimizing the unknown model and tree parameters by ML, the Bayesian approach instead integrates over this parameter space. Since this integration cannot be calculated explicitly, various flavors of the Markov chain Monte Carlo (MCMC) numerical integration method are used to approximate the posterior probability distribution. In recent years, the Bayesian approach in phylogenetics has become very popular and a variety of software tools have become available including MrBayes (Ronquist and Huelsenbeck, 2003), BEAST (Drummond and Rambaut, 2007), and PhyloBayes (Lartillot et al., 2009). As with ML, the accuracy of the results of Bayesian analysis depends not only on the realism of the substitution model but also on how thoroughly tree and parameter space is sampled during the MCMC procedure. Although the details depend on the specific MCMC implementations of the software tools, it is always important to run multiple independent "chains" of MCMC from different starting trees for as many iterations of MCMC as is practically possible and then to assess whether these multiple chains converge on the same posterior probability distribution for trees and parameters. Once the user has some confidence that convergence has been achieved (assessed by tools like Tracer; Rambaut and Drummond, 2007), then maximum posterior estimates of trees and posterior probabilities for trees or individual branches can be calculated from the posterior distribution.

There has been some debate over the relative merits of the Bayesian versus the ML approach. We have found that if the same phylogenetic model is used, extensive tree-sampling methods are employed and there is sufficient historical signal, that these two methods usually estimate very similar, if not identical, topologies. The major difference between these methods is usually in their associated statistical support measures for branching patterns, a subject addressed further below.

SELECTING AN APPROPRIATE PHYLOGENETIC MODEL

General Considerations

For ML distance, full ML, or Bayesian analyses, a model describing sequence evolution must be specified. This model comprises not only the **Q** matrix of rates of interchange between, and stationary frequencies of, bases or amino acids, but it will also often account for differences in the evolutionary rates at different sites, sometimes specify several classes of sites with different evolutionary dynamics or even more complex site-rate or site-type switching processes. For supermatrix phylogenomic analyses, there is the additional complexity of capturing differences in model parameters across different genes.

Substitution Rate Matrices

There are a variety of nucleotide substitution models that vary in the complexity of their rate matrix **Q**. For example, the simplest Jukes + Cantor (JC) substitution model assumes that all nucleotide bases in the sequences occur at equal frequencies and interchange at equal rates, while more complex models, such as Hasegawa-Kishino-Yano-1985 (HKY85), have adjustable parameters describing the transition : transversion rate ratios and equilibrium base compositions. The most complex, widely used reversible substitution model is the general time-reversible (GTR) model that allows every interchange its own rate parameter and all base frequencies are specified by separate parameters. However, it is not uncommon for the base compositions of genes to differ radically among species, violating the assumptions of GTR and simpler models. The log-determinant (LogDet) distance method (implemented, e.g., in PAUP*) is theoretically robust to this kind of heterogeneity (Gu and Li,

1996) and is frequently used for such data sets. Furthermore, several less frequently used, but very complex, models have been developed to allow base composition to shift over the tree in ML or Bayesian analysis, including those introduced by Yang and Roberts (1995), Galtier and Gouy (1998), and Foster (2004). Since these models are neither reversible nor stationary, the phylogenies examined must have a root position specified or estimated from the data. Further, the additional parameters of these models can make phylogeny estimation relatively slow. At the higher end of complexity, the Barry and Hartigan general discrete-time general Markov model (where each branch in the tree has its own estimated transition matrix) has also been shown to be useful in cases of compositional heterogeneity across the tree (Jayaswal et al., 2005), although the computational costs and the number of parameters to be estimated are prohibitive unless the numbers of species examined are small.

If protein-coding genes are analyzed at the nucleotide level, then it is important to somehow take into account the codon structure of the data. Complex models for codon evolution that specify rates of interchange between all 61 codons have been implemented, usually for the detection of positive selection given a known phylogeny (Yang, 2006). These models typically include a parameter (usually denoted ω) that is the ratio of the rate of nonsynonymous codon changes (i.e., amino acid changing substitutions) to synonymous changes. This parameter conveys the strength and direction of selection with $\omega > 1$ indicating positive selection, $\omega \approx 1$ indicating neutral evolution, and $\omega < 1$ indicating purifying selection. These models are most often fit on a predetermined gene phylogeny as estimation of phylogenies using these models is generally prohibitively computationally intensive for large data sets (Ren et al., 2005). For the estimation of phylogenies, codon-based data sets are sometimes partitioned into first, second, and third codon position site classes, then separate models (e.g., GTR) and rate parameters are fit for each of these site types.

Proteins are often more straightforwardly analyzed at the amino acid level by assuming a model of amino acid interchange (a 20×20 $\mathbf{Q}$ matrix). These rate matrices involve a large number of parameters, so to avoid problems with overfitting on single protein data sets (see below for a discussion of overfitting), fixed "empirical" matrices based on analyses of large databases of alignments are often used. Empirical amino acid models include the Dayhoff (PAM), Jones–Taylor–Thornton (JTT), Whelan and Goldman (WAG), and the recently developed Le and Gascuel (LG) matrix (see Whelan et al., 2001; Le and Gascuel, 2008 and references therein). Matrices for specific genome or protein types have also been developed (e.g., the chloroplast cpREV model; Adachi et al., 2000; or the reverse transcriptase model rtREV; Dimmic et al., 2002). Amino acid frequency parameters can be either derived from the databases from which the matrices were inferred or based on the data set under investigation (the latter models are denoted with an "F": e.g., the LG + F model). For phylogenomic supermatrices, there are sufficiently large numbers of sites that the amino acid version of the GTR model can be estimated from the data without problems associated with overfitting. Programs like RAxML (Stamatakis, 2006) and PAML (Yang, 1997) allow users to estimate these GTR amino acid models with a fixed phylogenetic tree derived, for example, from an analysis with one of the empirical matrices above. Estimation can be quite slow because of the large number of parameters involved. Accounting for changing amino acid compositions across sequences has received less attention than for base compositional heterogeneity. As with nucleotide data, LogDet distance approaches can also be used for amino acid data (Foster et al., 2009). In terms of models, the node-discrete composition heterogeneity (NCDH) model (Foster, 2004) and the break-point (BP) model (Blanquart and Lartillot, 2008) can be used in a Bayesian framework, although they can become very parameter rich and computationally intensive.

Site Rates and Classes

Individual sites in molecules evolve at different rates and this is usually modeled by a mixture model whereby the rates at sites come from a discrete rate distribution (e.g., the gamma [Γ] distribution), but the rate category assignment to sites is unknown (Yang, 2006). In the case of a gamma rate model, the shape of the rate distribution is governed by an adjustable parameter, α, that can be estimated from the data. For codon data, rates across sites can be modeled by estimation of separate rates for the first, second, and third codon positions in the data.

Attention has recently been directed at the phenomenon of heterotachy, whereby rates at sites can change either regularly or sporadically over the phylogeny, possibly as a result of drifting or shifting structural constraints in the molecules under study. Covarion-type models have been developed to capture regular site-rate shifting processes both for nucleotide (Galtier, 2001; Huelsenbeck, 2002) and amino acid data (Wang et al., 2009; Whelan et al., 2011), but the extremely large rate matrices they entail and additional parameters make them very slow for analyses of large data sets. Branch-length mixture models have also been developed (Kolaczkowski and Thornton, 2008; Zhou et al., 2007) that allow for more complex and irregular rate-shifting events, but they are, again, extremely slow.

Sites may vary not only in their rates but also in the kinds and rates of substitutions that may take place because of site-specific structural/functional constraints. For amino acid data, this has been addressed by allowing for models that utilize separate rate matrices for site classes. As with rates, often the site type is unknown for any particular position, so mixture models that compute weighted averages of the likelihood at sites over different site-class rate matrices are frequently used. For example, PhyloBayes implements a number of different versions of the CAT model that allow arbitrary numbers of site classes to be fit to the data during tree estimation, with the number of classes being a parameter that is explored by MCMC (Lartillot and Philippe, 2004). PhyloBayes (Lartillot et al., 2009) and a version of PhyML (Guindon et al., 2010) can also be used with fixed empirically estimated classes (so-called CATfix models with classes numbering 10–60; Le et al., 2008). A different mixture of empirically estimated site class frequency (cF) model is available in the ML program QmmRAxML (Wang et al., 2008).

Model Selection

Models should be selected to capture the important determinants of sequence evolution, but should not be so complex as to "overfit" the data; overfitting arises when the numbers of estimated model parameters approaches (or exceeds) the number of distinct patterns at aligned sites in the data set. Such overparameterized models may almost perfectly fit a given data set, but they will usually have no predictive accuracy and the variance in the phylogenetic estimates will be extremely large.

There are a number of statistical procedures for model selection. For a model series in which the simpler models are special cases of the more complex ones (e.g., "nested" models such as JC, HKY85, and GTR), likelihood ratio tests (LRTs) can be performed. For LRTs, two times the difference in the log-likelihoods of the fitted models is the test statistic, and significance can be assessed by comparing this to the chi-squared distribution, with degrees of freedom equal to the difference in the number of adjustable parameters in the models (Felsenstein, 2004; Yang, 2006). For cases where models are not nested (e.g., different empirical amino acid substitution matrices), measures such as the Akaike information criterion (AIC) or the Bayesian information criterion (BIC) can be used to choose the optimal model. For automated model selection using these methods, the software tool ModelTest (Posada, 2006) has been developed for nucleotide data and ProtTest (Darriba et al., 2011) for amino acid data.

The foregoing model selection methods are more typically used as part of an ML estimation framework. Bayesian model selection commonly involves comparing the marginal likelihoods (i.e., the integrated likelihood) of competing models to compute the Bayes factor (BF). Although simple harmonic mean estimates of marginal likelihoods are implemented in some software tools to compute BFs, this approach is not reliable and should be avoided (Newton and Raftery, 1994). Better approaches for BF estimation have recently been introduced that should be used where possible (Lartillot and Philippe, 2006; Xie et al., 2011). As an alternative, posterior-predictive simulation can also be performed for model selection as implemented in PhyloBayes (Lartillot et al., 2009) or P4 (http://www.bmnh.org/web_users/pf/p4.html). Reversible-jump MCMC procedures have also been developed (Huelsenbeck et al., 2004) to allow model selection concurrently with phylogeny estimation but are not yet widely available.

Modeling Multiple Gene Data

Individual genes often differ from each other in their rates of evolution in particular lineages of organisms or even in the dynamics of the substitutions that occur. If model selection methods like LRTs, AIC, or BIC suggest there is significant gene-to-gene variation in the evolutionary process, then this variation should, ideally, be taken into account in supermatrix analysis (Pupko et al., 2002). This can be achieved by "partitioned" analyses where various parameters of the evolutionary models are separately optimized (or, in Bayesian analysis, integrated over) for subsets of genes in the supermatrix. Several software tools such as RAxML and MrBayes explicitly allow for these kinds of partitioned analyses. Note, however, that if all parameters, including branch lengths, are evaluated separately for each partition, this can lead to problems with overparameterization and computational expense.

Even more problematic are cases of true historical incongruence in the phylogenies of genes/proteins in the supermatrix as a result of undiscovered paralogy, allelic sorting effects, or LGT. Rather than attempting to infer a phylogeny of an unholy mixture of historical signals, it is advisable to test for congruence between individual genes or protein phylogenies prior to analysis. Several methods have been published to test for congruence including the distance-based congruence among distance matrices (CADM) method (Campbell et al., 2011), a full likelihood ratio-based hierarchical congruence testing method (CONCATERPILLAR) (Leigh et al., 2008), and a much faster likelihood-based clustering method (CONCLUSTADOR) (Leigh et al., 2011b). A more comprehensive review and critique of these methods is provided by Leigh et al. (2011a).

ASSESSING STATISTICAL SIGNIFICANCE

It is important to know whether particular relationships in an estimated phylogenetic tree are well supported by the data or are merely a result of random "noise." By far, the most widely used measure of statistical support in phylogenetics is the bootstrap support value. Bootstrap support for a branch separating two sets of species (a "split" or bipartition in the tree) is the proportion of times that branch is observed when the aligned sites are resampled with replacement (usually 100–1000 times) to create data sets of the same length, which are then each analyzed using the phylogenetic method of choice. Although the bootstrap support value was initially proposed to be a statistical test for the hypothesis that the branch was *not* present in the true tree (1-bootstrap value = *P*-value of the test) (Felsenstein, 1985), doubts over its correct interpretation have lingered for decades (Felsenstein, 2004; Yang, 2006). Informally, bootstrap values for branches of >90% are often considered very strong support, with varying degrees of moderate strength accorded to values between 60% and 90%, and <50–60% generally indicating poor support. Recently, Susko has proved that the traditional bootstrap is not "first-order correct" as a statistical test (Susko, 2009) and has introduced methods to correct bootstrap values to allow them to be validly used in a hypothesis testing framework (Susko, 2010).

With the introduction of Bayesian analyses into phylogenetics, the posterior probability has become a widely reported support statistic that is usually calculated, like bootstrap support, for individual bipartitions in estimated phylogenies. Apparent discrepancies between bootstrap values and posterior probabilities in analyses have been vigorously debated; many studies have reported that posterior probabilities for branches tend to be larger than the corresponding bootstrap value (Douady et al., 2003; Suzuki et al., 2002), with the latter values tending to be more "trustable." Although the details are beyond the scope of this chapter, the difference is likely due to several factors including (i) the two measures having completely different statistical interpretations (Yang, 2006), (ii) the importance of convergence of MCMC procedures for obtaining accurate posterior probabilities, (iii) the use of overly simplistic phylogenetic models leading to spuriously high posterior probabilities (Huelsenbeck and Rannala, 2004), (iv) the sensitivity of posterior probabilities to branch-length prior distributions (Yang, 2008), and (v) the greater susceptibility of posterior probabilities to estimation biases (Susko, 2008). As a pragmatic approach, we suggest that researchers report both ML bootstrap values and Bayesian posterior probabilities and only draw firm conclusions when both measures strongly support the branch in question.

For technical reasons, in practice, bootstrap support values and posterior probabilities only allow the evaluation of support for one branch of a tree at a time. If we wish to consider different topologies differing by many branches as alternative hypotheses, then topology-testing frameworks must be used. These include generalized least squares confidence region methods for distance-based analysis (Susko, 2011) and ML topology tests such as the Kishino–Hasegawa test, the Shimodaira–Hasegawa test, the expected likelihood weights test, and the approximately unbiased test (see references in Yang, 2006; Shimodaira and Hasegawa, 2001). All of the latter methods are implemented in the program CONSEL (Shimodaira and Hasegawa, 2001). It is not widely appreciated that the *P*-values of many of these methods depend heavily on the numbers and types of trees included in the test and that it is best to include a large number of near-optimal "good" trees in the test in addition to the optimal tree and the alternative trees of interest. Good trees could include topologies returned during bootstrap analysis or a large number of top-scoring trees found during tree searching.

SYSTEMATIC ERROR AND DEALING PHYLOGENETIC ARTIFACTS

Estimated phylogenies may be incorrect not only because of lack of resolution and random noise but also because of misleading nonhistorical "signal." This systematic error can manifest as incorrect branching orders among sequences that are strongly supported by bootstrap support, posterior probability and/or topology tests. The most famous form of systematic error is the LBA phenomenon that, as discussed earlier, was first shown to be a problem for MP analysis under conditions when long branches are interspersed with short branches in the tree. Part of the theoretical justification for methods like ML distance, full ML, and Bayesian analysis was that these methods should be immune to such systematic error and should thus converge toward estimating the correct phylogenetic tree as more and more data are added to the analyses. However, this "statistical consistency" property hinges critically on whether the data under examination have evolved according to the process assumptions of the phylogenetic model (Yang, 2006). In other words, if the evolutionary process deviates significantly from the model used in phylogenetic inference, then ML distance, full ML, and Bayesian methods may not be consistent and systematic error can become a problem for estimation.

The LBA artifact and phylogenetic errors induced by compositional heterogeneity (e.g., high and low G + C base composition or changing amino acid composition) are the most common systematic errors. The following paragraphs focus mainly on LBA, which is probably the more widespread problem in comparisons of amino acid sequences.

Model Realism and LBA

A clear example of LBA manifested in early studies of eukaryote phylogeny that employed the small subunit ribosomal RNA (SSU rRNA) genes and elongation factor proteins, and relatively simple evolutionary models. The microsporidian parasites have extremely divergent sequences and tended to emerge as a deep branch in the eukaryotic tree, clustering with the long branch leading to the prokaryotic outgroups (Kamaishi et al., 1996; Van Keulen et al., 1993). As a result, many regarded the Microsporidia as an early eukaryote offshoot and are therefore crucial for understanding the evolution of the eukaryotic cell. However, it was later shown that the deeply branching position was in fact an LBA artifact; Microsporidia were in fact derived fungi (Edlind et al., 1996). The LBA effect became evident when increasingly realistic models of sequence evolution were used for phylogenetic reconstruction and support for the deep-branching position evaporated (Fischer and Palmer, 2005; Van de Peer et al., 2000). Phylogenomic analyses of Microspordia now clearly support their status as Fungi.

The importance of model realism has also been shown with regard to the performance of new site-class mixture models described above (e.g., CAT, CATfix, and cF). These models show strong potential to ameliorate LBA problems in both simulated and real phylogenomic data sets (Lartillot et al., 2007; Wang et al., 2008). It is important to note that these models are very computationally intensive and, for the very large data sets (>100 genes), it may be close to impossible to use an ML approach with current implementations. PhyloBayes implements many of these models in such a way as to avoid dramatically increasing the computational expense. However, our experience with these models and large phylogenomic supermatrices suggests that there are often difficulties in achieving convergence between multiple MCMC runs using the most parameter-rich versions of these models.

Data Filtering

Another way of combating LBA is to filter out the data that are contributing the most to the artifact. For example, fast-evolving sites may be removed from the data set prior to final analysis. In this case, sites in the alignment can be sorted from slowest evolving to fastest evolving, with the rate of evolution estimated through MP or ML approaches (Brinkmann and Philippe, 1999; Rodriguez-Ezpeleta et al., 2007b). A series of data sets is then generated by sequentially removing the fastest evolving sites. Each of these data sets is then reanalyzed, and change in the support for the "LBA topology" versus alternative topologies is evaluated.

In large phylogenomic analyses, it is possible to remove the most rapidly evolving sequences from the single gene alignments before concatenation. The branch length for each sequence in each of the single gene trees is estimated and a series of data sets is then generated by sequentially removing the longest branching of these sequences from the concatenated alignment. Each of the data sets is then analyzed, and change in support for different topologies is evaluated. Methods like this have been successfully used to suppress LBA in some large multigene analyses (Hampl et al., 2009).

Data Recoding

Another way of addressing both LBA and compositional bias is the recoding nucleotide or amino acid states into smaller numbers of states. In nucleotide data, sequences are typically recoded to R and Y (purines and pyrimidines; Phillips et al., 2004) so that the only substitutions considered are transversions. This is because the most common compositional biases are alterations in the G + C:A + T ratios that are mostly mediated by rapidly occurring transition substitutions. The rationale for recoding is similar in the case of suspected LBA because convergent substitutions that can worsen LBA effects may involve the most frequently occurring substitutions, and it is these changes where lack of fit in the evolutionary model is often most pronounced. In the case of protein sequences, amino acids are usually grouped into n categories based on chemical properties—amino acid interchange within a property-based class should be more rapid than the more radical between-class changes. Four or five categories have been the most commonly used, but it is possible to recode amino acids based on properties to any number of groups from 2 to 19 (Susko and Roger, 2007). At present, RAxML and PhyloBayes are the only widely used programs capable of dealing with models for the latter numbers of states.

Taxon Sampling

Another powerful way of addressing LBA is improving taxon sampling. The addition of short-branching taxa that are deep sister clades to the long-branching species can diminish the LBA artifact by breaking the long branches (Poe, 2003). Alternatively a long-branching species can sometimes be substituted by a short branching known relative in the data matrix. For example, in the deep eukaryote phylogeny shown in Figure 3.2, the position of the extremely long-branching diplomonads (e.g., *Giardia* and *Spironucleus*) remains contentious. Thus, to address concerns about the LBA affecting their position, we are currently sequencing transcriptomes of a number of recently discovered close relatives of diplomonads that in smaller-scale analyses appear to have much lower rates of evolution (Kolisko et al., 2010). By including these taxa in phylogenomic analyses to "break" the long diplomonad branches (or replace them), we hope to improve the resolution of deep eukaryotic phylogeny.

WHY PHYLOGENETIC SIGNALS OF DIFFERENT GENES CAN DIFFER AND WHY IT MATTERS

Ancestral gene duplication and loss events can cause confusion between paralogs and orthologs when data sets are being constructed, yielding apparent phylogenetic conflicts between genes in phylogenomic analyses. This phenomenon is discussed earlier in the chapter, so we will not elaborate on it further here.

Coalescent Effects

Incongruence can also arise when multiple alleles in ancestral populations were differentially fixed in descendant populations. This is most easily explained by referring to a recent phylogenomic analysis of primates. It is well-known that, among great apes, chimps are the closest relatives of humans. However, when Ebersberger and colleagues (2007) analyzed ~12,000 randomly sequenced genomic DNA regions from humans, chimpanzees, gorillas, orangutans, and rhesus monkeys that gave well-resolved trees, they found that 23% of these showed gorillas as the closest relatives of humans or chimps to the exclusion of the other. This discrepancy between DNA trees and species trees was caused by the coalescent process; the polymorphism from the ancestral population of these species was incompletely sorted, resulting in some parts of the human genome being more closely related to gorillas than to chimps. Note that this phenomenon is expected to mostly affect relationships between closely related species where consecutive speciation events were closely spaced in time. Even then, the signal for the correct species tree will often be dominant, like in the case of the hominids.

Gene Transfers between Organisms

Incongruence can also result from genes that have been transferred between organismal lineages (across species boundaries) either from direct transfer of one or several genes from an independent organism (LGT) or in the special context of an endosymbiotic organelle acquisition (EGT). Indeed it is becoming increasingly clear that the eukaryote genomes can contain considerable numbers of laterally transferred genes (Andersson et al., 2005, 2007). For example, it is probable that alpha-tubulin was laterally transferred between some eukaryotic lineages, perhaps from a relative of opisthokonts (animals, fungi, and relatives) to the clade that includes diplomonads and parabasalids, and to the free-living protist *Andalucia* (Simpson et al., 2008). This strongly affects the results of phylogenies of eukaryotes based on a few concatenated genes including tubulins, with improved phylogenetic estimation observed when alpha-tubulin was removed from the analysis (Hampl et al., 2005; Simpson et al., 2006). Individual LGTs of this type may be less problematic for large phylogenomic analyses including >100 genes, as the LGT signal may be overpowered by signal from the rest of the data set.

On the other hand, EGT turns out to be an important factor in large-scale phylogenomic analyses, especially when considering lineages with secondary plastids (plastids descended from eukaryotic algae whose "primary" plastids were, in turn, direct descendants of cyanobacteria). It has been well demonstrated that large numbers of genes were transferred from the endosymbiont to the host cell nucleus during both primary and secondary endosymbiosis (Lane and Archibald, 2008). In secondary endosymbiosis, this included eukaryotic genes from the endosymbiont nucleus, some of which may have replaced the original host orthologs. This means that the genomes of lineages that have (or had) secondary plastids might yield an admixture of two significant phylogenetic signals, one representing the host lineage and the other the lineage of the eukaryotic symbiont that became the plastid. It is proving very challenging to untangle the signal from endosymbiont-derived versus host-derived genes, and this may contribute to ongoing uncertainties in resolving the phylogenetic positions of some of the lineages with secondary plastids, especially cryptophytes and haptophytes (Baurain et al., 2010).

In some cases, phylogenomic analyses are actually targeted to discover genes that have been laterally transferred. By these analyses, we can learn to what degree adaptations (e.g., parasitism, anaerobism) are due to the modification of existing genetic capacity versus acquisition of novel genes. For example, Andersson et al. (2007) identified several genes in the fish intestinal parasite *Spironucleus salmonicida* (ruberythrin, A-type flavoprotein, and arginine deiminase) that were obtained via LGT and have the potential to protect the parasite against host innate immune systems (against nitric oxide and reactive oxygen species). Other examples include phylogenomic analyses that have identified LGTs of toxic secondary metabolite-producing genes between fungi (Slot and Rokas, 2011) and the acquisition of genes involved in plant parasitism by oomycetes from pathogenic fungi (Richards et al., 2006).

CONCLUSIONS

Phylogenomic inference is a rapidly changing field. Increasing computational power combined with the vast amounts of genomic information now available from inexpensive high throughput sequencing technologies have made a wide variety of "first-generation" phylogenomic analyses possible. However, as data sets get larger—both in terms of the numbers of taxa and the numbers of genes examined—problems with systematic errors in phylogenetic reconstructions can become more pronounced, and the use of more realistic and parameter-rich phylogenetic models has become essential. The computational complexity of increasingly sophisticated phylogenomic analyses based on complex models may ultimately be best tackled by the continued development of parallelized software tools that make use of new processing strategies (Suchard and Rambaut, 2009). The complexities of the process of genome evolution (coalescent allelic sorting effects, gene duplications and losses, and gene transfers between organisms) make distinguishing gene tree versus species tree conflicts even more challenging. The future will likely see the continued development of probabilistic phylogenomic methods (see Akerborg et al., 2009 for a recent attempt) that jointly model sequence evolutionary processes and genome dynamics (gene duplications, losses, LGT, etc.).

In any case, the revelation of the full complexities of genome evolution has revolutionized our understanding of the mechanisms of evolution, especially in the case of microbes. For example, phylogenomic analyses of prokaryotes have revealed the huge role of LGT in shaping prokaryotic evolution and indicate that the core of genes that is useful for reconstructing prokaryotic phylogeny may be very small indeed. In fact, the usefulness of the "tree of life" concept has been challenged in the case of prokaryotes (Doolittle, 1999). On the other hand, supermatrix analyses have shown that, apart from complications caused by EGT (see above), there is likely enough congruent signal for reconstruction of many of the deepest relationships among eukaryote supergroups (Burki et al., 2009; Hampl et al., 2009). Nevertheless, phylogenomic analyses have shown that LGT in eukaryotes (especially microbes) is surprisingly common, and we are only just now beginning to understand the role of this process in shaping the diversity of metabolisms, life histories, and adaptation of specific eukaryotic lineages to specific niches.

REFERENCES

Adachi J, Waddell PJ, Martin W, Hasegawa M. 2000. Plastid genome phylogeny and a model of amino acid substitution for proteins encoded by chloroplast DNA. Journal of Molecular Evolution 50: 348–358.

Akerborg O, Sennblad B, Arvestad L, Lagergren J. 2009. Simultaneous Bayesian gene tree reconstruction and reconciliation analysis. Proceedings of the National Academy of Sciences of the United States of America 106: 5714–5719.

Andersson JO, Sarchfield SW, Roger AJ. 2005. Gene transfers from Nanoarchaeota to an ancestor of diplomonads and parabasalids. Molecular Biology and Evolution 22: 85–90.

Andersson JO, Sjogren AM, Horner DS, Murphy CA, Dyal PL, Svard SG, Logsdon JM, Ragan MA, Hirt RP, Roger AJ. 2007. A genomic survey of the fish parasite *Spironucleus salmonicida* indicates genomic plasticity among diplomonads and significant lateral gene transfer in eukaryote genome evolution. BMC Genomics 8: 51–76.

Baurain D, Brinkman H, Petersen J, Rodriguez-Ezpeleta N, Stechmann A, Demoulin V, Roger AJ, Burger G, Lang BF, Philippe H. 2010. Phylogenomic evidence for separate acquisition of plastids in cryptophytes, haptophytes, and stramenopiles. Molecular Biology and Evolution 27: 1698–1709.

Blanquart S, Lartillot N. 2008. A site- and time-heterogeneous model of amino acid replacement. Molecular Biology and Evolution 25: 842–858.

Blouin C, Perry S, Lavell A, Susko E, Roger AJ. 2009. Reproducing the manual annotation of multiple sequence alignments using a SVM classifier. Bioinformatics (Oxford, England) 25: 3093–3098.

Bradley RK, Roberts A, Smoot M, Juvekar S, Do J, Dewey C, Holmes I, Pachter L. 2009. Fast statistical alignment. PLoS Computational Biology 5: 15.

Brinkmann H, Philippe H. 1999. Archaea sister group of bacteria? Indications from tree reconstruction artifacts in ancient phylogenies. Molecular Biology and Evolution 16: 817–825.

Burki F, Shalchian-Tabrizi K, Pawlowski J. 2008. Phylogenomics reveals a new "megagroup" including most photosynthetic eukaryotes. Biology Letters 4: 366–369.

Burki F, Inagaki Y, Brate J, Archibald JM, Keeling PJ, Cavalier-Smith T, Sakaguchi M, Hashimoto T, Horak A, Kumar S, Klaveness D, Jakobsen KS, Pawlowski J, Shalchian-Tabrizi K. 2009. Large-scale phylogenomic analyses reveal that two enigmatic protist lineages, Telonemia and Centroheliozoa, are related to photosynthetic chromalveolates. Genome Biology and Evolution 1: 231–238.

Campbell V, Legendre P, Lapointe FJ. 2011. The performance of the congruence among distance matrices (CADM) test in phylogenetic analysis. BMC Evolutionary Biology 11: 64.

Castresana J. 2000. Selection of conserved blocks from multiple alignments for their use in phylogenetic analysis. Molecular Biology and Evolution 17: 540–552.

Criscuolo A, Gribaldo S. 2010. BMGE (Block Mapping and Gathering with Entropy): A new software for selection of phylogenetic informative regions from multiple sequence alignments. BMC Evolutionary Biology 10: 210.

Darriba D, Taboada GL, Doallo R, Posada D. 2011. ProtTest 3: Fast selection of best-fit models of protein evolution. Bioinformatics (Oxford, England) 27: 1164–1165.

Desper R, Gascuel O. 2006. Getting a tree fast: Neighbor Joining, FastME, and distance-based methods. Curr Protoc Bioinformatics, Chapter 6, Unit 63.

Dimmic MW, Rest JS, Mindell DP, Goldstein RA. 2002. rtREV: An amino acid substitution matrix for inference of retrovirus and reverse transcriptase phylogeny. Journal of Molecular Evolution 55: 65–73.

Do CB, Brudno M, Batzoglou S. 2004. PROBCONS: Probabilistic consistency-based multiple alignment of amino acid sequences. In: *Proceeding of the Nineteenth National Conference on Artificial Intelligence and the Sixteenth Conference on Innovative Applications of Artificial Intelligence*. Menlo Pk, American Association of Artificial Intelligence.

Doolittle WF. 1999. Phylogenetic classification and the universal tree. Science 284: 2124–2128.

Douady CJ, Delsuc F, Boucher Y, Doolittle WF, Douzery EJ. 2003. Comparison of Bayesian and maximum likelihood bootstrap measures of phylogenetic reliability. Molecular Biology and Evolution 20: 248–254.

Drummond AJ, Rambaut A. 2007. BEAST: Bayesian evolutionary analysis by sampling trees. BMC Evolutionary Biology 7: 8.

Ebersberger I, Galgoczy P, Taudlen S, Taenzer S, Platzer M, Von Haeseler A. 2007. Mapping human genetic ancestry. Molecular Biology and Evolution 24: 2266–2276.

Eddy SR. 2009. A new generation of homology search tools based on probabilistic inference. Genome Informatics. International Conference on Genome Informatics 23: 205–211.

Edgar RC. 2004. MUSCLE: Multiple sequence alignment with high accuracy and high throughput. Nucleic Acids Research 32: 1792–1797.

Edlind TD, Li J, Visvesvara MH. 1996. Phylogenetic analysis of beta-tubulin sequences from amitochondrial protozoa. Molecular Phylogenetics and Evolution 5: 359–367.

Felsenstein J. 1978. Cases in which parsimony or compatibility methods will be positively misleading. Systematic Zoology 27: 401–410.

Felsenstein J. 1985. Confidence limits on phylogenies—An approach using the bootstrap. Evolution 39: 783–791.

Felsenstein J. 2004. *Inferring Phylogenies*. Sinauer Associates, Sunderland, MA.

Felsenstein J. 2005. PHYLIP (Phylogeny Inference Package) version 3.6. Distributed by the author, Department of Genome Sciences, University of Washington, Seattle.

Fischer WM, Palmer JD. 2005. Evidence from small-subunit ribosomal RNA sequences for a fungal origin of Microsporidia. Molecular Phylogenetics and Evolution 36: 606–622.

Foster PG. 2004. Modeling compositional heterogeneity. Systematic Biology 53: 485–495.

Foster PG, Cox CJ, Embley TM. 2009. The primary divisions of life: A phylogenomic approach employing composition-heterogeneous methods. Philosophical Transactions of the Royal Society of London. Series B, Biological Sciences 364: 2197–2207.

Frickey T, Lupas AN. 2004. PhyloGenie: Automated phylome generation and analysis. Nucleic Acids Research 32: 5231–5238.

Galtier N. 2001. Maximum-likelihood phylogenetic analysis under a covarion-like model. Molecular Biology and Evolution 18: 866–873.

Galtier N, Gouy M. 1998. Inferring pattern and process: Maximum-likelihood implementation of a nonhomogeneous model of DNA sequence evolution for phylogenetic analysis. Molecular Biology and Evolution 15: 871–879.

Gile GH, Faktorova D, Castlejohn CA, Burger G, Lang BF, Farmer MA, Lukes J, Keeling PJ. 2009. Distribution and phylogeny of EFL and EF-1 alpha in Euglenozoa suggest ancestral co-occurrence followed by differential loss. PLoS One 4: 9.

Gouret P, Thompson JD, Pontarotti P. 2009. PhyloPattern: Regular expressions to identify complex patterns in phylogenetic trees. BMC Bioinformatics 10: 12.

Gu X, Li WH. 1996. Bias-corrected paralinear and LogDet distances and tests of molecular clocks and phylogenies under nonstationary nucleotide frequencies. Molecular Biology and Evolution 13: 1375–1383.

Guindon S, Dufayard JF, Lefort V, Anisimova M, Hordijk W, Gascuel O. 2010. New algorithms and methods to estimate maximum-likelihood phylogenies: Assessing the performance of PhyML 3.0. Systematic Biology 59: 307–321.

Hampl V, Horner DS, Dyal P, Kulda J, Flegr J, Foster PG, Embley TM. 2005. Inference of the phylogenetic position of oxymonads based on nine genes: Support for Metamonada and Excavata. Molecular Biology and Evolution 22: 2508–2518.

Hampl V, Hug L, Leigh JW, Dacks JB, Lang BF, Simpson AGB, Roger AJ. 2009. Phylogenomic analyses support the monophyly of Excavata and resolve relationships among eukaryotic "supergroups." Proceedings of the National Academy of Sciences of the United States of America 106: 3859–3864.

Heath TA, Hedtke SM, Hillis DM. 2008. Taxon sampling and the accuracy of phylogenetic analyses. Journal of Systematics and Evolution 46: 239–257.

Holder MT, Zwickl DJ, Dessimoz C. 2008. Evaluating the robustness of phylogenetic methods to among-site variability in substitution processes. Philosophical Transactions of the Royal Society of London. Series B, Biological Sciences 363: 4013–4021.

Huelsenbeck J, Rannala B. 2004. Frequentist properties of Bayesian posterior probabilities of phylogenetic trees under simple and complex substitution models. Systematic Biology 53: 904–913.

Huelsenbeck JP. 2002. Testing a covariotide model of DNA substitution. Molecular Biology and Evolution 19: 698–707.

Huelsenbeck JP, Larget B, Alfaro ME. 2004. Bayesian phylogenetic model selection using reversible jump Markov chain Monte Carlo. Molecular Biology and Evolution 26: 1123.

Huson DH, Scornavacca C. 2011. A survey of combinatorial methods for phylogenetic networks. Genome Biology and Evolution 3: 23–35.

Inagaki Y, Simpson AGB, Dacks JB, Roger AJ. 2004. Phylogenetic artifacts can be caused by leucine, serine, and arginine codon usage heterogeneity: Dinoflagellate plastid origins as a case study. Systematic Biology 53: 582–593.

Jayaswal V, Jermiin LS, Robinson J. 2005. Estimation of phylogeny using a general Markov model. Evolutionary Bioinformatics 1: 62–80.

Jones MO, Koutsovoulos GD, Blaxter ML. 2011. iPhy: An integrated phylogenetic workbench for supermatrix analyses. BMC Bioinformatics Online 12: 30.

Kamaishi T, Hashimoto T, Nakamura Y, Nakamura F, Murata S, Okada N,K,O, Shimizu M, Hasegawa M. 1996. Protein phylogeny of translation elongation factor EF-1α suggests microsporidians are extremely ancient eukaryotes. Journal of Molecular Evolution 42: 257–263.

Kamikawa R, Sakaguchi M, Matsumoto T, Hashimoto T, Inagaki Y. 2010. Rooting for the root of elongation factor-like protein phylogeny. Molecular Phylogenetics and Evolution 56: 1082–1088.

Katoh K, Kei-Ichi K, Hiroyuki T, Takashi M. 2005. MAFFT version 5: Improvement in accuracy of multiple sequence alignment. Nucleic Acids Research 33: 511–518.

Kolaczkowski B, Thornton JW. 2008. A mixed branch length model of heterotachy improves phylogenetic accuracy. Molecular Biology and Evolution 25: 1054–1066.

Kolisko M, Silberman JD, Cepicka I, Yubuki N, Takishita K, Yabuki A, Leander BS, Inouye I, Inagaki Y, Roger AJ, Simpson AGB. 2010. A wide diversity of previously undetected free-living relatives of diplomonads isolated from marine/saline habitats. Environmental Microbiology 12: 2700–2710.

Landan G, Graur D. 2007. Heads or tails: A simple reliability check for multiple sequence alignments. Molecular Biology and Evolution 24: 1380–1383.

Lane CE, Archibald JM. 2008. The eukaryotic tree of life: Endosymbiosis takes its TOL. Trends in Ecology & Evolution 23: 268–275.

Larkin MA, Blackshields G, Brown NP, Chenna R, Mcgettigan PA, Mcwilliam H, Valentin F, Wallace IM, Wilm A, Lopez R, Thompson JD, Gibson TJ, Higgins DG. 2007. ClustalW and ClustalX version 2.0. Bioinformatics (Oxford, England) 23: 2947–2948.

Lartillot N, Philippe H. 2004. A Bayesian mixture model for across-site heterogeneities in the amino-acid replacement process. Molecular Biology and Evolution 21: 1095–1109.

Lartillot N, Philippe H. 2006. Computing Bayes factors using thermodynamic integration. Systematic Biology 55: 195–207.

Lartillot N, Brinkmann H, Philippe H. 2007. Suppression of long-branch attraction artefacts in the animal phylogeny using a site-heterogeneous model. BMC Evolutionary Biology 7(Suppl I): S4–S18.

Lartillot N, Lepage T, Blanquart S. 2009. PhyloBayes 3: A Bayesian software package for phylogenetic reconstruction and molecular dating. Bioinformatics (Oxford, England) 25: 2286–2288.

Le SQ, Gascuel O. 2008. An improved general amino acid replacement matrix. Molecular Biology and Evolution 25: 1307–1320.

Le SQ, Lartillot N, Gascuel O. 2008. Phylogenetic mixture models for proteins. Philosophical Transactions of the Royal Society of

London. Series B, Biological Sciences 363: 3965–3976.

Leigh JW, Susko E, Baumgartner M, Roger AJ. 2008. Testing congruence in phylogenomic analysis. Systematic Biology 57: 104–115.

Leigh JW, Lapointe F-J, Lopez P, BAPTESTE E. 2011a. Evaluating phylogenetic congruence in the post-genomic era. Genome Biology 3: 571–587.

Leigh JW, Schliep K, Lopez P, Bapteste E. 2011b. Let them fall where they may: Congruence analysis in massive, phylogenetically messy datasets. Molecular Biology and Evolution 28: 2773–2785.

Li L, Stoeckert CJ, Roos DS. 2003. OrthoMCL: Identification of ortholog groups for eukaryotic genomes. Genome Research 13: 2178–2189.

Linard B, Thompson JD, Poch O, Lecompte O. 2011. OrthoInspector: Comprehensive orthology analysis and visual exploration. BMC Bioinformatics 12: 11.

Mitreva M, Smant G, Helder J. 2009. Role of horizontal gene transfer in the evolution of plant parasitism among nematodes. Methods in Molecular Biology (Clifton, N.J.) 532: 517–535.

Moustafa A, Bhattacharya D. 2008. PhyloSort: A user-friendly phylogenetic sorting tool and its application to estimating the cyanobacterial contribution to the nuclear genome of Chlamydomonas. BMC Evolutionary Biology 8: 7.

Newton MA, Raftery AE. 1994. Approximate Bayesian inference with the weighted likelihood bootstrap. Journal of the Royal Statistical Society. Series B 56: 3–48.

Notredame C, Higgins DG, Heringa J. 2000. T-Coffee: A novel method for fast and accurate multiple sequence alignment. Journal of Molecular Biology 302: 205–217.

Ostlund G, Schmitt T, Forslund K, Kostler T, Messina DN, Roopra S, Frings O, Sonnhammer EL. 2011. InParanoid 7: New algorithms and tools for eukaryotic orthology analysis. Nucleic Acids Research 38: D196–D203.

Pei JM, Grishin NV. 2001. AL2CO: Calculation of positional conservation in a protein sequence alignment. Bioinformatics (Oxford, England) 17: 700–712.

Penn O, Privman E, Ashkenazy H, Landan G, Graur D, Pupko T. 2010. GUIDANCE: A web server for assessing alignment confidence scores. Nucleic Acids Research 38: W23–W28.

Philippe H, Lopez P, Brinkmann H, Budin K, Germot A, Laurent J, Moreira D, Muller M, Le Guyader H. 2000. Early-branching or fast-evolving eukaryotes? An answer based on slowly evolving positions. Proceedings of the Royal Society of London. Series B, Biological Sciences 267: 1213–1221.

Phillips MJ, Delsuc F, Penny D. 2004. Genome-scale phylogeny and the detection of systematic biases. Molecular Biology and Evolution 21: 1455–1458.

Poe S. 2003. Evaluation of the strategy of long-branch subdivision to improve the accuracy of phylogenetic methods. Systematic Biology 52: 423–428.

Poptsova MS, Gogarten JP. 2007. BranchClust: A phylogenetic algorithm for selecting gene families. BMC Bioinformatics 8: 120.

Posada D. 2006. ModelTest Server: A web-based tool for the statistical selection of models of nucleotide substitution online. Nucleic Acids Research 34: W700–W703.

Price MN, Dehal PS, Arkin AP. 2010. FastTree 2-approximately maximum-likelihood trees for large alignments. PLoS One 5: e9490.

Pupko T, Huchon D, Cao Y, Okada N, Hasegawa M. 2002. Combining multiple data sets in a likelihood analysis: Which models are the best? Molecular Biology and Evolution 19: 2294–2307.

Rambaut A, Drummond AJ. 2007. Tracer v1.4. Available from http://beast.bio.ed.ac.uk/Tracer

Rannala B, Yang ZH. 2008. Phylogenetic inference using whole genomes. Annual Review of Genomics and Human Genetics 9: 217–231.

Ren F, Tanaka H, Yang Z. 2005. An empirical examination of the utility of codon-substitution models in phylogeny reconstruction. Systematic Biology 54: 808–818.

Richards TA, Dacks JB, Jenkinson JM, Thornton CR, Talbot NJ. 2006. Evolution of filamentous plant pathogens: Gene exchange across eukaryotic kingdoms. Current Biology: CB 16: 1857–1864.

Rodriguez-Ezpeleta N, Brinkmann H, Burger G, Roger AJ, Gray MW, Philippe H, Lang BF. 2007a. Toward resolving the eukaryotic tree: The phylogenetic positions of jakobids and cercozoans. Current Biology: CB 17: 1420–1425.

Rodriguez-Ezpeleta N, Brinkmann H, Roure B, Lartillot N, Lang BF, Philippe H. 2007b. Detecting and overcoming systematic errors in genome-scale phylogenies. Systematic Biology 56: 389–399.

Ronquist F, Huelsenbeck JP. 2003. MrBayes 3: Bayesian phylogenetic inference under mixed models. Bioinformatics (Oxford, England) 19: 1572–1574.

Roshan U, Livesay DR. 2006. Probalign: Multiple sequence alignment using partition function posterior probabilities. Bioinformatics (Oxford, England) 22: 2715–2721.

Roure B, Rodriguez-Ezpeleta N, Philippe H. 2007. SCaFoS: A tool for selection, concatenation and fusion of sequences for phylogenomics. BMC Evolutionary Biology 7: 12.

Sawa G, Dicks J, Roberts IN. 2003. Current approaches to whole genome phylogenetic analysis. Briefings in Bioinformatics 4: 63–74.

Schreiber F, Pick K, Erpenbeck D, Worheide G, Morgenstern B. 2009. OrthoSelect: A protocol for selecting orthologous groups in phylogenomics. BMC Bioinformatics 10: 12.

Shimodaira H, Hasegawa M. 2001. CONSEL: For assessing the confidence of phylogenetic tree selection. Bioinformatics (Oxford, England) 17: 1246–1247.

Simpson AGB, Inagaki Y, Roger AJ. 2006. Comprehensive multigene phylogenies of excavate protists reveal the evolutionary positions of "primitive" eukaryotes. Molecular Biology and Evolution 23: 615–625.

Simpson AGB, Perley TA, Lara E. 2008. Lateral transfer of the gene for a widely used marker, alpha-tubulin, indicated by a multi-protein study of the phylogenetic position of *Andalucia* (Excavata). Molecular Phylogenetics and Evolution 47: 366–377.

Slot JC, Rokas A. 2011. Horizontal transfer of a large and highly toxic secondary metabolic gene cluster between fungi. Current Biology: CB 21: 134–139.

Spencer M, Sangaralingam A. 2009. A phylogenetic mixture model for gene family loss in parasitic bacteria. Molecular Biology and Evolution 26: 1901–1908.

Stamatakis A. 2006. RAxML-VI-HPC: Maximum likelihood-based phylogenetic analyses with thousands of taxa and mixed models. Bioinformatics (Oxford, England) 22: 2688–2690.

Suchard MA, Rambaut A. 2009. Many-core algorithms for statistical phylogenetics. Bioinformatics (Oxford, England) 25: 1370–1376.

Sun J, Jiang H, Flores R, Wen J. 2010. Gene duplication in the genome of parasitic *Giardia lamblia*. BMC Evolutionary Biology 10: 49.

Susko E. 2008. On the distributions of bootstrap support and posterior distributions for a star tree. Systematic Biology 57: 602–612.

Susko E. 2009. Bootstrap support is not first-order correct. Systematic Biology 58: 211–223.

Susko E. 2010. First-order correct bootstrap support adjustments for splits that allow hypothesis testing when using maximum likelihood estimation. Molecular Biology and Evolution 27: 1621–1629.

Susko E. 2011. Improved least squares topology testing and estimation. Systematic Biology 60: 668–675.

Susko E, Roger AJ. 2007. On reduced amino acid alphabets for phylogenetic inference. Molecular Biology and Evolution 24: 2139–2150.

Susko E, Inagaki Y, Roger AJ. 2004. On inconsistency of the neighbor-joining, least squares, and minimum evolution estimation when substitution processes are incorrectly modeled. Molecular Biology and Evolution 21: 1629–1642.

Suzuki Y, Glazko GV, Nei M. 2002. Overcredibility of molecular phylogenies obtained by Bayesian phylogenetics. Proceedings of the National Academy of Sciences of the United States of America 99: 16138–16143.

Swofford DL. 2002. *PAUP*. Phylogenetic Analysis Using Parsimony (* and Other Methods),*

*Version 4*B10*. Sinauer Associates, Sunderland, MA.

Tamura K, Dudley J, Nei M, Kumar S. 2007. MEGA4: Molecular evolutionary genetics analysis (MEGA) software version 4.0. Molecular Biology and Evolution 24: 1596–1599.

Thompson JD, Linard B, Lecompte O, Poch O. 2011. A comprehensive benchmark study of multiple sequence alignment methods: Current challenges and future perspectives. PLoS One 6: 14.

Trachana K, Larsson TA, Powell S, Chen WH, Doerks T, Muller J, Bork P. 2011. Orthology prediction methods: A quality assessment using curated protein families. BioEssays: News and Reviews in Molecular, Cellular and Developmental Biology 33: 769–780.

Van de Peer Y, Baldauf SL, Doolittle WF, Meyer A. 2000. An updated and comprehensive rRNA phylogeny of (crown) eukaryotes based on rate-calibrated evolutionary distances. Journal of Molecular Evolution 51: 565–576.

Van Keulen H, Gutell RR, Gates MA, Campbell SR, Erlandsen SL, Jarroll EL, Kulda J, Meyer EA. 1993. Unique phylogenetic position of Diplomonadida based on the complete small subunit ribosomal-RNA sequence of *Giardia ardae*, *G. muris*, *G. duodenalis* and *Hexamita* sp. The FASEB Journal: Official Publication of the Federation of American Societies for Experimental Biology 7: 223–231.

Wang HC, Li K, Susko E, Roger AJ. 2008. A class frequency mixture model that adjusts for site-specific amino acid frequencies and improves inference of protein phylogeny. BMC Evolutionary Biology 8: 331–344.

Wang HC, Susko E, Roger AJ. 2009. PROCOV: Maximum likelihood estimation of protein phylogeny under covarion models and site-specific covarion pattern analysis. BMC Evolutionary Biology 9: 225–238.

Wellner A, Lurie MN, Gophna U. 2007. Complexity, connectivity and duplicability as barriers to lateral gene transfer. Genome Biology 8: R156.

Wernersson R, Pedersen AG. 2003. RevTrans: Multiple alignment of coding DNA from aligned amino acid sequences. Nucleic Acids Research 31: 3537–3539.

Whelan S, Lio P, Goldman N. 2001. Molecular phylogenetics: State-of-the-art methods for looking into the past. Trends in Genetics: TIG 17: 262–272.

Whelan S, Blackburne BP, Spencer M. 2011. Phylogenetic substitution models for detecting heterotachy during plastid evolution. Molecular Biology and Evolution 28: 449–458.

Wu M, Eisen JA. 2008. A simple, fast, and accurate method of phylogenomic inference. Genome Biology 9: 34.

Xie W, Lewis PO, Fan Y, Kuo L, Chen MH. 2011. Improving marginal likelihood estimation for Bayesian phylogenetic model selection. Systematic Biology 60: 150–160.

Yang Z. 1997. PAML: A program package for phylogenetic analysis by maximum likelihood. Computer Applications in the Biosciences 13: 555–556.

Yang Z. 2006. *Computational Molecular Evolution*. Oxford University Press, Oxford, England.

Yang Z. 2008. Empirical evaluation of a prior for Bayesian phylogenetic inference. Philosophical Transactions of the Royal Society of London. Series B, Biological Sciences 363: 4031–4039.

Yang ZH, Roberts D. 1995. On the use of nucleic-acid sequences to infer early branchings in the tree of life. Molecular Biology and Evolution 12: 451–458.

Zhou Y, Rodrigue N, Lartillot N, Philippe H. 2007. Evaluation of the models handling heterotachy in phylogenetic inference. BMC Evolutionary Biology 7: 206.

Zwickl DJ. 2006. Genetic algorithm approaches for the phylogenetic analysis of large biological sequence datasets under the maximum likelihood criterion. PhD Dissertation, Austin, The University of Texas.

CHAPTER 4

PHYLOGENETICS AND EVOLUTION OF VIRULENCE IN THE KINGDOM FUNGI

MONICA A. GARCIA-SOLACHE and ARTURO CASADEVALL

GENERAL CONCEPTS ON FUNGAL EVOLUTION

Fungal biodiversity is one of the highest in the eukaryotic kingdoms, rivaling plants and animals in the number of predicted species. Although only about 70,000 have been formally described, the estimated number of fungal species is 1.5 million (Hawksworth, 1991, 2001; Lutzoni et al., 2004). Considering the great biodiversity of fungi, it is remarkable that only about 300 species are reported human pathogens, of which only a few cause relative frequent disease in humans (Taylor et al., 2001), compared to 15,000 described fungal pathogens of plants (Gonzalez-Fernandez et al., 2010) and 700 pathogenic fungi of invertebrates (Roy et al., 2006). In fact, the majority of fungal diseases in humans are caused by only a few species of *Candida* and dermatophytes. Hence, humans, as well as mammals in general, manifest strong resistance to the majority of fungi with pathogenic potential found in nature.

Fungi are part of the opisthokonts, a well-supported group composed of the Metazoa, the Choanozoa, the nucleariids, and the Fungi, as well as the less well-described groups Ichthyosporea and Capsaspora (Cavalier-Smith, 2002; Keeling et al., 2005; Shalchian-Tabrizi et al., 2008; Stechmann and Cavalier-Smith, 2002; Steenkamp et al., 2006). The nucleariids, a predatory clade of aflagellated amoebas with sharp, starlike pseudopodia, is a sister group to the Fungi (Fig. 4.1) (Liu et al., 2009b; Medina et al., 2003). Fungi are generally sparse in the fossil record, and this is especially true for unicellular fungi that constitute the overwhelming majority of pathogens. Fungal fossil history is incomplete, and the lack of a good record has complicated the calibration of molecular trees to obtain an accurate fungal tree of life (FTOL). Peterson et al. (2003) suggested a fungal identity for some Ediacaran biota fossils (635–442 million years ago [mya]). Butterfield (2005) described an even older putative fossilized fungus from the Proterozoic, some 1430 mya. However, the oldest fossils unambiguously considered to be a fungus are between 460 and 400 mya old (Heckman et al., 2001; Redecker et al., 2000; Remy et al., 1994).

Evolution of Virulence in Eukaryotic Microbes, First Edition. Edited by L. David Sibley, Barbara J. Howlett, and Joseph Heitman.

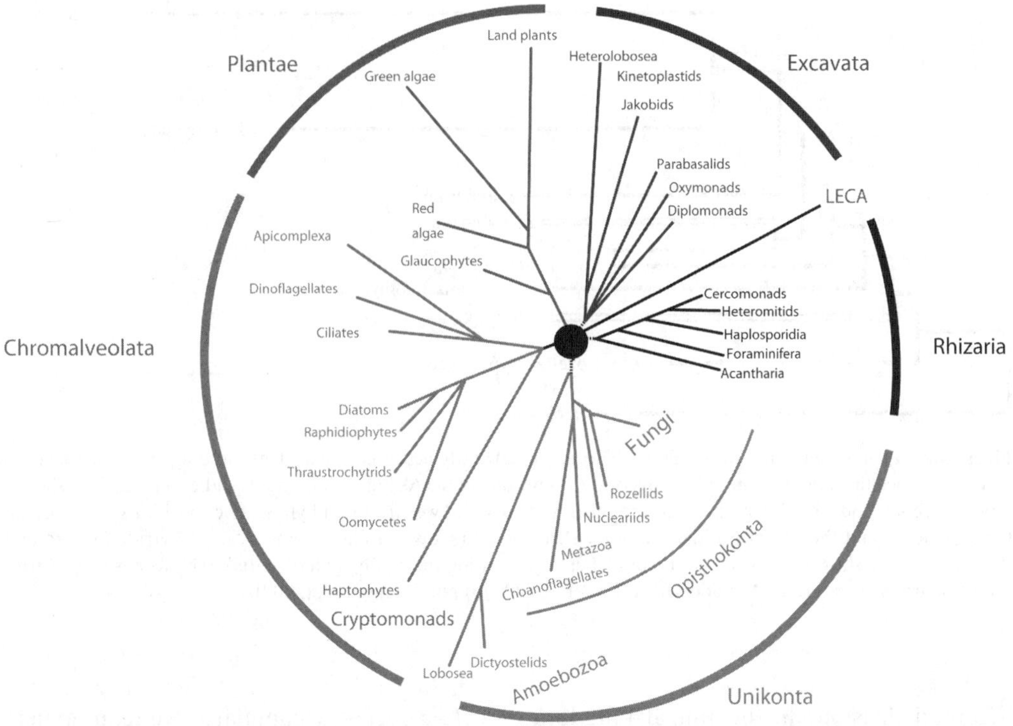

Figure 4.1 Eukaryote phylogenetic tree. Star phylogeny including the last eukaryote common ancestor (LECA). The root of the tree is not well resolved and is not shown here; the proposed relationships between the currently accepted five eukaryotic supergroups (Excavata, Rhizaria, Unikonta, Chromaleveolata, and Plantae) are shown. The monophyly of the plants and the unikonts is well supported, while others are not as well supported. Modified from Koonin (2010). See color insert.

In the absence of an extensive fossil record, most of our insights about fungal evolution are derived from the analysis of DNA data. Currently, based mostly on molecular data, fungi are considered to have diverged from the group leading to the Metazoa between 760 and 1500 mya (Hedges et al., 2004; Lucking et al., 2009; Lutzoni et al., 2004; Padovan et al., 2005; Taylor and Berbee, 2006; Wang et al., 1999). Fungal land colonization, a major event for land ecosystems, took place between 500 and 1000 mya (Berbee and Taylor, 2001; Heckman et al., 2001; Taylor and Osborn, 1996). It was close to this event that the divergence of the two largest fungal phyla, the Ascomycota and the Basidiomycota, occurred, which is described in detail in the next section.

FUNGAL PHYLOGENY

Historically, fungal systematics has been based largely on morphological and physiological characteristics, as well as nuances of sexual stages. However, within the last two decades, the widespread use of genetic data allowed revision of the FTOL, confirming the relationships of some groups and dramatically changing others (Fitzpatrick et al., 2006; James et al., 2006a; Wang et al., 2009). The continuously increasing use of multilocus data sets and more extensive taxon sampling has greatly helped us to understand the high level phylogeny of the kingdom and to clarify the relationships among its members; however, controversy regarding the rank of some groups and their position in the FTOL exists. Currently the

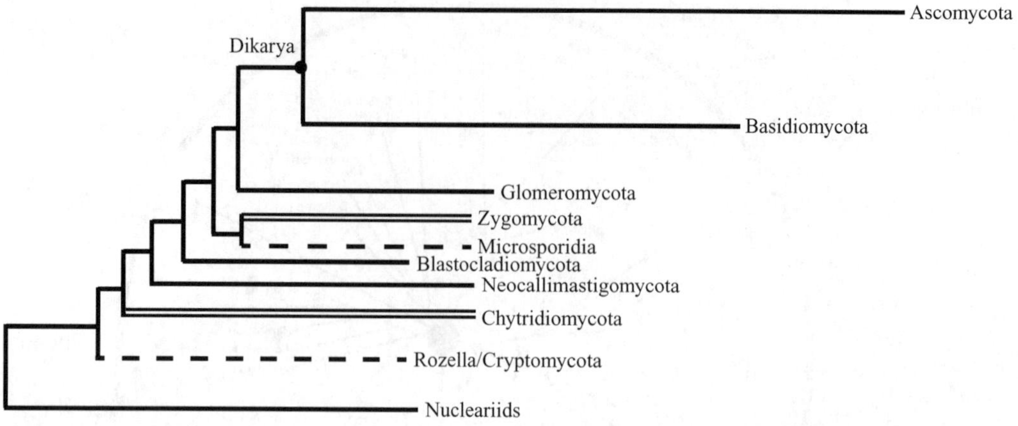

Figure 4.2 Phylogenetic tree of the Fungi. The seven currently accepted phyla (Chytridiomycota, Blastocladiomycota, Neocallimastigomycota, Zygomycota, Glomeromycota, Ascomycota, and Basidiomycota) are shown. Cryptomycota and the Microsporidia are also included; however, the phylogenetic positions of *Rozella*/Cryptomycota and the Microsporidia are not well resolved (shown in dashed lines). The Chytridiomycota and Zygomycota are shown with double line branches representing their polyphyletic nature. The tree is rooted with the nucleariids as outgroup. Modified from Liu et al. (2009b) and Wang et al. (2009).

accepted divisions in the fungal kingdom comprise seven phyla: the Chytridiomycota, Blastocladiomycota (previously included within the Chytridiomycota), Neocallimastigomycota, Zygomycota, Glomeromycota, Ascomycota, and Basidiomycota, with the two last forming the supraphylum (subkingdom) Dikarya (Fig. 4.2) (Hibbett et al., 2007). The Chytridiomycota and Zygomycota are not monophyletic assemblages (Hibbett et al., 2007; James et al., 2006a,b; Lutzoni et al., 2004; White et al., 2006), but the precise phylogenetic relationships within these groups are not entirely resolved and there are very few full genome sequences available (Wang et al., 2009). Recently, the endoparasitic genus *Rozella*, a chitin-lacking fungus formerly included in the Chytridiomycota, has been placed in a new, temporary clade known as the Cryptomycota, which includes a broad group of unculturable organisms sampled from the environment that seem to be in an early branching position within the fungal kingdom (Jones et al., 2011). As full genome sequences of zygomycete and chytridiomycete species accumulate, we can expect that the relationships in these two phyla will change, and some groups may possibly be elevated to the phylum category. Additionally, controversies exist, such as how many times the ancestral flagellum, which was present in the last common fungal ancestor, was lost during fungal evolution. The only fungi that bear flagellated spores are classified within the Chytridiomycota, but James et al. (2006a), using data for six genes in 200 fungal species, proposed at least four independent losses of the flagellum. In contrast, Liu et al. (2006) concluded that the flagellum loss occurred only once in evolution based on data from RNA polymerase II genes from 58 taxa and sampling the major groups of fungi, including Microsporidia.

Based mostly on morphological characteristics, it is considered that the most basal fungal lineages reside within the Chytridiomycota, as discussed in Lutzoni et al. (2004). The Neocallimastigomycota were formerly classified within the Chytridiomycota but have now ascended to phylum status (Hibbett et al., 2007); this group consists of anaerobic gut symbionts in ruminant and nonruminant vertebrates (Liggenstoffer et al., 2010).

The obligate endoparasitic Microsporidia, which for a long time were considered basal eukaryotes, are now regarded as bona fide fungi (Keeling et al., 2000). However, their precise phylogenetic affinities are not very well resolved due to their extremely reduced genome size and their high evolutionary rate (Keeling and Slamovits, 2004; Slamovits et al., 2004). There is some genetic evidence that Microsporidia diverged from a zygomycete ancestor, either within the Entomophthorales and Zoopagales or the Mucorales (Keeling, 2003; Lee et al., 2008; Thomarat et al., 2004); however, a more basal position in the FTOL (Hibbett et al., 2007; James et al., 2006a; Wang et al., 2009) or even a position as a sister taxon without a definite fungal identity (Liu et al., 2006) cannot be excluded.

The Glomeromycota, or arbuscular mycorrhyzea fungi, were previously considered to belong to the Zygomycota but is now elevated to phylum status. This group is the probable sister group to the Dikarya (Schuessler et al., 2001).

The Ascomycota and Basidiomycota are monophyletic sister taxa, which together form the supraphylum Dikarya. Depending on the data set used and the analytic tools used, the divergence time between the Ascomycota and the Basidiomycota has been calculated to be between 400 and 1000 mya (Heckman et al., 2001; Lutzoni et al., 2004; Taylor and Berbee, 2006), most likely shortly after land colonization occurred.

The Ascomycota is the biggest and most diverse group of fungi. This group contains the vast majority of pathogenic fungi, followed by the Zygomycota. In addition, human pathogenic fungi are also found within the Microsporidia and the Basidiomycota. There are no reports of chytrids being pathogenic to humans; however, a single species, the keratophilic fungus *Batrachochytrium dendrobatidis*, has caused a huge decline in amphibian populations worldwide and is responsible for extinction events in several species, as is thoroughly reviewed by Fisher et al. (2009). Chytrids are found predominantly in temperate aquatic habitats, perhaps retaining some of the traits that were present in the ancestors of modern-day fungi.

MECHANISMS TO GENERATE GENETIC DIVERSITY

Major mechanisms that organisms can use to generate genetic diversity include mutations, homologous recombination, gene inversions, deletions and insertions such as horizontal gene transfer (HGT), and duplications (for genes, chromosomes, or even whole genomes). The relative contribution to any given mechanism on pathogen adaptation is highly dependent on the group of organisms studied, but point mutations are regarded to be the foundation for generating most of the genetic variability in eukaryotes. Deletions are also important adaptive mechanisms in obligate parasites and endosymbionts, where genome reduction could be advantageous as many biosynthetic pathways are rendered redundant by their capacity for nutrient acquisition from the host. Dramatic genome reductions are present in the obligate intracellular parasitic Microsporidia. The human pathogenic microsporidian *Encephalitozoon intestinalis* has the smallest identified genome of any eukaryote, coding for about 1800 genes in a 2.3-Mbp genome (Corradi et al., 2010).

HGT is recognized as a major driving force in the evolution of bacterial populations, and the high degree of bacterial adaptation to diverse and changing environments has been associated with the ability to incorporate exogenous DNA into their genomes or extrachromosomically (de la

Cruz and Davies, 2000; Lawrence, 2002; Ochman et al., 2000). Sources of horizontally obtained DNA can be plasmids, phage-carried genes, or transposable elements. Some of the most medically relevant traits that pathogenic bacteria can acquire via HGT are antibiotic resistance (mostly through plasmids) and so-called pathogenicity islands, which are chromosomically encoded virulence genes that can transform an otherwise benign bacteria into a pathogenic one (Groisman and Ochman, 1996). These events can occur both within related strains and between very different species, or even transkingdom (Jain et al., 2002).

In fungi, HGT has been long suspected, but evidence of its occurrence has been elusive until recently, and its evolutionary implications are not fully understood. HGT has been reported to occur both via interkingdom acquisition from bacteria and by intrakingdom acquisition events (Hall and Dietrich, 2007; Hall et al., 2005; Rosewich and Kistler, 2000; Schmitt and Lumbsch, 2009). Large DNA data sets and powerful phylogenomic analyses have proved to be useful in the inference of ancient and rare HGT events, as shown by Richards et al. (2009), who demonstrated HGT between plants and fungi. Marcet-Houben and Gabaldon (2010) have proposed that HGT has played an important role on fungal genome evolution; these authors report that in a survey of 60 fungal genomes, putative HGT events could be tracked and extrapolated to most fungal clades. In agreement with these findings, in a recent report, McDonald and collaborators (2012) found that fungi, unlike all other eukaryotes, posses a methylammonium permease gene (MEPγ) that was acquired via HGT from a prokaryote donor early in fungal evolution. The rest of the eukaryotic kingdoms have an ammonium transporter (AMT) that forms a single clade. The authors propose that a fungal ancestor bearing the eukaryotic AMT acquired a MEPγ from bacteria eventually replacing the AMT gene. The horizontally acquired MEPγ is a synapomorphy of all extant fungi.

In human pathogenic fungi, there are only few reports suggesting the acquisition of genes via HGT. Sequence-based data from *Candida parapsilosis* suggest the acquisition of at least two genes from a bacterial donor: a proline racemase and a phenazine F protein (Fitzpatrick et al., 2008). Similarly, another *Candida* species, *Candida glabrata* also apparently acquired a racemase enzyme from a bacterial donor (Marcet-Houben and Gabaldon, 2010). Additionally, indirect evidence of HGT has been recently obtained for the plant pathogen *Fusarium oxysporum* involving genes implied in pathogenesis; the authors propose that the source of the donor genes was another *Fusarium* species (Ma et al., 2010). It remains to be determined the extent to which HGT has contributed to the origin and maintenance of pathogenic traits in fungi, and more research in this direction is necessary to fully understand the impact of HGT on fungal evolution.

The rapidly increasing number of sequenced fungal genomes will permit a better understanding of phylogenetic relationships, both at high level phylogenetic and species levels in the near future. Furthermore, the availability of complete sequenced genomes makes possible large comparative genetic studies (CGSs). CGSs in pathogenic fungi are still in the initial stages but are already generating a more comprehensive idea regarding the evolution of pathogenic lineages. Comparative studies have been performed in pathogenic aspergilli, *Candida*, *Cryptococcus*, *Coccidioides* (Butler et al., 2009; Diaz et al., 2005; D'Souza et al., 2011; Fedorova et al., 2008; Goulart et al., 2010; Neafsey et al., 2010; Sharpton et al., 2009). As result of these studies, many important discoveries have been produced. By analyzing eight different *Candida* genomes, Butler et al. (2009) found that genes involved in virulence and sexual recombination are very dynamic and are under a high rate of turnover. A comparison of *Coccidioides* and related pathogenic and nonpathogenic species in the Onygenales has shown that in this lineage, there was a reduction in genes involved in cellulose degradation, in accordance with a shift from being primarily associated with plants to be mostly associated with animals (Sharpton et al., 2009).

Larger CGSs will be extremely useful to analyze the genetic variations that are linked to virulence and pathogen emergence, as well as the relevant differences between closely related species of pathogenic versus nonpathogenic fungi.

Despite the increasing number of sequenced fungal genomes, there is a long lag in their annotation. Well-annotated complete genomes would be powerful tools to make inferences regarding metabolic pathways and lifestyle choices and to track and understand the driving forces behind the origin and maintenance of pathogenicity. Consequently, as fungal genome annotation progresses, we can expect that there will be a parallel increase in our understanding of the mechanisms responsible for conferring pathogenicity to fungi.

Another way to generate genetic diversity is via sexual reproduction. Meiotic recombination creates novel gene combinations, increasing variation and purging deleterious mutations at the population level. Sex in fungi is diverse and many different strategies are found across the kingdom; for a review and discussion about the role of sex in fungal evolution, see Logsdon and Heitman in this book and Zeyl (2009). In pathogenic fungi, evidence for sexual reproduction has been elusive and many pathogens were considered to be largely clonal. It was not until recently that population genetic studies and genomic analysis revealed that many fungi have population structures compatible with sexual reproduction and that the mating-type loci (*MAT*) were present and seemed functional (Butler, 2010; Butler et al., 2009; Fraser et al., 2007; Paoletti et al., 2005). In some pathogenic fungi, the sexual cycle has not been observed yet, like in the case of *Coccidioides*; however, the population structure, the presence of a nearly equal number of MAT isomorphs and the presence of gene introgression between the two *Coccidioides* species strongly suggest active sexual recombination (Mandel et al., 2007; Neafsey et al., 2010). In others, like *Aspergillus fumigatus*, the direct observation of the sexual cycle nicely complemented the results obtained from population genetic studies and genomic studies (O'orman et al., 2009), making a breakthrough in the understanding of this deadly pathogen.

Meiotic recombination cannot be only a source for new gene combination; in self-sex mating, it could help in segregating adaptive mutations and it has been proposed to be also a mutagenic process capable of contributing further to genetic variation (Heitman, 2010).

EVOLUTION OF FUNGAL PATHOGENS

Fungi can be pathogens of basically any other eukaryotic group, and the evolution of parasitism and pathogenic capability evolved many times in the fungal phylogeny. Pathogenic and parasitic fungi are distributed throughout the FTOL, but the majority of known plant, invertebrate, and vertebrate pathogenic/parasitic fungi belong to the Ascomycota; specifically, the subphylum Pezizomycotina contains most of the currently recognized pathogens. Plant pathogens are mostly found in three classes within the Pezizomycotina: the Leotiomycetes, Dothideomycetes, and Sordariomycetes. Entomopathogenic fungi are clustered in the order Hypocreales within the Sordariomycetes, and mammalian pathogens are clustered in the class Eurotiomycetes (Berbee, 2001). Several

Basidiomycete species are important plant pathogens and are concentrated in the subphylum Ustilagomycotina, including *Ustilago maydis*, *Tilletia*, and *Puccinia* (Begerow et al., 2006). The Entomophthoromycotina, which was previously classified within the Zygomycota, is an important group of invertebrate pathogens, mainly of insects. The Microsporidia are an important group of animal parasites, infesting insects, crustaceans, bryozoans, fish, and mammals, among other metazoans (Smith, 2009). Their phylogenetic affinities seem to be either within the Entomophthorales/Zoopagales or the Mucorales (Lee et al., 2008), two groups that contain several animal pathogens.

The interactions between and within species are among the most powerful evolutionary forces in existence. Unlike most mammalian fungal pathogens, which are incidental, plant and entomopathogenic fungi have close associations with their hosts. They are coevolving systems, meaning that each interacting partner is under selective pressure in response to its counterpart adaptations, producing in many cases very specific associations (Roy et al., 2006). It is outside of the scope of this chapter to analyze plant and invertebrate pathogens and their evolutionary histories, but the interested reader is referred to the following reviews: Berbee (2001), van der Does and Rep (2007), and Humber (2008).

Human pathogenic fungi are a phylogenetically diverse group that can be divided into two major categories based on their ecology (not reflecting phylogenetic relationships). The environmental fungi comprise organisms living mostly in soil, decaying matter, and vegetation. This group includes the dimorphic ascomycetes and the basidiomycete *Cryptococcus neoformans*. These fungi are not commensals and do not have an obvious need for an animal host. In contrast, other human-associated fungi are considered to be part of the human flora, like the ascomycete *Candida albicans* and related species or the basidiomycete *Malassezia furfur*. Both groups of pathogens have evolved under very different selective pressures of the different ecological niches in which they live.

The evolutionary history of human pathogenic fungi has not been thoroughly studied. Relatively little is known about their natural environment and the contribution of the different factors to the acquisition of pathogenicity. Environmental pressures unrelated to survival in a mammal/human host have selected for many of the traits that, in the presence of a suitable host, act as virulence factors. One fundamental characteristic required for all human pathogens is the capacity for heat tolerance, meaning the capacity to survive and ideally grow at the high temperature of the mammalian body (37°C). Consequently, only those fungi able survive at mammalian temperatures have the potential to be pathogens. It is not clear which are the major constraints for heat tolerance acquisition in fungi; the comparison of thermotolerant lineages versus nonthermotolerant lineages would shed light upon the evolution and selection of this trait in the fungal groups that harbor pathogens.

Soil-dwelling fungi live in complex ecosystems where they interact with a great variety of organisms like soil protozoa (amoebas, slime molds) and small invertebrates (nematodes). Predator–prey interactions play an important role in the selection of phenotypic traits that increase fitness for hostile environments. The relationships of the soil amoeba *Acanthamoeba castellanii* with different environmental pathogenic fungi have been studied, demonstrating that *A. castellanii* interacts with distantly related fungi such as *C. neoformans*, *Blastomyces dermatitidis*, *Sporothrix shenckii*, and *Histoplasma capsulatum* in similar ways (Steenbergen and Casadevall, 2003; Steenbergen et al., 2004). These fungi are capable of surviving and replicating in the presence of the amoeba, and in the case of the dimorphic fungi *B. dermatitidis*, *S. shenckii*, and *H. capsulatum*, the coincubation with *A. castellanii* induced a phase transition from yeast to hyphal growth at 37°C, a temperature that is normally associated with the yeast state. Interestingly, when *A. castellanii* was challenged with the domesticated yeast *Saccharomyces cerevisiae* or the human-adapted fungus *C. albicans*, the amoebas were able to kill them (Steenbergen et al., 2004). These studies demonstrated the importance of the predator–prey selective pressure for the maintenance of virulence traits in environmental fungi.

Microsporidia. The Microsporidia are a highly diverse group of obligate intracellular parasites. They are able to infect nearly all of the metazoan groups, from marine invertebrates to insects to mammals. There are about 150 described genera and 1200 species (Keeling and Fast, 2002); however, it is likely that the number of species described is only a small amount of the total number considering the undersampling of deep water fish and invertebrates and other not highly accessible environments. Six genera and about 14 species have been identified as agents for human disease (Franzen and Muller, 1999).

Zygomycetes. The Zygomycota is a polyphyletic phylum of the fungi. The traditional zygomycete orders are the Mucoromycotina, Kickxellomycotina, Zoopagomycotina, and Entomophthoromycotina, but Liu et al. (2009b) proposed that these groups should be considered *incertae sedis* until their proper relationships are determined. The Glomeromycota were formerly considered to be within the Zygomycota but now are in their own phylum (Schuessler et al., 2001). The few fully sequenced genomes make zygomycete phylogenetic relationships incomplete, and no accurate trees are yet available.

Most pathogenic zygomycetes belong to the Mucorales. The most common disease-causing agents are within the genera *Rhizopus*, *Rhizomucor*, *Lichtheimia* (*Absidia*), *Cunninghamella*, and *Mucor*. The Mucorales are thermotolerant molds that are present in organic matter (Richardson, 2009). Other medically important species are within the Entomophthorales, a group that contains many entomopathogenic species, but at least two are able to cause human disease: *Basidiobolus ranarum* and *Conidiobolus coronatus* (Restrepo, 1994).

Ascomycetes. As mentioned in the previous sections, the Ascomycota is the largest group of fungi, with over 64,000 species. It is divided into three major subphyla, each containing saprobes, plant pathogens, and human pathogens (Berbee, 2001). The Taphrinomycotina is the most basal group of the Ascomycota and includes the pathogenic genus *Pneumocystis*; the other two subphyla are the Saccharomycotina (Hemiascomycetes) and the Pezizomycotina (Euascomycetes) (Liu et al., 2009a). Within the Saccharomycotina, only the *Saccharomyces*/*Candida* clade is implicated in human disease (Berbee, 2001). In contrast, the Pezizomycotina subphylum contains the vast majority of human pathogenic fungi. Most of the human pathogenic ascomycetes are concentrated within two classes: the Eurotiomycetes and the Chaetothyriomycetes, which form a monophyletic clade. Even though it is widely accepted that the ability to be a human pathogen evolved more than once independently (Bowman et al., 1992, 1996), it is likely that inherent traits of the fungi within these classes facilitate the capability to grow and establish infection in mammals (Fig. 4.3).

Pneumocystis. The genus *Pneumocystis* is formed by obligate parasites of mammals that have high species-specific affinities (Aliouat-Denis et al., 2008). They belong to the Taphrinomycotina and form an order of their own, the Pneumocystidales (Eriksson, 1994). The current view of their evolution

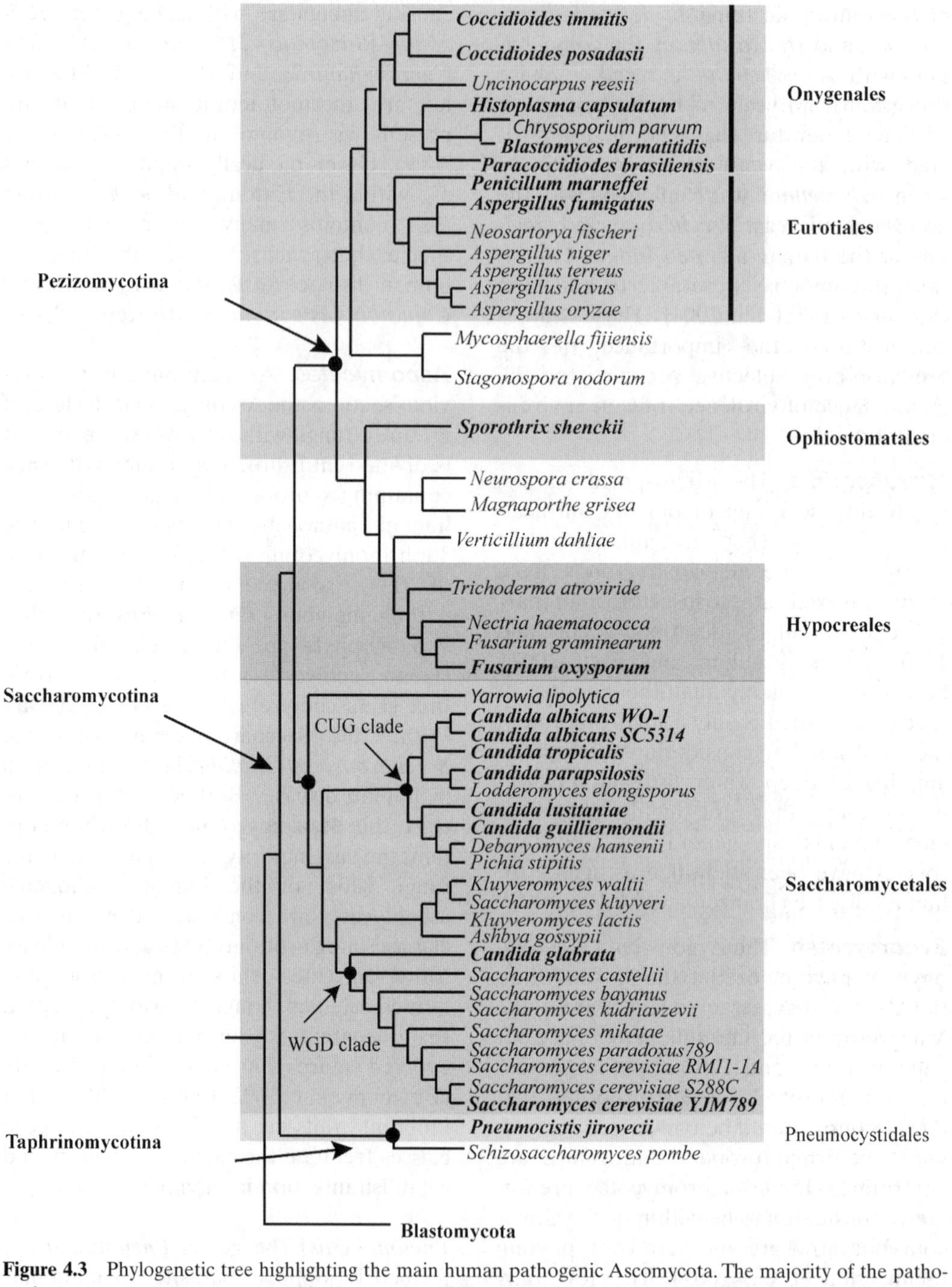

Figure 4.3 Phylogenetic tree highlighting the main human pathogenic Ascomycota. The majority of the pathogenic ascomycetes are concentrated in three orders: Onygenales, Eurotiales, and Saccharomycetales. Pathogenic species are shown in bold characters. The two major clades within the Saccharomycetales are shown; the CUG clade includes *Candida* species that translate the CUG codon into serine, and the other clade underwent a whole-genome duplication event and comprises *Saccharomyces*, which has emerged recently as an opportunistic pathogen. Modified from Wang et al. (2009). See color insert.

regards them as parasites that have coevolved with their hosts and that can colonize and multiply in the lungs of healthy individuals without necessarily causing disease (Chabe et al., 2004). Five species are formally recognized (*Pneumocystis carinii*, *Pneumocystis jirovecii*, *Pneumocystis wakefieldiae*, *Pneumocystis murina*, and *Pneumocystis oryctolagi*) (Cushion et al., 2004; Dei-Cas et al., 2006; Frenkel, 1999; Keely et al., 2004), but it is likely that a far greater number of species has yet to be described. In humans, only one species has been associated with disease, *P. jirovecii* (Durand-Joly et al., 2002). The evolutionary divergence time between the different *Pneumocystis* species and their mammalian hosts segregates in parallel, strongly suggesting that *Pneumocystis* speciation events occurred following the divergence of mammalian lineages and that we have ongoing coevolution with our *Pneumocystis* parasites (Demanche et al., 2001; Guillot et al., 2001; Keely et al., 2003).

Candida Species. Many of the yeasts in the *Candida/Saccharomyces* clade share a long evolutionary history with humans and other mammalian hosts. *C. albicans*, *Candida dubliniensis*, *Candida tropicalis*, *C. parapsilosis*, and *C. glabrata* have close associations with humans, and for these microbes, the change from commensal to pathogenic state largely reflects the interplay between the immune system and the fungus.

The majority of pathogenic *Candida* species are clustered in a monophyletic clade, but the genus *Candida* is not a monophyletic assemblage. Most of the pathogenic species form a monophyletic group characterized by the noncanonical coding change of the CUG codon from leucine into serine. This genetic change is calculated to have taken place about 175 mya (Massey et al., 2003), and in *C. albicans*, it has been proposed as a mechanism to generate phenotypic diversity by allowing codon ambiguity (some reversal from the acquired serine back to the ancestral leucine codon reading), increasing the potential proteome diversity of this organism (Gomes et al., 2007). Within the CUG clade are two different clusters: *C. albicans*, *C. dubliniensis*, *C. tropicalis*, *C. parapsilosis*, and *Lodderomyces elongisporus* form one group; *Candida lusitaniae*, *Candida guilliermondii*, and *Debaryomyces hansenii* form the other (Fitzpatrick et al., 2006). *C. glabrata* is more related to the baker's yeast *S. cerevisiae* than with the other *Candida* species. *S. cerevisiae* and *C. glabrata* are part of a clade that underwent a whole-genome duplication about 100–200 mya (Gordon et al., 2009), after the *Saccharomyces/Kluyveromyces* divergence (Wolfe and Shields, 1997).

The yeasts in the *Candida* clade (including all the species that use CUG_{ser}) have expanded gene families related to nutrient acquisition, cell wall-attached proteins, secretion, and transport. The expansion of gene families implicated in nutrient uptake and catabolic capabilities over a broad variety of carbon sources facilitates the growth in human hosts (van het Hoog et al., 2007). Some cell wall-associated proteins, amino acid permeases, lipases, proteases, and ferric reductases, among other gene families, have also expanded in *C. glabrata* but not in *S. cerevisiae*, suggesting a convergent evolutionary trend for useful traits in the context of host colonization (Butler et al., 2009).

Dimorphic Ascomycetes. The pathogenic ascomycetes *B. dermatitidis*, *Paracoccidiodes brasiliensis*, *H. capsulatum*, and *Coccidioides immitis/Coccidioides posadasii* are all phylogenetically related, forming part of a monophyletic clade in the order Onygenales within the class Eurotiomycetales, subphylum Pezizomycotina (Bowman et al., 1992; Kasuga et al., 2002; Sugiyama et al., 1999; Vilela et al., 2009). They belong to the Ajellomycetaceae, a family characterized as possessing keratinolytic activity (Untereiner et al., 2004).

In the same class but in different orders are *Sporothrix schenkii* and *Penicillum marneffei*. All of these species are thermally dimorphic, meaning that they are able to switch between mold and yeast phases in a temperature-controlled process. They exist as mold at 25°C, but at temperatures close to 37°C, morphogenesis is induced and they change to a yeast state.

B. dermatitidis and *H. capsulatum* are closely related, their telemorphs forming the genus *Ajellomyces* (Leclerc et al., 1994). Their estimated divergence time is between 32 and 138 mya (Kasuga et al., 2002). It is likely that both species originated on the American continent, and from there, *H. capsulatum* radiated some 3–13 mya to produce at least seven phylogenetic species with distinct geographic distributions (Kasuga et al., 2003). *P. brasiliensis* and its sister species *Lacazia loboi* (Vilela et al., 2009) are closely related with the *Ajellomyces* group and share many pathogenic traits with them. *P. brasiliensis* has a broad geographic extension being endemic to many countries in Central and South America; this broad distribution is reflected in a great genetic variation. *P. brasiliensis* is not a single species but forms a species complex (SC) that has at least three phylogenetic species that still have a certain degree of gene flow between them (Matute et al., 2006). Although not recognized as a major human pathogen, the noncultivable *L. loboi* can cause skin and subcutaneous tissue infections, producing highly deforming lesions, and it is also a widespread pathogen of marine mammals (Bermudez et al., 2009). *L. loboi* occurs in the American continent in tropical/subtropical humid forests, largely overlapping the area where *P. brasiliensis* occurs, and the existence of *L. loboi* as an independent species was demonstrated only relatively recently (Vilela et al., 2009).

Coccidioides has two recognized species, *C. immitis* and *C. posadasii*. According to the initial characterization of the two species, the divergence time between them was estimated to occur from 11.0 to 12.5 mya (Fisher et al., 2002). However, more recent evidence with an increased population sampling indicates the divergence occurred more recently, at about 5 mya (Sharpton et al., 2009). *C. immitis* and *C. posadasii* have only relatively recently been introduced to South America (9000–140,000 years ago) from a single population in Texas, probably carried by mammals or even by early Amerindians migrating south (Dignani et al., 2003; Fisher et al., 2001; Hector and Laniado-Laborin, 2005). The closest relative of *C. immitis* and *C. posadasii* is the nonpathogenic fungus *Uncinocarpus reesii*, which is itself heat tolerant (Fisher et al., 2002; Pan et al., 1994).

Considering the widespread heat tolerance found in these organisms, it is likely that their last common ancestor was heat tolerant, thermally dimorphic, and presumably potentially pathogenic. It has not been thoroughly studied to date whether thermotolerance is an ancestral trait of the Pezizomycotina subphylum, which is the group that includes all the previously mentioned dimorphic fungi. However, it seems likely that at least heat tolerance and pathogenic capability was present in the common ancestor of the Onygenales.

Aspergilli. The *Aspergillus* genus comprises a diverse group of filamentous fungi in the order Eurotiales, which are found primarily in soil and decaying vegetation and are remarkably resistant to harsh environmental conditions. The three major human pathogens *A. fumigatus*, *Aspergillus flavus*, and *Aspergillus terreus* are not members of a single clade, and all have more closely related species that are not pathogenic (Berbee, 2001).

The most prevalent pathogen among the *Aspergillus* spp. is *A. fumigatus*;

however, not much is known regarding its population structure or recent evolution. Environmental samples of *A. fumigatus* show apparently little genetic variation at the population level (Rydholm et al., 2006), while clinical isolates tend to have more genetic diversity, attributable to tandem duplications and point mutations (Balajee et al., 2007). It was recently demonstrated that *A. fumigatus* presents high interstrain variability; up to 1.4% difference in genome size and up to 2% unique genes have been observed in different pathogenic isolates (Fedorova et al., 2008). The difference in the genome content observed in the clinical isolate Af293 compared to isolate A1163 can be explained by a segmental duplication event that could occur under the strong selective pressure exerted by the host environment. This example illustrates a possible case of very rapid adaptation under extreme circumstances such as the human host environment. Based on their observations, Fedorova et al. (2008) proposed that duplication, diversification, and differential gene loss contribute to the origin of lineage-specific genes, and their discovery of variable genes concentrated in genomic islands is attributed to function as "designated gene dumps and, perhaps also as gene factories." Further studies, including environmental sampling, need to be conducted to fully validate this hypothesis.

Basidiomycetes. *C. neoformans SC.* The pathogenic *C. neoformans* is an SC that includes *C. neoformans* var. *neoformans* (serotype D), *C. neoformans* var. *grubii* (serotype A) (Franzot et al., 1998), and *Cryptococcus gattii* (serotypes B and C) (Fraser et al., 2005; Kwon-Chung et al., 2002); in turn, *C. gattii* comprises four recognized varieties (Bovers et al., 2008) that could be considered cryptic species.

The *C. neoformans* SC belongs to the *Filobasidiella* clade in the order Tremellales (Scorzetti et al., 2002). The *Filobasidiella* group comprises the saprophytic fungi *Filobasidiella depauperata*, the sibling species *Cryptococcus amylolentus* and *Tsuchiyaea wingfieldii*, and the mammalian pathogens *C. neoformans* and *C. gattii* (Findley et al., 2009).

The pathogenic cryptococci are likely to have evolved in a tropical or subtropical region, and molecular data suggest that it has only recently spread globally (Kavanaugh et al., 2006). The split of *C. neoformans* and *C. gattii* occurred about 37 mya (Xu et al., 2000). Recent data from population genetics studies suggest that *C. neoformans* originated in sub-Saharan Africa; the genetic diversity is greater in that region compared with other parts of the world (Litvintseva et al., 2003, 2007, 2011), where the populations are largely clonal and predominantly associated with pigeon guano (Littman and Borok, 1968). In contrast, in sub-Saharan Africa, cryptococcal populations are primarily associated with trees, like the Mopane tree, and wooden debris and can undergo sexual recombination (Botes et al., 2009; Litvintseva et al., 2011). *C. gattii* has been associated with *Eucalyptus* trees native from Australia (Ellis and Pfeiffer, 1990), and there is genetic evidence that the current Pacific Northwest outbreak originated from Australian strains (Fraser et al., 2005).

It has been proposed that the major virulence traits of *C. neoformans* were evolutionarily selected for survival in a hostile environment, such as pigeon guano or soil, where the fungus finds predators, toxic substances, and high temperature variation (Casadevall et al., 2003; Steenbergen and Casadevall, 2003). This hypothesis correlates with similar findings in unrelated fungi, such as the dimorphic ascomycetes (Steenbergen et al., 2004), with which *C. neoformans* shares some lifestyle characteristics. Moreover, cryptococcal virulence factors are not only important for fungal survival during interaction with mammalian macrophages but it has been shown that these factors also play a role in the

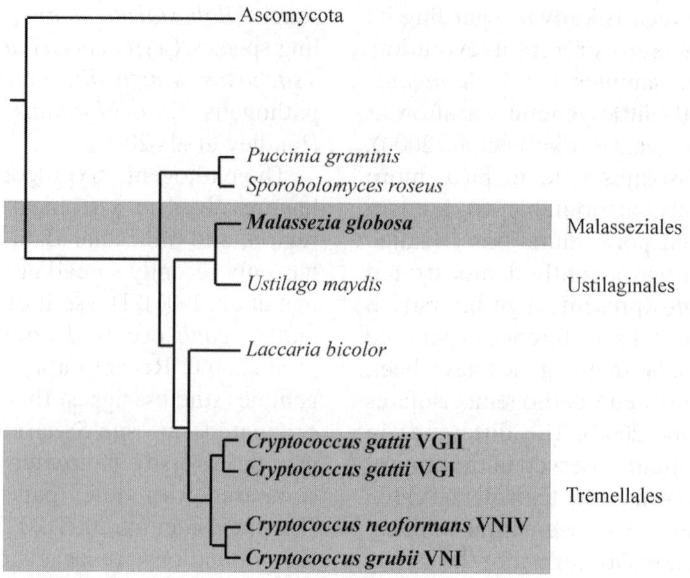

Figure 4.4 Phylogenetic tree of the human pathogenic Basidiomycota. Pathogenic species are shown in bold characters. Notice that the two human pathogens is this group, *Malassezia* and *Cryptococcus*, are not closely related. *Cryptococcus* comprises a species complex formed by at least three different species: *neoformans*, *grubii*, and *gattii*. Modified from Wang et al. (2009). See color insert.

survival strategy of *C. neoformans* against soil amoebae (Casadevall et al., 2003; Steenbergen and Casadevall, 2003; Steenbergen et al., 2001) and possibly other predators that form part of its natural environmental niche. A recent study has shown that a conserved mechanism triggered by phospholipids induces capsule enlargement in *C. neoformans* if the yeasts are coincubated with either mammalian macrophages or amoeba, suggesting that the regulation of capsule size has evolved, among other things, as a mechanism to avoid phagocytosis by potential predators (Chrisman et al., 2011).

The other basidiomycete associated with human disease is *Malassezia*. Unlike *C. neoformans*, which is an environmental organism, *Malassezia* spp. are skin commensals and have a close evolutionary history with that of humans and other mammals. They are the etiological agents of dandruff and seborrheic dermatitis but rarely cause systemic or deeper infections.

The genus *Malassezia* belongs to the order Malasseziales and is closely related with the plant pathogen *Ustilago* (Ustilaginales) (Begerow et al., 2006) (Fig. 4.4). It is proposed that due to adaptation and lifestyle change, *Malassezia* change their host preference from plants to humans. The genome analysis of *Malassezia globosa* reveals that certain gene families, such as extracellular hydrolases, including lipases, phospholipases, and aspartyl proteases, are expanded in comparison to *Ustilago*, which instead produces more enzymes involved in plant material digestion (Xu et al., 2007).

CONCLUDING REMARKS

In summary, we cannot generalize the evolutionary history of the pathogenic fungi. As outlined in the previous sections, human pathogens are scattered along almost the entire FTOL, with the exception of the Chytridiomycota. Hence, the capacity for human virulence among the fungi may have emerged numerous times in evolutionary history. With few exceptions, the fungal species able to cause disease in humans are accidental pathogens, and the traits that aid in the establishment and progression of disease have been selected to enhance fungal survival in their primary habitat. Despite this, different groups of fungi have followed different evolutionary histories, and the selection pressures exerted by their ecological niches (soil, guano, decomposing vegetation) have elicited in few organisms the capacity to cause disease in an otherwise unsolicited host due to convergent survival strategies, such as heat tolerance, iron scavenging capacity, the ability to use different carbon sources, toxin and enzyme secretion, evasion of phagocytosis, and other strategies. On the other hand, why closely related fungi that apparently are under very similar selective pressures find different survival strategies to cope, with some acquiring pathogenic capability and others remaining harmless saprobes, is a matter for further consideration.

REFERENCES

Aliouat-Denis C-M, Chabe M, Demanche C, Aliouat EM, Viscogliosi E, Guillot J, Delhaes L, Dei-Cas E. 2008. *Pneumocystis* species, co-evolution and pathogenic power. Infect Genet Evol 8(5): 708–726.

Balajee SA, Tay ST, Lasker BA, Hurst SF, Rooney AP. 2007. Characterization of a novel gene for strain typing reveals substructuring of *Aspergillus fumigatus* across North America. Eukaryot Cell 6(8): 1392–1399.

Begerow D, Stoll M, Bauer R. 2006. A phylogenetic hypothesis of Ustilaginomycotina based on multiple gene analyses and morphological data. Mycologia 98(6): 906–916.

Berbee ML. 2001. The phylogeny of plant and animal pathogens in the Ascomycota. Physiol Mol Plant Pathol 59(4): 165–187.

Berbee ML, Taylor G. 2001. Fungal molecular evolution: Gene trees and geologic time. In: McLaughlin DJ, Blackwell M, Spatafora JW, eds. *The Mycota Vol. VIIB, Systematics and Evolution*, Vol. VIIB. Springer, Berlin, pp. 229–246.

Bermudez L, Van Bressem M-F, Reyes-Jaimes O, Sayegh AJ, Paniz Mondolfi AE. 2009. Lobomycosis in man and lobomycosis-like disease in bottlenose dolphin, Venezuela. Emerg Infect Dis 15(8): 1301–1303.

Botes A, Boekhout T, Hagen F, Vismer H, Swart J, Botha A. 2009. Growth and mating of *Cryptococcus neoformans* var. *grubii* on woody debris. Microb Ecol 57(4): 757–765.

Bovers M, Hagen F, Kuramae EE, Boekhout T. 2008. Six monophyletic lineages identified within *Cryptococcus neoformans* and *Cryptococcus gattii* by multi-locus sequence typing. Fungal Genet Biol 45(4): 400–421.

Bowman BH, Taylor JW, White TJ. 1992. Molecular evolution of the fungi: Human pathogens. Mol Biol Evol 9(5): 893–904.

Bowman BH, White TJ, Taylor JW. 1996. Human pathogeneic fungi and their close nonpathogenic relatives. Mol Phylogenet Evol 6(1): 89–96.

Butler G. 2010. Fungal sex and pathogenesis. Clin Microbiol Rev 23(1): 140–159.

Butler G, Rasmussen MD, Lin MF, Santos MAS, Sakthikumar S, Munro CA, Rheinbay E, Grabherr M, Forche A, Reedy JL, Agrafioti I, Arnaud MB, Bates S, Brown AJP, Brunke S, Costanzo MC, Fitzpatrick DA, de Groot PWJ, Harris D, Hoyer LL, Hube B, Klis FM, Kodira C, Lennard N, Logue ME, Martin R, Neiman AM, Nikolaou E, Quail MA, Quinn J, Santos MC, Schmitzberger FF, Sherlock G, Shah P, Silverstein KAT, Skrzypek MS, Soll D, Staggs R, Stansfield I, Stumpf MPH, Sudbery PE, Srikantha T, Zeng Q, Berman J, Berriman M, Heitman J, Gow NAR, Lorenz MC, Birren BW, Kellis M, Cuomo CA. 2009. Evolution of pathogenicity and sexual reproduction in eight *Candida* genomes. Nature 459(7247): 657–662.

Butterfield NJ. 2005. Probable proterozoic fungi. Paleobiology 31(1): 165–182.

Casadevall A, Steenbergen JN, Nosanchuk JD. 2003. Ready made virulence and dual use virulence factors in pathogenic environmental fungi—The *Cryptococcus neoformans* paradigm. Curr Opin Microbiol 6(4): 332–337.

Cavalier-Smith T. 2002. The phagotrophic origin of eukaryotes and phylogenetic classification of protozoa. Int J Syst Evol Microbiol 52(2): 297–354.

Chabe M, Dei-Cas E, Creusy C, Fleurisse L, Respaldiza N, Camus D, Durand-Joly I. 2004. Immunocompetent hosts as a reservoir of *Pneumocystis* organisms: Histological and RT-PCR data demonstrate active replication. Eur J Clin Microbiol Infect Dis 23(2): 89–97.

Chrisman CJ, Albuquerque P, Guimaraes AJ, Nieves E, Casadevall A. 2011. Phospholipids trigger *Cryptococcus neoformans* capsular enlargement during interactions with amoebae and macrophages. PLoS Pathog 7(5): e1002047.

Corradi N, Pombert JF, Farinelli L, Didier ES, Keeling PJ. 2010. The complete sequence of the smallest known nuclear genome from the microsporidian *Encephalitozoon intestinalis*. Nat Commun 1:77. doi:10.1038/ncomms1082.

Cushion MT, Keely SP, Stringer JR. 2004. Molecular and phenotypic description of *Pneumocystis wakefieldiae* sp. nov., a new species in rats. Mycologia 96(3): 429–438.

Dei-Cas E, Chabe M, Moukhlis R, Durand-Joly I, Aliouat El M, Stringer JR, Cushion M, Noel C, de Hoog GS, Guillot J, Viscogliosi E. 2006. *Pneumocystis oryctolagi* sp. nov., an uncultured fungus causing pneumonia in rabbits at weaning: Review of current knowledge, and description of a new taxon on genotypic, phylogenetic and phenotypic bases. FEMS Microbiol Rev 30(6): 853–871.

de la Cruz F, Davies J. 2000. Horizontal gene transfer and the origin of species: Lessons from bacteria. Trends Microbiol 8(3): 128–133.

Demanche C, Berthelemy M, Petit T, Polack B, Wakefield AE, Dei-Cas E, Guillot J. 2001. Phylogeny of *Pneumocystis carinii* from 18 primate species confirms host specificity and suggests coevolution. J Clin Microbiol 39(6): 2126–2133.

Diaz MR, Boekhout T, Kiesling T, Fell JW. 2005. Comparative analysis of the intergenic spacer regions and population structure of the species complex of the pathogenic yeast *Cryptococcus neoformans*. FEMS Yeast Res 5(12): 1129–1140.

Dignani MC, Kiwan EN, Anaissie EJ. 2003. Hyalohyphomycoses. In: Anaissie E, McGinnis MR, Pfaller MA, eds. *Clinical Mycology*. Philadelphia: Churchill Livingston, pp. 309–324.

D'Souza CA, Kronstad JW, Taylor G, Warren R, Yuen M, Hu G, Jung WH, Sham A, Kidd SE, Tangen K, Lee N, Zeilmaker T, Sawkins J, McVicker G, Shah S, Gnerre S, Griggs A, Zeng Q, Bartlett K, Li W, Wang X, Heitman J, Stajich JE, Fraser JA, Meyer W, Carter D, Schein J, Krzywinski M, Kwon-Chung KJ, Varma A, Wang J, Brunham R, Fyfe M, Ouellette BFF, Siddiqui A, Marra M, Jones S, Holt R, Birren BW, Galagan JE, Cuomo CA. 2011. Genome variation in *Cryptococcus gattii*, an emerging pathogen of immunocompetent hosts. MBio 2(1): e00342–10.

Durand-Joly I, Aliouat El M, Recourt C, Guyot K, Francois N, Wauquier M, Camus D, Dei-Cas E. 2002. *Pneumocystis carinii* f. sp. *hominis* is not infectious for SCID mice. J Clin Microbiol 40(5): 1862–1865.

Ellis DH, Pfeiffer TJ. 1990. Natural habitat of *Cryptococcus neoformans* var. *gattii*. J Clin Microbiol 28(7): 1642–1644.

Eriksson OE. 1994. *Pneumocystis carinii*, a parasite in lungs of mammals, referred to a new family and order (Pneumocystidaceae, Pneumocystidales, Ascomycota). Systema Ascomycetum 13: 165–180.

Fedorova ND, Khaldi N, Joardar VS, Maiti R, Amedeo P, Anderson MJ, Crabtree J, Silva JC, Badger JH, Albarraq A, Angiuoli S, Bussey H, Bowyer P, Cotty PJ, Dyer PS, Egan A, Galens K, Fraser-Liggett CM, Haas BJ, Inman JM, Kent R, Lemieux S, Malavazi I, Orvis J, Roemer T, Ronning CM, Sundaram JP, Sutton G, Turner G, Venter JC, White OR, Whitty BR, Youngman P, Wolfe KH, Goldman GH, Wortman JR, Jiang B, Denning DW, Nierman WC. 2008. Genomic islands in the pathogenic filamentous fungus *Aspergillus fumigatus*. PLoS Genet 4(4): e1000046.

Findley K, Rodriguez-Carres M, Metin B, Kroiss J, Fonseca A, Vilgalys R, Heitman J. 2009. Phylogeny and phenotypic characterization of pathogenic *Cryptococcus* species and closely related saprobic taxa in the Tremellales. Eukaryot Cell 8(3): 353–361.

Fisher MC, Koenig GL, White TJ, San-Blas G, Negroni R, Alvarez IG, Wanke B, Taylor JW. 2001. Biogeographic range expansion into South America by *Coccidioides immitis* mirrors New World patterns of human migration. Proc Natl Acad Sci U S A 98(8): 4558–4562.

Fisher MC, Koenig GL, White TJ, Taylor JW. 2002. Molecular and phenotypic description of *Coccidioides posadasii* sp. nov., previously recognized as the non-California population of *Coccidioides immitis*. Mycologia 94(1): 73–84.

Fisher MC, Garner TWJ, Walker SF. 2009. Global emergence of *Batrachochytrium dendrobatidis* and amphibian chytridiomycosis in space, time, and host. Annu Rev Microbiol 63(1): 291–310.

Fitzpatrick D, Logue M, Stajich J, Butler G. 2006. A fungal phylogeny based on 42 complete genomes derived from supertree and combined gene analysis. BMC Evol Biol 6(1): 99.

Fitzpatrick D, Logue M, Butler G. 2008. Evidence of recent interkingdom horizontal gene transfer between bacteria and *Candida parapsilosis*. BMC Evol Biol 8(1): 181.

Franzen C, Muller A. 1999. Molecular techniques for detection, species differentiation, and phylogenetic analysis of Microsporidia. Clin Microbiol Rev 12(2): 243–285.

Franzot SP, Mukherjee J, Cherniak R, Chen LC, Hamdan JS, Casadevall A. 1998. Microevolution of a standard strain of *Cryptococcus neoformans* resulting in differences in virulence and other phenotypes. Infect Immun 66(1): 89–97.

Fraser JA, Giles SS, Wenink EC, Geunes-Boyer SG, Wright JR, Diezmann S, Allen A, Stajich JE, Dietrich FS, Perfect JR, Heitman J. 2005. Same-sex mating and the origin of the Vancouver Island *Cryptococcus gattii* outbreak. Nature 437(7063): 1360–1364.

Fraser JA, Stajich JE, Tarcha EJ, Cole GT, Inglis DO, Sil A, Heitman J. 2007. Evolution of the mating type locus: Insights gained from the dimorphic primary fungal pathogens *Histoplasma capsulatum, Coccidioides immitis*, and *Coccidioides posadasii*. Eukaryot Cell 6(4): 622–629.

Frenkel JK. 1999. *Pneumocystis* pneumonia, an immunodeficiency-dependent disease (IDD): A critical historical overview. J Eukaryot Microbiol 46(5): 89S–92S.

Gomes AC, Miranda I, Silva RM, Moura GR, Thomas B, Akoulitchev A, Santos MA. 2007. A genetic code alteration generates a proteome of high diversity in the human pathogen *Candida albicans*. Genome Biol 8(10): R206.

Gonzalez-Fernandez R, Prats E, Jorrin-Novo JV. 2010. Proteomics of plant pathogenic fungi. J Biomed Biotechnol 2010: 932527.

Gordon JL, Byrne KP, Wolfe KH. 2009. Additions, losses, and rearrangements on the evolutionary route from a reconstructed ancestor to the modern *Saccharomyces cerevisiae* genome. PLoS Genet 5(5): e1000485.

Goulart LC, Rosa E Silva LK, Chiapello L, Silveira C, Crestani J, Masih D, Vainstein MH. 2010. *Cryptococcus neoformans* and *Cryptococcus gattii* genes preferentially expressed during rat macrophage infection. Med Mycol 48(7): 932–941.

Groisman EA, Ochman H. 1996. Pathogenicity islands: Bacterial evolution in quantum leaps. Cell 87(5): 791–794.

Guillot J, Demanche C, Hugot JP, Berthelemy M, Wakefield AE, Dei-Cas E, Chermette R. 2001. Parallel phylogenies of *Pneumocystis* species and their mammalian hosts. J Eukaryot Microbiol 48(Suppl. 1S): 113S–115S.

Hall C, Dietrich FS. 2007. The reacquisition of biotin prototrophy in *Saccharomyces cerevisiae* involved horizontal gene transfer, gene duplication and gene clustering. Genetics 177(4): 2293–2307.

Hall C, Brachat S, Dietrich FS. 2005. Contribution of horizontal gene transfer to the evolution of *Saccharomyces cerevisiae*. Eukaryot Cell 4(6): 1102–1115.

Hawksworth DL. 1991. The fungal dimension of biodiversity: Magnitude, significance, and conservation. Mycol Res 95(6): 641–655.

Hawksworth DL. 2001. The magnitude of fungal diversity: The 1.5 million species estimate revisited. Mycol Res 105(12): 1422–1432.

Heckman DS, Geiser DM, Eidell BR, Stauffer RL, Kardos NL, Hedges SB. 2001. Molecular evidence for the early colonization of land by fungi and plants. Science 293(5532): 1129–1133.

Hector RF, Laniado-Laborin R. 2005. coccidioidomycosis, a fungal disease of the Americas. PLoS Med 2(1): e2.

Hedges SB, Blair JE, Venturi ML, Shoe JL. 2004. A molecular timescale of eukaryote evolution and the rise of complex multicellular life. BMC Evol Biol 4: 2.

Heitman J. 2010. Evolution of eukaryotic microbial pathogens via covert sexual reproduction. Cell Host Microbe 8(1): 86–99.

Hibbett DS, Binder M, Bischoff JF, Blackwell M, Cannon PF, Eriksson OE, Huhndorf S, James T, Kirk PM, Lücking R, Lumbsch TH, Lutzoni F, Matheny PB, McLaughlin DJ, Powell MJ, Redhead S, Schoch CL, Spatafora JW, Stalpers JA, Vilgalys R, Aime MC, Aptroot A, Bauer R, Begerow D, Benny GL, Castlebury LA, Crous PW, Dai Y-C, Gams W, Geiser DM, Griffith GW, Gueidan C, Hawksworth DL, Hestmark G, Hosaka K, Humber RA, Hyde KD, Ironside JE, Kõljalg U, Kurtzman CP, Larsson K-H, Lichtwardt R, Longcore J, Miadlikowska J, Miller A, Moncalvo J-M, Mozley-Standridge S, Oberwinkler F, Parmasto E, Reeb V, Rogers JD, Roux C, Ryvarden L, Sampaio JP, Schüssler A, Sugiyama J, Thorn RG, Tibell L, Untereiner WA, Walker C, Wang Z, Weir A, Weiss M, White MM, Winka K, Yao Y-J, Zhang N. 2007. A higher-level phylogenetic classification of the Fungi. Mycol Res 111(5): 509–547.

Humber RA. 2008. Evolution of entomopathogenicity in fungi. J Invertebr Pathol 98(3): 262–266.

Jain R, Rivera MC, Moore JE, Lake JA. 2002. Horizontal gene transfer in microbial genome evolution. Theor Popul Biol 61(4): 489–495.

James TY, Kauff F, Schoch CL, Matheny PB, Hofstetter V, Cox CJ, Celio G, Gueidan C, Fraker E, Miadlikowska J, Lumbsch HT, Rauhut A, Reeb V, Arnold AE, Amtoft A, Stajich JE, Hosaka K, Sung GH, Johnson D, O'ourke B, Crockett M, Binder M, Curtis JM, Slot JC, Wang Z, Wilson AW, Schussler A, Longcore JE, O'Donnell K, Mozley-Standridge S, Porter D, Letcher PM, Powell MJ, Taylor JW, White MM, Griffith GW, Davies DR, Humber RA, Morton JB, Sugiyama J, Rossman AY, Rogers JD, Pfister DH, Hewitt D, Hansen K, Hambleton S, Shoemaker RA, Kohlmeyer J, Volkmann-Kohlmeyer B, Spotts RA, Serdani M, Crous PW, Hughes KW, Matsuura K, Langer E, Langer G, Untereiner WA, Lucking R, Budel B, Geiser DM, Aptroot A, Diederich P, Schmitt I, Schultz M, Yahr R, Hibbett DS, Lutzoni F, McLaughlin DJ, Spatafora JW, Vilgalys R. 2006a. Reconstructing the early evolution of Fungi using a six-gene phylogeny. Nature 443(7113): 818–822.

James TY, Letcher PM, Longcore JE, Mozley-Standridge SE, Porter D, Powell MJ, Griffith GW, Vilgalys R. 2006b. A molecular phylogeny of the flagellated fungi (Chytridiomycota) and description of a new phylum (Blastocladiomycota). Mycologia 98(6): 860–871.

Jones MDM, Forn I, Gadelha C, Egan MJ, Bass D, Massana R, Richards TA. 2011. Discovery of novel intermediate forms redefines the fungal tree of life. Nature 474(7350): 200–203.

Kasuga T, White TJ, Taylor JW. 2002. Estimation of nucleotide substitution rates in Eurotiomycete fungi. Mol Biol Evol 19(12): 2318–2324.

Kasuga T, White TJ, Koenig G, McEwen J, Restrepo A, Castañeda E, Da Silva Lacaz C, Heins-Vaccari EM, De Freitas RS, Zancopé-Oliveira RM, Qin Z, Negroni R, Carter DA, Mikami Y, Tamura M, Taylor ML, Miller GF, Poonwan N, Taylor JW. 2003. Phylogeography of the fungal pathogen *Histoplasma capsulatum*. Mol Ecol 12(12): 3383–3401.

Kavanaugh LA, Fraser JA, Dietrich FS. 2006. Recent evolution of the human pathogen *Cryptococcus neoformans* by intervarietal transfer of a 14-gene fragment. Mol Biol Evol 23(10): 1879–1890.

Keeling PJ. 2003. Congruent evidence from alpha-tubulin and beta-tubulin gene phylogenies for a zygomycete origin of Microsporidia. Fungal Genet Biol 38(3): 298–309.

Keeling PJ, Fast NM. 2002. Microsporidia: Biology and evolution of highly reduced intracellular parasites. Annu Rev Microbiol 56: 93–116.

Keeling PJ, Slamovits CH. 2004. Simplicity and complexity of microsporidian genomes. Eukaryot Cell 3(6): 1363–1369.

Keeling PJ, Luker MA, Palmer JD. 2000. Evidence from beta-tubulin phylogeny that Microsporidia evolved from within the fungi. Mol Biol Evol 17(1): 23–31.

Keeling PJ, Burger G, Durnford DG, Lang BF, Lee RW, Pearlman RE, Roger AJ, Gray MW. 2005. The tree of eukaryotes. Trends Ecol Evol 20(12): 670–676.

Keely SP, Fischer JM, Stringer JR. 2003. Evolution and speciation of *Pneumocystis*. J Eukaryot Microbiol 50(Suppl.): 624–626.

Keely SP, Fischer JM, Cushion MT, Stringer JR. 2004. Phylogenetic identification of *Pneumocystis murina* sp. nov., a new species in laboratory mice. Microbiology 150(Pt 5): 1153–1165.

Koonin E. 2010. The origin and early evolution of eukaryotes in the light of phylogenomics. Genome Biol 11(5): 209.

Kwon-Chung KJ, Boekhout T, Fell J, Diaz M. 2002. Proposal to conserve the name *Cryptococcus gattii* against *C. hondurianus* and *C. bacillisporus* (Basidiomycota, Hymenomycetes, Tremellomycetidae). Taxon 51: 804–806.

Lawrence JG. 2002. Gene transfer in bacteria: Speciation without species? Theor Popul Biol 61(4): 449–460.

Leclerc MC, Philippe H, Gueho E. 1994. Phylogeny of dermatophytes and dimorphic fungi based on large subunit ribosomal RNA sequence comparisons. J Med Vet Mycol 32(5): 331–341.

Lee SC, Corradi N, Byrnes EJ 3rd, Torres-Martinez S, Dietrich FS, Keeling PJ, Heitman J. 2008. Microsporidia evolved from ancestral sexual fungi. Curr Biol 18(21): 1675–1679.

Liggenstoffer AS, Youssef NH, Couger MB, Elshahed MS. 2010. Phylogenetic diversity and community structure of anaerobic gut fungi (phylum Neocallimastigomycota) in ruminant and non-ruminant herbivores. ISME J 4(10): 1225–1235.

Littman ML, Borok R. 1968. Relation of the pigeon to cryptococcosis: Natural carrier state, heat resistance and survival of *Cryptococcus neoformans*. Mycopathologia 35(3): 329–345.

Litvintseva AP, Marra RE, Nielsen K, Heitman J, Vilgalys R, Mitchell TG. 2003. Evidence of sexual recombination among *Cryptococcus neoformans* serotype A isolates in sub-Saharan Africa. Eukaryot Cell 2(6): 1162–1168.

Litvintseva AP, Lin X, Templeton I, Heitman J, Mitchell TG. 2007. Many globally isolated AD hybrid strains of *Cryptococcus neoformans* originated in Africa. PLoS Pathog 3(8): e114.

Litvintseva AP, Carbone I, Rossouw J, Thakur R, Govender NP, Mitchell TG. 2011. Evidence that the human pathogenic fungus *Cryptococcus neoformans* var. *grubii* may have evolved in Africa. PLoS One 6(5): e19688.

Liu Y, Hodson M, Hall B. 2006. Loss of the flagellum happened only once in the fungal lineage: Phylogenetic structure of kingdom Fungi inferred from RNA polymerase II subunit genes. BMC Evol Biol 6(1): 74.

Liu Y, Leigh JW, Brinkmann H, Cushion MT, Rodriguez-Ezpeleta N, Philippe HE, Lang BF. 2009a. Phylogenomic analyses support the monophyly of Taphrinomycotina, including *Schizosaccharomyces* fission yeasts. Mol Biol Evol 26(1): 27–34.

Liu Y, Steenkamp E, Brinkmann H, Forget L, Philippe H, Lang BF. 2009b. Phylogenomic analyses predict sistergroup relationship of nucleariids and Fungi and paraphyly of zygomycetes with significant support. BMC Evol Biol 9(1): 272.

Lucking R, Huhndorf S, Pfister DH, Plata ER, Lumbsch HT. 2009. Fungi evolved right on track. Mycologia 101(6): 810–822.

Lutzoni F, Kauff F, Cox CJ, McLaughlin D, Celio G, Dentinger B, Padamsee M, Hibbett D, James TY, Baloch E, Grube M, Reeb V, Hofstetter V, Schoch C, Arnold AE, Miadlikowska J, Spatafora J, Johnson D, Hambleton S, Crockett M, Shoemaker R, Sung G-H, Lucking R, Lumbsch T, O'Donnell K, Binder M, Diederich P, Ertz D, Gueidan C, Hansen K, Harris RC, Hosaka K, Lim Y-W, Matheny B, Nishida H, Pfister D, Rogers J, Rossman A, Schmitt I, Sipman H, Stone J, Sugiyama J, Yahr R, Vilgalys R. 2004. Assembling the fungal tree of life: Progress, classification, and evolution of subcellular traits. Am J Bot 91(10): 1446–1480.

Ma L-J, van der Does HC, Borkovich KA, Coleman JJ, Daboussi M-J, Di Pietro A, Dufresne M, Freitag M, Grabherr M, Henrissat B, Houterman PM, Kang S, Shim W-B, Woloshuk C, Xie X, Xu J-R, Antoniw J, Baker SE, Bluhm BH, Breakspear A, Brown DW, Butchko RAE, Chapman S, Coulson R, Coutinho PM, Danchin EGJ, Diener A, Gale LR, Gardiner DM, Goff S, Hammond-Kosack KE, Hilburn K, Hua-Van A, Jonkers W, Kazan K, Kodira CD, Koehrsen M, Kumar L, Lee Y-H, Li L, Manners JM, Miranda-Saavedra D, Mukherjee M, Park G, Park J, Park S-Y, Proctor RH, Regev A, Ruiz-Roldan MC, Sain D, Sakthikumar S, Sykes S, Schwartz DC, Turgeon BG, Wapinski I, Yoder O, Young S, Zeng Q, Zhou S, Galagan J, Cuomo CA, Kistler HC, Rep M. 2010. Comparative genomics reveals mobile pathogenicity chromosomes in *Fusarium*. Nature 464(7287): 367–373.

Mandel MA, Barker BM, Kroken S, Rounsley SD, Orbach MJ. 2007. Genomic and population analyses of the mating type loci in *Coccidioides* species reveal evidence for sexual reproduction and gene acquisition. Eukaryot Cell 6(7): 1189–1199.

Marcet-Houben M, Gabaldon T. 2010. Acquisition of prokaryotic genes by fungal genomes. Trends Genet 26(1): 5–8.

Massey SE, Moura G, Beltrao P, Almeida R, Garey JR, Tuite MF, Santos MAS. 2003. Comparative evolutionary genomics unveils the molecular mechanism of reassignment of the CTG codon in *Candida* spp. Genome Res 13(4): 544–557.

Matute DR, McEwen JG, Puccia R, Montes BA, San-Blas G, Bagagli E, Rauscher JT, Restrepo A, Morais F, Nino-Vega G, Taylor JW. 2006. Cryptic speciation and recombination in the fungus *Paracoccidioides brasiliensis* as revealed by gene genealogies. Mol Biol Evol 23(1): 65–73.

McDonald T, Dietrich F, Lutzoni F. 2012. Multiple horizontal gene transfers of ammonium transporters/ammonia permeases from prokaryotes to eukaryotes: Toward a new functional and evolutionary classification. Mol Biol Evol 29(1): 51–60.

Medina M, Collins AG, Taylor JW, Valentine JW, Lipps JH, Amaral-Zettler L, Sogin ML. 2003. Phylogeny of Opisthokonta and the evolution of multicellularity and complexity in Fungi and Metazoa. Int J Astrobiology 2(3): 203–211.

Neafsey DE, Barker BM, Sharpton TJ, Stajich JE, Park DJ, Whiston E, Hung C-Y, McMahan C, White J, Sykes S, Heiman D, Young S, Zeng Q, Abouelleil A, Aftuck L, Bessette D, Brown A, FitzGerald M, Lui A, Macdonald JP, Priest M, Orbach MJ, Galgiani JN, Kirkland TN, Cole GT, Birren BW, Henn MR, Taylor JW, Rounsley SD. 2010. Population genomic sequencing of Coccidioides fungi reveals recent hybridization and transposon control. Genome Res 20(7): 938–946.

Ochman H, Lawrence JG, Groisman EA. 2000. Lateral gene transfer and the nature of bacterial innovation. Nature 405(6784): 299–304.

O'orman CM, Fuller HT, Dyer PS. 2009. Discovery of a sexual cycle in the opportunistic fungal pathogen *Aspergillus fumigatus*. Nature 457(7228): 471–474.

Padovan ACB, Sanson GFO, Brunstein A, Briones MRS. 2005. Fungi evolution revisited: Application of the penalized likelihood method to a Bayesian fungal phylogeny provides a new perspective on phylogenetic relationships and divergence dates of Ascomycota groups. J Mol Evol 60(6): 726–735.

Pan S, Sigler L, Cole GT. 1994. Evidence for a phylogenetic connection between *Coccidioides immitis* and *Uncinocarpus reesii* (Onygenaceae). Microbiology 140(Pt 6): 1481–1494.

Paoletti M, Rydholm C, Schwier EU, Anderson MJ, Szakacs G, Lutzoni F, Debeaupuis JP, Latge JP, Denning DW, Dyer PS. 2005. Evidence for sexuality in the opportunistic fungal pathogen *Aspergillus fumigatus*. Curr Biol 15(13): 1242–1248.

Peterson KJ, Waggoner B, Hagadorn JW. 2003. A fungal analog for Newfoundland Ediacaran fossils? Integr Comp Biol 43(1): 127–136.

Redecker D, Kodner R, Graham LE. 2000. Glomalean fungi from the Ordovician. Science 289(5486): 1920–1921.

Remy W, Taylor TN, Hass H, Kerp H. 1994. Four hundred-million-year-old vesicular arbuscular mycorrhizae. Proc Natl Acad Sci U S A 91(25): 11841–11843.

Restrepo A. 1994. Treatment of tropical mycoses. J Am Acad Dermatol 31(3 Pt 2): S91–S102.

Richards TA, Soanes DM, Foster PG, Leonard G, Thornton CR, Talbot NJ. 2009. Phylogenomic analysis demonstrates a pattern of rare and ancient horizontal gene transfer between plants and fungi. Plant Cell 21(7): 1897–1911.

Richardson M. 2009. The ecology of the Zygomycetes and its impact on environmental exposure. Clin Microbiol Infect 15(s5): 2–9.

Rosewich UL, Kistler HC. 2000. Role of horizontal gene transfer in the evolution of fungi. Annu Rev Phytopathol 38(1): 325–363.

Roy H, Steinkraus D, Eilenberg J, Hajek A, Pell J. 2006. Bizarre interactions and endgames: Entomopathogenic fungi and their arthropod hosts. Annu Rev Entomol 51: 331–357.

Rydholm C, Szakacs G, Lutzoni F. 2006. Low genetic variation and no detectable population structure in *Aspergillus fumigatus* compared to closely related *Neosartorya* species. Eukaryot Cell 5(4): 650–657.

Schmitt I, Lumbsch HT. 2009. Ancient horizontal gene transfer from bacteria enhances biosynthetic capabilities of fungi. PLoS One 4(2): e4437.

Schuessler A, Schwarzott D, Walker C. 2001. A new fungal phylum, the Glomeromycota: Phylogeny and evolution. Mycol Res 105(12): 1413–1421.

Scorzetti G, Fell JW, Fonseca A, Statzell-Tallman A. 2002. Systematics of basidiomycetous yeasts: A comparison of large subunit D1/D2 and internal transcribed spacer rDNA regions. FEMS Yeast Res 2(4): 495–517.

Shalchian-Tabrizi K, Minge MA, Espelund M, Orr R, Ruden T, Jakobsen KS, Cavalier-Smith T. 2008. Multigene phylogeny of Choanozoa and the origin of animals. PLoS One 3(5): e2098.

Sharpton TJ, Stajich JE, Rounsley SD, Gardner MJ, Wortman JR, Jordar VS, Maiti R, Kodira CD, Neafsey DE, Zeng Q, Hung C-Y, McMahan C, Muszewska A, Grynberg M, Mandel MA, Kellner EM, Barker BM, Galgiani JN, Orbach MJ, Kirkland TN, Cole GT, Henn MR, Birren BW, Taylor JW. 2009. Comparative genomic analyses of the human fungal pathogens *Coccidioides* and their relatives. Genome Res 19(10): 1722–1731.

Slamovits CH, Fast NM, Law JS, Keeling PJ. 2004. Genome compaction and stability in microsporidian intracellular parasites. Curr Biol 14(10): 891–896.

Smith JE. 2009. The ecology and evolution of microsporidian parasites. Parasitology 136(Special Issue 14): 1901–1914.

Stechmann A, Cavalier-Smith T. 2002. Rooting the Eukaryote tree by using a derived gene fusion. Science 297(5578): 89–91.

Steenbergen JN, Casadevall A. 2003. The origin and maintenance of virulence for the human pathogenic fungus *Cryptococcus neoformans*. Microbes Infect 5(7): 667–675.

Steenbergen JN, Shuman HA, Casadevall A. 2001. *Cryptococcus neoformans* interactions with amoebae suggest an explanation for its virulence and intracellular pathogenic strategy in macrophages. Proc Natl Acad Sci U S A 98(26): 15245–15250.

Steenbergen JN, Nosanchuk JD, Malliaris SD, Casadevall A. 2004. Interaction of *Blastomyces dermatitidis*, *Sporothrix schenckii*, and *Histoplasma capsulatum* with *Acanthamoeba castellanii*. Infect Immun 72(6): 3478–3488.

Steenkamp ET, Wright J, Baldauf SL. 2006. The protistan origins of animals and fungi. Mol Biol Evol 23(1): 93–106.

Sugiyama M, Ohara A, Mikawa T. 1999. Molecular phylogeny of onygenalean fungi based on small subunit ribosomal DNA (SSU rDNA) sequences. Mycoscience 40(3): 251–258.

Taylor JW, Berbee ML. 2006. Dating divergences in the fungal tree of life: Review and new analyses. Mycologia 98(6): 838–849.

Taylor LH, Latham SM, Woolhouse MEJ. 2001. Risk factors for human disease emergence. Philos Trans R Soc Lond B Biol Sci 356(1411): 983–989.

Taylor TN, Osborn JM. 1996. The importance of fungi in shaping the paleoecosystem. Rev Palaeobot Palynol 90(3–4): 249–262.

Thomarat F, Vivarès CP, Gouy M. 2004. Phylogenetic analysis of the complete genome sequence of *Encephalitozoon cuniculit*; supports the fungal origin of Microsporidia and reveals a high frequency of fast-evolving genes. J Mol Evol 59(6): 780–791.

Untereiner WA, Scott JA, Naveau FA, Sigler L, Bachewich J, Angus A. 2004. The Ajellomycetaceae, a new family of vertebrate-associated Onygenales. Mycologia 96(4): 812–821.

van der Does HC, Rep M. 2007. Virulence genes and the evolution of host specificity in plant-pathogenic fungi. Mol Plant Microbe Interact 20(10): 1175–1182.

van het Hoog M, Rast TJ, Martchenko M, Grindle S, Dignard D, Hogues H, Cuomo C, Berriman M, Scherer S, Magee BB, Whiteway M, Chibana H, Nantel A, Magee PT. 2007. Assembly of the *Candida albicans* genome into sixteen supercontigs aligned on the eight chromosomes. Genome Biol 8(4): R52.

Vilela R, Rosa PS, Belone AFF, Taylor JW, DiÛrio SM, Mendoza L. 2009. Molecular phylogeny of animal pathogen *Lacazia loboi* inferred from rDNA and DNA coding sequences. Mycol Res 113(8): 851–857.

Wang DY, Kumar S, Hedges SB. 1999. Divergence time estimates for the early history of animal phyla and the origin of plants, animals and fungi. Proc Biol Sci 266(1415): 163–171.

Wang H, Xu Z, Gao L, Hao B. 2009. A fungal phylogeny based on 82 complete genomes using the composition vector method. BMC Evol Biol 9(1): 195.

White MM, James TY, O'Donnell K, Cafaro MJ, Tanabe Y, Sugiyama J. 2006. Phylogeny of the Zygomycota based on nuclear ribosomal sequence data. Mycologia 98(6): 872–884.

Wolfe KH, Shields DC. 1997. Molecular evidence for an ancient duplication of the entire yeast genome. Nature 387(6634): 708–713.

Xu J, Vilgalys R, Mitchell TG. 2000. Multiple gene genealogies reveal recent dispersion and hybridization in the human pathogenic fungus *Cryptococcus neoformans*. Mol Ecol 9(10): 1471–1481.

Xu J, Saunders CW, Hu P, Grant RA, Boekhout T, Kuramae EE, Kronstad JW, DeAngelis YM, Reeder NL, Johnstone KR, Leland M, Fieno AM, Begley WM, Sun Y, Lacey MP, Chaudhary T, Keough T, Chu L, Sears R, Yuan B, Dawson TL. 2007. Dandruff-associated *Malassezia* genomes reveal convergent and divergent virulence traits shared with plant and human fungal pathogens. Proc Natl Acad Sci U S A 104(47): 18730–18735.

Zeyl C. 2009. The role of sex in fungal evolution. Curr Opin Microbiol 12(6): 592–598.

PART II

POPULATION GENETICS AND EVOLUTIONARY APPROACHES

PART II

CHAPTER 5

MALARIA: HOST RANGE, DIVERSITY, AND SPECIATION

ANANIAS A. ESCALANTE and FRANCISCO J. AYALA

Malaria has long been recognized as a scourge of humanity. Malaria is caused by parasitic protozoa belonging to the genus *Plasmodium*, a diverse group with a broad range of vertebrate hosts, including reptiles, birds, and mammals (Garnham, 1966). The known *Plasmodium* species found in mammals are mostly restricted to primates primarily from Africa and Southeast Asia, as well as a handful of *Plasmodium* species parasitic to African rodents. Four *Plasmodium* species are commonly found in humans, but two of them, *Plasmodium falciparum* and *Plasmodium vivax*, account for most of the malaria morbidity and mortality worldwide, with *P. falciparum* accounting for the largest share. Given the diversity of *Plasmodium* parasites found in humans and the importance of malaria as a major global health problem, it is of considerable interest to understand the evolutionary path through which malarial parasites colonized the human host.

The different species of primate malarias, including those parasitic to humans, have been traditionally identified based on morphological traits and life cycle characteristics, including the host that they infect. However, it is becoming clear that these traits are less reliable than once thought (Escalante et al., 1995, 1998; Martinsen et al., 2008; Perkins and Schall, 2002; Singh et al., 2004). There is compelling evidence indicating that host switches have been rather common among primate malarial parasites through their evolutionary history (Duval et al., 2010; Escalante et al., 1995; Krief et al., 2010; Liu et al., 2010; Prugnolle et al., 2010, 2011; Rich et al., 2009; Tazi and Ayala, 2011) and that such a dynamic has been important in the origin of human malarias. The public health implications of a previously unsuspected broad host range are just starting to be understood. As an example, the macaque parasite *Plasmodium knowlesi* has been recently found in humans (Singh et al., 2004). Although *P. knowlesi* seems to be a zoonosis, its discovery provides additional evidence that host specificity cannot any longer be assumed in primate malarias. Similar patterns are emerging among the African apes; for example, *P. falciparum* has been found in gorillas and chimps (Krief et al., 2010; Prugnolle et al., 2010). There is also evidence of host switches between humans

Evolution of Virulence in Eukaryotic Microbes, First Edition. Edited by L. David Sibley, Barbara J. Howlett, and Joseph Heitman.

and several New World monkeys, which share two genetically indistinguishable pairs of parasites. One pair involves *Plasmodium malariae* in humans and *Plasmodium brasilianum* in several species of South American nonhuman primates. The other pair involves *P. vivax* in humans and *Plasmodium simium* in nonhuman South American primates (Escalante et al., 1995; Tazi and Ayala, 2011). These examples indicate that "host specificity" in primate malarial parasites could be impacted, at least in part, by the local ecology rather than being simply determined by obligated associations of malarial parasites with given host species.

Here we review how our understanding of the evolution of primate malarias and of species parasitic to humans has changed in the last few years. Of particular importance has been the access to recent parasite samples from African apes (Duval et al., 2009, 2010; Krief et al., 2010; Liu et al., 2010; Ollomo et al., 2009; Prugnolle et al., 2010, 2011; Rich et al., 2009). The large and highly informative *Plasmodium* samples recently obtained provide previously unsuspected new perspectives on the evolution of ape (and other primate) malarias.

EARLY PHYLOGENETIC STUDIES

Ascertaining the origin of human malarias and the search for suitable animal models motivated early taxonomic and diversity investigations in nonhuman primate malarias (Coatney et al., 1971; Garnham, 1966). Formal molecular phylogenetic studies did not appear until the early 1990s, when a substantial amount of genetic data on housekeeping and antigen encoding genes became available for various *Plasmodium* species.

Initial molecular phylogenetic investigations focused on the origin of *P. falciparum* but provided important insight about the origin of other human malarias (Escalante and Ayala, 1994; Escalante et al., 1995; McCutchan et al., 1996; Waters et al., 1991). An important lesson learned from those early phylogenetic studies was that the four *Plasmodium* species parasitic to *Homo* arose independently as human pathogens. Independent origins were early on accepted for *P. falciparum* and *P. vivax*, but some investigators suggested that *Plasmodium ovale*, a parasite considered "*vivax*-like" given its biological characteristics (Coatney et al., 1971), could have been derived from *P. vivax*. Such hypothesis for the origin of *P. ovale* was ruled out by molecular data as described below (Qari et al., 1996). A second important result was that *P. falciparum* shared its most recent common ancestor with *Plasmodium reichenowi*, a parasite found in chimpanzees and gorillas (Escalante and Ayala, 1994; Escalante et al., 1995).

We have incorporated these early phylogenetic studies in Figure 5.1 using a Bayesian phylogeny that only includes those lineages known until about a decade ago. This phylogeny has been inferred using complete mitochondrial genomes. The mitochondrial cytochrome b gene is widely used in malaria molecular phylogenetic and population studies (Escalante et al., 1998; Perkins and Schall, 2002; Ricklefs and Fallon, 2002; Ricklefs et al., 2004), but the complete mitochondrial genome is more informative because it has more polymorphic sites (Cornejo and Escalante, 2006; Joy et al., 2003; Mu et al., 2005). In Figure 5.1, we have rooted the mammalian species of *Plasmodium* using avian and reptile malarial parasites as an outgroup (see Table 5.1 for a complete list of *Plasmodium* species and their hosts). It deserves notice that *P. falciparum* and *P. reichenowi* appear as a closely related pair that is phylogenetically distinct from all other mammalian parasites.

Regardless of the gene loci used, all early phylogenetic studies essentially reproduced the evolutionary relationships

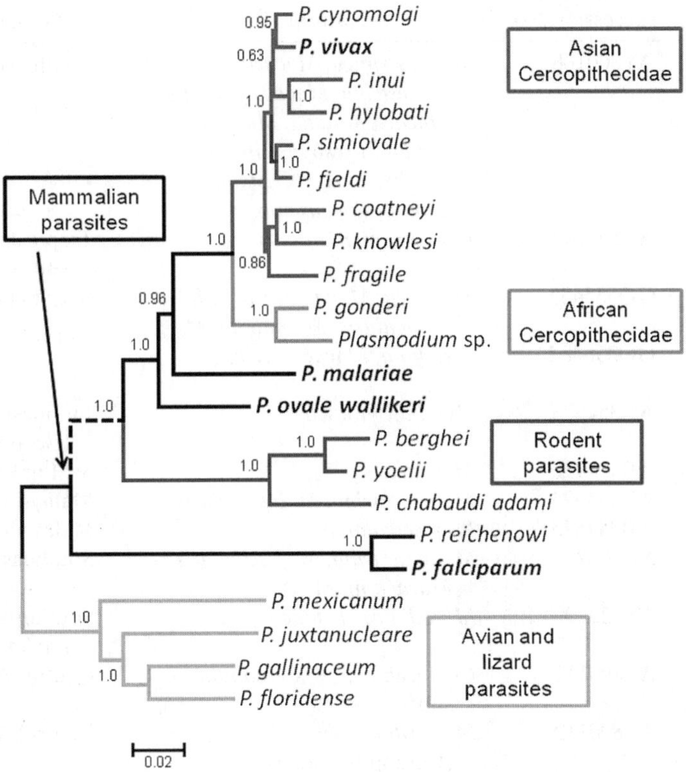

Figure 5.1 Bayesian phylogenetic tree of primate malarial parasites based on complete mitochondrial genomes. Avian and lizard parasites are used as an outgroup. The tree is inferred under the Hasegawa–Kishino–Yano (HKY) substitution model using MrBayes, version 3.1.2 (Ronquist and Huelsenbeck, 2003), with a Markov chain Monte Carlo (MCMC) of 3,000,000 generations, sampling trees each 100 generations. The first 10,000 sampled trees were discarded in the burning. Values above branches are posterior probabilities. A neighbor joining tree inferred under the Tamura and Nei substitution model with the software MEGA, version 4.0 (Tamura et al., 2007), was used as a prior. Branches including major parasite lineages or groups are indicated with different colors. Human malarial parasites are indicated in bold font. See color insert.

depicted in Figure 5.1. The four species of human malaria appeared distantly related to each other. Interestingly, primate malarias were subdivided into two distinct clades, one represented solely by the lineages of *P. falciparum* and *P. reichenowi*, and a separate grouping that included all other known primate malarial parasites found in New and Old World primates (Escalante and Ayala, 1994; Escalante et al., 1995, 1998; Perkins and Schall, 2002). Two parasites of New World primates, *P. simium* and *P. brasilianum*, were indistinguishable from the human parasites *P. vivax* and *P. malariae*, respectively (Escalante et al., 1995). In all cases, rodent malarias appeared as a monophyletic group, which, according to a recent proposal, may likely have originated from primate parasites (Martinsen et al., 2008).

Several questions were difficult to address at the time. Were *P. falciparum* and *P. reichenowi* parasites that had cospeciated with their hosts? Could *P. vivax* and *P. malariae* have originated from South American primates rather than being the ancestors of *P. simium* and *P. brasilianum*? These questions motivated a broad range

TABLE 5.1 ***Plasmodium* Species Included in Our Phylogenetic Analyses**

Species—Strain	GenBank Accession No.	Natural Host	Geographic Range
P. cynomolgi	AY800108	*Macaca sinica*, *Macaca nemestrina*, *Macaca fascicularis*, *Macaca mulatta*, *Macaca radiata*, *Presbytis entrellus*, *Trachypithecus cristatus*, *Semnopithecus* spp.	Southeast Asia
P. vivax	AY598140	*Homo sapiens*	Tropical, subtropical, and temperate regions
P. inui—Taiwan II	GQ355483	*M. sinica*, *M. nemestrina*, *M. fascicularis*, *M. mulatta*, *M. radiata*, *Macaca cyclopis*	South and East Asia
P. inui—leaf monkey II	GQ355482		
P. hylobati	AB354573	*Hylobati moloch*	Indonesia, Malaysia (Borneo)
P. simiovale	AB434920	*M. sinica*	Sri Lanka
P. fieldi—N-3	AB354574	*M. nemestrina*, *M. fascicularis*	Malaysia
P. coatneyi	AB354575	*M. fascicularis*	Malaysia, Philippines
P. knowlesi	NC_007232	*M. nemestrina*, *M. fascicularis*, *Macaca nigra*	Southeast Asia
P. fragile	AY722799	*M. radiata*, *M. sinica*	Southern India, Sri Lanka
P. gonderi	AY800111	*Cercocebus atys*, *Cercopithecus* spp.	Central Africa
Plasmodium sp.	AY800112	*Mandrillus sphinx* (Cercopithecidae)	Central Africa
P. malariae	AB354570	*H. sapiens*	Tropical, subtropical, and temperate regions
P. ovale wallikeri	HQ712053	*H. sapiens*	Tropical and subtropical regions
P. ovale curtisi	HQ712052	*H. sapiens*	Tropical and subtropical regions
Plasmodium sp. (A)	HQ712054	*Hapalemur griseus griseus*	Madagascar (eastern rainforest)
Plasmodium sp. (B)	HQ712055	*Varecia variegata*	
Plasmodium sp. (C)	HQ712056	*H. griseus griseus*	
Plasmodium sp. (D)	HQ712057	*Indri indri*	
P. berghei	AF014115	*Grammomys* sp.	Central Africa
P. yoelii	M29000	*Thamnomys* sp.	Africa
P. atheruri	HQ712051	*Atherurus africanus* (porcupine)	Central Africa
P. chabaudi	AF014116	*Thamnomys* sp.	Central Africa
P. billbrayi	GQ355468	*Pan troglodytes*	Uganda, Republic of the Congo
P. billcollinsi	GQ355479	*P. troglodytes*	Uganda, Republic of the Congo
P. reichenowi	NC-002235	*P. troglodytes*	Africa
P. reichenowi	GQ355476	*P. troglodytes*	Republic of the Congo
P. falciparum	AY282930	*H. sapiens*	Worldwide Tropical regions
P. gaboni	FJ895307	*P. troglodytes*	Gabon

TABLE 5.1 (*Continued*)

Species—Strain	GenBank Accession No.	Natural Host	Geographic Range
Plasmodium sp. (G1)	HM235308	*Gorilla* sp.	Republic of the Congo
Plasmodium sp. (G1)	HM235269	*Gorilla* sp.	Republic of the Congo
Plasmodium sp. (G2)	HM235307	*Gorilla* sp.	Cameroon
Plasmodium sp. (G3)	HM235294	*Gorilla* sp.	Central African Republic
Plasmodium sp. (C1)	HM235350	*P. troglodytes*	Cameroon
Plasmodium sp. (C1)	HM235388	*P. troglodytes*	Democratic Republic of the Congo
Plasmodium sp. (C2)	HM235319	*P. troglodytes*	Democratic Republic of the Congo
Plasmodium sp. (C2)	HM235349	*P. troglodytes*	Cameroon
Plasmodium sp. (C3)	HM235293	*P. troglodytes*	Cameroon
P. juxtanucleare	NC_008279	Galliformes and domestic birds	Tropical and subtropical regions
P. gallinaceum	NC_008288	Galliformes and Sphenisciformes	
P. floridense	NC_009961	*Anolis sagrei*	Caribbean
P. mexicanum	NC_009960	*Sceloporus occidentalis*	California

of molecular systematic investigations that were severely constrained by the limited access to new parasitic samples from primates. Most studies simply reproduced the same major clades over and over again, using a variety of arguments about the likelihood of one or other origin for pairs of species or groups of species. In spite of the rapid development of molecular methods, the progress in malaria evolutionary biology was handicapped for almost two decades by the limited information concerning the host range and distribution of malarial parasites. Indeed, until very recently, the malaria research community was using isolates that had been obtained during field studies carried out decades earlier. Recently, this state of affairs has changed drastically owing to the discovery of new malarial lineages in African apes, especially many that are closely related to *P. falciparum*. Accordingly, the malaria research community is at present engaged in actively searching for new *Plasmodium* samples that would allow us to reassess the host range of malarial parasites in primates and, by so doing, would help to resolve the origin of human malarias.

THE ORIGIN OF *P. FALCIPARUM*

An early view of *P. falciparum* evolution was that it had originated "recently" as a human parasite as the result of a host switch from an avian host. It was argued that such putative avian origin would explain its high virulence compared to other human malarias (Boyd, 1949). This recent-origin avian hypothesis appeared to be supported by several lines of evidence. The host range of avian malarias was considered broad since morphologically indistinguishable species were found in several avian hosts, even with worldwide distribution (Levine, 1988; Valkiūnas, 2005). Accordingly, host switches were considered common in avian malaria parasites, making it plausible that an avian parasite could have switched to humans. In addition, some avian malaria parasites shared morphological features with *P. falciparum* that other mammalian *Plasmodium* did not (e.g., falciform gametocytes), finally, avian species had similar A-T content in their genomes as *P. falciparum* (McCutchan et al., 1984).

The avian origin hypothesis motivated early molecular phylogenetic investigations (Escalante and Ayala, 1994; Escalante et al., 1995, 1998; Perkins and Schall, 2002; Waters et al., 1991), leading to an alternative model where *P. falciparum* did not originate from an avian host. Rather, it was shown genetically that its closest relative among the known *Plasmodium* species was a chimpanzee parasite, *P. reichenowi*. Whether or not this implied cospeciation with their host or a host switch was not solved at the time (Escalante and Ayala, 1994, 1995; Escalante et al., 1998; Rich et al., 1998). The coalescent theory of population genetics suggests that the ancestral parasite should have higher levels of adaptively neutral polymorphism than the derived parasite. Since only one isolate of *P. reichenowi* was available, this comparison could not be made. Cospeciation of *P. falciparum* and *P. reichenowi* with their respective hosts, humans and chimpanzees, was tentatively accepted as a working hypothesis. Estimates of neutral mutation rates led, in any case, to the conclusion that the world expansion of *P. falciparum* was very recent, of the order of a few thousand years, given the scarcity of neutral polymorphisms in a diversity of genes (Coluzzi, 1999; Conway et al., 2000; Joy et al., 2003; Rich and Ayala, 2003; Rich et al., 1998; Volkman et al, 2001). It was, nevertheless, hypothesized that *P. reichenowi*, as well as *P. falciparum*, may have had an avian origin (Wolfe et al., 2007). The hypothesis that *P. falciparum* may have originated as a result of a host switch from African apes, chimpanzees most likely, into humans was never ruled out (Escalante et al., 1998; Rich et al., 1998), but it was only very recently that putative new species found in African apes have shown that *P. falciparum* and *P. reichenowi* are closely related to several newly discovered ape parasites and that *P. falciparum* may have originated from ape parasites.

Figure 5.2 depicts a phylogeny based on partial sequences of the mitochondrial genome that includes data recently reported from numerous *Plasmodium* samples from chimpanzee and gorilla (Krief et al., 2010; Liu et al., 2010; Ollomo et al., 2009) and other primates (see Pacheco et al., 2011). This phylogeny does not include all available data (Duval et al., 2010; Kaiser et al., 2010; Prugnolle et al., 2010, 2011; Rich et al., 2009). However, including additional shorter mitochondrial sequences does not change in any significant way the phylogenetic relationships depicted in Figure 5.2. The overwhelming conclusion emerging from the recent ape samples is that the *P. falciparum–P. reichenowi* lineage is not isolated. Rather, it is part of a diverse group of malarial parasites that appear to have been long associated with the hominids and would belong to the *Laverania* subgenus, long ago named to take into account the distinct phylogeny of the *P. falciparum–P. reichenowi* lineage (Garnham, 1966). Another important conclusion is that there have been several host switches during the evolutionary history of the *Laverania* subgenus as evidenced by the fact that the *P. falciparum* lineage is intertwined with several chimpanzee and gorilla parasites and that several of the parasites are present in different host species.

The *Laverania* clade now includes several lineages that may reasonably be considered different species. Table 5.2 gives the genetic distances among five *Plasmodium* species, including *P. falciparum*, *P. reichenowi*, and three recently named ape species, as well as the genetic distances estimated for well-established rodent parasite species. The distance among the primate parasites

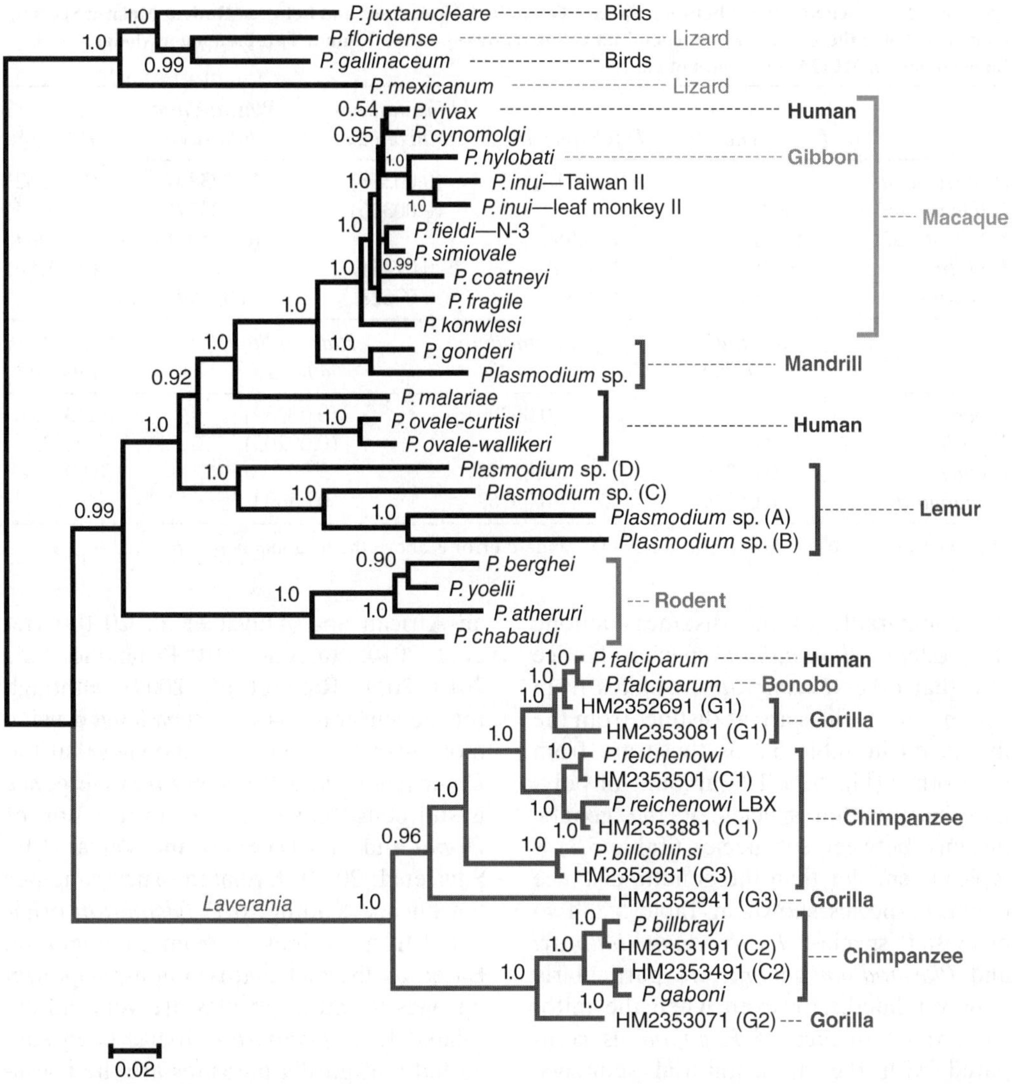

Figure 5.2 Bayesian phylogenetic tree using partial mitochondrial genomes, approximately 3400 bp in length. The alignment includes a partial cytochrome oxidase subunit 3 gene, the complete cytochrome oxidase subunit 1 gene, and a partial cytochrome b gene. The phylogeny incorporates several new lineages recently discovered in African apes (see text); those lineages are part of a monophyletic group (the *Laverania* subgenus), which also includes the human parasite *P. falciparum* and the chimpanzee parasite *P. reichenowi*. The phylogeny includes all known primate malarial parasites; avian and lizard parasites were used as an outgroup. The tree is inferred under the Hasegawa–Kishino–Yano (HKY) substitution model using MrBayes, version 3.1.2 (Ronquist and Huelsenbeck, 2003), with an MCMC of 6,000,000 generations, sampling trees each 100 generations. The first 20,000 sampled trees were discarded in the burning. Values above branches are posterior probabilities. A neighbor joining tree inferred under the Tamura and Nei substitution model with the software MEGA, version 4.0 (Tamura et al., 2007), was used as a prior. See color insert.

TABLE 5.2 Genetic Distance between Species Found in African Apes and between Rodent Malaria Species, Estimated Using the Complete Mitochondrial Genome, Based on the Kimura Two-Parameter Model as Implemented in MEGA 4 (Tamura et al., 2007)

	P. reichenowi	*P. falciparum*	*Plasmodium billcollinsi*	*Plasmodium billbrayi*	*P. gaboni*
P. reichenowi		(0.00218)	(0.00318)	(0.00383)	(0.00382)
P. falciparum	0.02380		(0.00316)	(0.00377)	(0.00384)
P. billcollinsi	0.05152	0.04956		(0.00403)	(0.00404)
P. billbrayi	0.07502	0.07502	0.07441		(0.00186)
P. gaboni	0.07623	0.07522	0.07623	0.01759	

	Plasmodium berghei	*Plasmodium yoelii*	*Plasmodium atheruri*	*Plasmodium chabaudi*
P. berghei		(0.00200)	(0.00239)	(0.00380)
P. yoelii	0.02169		(0.00203)	(0.00332)
P. atheruri	0.03350	0.02793		(0.00332)
P. chabaudi	0.05749	0.05267	0.05693	

The genetic distance is below the diagonal; the standard error is above the diagonal between brackets.

are comparable to the distances among the rodent *Plasmodium* species. Notice also that other *Plasmodium* isolates from chimps and gorillas are as distinct from the five listed in Table 5.2 as these are from each other (Fig. 5.2). The amount of polymorphism within species or the genetic distance between subspecies (Table 5.3) is typically smaller than the genetic distance between species shown in Table 5.2. Two proposed species, *Plasmodium billbrayi* and *Plasmodium gaboni*, are particularly closely related. However, when the mitochondrial sequence of *P. gaboni* is compared with the mitochondrial sequence from *P. billbrayi*, there are differences of sufficient importance (e.g., *P. gaboni* has a unique insert) that would support keeping these two lineages as two putative new species of the *Laverania* clade (Krief et al., 2010; Liu et al., 2010; Ollomo et al., 2009).

These new *Plasmodium* lineages in African apes indicate that the lineage of *P. falciparum* and *P. reichenowi* is part of a diverse group of malarial parasites of the Homininae subfamily. Taking into account the phylogeny that includes recent data, it seems likely that *P. falciparum* originated as a human parasite via a host switch from an African ape (Duval et al., 2010; Krief et al., 2010; Liu et al., 2010; Prugnolle et al., 2010, 2011; Rich et al., 2009), although this conclusion has been challenged using molecular clock analyses showing that the *P. reichenowi* and *P. falciparum* divergence is still consistent with the cospeciation of *Homo* and *Pan* (Hughes and Verra, 2010; Silva et al., 2010). A related issue is whether the lineage leading to *P. falciparum* originated from gorillas or from chimpanzees. Based on the fact that two non-*falciparum* lineages found in gorillas are very closely related to *P. falciparum*, it has been suggested that gorilla parasites are the ancestors of *P. falciparum* (Liu et al., 2010; Prugnolle et al., 2010), although most recent evidence suggests that an origin from chimpanzees is equally likely (Prugnolle et al., 2011). Moreover, we need to take into account that *P. falciparum* itself has been found in bonobos, chimpanzees, and gorillas (Duval et al., 2010; Krief et al., 2010; Prugnolle et al., 2011), and that some lineages reported in gorillas have been also found in chimpanzees (Duval et al., 2010). Figure 5.3 shows a minimum number of host switches required to explain the distribution of malarial parasites in humans

TABLE 5.3 Genetic Diversity within *Plasmodium* Species and Average Genetic Divergence between Subspecies of Mammalian *Plasmodium* Parasites

		Genetic Distance (d ± SE)				
Species	n	COXI	COXIII	CYTB	COXI + CYTB	Complete mtDNA
P. chabaudi chabaudi	7	0.0011 ± 0.0005	0.0017 ± 0.0009	0.0005 ± 0.0003	0.0009 ± 0.0003	0.0010 ± 0.0003
P. chabaudi adami	2	0.0035 ± 0.0016	0.0051 ± 0.0025	0.0035 ± 0.0017	0.0035 ± 0.0013	0.0030 ± 0.0007
P. ch. chabaudi–P. ch. adami	7 vs. 2	0.0051 ± 0.0017	0.0091 ± 0.0028	0.0056 ± 0.0019	0.0053 ± 0.0013	0.0048 ± 0.0007
P. falciparum	101	0.0001 ± 0.0001	0.0003 ± 0.0001	0.0005 ± 0.0003	0.0003 ± 0.0001	0.0003 ± 0.0001
P. ovale wallikeri–P. ovale curtisi	1 vs. 1	0.0141 ± 0.0031	0.0283 ± 0.0058	0.0140 ± 0.0035	0.0141 ± 0.0023	0.0152 ± 0.0016
P. vivax	110	0.0013 ± 0.0005	0.0007 ± 0.0005	0.0003 ± 0.0001	0.0009 ± 0.0003	0.0006 ± 0.0001
P. cynomolgi	12	0.0036 ± 0.0008	0.0033 ± 0.0011	0.0032 ± 0.0009	0.0034 ± 0.0006	0.0026 ± 0.0003
P. inui	14	0.0138 ± 0.0017	0.0179 ± 0.0029	0.0154 ± 0.0022	0.0145 ± 0.0015	0.0126 ± 0.0011
P. knowlesi (*Macaca* sp.–Human)	59	0.0013 ± 0.0004	0.0016 ± 0.0007	0.0008 ± 0.0004	0.0011 ± 0.0003	0.0009 ± 0.0002

Estimates are based on the complete mitochondrial genome as well as on some genes individually. All calculations use the Kimura two-parameter model as implemented in MEGA 4 (Tamura et al., 2007); n is the number of sequences analyzed. COXI, cytochrome c oxidase subunit I; COXIII, cytochrome c oxidase subunit III; CYTB, cytochrome b.

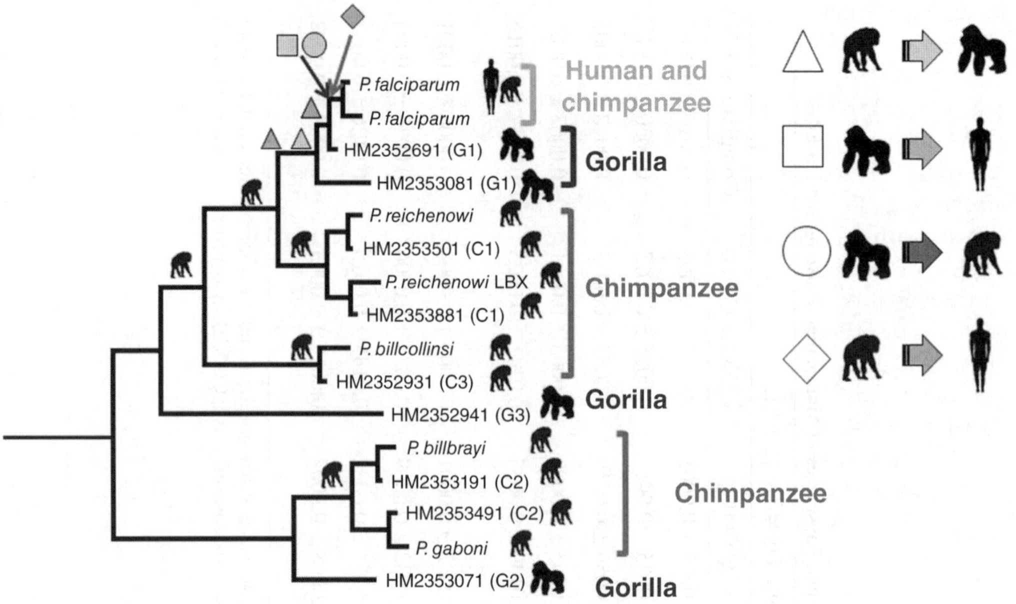

Figure 5.3 Phylogenetic tree of the malarial parasites found in African apes indicating whether lineages or species are found in *Pan* (chimpanzees), *Gorilla*, and/or humans. This phylogeny is part of the one depicted in Figure 5.2. Host switches are indicated with geometric symbols. Triangle: *Pan* to *Gorilla*; square: *Gorilla* to human; circle: *Gorilla* to *Pan*; and diamond: *Pan* to human. Some of these host switches may have occurred in the opposite direction or in both directions. The host switches required under the scenario of a *Gorilla* origin for *P. falciparum* are indicated in orange. The host switches required under the scenario that *P. falciparum* originated from a species parasitic to the genus *Pan* are indicated in green. See color insert.

and in African apes. Each geometric symbol represents one possible switch; host switches that assume a gorilla origin appear in yellow, while host switches that assume an ancestor in *Pan* are in green. In Figure 5.3, the number of host switches required is the same under both scenarios. Thus, it is plausible that gorillas are simply particularly susceptible to *falciparum*-like parasites and that chimpanzees were the original hosts from which both human and gorillas acquired their malarial parasites (Krief et al., 2010; Prugnolle et al., 2011; Rich et al., 2009).

THE ORIGIN OF *P. VIVAX*

Traditionally, the origin of primate malarial parasites, with the exception of *P. falciparum*, was placed in Southeast Asia. The proposal of an Asian origin was based on the high diversity of malarial parasites found in that region, where, moreover, many lineages of nonhuman primate parasites were identified as closely related to *P. vivax*, *P. ovale*, and *P. malariae*, based on life history traits. We now know that those similarities are convergent characteristics that emerged as a result of the radiation of nonhuman primate malarias in that region (Escalante et al., 1998, 2005). For instance, the length of periodicity is a convergent characteristic. The quartan parasites *Plasmodium inui* and *P. malariae* are not monophyletic (Figs. 5.1 and 5.2); *P. knowlesi*, the only quotidian primate parasite (24-hour erythrocytic cycle), shares this characteristic with several species parasitic to birds (Valkiūnas, 2005). Another example is the capacity for relapse. Parasites that share this characteristic do not form a monophyletic group (Escalante et al., 1998).

Nevertheless, in the case of *P. vivax*, an Asian origin was accepted because of its similarity to *Plasmodium cynomolgi* (Coatney et al., 1971; Garnham, 1966), a parasite with relatively broad host and geographic distributions. This conclusion was controversial owing to the high frequency in Africa of a Duffy negative blood group allele that does not allow for invasion of erythrocytes by *P. vivax* (Carter, 2003; Livingstone, 1984). Assuming that the Duffy blood group allele was selected to near fixation owing to exposure to *P. vivax*, its high frequency would indicate a long association of the African human population with this parasite. Considering that the extant *P. vivax* is by far less virulent than *P. falciparum*, such a strong selection by *P. vivax* in Africa would imply a more virulent parasite than the one we know today.

Molecular phylogenetic analyses have provided important information about these issues. First, there is an African origin for the clade that originated the current species observed in Southeast Asia and *P. vivax* (Fig. 5.2). It has been shown that the lineage leading to *P. vivax* was part of a parasite radiation, facilitated by host switches and a strong geographic structure, a process that took place in the region over the last 3–4 million years (Escalante et al., 1998, 2005). Further studies that include several antigens and housekeeping genes support the hypothesis that the closest species related to *P. vivax* are those found in Southeast Asian nonhuman primates (Mitsui et al., 2010; Pacheco et al., 2007, 2010; Sawai et al., 2010; Tanabe et al., 2007)

There is, however, a primate parasite lineage in the New World, *P. simium*, that is genetically indistinguishable from *P. vivax* (Escalante et al., 1995; Tazi and Ayala, 2011). The presence of *P. simium* in New World primates indicates that a host switch has taken place very recently, namely, after the colonization of the Americas some 15,000 years ago or, more likely, after the European colonizations that started some five centuries ago. We know that at least two independent host transfers have occurred (Tazi and Ayala, 2011). But the direction of the host switch (from humans to monkeys or monkeys to humans) cannot be established when only these two species are compared (Tazi and Ayala, 2011) because the scarcity of available isolates of *P. simium* does not allow comparing the relative amounts of neutral polymorphisms between *P. simium* and *P. vivax*. However, if we consider the phylogeny of nonhuman primate malarias, all the species closely related to the *P. vivax–P. simium* lineage are found in Old World primates. Thus, the phylogeny can be easily accounted for if we assume that the direction of the *P. simium*/*P. vivax* switch was from a Catarrhini (Old World primates) to a Platyrrhini primate (New World primates), making the origin of *P. simium* from *P. vivax* the most parsimonious hypothesis.

There is an increasing body of evidence indicating that *P. vivax* may exist at higher frequency in Africa than previously thought. There are several reports indicating that *P. vivax* can actually infect individuals that are Duffy negative (Culleton et al., 2009; Ménard et al., 2010; Ryan et al., 2006), challenging the notion that this parasite had been eliminated, or nearly so, from sub-Saharan Africa. *P. vivax* can also be found in chimpanzees (Kaiser et al., 2010; Krief et al., 2010), an observation consistent with the higher than previously assumed prevalence of *P. vivax* in African humans, as this may have facilitated a host switch from humans to chimps. Whereas these facts alone do not exclude the origin of *P. vivax* via a host switch from a nonhuman primate in Southeast Asia, there remain several issues that need further clarification. We have almost no data from malarial parasites in orangutans and gibbons. Parasite lineages from such hosts may confirm or challenge the current model about an Asian origin of primate parasites. It is also important to realize that there are relatively few studies about malarial parasites in African primates. Such lineages could challenge the Asian origin model for *P. vivax*.

In any case, it seems clear that *P. vivax* is capable of infecting New World monkeys as well as African apes, making it a highly flexible parasite in terms of suitable hosts. Such host plasticity may challenge its eradication if nonhuman primates could become reservoirs, thus potentially allowing the parasite to reinfect humans.

P. OVALE AND *P. MALARIAE*

The distribution of *P. ovale* and *P. malariae* and the limited information on closely related species in nonhuman primates make the origin of these species an unsolved puzzle. *P. ovale* is not found in the Americas and, until very recently, we had no information on any nonhuman parasite closely related to it. Previous hypotheses indicated that *P. ovale* might be related to *Plasmodium* species found in nonhuman primates from Southeast Asia, but such hypotheses were ruled out by phylogenetic studies (Qari et al., 1996). Nowadays, it is accepted that *P. ovale* includes two nonrecombinant forms or distinct lineages distributed in the Old World. The temporal and spatial persistence of these two lineages have been explained by the existence of two different species or subspecies (Sutherland et al., 2010). Recent studies have found that *P. ovale* (or at least a parasite indistinguishable from ovale on the basis of mitochondrial genes) is present in chimpanzees (Duval et al., 2009; Kaiser et al., 2010). Thus, like in the case of *P. vivax* and *P. falciparum*, *P. ovale* can be shared by humans and African apes. A distinct but closely related lineage to *P. ovale* has been found also in chimpanzees (Duval et al., 2009). This finding, if confirmed, suggests that *P. ovale* may have originated in Africa and may have expanded from it to the Old World. It seems interesting to note that this parasite is totally absent from the Americas, a phenomenon that could imply some ecological barrier such as unsuitable vectors.

In contrast to *P. ovale*, *P. malariae* is a parasite found worldwide. An identical species is found in new world primates, *P. brasilianum* (Escalante et al., 1995; Tazi and Ayala, 2011). A difference between the pair *P. simium*/*P. vivax* and *P. brasilianum*/*P. malariae* is that the distribution of *P. brasilianum* in South America is much more geographically widespread than the one documented for *P. simium*. For example, *P. brasilianum* has been found from Panama to the north of Brazil (Coatney et al., 1971). In addition, *P. brasilianum* has been found in several hosts of New World primates that are classified in eight different genera (Coatney et al., 1971). Thus, in the case of *P. malariae* and *P. brasilianum*, any model proposed for their origin should account not for one host switch but for a number of them, if the transfer happened from humans to monkeys. Unfortunately, there is a lack of molecular data from isolates found in nonhuman hosts so that it seems impossible to ascertain the complex dynamics between *P. brasilianum* and South American primates (Tazi and Ayala, 2011). The situation is further complicated by the fact that an identical parasite to *P. malariae*, in terms of mitochondrial markers, has been found in chimpanzees (Hayakawa et al., 2009; Kaiser et al., 2010; Krief et al., 2010). The classical view, based solely on a putative *malariae*-like species called *Plasmodium rodhaini*, was that this parasite originated in Africa from African apes. The fact that *P. malariae* has been found in chimpanzees could be taken as evidence for an African origin (Krief et al., 2010). However, it is clearly premature to reach such a conclusion in the absence of more data from New World primate parasites (Tazi and Ayala, 2011). What is conclusive from the available evidence is that *P. malariae* is a highly resilient parasite capable of infecting a broad range of hosts.

HOST SWITCHES: PUBLIC HEALTH IMPLICATIONS

Host switches have been common during the evolutionary history of primate malarias and have played an important role in the origin of *Plasmodium* species parasitic to humans. Of special interest is that African apes can sustain infections by all known human malarial parasites. The public health import of these findings deserves consideration.

Host switches are often investigated at the species level by using phylogenetic approaches, which provide a good approximation of the "big picture" toward identifying the origin of parasite lineages (e.g., Hafner et al., 1994). Phylogenetic approaches have been proven very useful in fast-evolving parasites (high rate of evolution in the parasites/pathogens relative to their hosts), such as in the case of viruses (e.g., Biek et al., 2007), where there is an evolutionary timescale in the parasite and an ecological timescale in the host. The situation is different in the case of eukaryotic parasites, such as *Plasmodium*. The observation of host switches using relatively slow-evolving genes does not preclude that parasite populations could be structured in different host populations in space; that is, at an ecological timescale for the host, parasite lineages may be ecologically isolated in different hosts so that alternative hosts are not actual reservoirs of the disease. If host switches are relatively infrequent between two species of hosts (low relative number of parasite migrants between two host species compared to the parasite's effective population sizes in each host species), we should expect some "host species structure" in the parasite populations, that is, some level of divergence between the parasite subpopulations found in each host species. Thus, if intrahost species transmission (parasite replication within populations of the same host species) is clearly favored over interhost species transmission (host switches) then, the host population structure will drive the parasite divergence and we should expect parasite lineages to drift apart in different hosts even in relative sympatry (Krief et al., 2010). There are of course more complex dynamics than the one just described; however, the underlying fact is that parasite populations structured in different species of host could be maintained by the local ecology simply by host segregation (differences in behavior or in vectors), the spatial host distributions, and/or natural selection favoring intrahost species transmission over interhost species transmission. Assessing the relative importance of such factors is necessary to understand the importance of African apes as potential reservoirs for human malarias or in order to assess the risk of African apes to become reservoirs if the ecological conditions change.

Empirical evidence still needs to be further gathered in the case of primate malarias, but interesting patterns are emerging from avian malarias that suggest such host structured populations. There are a rising number of parasite lineages that have been discovered by the extensive use of the mitochondrial cytochrome b (Ricklefs and Fallon, 2002; Ricklefs et al., 2004). Whether these molecular lineages are indeed different species or could be considered an intraspecific variation is still a matter of resolution (Martinsen et al., 2006; Ricklefs and Fallon, 2002; Ricklefs et al., 2004). Nevertheless, whereas there is evidence for several host switches, recent molecular studies have provided evidence for some host specificity—that is, host structured population—in avian malarial parasites (Ricklefs et al., 2004). This complex dynamics illustrates the need for addressing population-level processes when studying parasite host range, something that is still pending in most multihost malarial parasites. Finally, an important factor to consider when assessing the putative role of African apes as reservoirs for human malaria is the ecology of suitable vectors. Recent studies of *P. knowlesi* have shown that the behavior and distribution of the vectors seem to be the limiting factor maintaining this parasite as a zoonosis (Tan et al., 2008). Without information about the vectors, any assessment on the importance of African apes as human malaria reservoirs will be incomplete.

MOLECULAR BASIS FOR THE HUMAN MALARIA TOLL

The understanding of the origin of human malarias, specially the origin of *P. falciparum*, has changed dramatically in the last few years. Whereas these findings may have unraveled the evolutionary history of *P. falciparum*, they also offer unique opportunities for increasing the power and impact that comparative genomics could have in malaria research in the context of the most malignant form of the disease. Until recently, the *P. falciparum* genomic diversity was assessed by comparing it with the only isolate of *P. reichenowi* previously available to the scientific community (Jeffares et al., 2007; Weedall et al., 2008). The information from a single genome used as an outgroup was hampering our capacity to fully understand the observed polymorphism of *P. falciparum* and the potential molecular adaptations that make this human parasite successful in terms of its high reproductive rate.

The information that can be gathered from comparative genomic approaches among *P. falciparum* and related species will be of great value for identifying proteins that are relevant to invading the human red blood cells and that can be used as vaccine targets, for improving our understanding of the functional basis of *falciparum* pathogenesis in humans (a phenomenon that has not been observed in other nonhuman primate malarias), and for assessing the rate and mode of evolution of the *falciparum* genome. Such information will be an important contribution from evolutionary biologists to the global effort against human malaria. Martin et al. (2005; see also Varki and Gagneux, 2009) have made important advances in this respect by showing that (i) the cell surface glycans are different in humans and in apes, (ii) that this difference is due to a single human mutation, and (iii) that it accounts for the greater malignancy of *P. falciparum* for humans than for apes. Much additional research along the same lines is called for.

Studying in detail the evolution of the *P. falciparum* clade will require an extensive sampling of parasites from African apes, sampling that is difficult due to ethical concerns and regulations limiting research on apes and the difficult accessibility to wild apes. In addition, culturing these *falciparum*-like parasites as viable isolates in order to use them for comparative studies may prove difficult. Nevertheless, it is worth considering that such studies need not expose African apes to significant risks. Establishing a partnership between the malaria and primatology research communities would facilitate ascertaining whether and how these *falciparum*-like parasites may have been "among us" for several million years, and that *falciparum* pathogenesis may be a modern event associated with increasing numbers of particularly susceptible hosts (humans) among suitable vectors, as well as the concomitant selective processes (Martin et al., 2005; Varki and Gagneux, 2009).

REFERENCES

Biek R, Henderson JC, Waller LA, Rupprecht CE, Real LA. 2007. A high-resolution genetic signature of demographic and spatial expansion in epizootic rabies virus. Proc Natl Acad Sci U S A 104: 7993–7998.

Boyd MF. 1949. *Malariology: A Comprehensive Survey of All Aspects of This Group of Diseases from a Global Standpoint.* Saunders, Philadelphia, PA.

Carter R. 2003. Speculations on the origins of *Plasmodium vivax* malaria. Trends Parasitol 19: 214–219.

Coatney RG, Collins WE, Warren M, Contacos PG. 1971. *The Primate Malaria.* U.S. Government Printing Office, Washington, DC.

Coluzzi M. 1999. The clay feet of the malaria giant and its African roots: Hypotheses and inferences about the origin, spread and control of *Plasmodium falciparum.* Parassitologia 41: 277–283.

Conway DJ, Fanello C, Lloyd JM, Al-Joubori BM, Baloch AH, Somanath SD, Roper C, Oduola AM, Mulder B, Povoa MM, Singh B, Thomas AW. 2000. Origin of *Plasmodium falciparum* malaria is traced by mitochondrial DNA. Mol Biochem Parasitol 111: 163–171.

Cornejo OE, Escalante AA. 2006. The origin and age of *Plasmodium vivax.* Trends Parasitol 22: 558–563.

Culleton R, Ndounga M, Zeyrek FY, Coban C, Casimiro PN, Takeo S, Tsuboi T, Yadava A, Carter R, Tanabe K. 2009. Evidence for the transmission of *Plasmodium vivax* in the Republic of the Congo, West Central Africa. J Infect Dis 200: 1465–1469.

Duval L, Nerrienet E, Rousset D, Sadeuh Mba SA, Houze S, Fourment M, Le Bras J, Robert V, Ariey F. 2009. Chimpanzee malaria parasites related to *Plasmodium ovale* in Africa. PLoS One 4: e5520.

Duval L, Fourment M, Nerrienet E, Rousset D, Sadeuh SA, Goodman SM, Andriaholinirina NV, Randrianarivelojosia M, Paul RE, Robert V, Ayala FJ, Ariey F. 2010. African apes as reservoirs of *Plasmodium falciparum* and the origin and diversification of the *Laverania* subgenus. Proc Natl Acad Sci U S A 107: 10561–10566.

Escalante AA, Ayala FJ. 1994. Phylogeny of the malarial genus *Plasmodium*, derived from rRNA gene sequences. Proc Natl Acad Sci U S A 91: 11373–11377.

Escalante AA, Ayala FJ. 1995. Origin of *Plasmodium* and other Apicomplexa based on rRNA genes. Proc Natl Acad Sci U S A 92: 5793–5797.

Escalante AA, Barrio E, Ayala FJ. 1995. Evolutionary origin of human and primate malarias: Evidence from the circumsporozoite protein gene. Mol Biol Evol 12: 616–626.

Escalante AA, Freeland DE, Collins WE, Lal AA. 1998. The evolution of primate malaria parasites based on the gene encoding cytochrome b from the linear mitochondrial genome. Proc Natl Acad Sci U S A 95: 8124–8129.

Escalante AA, Cornejo OE, Freeland DE, Poe AC, Durrego E, Collins WE, Lal AA. 2005. A monkey's tale: The origin of *Plasmodium vivax* as a human malaria parasite. Proc Natl Acad Sci U S A 102: 1980–1985.

Garnham PCC. 1966. *Malaria Parasites and Other Haemosporidia.* Blackwell Scientific Publication, Oxford.

Hafner MS, Sudman PD, Villablanca FX, Spradling TA, Demastes JW, Nadler SA. 1994. Disparate rates of molecular evolution in cospeciating hosts and parasites. Science 265: 1087–1090.

Hayakawa T, Arisue N, Udono T, Hirai H, Sattabongkot J, Toyama T, Tsuboi T, Horii T, Tanabe K. 2009. Identification of *Plasmodium malariae*, a human malaria parasite, in imported chimpanzees. PLoS One 4: e7412.

Hughes AL, Verra F. 2010. Malaria parasite sequences from chimpanzee support the co-speciation hypothesis for the origin of virulent human malaria (*Plasmodium falciparum*). Mol Phylogenet Evol 57: 135–143.

Jeffares DC, Pain A, Berry A, Cox AV, Stalker J, Ingle CE, Thomas A, Quail MA, Siebenthall K, Uhlemann AC, Kyes S, Krishna S, Newbold C, Dermitzakis ET, Berriman M. 2007. Genome variation and evolution of the malaria parasite *Plasmodium falciparum.* Nat Genet 39: 120–125.

Joy DA, Feng X, Mu J, Furuya T, Chotivanich K, Krettli AU, Ho M, Wang A, White NJ, Suh E, Beerli P, Su XZ. 2003. Early origin and recent expansion of *Plasmodium falciparum.* Science 300: 318–321.

Kaiser M, Löwa A, Ulrich M, Ellerbrok H, Goffe AS, Blasse A, Zommers Z,

Couacy-Hymann E, Babweteera F, Zuberbühler K, Metzger S, Geidel S, Boesch C, Gillespie TR, Leendertz FH. 2010. Wild chimpanzees infected with 5 *Plasmodium* species. Emerg Infect Dis 16: 1956–1959.

Krief S, Escalante AA, Pacheco MA, Mugisha L, André C, Halbwax M, Fischer A, Krief JM, Kasenene JM, Crandfield M, Cornejo OE, Chavatte JM, Lin C, Letourneur F, Grüner AC, McCutchan TF, Rénia L, Snounou G. 2010. On the diversity of malaria parasites in African apes and the origin of *Plasmodium falciparum* from Bonobos. PLoS Pathog 6: e1000765.

Levine ND. 1988. *The Protozoan Phylum Apicomplexa.* CRC Press, Boca Raton, FL.

Liu W, Li Y, Learn GH, Rudicell RS, Robertson JD, Keele BF, Ndjango JB, Sanz CM, Morgan DB, Locatelli S, Gonder MK, Kranzusch PJ, Walsh PD, Delaporte E, Mpoudi-Ngole E, Georgiev AV, Muller MN, Shaw GM, Peeters M, Sharp PM, Rayner JC, Hahn BH. 2010. Origin of the human malaria parasite *Plasmodium falciparum* in gorillas. Nature 467: 420–425.

Livingstone FB. 1984. The Duffy blood groups, vivax malaria, and malaria selection in human populations: A review. Hum Biol 56: 413–425.

Martin MJ, Rayner JC, Gagneux P, Barnwell JW, Varki A. 2005. Evolution of human-chimpanzee differences in malaria susceptibility: Relationship to human genetic loss of N-glycolylneuraminic acid. Proc Natl Acad Sci U S A 102: 12819–12824.

Martinsen ES, Paperna I, Schall JJ. 2006. Morphological versus molecular identification of avian Haemosporidia: An exploration of three species concepts. Parasitology 133: 279–288.

Martinsen ES, Perkins SL, Schall JJ. 2008. A three-genome phylogeny of malaria parasites (*Plasmodium* and closely related genera): Evolution of life-history traits and host switches. Mol Phylogenet Evol 47: 261–273.

McCutchan TF, Dame JB, Miller LH, Barnwell J. 1984. Evolutionary relatedness of *Plasmodium* species as determined by the structure of DNA. Science 225: 808–811.

McCutchan TF, Kissinger JC, Touray MG, Rogers MJ, Li J, Sullivan M, Braga EM, Krettli AU, Miller LH. 1996. Comparison of circumsporozoite proteins from avian and mammalian malarias: Biological and phylogenetic implications. Proc Natl Acad Sci U S A 93: 11889–11894.

Ménard D, Barnadas C, Bouchier C, Henry-Halldin C, Gray LR, Ratsimbasoa A, Thonier V, Carod JF, Domarle O, Colin Y, Bertrand O, Picot J, King CL, Grimberg BT, Mercereau-Puijalon O, Zimmerman PA. 2010. *Plasmodium vivax* clinical malaria is commonly observed in Duffy-negative Malagasy people. Proc Natl Acad Sci U S A 107: 5967–5971.

Mitsui H, Arisue N, Sakihama N, Inagaki Y, Horii T, Hasegawa M, Tanabe K, Hashimoto T. 2010. Phylogeny of Asian primate malaria parasites inferred from apicoplast genome-encoded genes with special emphasis on the positions of *Plasmodium vivax* and *P. fragile*. Gene 450: 32–38.

Mu J, Joy DA, Duan J, Huang Y, Carlton J, Walker J, Barnwell J, Beerli P, Charleston MA, Pybus OG, Su XZ. 2005. Host switch leads to emergence of *Plasmodium vivax* malaria in humans. Mol Biol Evol 22: 1686–1693.

Ollomo B, Durand P, Prugnolle F, Douzery E, Arnathau C, Nkoghe D, Leroy E, Renaud F. 2009. A new malaria agent in African hominids. PLoS Pathog 5: e1000446.

Pacheco MA, Poe AC, Collins WE, Lal AA, Tanabe K, Kariuki SK, Udhayakumar V, Escalante AA. 2007. A comparative study of the genetic diversity of the 42 kDa fragment of the merozoite surface protein 1 in *Plasmodium falciparum* and *P. vivax*. Infect Genet Evol 17: 180–187.

Pacheco MA, Ryan EM, Poe AC, Basco L, Udhayakumar V, Collins WE, Escalante AA. 2010. Evidence for negative selection on the gene encoding rhoptry-associated protein 1 (RAP-1) in *Plasmodium* spp. Infect Genet Evol 10: 655–661.

Pacheco MA, Battistuzzi FU, Junge RE, Cornejo OE, Williams CV, Landau I, Rabetafika L, Snounou G, Jones-Engel L, Escalante AA. 2011. Timing the origin of human malarias: The lemur puzzle. BMC Evol Biol 11: 299.

Perkins SL, Schall JJ. 2002. A molecular phylogeny of malarial parasites recovered from cytochrome b gene sequences. J Parasitol 88: 972–978.

Prugnolle F, Durand P, Neel C, Ollomo B, Ayala FJ, Arnathau C, Etienne L, Mpoudi-Ngole E, Nkoghe D, Leroy E, Delaporte E, Peeters M, Renaud F. 2010. African great apes are natural hosts of multiple related malaria species, including *Plasmodium falciparum*. Proc Natl Acad Sci U S A 107: 1458–1463.

Prugnolle F, Ollomo B, Durand P, Yalcindag E, Arnathau C, Elguero E, Berry A, Pourrut X, Gonzalez J-P, Nkoghe D, Verrier D, Leroy E, Ayala FJ, Renaud F. 2011. African monkeys are infected by *Plasmodium falciparum* primate-specific strains. Proc Natl Acad Sci U S A 108: 11948–11953.

Qari SH, Shi YP, Pieniazek NJ, Collins WE, Lal AA. 1996. Phylogenetic relationship among the malaria parasites based on small subunit rRNA gene sequences: Monophyletic nature of the human malaria parasite, *Plasmodium falciparum*. Mol Phylogenet Evol 6: 157–165.

Rich SM, Ayala FJ. 2003. Progress in malaria research: The case for phylogenetics. In: Littlewood DTJ, ed. *Advances in Parasitology. The Evolution of Parasitism a Phylogenetic Perspective*, Vol. 54. Elsevier/Academic Press, Amsterdam, pp. 255–280.

Rich SM, Licht MC, Hudson RR, Ayala FJ. 1998. Malaria's eve: Evidence of a recent population bottleneck throughout the world populations of *Plasmodium falciparum*. Proc Natl Acad Sci U S A 95: 4425–4430.

Rich SM, Leendertz FH, Xu G, LeBreton M, Djoko CF, Aminake MN, Takang EE, Diffo JL, Pike BL, Rosenthal BM, Formenty P, Boesch C, Ayala FJ, Wolfe ND. 2009. The origin of malignant malaria. Proc Natl Acad Sci U S A 106: 14902–14907.

Ricklefs RE, Fallon SM. 2002. Diversification and host switching in avian malaria parasites. Proc R Soc London Ser B 269: 885–892.

Ricklefs RE, Fallon SM, Bermingham E. 2004. Evolutionary relationships, cospeciation, and host switching in avian malaria parasites. Syst Biol 53: 111–119.

Ronquist F, Huelsenbeck JP. 2003. MrBayes 3: Bayesian phylogenetic inference under mixed models. Bioinformatics 19: 1572–1574.

Ryan JR, Stoute JA, Amon J, Dunton RF, Mtalib R, Koros J, Owour B, Luckhart S, Wirtz RA, Barnwell JW, Rosenberg R. 2006. Evidence for transmission of *Plasmodium vivax* among a duffy antigen negative population in Western Kenya. Am J Trop Med Hyg 75: 575–581.

Sawai H, Otani H, Arisue N, Palacpac N, De Oliveira Martins L, Pathirana S, Handunnetti S, Kawai S, Kishino H, Horii T, Tanabe K. 2010. Lineage-specific positive selection at the merozoite surface protein 1 (msp1) locus of *Plasmodium vivax* and related simian malaria parasites. BMC Evol Biol 10: 52.

Silva JC, Egan A, Friedman R, Munro JB, Carlton JM, Hughes AL. 2010. Genome sequences reveal divergence times of malaria parasite lineages. Parasitology 138: 1–13.

Singh B, Kim Sung L, Matusop A, Radhakrishnan A, Shamsul SS, Cox-Singh J, Thomas A, Conway DJ. 2004. A large focus of naturally acquired *Plasmodium knowlesi* infections in human beings. Lancet 363: 1017–1024.

Sutherland CJ, Tanomsing N, Nolder D, Oguike M, Jennison C, Pukrittayakamee S, Dolecek C, Hien TT, Do Rosário VE, Arez AP, Pinto J, Michon P, Escalante AA, Nosten F, Burke M, Lee R, Blaze M, Otto TD, Barnwell JW, Pain A, Williams J, White NJ, Day NP, Snounou G, Lockhart PJ, Chiodini PL, Imwong M, Polley SD. 2010. Two nonrecombining sympatric forms of the human malaria parasite *Plasmodium ovale* occur globally. J Infect Dis 201: 1544–1550.

Tamura K, Dudley J, Nei M, Kumar S. 2007. MEGA4: Molecular Evolutionary Genetics Analysis (MEGA) software version 4.0. Mol Biol Evol 24: 1596–1599.

Tan CH, Vythilingam I, Matusop A, Chan ST, Singh B. 2008. Bionomics of Anopheles latens in Kapit, Sarawak, Malaysian Borneo in relation to the transmission of zoonotic simian malaria parasite *Plasmodium knowlesi*. Malar J 7: 52.

Tanabe K, Escalante A, Sakihama N, Honda M, Arisue N, Horii T, Culleton R, Hayakawa T, Hashimoto T, Longacre S, Pathirana S,

Handunnetti S, Kishino H. 2007. Recent independent evolution of msp1 polymorphism in *Plasmodium vivax* and related simian malaria parasites. Mol Biochem Parasitol 156: 74–79.

Tazi L, Ayala FJ. 2011. Unresolved direction of host transfer of *Plasmodium vivax* v. *P. simium* and *P. malariae* v. *P. brasilianum*. Infect Genet Evol 11: 209–221.

Valkiūnas G. 2005. *Avian Malaria Parasites and Other Haemosporina*. CRC Press, Boca Raton, FL.

Varki A, Gagneux P. 2009. Human-specific evolution of sialic acid targets: Explaining the malignant malaria mystery? Proc Natl Acad Sci U S A 106: 14739–14740.

Volkman SK, et al. 2001. Recent origin of *Plasmodium falciparum* from a single progenitor. Science 293: 482–484.

Waters AP, Higgins DG, McCutchan TF. 1991. *Plasmodium falciparum* appears to have arisen as a result of lateral transfer between avian and human hosts. Proc Natl Acad Sci U S A 88: 3140–3144.

Weedall GD, Polley SD, Conway DJ. 2008. Gene-specific signatures of elevated non-synonymous substitution rates correlate poorly across the *Plasmodium* genus. PLoS One 3: e2281.

Wolfe ND, Dunavan CP, Diamond J. 2007. Origins of major human infectious diseases. Nature 447: 279–283.

CHAPTER 6

FROM POPULATION GENOMICS TO ELUCIDATED TRAITS IN *PLASMODIUM FALCIPARUM*

Population Genomics, Genetic Diversity, and Association in Malaria

SARAH K. VOLKMAN, DANIEL E. NEAFSEY, STEPHEN F. SCHAFFNER, PARDIS C. SABETI, and DYANN F. WIRTH

The 2010 World Health Organization (WHO) World Malaria Report estimates that approximately 800,000 people are killed from the disease each year, with the highest mortality occurring among children in sub-Saharan Africa (Snow et al., 2005). Genetic evidence suggests that malaria is not only a contemporary scourge but in fact has been among the strongest selection pressures acting to change the human genome over at least the last 10,000 years (Kwiatkowski, 2005). Variations in hemoglobin (*HBB*), Duffy (*FY*), *CD36*, and numerous other genes have emerged and spread through human populations to confer protection from this deadly disease. Malaria parasites have responded in kind to these developments, evolving new mechanisms to overcome host immunity and to acquire resistance to a multiplicity of different drug compounds. Achieving a better understanding of the human and malaria genomes and how each continues to evolve in response to disease, immune response, and therapy may enable us to finally tip the host–pathogen evolutionary arms race in our favor. Recent research efforts have generated a map of genetic diversity across the *Plasmodium falciparum* genome, sampling hundreds of parasites from diverse geographic settings. These genomic data provide great opportunity for the discovery of signatures of selection in the parasite and usher in a new era of pathogen research with the promise of a faster therapeutic development pipeline (Jeffares et al., 2007; Mu et al., 2007; Volkman et al., 2007). In this chapter, we will examine the research opportunities afforded by genomic analysis, discuss population structure and implications for genomic analysis, review strategies to identify signatures of selection using comparisons of diversity and divergence, and emphasize association studies for the identification of loci connected to important clinical phenotypes. We will

Evolution of Virulence in Eukaryotic Microbes, First Edition. Edited by L. David Sibley, Barbara J. Howlett, and Joseph Heitman.

demonstrate how population genetic approaches have elucidated traits in *P. falciparum* and promise to describe clinically important characteristics that may be leveraged toward the elimination or eradication of malaria.

Discovery and characterization of *P. falciparum* genetic diversity has accelerated in recent years with the advancement of technologies and strategies that take advantage of the rapid evolution of the parasite genome in response to control and eradication measures. Since the first malaria genome was sequenced in 2002 (Gardner et al., 2002), over 60,000 unique single-nucleotide polymorphisms (SNPs) have been identified by concerted sequencing efforts (Jeffares et al., 2007; Mu et al., 2007; Volkman et al., 2007), and several genomic tiling arrays (Carret et al., 2005; Dharia et al., 2009; Jiang et al., 2008; Kidgell et al., 2006; Tan et al., 2009) and low density SNP arrays (Mu et al., 2010; Neafsey et al., 2008) have been developed to query this genetic variation. Recently, the first malaria genome-wide association study (GWAS) was published (Mu et al., 2010), in which 189 drug-phenotyped parasites from Asia, Africa, and the Americas were genotyped using a low density array (3257 SNPs); that study identified loci under positive selection and found several novel drug resistance candidates. Subsequent studies using GWAS have not only identified additional candidate drug resistance genes but have also performed functional validation of at least one of these loci to demonstrate the utility of using GWAS approaches to identify novel genetic loci that have a role in modulating responses to antimalarial compounds (Van Tyne et al., 2011).

As the cost of genome sequencing continues to fall, association studies between genotype and phenotype promise to become a powerful and widely deployed approach to better understand mechanisms of disease as well as therapy, and to inform elimination and eradication strategy efforts that are now under way. We are now moving from sampling parasite genomes to sequencing full genomes as a consequence of technological advancements that make it possible to inexpensively sequence hundreds of parasite samples, including those directly obtained from patients. These genome sequence data allow both the analysis of genetic variation across the entire genome from geographically diverse parasites, and the use of population genomic approaches such as GWAS to identify novel loci associated with relevant clinical phenotypes such as drug resistance. For the successful application of genomic sequence information to identify relevant genetic variants that compromise intervention strategies, we must know about population structure and variation both within (diversity) and between (divergence) populations of parasites. And, to ultimately leverage genomic data to identify genetic variants associated with key clinical traits, we will need accurate classification of appropriate parasite phenotypes.

POPULATION STRUCTURE

Central to understanding genetic variation in a species is determining population structure, that is, how allele frequencies vary between different populations within the species. Population structure can give insights into the history of the species; populations diverge genetically because of neutral genetic drift, and the profile of divergence can reveal information about the past demography of parasite populations, as well as the amount of contemporary gene flow among them. Because different populations typically encounter different environments and differing selection pressures, specific differences between populations can reveal how natural selection has been operating on the species. Population structure is of particular interest in *P. falciparum* because it is widely distributed geographically and is an obligate parasite whose insect host species varies geographically.

Understanding population structure is also of considerable practical importance when trying to detect associations between genotype and phenotype: If the phenotype of interest differs between two populations, any alleles that occur with different frequencies will be spuriously associated with the phenotype even if the frequency differences are purely random. Knowledge of population structure, then, can be used to inform sample selection for association studies or, if necessary, to correct for it.

A number of techniques and data sets have been used to characterize population structure in *P. falciparum*, with generally concordant results. Figure 6.1a shows the results of a principal component analysis (PCA) of a worldwide sample of strains. The largest-scale structure reflects the historical spread of the species from its origin in Africa, with one major branch populating Asia and another the Americas (Mu et al., 2005; Neafsey et al., 2008). There is also considerable structure apparent within continents, and, in some cases, within a single country (e.g., Brazil). Measuring divergence using the F_{ST} statistic, one study (Neafsey et al., 2008) showed the largest difference to be between Asian and American populations ($F_{ST} = 0.431$), with smaller differences between Africa and the Americas (0.306), and between Africa and Asia (0.236); within Africa, samples from Senegal and from Malawi yielded an F_{ST} of 0.181. (Refer to Chapter 1 for a discussion of F_{ST} and how it is used for measuring divergence between populations.)

LINKAGE DISEQUILIBRIUM (LD)

Another important characteristic of genetic variation in a population is the degree to which alleles at neighboring variant sites are correlated. This correlation (LD) arises because an allele always starts out as new mutation on a particular chromosome and therefore is completely correlated with alleles in neighboring genes around it. It is broken down over time by recombination, which can move the allele to different chromosomal backgrounds. The extent of LD is again a practical issue in designing an association study: If LD is high and extends over many markers as it does in humans, only a limited subset of markers need be interrogated to find evidence for association since correlated alleles can serve as partial proxies for one another. This approach has been the basis for the hundreds of disease-associated loci found in humans, using relatively low density genotyping arrays.

In the case of *P. falciparum*, there is in fact very little LD, and what there is extends only a short distance. On average, LD in African populations is almost undetectable at distances greater than 1 kb (Mu et al., 2005). In non-African populations, malaria transmission, and with it the effective recombination rate (detectable recombination only occurs when multiple distinct strains infect the same mosquito host simultaneously), is typically lower. As a result, LD is somewhat greater outside Africa. Population bottlenecks in the founding of non-African *P. falciparum* populations may also contribute to greater LD. Even in Brazil, however, which has the largest extent of LD observed to date, useful LD ($r^2 > 0.5$) extends only ~10 kb.

The near absence of LD implies that any GWAS will have to interrogate a large fraction of variant sites directly, either through dense genotyping or through full sequencing, to have any power—although it does also mean that it will be easy to narrow down the causal locus when an association is detected. An exception occurs in cases of recent positive selection, where the selected allele sweeps the chromosomal region around it to higher frequency along with it, increasing LD locally and making detection possible with lower density genotyping

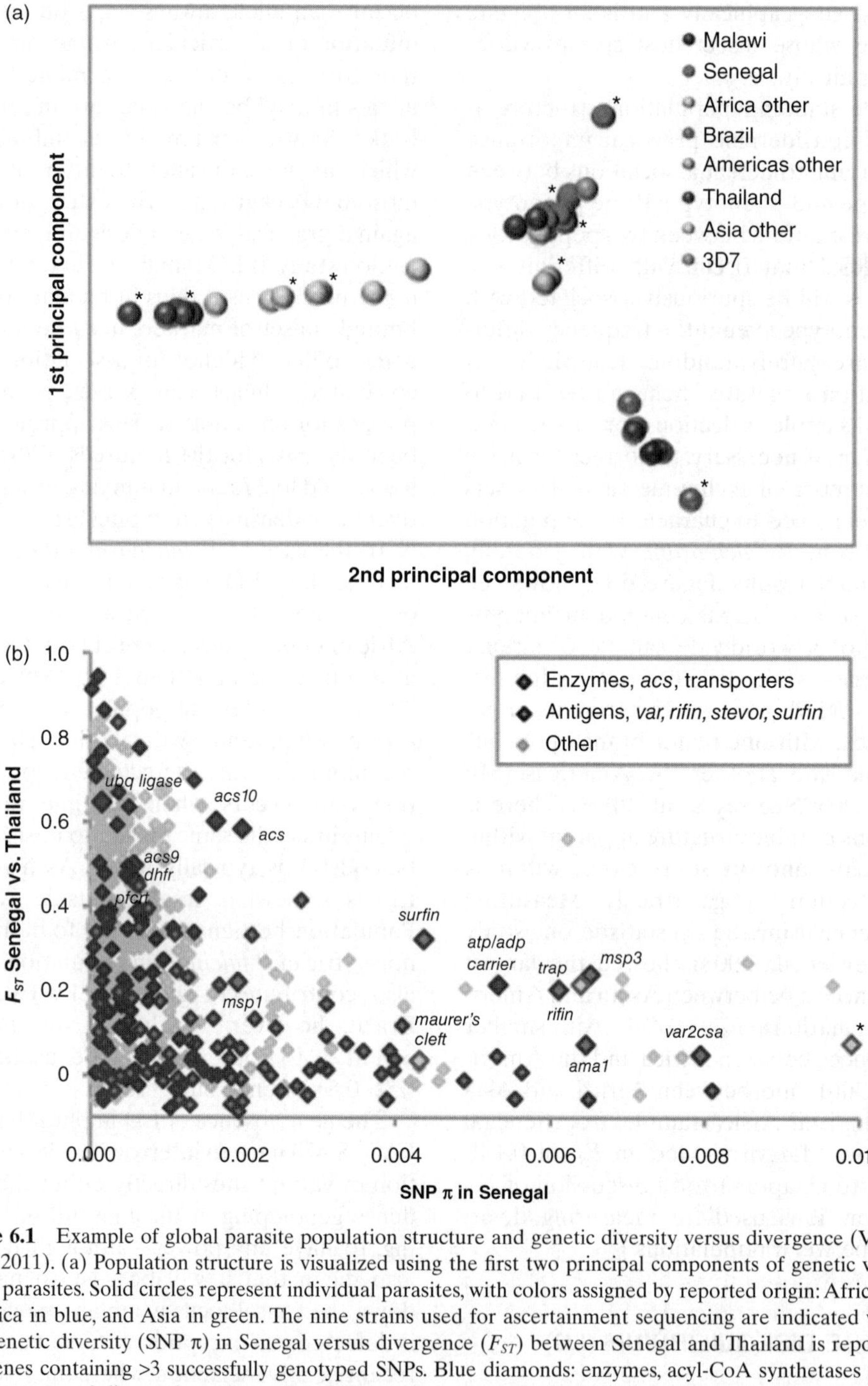

Figure 6.1 Example of global parasite population structure and genetic diversity versus divergence (Van Tyne et al., 2011). (a) Population structure is visualized using the first two principal components of genetic variation for 57 parasites. Solid circles represent individual parasites, with colors assigned by reported origin: Africa in red, America in blue, and Asia in green. The nine strains used for ascertainment sequencing are indicated with (*). (b) Genetic diversity (SNP π) in Senegal versus divergence (F_{ST}) between Senegal and Thailand is reported for 688 genes containing >3 successfully genotyped SNPs. Blue diamonds: enzymes, acyl-CoA synthetases (*acs*), or transporters; red diamonds: antigens, *vars*, *rifins*, *stevors*, or *surfins*; gray diamonds: all other genes. Gene IDs (PlasmoDB.org) for highlighted genes are listed in table S7 from Van Tyne et al. (2011). A gene with an unknown function is flagged with an asterisk (*) to indicate that SNP π is off-scale (0.014). See color insert.

arrays. It is not surprising, therefore, that the first attempts at whole-genome association studies in *P. falciparum*, and the first positive results, have been for loci associated with drug resistance (Mu et al., 2010; Van Tyne et al., 2011).

DIVERSITY AND DIVERGENCE

Comparisons of genetic diversity in sets of genes belonging to different functional categories reveal stark differences between evolutionarily dynamic and stable categories. For example, categories containing surface molecules involved in cytoadherence and antigenic variation (such as those encoded by the *var*, *rifin*, and *stevor* genes) had the highest degree of diversity, while categories containing genes related to mitochondrial function and electron transport chain showed low diversity (Jeffares et al., 2007; Mu et al., 2007; Volkman et al., 2007). Given that so many of the rapidly evolving and highly diverse genes are known antigens, many of the yet unclassified genes identified with high diversity may also be antigens and could therefore make promising vaccine candidates. Perhaps even more promising are the small number of molecules that are likely antigenic, which have little or no genetic diversity, as these may make stable targets for future vaccines. One strategy that has been explored to identify which of these rapidly evolving and highly diverse genes are antigenic was taken by Mu et al. (2007), who expressed the 65 most highly diverse genes using an *Escherichia coli* cell-free system and asked whether immune sera from humans exposed to malaria would contain antibodies that recognized these gene products. They found 11 of the 65 gene products were recognized by pooled sera and identified 7 previously unknown antigens. These results imply that diverse regions of the parasite may encode proteins that are antigenic in humans and thus may be useful to consider for vaccine development.

In addition to using genomic variation to understand global population structure, a comparison of the profile of divergence among populations with genomic diversity *within* populations can inform our understanding of parasite population biology and selective pressures operating on various classes of genes. The population genetic statistic "π" is most frequently used to quantify genetic variability within populations. The π statistic is defined as the mean rate of pairwise differences among all isolates within a population sample. It may be defined on an individual gene basis or calculated across all of the coding and noncoding sequences of an entire genome. Three major studies published concurrently in the January 2007 issue of *Nature Genetics* (Jeffares et al., 2007; Mu et al., 2007; Volkman et al., 2007), involving groups from the Wellcome Trust Sanger Institute, the Broad Institute and Harvard School of Public Health, and the National Institutes of Health, constituted the first step toward an effort to develop a genome-wide diversity map of *P. falciparum*. The groups collectively identified ~65,000 SNPs (single-nucleotide sites at which individuals in a population may differ) by sequencing 20 worldwide *P. falciparum* lines in comparison to the well-characterized 3D7 reference clone (Gardner et al., 2002). The 20 lines were each sequenced to varying extents, and overall provide basic information about parasite population structure and diversity.

Genome-wide diversity mapping in *P. falciparum* in two studies using whole-genome shotgun sequencing approaches gave similar results, both estimating a nucleotide diversity (π) of 1×10^{-3} (Jeffares et al., 2007; Volkman et al., 2007) for the global *P. falciparum* population. This means that two randomly chosen parasite samples would differ, on average, at 1 in 1000 nucleotide positions. The study of directed polymerase chain reaction (PCR)-based sequencing within protein-coding genes found lower diversity of 5×10^{-4}, likely indicative of greater functional constraints within coding regions (Mu et al., 2007). All groups found considerable variability in genetic diversity among chromosomes and within regions of individual chromosomes. For example, regions of the genome containing *var* genes have extensive diversity, with notable increases in diversity not just in the *var* genes but in surrounding regions as well. One group additionally sequenced *Plasmodium reichenowi*, a chimpanzee parasite that is the most closely related known species to *P. falciparum*. Their sequencing effort, covering roughly 42% of the *P. reichenowi* genome, identified 216,619 fixed differences between the *P. falciparum* reference strain 3D7 and *P. reichenowi*, suggesting roughly 10-fold greater interspecies divergence than within-species polymorphism (Jeffares et al., 2007).

By analyzing diversity (π) on an individual gene level and comparing diversity within a geographically defined population to divergence among populations, the role of selection of sculpting genomic variation in the *P. falciparum* genome becomes evident (Neafsey et al., 2008). Under neutral conditions, diversity within populations is expected to scale positively with divergence between populations under the assumption that a certain fraction of polymorphisms within populations will ultimately go to fixation locally. However, as Figure 6.1b illustrates, there are many individual genes that do not conform to this expectation. Most notably, genes undergoing strong pressure to evade the human adaptive immune response exhibit a signal of balancing selection, characterized by high diversity within the Senegal population and low divergence between the Senegal and Thailand populations. Frequency-dependent balancing selection prevents or retards the rate at which polymorphisms rise to high frequencies and fix within populations, resulting in the accrual of large amounts of diversity within populations and a paucity of divergent positions between population samples at selected loci. A number of studies have shown that regions of the *P. falciparum* genome that are highly polymorphic and appear to be under balancing selection encode antigens that are recognized by the human immune system (Mu et al., 2007). A number of well-studied antigens are found in the region of the Figure 6.1b plot exhibiting high diversity but divergence, but many unstudied loci also reside in that region, suggesting that this analysis approach could be useful in identifying previously unknown immunogenic loci that could be of use in identifying vaccine candidates or in better understanding the mechanisms of host immune evasion by the parasite.

Directional natural selection would be expected to result in the opposite pattern: low diversity within a population due to the effects of selective sweeps and high divergence between populations due to the accrual of selection-driven fixations. Large samples of fully sequenced genomes would be required to confidently identify genes exhibiting a signal of directional selection by this approach, but at the genome-wide level, there is a curious discordance of divergence profile between nonsynonymous and silent SNPs that is perhaps indicative of directional selection (Neafsey et al., 2008). Higher F_{ST} values are observed for nonsynonymous SNPs than for silent SNPs in all population comparisons except Africa versus the Americas (which exhibited highly similar values for both classes). This suggests a role for natural selection in the ongoing genetic differentiation of malaria populations in disparate geographic locales, possibly due to differences in local human host or vector biology, or variability in the nature and intensity of malaria control measures.

GWASs

An analysis of the *P. falciparum* genome reveals that approximately half of the genome is composed of genes that have no known functional homolog outside of the genus. Genomic diversity data can, in some cases, help to elucidate the general function of these mystery genes, for example, exported or surface-localized proteins associated with immune escape. Hidden among the thousands of mystery genes are likely dozens of potential drug and vaccine targets, as well as many genetic loci that could provide key insights into malaria biology and pathogenesis. The key advance in the coming years will be developing and applying strategies to find those rare therapeutically valuable genes in the malaria genome haystack to pinpoint key fronts in the evolutionary arms race.

This process is done agnostic to hypothesis but uses the principle that variants associated with that phenotype will be significantly enriched in the parasite group that harbors the trait of interest. This process involves the principles of forward genetics while combining deep sequencing with population genetic analysis. Identification of candidate variants requires confirmation and validation to demonstrate their role in that phenotype. These variants can then provide information about disease mechanisms and may provide useful information for the development of better strategies to detect, treat, or prevent malaria infection.

Genomic sequencing provides rich information about the genetic variation across the parasite genome. The decrease in sequencing costs coupled with the small genome size has contributed to the marked increase in full genomic sequencing being done by a number of groups. Because clinical samples contain a disproportionately large amount (100-fold or greater) of human genetic material, strategies to either deplete the human white blood cells or to isolate the parasite genetic material from the mixture are critical to keeping these costs feasible for a large number of samples. *P. falciparum* harbors certain attributes that also increase the power of identifying genetic variants associated with a key clinical phenotype. First, haploidy makes it unnecessary to disentangle the proper phasing of contiguous polymorphic markers, as must occur when genotyping a diploid sample. Haploidy greatly reduces ambiguity about exactly which allele at a polymorphic locus is associated with a trait of interest. Second, genes that induce a phenotype of interest are easier to identify because there is no issue of phenotypic dominance in a haploid. Mutations that would ordinarily cause recessive phenotypes in a diploid, confounding an association study through disrupting a 1:1 correspondence between genotype and phenotype, are plainly expressed in a haploid and are therefore much easier to detect. Finally, because of the increased density of markers as well as the short LD across in the *P. falciparum* genome in general, we are likely to identify genetic variants that directly correspond to the phenotype of interest. As it happens, 1 kb is the average size of a gene in *P. falciparum*. This means that genotype–phenotype association studies can narrow an association not only to a region of chromosome but also to an actual gene. This is particularly so in Africa where LD is short. In other words, the population structure of the malaria parasite, in general, exemplifies the population geneticist's ideal situation: an organism in which selection occurs almost exclusively in the haploid phase of the life cycle but in which there is a brief diploid phase with so much recombination that there is little LD.

Drug resistance is a phenotype well suited for GWAS because it is expected to be caused by common alleles of large effect at few genomic loci (Hayton and Su, 2008). If this is the case, associations will be much easier to detect than in a typical human GWAS, in which the phenotype is caused by alleles at many loci that are rare and/or of small effect. The increased LD caused by recent selection for drug resistance counteracts the loss of power that comes from short LD, small sample size, and the temporal and geographic stratification of the parasite population that we examined. Thus, despite the potential limitations, we were able to detect a known drug resistance locus (*pfcrt*), observed little *P*-value inflation in our GWAS data, and identified a number of genome-wide significant loci associated with drug resistance (Van Tyne et al., 2011). Part of this success was likely due to specific tests we used to account for population structure, as described below.

APPLICATION OF GWAS TO DRUG RESISTANCE

To demonstrate how a GWAS approach can be used for discovery, we will describe some of our recent work (Van Tyne et al., 2011) using this approach to identify a novel drug resistance locus in *P. falciparum*. This same approach can be applied to many other clinically relevant phenotypes including host response, cytoadherence, invasion, and gametocyte formation. The first step is to carefully define the phenotype of interest for which one would like to identify causal genetic variants. In our study, we measured drug response and determined the effective concentration of 50% (EC_{50}) to a number of known antimalarial drugs for a set of 50 parasites that had been culture adapted. These phenotype data were necessarily reproducible and covered a range of values across the population that provided a distribution for both sensitive and resistant parasites. Depending upon the range and distribution of EC_{50} values for a given compound across the parasite population set, one can also focus on parasites that represent the extremes of the distribution, that is, the most highly resistant and the most sensitive parasites from the population. As was discussed in the section on population structure, beginning with a homogeneous parasite population is ideal to avoid confounding the results with genetic variants that differ in the population but are unrelated to the phenotype under investigation.

There are a number of statistical tests that can be applied to identify genetic variants that are associated with a given phenotype. For our analysis (Van Tyne et al., 2011), we used the mixed-model association (EMMA) and haplotype likelihood ratio (HLR) tests. EMMA identifies quantitative trait associations in individuals with complex population structures and hidden relatedness, which was ideal for our global set of parasites evaluated in this initial study. EMMA is particularly applicable for small and structured sample sets and has been shown to outperform both PCA-based and λ_{GC}-based correction approaches in highly inbred and structured mouse, maize, and *Arabidopsis* populations (Kang et al., 2008). The HLR test is a multimarker test designed to detect the association of a single haplotype with a phenotype and is particularly powerful when the associated haplotype has experienced recent strong selection (and is therefore long) and occurs on a low LD background (Lindblad-Toh et al., 2005); it is therefore particularly appropriate for a study seeking loci under recent drug selection. When used together, these two complementary approaches provide a highly sensitive screen for association signals within the *P. falciparum* genome.

The well-characterized chloroquine resistance locus, *pfcrt*, served as a positive control for this study, and as expected, we found evidence for association with resistance to chloroquine using both tests, consistent with previous studies (Mu et al., 2010). Applying the same tests to the other

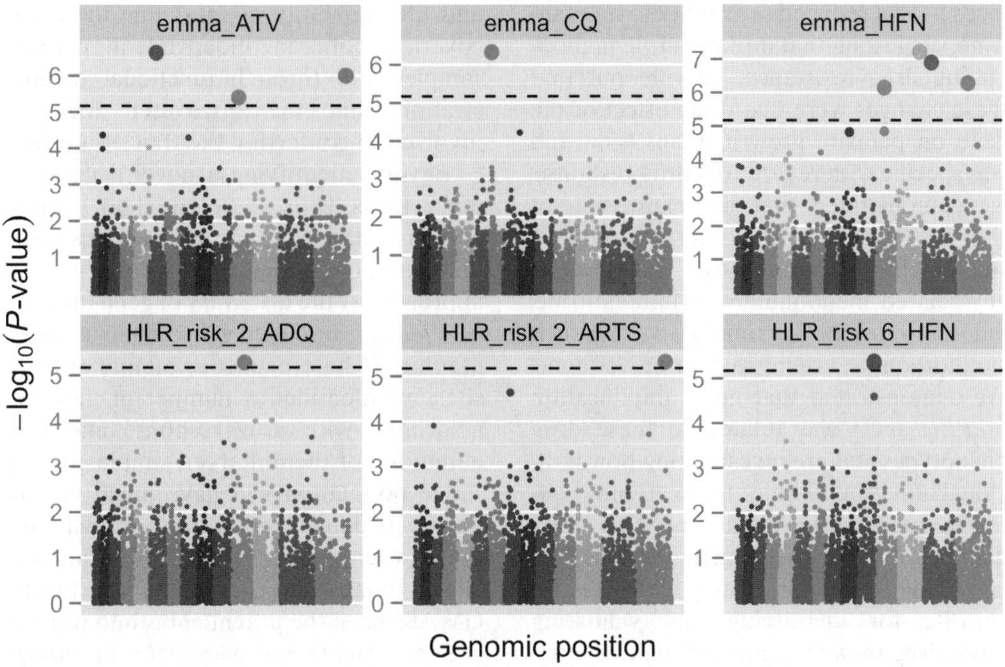

Figure 6.2 Genome-wide association study (GWAS) results (Van Tyne et al., 2011). Genome-wide significant associations were found for 5 antimalarials out of 13 tested using EMMA and HLR tests. They include *pfcrt* (chromosome 7) associated with chloroquine resistance and 11 novel associations with resistance to several drugs. See color insert.

drug phenotypes, we detected numerous novel loci showing significant associations with drug resistance (Fig. 6.2), with 11 novel loci achieving genome-wide significance for association with resistance to five different drugs: amodiaquine, artemisinin, atovaquone, chloroquine, and halofantrine. In most cases, the short extent of LD allowed localization to individual genes. Among the loci identified were various transporters and membrane proteins, as well as five conserved genes with unknown function (Van Tyne et al., 2011).

ELUCIDATED TRAIT FOR DRUG RESISTANCE DISCOVERED USING GWAS

Demonstrating that a signal of association actually reflects a causal molecular process requires functional testing and validation of the candidate locus, both because of concerns about power and reproducibility of genetic association tests and because even a robust statistical correlation need not imply biological causation. To confirm the ability of GWAS to identify functionally relevant candidates, we investigated one of our association findings, *PF10_0355*, in greater depth (Van Tyne et al., 2011). This gene contains multiple SNPs associated with halofantrine resistance and encodes a putative erythrocyte membrane protein (http://plasmodb.org/plasmo) characterized by high genetic diversity.

We set out to determine the role of *PF10_0355* in halofantrine resistance by introducing this genetic locus into halofantrine-sensitive parasites using transfection approaches. These studies showed that parasites that overexpress the *PF10_0355* locus have a reduced sensitivity to halofantrine, suggesting that it plays a role in modulating drug resistance. To rule out that these changes were due to an effect of the gene on parasite growth or invasion and was specific with regard to a drug response, we demonstrated that overexpression of the *PF10_0355* gene conferred reduced susceptibility to halofantrine and structurally related mefloquine and lumefantrine, but not to structurally unrelated drugs such as chloroquine, artemisinin, or atovaquone. We demonstrated that gene amplification of *PF10_0355* was related to these drug phenotypes and are investigating how individual mutations within this locus may contribute to this drug phenotype. This study demonstrates the feasibility of coupling GWAS and functional testing in the malaria parasite for identifying and validating novel drug resistance loci through the coupling of GWAS, and it illustrates the power of GWAS to find functionally important alleles.

Natural selection and genome-wide association methods are complementary approaches to probe the genetic basis of adaptation in *P. falciparum*: Scanning for selected loci permits an unbiased search for unknown adaptive changes but provides little information about the processes at work, while GWAS gives a focused look at one easily identified (and clinically critical) adaptive phenotype. Results from both approaches open up new avenues for study as we seek to understand the biological significance of the findings. In this chapter, we have focused on how understanding population structure and genetic diversity is important to conclusions one can draw about associations between genetic variants and phenotypes of interest.

Challenges for malaria to perform association studies will be obtaining sufficient numbers of parasite samples with carefully defined phenotypes that are reproducible; obtaining appropriate samples that can be sequenced to carry out association studies; and the capability of validating these discoveries both in independent parasite samples and through functional demonstration that the discovered loci are involved in conferring the trait of interest.

Beyond identifying a novel drug resistance locus, with its potential to provide new insights into drug response, this study illustrates the general utility of a GWAS approach for the discovery of gene function in *P. falciparum*. Even with a small and geographically heterogeneous sample of parasites, we identified a number of new loci associated with drug response and validated one of them. Larger samples from a single population will have much greater power to detect additional loci, including those where multiple and low frequency alleles contribute to resistance. Future GWASs have the potential both to provide greater insights into basic parasite biology and to identify biomarkers for drug resistance and other clinically relevant phenotypes like acquired protection, pathogenesis, and placental malaria.

Future GWAS will be able to counteract the loss of power caused by low LD, either by focusing on parasite populations with reduced outcrossing rates or by studying cases of very strong selective pressure, but this issue will soon become moot as the declining cost of whole-genome sequencing makes it practical to assay every nucleotide in the genome on a routine basis. Culture-adapted parasites, which are amenable to robust and reproducible phenotypic characterization, retain an important role, but their limitations—the potential for artifactual mutations during adaptation and for a biased selection of a subset of parasites within a given infection—mean that genetic changes identified using them require both functional validation and demonstration that the changes are important during natural infection. As direct sequencing of clinical isolates with demonstrable clinical phenotypes like *ex vivo* drug response or invasion properties becomes increasingly feasible, sequencing will enable us to directly identify genetic changes in the parasite associated with clinically relevant phenotypes. In the years ahead, genome analysis of *P. falciparum* has the potential to identify genetic loci associated with many phenotypes and to enhance our understanding of the biology of this important human pathogen as well as inform the development of diagnostic and surveillance tools for malaria eradication.

SUMMARY POINTS

Genomic diversity mapping has immediate application to the design of therapeutics. The parasite's capacity for evolutionary adaption, mediated in part by genomic diversity, is one of the major impediments to the design of therapeutics for malaria (Good et al., 2004). A comprehensive understanding of the properties of the parasite's genomic variation, however, will help to transform it from a hindrance into a useful research tool. Hot spots of variation and unusual patterns of LD among SNPs will point to promising new targets for drug and vaccine design. Novel drug and vaccine targets will also be generated by association studies that allow investigators to identify the genetic basis of many clinical and laboratory phenotypes. Awareness of the full breadth of malaria diversity will enable efficacy testing of therapeutics across as genetically diverse a set of populations as possible, and knowledge of gene flow patterns will inform rational application of existing and future therapeutics to minimize the risk of engendering resistance.

The utility of malaria diversity data can be multiplied by combining it with comparable human data generated by the HapMap project (2005). Few evolutionary relationships in nature are as close as that between an obligate parasite and its host. Analyses of human genomic diversity have already revealed the surprising extent to which malaria has impacted recent human evolution (Kwiatkowski, 2005). The human immune system has had a reciprocal impact on the malaria genome, inducing high genetic variation or rapid evolution at many loci through an "evolutionary arms race" (Hastings and Donnelly, 2005; Roy et al., 2006). By leveraging the malaria diversity map with the human HapMap and analyzing how specific human and malaria gene variants intermesh across populations, we will have a much clearer picture of the conflict and will be able to identify particularly vulnerable targets in the parasite. This kind of analysis has the potential to illuminate the basis of variation in virulence by identifying particular combinations of host and parasite genotypes that result in mild versus severe infection. For example, a patient with a certain MHC profile, or carrying the resistance-conferring sickle cell or G6PD variants, might be expected to progress more slowly if at all to a state of severe anemia or parasitemia, whereas a patient exhibiting a particularly susceptible MHC profile could be expected to experience severe parasitemia and could therefore receive an aggressively tailored treatment regimen.

Tracking the clinical outcome of disease progression and treatment response for various host and parasite genotypic combinations across large numbers of patients could ultimately allow researchers to control for genetic variation effectively in analyses aimed at investigating the importance of other factors on disease progression. For example, the importance of past infection history, comorbidity, age, gender, and many other factors could be more effectively ascertained after controlling for genetic variation in host and parasite. Once a suitable genotyping platform becomes widely available, the data to create such controls will begin accumulating rapidly.

In addition to casting new light on host–parasite interactions, a detailed genomic diversity map will facilitate investigations into parasite–parasite interactions, such as longitudinal studies within patients of the competitive dynamics of multiple infections, or differences among samples collected from regions with high versus low incidences of malaria infection. Variation in the rates of transmission, sexual outcrossing, gene flow, and local population diversity among different malaria-endemic regions could create clinically important genetic heterogeneity in the parasite population that is not detectable through traditional, limited surveys of parasite genetic variation. Further, high resolution genotyping across the *P. falciparum* genome will identify whether certain strains in the global population are particularly inclined to cause certain severe clinical outcomes, such as respiratory disease or cerebral or placental malaria, and yield markers to assist in the diagnosis of such infections.

The rapidly falling cost of DNA sequencing and genotyping technologies means that we will soon have genome diversity maps for a host of human pathogens and parasites. *P. falciparum* will serve as a prototype for population genomic analysis of eukaryotic pathogens, as many of the analytical approaches being developed to mine the *P. falciparum* data will find application in these other infectious organisms. This approach has the general potential to (i) provide inexpensive, detailed diagnostic tools to clinicians and researchers; (ii) accelerate therapeutic development by supplying new targets for vaccines and drugs and speeding basic biological research; and (iii) counter the evolution of resistance to therapeutics through a better understanding of pathogen population structure and evolution.

The development of technologies and methodologies for generating and analyzing genetic diversity data may ultimately have a large impact on human health in the infectious disease arena if the malaria model is extended to other pathogens. Pathogen genomic diversity maps can be only fully exploited with enhanced collaboration between clinicians and researchers to enable broad epidemiological sampling, with a worldwide community working together to adopt new paradigms for infectious disease research that recognize the value of pooled resources and large-scale initiatives for achieving ambitious common goals.

REFERENCES

Carret CK, et al. 2005. Microarray-based comparative genomic analyses of the human malaria parasite *Plasmodium falciparum* using Affymetrix arrays. Mol Biochem Parasitol 144: 177–186.

Dharia NV, et al. 2009. Use of high-density tiling microarrays to identify mutations globally and elucidate mechanisms of drug resistance in *Plasmodium falciparum*. Genome Biol 10: R21.

Gardner MJ, et al. 2002. Genome sequence of the human malaria parasite *Plasmodium falciparum*. Nature 419(6906): 498–511.

Good MF, et al. 2004. The immunological challenge to developing a vaccine to the blood stages of malaria parasites. Immunol Rev 201: 254–267.

Hastings IM, Donnelly MJ. 2005. The impact of antimalarial drug resistance mutations on parasite fitness, and its implications for the evolution of resistance. Drug Resist Updat 8(1–2): 43–50.

Hayton K, Su XZ. 2008. Drug resistance and genetic mapping in *Plasmodium falciparum*. Curr Genet 54(5): 223–239.

International HapMap Consortium. 2005. A haplotype map of the human genome. Nature 437(7063): 1299–1320.

Jeffares DC, et al. 2007. Genome variation and evolution of the malaria parasite *Plasmodium falciparum*. Nat Genet 39: 120–125.

Jiang H, et al. 2008. Detection of genome-wide polymorphisms in the AT-rich *Plasmodium falciparum* genome using a high-density microarray. BMC Genomics 9: 398.

Kang HM, et al. 2008. Efficient control of population structure in model organism association mapping. Genetics 178(3): 1709–1723.

Kidgell C, et al. 2006. A systematic map of genetic variation in *Plasmodium falciparum*. PLoS Pathog 2: e57.

Kwiatkowski DP. 2005. How malaria has affected the human genome and what human genetics can teach us about malaria. Am J Hum Genet 77(2): 171–192.

Lindblad-Toh K, et al. 2005. Genome sequence, comparative analysis and haplotype structure of the domestic dog. Nature 438(7069): 803–819.

Mu J, et al. 2005. Recombination hotspots and population structure in *Plasmodium falciparum*. PLoS Biol 3: e355.

Mu J, et al. 2007. Genome-wide variation and identification of vaccine targets in the *Plasmodium falciparum* genome. Nat Genet 39: 126–130.

Mu J, et al. 2010. *Plasmodium falciparum* genome-wide scans for positive selection, recombination hot spots and resistance to antimalarial drugs. Nat Genet 42(3): 268–271.

Neafsey DE, et al. 2008. Genome-wide SNP genotyping highlights the role of natural selection in *Plasmodium falciparum* population divergence. Genome Biol 9: R171.

Roy SW, Ferreira MU, Hartl DL. 2006. Evolution of allelic dimorphism in malarial surface antigens. Heredity 100(2): 103–110.

Snow RW, et al. 2005. The global distribution of clinical episodes of *Plasmodium falciparum* malaria. Nature 434(7030): 214–217.

Tan JC, et al. 2009. Optimizing comparative genomic hybridization probes for genotyping and SNP detection in *Plasmodium falciparum*. Genomics 93(6): 543–550.

Van Tyne D, et al. 2011. Identification and functional validation of the novel antimalarial resistance locus PF10_0355 in *Plasmodium falciparum*. PLoS Genet 7(4): e1001383.

Volkman SK, et al. 2007. A genome-wide map of diversity in *Plasmodium falciparum*. Nat Genet 39(1): 113–119.

CHAPTER 7

SELECTIVE SWEEPS IN HUMAN MALARIA PARASITES

XIN-ZHUAN SU and JOHN C. WOOTTON

INTRODUCTION

Malaria is still killing nearly a million people a year, mostly children in Africa (WHO, 2008). The parasites that cause human malaria consist of five *Plasmodium* species, comprising *Plasmodium falciparum*, *Plasmodium vivax*, *Plasmodium malariae*, *Plasmodium ovale*, and the nonhuman primate parasite *Plasmodium knowlesi* that has also been reported to infect humans (Singh et al., 2004). The most deadly parasite, *P. falciparum*, is also the species that has received the most attention including drug and vaccine development. It is also the species for which *in vitro* culture is available. Due to the application of antimalarial drugs, drug-resistant parasites are widespread, including resistance to sulfadoxine/pyrimethamine (SP), chloroquine (CQ), and potentially, artemisinin (ART) (Dondorp et al., 2009; Hayton and Su, 2008).

The human malaria parasites are transmitted by *anopheline* mosquitoes and have a complex life cycle that includes the obligatory sexual cycle in the mosquitoes and two types of asexual reproduction in the human host known as schizogony. A parasite sexual stage called a gametocyte develops in a minority of infected red blood cells (RBCs) and enters the mosquito midgut after a blood meal. In the midgut, the haploid male and female gametocytes quickly produce gametes that fuse to form a diploid zygote called an ookinete. By a morphogenetic process that includes meiosis, the ookinete develops via the oocyst stage to produce haploid sporozoites that then migrate to the salivary gland of the mosquito. When a mosquito bites again, the sporozoite enters the human host, where it rapidly targets the liver and replicates in hepatocytes. Merozoites are released into the bloodstream and infect RBCs. The parasites can replicate and grow in RBCs as an asexual cycle, causing disease symptoms. For yet unknown reasons, some parasites may switch to sexual gametocyte development, thus completing the life cycle.

In malaria-endemic regions with moderate to high entomological infection rates, the blood of an infected individual typically carries a mixture of different parasite genotypes. This creates a theater for in-host evolutionary dynamics in which selection pressures—such as drugs, host specificity, or immunity—will favor advantageous alleles

Evolution of Virulence in Eukaryotic Microbes, First Edition. Edited by L. David Sibley, Barbara J. Howlett, and Joseph Heitman.

that influence parasite growth, survival, and transmission. Consequently, in mosquitoes, mating between favored genotypes becomes more probable, and progeny parasites carrying the allele of advantage will be further selected in subsequent infections when the same selection pressures persist. In a relatively small number of generations, this process can approach fixation with the majority of parasites in a population carrying a single allele at the selected locus, with concomitant removal of less favored alleles. If the selective conditions are widespread, the allele can spread to various geographic regions. Such evolutionary processes define the characteristics of selective sweeps in malaria parasites, the topic of this chapter. Indeed, the rapid spread of drug-resistant parasites across continents, following the extensive application of CQ or SP, provides exceptionally clear-cut examples of selective sweeps acting on alleles in their chromosomal context as conceived by the genetic hitchhiking theory (discussed in more detail below). We will focus our discussions on data and discoveries from *P. falciparum*, on which the majority of the studies have been done, with occasional reference to other species such as *P. vivax*.

TYPES OF SELECTION

The population genetic concept of a selective sweep (i.e., the definition of the term that determines the scope of this chapter) requires that Darwinian selection is manifest at the level of a substantial population of organisms and considers extended effects on the genome including the chromosomal context of any individual gene or genes under selection. For the more narrowly defined levels of selection operating on a single gene, several types are recognized, notably positive selection that is the usual underlying drive of a selective sweep. Under positive selection, the allele frequency for a genetic locus will shift in one advantageous direction. For a sweep to occur, a particular directional selective pressure needs to be consistently sustained in space and time so that an advantageous allele increases relatively rapidly in frequency and may eventually reach fixation in a population; however, positive selection is not in itself sufficient for a sweep, as selective pressures can act in different directions over time or on various subpopulations, resulting in multiple versions of a gene being maintained in the gene pool at frequencies well above those expected from de novo mutation and random genetic drift. In haploid organisms such as malaria parasites, this type of multidirectional selection may result in high frequencies of two or more favored alleles at a given locus within a single population and, in this respect, resembles balancing selections in higher diploid organisms. A difference, however, is that heterozygote advantage, which is a familiar cause of balancing selection in diploids when the heterozygotes for the alleles under consideration have a higher adaptive value than the homozygotes, could not apply to the parasites except at the transient premeiotic zygote stage. A third type of selection is negative or purifying selection, a process of selective decrease in frequency or removal of deleterious alleles while neutral or nearly neutral alleles are concomitantly maintained by genetic drift as the population stabilizes on a particular trait value. Because deleterious mutations relentlessly arise de novo, such purifying selection, like genetic drift, is recognized as one of the principal continuous forces of long-term genome evolution and is called background selection in that context.

Malaria parasites will likely experience many strong selective pressures within the human and mosquito hosts. For example, the human immune response strongly selects against parasites carrying epitopes targeted by preexisting immune reactions, such as those that may be present on merozoite surface proteins or parasite antigens expressed on the surface of infected RBCs. Given mixed genotype infections, such in-host immune selection can favor the rapid evolution of multiple alleles encoding mosaics of different non-cross-reactive amino acids on these proteins. These rapid evolutionary dynamics can resemble the effects of balancing selection or may also lead to more sustained positive directional selection if some of the new alleles are beneficial in providing parasite strategies to escape host immune response. In contrast, during the evolution of resistance to antimalarial drugs, the interplay of positive directional selection and deleterious fitness effects appears to have occurred. Under continuous drug pressure, mutations may be strongly favored in drug targets or in molecules transporting the drug into, within, or out of the parasite; however, the mutations that confer a chloroquine-resistant (CQR) phenotype in the crucial *P. falciparum* CQ resistance transporter (*pfcrt*) gene (Fidock et al., 2000) are evidently deleterious in the absence of CQ. Removal of CQ pressure in endemic regions has resulted in the rapid return of chloroquine-sensitive (CQS) *pfcrt* alleles in the populations (Laufer et al., 2006; Wang et al., 2005), suggesting that parasites carrying the ancestral *pfcrt* alleles have substantially greater genetic fitness when there is no CQ pressure.

HITCHHIKING

The classical hallmark of a selective sweep is the reduction or elimination of genetic variability in the chromosomal haplotype surrounding a mutation that has experienced recent and strong positive selection. The term genetic hitchhiking (Smith and Haigh, 1974) refers to the process by which flanking genomic segments are dragged toward fixation along with the advantageous allele or are removed with the less fit alleles. To observe hitchhiking, extended haplotypes, physically linked to the gene of interest, are determined for a sample of individuals from a population. Typically, this requires a sufficiently dense chromosomal coverage of neutral or nearly neutral polymorphic molecular markers such as microsatellites (MSs) or single-nucleotide polymorphisms (SNPs) that are functionally unrelated to the gene under selection. Given sufficiently high meiotic recombination rates, crossovers may have separated some of the linked markers from the selected gene in some individuals, with the result that the strength of hitchhiking tends to decrease with distance from the selected locus. For a selective sweep in a population of haploid organisms such as malaria parasites, such linkage decay with physical distance can be thought of as depending on just three parameters: (i) the intrinsic frequency of crossovers per meiosis, (ii) the number of sexual generations since the original favorable founder allele originated in the population by mutation or introgression, and (iii) the proportion of inbreeding in the population.

It is important to note that hitchhiking is the subject of a long-standing controversy in higher organism population genetics (e.g., human, insects, plants), because the reduction of genetic variability in an extended chromosomal segment does not in itself prove that a locus in that segment has undergone a recent rapid selective sweep (reviewed by Stephan, 2010). In addition to the original hitchhiking mechanisms based on positive directional selection (Kaplan et al., 1989; Ota and Kimura, 1975), an alternative mechanism based on a more long-term background selection against linked deleterious alleles can explain the observed patterns of reduced variability equally well, as first demonstrated by Charlesworth et al. (1993). Consequently, positive directional selective sweeps are hard to verify in higher eukaryotes especially because the strength of selection typically inferred is relatively weak, and the majority of such cases remain unresolved or controversial; however, the selective sweeps that we review here involving drug resistance in *P. falciparum* cannot be explained by alternative background selection mechanisms. Very powerful selective pressures favoring drug-resistant parasites over their wild-type, drug-sensitive ancestors have been abundantly demonstrated by physiological and parasitological observations both *in vivo* and with cultured parasites. Moreover, the strength of such selection is quantitatively consistent with the high directional selection coefficients inferred using hitchhiking theory from the observed chromosomal segments of reduced variability, and the rapid fixation rate inferred is consistent with the known history of transcontinental spread of drug-resistant parasites. Thus, the malaria parasite systems provide unusually clear exemplars of genetic hitchhiking and selective sweeps, with additional advantages for genetic studies of having haploid inheritance, a high genome-wide density of polymorphic markers, and relatively uniform meiotic crossover frequencies.

COMMON GENETIC VARIATIONS IN MALARIA PARASITES

Genetic variation can be displayed in different forms such as SNP, deletion and insertion (including micro-and minisatellites), and copy number variation (CNV), as well as epigenetic modifications. The genomes of many malaria parasites including *P. falciparum*, *P. vivax*, and *Plasmodium yoelii* have been shown to be quite polymorphic with large numbers of SNPs, MSs, and CNVs (Dharia et al., 2010; Feng et al., 2003; Jeffares et al., 2007; Jiang et al., 2008; Kidgell et al., 2006; Li et al., 2009; Mu et al., 2007; Su and Wellems, 1996; Su et al., 1999; Volkman et al., 2007). Because the *P. falciparum* parasite has an AT-rich genome with an average AT content of ~82% (Gardner et al., 2002), AT-rich repetitive sequences or MSs are abundant in the parasite genome (Su and Wellems, 1996) and are highly polymorphic. In one study, an average of ~15 alleles per MS were obtained from 342 MSs over 87 *P. falciparum* isolates collected worldwide (Wootton et al., 2002). The presence of multiple alleles per MS is in contrast with SNPs that generally have two alleles, although more than two SNP alleles have been observed in genes under strong selection such as the *pfcrt* gene conferring resistance to CQ and the gene encoding surface antigen *P. falciparum* apical membrane antigen 1 (PfAMA1) (Cooper et al., 2005, 2007; Duan et al., 2008). The presence of multiple alleles enables relatively reliable estimates of heterozygosity for each MS, which can be used in turn for the detection of selection. Another advantage of MSs is that the majority of polymorphic MSs are present in the noncoding regions. If we assume that there will be less impact on gene function when a mutation occurs in a noncoding region (more neutral), then estimates using MSs may more closely reflect a neutral evolutionary process. Finally, many factors have been shown to influence the extent of MS stability, including repeat unit length such as (TA) versus (TAA), length of an MS or number of repeat units, base composition, and the degree of "perfection" of the repeated MS (perfect repeat of single repeat type or with more than one repeat units) (Boyer et al., 2008; Chung et al., 2010). Generally, longer MSs mutate more frequently than shorter ones, and pure repeats are less stable than mixed repeats. Therefore, each MS has its own mutation rate that is determined by the repeat types and the length of a repeat.

A second type polymorphism in malaria parasites is SNP. Although there have been debates on whether *P. falciparum* has a clonal or diverse population (Hughes and Verra, 2001; Rich et al., 1998; Su et al., 2003; Tibayrenc et al., 1990; Volkman et al., 2001), the genome has recently been shown to be quite polymorphic, with a frequency of ~1 SNP every 500–900 bp (Mu et al., 2002, 2007; Volkman et al., 2007). Despite the different views on the age and evolutionary history of *P. falciparum*, one interesting observation of nucleotide substitution is the high frequency of nonsynonymous substitution in the parasite genome. There were more than twice as many nonsynonymous single-nucleotide polymorphisms (nsSNPs) as synonymous single-nucleotide polymorphisms (sSNPs) in the coding regions of *P. falciparum*, in sharp contrast to similar frequencies of sSNPs and nsSNPs—or slightly more sSNPs—in the genomes of human and other organisms (Cargill et al., 1999; Mu et al., 2002). Selections from the hosts and antimalarial drugs may play a role in the high frequency of nonsynonymous substitution, which can be used to detect the "signature" of drug selection.

A third type of variation in the parasite genome is CNV. CNV and its role in parasite phenotypic variation have received more attention recently. Regions of deletion or amplification have been detected in the *P. falciparum* genome using microarray or quantitative polymerase chain reaction (PCR) (Jiang et al., 2008; Kidgell et al., 2006; Ribacke et al., 2007). Some CNVs have been associated with drug selection. One example is the association of a higher copy number of a gene encoding a homolog of the human P-glycoprotein or *P. falciparum* multiple drug resistance 1 (PfMDR1) and responses to antimalarial drugs mefloquine, ART, and others (Chavchich et al., 2010; Cowman et al., 1994; Price et al., 2004). Manipulation of the copy number alters the response to multiple drugs (Sidhu et al., 2006). Another example of gene amplification is the *P. falciparum* GTP-cyclohydrolase I (*gch1*) in the folate biosynthesis pathway. The gene was shown to be under selection of folate drugs, and the CNV of the gene was associated with parasite response to folate drugs (Nair et al., 2008). Screening for CNVs may provide a method for locating genes under selection, and, in such cases, the adaptation may be a consequence of CNV-associated differences in gene expression levels rather than amino acid changes in the encoded proteins.

DRUG SELECTIVE SWEEPS IN *P. FALCIPARUM*

Numerous methods, not reviewed here, have been used in population genetic studies of higher eukaryotic organisms to infer deviations from neutral expectation, the existence of selection, or the occurrence of selective sweeps; however, in the malaria community, following the development of genetic tools and molecular markers for *P. falciparum* in the 1990s, the most popular method used to detect the signature of a selective sweep has been to analyze the MS or SNP diversity flanking a known drug selection target, applying the analytical principles outlined in previous sections of this chapter.

CQ Selective Sweeps

The first such study was by Wootton et al. (2002), in which 342 polymorphic MS markers were typed genome-wide across all 14 nuclear chromosomes from 87 worldwide *P. falciparum* isolates that included both CQR and CQS parasites collected from different malaria-endemic regions in Africa, Asia, South and Central America, and Papua New Guinea (PNG). The genome-wide MS haplotypes grouped the parasites into clusters reflecting their continental origins, with very high and significant allele sharing among the American parasites, a lower but significant level of allele sharing among the Asian parasites, and lack of allele sharing among the African parasites (Wootton et al., 2002). Substantial differences in allelic diversity were also

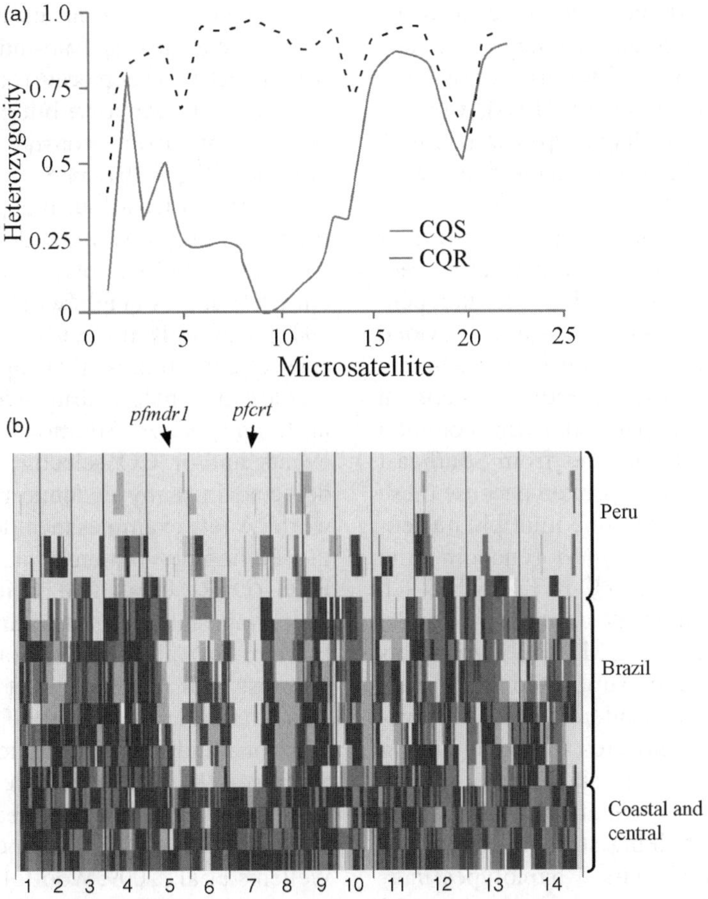

Figure 7.1 Detection of drug selective sweeps by searching for loci with reduced microsatellite (MS) diversity. (a) Reduction in MS diversity in chloroquine-resistant (CQR) parasites from Southeast Asia and Africa. The red solid line is from CQR parasites, and the black dashed line is from chloroquine-sensitive (CQS) parasites. The figure is redrawn from data published in Wotton et al. 2002. *Nature* 418: 320–323. (b) MS allele sharing shows reduced MS diversity among parasites from America. The reference parasite genome from Peru at the top of the figure is highly similar to other parasites from Peru (single cyan color). There are regions of reduced diversity on chromosomes 5 and 7, surrounding the *pfmdr1* and *pfcrt* genes, respectively, in parasites from the Brazil Amazon region, whereas the parasites from Columbia and Honduras (coastal and central) are diverse and are sensitive to chloroquine. The darker blue, red, and yellow colors indicate increasing mutational distance of MS markers from the reference parasite genome. See color insert.

observed among and across the chromosomes, suggesting recent directional selection superimposed on the patterns of geographic diversity, especially at a single region of chromosome 7 that encompassed *pfcrt*, the principal known determinant of the CQR phenotype (Fidock et al., 2000). A dramatic reduction in heterozygosity in the markers flanking *pfcrt* in the CQR parasites, but not the CQS parasites, was found, revealing a "selection valley" in the CQR parasites (Fig. 7.1a). The classical hallmarks of a selective sweep were notable in this region, namely, strongly reduced allelic diversity associated with the specific phenotype (CQR) and the presence of a specific allele (of *pfcrt*) known to confer that phenotype. The gradual decrease in MS diversity toward the *pfcrt* gene occurred over a chromosomal segment spanning >200 kb among the CQR parasites (a decay in linkage disequilibrium [LD]), whereas highly diverse MS haplotypes were found among the CQS parasites in the same flanking regions.

A comparison of *pfcrt* alleles and their flanking MS haplotypes from different geographic locations showed several independent CQ founder mutations and provided the first convincing evidence of large-scale cross-continent CQ selective sweeps in *P. falciparum*. In particular, the common origins for CQR parasites from Southeast Asia and Africa were proven because these parasites shared the same multiple nucleotide substitutions in *pfcrt* (encoding the amino acid pattern C-IET-H-S-E-S-T-I. These are single-letter amino acid codes) and the same flanking MS haplotypes. This molecular evidence supported the earlier conjecture from epidemiologic observations that CQR parasites spread by human travel from Southeast Asia to Africa (Payne, 1987). Moreover, calculations of the rate of decline of hitchhiking strength across the extended chromosome 7 haplotypes indicated that selection to near fixation had occurred in 20–80 parasite generations. This is consistent with the known history and spread of CQR parasites, as CQR was first reported in Southeast Asia along the Thai–Cambodia border in the late 1950s, in the Amazon region in the early 1960s, and then in East Africa in the late 1970s (Payne, 1987). Subsequently, the CQR parasites have spread to many countries in Asia and throughout South America. Regarding parasites from the Americas, allele-sharing analysis of genome-wide MS haplotypes was used to compare isolates from the Peruvian and Brazilian Amazon regions with isolates from coastal South and Central America. The Amazonian parasites showed reduced heterozygosity surrounding two loci known to be involved in drug resistance—namely, the drug transporter genes *pfmdr1* on chromosome 5 and *pfcrt* on chromosome 7—indicating that a drug selective sweep occurred among parasites from the Amazon but not in parasite populations from coastal and central America (Fig. 7.1b).

Overall, at least four independent CQR founder mutations were detected in this study using MS haplotypes flanking different *pfcrt* alleles (Fig. 7.2) (Wootton et al., 2002). The study also set the stage for many subsequent studies that applied similar methods to analyze drug selective sweeps in *P. falciparum*. Additional CQ founder events and/or CQ selective sweeps were detected in many endemic regions of the world. A few examples included those from the Philippines (Chen et al., 2003, 2005), PNG (DaRe et al., 2007; Mehlotra et al., 2001; Nagesha et al., 2003), India (Mixson-Hayden et al., 2010), South America (Cortese et al., 2002; Vieira et al., 2004), and Africa (Laufer et al., 2010). Signatures of additional independent CQ founder events and more than 20 nucleotide substitutions in the *pfcrt* have now been recognized (Fig. 7.2) (Chen et al., 2005; Cooper et al., 2005; Wellems et al., 2009; Wootton et al., 2002); however, parasites with a same *pfcrt* allele but having polymorphic MSs in the introns of the gene have been observed (Das and Dash, 2007; Mittra et al., 2006; Vinayak et al., 2006), raising some questions on recent CQ selective sweeps in some specific regions. The observations could be results of highly unstable MSs and/or recombination in high transmission regions.

Mutations in selected genes are often deleterious, and parasites with mutated genes may not compete well with parasites having wild-type alleles. Removal of selection pressure may allow the return of the parasites with wild-type alleles. Indeed, increased frequencies of parasites with wild-type *pfcrt* have been reported in Africa and China after the cessation of CQ use (Laufer et al., 2006; Wang et al., 2005). The increase in the ancestral CQS allele

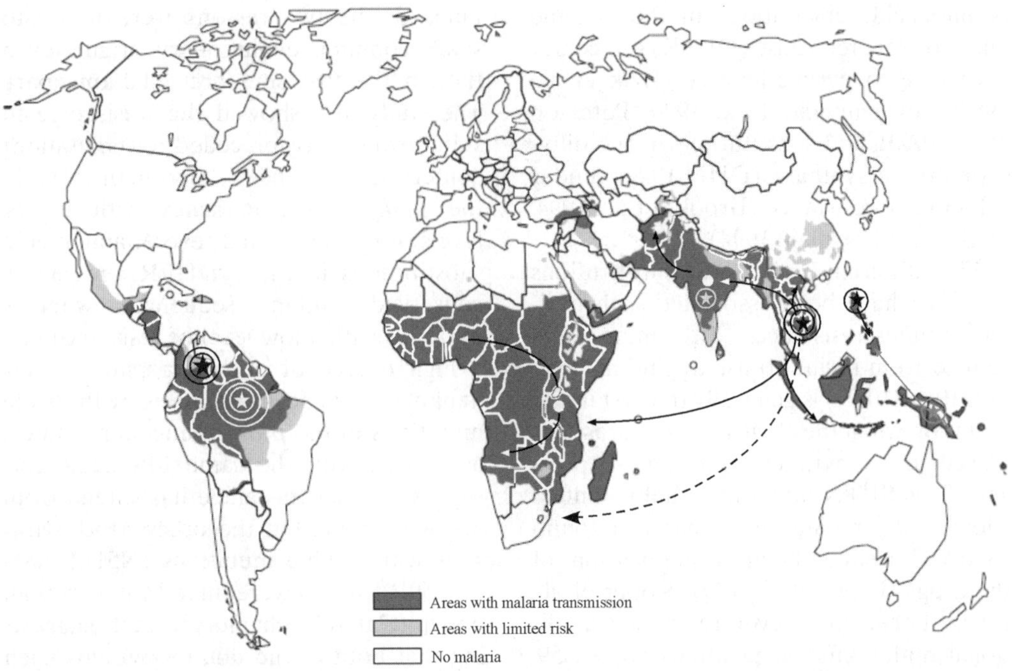

Figure 7.2 Chloroquine resistance (CQ) founder mutations and drug selective sweeps in *Plasmodium falciparum*. At least five independent CQ founder events, indicated by the stars of different colors, have been detected through haplotype analysis of *pfcrt* alleles and their flanking microsatellites (MSs). Each star represents one site where unique *pfcrt* and/or flanking MS haplotypes have been observed, although selective sweeps and the number of founder mutations in India are still being debated. The dashed line represents a sweep of high level resistant *pfdhfr* allele from Southeast Asia to Africa. The colored areas of the world map represent malaria-endemic regions with two levels of transmission risk as defined on the World Health Organization (WHO) Web site (http://www.who.int/malaria/en). See color insert.

frequency is likely from the re-expansion of the minority of drug-susceptible parasites that survive in the population (Laufer et al., 2010); however, this reversal to CQS has not been observed in South America. One explanation for the lack of reversal in South America is that the parasites there have had compensatory mutations that might stabilize the initial substitutions, specifically mutations in *pfmdr1* (Sa et al., 2009). Another possibility is that the mutant *pfcrt* allele in America has fewer substitutions (four) and may be more "stable" than the Asian *pfcrt* allele. The specific *pfcrt* and *pfmdr1* alleles seen in the resistant South American parasites are believed to be the results of double selection by CQ and amodiaquine (Sa et al., 2009), which is consistent with the observation of simultaneous selection on *pfcrt* and *pfmdr1* (Fig. 7.1b).

SP Selective Sweeps

Another antimalarial therapy that has been extensively used to treat malaria infection is a drug combination of sulfadoxine and pyrimethamine (SP). Similar to CQ, resistance to SP has been reported worldwide; highly resistant parasites have spread from Southeast Asia and South America in the 1970s–1980s and to Africa in the late 1990s (Mita, 2010; Roper et al., 2003, 2004). The resistance to SP is due to the mutations in two enzymes in the folate synthesis pathway.

Amino acid substitutions in the enzyme dihydrofolate reductase (DHFR) can confer resistance to pyrimethamine (Bzik et al., 1987; Cowman and Lew, 1990; Peterson et al., 1990), and substitutions in the dihydropteroate synthase (DHPS) can render sulfadoxine resistance (Brooks et al., 1994; Triglia and Cowman, 1994).

Five different amino acid substitutions in *pfdhfr* have been associated with pyrimethamine resistance. An amino acid change from serine to asparagine at position 108 (S108N) is generally the first initial mutation when the *P. falciparum* parasite is placed under pyrimethamine or SP pressure. This S108N mutation usually confers a low level of resistance and has been found to occur frequently after application of the drugs (Plowe et al., 1997; Roper et al., 2003). For higher level resistance, additional mutation(s) at positions 50, 51, 59, and 164 are necessary (Foote et al., 1990; Peterson et al., 1988; Plowe, 2001), with the highest level of resistance in parasites having a four-amino acid substitution (N51I + C59R + S108N + I164R) in the PfDHFR. Similar to PfDHFR, an amino acid substitution at position 437 (A437G) of PfDHPS is often the initial mutation for sulfadoxine resistance, and additional mutations at positions 436, 540, 581, and 613 are observed for higher levels of resistance to sulfadoxine (Triglia and Cowman, 1994; Triglia et al., 1997).

The first systematic study of evolutionary origin of the resistance to SP also employed a strategy similar to that used by Wootton et al., that is, comparing *pfdhrf*/*pfdhps* genes and their flanking MS haplotypes between parasites with different levels of resistance to SP (Roper et al., 2003). In this study, parasites collected from KwaZulu-Natal, South Africa, over a 5-year period were analyzed for polymorphisms in the *pfdhfr* and *pfdhps* genes and in MSs from sequences flanking the genes, and the results were compared to those from the Kilimanjaro region of northern Tanzania. The allelic determinants of resistance in the two regions were found to share common evolutionary origins even though the two sites are 4000 km apart. The study also showed that resistance to sulfadoxine was preceded by mutations conferring resistance to pyrimethamine in the *pfdhfr* gene. Parasites with higher levels of resistance had several amino acid substitutions in the PfDHFR and highly conserved flanking sequences, whereas parasites with a low level of resistance had a high degree of polymorphism in their flanking sequences. Parasites with triple mutations in the *pfdhfr* gene were always associated with the same MS haplotype, suggesting that this allele has spread from a single origin. On the other hand, parasites with double mutations (N51I/S108N or C59R/S108N) were linked to more than one unrelated MS haplotype each, suggesting that both of the double mutants each have arisen more than once. By contrast, highly polymorphic flanking sequences were found in parasites with sensitive *pfdhfr* alleles.

The linkage of the *pfdhfr* triple mutant (N51I, C59R, S108N), corresponding to the highest level of pyrimethamine resistance, to the same MS haplotype indicates a single founder event. The different flanking MS haplotypes in the double-mutant parasites also suggest that none of the double mutants is an ancestor of this triple mutant. Additionally, the frequency of the *pfdhfr* triple mutant was found to increase with time over the 5-year sampling period, displacing a double-mutant allele. These results suggest that selective sweeps are ongoing in the parasite populations, and a high level of resistance to pyrimethamine did not arise de novo as frequently as earlier thought (whereas the low level resistance with a single mutation did indeed occur multiple times).

In a follow-up study, the triple-mutant *pfdhfr* found in South Africa was subsequently shown to have originated from Southeast Asia. Parasites with triple mutations in the *pfdhfr* gene from Southeast

Asia and Africa were genotyped with eight MS markers, and the results showed that a predominant five-locus MS haplotype associated with the triple-mutant *pfdhfr* in Africa was identical to the *pfdhfr* alleles carrying two to four mutations in Southeast Asia. The results suggest a cross-continent introgression and selective sweep parallel to that of a CQ selective sweep (Fig. 7.2) (Roper et al., 2004).

In a separate study in Southeast Asia, Nair et al. genotyped 33 MS markers on chromosome 4 (*pfdhfr* is located on chromosome 4) in 61 parasites from a location on the Thailand/Myanmar border (Nair et al., 2003). Minimal MS length variation in a 12-kb region flanking the *pfdhfr* gene was observed in resistant parasites, and the same or similar MS haplotypes flanking the resistant *pfdhfr* alleles were found in parasites collected from 11 locations. Based on a dramatic selection valley observed, the authors concluded that "all extant resistant dhfr alleles observed have a single origin and that variation around dhfr has been purged as a result of a single selective event." Again, the results showed strong selective sweeps involving *pfdhfr* genes in Southeast Asia.

In South America, drug-resistant SP mutants were fixed in *P. falciparum* populations (McCollum et al., 2007). In one study, analysis of *pfdhfr* and its flanking MS haplotypes of malaria parasites from five regions of the South American Amazon (Colombia, Venezuela, Brazil, Peru, and Bolivia) showed that the mutations conferring mid- and high level resistance to SP have a common origin (Cortese et al., 2002). In another study, the genes encoding *pfdhfr* and *pfdhps* and MS markers flanking the two loci were examined for 97 samples collected from Bolivar State, Venezuela. Two genotypes of triple mutants in the genes, *pfdhfr* (50R/51I/108N) and *pfdhps* (437G/540E/581G) (90.7%), and two double mutants, *pfdhfr* (51I/108N) and *pfdhps* (437G/581G) (9.3%), were found. All the *pfdhfr* and *pfdhps* alleles carried a single MS haplotype that was different from those of Africa. The data suggested a single origin for both *pfdhfr* and *pfdhps* SP-resistant alleles and the presence of a "hitchhiking effect" due to selection by SP use (McCollum et al., 2007).

Similar results and conclusions of drug selective sweeps were also reported in India (Lumb et al., 2009). One hundred ninety *P. falciparum* parasite isolates were collected from five different geographic regions of India (Uttar Pradesh, Madhya Pradesh, Assam, Orissa, and Andaman and Nicobar Islands) and were genotyped with MS markers. A significant reduction in genetic diversity in the ±20-kb vicinity of the mutant *pfdhfr* alleles was observed, suggesting a drug selective sweep. The reduction in genetic diversity was more dramatic around quadruple *pfdhfr* alleles than around double- and single-mutant alleles (Lumb et al., 2009). Additionally, all the resistant *pfdhfr* alleles share a single MS haplotype similar to that in Thailand *pfdhfr* mutants, suggesting potential common origins of SP-resistant parasites in Thailand and India.

All these studies establish a common theme: The origins of high levels (triple or quadruple mutants) of resistance to SP are limited, and drug selective sweeps are always involved, including large-scale cross-continent sweeps; however, evidence of multiple origins of high level resistance to SP has also been reported in Africa. Using the same method of direct sequencing of PCR products and typing MS markers flanking the *pfdhfr* gene, McCollum et al. examined 479 *P. falciparum* samples from western Kenya and showed multiple lineages for the triple-mutant *pfdhfr* genotypes in the parasite samples (McCollum et al., 2006). The 51I/108N/164L genotype was associated with at least four different MS haplotypes, which also suggests that recombination, rather than the accumulation of simple point mutations, may have played a role in the evolution of specific genotypes.

Because SP contains two drugs, the parasites are under the pressure of both drugs at the same time, including sulfadoxine. At least two different triple-mutant *pfdhps* alleles (A-G-E-A-A and S-G-E-G-A) have been reported in Southeast Asia (Ahmed et al., 2006; Khim et al., 2005), but only a single triple-mutant allele S-G-E-G-A was described in South America (Cortese et al., 2002; Kublin et al., 1998; McCollum et al., 2007). In a recent study, the origins and lineages of the highly resistant *pfdhps* alleles in Southeast Asia were investigated (Vinayak et al., 2010). Two hundred thirty-four *P. falciparum* DNA samples from Cambodia were sequenced to confirm the *pfdhps* substitutions S436A/F, A437G, K540E, A581G, and A613S/T that are implicated in sulfadoxine resistance. Additionally, 10 MSs flanking the *pfdhps* were typed to determine the genetic backgrounds of various alleles. The two well-known, highly resistant triple-mutant *pfdhps* alleles (S-G-E-G-A and A-G-E-A-A) and a new triple-mutant allele (S-G-N-G-A) were found among the parasites. At least three independent origins for the double mutants and one for each of the triple-mutant S-G-K-G-A, A-G-K-A-A, and S-G-E-A-A alleles were detected. The data also suggest that the triple-mutant allele S-G-E-G-A and the novel allele S-G-N-G-A have a common origin on the S-G-K-G-A background, whereas the A-G-E-A-A triple mutant is derived from its precursor A-G-K-A-A on multiple genetic backgrounds. The MS haplotypes flanking the *pfdhps* alleles from Cambodia, Kenya, Cameroon, and Venezuela also suggest an independent origin of sulfadoxine-resistant alleles in each of these regions (Vinayak et al., 2010).

Selection Leading to CNV

Amino acid substitutions in PfMDR1 and gene copy variation have been implicated in resistance to many antimalarial drugs including CQ, mefloquine, and ART (Cowman et al., 1994; Hayton and Su, 2004; Price et al., 1997; Sa et al., 2009). A "soft" (or weak) selective sweep, leading to increases in copy number but not to fixation of any single CNV in the population, has been reported in Southeast Asia for the genomic region containing the *pfmdr1* gene on chromosome 5 (Nair et al., 2007). Five to fifteen independent amplification events in the *pfmdr1* locus were detected using MS markers flanking the *pfmdr1* and real-time PCR to determine the copy number. MS diversity was reduced in a 170- to 250-kb (10–15 cM) region of the *pfmdr1* locus, suggesting hitchhiking and rapid recent spread of the selected chromosomes, although the reduction of diversity in the flanking regions was not very strong. Based on this "broad signature" and limited reduction in diversity, the authors suggested that "soft sweeps are common in nature and the statistical methods based on diversity reduction may be inefficient at detecting evidence for selection in genome-wide marker screens."

Another example of selection leading to CNV is the first gene in the *Plasmodium* folate biosynthesis pathway, *gch1* (Nair et al., 2008). In this study, the copy number of *pfgch1* was shown to be an adaptive response to selection by antifolate drugs that target enzymes downstream in this pathway. The CNV of *pfgch1* in parasites from Thailand that have been through strong historical antifolate selection were compared to those from neighboring Laos that had weak antifolate selection. Only 2% of chromosomes had amplified *pfgch-1*copy number in Laos, while 72% of the Thai parasites had multiple copies (2–11). Reduced MS diversity and increased LD in a 900-kb region flanking the *pfgch1* in the Thai parasites were observed, indicating rapid recent spread of chromosomes carrying multiple copies of *pfgch1*. Additionally, this CNV was also associated with the *pfdhfr* 164L allele, suggesting functional association between genes located on different chromosomes.

Selection on mtDNA Cytochrome b (*pfcytb*) Gene

Resistance to atovaquone is conferred by nucleotide substitutions at position 4296 (a4296g or a4296c) in the *pfcytb* gene located on the 6-kb mitochondrial (mt) genome (Kessl et al., 2005; Korsinczky et al., 2000). The deployment of the antimalarial drug combination atovaquone–proguanil may select for mutations that can confer resistance to both drugs. Indeed, mutations at *pfcybt* at the 4296 position conferring resistance to atovaquone–proguanil have been reported (Musset et al., 2007). By comparing the mitochondrial sequences and MS genotypes of parasite isolates before treatment and at the day of failure for each uncured patient, Musset et al. observed repeated occurrences of the mutations at position 4296 of the *pfcytb* in distinct mitochondrial haplotypes, suggesting multiple origins of the selected alleles (Musset et al., 2007). The observation is similar to low level resistance to SP, where a single-nucleotide substitution can contribute to resistance and lack of (or weak) selective sweeps.

Detection of Genome-Wide Selection of Unknown Sources

Recently, the development of genome-wide SNPs from multiple field isolates has allowed the identification of genomic loci under selection. Based on the principle of selection and polymorphic nature of genes encoding malaria antigens and genes conferring drug-resistance, genomic loci with signatures of selection under nonspecified selection forces have been identified (Mu et al., 2007, 2010; Volkman et al., 2007). In a recent study, genome-wide maps of selection for parasite populations from Asia, Africa, and America were developed, and multiple loci under significant positive selection were detected (Mu et al., 2010). Some obvious examples include the locus on chromosome 7 containing *pfcrt*, a locus on chromosome 11 containing the gene encoding *P. falciparum* apical membrane antigen 1 (*pfama-1*), and a locus on chromosome 13. As discussed above, the *pfcrt* gene is under CQ selection, and *pfama-1* is a target of host immune response (or under immune selection), although *pfama-1* is more likely under balancing selection (Polley et al., 2003). The PF13_0271 gene on chromosome 13 encodes an ATP-binding cassette (ABC) transporter and can be involved in transporting drugs. This gene is a good candidate for drug selection because ABC transporters are well-known to be involved in drug resistance (Dean, 2005).

Possible signatures for selective sweeps that drive specific alleles to fixation in one continental population but remain polymorphic in others were also detected using cross-population extended haplotype homozygosity (Pickrell et al., 2009; Sabeti et al., 2007). In a recent study, a locus on chromosome 5 containing a gene (PFE1445c) that encodes a *Plasmodium* conserved protein was found to be under differential selection between populations, in addition to the *pfcrt* locus (Mu et al., 2010). Further, several large extended haplotypes (519,126–922,368 bp on chromosome 7; 831,749–925,515 bp on chromosome 8; and 319,075–495,408 bp on chromosome 9) between Asian and American populations also had signals of selective sweeps. A total of 11 genes apparently under significant selection were detected using three different methods for detecting signatures of selection (Mu et al., 2010). More recently, the same strategy was used to analyze 44 worldwide *P. falciparum* isolates and showed that 15 genes, including *pfcrt* and *pfdhfr*, had undergone recent positive selection (Van Tyne et al., 2011) This study also reported 11 candidate loci associated with variation in sensitivity to antimalarial drugs, including *pfcrt* and a newly implicated gene, PF10_0355, in which CNV-associated overexpression was correlated with decreased sensitivity to halofantrine, mefloquine, and lumefantrine. Further studies are necessary to link the loci under selection to their selection agents.

Lack of Selective Sweep Signatures in *P. vivax*

Many antimalarial drugs used to treat *P. falciparum* infection have also been applied to treat *P. vivax* infection. Although mutations in the *pfcrt* gene have been shown to confer resistance to CQ in *P. faciparum*, polymorphism in the *vivax* homolog of *pfcrt* does not appear to be associated with CQ resistance (Nomura et al., 2001). Therefore, the search for a selection signature based on CQ selection valley at the locus containing the homolog of *pfcrt* is not possible at this time.

Similar to the observations in *P. falciparum*, mutations in the *pvdhfr* and *pvdhps* have been reported from many endemic regions and have been shown to be associated with resistance to SP (Auliff et al., 2006; Imwong et al., 2001; Lu et al., 2010; Zakeri et al., 2009). In one study, the *pvdhfr* coding regions, 792 bp upstream and 683 bp downstream of the gene, were amplified and sequenced from 137 isolates collected from Colombia, India, Indonesia, PNG, Sri Lanka, Thailand, and Vanuatu (Hawkins et al., 2008). A repeat motif located 2.6 kb upstream of *pvdhfr* was also sequenced from 75 of 137 patient isolates. A double mutant (58R/117N) of the *pvdhfr* was found to evolve from several origins, and triple (58R/61M/117T) and three quadruple (57L/61M/117T/173F, 57I/58R/61M/117T, and 57L/58R/61M/117T) mutant alleles were found to have at least three independent origins in Thailand, Indonesia, and PNG/Vanuatu. The results suggest multiple founding events for high level SP resistance, and the scale of selective sweep in *P. vivax* is smaller than that of *P. falciparum*. Similarly, analysis of four MS markers at −230.54, −38.83, +6.15, and +283.28 kb from the *pvdhfr* gene showed highly polymorphic MS markers in 110 Indian *P. vivax* isolates; no association between *pvdhfr* mutations and the flanking MS alleles was found (Alam et al., 2007). These observations are in contrast to those found in *P. falciaprum*, where markers around sensitive *pfdhfr* alleles have highly polymorphic MS markers, whereas those around resistant alleles have reduced diversity leading to strong LD. Although the reasons for this lack of association are not clear, several possibilities could contribute to this observation, including high MS mutation rates, high recombination rates in the *P. vivax* populations, multiple founder mutations, lower entomological transmission rates, or effects of the local parasite population structure.

CONCLUSION

Selective sweeps, particularly drug selective sweeps, have occurred extensively in populations of *P. falciparum*. Sustained selection over time and over different geographic regions has imposed the conditions that result in a unique haplotype surrounding a selection target, fulfilling the strict criteria of genetic hitchhiking. Candidates for other targets of selection can also be detected using other criteria suggestive of selection signatures. Because MS loci provide highly diverse neutral markers with multiple alleles in parasite populations and dense coverage of the genome, they are very useful for detecting signatures of selection, selective sweeps, and the foci of founder mutations. Using these methods, the malaria community has now assembled a very detailed picture of the multiple origins and rapid spread of *P. falciparum* genotypes that confer resistance to the widely applied CQ and SP drug regimens. The parasite populations clearly have a great capacity to evolve new, complex, multisite mutations and genotypes of sufficient genetic fitness to spread and overwhelm such drug regimens across continents. A lesson for antimalarial drug research and development and deployment policies in the future is to consider how such pervasively directional drug selection pressures on the parasites can be minimized, perhaps by designing various drug combinations that impose opposite selective pressures on the physiology and evolving genomes of the parasites. Similar considerations apply to the development of antiparasite vaccines or transmission-blocking vaccines. Sustainable vaccine strategies will likely require a diversity of different parasite molecular targets in order to minimize widespread directional selection on any single parasite target of immunity and to stem the spread of vaccine-resistant parasites. Developing suitable vaccine combinations, however, is very challenging, partly because of gaps in the current understanding of naturally acquired immunity to malaria and of the interplay of immune protection, which is commonly incomplete or short-lasting, and parasite transmission. Finally, these investigations emphasize how signatures of selection in parasite populations can be explored in the future in fundamental research for identifying and mapping genes conferring drug resistance (positive directional selection), genes encoding immune targets (balancing selection), and genes playing important roles in parasite transmission (directional or balancing selections) through the mosquito and human hosts.

ACKNOWLEDGMENTS

This work was supported by the Intramural Research Program of the Division of Intramural Research, National Institute of Allergy and Infectious Diseases (NIAID), National Institutes of Health, and by the National Center for Biotechnology Information, National Library of Medicine, National Institutes of Health. We thank NIAID intramural editor Brenda Rae Marshall for assistance.

REFERENCES

Ahmed A, Lumb V, Das MK, Dev V, Wajihullah, Sharma YD. 2006. Prevalence of mutations associated with higher levels of sulfadoxine-pyrimethamine resistance in *Plasmodium falciparum* isolates from Car Nicobar Island and Assam, India. Antimicrob Agents Chemother 50: 3934–3938.

Alam MT, Agarwal R, Sharma YD. 2007. Extensive heterozygosity at four microsatellite loci flanking *Plasmodium vivax* dihydrofolate reductase gene. Mol Biochem Parasitol 153: 178–185.

Auliff A, Wilson DW, Russell B, Gao Q, Chen N, et al. 2006. Amino acid mutations in *Plasmodium vivax* DHFR and DHPS from several geographical regions and susceptibility to antifolate drugs. Am J Trop Med Hyg 75: 617–621.

Boyer JC, Hawk JD, Stefanovic L, Farber RA. 2008. Sequence-dependent effect of interruptions on microsatellite mutation rate in mismatch repair-deficient human cells. Mutat Res 640: 89–96.

Brooks DR, Wang P, Read M, Watkins WM, Sims PF, et al. 1994. Sequence variation of the hydroxymethyldihydropterin pyrophosphokinase: Dihydropteroate synthase gene in lines of the human malaria parasite, *Plasmodium falciparum*, with differing resistance to sulfadoxine. Eur J Biochem 224: 397–405.

Bzik DJ, Li WB, Horii T, Inselburg J. 1987. Molecular cloning and sequence analysis of the *Plasmodium falciparum* dihydrofolate reductase-thymidylate synthase gene. Proc Natl Acad Sci U S A 84: 8360–8364.

Cargill M, Altshuler D, Ireland J, Sklar P, Ardlie K, et al. 1999. Characterization of single-nucleotide polymorphisms in coding regions of human genes. Nat Genet 22: 231–238.

Charlesworth B, Morgan MT, Charlesworth D. 1993. The effect of deleterious mutations on neutral molecular variation. Genetics 134: 1289–1303.

Chavchich M, Gerena L, Peters J, Chen N, Cheng Q, et al. 2010. Role of pfmdr1 amplification and expression in induction of resistance to artemisinin derivatives in *Plasmodium falciparum*. Antimicrob Agents Chemother 54: 2455–2464.

Chen N, Kyle DE, Pasay C, Fowler EV, Baker J, et al. 2003. pfcrt allelic types with two novel amino acid mutations in chloroquine-resistant *Plasmodium falciparum* isolates from the Philippines. Antimicrob Agents Chemother 47: 3500–3505.

Chen N, Wilson DW, Pasay C, Bell D, Martin LB, et al. 2005. Origin and dissemination of chloroquine-resistant *Plasmodium falciparum* with mutant pfcrt alleles in the Philippines. Antimicrob Agents Chemother 49: 2102–2105.

Chung H, Lopez CG, Holmstrom J, Young DJ, Lai JF, et al. 2010. Both microsatellite length and sequence context determine frameshift mutation rates in defective DNA mismatch repair. Hum Mol Genet 19: 2638–2647.

Cooper RA, Hartwig CL, Ferdig MT. 2005. pfcrt is more than the *Plasmodium falciparum* chloroquine resistance gene: A functional and evolutionary perspective. Acta Trop 94: 170–180.

Cooper RA, Lane KD, Deng B, Mu J, Patel JJ, et al. 2007. Mutations in transmembrane domains 1, 4 and 9 of the *Plasmodium falciparum* chloroquine resistance transporter alter susceptibility to chloroquine, quinine and quinidine. Mol Microbiol 63: 270–282.

Cortese JF, Caraballo A, Contreras CE, Plowe CV. 2002. Origin and dissemination of *Plasmodium falciparum* drug-resistance mutations in South America. J Infect Dis 186: 999–1006.

Cowman AF, Lew AM. 1990. Chromosomal rearrangements and point mutations in the DHFR-TS gene of *Plasmodium chabaudi* under antifolate selection. Mol Biochem Parasitol 42: 21–29.

Cowman AF, Galatis D, Thompson JK. 1994. Selection for mefloquine resistance in *Plasmodium falciparum* is linked to amplification of the pfmdr1 gene and cross-resistance to halofantrine and quinine. Proc Natl Acad Sci U S A 91: 1143–1147.

DaRe JT, Mehlotra RK, Michon P, Mueller I, Reeder J, et al. 2007. Microsatellite polymorphism within pfcrt provides evidence of continuing evolution of chloroquine-resistant alleles in Papua New Guinea. Malar J 6: 34.

Das A, Dash AP. 2007. Evolutionary paradigm of chloroquine-resistant malaria in India. Trends Parasitol 23: 132–135.

Dean M. 2005. The genetics of ATP-binding cassette transporters. Methods Enzymol 400: 409–429.

Dharia NV, Bright AT, Westenberger SJ, Barnes SW, Batalov S, et al. 2010. Whole-genome sequencing and microarray analysis of ex vivo *Plasmodium vivax* reveal selective pressure on putative drug resistance genes. Proc Natl Acad Sci U S A 107: 20045–20050.

Dondorp AM, Nosten F, Yi P, Das D, Phyo AP, et al. 2009. Artemisinin resistance in *Plasmodium falciparum* malaria. N Engl J Med 361: 455–467.

Duan J, Mu J, Thera MA, Joy D, Kosakovsky Pond SL, et al. 2008. Population structure of the genes encoding the polymorphic *Plasmodium falciparum* apical membrane antigen 1: Implications for vaccine design. Proc Natl Acad Sci U S A 105: 7857–7862.

Feng X, Carlton JM, Joy DA, Mu J, Furuya T et al. 2003. Single-nucleotide polymorphisms and genome diversity in *Plasmodium vivax*. Proc Natl Acad Sci U S A 100: 8502–8507.

Fidock DA, Nomura T, Talley AK, Cooper RA, Dzekunov SM, et al. 2000. Mutations in the *P. falciparum* digestive vacuole transmembrane protein PfCRT and evidence for their role in chloroquine resistance. Mol Cell 6: 861–871.

Foote SJ, Galatis D, Cowman AF. 1990. Amino acids in the dihydrofolate reductase-thymidylate synthase gene of *Plasmodium falciparum* involved in cycloguanil resistance differ from those involved in pyrimethamine resistance. Proc Natl Acad Sci U S A 87: 3014–3017.

Gardner MJ, Hall N, Fung E, White O, Berriman M, et al. 2002. Genome sequence of the human malaria parasite *Plasmodium falciparum*. Nature 419: 498–511.

Hawkins VN, Auliff A, Prajapati SK, Rungsihirunrat K, Hapuarachchi HC, et al. 2008. Multiple origins of resistance-conferring mutations in *Plasmodium vivax* dihydrofolate reductase. Malar J 7: 72.

Hayton K, Su X-Z. 2004. Genetic and biochemical aspects of drug resistance in malaria parasites. Curr Drug Targets Infect Disord 4: 1–10.

Hayton K, Su XZ. 2008. Drug resistance and genetic mapping in *Plasmodium falciparum*. Curr Genet 54: 223–239.

Hughes AL, Verra F. 2001. Very large long-term effective population size in the virulent human malaria parasite *Plasmodium falciparum*. Proc R Soc Lond B Biol Sci 268: 1855–1860.

Imwong M, Pukrittakayamee S, Looareesuwan S, Pasvol G, Poirreiz J, et al. 2001. Association of genetic mutations in *Plasmodium vivax* dhfr with resistance to sulfadoxine-pyrimethamine: Geographical and clinical correlates. Antimicrob Agents Chemother 45: 3122–3127.

Jeffares DC, Pain A, Berry A, Cox AV, Stalker J, et al. 2007. Genome variation and evolution of the malaria parasite *Plasmodium falciparum*. Nat Genet 39: 120–125.

Jiang H, Yi M, Mu J, Zhang L, Ivens A, et al. 2008. Detection of genome wide polymorphisms in the AT rich *Plasmodium falciparum* genome using a high density microarray. BMC Genomics 9: 398.

Kaplan NL, Hudson RR, Langley CH. 1989. The "hitchhiking effect" revisited. Genetics 123: 887–899.

Kessl JJ, Ha KH, Merritt AK, Lange BB, Hill P, et al. 2005. Cytochrome b mutations that modify the ubiquinol-binding pocket of the cytochrome bc1 complex and confer antimalarial drug resistance in *Saccharomyces cerevisiae*. J Biol Chem 280: 17142–17148.

Khim N, Bouchier C, Ekala MT, Incardona S, Lim P, et al. 2005. Countrywide survey shows very high prevalence of *Plasmodium falciparum* multilocus resistance genotypes in Cambodia. Antimicrob Agents Chemother 49: 3147–3152.

Kidgell C, Volkman SK, Daily J, Borevitz JO, Plouffe D, et al. 2006. A systematic map of genetic variation in *Plasmodium falciparum*. PLoS Pathog 2: e57.

Korsinczky M, Chen N, Kotecka B, Saul A, Rieckmann K, et al. 2000. Mutations in *Plasmodium falciparum* cytochrome b that are associated with atovaquone resistance are located at a putative drug-binding site. Antimicrob Agents Chemother 44: 2100–2108.

Kublin JG, Witzig RS, Shankar AH, Zurita JQ, Gilman RH, et al. 1998. Molecular assays for surveillance of antifolate-resistant malaria. Lancet 351: 1629–1630.

Laufer MK, Thesing PC, Eddington ND, Masonga R, Dzinjalamala FK, et al. 2006. Return of chloroquine antimalarial efficacy in Malawi. N Engl J Med 355: 1959–1966.

Laufer MK, Takala-Harrison S, Dzinjalamala FK, Stine OC, Taylor TE, et al. 2010. Return of chloroquine-susceptible *falciparum* malaria in Malawi was a reexpansion of diverse susceptible parasites. J Infect Dis 202: 801–808.

Li J, Zhang Y, Liu S, Hong L, Sullivan M, et al. 2009. Hundreds of microsatellites for genotyping *Plasmodium yoelii* parasites. Mol Biochem Parasitol 166: 153–158.

Lu F, Lim CS, Nam DH, Kim K, Lin K, et al. 2010. Mutations in the antifolate-resistance-associated genes dihydrofolate reductase and dihydropteroate synthase in *Plasmodium vivax* isolates from malaria-endemic countries. Am J Trop Med Hyg 83: 474–479.

Lumb V, Das MK, Singh N, Dev V, Wajihullah, Sharma YD. 2009. Characteristics of genetic hitchhiking around dihydrofolate reductase gene associated with pyrimethamine resistance in *Plasmodium falciparum* isolates from India. Antimicrob Agents Chemother 53: 5173–5180.

McCollum AM, Poe AC, Hamel M, Huber C, Zhou Z, et al. 2006. Antifolate resistance in *Plasmodium falciparum*: Multiple origins and identification of novel dhfr alleles. J Infect Dis 194: 189–197.

McCollum AM, Mueller K, Villegas L, Udhayakumar V, Escalante AA. 2007. Common origin and fixation of *Plasmodium falciparum* dhfr and dhps mutations associated with sulfadoxine-pyrimethamine resistance in a low-transmission area in South America.

Antimicrob Agents Chemother 51: 2085–2091.

Mehlotra RK, Fujioka H, Roepe PD, Janneh O, Ursos LM, et al. 2001. Evolution of a unique *Plasmodium falciparum* chloroquine-resistance phenotype in association with pfcrt polymorphism in Papua New Guinea and South America. Proc Natl Acad Sci U S A 98: 12689–12694.

Mita T. 2010. Origins and spread of pfdhfr mutant alleles in *Plasmodium falciparum*. Acta Trop 114: 166–170.

Mittra P, Vinayak S, Chandawat H, Das MK, Singh N, et al. 2006. Progressive increase in point mutations associated with chloroquine resistance in *Plasmodium falciparum* isolates from India. J Infect Dis 193: 1304–1312.

Mixson-Hayden T, Jain V, McCollum AM, Poe A, Nagpal AC, et al. 2010. Evidence of selective sweeps in genes conferring resistance to chloroquine and pyrimethamine in *Plasmodium falciparum* isolates in India. Antimicrob Agents Chemother 54: 997–1006.

Mu J, Duan J, Makova KD, Joy DA, Huynh CQ, et al. 2002. Chromosome-wide SNPs reveal an ancient origin for *Plasmodium falciparum*. Nature 418: 323–326.

Mu J, Awadalla P, Duan J, McGee KM, Keebler J, et al. 2007. Genome-wide variation and identification of vaccine targets in the *Plasmodium falciparum* genome. Nat Genet 39: 126–130.

Mu J, Myers RA, Jiang H, Liu S, Ricklefs S, et al. 2010. Plasmodium falciparum genome-wide scans for positive selection, recombination hot spots and resistance to antimalarial drugs. Nat Genet 42: 268–271.

Musset L, Le Bras J, Clain J. 2007. Parallel evolution of adaptive mutations in *Plasmodium falciparum* mitochondrial DNA during atovaquone-proguanil treatment. Mol Biol Evol 24: 1582–1585.

Nagesha HS, Casey GJ, Rieckmann KH, Fryauff DJ, Laksana BS, et al. 2003. New haplotypes of the *Plasmodium falciparum* chloroquine resistance transporter (pfcrt) gene among chloroquine-resistant parasite isolates. Am J Trop Med Hyg 68: 398–402.

Nair S, Williams JT, Brockman A, Paiphun L, Mayxay M, et al. 2003. A selective sweep driven by pyrimethamine treatment in southeast asian malaria parasites. Mol Biol Evol 20: 1526–1536.

Nair S, Nash D, Sudimack D, Jaidee A, Barends M, et al. 2007. Recurrent gene amplification and soft selective sweeps during evolution of multidrug resistance in malaria parasites. Mol Biol Evol 24: 562–573.

Nair S, Miller B, Barends M, Jaidee A, Patel J, et al. 2008. Adaptive copy number evolution in malaria parasites. PLoS Genet 4: e1000243.

Nomura T, Carlton JM, Baird JK, Del Portillo HA, Fryauff DJ, et al. 2001. Evidence for different mechanisms of chloroquine resistance in 2 *Plasmodium* species that cause human malaria. J Infect Dis 183: 1653–1661.

Ota T, Kimura M. 1975. The effect of selected linked locus on heterozygosity of neutral alleles (the hitch-hiking effect). Genet Res 25: 313–326.

Payne D. 1987. Spread of chloroquine resistance in *Plasmodium falciparum*. Parasitol Today 3: 241–246.

Peterson DS, Walliker D, Wellems TE. 1988. Evidence that a point mutation in dihydrofolate reductase-thymidylate synthase confers resistance to pyrimethamine in falciparum malaria. Proc Natl Acad Sci U S A 85: 9114–9118.

Peterson DS, Milhous WK, Wellems TE. 1990. Molecular basis of differential resistance to cycloguanil and pyrimethamine in *Plasmodium falciparum* malaria. Proc Natl Acad Sci U S A 87: 3018–3022.

Pickrell JK, Coop G, Novembre J, Kudaravalli S, Li JZ, et al. 2009. Signals of recent positive selection in a worldwide sample of human populations. Genome Res 19: 826–837.

Plowe CV. 2001. Folate antagonists and mechanisms of resistance. In: Rosenthal PJ, ed. *Antimalarial Chemotherapy*. Humana Press, Totowa, NJ, pp. 173–190.

Plowe CV, Cortese JF, Djimde A, Nwanyanwu OC, Watkins WM, et al. 1997. Mutations in *Plasmodium falciparum* dihydrofolate reductase and dihydropteroate synthase and epidemiologic patterns of pyrimethamine-sulfadoxine use and resistance. J Infect Dis 176: 1590–1596.

Polley SD, Chokejindachai W, Conway DJ. 2003. Allele frequency-based analyses robustly map sequence sites under balancing selection in a malaria vaccine candidate antigen. Genetics 165: 555–561.

Price R, Robinson G, Brockman A, Cowman A, Krishna S. 1997. Assessment of pfmdr 1 gene copy number by tandem competitive polymerase chain reaction. Mol Biochem Parasitol 85: 161–169.

Price RN, Uhlemann AC, Brockman A, McGready R, Ashley E, et al. 2004. Mefloquine resistance in *Plasmodium falciparum* and increased pfmdr1 gene copy number. Lancet 364: 438–447.

Ribacke U, Mok BW, Wirta V, Normark J, Lundeberg J, et al. 2007. Genome wide gene amplifications and deletions in *Plasmodium falciparum*. Mol Biochem Parasitol 155: 33–44.

Rich SM, Licht MC, Hudson RR, Ayala FJ. 1998. Malaria's eve: Evidence of a recent population bottleneck throughout the world populations of *Plasmodium falciparum*. Proc Natl Acad Sci U S A 95: 4425–4430.

Roper C, Pearce R, Bredenkamp B, Gumede J, Drakeley C, et al. 2003. Antifolate antimalarial resistance in Southeast Africa: A population-based analysis. Lancet 361: 1174–1181.

Roper C, Pearce R, Nair S, Sharp B, Nosten F, et al. 2004. Intercontinental spread of pyrimethamine-resistant malaria. Science 305: 1124.

Sa JM, Twu O, Hayton K, Reyes S, Fay MP, et al. 2009. Geographic patterns of *Plasmodium falciparum* drug resistance distinguished by differential responses to amodiaquine and chloroquine. Proc Natl Acad Sci U S A 106: 18883–18889.

Sabeti PC, Varilly P, Fry B, Lohmueller J, Hostetter E, et al. 2007. Genome-wide detection and characterization of positive selection in human populations. Nature 449: 913–918.

Sidhu AB, Uhlemann AC, Valderramos SG, Valderramos JC, Krishna S, et al. 2006. Decreasing pfmdr1 copy number in *Plasmodium falciparum* malaria heightens susceptibility to mefloquine, lumefantrine, halofantrine, quinine, and artemisinin. J Infect Dis 194: 528–535.

Singh B, Kim Sung L, Matusop A, Radhakrishnan A, Shamsul SS, et al. 2004. A large focus of naturally acquired *Plasmodium knowlesi* infections in human beings. Lancet 363: 1017–1024.

Smith JM, Haigh J. 1974. The hitch-hiking effect of a favourable gene. Genet Res 23: 23–35.

Stephan W. 2010. Genetic hitchhiking versus background selection: The controversy and its implications. Philos Trans R Soc Lond B Biol Sci 365: 1245–1253.

Su X-Z, Wellems TE. 1996. Toward a high-resolution *Plasmodium falciparum* linkage map: Polymorphic markers from hundreds of simple sequence repeats. Genomics 33: 430–444.

Su X-Z, Ferdig MT, Huang Y, Huynh CQ, Liu A, et al. 1999. A genetic map and recombination parameters of the human malaria parasite *Plasmodium falciparum*. Science 286: 1351–1353.

Su X-Z, Mu J, Joy DA. 2003. The "Malaria's eve" hypothesis and the debate concerning the origin of the human malaria parasite *Plasmodium falciparum*. Microbes Infect 5: 891–896.

Tibayrenc M, Kjellberg F, Ayala FJ. 1990. A clonal theory of parasitic protozoa: The population structures of *Entamoeba*, *Giardia*, *Leishmania*, *Naegleria*, *Plasmodium*, *Trichomonas*, and *Trypanosoma* and their medical and taxonomical consequences [published erratum appears in Proc Natl Acad Sci U S A 1990 Oct;87(20):8185]. Proc Natl Acad Sci U S A 87: 2414–2418.

Triglia T, Cowman AF. 1994. Primary structure and expression of the dihydropteroate synthetase gene of *Plasmodium falciparum*. Proc Natl Acad Sci U S A 91: 7149–7153.

Triglia T, Menting JG, Wilson C, Cowman AF. 1997. Mutations in dihydropteroate synthase are responsible for sulfone and sulfonamide resistance in *Plasmodium falciparum*. Proc Natl Acad Sci U S A 94: 13944–13949.

Van Tyne D, Park DJ, Schaffner SF, Neafsey DE, Angelino E, et al. 2011. Identification and functional validation of the novel antimalarial resistance locus PF10_0355 in *Plasmodium falciparum*. PLoS Genet 7: e1001383.

Vieira PP, Ferreira MU, Alecrim MG, Alecrim WD, da Silva LH, et al. 2004. pfcrt polymorphism and the spread of chloroquine resistance in *Plasmodium falciparum* populations across the Amazon Basin. J Infect Dis 190: 417–424.

Vinayak S, Mittra P, Sharma YD. 2006. Wide variation in microsatellite sequences within each Pfcrt mutant haplotype. Mol Biochem Parasitol 147: 101–108.

Vinayak S, Alam MT, Mixson-Hayden T, McCollum AM, Sem R, et al. 2010. Origin and evolution of sulfadoxine resistant *Plasmodium falciparum*. PLoS Pathog 6: e1000830.

Volkman SK, Barry AE, Lyons EJ, Nielsen KM, Thomas SM, et al. 2001. Recent origin of *Plasmodium falciparum* from a single progenitor. Science 293: 482–484.

Volkman SK, Sabeti PC, DeCaprio D, Neafsey DE, Schaffner SF, et al. 2007. A genome-wide map of diversity in *Plasmodium falciparum*. Nat Genet 39: 113–119.

Wang X, Mu J, Li G, Chen P, Guo X, et al. 2005. Decreased prevalence of the *Plasmodium falciparum* chloroquine resistance transporter 76T marker associated with cessation of chloroquine use against *P. falciparum* malaria in Hainan, People's Republic of China. Am J Trop Med Hyg 72: 410–414.

Wellems TE, Hayton K, Fairhurst RM. 2009. The impact of malaria parasitism: From corpuscles to communities. J Clin Invest 119: 2496–2505.

WHO. 2008. World Malaria Report 2008. http://www.who.int/malaria/wmr2008/malaria2008.pdf

Wootton JC, Feng X, Ferdig MT, Cooper RA, Mu J, et al. 2002. Genetic diversity and chloroquine selective sweeps in *Plasmodium falciparum*. Nature 418: 320–323.

Zakeri S, Motmaen SR, Afsharpad M, Djadid ND. 2009. Molecular characterization of antifolates resistance-associated genes, (dhfr and dhps) in *Plasmodium vivax* isolates from the Middle East. Malar J 8: 20.

CHAPTER 8

EVOLUTION OF DRUG RESISTANCE IN FUNGI

JESSICA A. HILL,* SAMANTHA J. HOOT,* THEODORE C. WHITE, and LEAH E. COWEN

Adaptation in response to selective pressure is a ubiquitous phenomenon. Antifungal drugs exert strong selection for resistance in fungi, such that natural selection favors cells that have acquired mutations or epigenetic changes that permit proliferation in the presence of the antifungal. The rate at which resistance evolves depends on population parameters like mutation rate and population size, and resistance is frequently found in fungal pathogens isolated in the clinic (Cowen, 2008).

Several factors are implicated in the emergence of drug resistance in fungi as a major clinical issue. These include the increase in the number of immunocompromised individuals due to the HIV/AIDS epidemic and advances in medical care that leave patients immunosuppressed. This population is susceptible to infection, particularly by opportunistic fungi such as *Candida albicans* (Pfaller and Diekema, 2007). However, life-threatening fungal infection is not limited to the immunocompromised, exemplified by the high incidence of *Cryptococcus gattii* infection in healthy hosts (Byrnes et al., 2010). Increased drug resistance is also selected for with the preemptive use of antifungals for high risk patients, and with agricultural deployment of antifungals (Mortensen et al., 2010; Verweij et al., 2009). For example, environmental isolates of azole-resistant *Aspergillus fumigatus* cluster with azole-resistant *A. fumigatus* clinical isolates, and away from susceptible isolates, suggesting that human hosts are being colonized with isolates that have environmentally acquired azole resistance (Snelders et al., 2009). Additionally, with widespread antifungal drug deployment, the prevalence of intrinsically resistant fungi is increasing (Pfaller and Diekema, 2004).

Resistance to antifungals evolves at a rate faster than new antifungals are being developed. The most significant challenge in developing new antifungals is avoiding host toxicity while effectively killing the pathogen. The close evolutionary relationship between fungal pathogens and their human hosts limits the number of fungal-specific drug targets available (Baldauf et al., 2000). The most commonly used classes of antifungals, the azoles and the

*These authors contributed equally to this work.

Evolution of Virulence in Eukaryotic Microbes, First Edition. Edited by L. David Sibley, Barbara J. Howlett, and Joseph Heitman.

echinocandins, target the biosynthesis of the fungal-specific cell membrane sterol ergosterol and the fungal cell wall, respectively (Cowen and Steinbach, 2008). These limitations in drug development underscore the importance of determining how drug resistance evolves in order to predict and prevent it.

Antifungal drug resistance is tremendously costly, in terms of both human mortality (McNeil et al., 2001) and financial burden (Dasbach et al., 2000; Morgan et al., 2005). By understanding how drug resistance evolves, combating fungal infections will be greatly facilitated.

THE MAJOR ANTIFUNGAL CLASSES AND THEIR MODE OF ACTION

The number of fungal-specific drug targets is limited due to conservation with the human host. The majority of antifungals that are clinically deployed target the fungal-specific sterol ergosterol, or its biosynthesis, or the fungal cell wall (Cowen, 2008; Shapiro et al., 2011). The utility of these antifungals is compromised by their often fungistatic (antiproliferative) nature. Reservoirs of cells tolerant to fungistatic drugs are poised for the evolution of resistance. This section will focus on antifungals currently available for the treatment of the most prevalent fungal infections.

Ergosterol Biosynthesis Inhibitors (EBIs)

The EBIs are a large class of antifungals that function by targeting enzymes involved in the biosynthesis of ergosterol, the bulk sterol of fungal membranes, from squalene. The genes in this pathway encode products that act in the following order: Erg1, Erg7, Erg11, Erg24, Erg25, Erg26, Erg27, Erg6, Erg2, Erg3, Erg5, and Erg4 (White et al., 1998).

The most widely deployed class of antifungals, the azoles, are EBIs. They inhibit fungal growth by binding to the P450 enzyme α-14-lanosterol demethylase (encoded by *ERG11*), thus blocking the production of ergosterol and leading to the accumulation of methylated sterol intermediates in the cell membrane, which alters membrane fluidity and inhibits growth (Cowen and Steinbach, 2008). The azoles include imidazoles and newer triazoles. Imidazoles suffer from toxicity and bioavailability issues and are only available in topical, over-the-counter formulations for vaginal or skin infections (Ostrosky-Zeichner et al., 2010). Triazoles must be prescribed and are available in oral or injection formats. The four triazoles currently available clinically are fluconazole, itraconazole, posaconazole, and voriconazole. Azoles are effective for treating diverse fungi, including species of *Candida*, *Cryptococcus*, and *Aspergillus*, as well as dimorphic and dermatophyte fungi (Shapiro et al., 2011). Posaconazole is also effective against zygomycete fungi (Alastruey-Izquierdo et al., 2009). Due to the fungistatic nature of azoles against yeasts like *Candida*, the development of clinical resistance has become a significant issue.

Other commonly used classes of EBIs are the allylamines and morpholines. The allylamines, including terbinafine, and the thiocarbamates, including tolnaftate, act by inhibiting Erg1; they are available in over-the-counter topical preparations. In contrast, morpholines such as fenpropimorph are commonly used in agricultural settings to target fungal pathogens of plants (Zocco et al., 2011). Morpholines target two enzymes in the ergosterol biosynthetic pathway, C-8 sterol isomerase (encoded by *ERG2*) and C-14 sterol reductase (encoded by *ERG24*).

Echinocandins

The newest class of antifungals is the echinocandins. These drugs disrupt cell wall biosynthesis by inhibiting the enzyme β-(1,3)-glucan synthase, the subunits of which are encoded by *FKS1* and *FKS2* (Cowen and Steinbach, 2008). The echinocandins are broadly effective against many fungi including *Candida* and *Aspergillus* species, but not against *Cryptococcus* or Zygomycetes. Unlike azoles, echinocandins are generally fungicidal against yeasts and resistance is not yet widespread, although numerous clinical isolates with elevated resistance have been identified (Shapiro et al., 2011). The echinocandins have little to no toxicity to the host, and because they are fungicidal, they are becoming the drug of choice for the treatment of invasive fungal infections.

Polyenes

The polyenes have been used in the clinic for the past 50 years. They bind to ergosterol in fungal cell membranes, resulting in the formation of pores that allow small molecules like ions to diffuse across the membrane, ultimately resulting in cell death (White et al., 1998). The most commonly deployed polyenes are amphotericin B and nystatin. Amphotericin B is effective against *Candida*, *Cryptococcus*, and *Aspergillus* infections and has been previously considered the gold standard for the treatment of invasive fungal infections; it is highly insoluble and is typically delivered in liposomal formulations to increase solubility when given intravenously. Therapeutic use of amphotericin B is complicated by significant liver toxicity and, as such, it is not typically used to treat chronic fungal infections or for prophylaxis. Nystatin is also nephrotoxic and highly insoluble; it is typically given only in lozenge format to treat oropharyngeal candidiasis. Due to their limited use and fungicidal nature, acquired resistance to polyenes is not widespread, although it has been documented in clinical cases.

Other Antifungals

There are several other antifungals that are clinically employed or currently in development. The classic antifungal 5-flucytosine inhibits fungal nucleic acid biosynthesis, but its effect is fungistatic and resistance develops readily (White et al., 1998). 5-Flucytosine is therefore typically only used in combination with other antifungals. Emerging antifungals include sordarins and nikkomycin Z. Sordarins are semisynthetic natural products that had shown antifungal activity as early as the 1970s (Ostrosky-Zeichner et al., 2010). They function as antifungals by inhibiting fungal elongation factor 2 during protein biosynthesis. The sordarin derivative FR290581 has demonstrated robust antifungal activity against *C. albicans* and is currently in development. Nikkomycin Z interferes with cell wall biosynthesis by competitively inhibiting chitin synthases. Phase I clinical trials of nikkomycin Z are currently under way (Ostrosky-Zeichner et al., 2010).

ADAPTIVE MECHANISMS

Acquired resistance to antifungals can occur through several mechanisms. The canonical resistance mechanisms are drug target alteration or overexpression, reduction of intracellular drug accumulation, and upregulation of stress response pathways (Fig. 8.1). These resistance mechanisms need not occur in isolation; in fact, clinical isolates frequently exhibit several mechanisms of resistance. This section focuses on genetically stable resistance mechanisms.

Alteration of Drug Target

The most direct mechanism to increase resistance to an antimicrobial is by mutation of the drug target that reduces drug binding. One example of this is alteration of *ERG11* encoding the azole target enzyme in *C. albicans*. In a well-characterized series

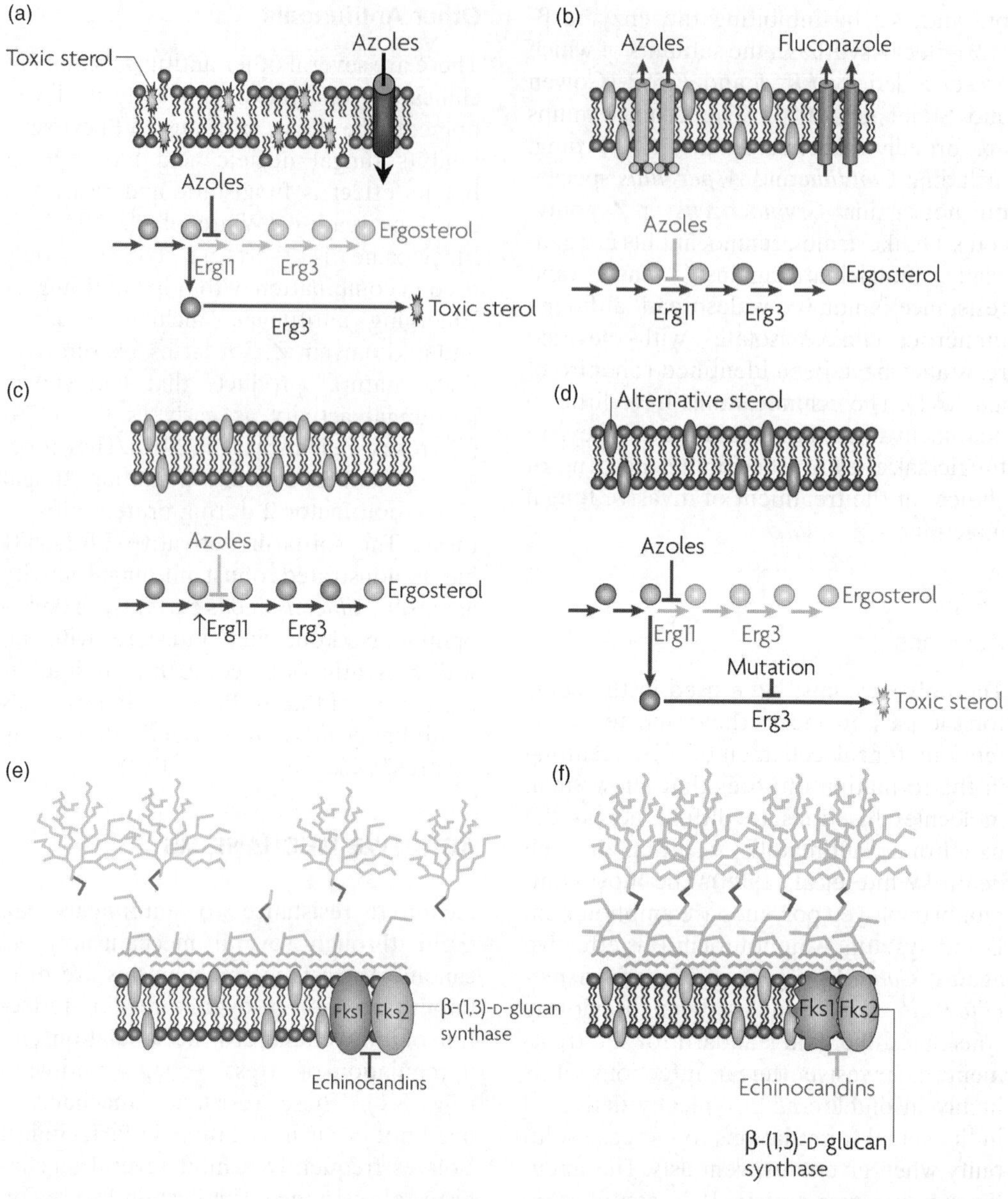

Figure 8.1 Mode of action of azoles and echinocandins and resistance mechanisms. (a) Azoles enter the cell by facilitated diffusion and inhibit Erg11, blocking the production of ergosterol and resulting in the accumulation of a toxic sterol intermediate, which results in cell membrane stress. Resistance to azoles can arise by (b) upregulation of two classes of efflux pumps that remove the drug from the cell; (c) mutation or overexpression of *ERG11*, which minimizes the impact of the drug on the target; or (d) loss-of-function mutation of *ERG3*, which blocks the accumulation of a toxic sterol that is otherwise produced when Erg11 is inhibited by azoles. (e) Echinocandins inhibit β-(1,3)-glucan synthase (the catalytic subunit is encoded by *FKS1* and *FKS2* in *S. cerevisiae*) and thus disrupt cell wall integrity. (f) Resistance to echinocandins can arise by mutations in *FKS1* or *FKS2* that minimize the impact of the drug on the target. Adapted by permission from Macmillan Publishers Ltd: Nature Publishing Group, Nature Reviews Microbiology (Cowen, 2008), © 2008. See color insert.

of matched susceptible and resistant isolates, an R467K amino acid substitution was identified that conferred azole resistance (White, 1997b). This mutation reduces the susceptibility of the enzyme to azoles, likely through decreasing the azole binding affinity (Lamb et al., 2000). Numerous other *ERG11* mutations have been identified within several mutational "hot spots" that result in amino acid substitutions that are in or adjacent to the active site. Elevated resistance is often associated with subsequent loss of heterozygosity that results in two mutated alleles of the gene.

Echinocandin resistance also occurs by target alteration. In echinocandin-resistant isolates of *C. albicans*, mutations in the target genes *FKS1* and *FKS2* (also referred to as *GSL1* and *GSC1*) have been found, leading to alteration of the glucan synthase (Munro, 2010). Similar mutations in *FKS1* and *FKS2* contributing to echinocandin resistance have been identified in *Candida glabrata* (Zimbeck et al., 2010). Elevated echinocandin resistance in *C. albicans* has been associated with loss of heterozygosity of *FKS1*, leaving two mutant alleles, as with *ERG11* and azole resistance (Niimi et al., 2010).

Overexpression of Drug Target

Increased expression of the drug target can also confer resistance. For example, overexpression of *ERG11* has been documented in azole-resistant *C. albicans* clinical isolates. *ERG11* overexpression can occur by gain-of-function mutations in the transcriptional activator encoded by *UPC2*, leading to overexpression of ergosterol biosynthesis genes including *ERG11* (Dunkel et al., 2008b; Hoot et al., 2011). Multiple studies identified mutations in the same C-terminal region of *UPC2* in azole-resistant *C. albicans* clinical isolates (Dunkel et al., 2008b; Heilmann et al., 2010; Hoot et al., 2011). In a clinical isolate containing *UPC2* A643T, loss of heterozygosity occurred, rendering the mutation homozygous and further elevating resistance. That these amino acid substitutions in diverse isolates arise in a restricted region of the protein suggests that this region has an essential regulatory function.

Recently, two *UPC2* homologs were identified in *C. glabrata* that, like in *C. albicans*, regulate the response to azoles and hypoxia (Nagi et al., 2011). However, *UPC2* homologs are limited to these and related species. In other ascomycetes and in basidiomycetes such as *Cryptococcus neoformans*, ergosterol biosynthesis is regulated at the transcriptional level by a sterol regulatory element-binding protein (SREBP)-like transcription factor Sre1 (Bien and Espenshade, 2010). This protein has functional homology to the mammalian cholesterol regulatory transcription factor SREBP. In *A. fumigatus*, a similar SREBP homolog called SrbA regulates sterol biosynthesis (Willger et al., 2008). These species also contain homologs of some of the other proteins that regulate the SREBP pathway, such as the proteases or chaperones involved in regulating SREBP proteolytic cleavage (Bien and Espenshade, 2010). In both *C. neoformans* and *A. fumigatus*, the SREBP-like proteins are required for sterol regulation in response to hypoxia and antifungal drugs and are also implicated in virulence in animal models.

Increased Drug Efflux

Reduction of intracellular drug concentration by active drug efflux can also lead to antifungal drug resistance, especially for the azoles, and is well studied in *C. albicans* (White et al., 1998). In *C. albicans*, azole efflux is mediated by overexpression of the ATP-dependent pumps Cdr1 and Cdr2, or the major facilitator pump Mdr1. Tac1 is the central transcriptional activator of *CDR1* and *CDR2* expression in *C. albicans*. *TAC1* gain-of-function mutations have been identified and are often homozygous in *CDR1* overexpressing clinical isolates (Coste et al., 2004, 2006). Another *C. albicans* transcription factor, Ndt80, has been implicated in *CDR1* transcriptional activation (Wang et al., 2006). Interestingly, *CDR2* overexpression can confer a slight decrease in susceptibility to echinocandins, however, only on solid agar media (Schuetzer-Muehlbauer et al., 2003; Silver et al., 2008). The major transcriptional activator of *MDR1* expression in *C. albicans* is Mrr1; *MRR1* mutations can lead to overexpression of *MDR1* (Dunkel et al., 2008a). However, unlike Upc2, neither Tac1 nor Mrr1 are activated in response to azoles.

Increased efflux of antifungals is a major mechanism of resistance in other fungal pathogens as well. In *C. neoformans*, overexpression of the efflux pumps encoded by *AFR1* and *MDR1* confers a decrease in azole susceptibility (Heilmann et al., 2010). *A. fumigatus* also encodes a major facilitator gene, *MDR1*, but expression of this gene in *Saccharomyces cerevisiae* does not impart resistance to azoles but only to the drug cilofungin, which is not used clinically (Heilmann et al., 2010). An ABC transporter encoding gene in *A. fumigatus*, *atrF*, was shown to be induced at the transcriptional level by azoles, and overexpression of this gene was correlated with azole resistance (Sionov et al., 2010).

Reduced Drug Import

It was previously assumed that azoles enter the cell through passive diffusion in most fungi (Prasad et al., 2006), and as such, alteration of drug importers did not seem a likely candidate resistance mechanism. Recent work has shown, however, that azoles are imported into fungal cells via an energy-independent facilitated diffusion mechanism (Mansfield et al., 2010). Some azole-resistant clinical isolates had reduced fluconazole import compared to the majority of clinical isolates, although at the moment, this remains a correlation. Facilitated diffusion of azoles suggests that import likely occurs through a specific protein, although this importer has yet to be identified. Identification of the azole importer and characterization of this importer in azole-resistant isolates will determine whether alteration of import can contribute to resistance.

Genomic Plasticity

Aneuploidy can facilitate increased resistance to antifungals. While aneuploidy itself is not known to confer resistance, the increased dosage of specific genes on an extra chromosome can contribute to resistance. In *C. albicans* clinical isolates, evidence of genomic plasticity contributing to resistance is well documented. A survey of 70 azole-resistant and susceptible isolates revealed that resistant isolates had a preponderance of aneuploidies, the most common of which involved chromosome 5, often with duplications of the left chromosome arm. Recombination events occurring at a common breakpoint in repetitive sequences flanking the centromere led to the production of an isochromosome with two left arms of chromosome 5, termed i(5L) (Selmecki et al., 2006). Functional analyses confirmed that in i(5L)-containing strains, azole resistance is due to duplication of both *ERG11* and *TAC1* alleles that reside on the left arm of chromosome 5 (Selmecki et al., 2008). *C. neoformans* also adapts to azoles by duplication of specific chromosomes, including chromosome 1, which harbors *ERG11* as well as the major azole transporter gene *AFR1* (Sionov et al., 2010).

Upregulation of Stress Response Pathways

Stress response pathways play a critical role in mediating both basal tolerance and resistance to diverse antifungals. For example, the molecular chaperone Hsp90 regulates crucial responses to both azoles and echinocandins in diverse fungi (Cowen, 2008; Cowen and Lindquist, 2005; Cowen et al., 2009; Singh et al., 2009). In *C. albicans*, Hsp90 regulates drug resistance by stabilizing key regulators of cellular stress responses, including the protein phosphatase calcineurin and the terminal mitogen-activated protein kinase in the Pkc1 cell wall integrity signaling pathway, Mkc1 (LaFayette et al., 2010; Singh et al., 2009). The cell wall integrity pathway regulates not only resistance to drugs that target the cell wall but also resistance to drugs that target the cell membrane (LaFayette et al., 2010). Drugs that inhibit Hsp90, calcineurin, or Pkc1 reduce azole and echinocandin resistance of isolates that evolved resistance in a human host (Cruz et al., 2002; LaFayette et al., 2010; Singh et al., 2009). While upregulation of stress response pathways has not yet been identified as a resistance mechanism, it is clear that these signaling pathways are key components of the cellular circuitry regulating drug resistance.

NATURAL VARIATION IN RESISTANCE TO ANTIFUNGAL DRUGS

Fungal pathogens are composed of diverse members of the fungal kingdom. Variation in response to antifungals exists between both distantly and closely related species, as well as between strains within a species and even between cells within a single strain. Pathogens can differ in their inherent ability to survive and reproduce despite the detrimental presence of a drug, often referred to as tolerance. Pathogens can also acquire specific mechanisms that reduce the inhibitory effect of the drug, referred to as resistance. High levels of basal tolerance can enable the evolution of resistance by maintaining viability in the context of strong selective pressure. The vast differences among species and strains in response to specific antifungals highlight the importance of selecting the appropriate antifungal to administer during clinical treatment.

This section will discuss the differences in resistance and tolerance between and within the major pathogenic fungal species. Differences in resistance within a species will be covered and special phenotypic states conferring elevated resistance will be examined.

Variation between Species

The major pathogenic fungi, *Candida* species, *A. fumigatus*, and *C. neoformans*, are very distantly related and vary dramatically in their ability to tolerate and acquire resistance to antifungals (Cowen and Steinbach, 2008; Shapiro et al., 2011). For example, fluconazole is one of the most frequently deployed antifungals in treating *Candida* infections, yet it is completely ineffective against *A. fumigatus* (Marichal and Vanden Bossche, 1995). While much remains to be understood about its intrinsic tolerance, *A. fumigatus* can be rendered susceptible to azoles by the deletion of the transcriptional regulator of sterol production, *srbA* (Willger et al., 2008). *A. fumigatus* is susceptible to other azoles, such as itraconazole and posaconazole (Pfaller et al., 2011b). Tolerance to azoles in *A. fumigatus* can be abrogated by the addition of an inhibitor of Hsp90 or calcineurin (Cowen et al., 2009). While *C. neoformans* is inherently susceptible to azoles, it shows remarkable tolerance to the echinocandins (Abruzzo et al., 1997; Krishnarao and Galgiani, 1997). How *C. neoformans* is able to tolerate echinocandin concentrations that are typically inhibitory is unclear, although it is known that *C. neoformans* β-(1,3)-glucan synthase (the echinocandin target) is inhibited by echinocandins (Thompson et al., 1999). Echinocandin tolerance in *C. neoformans* is abrogated by the inhibition of calcineurin (Del Poeta et al., 2000). The echinocandins are fungicidal against *C. albicans* and fungistatic against *A. fumigatus* (Denning, 2003).

Variation in tolerance also occurs between closely related fungal pathogens. *Candida krusei* and *C. glabrata*, for example, have lower intrinsic susceptibility to fluconazole (Blot et al., 2006). In *C. krusei*, this is mainly due to a low affinity for Erg11 (the target of azoles) (Lamping et al., 2009). Other azoles, however, can be effective in the treatment of *C. krusei* infection (Hoffman et al., 2000). A recent study assaying the epidemiological cutoff value of voriconazole for over 16,000 strains of *Candida* species found reduced susceptibility in *Candida guilliermondii*, *C. glabrata*, and *C. krusei* and greater susceptibility in *C. albicans*, *Candida tropicalis*, *Candida parapsilosis*, *Candida kefyr*, and *Candida lusitaniae* (Pfaller et al., 2011a). Unlike other *Candida* species, *C. lusitaniae* and *C. guilliermondii* are intrinsically resistant to amphotericin B. *Candida* species also vary in their sensitivity to echinocandins. *C. parapsilosis*, for example, shows greater intrinsic resistance to caspofungin and its incidence in the clinic increases with the application of caspofungin to treat candidemia, while incidence of the more susceptible *C. glabrata* and *C. tropicalis* decreases (Arendrup, 2010; Forrest et al., 2008).

Variation exists between *Aspergillus* species in their susceptibility to the polyene amphotericin B. Testing hundreds of clinical and environmental *Aspergillus* isolates revealed higher levels of amphotericin B tolerance in *Aspergillus flavus* and *Aspergillus terreus*, while *A. fumigatus* and *Aspergillus glaucus* were more susceptible (Araujo et al., 2007). It has been speculated that the intrinsic amphotericin B resistance of *A. terreus* is due to increased catalase production, which limits oxidative damage (Blum et al., 2008). Additionally, most *Aspergillus* species are susceptible to echinocandins, but there is reduced susceptibility in select *Aspergillus nidulans* isolates (Araujo et al., 2007).

The mechanisms of resistance an organism acquires are influenced by the genetic architecture of tolerance. While most fungal pathogens ultimately evolve antifungal resistance by several mechanisms, the most prevalent mechanisms may vary between species whose intrinsic resistance varies. For example, resistance of *A. fumigatus* to azoles is often due to alteration in the drug target *cyp51A*, the *A. fumigatus* equivalent of *ERG11* (Balashov et al., 2005). *C. neoformans* infrequently evolves resistance to azoles and does so largely by increased azole efflux or alteration in the drug target (Pfaller et al., 2005; Venkateswarlu et al., 1997), or via heteroresistance and aneuploidy for chromosome 1 (Sionov et al., 2010). Similarly, increased azole efflux and target alteration are common mechanisms of azole resistance in *C. albicans* (Shapiro et al., 2011). With respect to echinocandins, the most frequently encountered resistance mechanism in *C. albicans* clinical isolates is mutation in the drug target gene *FKS1* (Park et al., 2005). However, *A. fumigatus* echinocandin resistance is mostly facilitated through stress response pathways and rarely by *FKS1* mutations (Gardiner et al., 2005).

Mechanisms of resistance can vary between closely related species as well, like the model yeast *S. cerevisiae* and *C. glabrata*, which is more closely related to *S. cerevisiae* than to other *Candida* species. *S. cerevisiae* is able to acquire fluconazole resistance with the addition of exogenous ergosterol by importing the sterol (Kuo et al., 2010). However, *C. glabrata* lacks sterol import genes and typically becomes fluconazole resistant by increased efflux (Kuo et al., 2010).

Variation between Strains

Natural variation within a species can account for differing levels of tolerance among strains. For example, *erg3*Δ-mediated azole resistance in *S. cerevisiae* varies between strains W303 and BY4741/2 in its dependence on downstream effectors of calcineurin, Crz1, Hph1, and Hph2, in a manner that is dependent on nutrient signaling (Cowen et al., 2006; Robbins et al., 2010).

Clinically isolated *Candida* species have a broad diversity of genotypes. Four major clades of *C. albicans* strains have been identified by multilocus sequence typing (MLST) and DNA fingerprinting (MacCallum et al., 2009; Odds et al., 2007). Nearly all 5-flucytosine-resistant samples are members of the first clade, which has acquired a mutation that confers resistance (Tavanti et al., 2005). Likewise, one *Candida dubliniensis* clade has been associated with 5-flucytosine resistance (Al Mosaid et al., 2005). Members of the *C. albicans* fourth clade demonstrate increased tolerance to amphotericin B (Blignaut et al., 2005). However, no correlation between azole resistance and multilocus genotype was found in a set of *C. albicans* clinical isolates, indicating that the azole resistance was too transient to become associated with a neutral marker (Cowen et al., 1999). Similarly, five *C. glabrata* clades have been identified and the isolates with the highest levels of azole resistance were distributed across clades (Dodgson et al., 2003). Very little is known about the impact of population structure on resistance in the major fungal pathogens besides *Candida* species.

Variation within a Population

Variation in antifungal resistance occurs not only between and within fungal species but also within a population of cells. Unlike the predominantly stable variation in resistance found between strains and species, variation within a population of cells is often more transient.

Heteroresistance is an intriguing example of variation in azole resistance in *C. neoformans* that is commonly documented in clinical isolates and in both serotypes A and D, as well as in *C. gattii* (Sionov et al., 2009). Heteroresistance is defined as the emergence of a subset of azole-resistant cells in addition to azole-susceptible cells from a susceptible progenitor. This subpopulation gains azole resistance in a stepwise manner and loses this resistance with extended passage in azole-free media (Sionov et al., 2009). Recently, comparative genomic hybridization experiments determined that the azole-resistant cells are disomic, most commonly for chromosome 1 (Sionov et al., 2010). The duplicated chromosomes contain genes important for resistance, such as the azole target *ERG11* and azole transporter *AFR1*, and azole resistance was shown to correlate with the number of disomic chromosomes. Heteroresistance is also observed in strains of *C. albicans* (Marr et al., 2001).

There are several other distinct ways in which variation in resistance within a population has been described. Heterogeneous resistance is a broad term for a phenomenon in fungi and bacteria, which involves subpopulations of resistant cells that emerge at a low frequency (10^{-1} to 10^{-4}) from a larger susceptible population; colonies formed by the resistant cells again give rise to the same distribution with only a few resistant cells among a larger susceptible population (White et al., 2004). Another related phenomenon is high frequency azole resistance, in which some strains produce azole-resistant colonies at a frequency that is higher than a typical mutation rate (White et al., 2004). In contrast to heterogeneous resistance, the resistant colonies that emerge in the high frequency azole resistance context have stable resistance phenotypes, and it is thought that altered mitochondrial function might play a role in this phenomenon.

Biofilms are a clinically important example of heterogeneity in drug resistance. Fungal biofilms are complex surface-associated communities surrounded by an extracellular matrix (Blankenship and Mitchell, 2006). Besides having important virulence implications, biofilms demonstrate a level of antifungal resistance that is unprecedented in planktonic cells. *C. neoformans* biofilms, for example, have significantly higher resistance to amphotericin B and the echinocandin caspofungin than planktonic cells (Martinez and Casadevall, 2006). *C. albicans* biofilms are extremely resistant to azoles (d'Enfert, 2006). There are several mechanisms that contribute to the resistance of *C. albicans* biofilms, including increased levels of β-(1,3)-glucan (Nett et al., 2007), but the most important may be the advent of persister cells. Biofilms offer refuge from antifungal drugs and some host immune response, which facilitates the development of persister cells that are able to tolerate high drug concentrations. Persisters are a phenotypic variant that is only able to arise from biofilms (LaFleur et al., 2006). Persister cells are frequently found in bacterial biofilms; while bacterial persisters are quiescent, it is unclear whether fungal persisters are quiescent as well (LaFleur et al., 2006). Persisters are thought to be a major factor affecting the recalcitrance of fungal infection to treatments.

EVOLUTION OF DRUG RESISTANCE IN THE HUMAN HOST

One of the best ways to study how resistance to antifungals evolves in the human host is to examine fungal specimens that have been isolated from an infected individual over time. The most clinically relevant mechanisms of resistance can be elucidated and ultimately targeted to abrogate resistance. However, exclusively examining clinical isolates can be limiting due to the small sample sizes and the inability to control population parameters such as the number of generations, effective population size, and the genotype of the initial susceptible strain. This makes it challenging to determine what selective pressures directly contribute to the evolution of resistance. For these reasons, *in vivo* evolution of antifungal resistance studies is complemented by *in vitro* experimental evolution counterparts (Fig. 8.2), as discussed in a later section. This section focuses on studies of the evolution of drug resistance in the human host.

Population Dynamics of Fungi in the Human Host

When an antifungal-resistant fungal strain arises in an infected patient, it can be due to resistance mutations occurring in the initial strain or due to replacement with a resistant strain or species (reviewed in White et al., 1998). Replacement with a resistant species can occur due to intrinsic differences in drug susceptibility between related species and due to the fact that many fungi are environmentally ubiquitous. For example, it is well documented that during treatment, a patient with an initially azole-susceptible *C. albicans* strain may develop a resistant infection due to acquisition of a species that is intrinsically less susceptible to azoles, such as *C. glabrata*. It is also possible that a patient can acquire a resistant strain of the same species, and multiple studies have focused on strain replacement in the clinical setting; between 20–33% of cases of resistance that developed during treatment were due to strain replacement. To distinguish clonal relationships from strain replacement requires the susceptible and resistant isolates from the patient and involves typing methods such as restriction fragment length polymorphism (RFLP) analysis or MLST, which both take advantage of genotypic variability between strains of the same species. Clonal relationships among isolates would suggest that resistance emerged in the lineage due to the acquisition of resistance mutations.

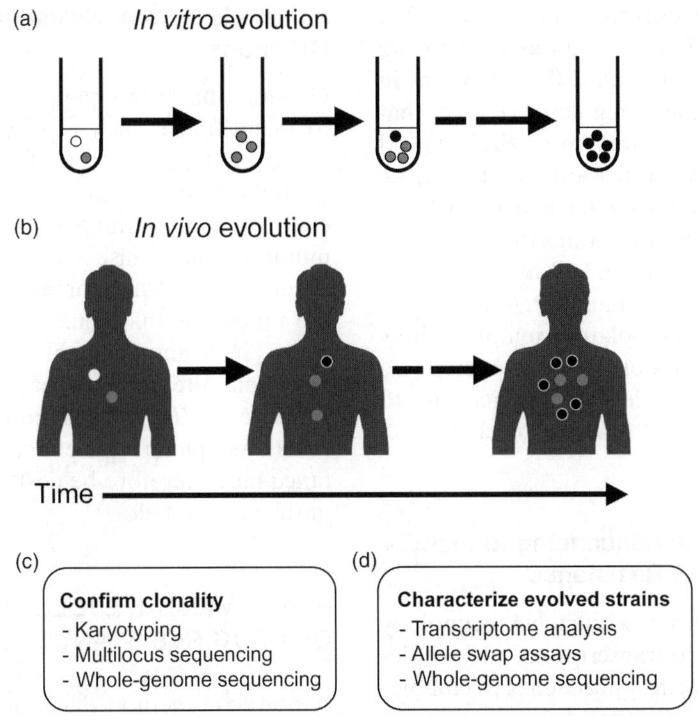

Figure 8.2 Studying the evolution of resistance to antifungals. The evolution of resistance to antifungals can be monitored in real time and the mechanisms of resistance can be dissected, in both *in vitro* and *in vivo* evolved strains. (a) Propagating a strain in the presence of antifungal selects for cells that have acquired a mutation that permits growth in the presence of the drug (gray cells). With continued propagation, additional mutations can accumulate (black cells) that confer resistance, abrogate the cost of resistance, or have no beneficial effect on their own ("hitchhiking" mutations). (b) Resistance evolves over time in the human host undergoing antifungal treatment. Serial isolates are taken from the host and their resistance to the administered antifungal can be assayed. Frequently, multiple mutations conferring resistance are found in clinical isolates (gray and black cells). The image of the human form was generously provided by Carolyn Thomas. (c) The clonality of evolved strains must be confirmed in order to determine how frequently beneficial mutations arose. Clinical isolates can represent several strains that the host was infected with. (d) Resistance mechanisms can be determined in evolved strains by several methods.

Progressive Drug Resistance

Matched sets of clinical isolates from fungal infections have informed the field a great deal as to how resistance mutations arise during the course of antifungal treatment. One of the best-characterized sets of isolates illustrates the concept of progressive drug resistance, or resistance that increases over time as sequential isolates accumulate multiple resistance mutations (White, 1997a). This set of 17 isolates of *C. albicans* was collected over 2 years from an HIV-infected patient who was receiving azole treatment for recurrent oropharyngeal candidiasis. The initial isolate in the series was highly susceptible to fluconazole treatment but by the end of the 2-year period had acquired a level of resistance approximately 200-fold higher than that of the original isolate. This set of isolates was extensively characterized to identify the molecular mechanisms of azole resistance (reviewed in Anderson, 2005; White et al., 1998). Initially, there was a selection of a strain variant, as detected by single-nucleotide polymorphisms. This was followed by overexpression of the efflux pump gene *MDR1*, mutations in the drug target *ERG11*, the A643V mutation in the transcriptional regulator *UPC2* that resulted in overexpression of *ERG11*, and ultimately, overexpression of the drug efflux pump genes *CDR1* and *CDR2* due to mutations in the transcription factor *TAC1* (Hoot et al., 2011; White, 1997a–c). The progressive accumulation of resistance mutations in these isolates exemplifies how resistance often develops in a stepwise fashion, with high level resistance due to the combination of multiple mechanisms.

Whole-Genome Sequencing to Identify Mechanisms of Resistance

Recent advances in sequencing technology allow for whole-transcriptome or whole-genome sequencing approaches to complement studies of antifungal drug resistance. For example, it would be feasible to perform whole-genome sequencing (DNA-seq) to identify acquired molecular mechanisms of drug resistance in clinical isolates, as well as to identify novel resistance mechanisms if matched sets of resistant and susceptible isolates are available. This approach is contingent on matched sets because of tremendous variation in genotypes among strains that are not clonal. Indeed, even among clonal isolates from the same patient, there is a baseline level of variation in genotype in the fungal population that would complicate determining functionally relevant mutations. In addition to genome DNA-seq, RNA-seq technology could enable the identification of genes that are specifically overexpressed in drug-resistant isolates and could thereby illuminate mechanisms of resistance. The most reproducible setting for genome-scale analyses would be *in vitro* experimental evolution of drug resistance in the laboratory, where the initial population and selection conditions can be carefully controlled and reproduced.

Fitness Effects of Resistance Mutations

Genetic alterations that confer resistance to an antifungal, such as point mutations in the drug target or specific aneuploidies, confer a selective advantage in the presence of the antifungal. However, these mutations may cause a fitness defect in the absence of a drug. For example, some of the mutations that confer azole resistance occur in transcriptional regulators and result in altered expression of multiple genes (e.g., *UPC2*, *TAC1*, and *MRR1* mutations). The pleiotropic effects of drug resistance may therefore be costly, as discussed in the section below.

EXPERIMENTAL EVOLUTION OF DRUG RESISTANCE

A powerful method that is gaining prominence in the study of antifungal resistance is experimental evolution. Experimental evolution generally refers to a population initiated from a single progenitor that is propagated for several generations, such that new traits can evolve and can be monitored in real time (Elena and Lenski, 2003). This broadly appealing approach is used to model theoretical evolutionary questions as well as more pragmatic issues in strain development (Fig. 8.2).

This section will describe the contribution experimental evolution has made to our understanding of how antifungal resistance occurs. Comparisons to more exhaustive studies in bacteria and the model yeast *S. cerevisiae* will also be discussed.

Benefits and Limitations of Experimental Evolution

In vitro evolution experiments can provide a simple model for how drug resistance is generated in the host. The researcher can manipulate parameters and thus more explicitly determine the factors that give rise to drug resistance. Another benefit is that these experiments are easily replicated so the sample size can be much greater than what can be expected from retrieving clinical isolates from patients. Experimental evolution can be performed with strains where the genome is sequenced, such that mutations conferring resistance can be more readily identified. However, experimental evolution cannot capture the complexity of growth in the human host. In the human body, organisms experience diverse challenges including variation in microenvironments, nutrient limitation, spatial structure, and competition with other unrelated pathogens. Despite these limitations, there has been much success in the *in vitro* evolution of lineages that exhibit mechanisms of resistance found in clinical isolates (Cowen et al., 2000; MacLean et al., 2010; Scully and Bidochka, 2005).

Experimental Evolution Methodology

Employing experimental evolution to study the evolution of drug resistance in fungi involves propagating cells in the presence of an antifungal, which provides the selective pressure for the evolution of resistance. Typically, cells are passaged by serial dilution of a stationary-phase culture to a fresh medium containing a concentration of antifungal that is inhibitory but not lethal (Dunham, 2010). It is crucial that the population size at each transfer is great enough that any beneficial mutations (conferring resistance) that have occurred are not lost to random drift (Elena and Lenski, 2003). This process is repeated until a sufficient number of generations have occurred for substantial changes to accrue in the population, conferring resistance. Occasionally, propagation occurs on a solid medium containing an antifungal. In this instance, cells are scraped or washed off the plate to ensure a sufficient population size and are then transferred to a new plate.

Experimental Evolution Studies in Fungi

There are three major factors important in the evolution of drug resistance: the mutations conferring resistance, the cost of maintaining resistance in terms of reduced reproduction (fitness cost), and the nature of mutations that can mitigate this cost of resistance (compensatory mutations). These parameters have important treatment implications: The deployment of antifungals selects for resistance; if resistance is costly, then discontinuing treatment may allow susceptible cells to outcompete their resistant counterparts, but if the cost of resistance is reduced by compensatory mutations, the resistant population will not be eliminated by selection.

C. albicans. The vast majority of experimental evolution studies with fungal pathogens have been performed with *C. albicans*. *C. albicans* replicate populations propagated for 330 generations in the presence of fluconazole evolved resistance to fluconazole, as well as cross resistance to ketoconazole and itraconazole (Cowen et al., 2000). The expression levels of four genes known to be important in azole resistance in clinical isolates (*ERG11*, *CDR1*, *CDR2*, and *MDR1*) varied among the populations, indicating the occurrence of different mechanisms of resistance. Subsequent studies monitoring the expression of 5000 open reading frames in a subset of these evolved lineages identified three different expression profiles that matched expression patterns found in clinical isolates (Cowen et al., 2002), further underscoring the utility of using *in vitro* evolution as a model for *in vivo* evolution of drug resistance. Aneuploidy also arose frequently in these populations and conferred resistance by increased gene dosage (Selmecki et al., 2006). For example, an isochromosome formed by the duplication of the left arm of chromosome 5 increased resistance through the elevated expression of two genes: *ERG11*, encoding the drug target, and *TAC1*, encoding a transcription factor that regulates expression of the ABC transporter genes *CDR1* and *CDR2* (Selmecki et al., 2008).

Studies examining the fitness effects of resistance mutations in human pathogenic fungi have been performed exclusively in *C. albicans* to date. Fitness was measured by competing evolved strains against their progenitor in both selective (with antifungal) and nonselective conditions and by measuring doubling times. Despite a predicted cost of resistance, the fitness of drug-resistant populations in a nonselective medium was just as often significantly better than the progenitor as it was worse (Cowen et al., 2001). In some populations, the chromosome 5 isochromosome provided a fitness advantage in both the presence and absence of fluconazole (Selmecki et al., 2009). This suggests that resistance need not be costly, and if it is, compensatory mutations occur readily enough to ameliorate this cost. Furthermore, any cost of resistance in these lineages was abrogated with additional evolution, indicating compensatory mutations occurred. Mutations reducing the cost of resistance remain enigmatic in *C. albicans*. The nature of compensatory mutations is best understood in bacteria, where compensatory mutations occur in several distinct ways, such as amplification of a related pathway when resistance has been caused by loss of function (Andersson and Hughes, 2010; MacLean et al., 2010).

A study was performed evolving strains of *C. albicans* in test tubes as well as in a mouse model host (Forche et al., 2009). Passage in the mouse host led to different population dynamics from *in vitro* evolution, including slower growth rates as well as increased genotypic and phenotypic variations in the mouse-evolved strains. Future studies could focus on *in vitro* conditions that may be more relevant to evolution in the human host. For example, bacterial evolution experiments performed with spatial structure revealed that structure inhibits the spread of a beneficial mutation relative to homogenized cultures (Habets et al., 2007).

S. cerevisiae. *S. cerevisiae* shares many aspects of its resistance to azoles with *C. albicans* and provides a tractable system with which to study resistance mechanisms. *S. cerevisiae* strains that were exposed to a single high dose of fluconazole evolved a different class of resistance mechanism from those that had been propagated for 400 generations in the presence of stepwise increased concentrations of azole. Specifically, the former method reproducibly selects for loss-of-function mutations in *ERG3*, a mode of resistance dependent on the stress response protein calcineurin and its chaperone Hsp90, whereas the latter predominantly selects for overexpression of efflux pumps mediated by mutations in the transcription factors *PDR1* and *PDR3* (Anderson et al., 2003). These mechanisms of resistance favored by different modes of selection are strongly antagonist, such that hybrids with both resistance mechanisms have impaired growth at certain concentrations of fluconazole compared to either parent (Anderson et al., 2006).

The tractability of *S. cerevisiae* enables experiments to determine the impact of genetic perturbations or ploidy on the evolution of drug resistance. In a study that examined the impact of gene deletion, *pdr16Δ* mutants repeatedly went extinct during selection with fluconazole, while all other deletion mutants evolved resistance (Anderson et al., 2009). This suggests that *PDR16*, a gene involved in lipid metabolism, is important for the evolution of fluconazole resistance. A separate study using haploid and diploid strains showed that different ploidies favored different resistance mechanisms (Anderson et al., 2004). For example, mutations in *PDR1* and *PDR3* were favored in diploids due to the larger mutational target (with twice the number of genes as haploids), whereas recessive, loss-of-function mutations were more frequently found in haploid lineages. This indicates that ploidy influences the mutations that are favored during the evolution of drug resistance.

Other Fungi. A more limited number of experimental evolution studies have addressed drug resistance in other fungal pathogens of humans. In one study, *A. fumigatus* was propagated in the presence of itraconazole (da Silva Ferreira et al., 2004). Resistance readily arose and was variable between replicate populations, indicating different mutations conferring resistance. Resistance mechanisms included overexpression of efflux pumps and mutations in the drug targets, *cyp51A* and *cyp51B*. Similar to clinical isolates, multiple resistance mechanisms accrued within the same population. No experimental evolution studies have addressed the evolution of drug resistance in *Cryptococcus* spp., although the rate of loss of sex has been estimated by experimental evolution in *C. neoformans* (Xu, 2002). *A. nidulans* is an emerging model organism for experimental studies of beneficial mutations due to the simplicity of detecting mutations that have arisen by changes in the spatial structure of its growth (Schoustra et al., 2009).

THWARTING THE EVOLUTION OF DRUG RESISTANCE

Combating the evolution of resistance to antifungals is imperative, given the high mortality rates and economic burden imposed by fungal drug resistance. There are three main strategies used to counter the development of antifungal resistance: combination therapies with currently deployed antifungals, identifying and targeting cellular regulators that potentiate resistance, and the discovery of novel fungal-specific drug targets. This section will address each of these strategies.

Combination Therapies

Combination therapies have long been known to be more effective than single drug therapy at limiting the development of resistance. Typically, combination therapy employs two antifungals that target more than one pathway (White et al., 1998). The synergistic effect of appropriately selected antifungals in combination will more substantially reduce the fungal population size, therefore reducing the genetic space over which mutations can arise. Furthermore, in order for resistance to occur, mutations would need to arise that permit fungal growth in the presence of both drugs.

One example of combination therapy is the use of 5-flucytosine and amphotericin B to treat fungal infections. This drug combination is commonly used in the treatment of cryptococcal meningitis (Bennett et al., 1979). Resistance to 5-flucytosine develops readily, but its effectiveness can be prolonged when used in combination with amphotericin B (White et al., 1998). This combination therapy is complicated, however, due to the nephrotoxicity of amphotericin B. Notably, 5-flucytosine can also be used in combination with azole therapy. Other combinations that show promise *in vitro* and in animal models are Hsp90 inhibitors with azoles or echinocandins in the treatment of *C. albicans* or *A. fumigatus* infections (Cowen et al., 2009). As with Hsp90 inhibitors, calcineurin inhibitors enhance antifungal efficacy and block the emergence of drug resistance, motivating the search for nonimmunosuppressive agents that retain antifungal activity (Steinbach et al., 2007).

Development of New Antifungals

There are relatively few drugs available to treat fungal infections and resistance to these antifungals has had a significant clinical impact, such that the development of new antifungals will be essential to keep the upper hand in the battle against these pervasive pathogens. Generally, new antifungals have an average of 10 years after the start of clinical use before resistance becomes common. This necessitates the perpetual identification of new drug targets and small molecules to inhibit those targets.

The development of new antifungals is complicated by the evolutionary relatedness of these eukaryotic pathogens to humans. To avoid host toxicity, differences between fungi and humans must be exploited to develop new drugs; however, new targets remain somewhat elusive. One set of candidates is the fungal-specific zinc cluster transcription factors, many of which regulate aspects of fungal drug resistance (such as Upc2, Tac1, and Mrr1). Efforts to identify drugs that inhibit these proteins have not yet been successful. The search for new fungal-specific drug targets will be an ongoing area of research for the control of fungal infections.

Targeting Regulators of Evolution of Drug Resistance

Traditional antifungals target fungal pathways required for proliferation or viability. Recent work has focused on inhibiting cellular factors that are required for the evolution of resistance but do not directly confer resistance. For example, the molecular chaperone Hsp90 has been shown to be required for resistance or tolerance to echinocandins and azoles in *Candida* and *Aspergillus* species, as previously described in this chapter. Hsp90 inhibitors are in clinical development as anticancer agents and have recently been tested to assess their utility in combination therapy with azoles or echinocandins in the wax moth *Galleria mellonella* and mouse models of infection (Cowen et al., 2009). While effective at enhancing the efficacy of antifungals in the wax moth larval model of infection, toxicity to the mouse in the context of an acute fungal infection compromised the therapeutic utility. Fungal-specific Hsp90 inhibitors would therefore be an extremely valuable tool to prevent the development of resistance *in vivo* and extend the clinical usefulness of antifungal drugs. Hsp90 holds great promise as a therapeutic target for extending the clinical efficacy of existing drugs to which antifungal resistance has already been documented.

An alternative strategy is to target a process required for fungi to undergo aneuploidy or loss of heterozygosity. Numerous examples of drug resistance mutations in clinical isolates are followed by loss of heterozygosity events in diploid *C. albicans*. Additionally, whole or segmental aneuploidies that amplify resistance genes also contribute to the evolution of antifungal resistance. Molecules that disrupt the ability of fungi to undergo aneuploidy or loss of heterozygosity could extend the life span of existing antifungals, providing that such molecules have low toxicity to mammalian cells (Selmecki et al., 2006). As we learn more about the mechanisms that contribute to chromosome rearrangements during the evolution of drug resistance, we will better be able to target regulators of this process.

HOST MODEL SYSTEMS FOR DRUG SCREENING

In an experimental context, the efficacy of an antifungal drug or an antifungal drug combination is primarily assessed by *in vitro* methods such as minimum inhibitory concentration assays. These experiments provide a quantitative measure of the impact of drugs on the pathogen's ability to survive and proliferate, but they fail in two respects: They do not recapitulate the complexity of the host, and they neglect an important aspect of drug resistance, the host response. *In vitro* experiments are therefore often validated in an experimental host model system. Aspects of host response to fungal pathogens are conserved between humans and several model hosts. The researcher can assess the impact of immune response on drug tolerance and can determine if host toxicity occurs at drug concentrations that inhibit the fungus. Drug screening in host model systems can provide more biologically relevant assays.

While the majority of host model systems are used for assaying virulence of fungal pathogens, this section will focus on the use of model systems in validating fungicidal activity and screening for novel drugs or drug combinations. The focus will be on three most commonly used model host systems: wax moth larvae, nematodes, and mice, which all offer potential not only for drug screening but also for monitoring the evolution of drug resistance.

Wax Moth Model

The greater wax moth, *G. mellonella*, is a relatively new host model. Its advantages include its ability to thrive between 25 and 37°C, the ease of infection administration, the ability to readily distinguish between live (off-white, active) and killed (dark, nonmotile) larvae, as well as the presence of an innate immune system (Fuchs et al., 2010). In fact, a correlation has been found between pathogenicity in mouse and *G. mellonella* models of infectivity (Brennan et al., 2002). Most commonly, the larvae are infected by injection in a rear proleg of the hemocoel (Fuchs et al., 2010), although topical application and ingestion have been noted as alternate methods of infection (Fedhila et al., 2010; Scully and Bidochka, 2005). Injection has the advantage of controlling the amount of inoculum (Cotter et al., 2000).

G. mellonella larvae have been used as a model system to examine antifungal drug efficacy and virulence of *Candida* spp., *C. neoformans*, and *Aspergillus* spp. (Cotter et al., 2000; Mylonakis et al., 2005). In a study of antifungals used to treat *C. neoformans*, assays in *G. mellonella* showed that monotherapies of fluconazole, flucytosine, and amphotericin B, as well as combination therapy of amphotericin B and flucytosine, enhanced larval survival (Mylonakis et al., 2005). *G. mellonella* larvae have been used to assess the efficacy of treating *C. albicans* and *A. fumigatus* infections with a combination of an Hsp90 inhibitor and an antifungal. *G. mellonella* larvae infected with a lethal *C. albicans* fungal load were rescued by combination therapy with an Hsp90 inhibitor and fluconazole (Cowen et al., 2009). Inhibiting Hsp90 in combination with caspofungin reduced killing of the wax moth larvae by *A. fumigatus* (Cowen et al., 2009). A recent study has shown that preemptive injection of caspofungin into the larval gut primes the organism's immune response such that general infections caused by *C. albicans* as well as *Staphylococcus aureus* are minimized (Kelly and Kavanagh, 2011).

Nematode Model

Like other invertebrate models of disease, the nematode *Caenorhabditis elegans* offers advantages like cost-effectiveness and conservation of components of the immune system (Sifri et al., 2005). In particular, *C. elegans* is a burgeoning system for the discovery of new antifungals or antifungal combinations because it is highly amenable to high throughput screening (Okoli et al., 2009). Another advantage specific to *C. elegans* is its genetic tractability as a host. Typically, screens are performed using a *glp-4;sek-1* double mutant; the *glp-4* mutation renders the nematodes unable to reproduce at restrictive temperatures and the *sek-1* mutation renders them more susceptible to fungal pathogens (Beanan and Strome, 1992; Kim et al., 2002). Finally, infection can be easily monitored in the translucent nematode by using fluorescently labeled pathogens (Breger et al., 2007).

While *C. elegans* has been used as a model host to study the virulence of *C. neoformans* and *Candida* spp., to date, antifungal assessment studies have been performed using *Candida* spp. Infection of *C. elegans* with *C. albicans* involves the transfer of worms at the L4 developmental stage from a plate of *E. coli* (their standard food source) to one of the pathogen; infection can also be achieved in a liquid environment. Upon ingestion, *C. albicans* is able to cause infection in the nematode intestine, ultimately forming hyphae that break through the cuticle (Breger et al., 2007). Screening of 1266 compounds for anti-*Candida* activity in *C. elegans* identified 72 compounds that increased the duration of nematode survival (Breger et al., 2007). A similar screen of an additional 3228 compounds identified 19 distinct antifungal compounds, validated known antifungals, and identified compounds toxic to *C. elegans* (Okoli et al., 2009).

Mouse Model

The standard model for assessing antifungal efficacy against fungal infection is the mouse model system. This is the most commonly used vertebrate system and offers the greatest level of conservation with the human host out of the models mentioned above. Mice, for example, have a homolog of human Dectin-1, the myeloid receptor of beta-glucans that has been shown to be necessary for the recognition of fungi like *C. albicans* and *A. fumigatus* (Steele et al., 2005; Taylor et al., 2007). Antifungal efficacy that has been assayed in *C. elegans* or *G. mellonella* is typically verified in the mouse, and discrepancies can occur. For example, *C. albicans* fungal burden was reduced with the combination of Hsp90 inhibitor and azole in the *G. mellonella* host, yet the murine host was not able to tolerate pharmacological inhibition of Hsp90 in the context of an acute fungal infection (Cowen et al., 2009). This underscores the importance of validating findings in the vertebrate model. Some of the challenges of working with a mouse model include expense and ethical concerns.

Most potential antifungals are assayed for effectiveness in the murine host. For *C. albicans*, the pathogen is typically administered by tail vein injection, which effectively mimics systemic fungal infection (MacCallum and Odds, 2005). Virulence is assayed by determining the fungal burden of certain organs, most commonly the kidneys, where fungal infection is established (Papadimitriou and Ashman, 1986). For *A. fumigatus*, a more relevant mode of pathogen delivery is inhalation, and fungal burden can be monitored in the lung (Steinbach et al., 2004). Subsequent to infection, antifungals are delivered, typically by intraperitoneal injection. Mouse models have been used for assays of antifungal efficacy against many fungal species, including determining the antifungal efficacy of echinocandins and the impact of mutations associated with echinocandin resistance in *Candida* spp. (Park et al., 2005), identifying new antifungals against *C. albicans* (Epp et al., 2010), and assaying a new-generation azole, posaconazole, for antifungal activity against *Aspergillus* spp., *C. neoformans*, and *C. albicans* (Groll and Walsh, 2006). Other examples of vertebrate hosts include guinea pigs (MacCallum et al., 2005), rats (Marchetti et al., 2000), and rabbits (Chemlal et al., 1996). Rat and mouse venous catheter models have proved of particular relevance for monitoring drug efficacies against biofilms, which are complex communities surrounded by matrix that form on specific surfaces such as catheters and are highly resistant to antifungal drugs (Andes et al., 2004; Lazzell et al., 2009).

CONCLUSION

The evolution of drug resistance in fungal pathogens remains a critical concern for human health. Many paradigms that have emerged for the evolution of drug resistance in fungal pathogens of humans also pertain to the evolution of fungicide resistance in fungal pathogens of plants, which have a profound impact on agriculture. That resistance has evolved to a diverse repertoire of antifungals provides a poignant illustration of the basic principle that resistance will ultimately evolve to any new antifungal or a combination of antifungals used for treatment. Viewing antifungal resistance as an evolutionary problem may help us better understand how to combat it.

REFERENCES

Abruzzo GK, Flattery AM, Gill CJ, Kong L, Smith JG, Pikounis VB, Balkovec JM, Bouffard AF, Dropinski JF, Rosen H, Kropp H, Bartizal K. 1997. Evaluation of the echinocandin antifungal MK-0991 (L-743,872): Efficacies in mouse models of disseminated aspergillosis, candidiasis, and cryptococcosis. Antimicrob Agents Chemother 41: 2333–2338.

Alastruey-Izquierdo A, Castelli MV, Cuesta I, Zaragoza O, Monzon A, Mellado E, Rodriguez-Tudela JL. 2009. *In vitro* activity of antifungals against Zygomycetes. Clin Microbiol Infect 15(Suppl. 5): 71–76.

Al Mosaid A, Sullivan DJ, Polacheck I, Shaheen FA, Soliman O, Al Hedaithy S, Al Thawad S, Kabadaya M, Coleman DC. 2005. Novel 5-flucytosine-resistant clade of *Candida dubliniensis* from Saudi Arabia and Egypt identified by Cd25 fingerprinting. J Clin Microbiol 43: 4026–4036.

Anderson JB. 2005. Evolution of antifungal-drug resistance: Mechanisms and pathogen fitness. Nat Rev Microbiol 3: 547–556.

Anderson JB, Sirjusingh C, Parsons AB, Boone C, Wickens C, Cowen LE, Kohn LM. 2003. Mode of selection and experimental evolution of antifungal drug resistance in *Saccharomyces cerevisiae*. Genetics 163: 1287–1298.

Anderson JB, Sirjusingh C, Ricker N. 2004. Haploidy, diploidy and evolution of antifungal drug resistance in *Saccharomyces cerevisiae*. Genetics 168: 1915–1923.

Anderson JB, Ricker N, Sirjusingh C. 2006. Antagonism between two mechanisms of antifungal drug resistance. Eukaryot Cell 5: 1243–1251.

Anderson JB, Sirjusingh C, Syed N, Lafayette S. 2009. Gene expression and evolution of antifungal drug resistance. Antimicrob Agents Chemother 53: 1931–1936.

Andersson DI, Hughes D. 2010. Antibiotic resistance and its cost: Is it possible to reverse resistance? Nat Rev Microbiol 8: 260–271.

Andes D, Nett J, Oschel P, Albrecht R, Marchillo K, Pitula A. 2004. Development and characterization of an in vivo central venous catheter *Candida albicans* biofilm model. Infect Immun 72: 6023–6031.

Araujo R, Pina-Vaz C, Rodrigues AG. 2007. Susceptibility of environmental versus clinical strains of pathogenic *Aspergillus*. Int J Antimicrob Agents 29: 108–111.

Arendrup MC. 2010. Epidemiology of invasive candidiasis. Curr Opin Crit Care 16: 445–452.

Balashov SV, Gardiner R, Park S, Perlin DS. 2005. Rapid, high-throughput, multiplex, real-time PCR for identification of mutations in the *cyp51A* gene of *Aspergillus fumigatus* that confer resistance to itraconazole. J Clin Microbiol 43: 214–222.

Baldauf SL, Roger AJ, Wenk-Siefert I, Doolittle WF. 2000. A kingdom-level phylogeny of eukaryotes based on combined protein data. Science 290: 972–977.

Beanan MJ, Strome S. 1992. Characterization of a germ-line proliferation mutation in *C. elegans*. Development 116: 755–766.

Bennett JE, Dismukes WE, Duma RJ, Medoff G, Sande MA, Gallis H, Leonard J, Fields BT, Bradshaw M, Haywood H, McGee ZA, Cate TR, Cobbs CG, Warner JF, Alling DW. 1979. A comparison of amphotericin B alone and combined with flucytosine in the treatment of cryptoccal meningitis. N Engl J Med 301: 126–131.

Bien CM, Espenshade PJ. 2010. Sterol regulatory element binding proteins in fungi: Hypoxic transcription factors linked to pathogenesis. Eukaryot Cell 9: 352–359.

Blankenship JR, Mitchell AP. 2006. How to build a biofilm: A fungal perspective. Curr Opin Microbiol 9: 588–594.

Blignaut E, Molepo J, Pujol C, Soll DR, Pfaller MA. 2005. Clade-related amphotericin B resistance among South African *Candida albicans* isolates. Diagn Microbiol Infect Dis 53: 29–31.

Blot S, Janssens R, Claeys G, Hoste E, Buyle F, De Waele JJ, Peleman R, Vogelaers D, Vandewoude K. 2006. Effect of fluconazole consumption on long-term trends in candidal ecology. J Antimicrob Chemother 58: 474–477.

Blum G, Perkhofer S, Haas H, Schrettl M, Wurzner R, Dierich MP, Lass-Florl C. 2008. Potential basis for amphotericin B resistance in *Aspergillus terreus*. Antimicrob Agents Chemother 52: 1553–1555.

Breger J, Fuchs BB, Aperis G, Moy TI, Ausubel FM, Mylonakis E. 2007. Antifungal chemical compounds identified using a *C. elegans* pathogenicity assay. PLoS Pathog 3: e18.

Brennan M, Thomas DY, Whiteway M, Kavanagh K. 2002. Correlation between virulence of *Candida albicans* mutants in mice and *Galleria mellonella* larvae. FEMS Immunol Med Microbiol 34: 153–157.

Byrnes EJ, III, Li W, Lewit Y, Ma H, Voelz K, Ren P, Carter DA, Chaturvedi V, Bildfell RJ, May RC, Heitman J. 2010. Emergence and pathogenicity of highly virulent *Cryptococcus gattii* genotypes in the northwest United States. PLoS Pathog 6: e1000850.

Chemlal K, Saint-Julien L, Joly V, Farinotti R, Seta N, Yeni P, Carbon C. 1996. Comparison of fluconazole and amphotericin B for treatment of experimental *Candida albicans* endocarditis in rabbits. Antimicrob Agents Chemother 40: 263–266.

Coste A, Turner V, Ischer F, Morschhauser J, Forche A, Selmecki A, Berman J, Bille J, Sanglard D. 2006. A mutation in Tac1p, a transcription factor regulating *CDR1* and *CDR2*, is coupled with loss of heterozygosity at chromosome 5 to mediate antifungal resistance in *Candida albicans*. Genetics 172: 2139–2156.

Coste AT, Karababa M, Ischer F, Bille J, Sanglard D. 2004. TAC1, transcriptional activator of CDR genes, is a new transcription factor involved in the regulation of *Candida albicans* ABC transporters *CDR1* and *CDR2*. Eukaryot Cell 3: 1639–1652.

Cotter G, Doyle S, Kavanagh K. 2000. Development of an insect model for the in vivo pathogenicity testing of yeasts. FEMS Immunol Med Microbiol 27: 163–169.

Cowen LE. 2008. The evolution of fungal drug resistance: Modulating the trajectory from genotype to phenotype. Nat Rev Microbiol 6: 187–198.

Cowen LE, Lindquist S. 2005. Hsp90 potentiates the rapid evolution of new traits: Drug resistance in diverse fungi. Science 309: 2185–2189.

Cowen LE, Steinbach WJ. 2008. Stress, drugs, and evolution: The role of cellular signaling in fungal drug resistance. Eukaryot Cell 7: 747–764.

Cowen LE, Sirjusingh C, Summerbell RC, Walmsley S, Richardson S, Kohn LM, Anderson JB. 1999. Multilocus genotypes and DNA fingerprints do not predict variation in azole resistance among clinical isolates of *Candida albicans*. Antimicrob Agents Chemother 43: 2930–2938.

Cowen LE, Sanglard D, Calabrese D, Sirjusingh C, Anderson JB, Kohn LM. 2000. Evolution of drug resistance in experimental populations of *Candida albicans*. J Bacteriol 182: 1515–1522.

Cowen LE, Kohn LM, Anderson JB. 2001. Divergence in fitness and evolution of drug resistance in experimental populations of *Candida albicans*. J Bacteriol 183: 2971–2978.

Cowen LE, Nantel A, Whiteway MS, Thomas DY, Tessier DC, Kohn LM, Anderson JB. 2002. Population genomics of drug resistance in *Candida albicans*. Proc Natl Acad Sci U S A 99: 9284–9289.

Cowen LE, Carpenter AE, Matangkasombut O, Fink GR, Lindquist S. 2006. Genetic architecture of Hsp90-dependent drug resistance. Eukaryot Cell 5: 2184–2188.

Cowen LE, Singh SD, Kohler JR, Collins C, Zaas AK, Schell WA, Aziz H, Mylonakis E, Perfect JR, Whitesell L, Lindquist S. 2009. Harnessing Hsp90 function as a powerful, broadly effective therapeutic strategy for fungal infectious disease. Proc Natl Acad Sci U S A 106: 2818–2823.

Cruz MC, Goldstein AL, Blankenship JR, Del Poeta M, Davis D, Cardenas ME, Perfect JR, McCusker JH, Heitman J. 2002. Calcineurin is essential for survival during membrane stress in *Candida albicans*. EMBO J 21: 546–559.

Dasbach EJ, Davies GM, Teutsch SM. 2000. Burden of aspergillosis-related hospitalizations in the United States. Clin Infect Dis 31: 1524–1528.

Da Silva Ferreira ME, Capellaro JL, Dos Reis Marques E, Malavazi I, Perlin D, Park S, Anderson JB, Colombo AL, Arthington-Skaggs BA, Goldman MH, Goldman GH. 2004. *In vitro* evolution of itraconazole resistance in *Aspergillus fumigatus* involves multiple mechanisms of resistance. Antimicrob Agents Chemother 48: 4405–4413.

Del Poeta M, Cruz MC, Cardenas ME, Perfect JR, Heitman J. 2000. Synergistic antifungal activities of bafilomycin A(1), fluconazole, and the pneumocandin MK-0991/caspofungin acetate (L-743,873) with calcineurin inhibitors FK506 and L-685,818 against *Cryptococcus neoformans*. Antimicrob Agents Chemother 44: 739–746.

d'Enfert C. 2006. Biofilms and their role in the resistance of pathogenic *Candida* to antifungal agents. Curr Drug Targets 7: 465–470.

Denning DW. 2003. Echinocandin antifungal drugs. Lancet 362: 1142–1151.

Dodgson AR, Pujol C, Denning DW, Soll DR, Fox AJ. 2003. Multilocus sequence typing of *Candida glabrata* reveals geographically enriched clades. J Clin Microbiol 41: 5709–5717.

Dunham MJ. 2010. Experimental evolution in yeast: A practical guide. Methods Enzymol 470: 487–507.

Dunkel N, Blass J, Rogers PD, Morschhauser J. 2008a. Mutations in the multi-drug resistance regulator *MRR1*, followed by loss of heterozygosity, are the main cause of *MDR1* overexpression in fluconazole-resistant *Candida albicans* strains. Mol Microbiol 69: 827–840.

Dunkel N, Liu TT, Barker KS, Homayouni R, Morschhauser J, Rogers PD. 2008b. A gain-of-function mutation in the transcription factor Upc2p causes upregulation of ergosterol biosynthesis genes and increased fluconazole resistance in a clinical *Candida albicans* isolate. Eukaryot Cell 7: 1180–1190.

Elena SF, Lenski RE. 2003. Evolution experiments with microorganisms: The dynamics and genetic bases of adaptation. Nat Rev Genet 4: 457–469.

Epp E, Vanier G, Harcus D, Lee AY, Jansen G, Hallett M, Sheppard DC, Thomas DY, Munro CA, Mullick A, Whiteway M. 2010. Reverse genetics in *Candida albicans* predicts ARF cycling is essential for drug resistance and virulence. PLoS Pathog 6: e1000753.

Fedhila S, Buisson C, Dussurget O, Serror P, Glomski IJ, Liehl P, Lereclus D, Nielsen-LeRoux C. 2010. Comparative analysis of the virulence of invertebrate and mammalian pathogenic bacteria in the oral insect infection model *Galleria mellonella*. J Invertebr Pathol 103: 24–29.

Forche A, Magee PT, Selmecki A, Berman J, May G. 2009. Evolution in *Candida albicans* populations during a single passage through a mouse host. Genetics 182: 799–811.

Forrest GN, Weekes E, Johnson JK. 2008. Increasing incidence of *Candida parapsilosis* candidemia with caspofungin usage. J Infect 56: 126–129.

Fuchs BB, O'Brien E, Khoury JB, Mylonakis E. 2010. Methods for using *Galleria mellonella* as a model host to study fungal pathogenesis. Virulence 1: 475–482.

Gardiner RE, Souteropoulos P, Park S, Perlin DS. 2005. Characterization of *Aspergillus fumigatus* mutants with reduced susceptibility to caspofungin. Med Mycol 43(Suppl. 1): S299–S305.

Groll AH, Walsh TJ. 2006. Antifungal efficacy and pharmacodynamics of posaconazole in experimental models of invasive fungal infections. Mycoses 49(Suppl. 1): 7–16.

Habets MG, Czaran T, Hoekstra RF, De Visser JA. 2007. Spatial structure inhibits the rate of invasion of beneficial mutations in asexual populations. Proc Biol Sci 274: 2139–2143.

Heilmann CJ, Schneider S, Barker KS, Rogers PD, Morschhauser J. 2010. An A643T mutation in the transcription factor Upc2p causes constitutive *ERG11* upregulation and increased fluconazole resistance in *Candida albicans*. Antimicrob Agents Chemother 54: 353–359.

Hoffman HL, Ernst EJ, Klepser ME. 2000. Novel triazole antifungal agents. Expert Opin Investig Drugs 9: 593–605.

Hoot SJ, Smith AR, Brown RP, White TC. 2011. An A643V amino acid substitution in Upc2p contributes to azole resistance in well-characterized clinical isolates of *Candida albicans*. Antimicrob Agents Chemother 55: 940–942.

Kelly J, Kavanagh K. 2011. Caspofungin primes the immune response of the larvae of *Galleria mellonella* and induces a non-specific antimicrobial response. J Med Microbiol 60: 189–196.

Kim DH, Feinbaum R, Alloing G, Emerson FE, Garsin DA, Inoue H, Tanaka-Hino M, Hisamoto N, Matsumoto K, Tan MW, Ausubel FM. 2002. A conserved p38 MAP kinase pathway in *Caenorhabditis elegans* innate immunity. Science 297: 623–626.

Krishnarao TV, Galgiani JN. 1997. Comparison of the in vitro activities of the echinocandin LY303366, the pneumocandin MK-0991, and fluconazole against *Candida* species and *Cryptococcus neoformans*. Antimicrob Agents Chemother 41: 1957–1960.

Kuo D, Tan K, Zinman G, Ravasi T, Bar-Joseph Z, Ideker T. 2010. Evolutionary divergence in the fungal response to fluconazole revealed by soft clustering. Genome Biol 11: R77.

LaFayette SL, Collins C, Zaas AK, Schell WA, Betancourt-Quiroz M, Gunatilaka AA, Perfect JR, Cowen LE. 2010. PKC signaling regulates drug resistance of the fungal pathogen *Candida albicans* via circuitry comprised of Mkc1, calcineurin, and Hsp90. PLoS Pathog 6: e1001069.

LaFleur MD, Kumamoto CA, Lewis K. 2006. *Candida albicans* biofilms produce antifungal-tolerant persister cells. Antimicrob Agents Chemother 50: 3839–3846.

Lamb DC, Kelly DE, White TC, Kelly SL. 2000. The R467K amino acid substitution in *Candida albicans* sterol 14alpha-demethylase causes drug resistance through reduced affinity. Antimicrob Agents Chemother 44: 63–67.

Lamping E, Ranchod A, Nakamura K, Tyndall JD, Niimi K, Holmes AR, Niimi M, Cannon RD. 2009. Abc1p is a multidrug efflux transporter that tips the balance in favor of innate azole resistance in *Candida krusei*. Antimicrob Agents Chemother 53: 354–369.

Lazzell AL, Chaturvedi AK, Pierce CG, Prasad D, Uppuluri P, Lopez-Ribot JL. 2009. Treatment and prevention of *Candida albicans* biofilms with caspofungin in a novel central venous catheter murine model of candidiasis. J Antimicrob Chemother 64: 567–570.

MacCallum DM, Odds FC. 2005. Temporal events in the intravenous challenge model for experimental *Candida albicans* infections in female mice. Mycoses 48: 151–161.

MacCallum DM, Whyte JA, Odds FC. 2005. Efficacy of caspofungin and voriconazole combinations in experimental aspergillosis. Antimicrob Agents Chemother 49: 3697–3701.

MacCallum DM, Castillo L, Nather K, Munro CA, Brown AJ, Gow NA, Odds FC. 2009. Property differences among the four major *Candida albicans* strain clades. Eukaryot Cell 8: 373–387.

MacLean RC, Hall AR, Perron GG, Buckling A. 2010. The population genetics of antibiotic resistance: Integrating molecular mechanisms and treatment contexts. Nat Rev Genet 11: 405–414.

Mansfield BE, Oltean HN, Oliver BG, Hoot SJ, Leyde SE, Hedstrom L, White TC. 2010. Azole drugs are imported by facilitated diffusion in *Candida albicans* and other pathogenic fungi. PLoS Pathog 6: e1001126.

Marchetti O, Entenza JM, Sanglard D, Bille J, Glauser MP, Moreillon P. 2000. Fluconazole plus cyclosporine: A fungicidal combination effective against experimental endocarditis due to *Candida albicans*. Antimicrob Agents Chemother 44: 2932–2938.

Marichal P, Vanden Bossche H. 1995. Mechanisms of resistance to azole antifungals. Acta Biochim Pol 42: 509–516.

Marr KA, Lyons CN, Ha K, Rustad TR, White TC. 2001. Inducible azole resistance associated with a heterogeneous phenotype in

Candida albicans. Antimicrob Agents Chemother 45: 52–59.

Martinez LR, Casadevall A. 2006. Susceptibility of *Cryptococcus neoformans* biofilms to antifungal agents in vitro. Antimicrob Agents Chemother 50: 1021–1033.

McNeil MM, Nash SL, Hajjeh RA, Phelan MA, Conn LA, Plikaytis BD, Warnock DW. 2001. Trends in mortality due to invasive mycotic diseases in the United States, 1980–1997. Clin Infect Dis 33: 641–647.

Morgan J, Meltzer MI, Plikaytis BD, Sofair AN, Huie-White S, Wilcox S, Harrison LH, Seaberg EC, Hajjeh RA, Teutsch SM. 2005. Excess mortality, hospital stay, and cost due to candidemia: A case-control study using data from population-based candidemia surveillance. Infect Control Hosp Epidemiol 26: 540–547.

Mortensen KL, Mellado E, Lass-Florl C, Rodriguez-Tudela JL, Johansen HK, Arendrup MC. 2010. Environmental study of azole-resistant *Aspergillus fumigatus* and other aspergilli in Austria, Denmark, and Spain. Antimicrob Agents Chemother 54: 4545–4549.

Munro CA. 2010. Fungal echinocandin resistance. F1000 Biol Rep 2: 66.

Mylonakis E, Moreno R, El Khoury JB, Idnurm A, Heitman J, Calderwood SB, Ausubel FM, Diener A. 2005. *Galleria mellonella* as a model system to study *Cryptococcus neoformans* pathogenesis. Infect Immun 73: 3842–3850.

Nagi M, Nakayama H, Tanabe K, Bard M, Aoyama T, Okano M, Higashi S, Ueno K, Chibana H, Niimi M, Yamagoe S, Umeyama T, Kajiwara S, Ohno H, Miyazaki Y. 2011. Transcription factors *CgUPC2A* and *CgUPC2B* regulate ergosterol biosynthetic genes in *Candida glabrata*. Genes Cells 16: 80–89.

Nett J, Lincoln L, Marchillo K, Massey R, Holoyda K, Hoff B, VanHandel M, Andes D. 2007. Putative role of beta-1,3 glucans in *Candida albicans* biofilm resistance. Antimicrob Agents Chemother 51: 510–520.

Niimi K, Monk BC, Hirai A, Hatakenaka K, Umeyama T, Lamping E, Maki K, Tanabe K, Kamimura T, Ikeda F, Uehara Y, Kano R, Hasegawa A, Cannon RD, Niimi M. 2010. Clinically significant micafungin resistance in *Candida albicans* involves modification of a glucan synthase catalytic subunit *GSC1 (FKS1)* allele followed by loss of heterozygosity. J Antimicrob Chemother 65: 842–852.

Odds FC, Bougnoux ME, Shaw DJ, Bain JM, Davidson AD, Diogo D, Jacobsen MD, Lecomte M, Li SY, Tavanti A, Maiden MC, Gow NA, d'Enfert C. 2007. Molecular phylogenetics of *Candida albicans*. Eukaryot Cell 6: 1041–1052.

Okoli I, Coleman JJ, Tampakakis E, An WF, Holson E, Wagner F, Conery AL, Larkins-Ford J, Wu G, Stern A, Ausubel FM, Mylonakis E. 2009. Identification of antifungal compounds active against *Candida albicans* using an improved high-throughput *Caenorhabditis elegans* assay. PLoS One 4: e7025.

Ostrosky-Zeichner L, Casadevall A, Galgiani JN, Odds FC, Rex JH. 2010. An insight into the antifungal pipeline: Selected new molecules and beyond. Nat Rev Drug Discov 9: 719–727.

Papadimitriou JM, Ashman RB. 1986. The pathogenesis of acute systemic candidiasis in a susceptible inbred mouse strain. J Pathol 150: 257–265.

Park S, Kelly R, Kahn JN, Robles J, Hsu MJ, Register E, Li W, Vyas V, Fan H, Abruzzo G, Flattery A, Gill C, Chrebet G, Parent SA, Kurtz M, Teppler H, Douglas CM, Perlin DS. 2005. Specific substitutions in the echinocandin target Fks1p account for reduced susceptibility of rare laboratory and clinical *Candida* sp. isolates. Antimicrob Agents Chemother 49: 3264–3273.

Pfaller MA, Diekema DJ. 2004. Rare and emerging opportunistic fungal pathogens: Concern for resistance beyond *Candida albicans* and *Aspergillus fumigatus*. J Clin Microbiol 42: 4419–4431.

Pfaller MA, Diekema DJ. 2007. Epidemiology of invasive candidiasis: A persistent public health problem. Clin Microbiol Rev 20: 133–163.

Pfaller MA, Messer SA, Boyken L, Rice C, Tendolkar S, Hollis RJ, Doern GV, Diekema DJ. 2005. Global trends in the antifungal susceptibility of *Cryptococcus neoformans* (1990 to 2004). J Clin Microbiol 43: 2163–2167.

Pfaller MA, Boyken L, Hollis RJ, Kroeger J, Messer SA, Tendolkar S, Diekema DJ. 2011a. Wild-type MIC distributions and epidemiological cutoff values for posaconazole and voriconazole and *Candida* spp. as determined by 24-hour CLSI broth microdilution. J Clin Microbiol 49: 630–637.

Pfaller MA, Castanheira M, Messer SA, Moet GJ, Jones RN. 2011b. Echinocandin and triazole antifungal susceptibility profiles for *Candida* spp., *Cryptococcus neoformans*, and *Aspergillus fumigatus*: Application of new CLSI clinical breakpoints and epidemiologic cutoff values to characterize resistance in the SENTRY Antimicrobial Surveillance Program (2009). Diagn Microbiol Infect Dis 69: 45–50.

Prasad T, Chandra A, Mukhopadhyay CK, Prasad R. 2006. Unexpected link between iron and drug resistance of *Candida* spp.: Iron depletion enhances membrane fluidity and drug diffusion, leading to drug-susceptible cells. Antimicrob Agents Chemother 50: 3597–3606.

Robbins N, Collins C, Morhayim J, Cowen LE. 2010. Metabolic control of antifungal drug resistance. Fungal Genet Biol 47: 81–93.

Schoustra SE, Bataillon T, Gifford DR, Kassen R. 2009. The properties of adaptive walks in evolving populations of fungus. PLoS Biol 7: e1000250.

Schuetzer-Muehlbauer M, Willinger B, Krapf G, Enzinger S, Presterl E, Kuchler K. 2003. The *Candida albicans* Cdr2p ATP-binding cassette (ABC) transporter confers resistance to caspofungin. Mol Microbiol 48: 225–235.

Scully LR, Bidochka MJ. 2005. Serial passage of the opportunistic pathogen *Aspergillus flavus* through an insect host yields decreased saprobic capacity. Can J Microbiol 51: 185–189.

Selmecki A, Forche A, Berman J. 2006. Aneuploidy and isochromosome formation in drug-resistant *Candida albicans*. Science 313: 367–370.

Selmecki A, Gerami-Nejad M, Paulson C, Forche A, Berman J. 2008. An isochromosome confers drug resistance in vivo by amplification of two genes, *ERG11* and *TAC1*. Mol Microbiol 68: 624–641.

Selmecki AM, Dulmage K, Cowen LE, Anderson JB, Berman J. 2009. Acquisition of aneuploidy provides increased fitness during the evolution of antifungal drug resistance. PLoS Genet 5: e1000705.

Shapiro RS, Robbins N, Cowen LE. 2011. Regulatory circuitry governing fungal development, drug resistance, and disease. Microbiol Mol Biol Rev 75: 213–267.

Sifri CD, Begun J, Ausubel FM. 2005. The worm has turned—Microbial virulence modeled in *Caenorhabditis elegans*. Trends Microbiol 13: 119–127.

Silver PM, Oliver BG, White TC. 2008. Characterization of caspofungin susceptibilities by broth and agar in *Candida albicans* clinical isolates with characterized mechanisms of azole resistance. Med Mycol 46: 231–239.

Singh SD, Robbins N, Zaas AK, Schell WA, Perfect JR, Cowen LE. 2009. Hsp90 governs echinocandin resistance in the pathogenic yeast *Candida albicans* via calcineurin. PLoS Pathog 5: e1000532.

Sionov E, Chang YC, Garraffo HM, Kwon-Chung KJ. 2009. Heteroresistance to fluconazole in *Cryptococcus neoformans* is intrinsic and associated with virulence. Antimicrob Agents Chemother 53: 2804–2815.

Sionov E, Lee H, Chang YC, Kwon-Chung KJ. 2010. *Cryptococcus neoformans* overcomes stress of azole drugs by formation of disomy in specific multiple chromosomes. PLoS Pathog 6: e1000848.

Snelders E, Huis In 't Veld RA, Rijs AJ, Kema GH, Melchers WJ, Verweij PE. 2009. Possible environmental origin of resistance of *Aspergillus fumigatus* to medical triazoles. Appl Environ Microbiol 75: 4053–4057.

Steele C, Rapaka RR, Metz A, Pop SM, Williams DL, Gordon S, Kolls JK, Brown GD. 2005. The beta-glucan receptor dectin-1 recognizes specific morphologies of *Aspergillus fumigatus*. PLoS Pathog 1: e42.

Steinbach WJ, Benjamin DK, Jr., Trasi SA, Miller JL, Schell WA, Zaas AK, Foster WM, Perfect JR. 2004. Value of an inhalational model of invasive aspergillosis. Med Mycol 42: 417–425.

Steinbach WJ, Reedy JL, Cramer RA, Jr., Perfect JR, Heitman J. 2007. Harnessing calcineurin as a novel anti-infective agent against invasive fungal infections. Nat Rev Microbiol 5: 418–430.

Tavanti A, Davidson AD, Fordyce MJ, Gow NA, Maiden MC, Odds FC. 2005. Population structure and properties of *Candida albicans*, as determined by multilocus sequence typing. J Clin Microbiol 43: 5601–5613.

Taylor PR, Tsoni SV, Willment JA, Dennehy KM, Rosas M, Findon H, Haynes K, Steele C, Botto M, Gordon S, Brown GD. 2007. Dectin-1 is required for beta-glucan recognition and control of fungal infection. Nat Immunol 8: 31–38.

Thompson JR, Douglas CM, Li W, Jue CK, Pramanik B, Yuan X, Rude TH, Toffaletti DL, Perfect JR, Kurtz M. 1999. A glucan synthase *FKS1* homolog in *Cryptococcus neoformans* is single copy and encodes an essential function. J Bacteriol 181: 444–453.

Venkateswarlu K, Taylor M, Manning NJ, Rinaldi MG, Kelly SL. 1997. Fluconazole tolerance in clinical isolates of *Cryptococcus neoformans*. Antimicrob Agents Chemother 41: 748–751.

Verweij PE, Snelders E, Kema GH, Mellado E, Melchers WJ. 2009. Azole resistance in *Aspergillus fumigatus*: A side-effect of environmental fungicide use? Lancet Infect Dis 9: 789–795.

Wang JS, Yang YL, Wu CJ, Ouyang KJ, Tseng KY, Chen CG, Wang H, Lo HJ. 2006. The DNA-binding domain of CaNdt80p is required to activate *CDR1* involved in drug resistance in *Candida albicans*. J Med Microbiol 55: 1403–1411.

White TC. 1997a. Increased mRNA levels of *ERG16*, *CDR*, and *MDR1* correlate with increases in azole resistance in *Candida albicans* isolates from a patient infected with human immunodeficiency virus. Antimicrob Agents Chemother 41: 1482–1487.

White TC. 1997b. The presence of an R467K amino acid substitution and loss of allelic variation correlate with an azole-resistant lanosterol 14alpha demethylase in *Candida albicans*. Antimicrob Agents Chemother 41: 1488–1494.

White TC, Pfaller MA, Rinaldi MG, Smith J, Redding SW. 1997c. Stable azole drug resistance associated with a substrain of *Candida albicans* from an HIV-infected patient. Oral Dis 3(Suppl. 1): S102–S109.

White TC, Marr KA, Bowden RA. 1998. Clinical, cellular, and molecular factors that contribute to antifungal drug resistance. Clin Microbiol Rev 11: 382–402.

White TC, Harry J, Oliver BG. 2004. Antifungal drug resistance: Pumps and permutations. In: Domer JE, Kobyashi GS, eds. *Human Fungal Pathogens*. Springer, Heidelberg.

Willger SD, Puttikamonkul S, Kim KH, Burritt JB, Grahl N, Metzler LJ, Barbuch R, Bard M, Lawrence CB, Cramer RA, Jr. 2008. A sterol-regulatory element binding protein is required for cell polarity, hypoxia adaptation, azole drug resistance, and virulence in *Aspergillus fumigatus*. PLoS Pathog 4: e1000200.

Xu J. 2002. Estimating the spontaneous mutation rate of loss of sex in the human pathogenic fungus *Cryptococcus neoformans*. Genetics 162: 1157–1167.

Zimbeck AJ, Iqbal N, Ahlquist AM, Farley MM, Harrison LH, Chiller T, Lockhart SR. 2010. *FKS* mutations and elevated echinocandin MIC values among *Candida glabrata* isolates from U.S. population-based surveillance. Antimicrob Agents Chemother 54: 5042–5047.

Zocco D, Van Aarle IM, Oger E, Lanfranco L, Declerck S. 2011. Fenpropimorph and fenhexamid impact phosphorus translocation by arbuscular mycorrhizal fungi. Mycorrhiza 21: 363–374.

CHAPTER 9

DISCOVERY OF EXTANT SEXUAL CYCLES IN HUMAN PATHOGENIC FUNGI AND THEIR ROLES IN THE GENERATION OF DIVERSITY AND VIRULENCE

RICHARD J. BENNETT and KIRSTEN NIELSEN

Several of the most prominent fungal pathogens were originally believed to be exclusively asexual. Genomic studies first revealed the potential for extant sexual cycles in several of these pathogenic species and have since been validated experimentally. Sexual (or parasexual) reproduction has now been established for three of the most common fungal pathogens: *Cryptococcus neoformans*, *Aspergillus fumigatus*, and *Candida albicans*. Population genetic studies in both *C. neoformans* and *C. albicans* suggest these organisms exhibit primarily clonal modes of reproduction with limited evidence for genetic recombination, indicating that sexual reproduction is rare. This has led to the hypothesis that limiting sexual reproduction generates clonal populations that are well adapted for life in the host while still retaining the potential for mating and generation of recombinant strains (Heitman, 2006; Nielsen and Heitman, 2007).

Another feature common to both *C. neoformans* and *C. albicans* is the possibility of undergoing same-sex mating (homothallism) in addition to conventional opposite-sex mating (heterothallism). In the case of *C. neoformans*, the homothallic mode of reproduction appears to be predominant as the vast majority of natural isolates are of the α mating type. Thus, homothallism may represent an important mechanism whereby fungal pathogens generate forms with increased pathogenicity and could be particularly important for species where individual populations are unlikely to come into contact with cells of the opposite mating type. Homothallic mating also has important ramifications for sexual programs in other fungi due to the more limited repertoire of signals necessary for this form of mating.

Finally, in this chapter, we note that sexual reproduction has been characterized for several additional human pathogens with interesting implications for pathogenesis. It is therefore a relevant question for pathogens without known sexual cycles whether the conditions for sex have yet to be identified in these species or whether sex has recently been lost as potentially

Evolution of Virulence in Eukaryotic Microbes, First Edition. Edited by L. David Sibley, Barbara J. Howlett, and Joseph Heitman.

evident in the degeneration of the mating locus in *Candida parapsilosis*.

MATING IN *C. NEOFORMANS*

C. neoformans is a ubiquitous basidiomycete yeast and a leading cause of fungal meningoencephalitis in immunosuppressed hosts (Park et al., 2009, 2011). It is found worldwide and is associated with soil, trees, and pigeon guano in the environment. *C. neoformans* isolates are typically divided into three serotypes based on their capsular aggregation reaction and genome sequence. These are *C. neoformans* var. *grubii* serotype A (VNI, VNII, and VNB clades), var. *neoformans* serotype D (VNIV clade), and serotype AD hybrids, which are often diploid or aneuploid (VNIII clade) (Belay et al., 1996). There is also a closely related sibling species, *Cryptococcus gattii*, which itself is a cause of *Cryptococcus* infections and contains B and C serotypes (VGI–IV clades) (Kwon-Chung et al., 2011). Unlike *C. neoformans*, which primarily causes infections in immunocompromised hosts, infections with *C. gattii* can occur in immunocompetent, apparently healthy individuals.

A conventional sexual cycle in *C. neoformans* was discovered in the mid-1970s whereby mating was demonstrated between **a** and α cell types (Kwon-Chung, 1975, 1976a,b). Cell mating type is determined by genes encoded at the mating-type (*MAT*) locus, in a manner analogous to the sex-determining regions of other eukaryotes. Mating in many basidiomycetes is regulated by a tetrapolar mating locus in which pheromones and their receptors are present at one locus and transcription factors regulating mating are at a second unlinked locus (Hull and Heitman, 2002). This produces a system in which four loci act in concert to regulate mating, as two distinct loci are located in each mating partner. In *C. neoformans* and *C. gattii*, however, mating is bipolar, as the pheromone/receptor locus and transcription factor locus have become fused into one larger *MAT* locus that contains at least 23 genes and is more than 100 kb in length (Lengeler et al., 2002). A model describing how this locus evolved from a tetrapolar mating ancestor has been proposed, tested, and validated (Fraser et al., 2005b; Hsueh and Heitman, 2008; Hsueh et al., 2006, 2008; Metin et al., 2010).

Cell identity of *C. neoformans* cells is determined by a combination of pheromones, pheromone receptors, and homeodomain proteins encoded at the *MAT* locus. Gene exchange experiments with the *MAT*α homeodomain protein Sxi1α first showed that Sxi1α is sufficient to confer α identity and to change the behavior of a *MAT***a** cell into a *MAT***a**/α cell (Hull et al., 2002, 2005). Similar gene swapping of pheromone and pheromone receptor genes between *MAT***a** and *MAT*α cells subsequently demonstrated that these genes are both necessary and sufficient to confer haploid cell identity (Stanton et al., 2010). Thus, an engineered α cell expressing the pheromone and pheromone receptor from the *MAT***a** locus (in place of the endogenous pheromone/receptor) mated as an **a** cell, including undergoing fusion with an α cell. However, such crosses did not complete the sexual program unless both homeodomain transcription factors, Sxi1α and Sxi2**a**, were present (Stanton et al., 2010). In heterothallic *C. neoformans* mating, **a** and α cells fuse to form dikaryotic filaments that eventually undergo nuclear fusion and meiosis, generating sexual basidiospores (Kruzel and Hull, 2010; Lin and Heitman, 2006). While cell fusion is controlled via pheromone sensing (Stanton et al., 2010), Sxi1α and Sxi2**a** homeodomain proteins are essential for postfusion events. The exact role of Sxi1α and Sxi2**a** in mediating sexual development is yet to be elucidated, although one important target of Sxi1α/Sxi2**a** function appears to be Clp1 (clampless-1), which is a conserved regulator of development in basidiomycetes and is necessary for dikaryon formation in *C. neoformans* (Ekena et al., 2008). Two additional transcription factors, Mat2 and Znf2, are also important for mating in *C. neoformans* (Lin et al., 2010). Mat2 is a high mobility group (HMG) box transcription factor that regulates pheromone expression and subsequent cell fusion events. In contrast, the zinc finger protein Znf2 is essential for filamentation during the mating process.

The mating type of *C. neoformans* isolates has been directly linked to pathogenesis, as α cells from serotype D were shown to be more pathogenic than congenic **a** cells when compared in a murine model of infection (Kwon-Chung et al., 1992). Subsequent experiments revealed that the genetic background is critical for the influence of the mating locus on virulence and that the *MAT* locus interacts with genes from outside the locus to affect pathogenesis (Nielsen et al., 2003, 2005b). In addition, the overwhelming majority of *C. neoformans* environmental and clinical isolates are *MAT*α strains, although *MAT***a** strains of the predominant pathogenic form (serotype A var. *grubii*) have been identified (Lengeler et al., 2000). Interestingly, the majority of var. *grubii MAT***a** strains have been observed in sub-Saharan Africa where there is evidence for recent or ongoing sexual recombination taking place between **a** and α isolates (Litvintseva et al., 2003, 2005, 2011; Simwami et al., 2011).

SAME-SEX MATING IN *CRYPTOCOCCUS*

The predominance of *MAT*α strains in natural populations of *C. neoformans* was originally thought to preclude sexual reproduction. This dogma changed with the discovery that α cells exhibiting monokaryotic fruiting are actually undergoing a novel same-sex mode of diploidization and meiosis (Lin et al., 2005). Monokaryotic fruiting refers to the ability of α cells to form filaments and sporulate, and this developmental program was originally assumed to be asexual and mitotic (Wickes et al., 1996). Lin and colleagues revealed that fruiting represents a novel α–α homothallic sexual program that culminates with the production of recombinant haploid spores. Similar to meiosis in other eukaryotes, recombination during fruiting requires the Spo11 and Dmc1 proteins, which are necessary for the formation and repair of DNA double-strand breaks (Lin et al., 2005).

The discovery of same-sex mating in *C. neoformans* provided an explanation as to how recombinant strains could arise within a unisexual population. Examination of the population structure of *Cryptococcus* confirms that a low level of recombination is evident among natural isolates (Campbell et al., 2005; Litvintseva et al., 2003, 2005; Saul et al., 2008), and in the absence of *MAT***a** isolates, a same-sex mode of sexual reproduction is the most likely explanation. Further support for same-sex mating in *C. neoformans* var. *grubii* serotype A (the predominant pathogenic form) has come from an analysis of natural isolates showing the existence of α/α diploids that are intermediates in the unisexual mating cycle (Bui et al., 2008; Lin et al., 2009), including hybrid diploids that were produced by fusion between two distinct α parental cells (Lin et al., 2009).

Additional evidence for same-sex mating has accumulated from an ongoing *C. gattii* outbreak that initiated on Vancouver Island, Canada. Evidence suggests that the major genotype, VGIIa, is the product of a same-sex mating event between two closely related α isolates of the VGII type (Fraser et al., 2005a). The outbreak has since expanded into the Pacific Northwest of the United States, and a novel variant (VGIIc) has been uncovered that is also hypervirulent in animal models and cultured cells (Byrnes and Heitman, 2009; Byrnes et al., 2010). High mortality rates (>25%) are associated with these infections, and same-sex mating may be directly responsible for generating the infectious spores that are inhaled into the lung (Bartlett et al., 2008; Datta et al., 2009; D'Souza et al., 2011).

Some *MAT***a** isolates of *C. neoformans* have also been shown to be competent for monokaryotic fruiting (Tscharke et al., 2003), and it was therefore unclear as to the role of the *MAT* locus in directing fruiting and unisexual mating. Subsequent studies utilized quantitative trait locus (QTL) mapping to demonstrate that while both **a** and α strains can undergo fruiting, the *MAT*α locus acts to enhance hyphal growth during fruiting (Lin et al., 2006). The domination of α isolates in nature may therefore be due to their increased ability to undergo monokaryotic fruiting and same-sex reproduction. It has also been shown that *C. neoformans* var. *grubii* α isolates are better able to disseminate through the central nervous system than congenic **a** isolates during coinfection with opposite mating-type strains, indicating that the two cell types may have distinct interactions with the mammalian host (Nielsen et al., 2005a).

THE ADVANTAGES OF UNISEXUAL MATING IN *CRYPTOCOCCUS* SPECIES

Why might *C. neoformans* choose to undergo same-sex mating when an efficient mechanism for opposite-sex mating exists? Transitions in the mode of sexual reproduction have been documented in higher eukaryotes, including both plants and animals. For example, the switch from outcrossing to selfing is a common theme in many flowering plants and is thought to be beneficial for colonization of new territories where interactions between sexes would be relatively rare (Barrett, 2002; Tang et al., 2007). Similar evolutionary pressures may affect fungal development, where an isolate could find itself secluded from opposite-sex partners, and therefore the ability to mate with a clonal partner of the same sex could be highly beneficial.

In conventional mating, outcrossing is thought to be important as it is accompanied by extensive genetic recombination and results in variant strains. In contrast, many pathogenic fungi are often asexual or have limited access to their sexual cycles (Heitman, 2006; Nielsen and Heitman, 2007). It may be that there is a fitness advantage to restricted sexual reproduction in pathogenic species as this limits the rearrangement of alleles that are already highly adapted for growth in the host. This hypothesis may not be restricted to pathogens; even in the model yeast *Saccharomyces cerevisiae*, evidence suggests that natural strains reproduce predominantly via clonal propagation (Ezov et al., 2006, 2010). Furthermore, natural *S. cerevisiae* strains can be *Ho*$^{+}$ and self-fertile due to mating-type switching, or *ho*$^{-}$ and thus exclusively cross-fertile. These studies suggest that both inbreeding and outbreeding modes of reproduction may be advantageous under the right circumstances.

In the case of *C. neoformans*, several hypotheses have been proposed for the utility of same-sex mating (Heitman, 2010; Lee et al., 2010). For example, mating between two identical haploid isolates generates a diploid form that may represent a preferential state for undergoing mutagenesis. Diploid cells, unlike haploid cells, can accumulate recessive lethal alleles that individually are harmful but together can form advantageous combinations. This argues that same-sex reproduction is not mixing preexisting diversity from two different parental strains as in conventional outbreeding, but instead generates new diversity from identical parents. In support of this model, diploid strains of *Aspergillus nidulans* were shown to reach a higher fitness level than isogenic haploid strains, despite reverting to haploidy during the course of the experiment. The increased fitness levels of diploids were due to beneficial recessive mutations that accumulated in the diploid (Schoustra et al., 2007). Other advantages of same-sex mating include purging the nuclear genome of deleterious mutations and removing defective mitochondrial genomes due to uniparental inheritance of the mitochondrial genome.

Another possibility is that same-sex mating may actually promote more genomic rearrangements than opposite-sex mating due to the activation of transposons. *C. neoformans* was recently shown to have a sex-induced RNAi genomic defense system, referred to as sex-induced silencing (SIS) (Wang et al., 2010). SIS gene silencing occurs at a much higher frequency (250-fold) during opposite-sex mating compared with vegetative growth, thereby limiting the expression of a group of retrotransposons (Ni et al., 2011). Interestingly, phenotypic and genotypic diversity is readily observed in same-sex mating, indicating SIS activity is limited under these conditions (M. Ni and J. Heitman, unpublished data). These studies suggest that same-sex mating may promote genetic changes and strain evolution even in the absence of an opposite-sex mating partner while still retaining traits necessary for pathogenesis.

MATING AND PATHOGENESIS IN *C. NEOFORMANS*

As already discussed above, there are several well-established connections between mating type, sexual reproduction and virulence in *Cryptococcus* (Fig. 9.1) This includes the fact that α isolates exhibit

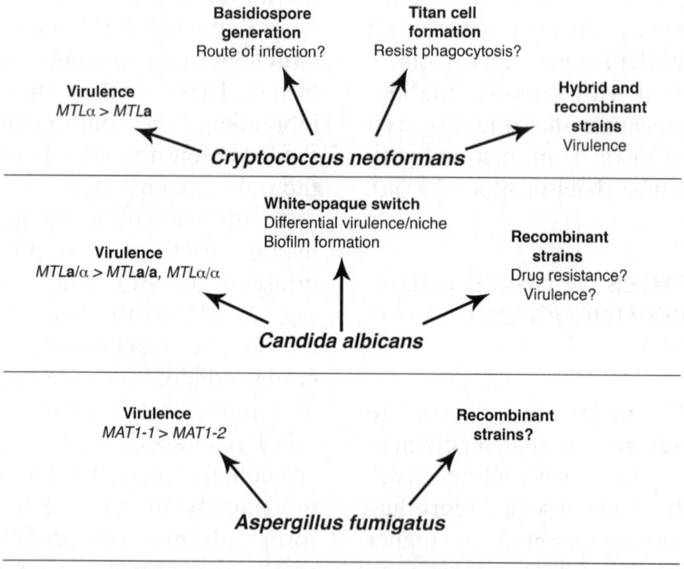

Figure 9.1 Overview of links between pathogenesis and sexual reproduction in the three major human fungal pathogens. In *C. neoformans*, strain mating type, basidiospore generation, titan cell formation, and recombinant strains resulting from sexual mating are all connected with infection. In *C. albicans*, the mating type of the strain influences virulence, as does phenotypic switching between white and opaque forms and potentially genetic recombination during parasexual reproduction. In *A. fumigatus*, mating type can affect strain virulence as could recombination between strains during heterothallic mating.

greater virulence than congenic **a** strains and that sexual reproduction can generate recombinant forms of the species that are hypervirulent in the mammalian host, as seen in the Vancouver Island outbreak. Recent evidence also provides strong support for sexual spores as the infectious particle. Most notably, spores are readily phagocytosed by alveolar macrophages and produce high mortality rates in a murine inhalation model of infection (Giles et al., 2009; Sukroongreung et al., 1998; Velagapudi et al., 2009).

Three outcomes can follow the inhalation of spores. First, the initial pulmonary infection can be cleared by the host immune system. Second, a latent asymptomatic infection within granulomas can develop. Finally, a robust pulmonary infection that disseminates through the bloodstream can develop either from the primary exposure or upon reactivation of the latent infection. In the case of *C. neoformans*, infectious particles do not usually cause disease unless the host is immunocompromised, in which case the infection spreads to the central nervous system and progresses to meningoencephalitis. The latter is uniformly fatal without drug intervention (Driver et al., 1995). One recent study that was successful in obtaining high amounts of purified spores from an **a**–α mating cross demonstrated that spores were also more resistant than yeast cells to a variety of environmental stresses (Botts et al., 2009). The ability to readily obtain purified spores is likely to accelerate the understanding of these structures in infection and pathogenesis.

Recent experiments have indicated that pheromone signaling between mating cells can also influence *C. neoformans* cell morphology and virulence. During mating of **a** and α cells, pheromones released by these two cell types are recognized by sex-specific receptors on the opposite mating type. Thus, **a** pheromone is bound by Ste3α on the α cell, and α pheromone is bound by Ste3**a** on the **a** cell (Davidson et al., 2003; McClelland et al., 2004; Nielsen and Heitman, 2007). Pheromone signaling ultimately leads to fusion of **a** and α cells, formation of a dikaryon, and filamentation during *in vitro* mating. Curiously, simultaneous infection of the host with both **a** and α cells leads to increased formation of enlarged cells, or titan cells, by *MAT***a** mating types (Okagaki et al., 2010). The enhanced formation of titan cells is dependent on pheromone signaling as deletion of Ste3**a** (the receptor for α pheromone) reduced *MAT***a** titan formation to basal levels. Recent studies have further shown that Ste3**a** interacts with the G-protein α subunit, Gpa1, and signals via the protein kinase A (PKA) pathway to promote titan cell formation (Okagaki et al., 2011). Titan cells are also often seen in clinical isolates from humans (Cruickshank et al., 1973; Love et al., 1985) and are resistant to phagocytosis by host macrophages (Okagaki et al., 2010; Zaragoza et al., 2010). Titan cells are polyploid, containing four or eight copies of the genome within a single nucleus, suggesting alterations in the cell cycle (Fig. 9.2; Okagaki et al., 2010). Curiously, only **a** cells and not α cells showed increased formation of titan cells during coinfection by **a**/α mixes. Thus, the observation that titan cell formation can be differentially regulated in a mating-type-specific manner may have implications for virulence of the two mating types.

MATING IN *ASPERGILLUS* SPECIES

Aspergillus species are Ascomycetes that are part of the order Eurotiales. The most common human pathogen from this genus is *A. fumigatus* (sexual state *Neosartorya fumigata*), which is found among decaying materials in the soil (Latge, 1999). *A. fumigatus* can cause infection of the lungs, particularly in immunocompromised individuals, but also as a primary pathogen. The outcome of an infection is a balance

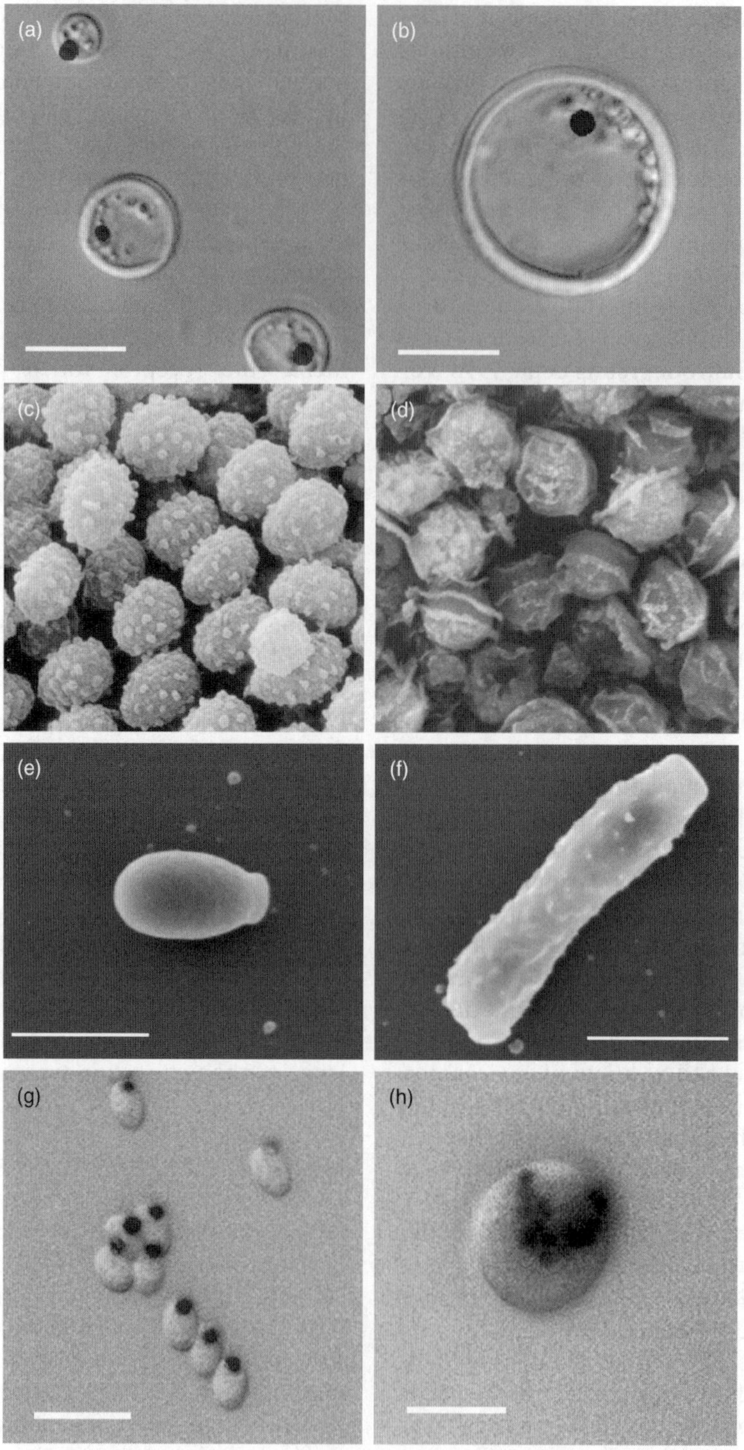
(a)
(b)
(c)
(d)
(e)
(f)
(g)
(h)

Figure 9.2 Sexual reproduction is associated with morphological diversity in human fungal pathogens. (a) Cryptococcal cells are normally 5–10 μm in diameter and have a single haploid nucleus. (b) Coinfection of mice with both mating types simultaneously results in increased production of large, polyploidy titan cells by *C. neoformans*. (c) *A. fumigatus* asexual conidia are produced under a variety of environmental conditions. (d) Sexual mating in *A. fumigatus* takes 1.5–6.0 months and generates sexual ascospores with two ornamental crests. (e) *C. albicans* white cells are unable to mate. (f) Opaque cells are the mating-competent cell morphology in *C. albicans*. (g) *M. circinelloides* (+) sporangiospores are small, uninucleate and have low virulence. (h) The larger multinucleate (–) sporangiospores of *M. circinelloides* are highly virulent in both insect and mouse models of disease. Scale bars in (a), (b), (g), and (h) = 10 μm. Scale bar in (d) = 4 μm. Scale bars in (e) and (f) = 3 μm. Punctate staining of the nuclei is indicated in panels (a), (b), (g), and (h). Images courtesy of Laura H. Okagaki and Kirsten Nielsen (a,b); Jean-Paul Latgé (c); Céline O'Gorman, Hubert Fuller, and Paul Dyer (d); Matthew Hirakawa and Richard Bennett (e,f); and Soo Chan Lee and Joseph Heitman (g,h). See color insert.

between the inoculum, the virulence of the strain, and the strength of the host defense (Hohl and Feldmesser, 2007; Latge, 1999). *A. fumigatus* primarily reproduces via asexual spores (conidia) that are easily dispersed and can grow under a wide variety of environmental conditions (Fig. 9.2) It is these conidiospores (or aspergillum) that define the genus morphologically (Dyer, 2007; Latge, 1999).

A. fumigatus was long believed to be an obligately asexual fungus, although this view was challenged when the sequenced genome revealed the presence of multiple genes associated with sexual reproduction (Nierman et al., 2005; Paoletti et al., 2005; Poggeler, 2002). Further analysis of *A. fumigatus* isolates showed that both mating types are present in the environment in a 1:1 ratio (Dyer and Paoletti, 2005; Paoletti et al., 2005). While strain diversity is low, evidence for recombination exists, supportive of sexual reproduction in the wild (Bain et al., 2007; Pringle et al., 2005; Rydholm et al., 2006; Varga and Toth, 2003). Due to the lack of apparent sexuality in *A. fumigatus*, many aspects of the life cycle have been studied in the model system *A. nidulans* (sexual state *Emericella nidulans*). In this species, mating is homothallic; there are no differentiated mating types and cells are able to mate with identical sister cells (Dyer, 2007). Formation of sexual ascospores occurs in fruiting bodies known as cleistotheicia and is induced by growth in darkness under limited aeration, perhaps resembling conditions in the soil (Pontecorvo et al., 1953).

In *A. nidulans*, homothallic mating is regulated by the concerted action of two mating-type genes, *MAT1-1* and *MAT1-2* (Paoletti et al., 2007). These genes encode an alpha-box domain protein (*MAT1-1*) and an HMG domain protein (*MAT1-2*). Both of these regulatory factors are encoded at the same *MAT* locus in *A. nidulans*, while in outcrossing (heterothallic) species, these factors are encoded by different mating partners, thereby ensuring that only compatible cells undergo mating and cell fusion. Expression of both factors in a single cell type allows self-mating of *A. nidulans*, culminating in the formation of cleistothecia. In fact, overexpression of *MAT1-1* and *MAT1-2* genes induced *A. nidulans* to undergo sexual development under conditions that normally do not favor sexual reproduction (Paoletti et al., 2007). These studies revealed that homothallic and heterothallic species can evolve from one another by simple rewiring of *MAT* regulation. There is still an active discussion as to which form of mating was present in the ancestral state that gave rise to these lineages, but it is now thought that homothallism arose out of a heterothallic ancestor rather than vice versa (Butler, 2010; Dyer, 2007; Scazzocchio, 2006).

DISCOVERY OF A SEXUAL CYCLE IN *A. FUMIGATUS*

Preliminary analysis of the *A. fumigatus* genome sequence revealed the presence of a mating-type gene (*MAT1-2*, encoding an HMG DNA-binding domain), a pheromone gene (related to *S. cerevisiae* α pheromone), and two pheromone receptor genes (analogous to **a**-pheromone and α-pheromone receptors in *S. cerevisiae*) (Poggeler, 2002). Paoletti et al. subsequently identified the *MAT1-1* gene encoding an alpha-domain protein and showed that natural populations of *A. fumigatus* were an equal distribution of *MAT1-1* and *MAT1-2* strains, as expected of an obligate heterothallic (outbreeding) fungus with an extant sexual cycle (Paoletti et al., 2005). These authors further showed that the mating-type, pheromone, and pheromone receptor genes were actively expressed during mycelial growth.

The functionality of *A. fumigatus MAT*-encoded proteins was directly demonstrated by complementation assays in *A. nidulans*. Interestingly, the *A. fumigatus MAT1-2* gene could complement the *A. nidulans* gene deletion if expressed under the *A. nidulans* promoter but not if expressed under its native promoter (Pyrzak et al., 2008). These experiments revealed that the *A. fumigatus MAT1-2*-encoded protein was active, but that transcriptional regulation of the gene had diverged since *A. fumigatus* and *A. nidulans* last shared a common ancestor (Pyrzak et al., 2008). Similar analyses showed the *A. fumigatus MAT1-1* gene (and *nsdD* gene, a positive regulator of sex) were functional in promoting *A. nidulans* sexual development (Grosse and Krappmann, 2008).

While genomic and gene-swapping experiments hinted at the potential for sex in *A. fumigatus*, it was not until 2009 that productive mating events were obtained for this species. In the end, it took a valiant experiment in which *MAT1-1* and *MAT1-2* strains were crossed on parafilm-sealed, oatmeal agar plates for 6 months in the dark (O'Gorman et al., 2009). Mature cleistothecia were observed along the junctions of intersecting colonies of opposite mating type, either singly or in small clusters of two to five. Disruption of the cleistothecia released multiple yellow- to greenish-white ascospores with two ornamental crests (Fig. 9.2). These ascopores were shown to germinate on malt extract agar producing typical *A. fumigatus* colonies containing a number of asexual conidia (O'Gorman et al., 2009). The sexual cycle in *A. fumigatus* was also shown to produce recombinant progeny as analysis of multiple genetic markers revealed independent segregation of these markers (O'Gorman et al., 2009). The sexual teleomorph of *A. fumigatus* was named *N. fumigata*; however, it should be noted that the use of a dual nomenclature for sexual and asexual forms has raised concerns about how to avoid confusion over species with alternative names (Hawksworth, 2010). (In this chapter, the anamorph names are used to indicate both the sexual and asexual states.) The limited conditions and length of incubation period also beg the question as to where and how *A. fumigatus* strains might undergo sexual development in nature and whether this species may be evolving into an asexual organism (Dyer and Paoletti, 2005; Kwon-Chung and Sugui, 2009).

The first *A. fumigatus* mating experiments were performed with a set of environmental isolates that had been collected by air sampling in Dublin, Ireland (O'Gorman et al., 2009). Even among the Dublin isolates, fertility was highly variable, with 80% of the crosses producing fewer than 60 cleistothecia and only 8% producing more than 100 cleistothecia, whereas other sexual *Aspergilli* produce cleistothecia too numerous to count (Kwon-Chung and Sugui, 2009; O'Gorman et al., 2009). A subsequent study extended these findings by showing that clinical isolates of *A.*

fumigatus could mate and form cleistothecia when cocultured on oatmeal agar for 6 weeks at 30°C (Szewczyk and Krappmann, 2010). These isolates were from patients that ultimately died from invasive aspergillosis (Nierman et al., 2005; Staib et al., 1980). Further coincubation of the clinical isolates for 3–4 months resulted in abundant formation of fruiting bodies containing multiple ascospores. Again, recombination between isolates was demonstrated and was consistent with random assortment of genetic markers (Szewczyk and Krappmann, 2010). Szewczyk and Krappmann also addressed the potential role of *MAT1-1* and *MAT1-2* genes in mating and showed that *A. fumigatus* mutants lacking either of these genes were prevented from undergoing sexual development. The *MAT* alleles regulated expression of *A. fumigatus* pheromone and pheromone receptor genes, consistent with their role as master regulators of mating in heterothallic fungi (Szewczyk and Krappmann, 2010).

Along with the discovery of sex in *A. fumigatus*, researchers have recently uncovered sexual programs in two other *Aspergillus* species. *Aspergillus flavus* and *Aspergillus parasiticus* are two of the primary producers of aflatoxin, a mycotoxin that can contaminate commercially important grain and nut crops. Aflatoxin is both toxic and carcinogenic and is a potential risk factor for hepatocellular carcinoma, particularly in Africa and Asia (Groopman and Kensler, 2005). *A. flavus* is also a significant human pathogen, second only to *A. fumigatus* as a cause of human invasive and noninvasive aspergillosis (Hedayati et al., 2007; Morgan et al., 2005). Similar to *A. fumigatus*, *A. flavus* and *A. parasiticus* species were shown to contain *MAT1-1* and *MAT1-2* idiomorphs, with both idiomorphs present in equal abundance in the population, suggestive of a sexually recombining population (Dyer, 2007; Ramirez-Prado et al., 2008). Subsequent experiments discovered that *A. flavus* and *A. parasiticus* both undergo heterothallic mating culminating in cleistothecia formation within stomata, revealing extant sexual cycles (Horn et al., 2009a,b).

MATING AND DISEASE IN *ASPERGILLUS*

What are the ramifications of sexual reproduction in *A. fumigatus* for the disease in humans? First, from the broader viewpoint of fungal pathogenesis, the identification of a mating cycle adds *A. fumigatus* to the growing list of species for which extant sexual programs have been identified. It is therefore apparent that the ability to undergo sexual reproduction, even if rare, is the norm rather than the exception for human fungal pathogens. Sexual cycles are likely to be discovered in other "fungi imperfecti" as genomic and genetic tools to dissect these species become available. Second, sexual reproduction is associated with the generation of recombinant forms, and it remains to be seen if progenies exhibit either increased virulence or enhanced resistance to antifungal drugs. Third, the sexual program could allow classical genetic studies on *A. fumigatus*. This includes the ability to cross strains of interest or the identification of quantitative traits associated with virulence or drug resistance. In this case, the lengthy incubation period acts as a barrier to genetic manipulations, and experiments to address this would be beneficial.

It is also worth noting that the sexual mating types of *A. fumigatus* differ in their virulence. Alvarez-Perez et al. demonstrated a significant association between the *MAT1-1* genotype and invasive growth in patients (Alvarez-Perez et al., 2010). In addition, they observed a significant association between the *MAT1-1* genotype and a clinical origin for the strain (vs. an environmental origin). More recently, the virulence of *MAT1-1* and *MAT1-2* strains was directly compared in a *Galleria mellonella* insect model of infection. Conidia from both clinical and environmental isolates of *A. fumigatus* were compared, and *MAT1-1* strains were found to be more virulent than *MAT1-2* strains (Cheema and Christians, 2011). Surprisingly, in their analysis of 20 isolates, they also found environmental strains were more virulent than clinical isolates. Extending these observations to a greater number of strains and identifying the source of these differences are important research topics for the future.

MATING IN *CANDIDA* SPECIES

Historically, *Candida* species were defined as yeasts that could form pseudohyphae or true hyphae but did not display evidence of sexual reproduction or sporulation. This taxonomy included a large number of species (>150), which phylogenetic analysis has subsequently revealed to be highly diverse (Reedy and Heitman, 2007). Conveniently, many pathogenic species of *Candida* group together in a single clade within the hemiascomycetes, known as the *Candida* clade. This clade is also referred to as the CTG clade, as member species use the CTG codon to encode for serine rather than leucine, a violation of the universal genetic code (Butler et al., 2009).

Candida species are the leading cause of fungal bloodstream infections in the United States, and the fourth leading cause of nosocomial infections overall (after coagulase-negative staphylococci, *Staphylococcus aureus*, and enterococci) (Wisplinghoff et al., 2004). The leading protagonist is *C. albicans*, which is found inhabiting the gastrointestinal tract of at least 70% of the population, as well as other commensal sites in the human body (Ruhnke and Maschmeyer, 2002). Unlike *Cryptococcus* and *Aspergillus* species, there does not appear to be a clear environmental niche for *C. albicans*, and it is found closely associated with its animal or avian hosts (Edelmann et al., 2005; Jacobsen et al., 2008a; Wrobel et al., 2008).

Like all *Candida* species, *C. albicans* was originally labeled as an obligate asexual yeast. However, this view was challenged when Hull and Johnson revealed the presence of a mating-type-like (*MTL*) locus that resembled the mating-type (*MAT*) locus of the model hemiascomycete, *S. cerevisiae* (Hull and Johnson, 1999). Subsequent experiments in the Johnson laboratory confirmed that *MTL***a** and *MTL*α diploids could mate during infection of a mammalian host to generate tetraploid products, while independent studies by the Magee group demonstrated mating on laboratory media *in vitro* (Hull et al., 2000; Magee and Magee, 2000). The efficiency of mating in both systems was extremely low, questioning the relevance of mating to *C. albicans* biology. Subsequently, an exciting breakthrough was made in the discovery of a direct connection between mating and white-opaque phenotypic switching. The white-opaque switch was first described by Soll and coworkers more than 20 years ago (Slutsky et al., 1987). In this work, *C. albicans* was shown to undergo a reversible and heritable switch between cells in the white state (round cells that form dome-shaped colonies) and cells in the opaque state (elongated cells that form flatter, darker colonies) (Fig. 9.2). Miller and Johnson revealed that the white-opaque switch was regulated by genes encoded at the *MTL*, so that only *MTL***a** or *MTL*α strains (and not *MTL***a**/α strains) could switch to the opaque state (Miller and Johnson, 2002). They further demonstrated that opaque cells could undergo efficient mating in contrast to the low levels of mating observed between white cells. These seminal studies opened up the field of *C. albicans* mating to detailed investigation.

THE WHITE-OPAQUE PHENOTYPIC SWITCH

The transition between white and opaque states is unique to *C. albicans* (and its sister species *Candida dubliniensis*) and may represent an adaptation to the commensal lifestyle (Johnson, 2003). The molecular mechanism underlying the white-opaque switch has begun to be elucidated, with the master regulator of the opaque state being the transcription factor Wor1. Expression of Wor1 is low in white cells, whereas expression of Wor1 is elevated in opaque cells, and positive feedback of Wor1 on its own promoter helps stabilize high expression (Huang et al., 2006; Srikantha et al., 2006; Zordan et al., 2006, 2007). A heterodimeric complex between **a**1 and α2 proteins inhibits Wor1 expression, thereby explaining why **a**/α cells are prevented from switching to opaque (Miller and Johnson, 2002). Approximately 3–9% of natural isolates are homozygous at the mating locus (i.e., *MTL***a**/**a** or *MTL*α/α) and are thus competent to undergo the white-opaque switch (Lockhart et al., 2002; Odds et al., 2007). At least three additional transcription factors regulate white-opaque stability with Wor1. The Efg1 protein is antagonistic to Wor1 and promotes formation of the white state, while Wor2 and Czf1 promote formation of the opaque state via positive feedback, either directly or indirectly, on Wor1 expression (Zordan et al., 2007).

Homologs of the Wor1 transcription factor have been found in all fungal lineages, yet the protein has no sequence or structural similarity to known DNA-binding proteins. Lohse et al. recently purified the conserved N-terminal region of Wor1 and established that it acts as a novel sequence-specific DNA-binding protein *in vitro* (Lohse et al., 2010). Wor1 homologs in several other fungal species have now been investigated, including Ryp1 in *Histoplasma capsulatum*, which was shown to regulate the yeast–mycelial transition, and Sge1 in *Fusarium oxysporum*, which controls expression of genes required for growth in the plant host (Michielse et al., 2009; Nguyen and Sil, 2008). These studies suggest a wide variety of functions for Wor1 homologs in diverse fungal species.

A number of environmental conditions have been shown to impact the white-opaque switch in *C. albicans*. For example, elevated CO_2 concentrations, anaerobic growth, genotoxic stress, oxidative stress, and N-acetylglucosamine (a monosaccharide produced by gastrointestinal bacteria) have all been shown to induce switching from white to opaque (Alby and Bennett, 2009; Dumitru et al., 2007; Huang et al., 2009, 2010; Kolotila and Diamond, 1990; Ramirez-Zavala et al., 2008). Some of these conditions may mimic those encountered in the mammalian host and, in support of this, white-opaque switching and mating has been observed in the anaerobic environment of the gastrointestinal tract (Dumitru et al., 2007; Ramirez-Zavala et al., 2008). Mating has also been observed on the skin (Lachke et al., 2003) and mating products have been recovered from internal organs following systemic infection (Hull et al., 2000), so the preferred *in vivo* niche for white-opaque switching and mating in *C. albicans* has yet to be determined.

Another intriguing aspect of the white-opaque switch is that while opaque cells are the mating-competent form, white cells can mount their own unique response to pheromone. Thus, while pheromone-treated opaque cells upregulate several hundred genes, including those necessary for mating, pheromone-treated white cells upregulate a smaller set of genes that participate in biofilm formation (Daniels et al., 2006). It is hypothesized that the response to pheromone by white cells generates a matrix within which pheromone gradients are maintained, thereby enabling opaque cells to accurately locate one another in the host (Daniels et al., 2006). In both white and opaque cells, pheromone sensing involves the same mitogen-activated protein kinase (MAPK) cascade. This pathway activates the Ste12/Cph1 transcription factor in opaque cells, whereas in white cells, it activates the Tec1 transcription factor (Sahni et al., 2010). The MAPK signaling pathway is therefore able to target two distinct transcription factors in white and opaque cells, although the mechanism by which this occurs has yet to be determined.

C. ALBICANS SAME-SEX MATING AND THE PARASEXUAL CYCLE

Pheromone signaling between opaque **a** and α cells drives conventional heterothallic mating in *C. albicans* as **a** cells respond to α pheromone secreted by α cells and vice versa (Bennett and Johnson, 2003; Lockhart et al., 2003; Panwar et al., 2003). Recent experiments by Alby et al. reveal that *C. albicans* cells can also undergo a homothallic mode of mating, during which same-sex fusion of cells occurs (Alby et al., 2009). Surprisingly, *C. albicans* **a** cells were shown to express not only the canonical **a** pheromone but also α pheromone; α pheromone was prevented from activating self-mating in these cells by the action of Bar1, an aspartyl protease that cleaves α pheromone (Schaefer et al., 2007). However, in the absence of Bar1, autocrine signaling due to secreted α pheromone caused **a** cells to mate efficiently with other **a** cells. Same-sex mating was also observed in "ménage à trois" matings, in which **a** cells were mixed with α cells. In this case, the pheromone produced by α cells was sufficient to drive same-sex mating of **a** cells and vice versa (Alby et al., 2009).

The critical role of pheromones in directing *C. albicans* mating has been confirmed by recent studies utilizing synthetic pheromones. Synthetic pheromone alone was shown to be sufficient to induce same-sex mating of opaque cells and biofilm formation by white cells (Alby and Bennett, 2011). Significantly, not only *C. albicans* pheromone but synthetic pheromones from several other *Candida* species also induced these responses in opaque and white cells. These findings broaden the potential conditions under which sexual reproduction may occur and suggest that interspecies signaling in mixed communities of *Candida* species could drive mating and sex-related processes.

In most sexual fungi, completion of the mating program occurs via meiosis, which halves the genetic content of the cell and produces recombinant spores. In *C. albicans*, a conventional meiosis has not been observed, and tetraploid cells (either from **a–a** or **a**–α mating) undergo the reduction back to diploid via a mechanism of stochastic chromosome loss (Bennett and Johnson, 2003). Many of the products of chromosome loss are aneuploid cells, and there is no evidence for ascospore formation. *C. albicans* tetraploids also underwent chromosome loss during bloodstream infection of a mammalian host (Ibrahim et al., 2005). The parasexual mechanism of chromosome loss may have direct benefits for the lifestyle of *C. albicans* as sexual spores are likely to be antigenic and hence are easily targeted by the host immune system (Heitman, 2006). Despite lacking the hallmarks of sexual sporulation, the products of the parasexual mating cycle included several that had undergone recombination between chromosome homologs (Forche et al., 2008). Furthermore, deletion of the meiosis-specific gene *SPO11* resulted in progeny that no longer exhibited interhomolog recombination (Forche et al., 2008). It therefore remains to be seen if *C. albicans* undergoes a variant form of meiosis or whether parasexual chromosome loss is due to mitotic chromosome nondisjunction events.

SEXUAL REPRODUCTION IN OTHER *CANDIDA* SPECIES

Among the *Candida* clade, several other pathogenic species have been shown to undergo sexual reproduction. For example, *Candida lusitaniae* is a haploid species that is an emerging human pathogen (Francois et al., 2001; Gargeya et al., 1990) and has a complete sexual cycle culminating in meiosis and sporulation. Reedy et al. demonstrated that *C. lusitaniae* undergoes Spo11-mediated meiotic recombination despite lacking many of the genes that are integral to meiosis in other fungal species (Reedy et al., 2009). Meiosis in *C. lusitaniae* was imprecise, with approximately one-third of the products exhibiting aneuploidy, in a manner reminiscent of *C. albicans* parasexual progeny. The extent of aneuploidy in *C. lusitaniae* meiosis was similar to that in human oogenesis (7–35%) (Hunt and Hassold, 2008), indicating that aneuploidy is a common by-product of meiosis in highly diverse eukaryotic species. The *C. lusitaniae* studies were also significant as they revealed that genomic analysis alone cannot accurately predict whether a species has an extant sexual cycle (Butler et al., 2009; Reedy et al., 2009; Schurko and Logsdon, 2008).

Candida tropicalis and *C. parapsilosis* are two diploid species that are both major human pathogens and close relatives of *C. albicans*, but neither species has yet been observed to mate. *C. parapsilosis* isolates were shown to be exclusively of the *MTL***a**/**a** mating type where the *MTL***a**1 gene is a pseudogene, suggesting that the *MTL* locus is degenerating and sexual reproduction may no longer be possible in this species (Logue et al., 2005; Sai et al., 2011). Two additional species, *Candida orthopsilosis* and *Candida metapsilosis*, which were formerly known as *C. parapsilosis* group II and group III, respectively, have now been analyzed. All *C. metapsilosis* isolates analyzed to date are *MTL*α/α strains, while *C. orthopsilosis* consists of a mixture of *MTL***a**/**a**, *MTL*α/α, and *MTL***a**/α strains (Sai et al., 2011). Despite the fact that both mating types of *C. orthopsilosis* exist, no mating between these isolates has been detected. It remains to be seen if the appropriate conditions for mating have yet to be identified for these species, as population genetics suggests that recombination is occurring among isolates of *C. tropicalis*, *C. orthopsilosis*, and *C. metapsilosis* (Hensgens et al., 2009; Jacobsen et al., 2008b; Tavanti et al., 2007).

Another *Candida* clade species of note is *Lodderomyces elongisporus*. This species causes rare infections in the clinic, and analysis of its genome sequence revealed it lacks an *MTL* locus (Butler et al., 2009; Lockhart et al., 2008). Despite the absence of this locus, it is reported to be a homothallic yeast able to generate sexual ascospores. Curiously, the sequenced *L. elongisporus* genome contains homologs of α pheromone and α-pheromone receptor, but not **a** pheromone or **a**-pheromone receptor (Butler et al., 2009). While the mechanism of mating and spore formation has not been identified, it is speculated that it could utilize an autocrine pheromone mechanism similar to that observed in self-mating strains of *C. albicans* (Alby et al., 2009).

MATING AND PATHOGENESIS IN *CANDIDA*

Mating in pathogenic *Candida* species appears to be rare as populations are generally clonal with only limited evidence for genetic exchange (Graser et al., 1996; Tibayrenc, 1997). The majority of *C. albicans* isolates are diploid *MTL***a**/α strains, which could preclude mating unless the *MTL* chromosome is lost and the remaining chromosome homolog reduplicated, or mitotic recombination at the mating locus occurs, generating **a**/**a** or α/α cells (Wu et al., 2005, 2007). Alternatively, *MTL* expression could be regulated by environmental factors, thereby allowing *MTL***a**/α cells to be behave phenotypically as **a** or α cells (Pendrak et al., 2004).

White and opaque cells of *C. albicans* exhibit very different characteristics during interaction with the host. These two states differentially express between 400 and 1300 genes (Lan et al., 2002; Tsong et al., 2003; Tuch et al., 2010), and while white cells are more virulent in a systemic model of infection, opaque cells are more effective at colonizing the skin (Kvaal et al., 1999; Lachke et al., 2003). White cells, but not opaque cells, also secrete a chemoattractant that is detected by leukocytes (Geiger et al., 2004) and, perhaps as a result, white cells are phagocytosed by macrophages more efficiently than opaque cells (Lohse and Johnson, 2008). As discussed above, pheromone signaling in white cells can also enhance biofilm formation and is dependent on the canonical MAPK pheromone-signaling cascade. It remains to be seen how this mechanism contributes to biofilms during device-associated infections, a common cause of clinical infections (Douglas, 2002; Nobile and Mitchell, 2006). Overall, the differential behavior of white and opaque cells suggests that these two forms are specialized for growth in different niches *in vivo*, although understanding the role of the two phenotypic states during infection is rudimentary at best (Lohse and Johnson, 2009; Morschhauser, 2010; Soll, 2009).

C. albicans **a**/**a** and α/α isolates exhibit slightly decreased virulence compared to **a**/α isolates, and loss of heterozygosity of non-*MTL* genes on the same chromosome may lead to a further decrease in virulence (Ibrahim et al., 2005; Wu et al., 2007). Tetraploid strains from diploid–diploid matings also show decreased virulence compared to diploid strains (Ibrahim et al., 2005). Recombinant progenies from the parasexual mating cycle have not been tested for virulence but exhibit a wide variety of phenotypes and are therefore likely to differ in their ability to infect the host (Forche et al., 2008). Given that parasexual progeny are often aneuploidy, it will also be interesting to determine their resistance to antifungal agents as chromosomal aneuploidies are directly implicated in causing increased drug resistance (Selmecki et al., 2006, 2008).

SEX IN OTHER HUMAN FUNGAL PATHOGENS

Genomes of multiple human pathogenic fungi have now been sequenced. With the exception of the rare pathogen *L. elongisporus*, all of the genomes have revealed the presence of mating-type loci and in some cases have provided new insight into the role of mating in pathogenesis (Heitman, 2006; Lee et al., 2010; Nielsen and Heitman, 2007). For example, genome sequencing has led to analysis of the mating-type loci in the pathogenic zygomycete *Mucor circinelloides* and the identification of differences in spore size and infectivity in animal models (Li et al., 2011).

In *Mucor*, the (+) sex locus encodes SexP, and the (−) sex locus encodes SexM, the HMG domain sex determination proteins. The (−) mating type produces larger sporangiospores (asexual spores) than the (+) mating type (Fig. 9.2). The larger sporangiospores are also more virulent in both wax moth and diabetic murine models and are able to germinate inside and lyse macrophages. Whereas *C. neoformans* titan cells exhibit increased ploidy within a single nucleus, the (−) mating-type sporangiospores are polyploid and contain multiple nuclei (Fig. 9.2). Interestingly, *sexMΔ* mutants still produce larger, virulent sporangiospores, suggesting a complex mechanism of mating-type-specific spore size regulation.

Not only do these studies show similarity to the size dimorphism observed during *C. neoformans* infection but they are also reminiscent of early work on *H. capsulatum*. Work done in the 1970s showed that both mating types (+ and –) of *H. capsulatum* are prevalent in the environment, but clinical infections are predominantly caused by the – mating type (Kwon-Chung et al., 1974, 1984). In contrast, both mating types were equally virulent in mice when infected with yeast cells. The molecular basis for the increased infectivity of – mating-type cells of *H. capsulatum* in humans is unknown, but spore size dimorphism as now reported in *Mucor* is one possibility.

SEX AND FUNGAL PATHOGENS: A BROADER VIEW

What have we learned from studying sex in human fungal pathogens? First of all, sexual reproduction is apparently playing a paradoxical role in these species. On one hand, heterothallic (opposite-sex) mating is limited in these fungi, perhaps indicative that extensive outcrossing and genetic recombination can be detrimental to these species. On the other hand, several fungal pathogens have developed elaborate mechanisms to promote inbreeding, including same-sex mating systems that may allow rapid strain evolution and recombination but without the reassortment of genes that are critical for pathogenesis. In addition to the anticipated roles of sex in generating recombinant strains, other more cryptic roles are now emerging. This includes a potential role for the sexual machinery in biofilm formation, drug resistance, and even specialized changes in cell morphology associated with pathogenesis. Additional studies will further reveal the exciting, yet complex, role of sex and sex-related processes in the evolution and virulence of the human fungal pathogens.

ACKNOWLEDGMENTS

This work was supported by the NIH (AI081560 and AI081704 grants to R.J.B.; AI080275, AI070152, and AI089244 to K.N.) and the NSF (MCB1021120 to R.J.B.). R.J.B. holds an investigator in the Pathogenesis of Infectious Disease Award from the Burroughs Wellcome Fund. The authors also thank Hubert Fuller, Joseph Heitman, Matthew Hirakawa, Jean Paul Latgé, Soo Chan Lee, Celine O'Gorman, and Laura Okagaki for images used in this chapter.

REFERENCES

Alby K, Bennett RJ. 2009. Stress-induced phenotypic switching in *Candida albicans*. Mol Biol Cell 20: 3178–3191.

Alby K, Bennett RJ. 2011. Interspecies pheromone signaling promotes biofilm formation and same-sex mating in *Candida albicans*. Proc Natl Acad Sci U S A 108: 2510–2515.

Alby K, Schaefer D, Bennett RJ. 2009. Homothallic and heterothallic mating in the opportunistic pathogen *Candida albicans*. Nature 460: 890–893.

Alvarez-Perez S, Blanco JL, Alba P, Garcia ME. 2010. Mating type and invasiveness are significantly associated in *Aspergillus fumigatus*. Med Mycol 48: 273–277.

Bain JM, Tavanti A, Davidson AD, Jacobsen MD, Shaw D, Gow NA, Odds FC. 2007. Multilocus sequence typing of the pathogenic fungus *Aspergillus fumigatus*. J Clin Microbiol 45: 1469–1477.

Barrett SC. 2002. The evolution of plant sexual diversity. Nat Rev Genet 3: 274–284.

Bartlett KH, Kidd SE, Kronstad JW. 2008. The emergence of *Cryptococcus gattii* in British Columbia and the Pacific Northwest. Curr Infect Dis Rep 10: 58–65.

Belay T, Cherniak R, O'Neill EB, Kozel TR. 1996. Serotyping of *Cryptococcus neoformans* by dot enzyme assay. J Clin Microbiol 34: 466–470.

Bennett RJ, Johnson AD. 2003. Completion of a parasexual cycle in *Candida albicans* by induced chromosome loss in tetraploid strains. EMBO J 22: 2505–2515.

Botts MR, Giles SS, Gates MA, Kozel TR, Hull CM. 2009. Isolation and characterization of *Cryptococcus neoformans* spores reveal a critical role for capsule biosynthesis genes in spore biogenesis. Eukaryot Cell 8: 595–605.

Bui T, Lin X, Malik R, Heitman J, Carter D. 2008. Isolates of *Cryptococcus neoformans* from infected animals reveal genetic exchange in unisexual, alpha mating type populations. Eukaryot Cell 7: 1771–1780.

Butler G. 2010. Fungal sex and pathogenesis. Clin Microbiol Rev 23: 140–159.

Butler G, Rasmussen MD, Lin MF, Santos MA, Sakthikumar S, Munro CA, Rheinbay E, Grabherr M, Forche A, Reedy JL, Agrafioti I, Arnaud MB, Bates S, Brown AJ, Brunke S, Costanzo MC, Fitzpatrick DA, De Groot PW, Harris D, Hoyer LL, Hube B, Klis FM, Kodira C, Lennard N, Logue ME, Martin R, Neiman AM, Nikolaou E, Quail MA, Quinn J, Santos MC, Schmitzberger FF, Sherlock G, Shah P, Silverstein KA, Skrzypek MS, Soll D, Staggs R, Stansfield I, Stumpf MP, Sudbery PE, Srikantha T, Zeng Q, Berman J, Berriman M, Heitman J, Gow NA, Lorenz MC, Birren BW, Kellis M, Cuomo CA. 2009. Evolution of pathogenicity and sexual reproduction in eight *Candida* genomes. Nature 459: 657–662.

Byrnes EJ, Heitman J. 2009. *Cryptococcus gattii* outbreak expands into the Northwestern United States with fatal consequences. F1000 Biol Rep 1: 62.

Byrnes EJ, 3rd, Li W, Lewit Y, Ma H, Voelz K, Ren P, Carter DA, Chaturvedi V, Bildfell RJ, May RC, Heitman J. 2010. Emergence and pathogenicity of highly virulent *Cryptococcus gattii* genotypes in the northwest United States. PLoS Pathog 6: e1000850.

Campbell LT, Currie BJ, Krockenberger M, Malik R, Meyer W, Heitman J, Carter D. 2005. Clonality and recombination in genetically differentiated subgroups of *Cryptococcus gattii*. Eukaryot Cell 4: 1403–1409.

Cheema MS, Christians JK. 2011. Virulence in an insect model differs between mating types in *Aspergillus fumigatus*. Med Mycol 49: 202–207.

Cruickshank JG, Cavill R, Jelbert M. 1973. *Cryptococcus neoformans* of unusual morphology. Appl Microbiol 25: 309–312.

Daniels KJ, Srikantha T, Lockhart SR, Pujol C, Soll DR. 2006. Opaque cells signal white cells to form biofilms in *Candida albicans*. EMBO J 25: 2240–2252.

Datta K, Bartlett KH, Baer R, Byrnes E, Galanis E, Heitman J, Hoang L, Leslie MJ, Macdougall L, Magill SS, Morshed MG, Marr KA. 2009. Spread of *Cryptococcus gattii* into Pacific Northwest region of the United States. Emerg Infect Dis 15: 1185–1191.

Davidson RC, Nichols CB, Cox GM, Perfect JR, Heitman J. 2003. A MAP kinase cascade composed of cell type specific and non-specific elements controls mating and differentiation of the fungal pathogen *Cryptococcus neoformans*. Mol Microbiol 49: 469–485.

Douglas LJ. 2002. Medical importance of biofilms in *Candida* infections. Rev Iberoam Micol 19: 139–143.

Driver JA, Saunders CA, Heinze-Lacey B, Sugar AM. 1995. Cryptococcal pneumonia in AIDS: Is cryptococcal meningitis preceded by clinically recognizable pneumonia? J Acquir Immune Defic Syndr Hum Retrovirol 9: 168–171.

D'Souza CA, Kronstad JW, Taylor G, Warren R, Yuen M, Hu G, Jung WH, Sham A, Kidd SE, Tangen K, Lee N, Zeilmaker T, Sawkins J, Mcvicker G, Shah S, Gnerre S, Griggs A, Zeng Q, Bartlett K, Li W, Wang X, Heitman J, Stajich JE, Fraser JA, Meyer W, Carter D, Schein J, Krzywinski M, Kwon-Chung KJ, Varma A, Wang J, Brunham R, Fyfe M, Ouellette BF, Siddiqui A, Marra M, Jones S, Holt R, Birren BW, Galagan JE, Cuomo CA. 2011. Genome variation in *Cryptococcus gattii*, an emerging pathogen of immunocompetent hosts. MBio 2: e00342–10.

Dumitru R, Navarathna DH, Semighini CP, Elowsky CG, Dumitru RV, Dignard D, Whiteway M, Atkin AL, Nickerson KW. 2007. In vivo and in vitro anaerobic mating in *Candida albicans*. Eukaryot Cell 6: 465–472.

Dyer PS. 2007. Sexual reproduction and significance of *MAT* in the *Aspergilli*. In: Heitman J, ed. *Sex in Fungi*. ASM Press, Washington, DC.

Dyer PS, Paoletti M. 2005. Reproduction in *Aspergillus fumigatus*: Sexuality in a supposedly asexual species? Med Mycol 43(Suppl. 1): S7–S14.

Edelmann A, Kruger M, Schmid J. 2005. Genetic relationship between human and animal isolates of *Candida albicans*. J Clin Microbiol 43: 6164–6166.

Ekena JL, Stanton BC, Schiebe-Owens JA, Hull CM. 2008. Sexual development in *Cryptococcus neoformans* requires CLP1, a target of the homeodomain transcription factors Sxi1alpha and Sxi2a. Eukaryot Cell 7: 49–57.

Ezov TK, Boger-Nadjar E, Frenkel Z, Katsperovski I, Kemeny S, Nevo E, Korol A, Kashi Y. 2006. Molecular-genetic biodiversity in a natural population of the yeast *Saccharomyces cerevisiae* from "Evolution Canyon": Microsatellite polymorphism, ploidy and controversial sexual status. Genetics 174: 1455–1468.

Ezov TK, Chang SL, Frenkel Z, Segre AV, Bahalul M, Murray AW, Leu JY, Korol A, Kashi Y. 2010. Heterothallism in *Saccharomyces cerevisiae* isolates from nature: Effect of HO locus on the mode of reproduction. Mol Ecol 19: 121–131.

Forche A, Alby K, Schaefer D, Johnson AD, Berman J, Bennett RJ. 2008. The parasexual cycle in *Candida albicans* provides an alternative pathway to meiosis for the formation of recombinant strains. PLoS Biol 6: e110.

Francois F, Noel T, Pepin R, Brulfert A, Chastin C, Favel A, Villard J. 2001. Alternative identification test relying upon sexual reproductive abilities of *Candida lusitaniae* strains isolated from hospitalized patients. J Clin Microbiol 39: 3906–3914.

Fraser JA, Giles SS, Wenink EC, Geunes-Boyer SG, Wright JR, Diezmann S, Allen A, Stajich JE, Dietrich FS, Perfect JR, Heitman J. 2005a. Same-sex mating and the origin of the Vancouver Island *Cryptococcus gattii* outbreak. Nature 437: 1360–1364.

Fraser JA, Huang JC, Pukkila-Worley R, Alspaugh JA, Mitchell TG, Heitman J. 2005b. Chromosomal translocation and segmental duplication in *Cryptococcus neoformans*. Eukaryot Cell 4: 401–406.

Gargeya IB, Pruitt WR, Simmons RB, Meyer SA, Ahearn DG. 1990. Occurrence of *Clavispora lusitaniae*, the teleomorph of *Candida lusitaniae*, among clinical isolates. J Clin Microbiol 28: 2224–2227.

Geiger J, Wessels D, Lockhart SR, Soll DR. 2004. Release of a potent polymorphonuclear leukocyte chemoattractant is regulated by white-opaque switching in *Candida albicans*. Infect Immun 72: 667–677.

Giles SS, Dagenais TR, Botts MR, Keller NP, Hull CM. 2009. Elucidating the pathogenesis of spores from the human fungal pathogen *Cryptococcus neoformans*. Infect Immun 77: 3491–3500.

Graser Y, Volovsek M, Arrington J, Schonian G, Presber W, Mitchell TG, Vilgalys R. 1996. Molecular markers reveal that population structure of the human pathogen *Candida albicans* exhibits both clonality and recombination. Proc Natl Acad Sci U S A 93: 12473–12477.

Groopman JD, Kensler TW. 2005. Role of metabolism and viruses in aflatoxin-induced liver cancer. Toxicol Appl Pharmacol 206: 131–137.

Grosse V, Krappmann S. 2008. The asexual pathogen *Aspergillus fumigatus* expresses functional determinants of *Aspergillus nidulans* sexual development. Eukaryot Cell 7: 1724–1732.

Hawksworth DL. 2010. Naming *Aspergillus* species: Progress towards one name for each species. Med Mycol 49(Suppl. 1): S70–S76.

Hedayati MT, Pasqualotto AC, Warn PA, Bowyer P, Denning DW. 2007. *Aspergillus flavus*: Human pathogen, allergen and mycotoxin producer. Microbiology 153: 1677–1692.

Heitman J. 2006. Sexual reproduction and the evolution of microbial pathogens. Curr Biol 16: R711–R725.

Heitman J. 2010. Evolution of eukaryotic microbial pathogens via covert sexual reproduction. Cell Host Microbe 8: 86–99.

Hensgens LA, Tavanti A, Mogavero S, Ghelardi E, Senesi S. 2009. AFLP genotyping of *Candida metapsilosis* clinical isolates: Evidence for recombination. Fungal Genet Biol 46: 750–758.

Hohl TM, Feldmesser M. 2007. *Aspergillus fumigatus*: Principles of pathogenesis and host defense. Eukaryot Cell 6: 1953–1963.

Horn BW, Moore GG, Carbone I. 2009a. Sexual reproduction in *Aspergillus flavus*. Mycologia 101: 423–429.

Horn BW, Ramirez-Prado JH, Carbone I. 2009b. The sexual state of *Aspergillus parasiticus*. Mycologia 101: 275–280.

Hsueh YP, Heitman J. 2008. Orchestration of sexual reproduction and virulence by the fungal mating-type locus. Curr Opin Microbiol 11: 517–524.

Hsueh YP, Idnurm A, Heitman J. 2006. Recombination hotspots flank the *Cryptococcus* mating-type locus: Implications for the evolution of a fungal sex chromosome. PLoS Genet 2: e184.

Hsueh YP, Fraser JA, Heitman J. 2008. Transitions in sexuality: Recapitulation of an ancestral tri- and tetrapolar mating system in *Cryptococcus neoformans*. Eukaryot Cell 7: 1847–1855.

Huang G, Wang H, Chou S, Nie X, Chen J, Liu H. 2006. Bistable expression of WOR1, a master regulator of white-opaque switching in *Candida albicans*. Proc Natl Acad Sci U S A 103: 12813–12818.

Huang G, Srikantha T, Sahni N, Yi S, Soll DR. 2009. CO(2) regulates white-to-opaque switching in *Candida albicans*. Curr Biol 19: 330–334.

Huang G, Yi S, Sahni N, Daniels KJ, Srikantha T, Soll DR. 2010. N-acetylglucosamine induces white to opaque switching, a mating prerequisite in *Candida albicans*. PLoS Pathog 6: e1000806.

Hull CM, Heitman J. 2002. Genetics of *Cryptococcus neoformans*. Annu Rev Genet 36: 557–615.

Hull CM, Johnson AD. 1999. Identification of a mating type-like locus in the asexual pathogenic yeast *Candida albicans*. Science 285: 1271–1275.

Hull CM, Raisner RM, Johnson AD. 2000. Evidence for mating of the "asexual" yeast *Candida albicans* in a mammalian host. Science 289: 307–310.

Hull CM, Davidson RC, Heitman J. 2002. Cell identity and sexual development in *Cryptococcus neoformans* are controlled by the mating-type-specific homeodomain protein Sxi1alpha. Genes Dev 16: 3046–3060.

Hull CM, Boily MJ, Heitman J. 2005. Sex-specific homeodomain proteins Sxi1alpha and Sxi2a coordinately regulate sexual development in *Cryptococcus neoformans*. Eukaryot Cell 4: 526–535.

Hunt PA, Hassold TJ. 2008. Human female meiosis: What makes a good egg go bad? Trends Genet 24: 86–93.

Ibrahim AS, Magee BB, Sheppard DC, Yang M, Kauffman S, Becker J, Edwards JE, Jr., Magee PT. 2005. Effects of ploidy and mating type on virulence of *Candida albicans*. Infect Immun 73: 7366–7374.

Jacobsen MD, Bougnoux ME, D'Enfert C, Odds FC. 2008a. Multilocus sequence typing of *Candida albicans* isolates from animals. Res Microbiol 159: 436–440.

Jacobsen MD, Davidson AD, Li SY, Shaw DJ, Gow NA, Odds FC. 2008b. Molecular phylogenetic analysis of *Candida tropicalis* isolates by multi-locus sequence typing. Fungal Genet Biol 45: 1040–1042.

Johnson A. 2003. The biology of mating in *Candida albicans*. Nat Rev Microbiol 1: 106–116.

Kolotila MP, Diamond RD. 1990. Effects of neutrophils and in vitro oxidants on survival and phenotypic switching of *Candida albicans* WO-1. Infect Immun 58: 1174–1179.

Kruzel EK, Hull CM. 2010. Establishing an unusual cell type: How to make a dikaryon. Curr Opin Microbiol 13: 706–711.

Kvaal C, Lachke SA, Srikantha T, Daniels K, Mccoy J, Soll DR. 1999. Misexpression of the opaque-phase-specific gene *PEP1 (SAP1)* in the white phase of *Candida albicans* confers increased virulence in a mouse model of cutaneous infection. Infect Immun 67: 6652–6662.

Kwon-Chung KJ. 1975. A new genus, *Filobasidiella*, the perfect state of *Cryptococcus neoformans*. Mycologia 67: 1197–1200.

Kwon-Chung KJ. 1976a. A new species of *Filobasidiella*, the sexual state of *Cryptococcus neoformans* B and C serotypes. Mycologia 68: 943–946.

Kwon-Chung KJ. 1976b. Morphogenesis of *Filobasidiella neoformans*, the sexual state of *Cryptococcus neoformans*. Mycologia 68: 821–833.

Kwon-Chung KJ, Sugui JA. 2009. Sexual reproduction in *Aspergillus* species of medical or economical importance: Why so fastidious? Trends Microbiol 17: 481–487.

Kwon-Chung KJ, Weeks RJ, Larsh HW. 1974. Studies on *Emmonsiella capsulata* (*Histoplasma capsulatum*). II. Distribution of the two mating types in 13 endemic states of the United States. Am J Epidemiol 99: 44–49.

Kwon-Chung KJ, Bartlett MS, Wheat LJ. 1984. Distribution of the two mating types among *Histoplasma capsulatum* isolates obtained from an urban histoplasmosis outbreak. Sabouraudia 22: 155–157.

Kwon-Chung KJ, Edman JC, Wickes BL. 1992. Genetic association of mating types and virulence in *Cryptococcus neoformans*. Infect Immun 60: 602–605.

Kwon-Chung KJ, Boekhout T, Wickes BL, Fell JW. 2011. Systematics of the genus *Cryptococcus* and its type species *C. neoformans*. In: Heitman J, Kozel TR, Kwon-Chung KJ, Perfect JR, Casadevall A, eds. *Cryptococcus: From Human Pathogen to Model Yeast*. ASM Press, Washington, DC.

Lachke SA, Lockhart SR, Daniels KJ, Soll DR. 2003. Skin facilitates *Candida albicans* mating. Infect Immun 71: 4970–4976.

Lan CY, Newport G, Murillo LA, Jones T, Scherer S, Davis RW, Agabian N. 2002. Metabolic specialization associated with phenotypic switching in *Candida albicans*. Proc Natl Acad Sci U S A 99: 14907–14912.

Latge JP. 1999. *Aspergillus fumigatus* and aspergillosis. Clin Microbiol Rev 12: 310–350.

Lee SC, Ni M, Li W, Shertz C, Heitman J. 2010. The evolution of sex: A perspective from the fungal kingdom. Microbiol Mol Biol Rev 74: 298–340.

Lengeler KB, Wang P, Cox GM, Perfect JR, Heitman J. 2000. Identification of the MATa mating-type locus of *Cryptococcus neoformans* reveals a serotype A MATa strain thought to have been extinct. Proc Natl Acad Sci U S A 97: 14455–14460.

Lengeler KB, Fox DS, Fraser JA, Allen A, Forrester K, Dietrich FS, Heitman J. 2002. Mating-type locus of *Cryptococcus neoformans*: A step in the evolution of sex chromosomes. Eukaryot Cell 1: 704–718.

Li CH, Cervantes M, Springer DJ, Boekhout T, Ruiz-Vasquez RM, Torres-Martinez SR, Heitman J, Lee SC. 2011. Sporangiospore size dimorphism is linked to virulence of *Mucor circinelloides*. PLoS Pathog 7: e1002086.

Lin X, Heitman J. 2006. The biology of the *Cryptococcus neoformans* species complex. Annu Rev Microbiol 60: 69–105.

Lin X, Hull CM, Heitman J. 2005. Sexual reproduction between partners of the same mating type in *Cryptococcus neoformans*. Nature 434: 1017–1021.

Lin X, Huang JC, Mitchell TG, Heitman J. 2006. Virulence attributes and hyphal growth of *C. neoformans* are quantitative traits and the *MATalpha* allele enhances filamentation. PLoS Genet 2: e187.

Lin X, Patel S, Litvintseva AP, Floyd A, Mitchell TG, Heitman J. 2009. Diploids in the *Cryptococcus neoformans* serotype A population homozygous for the alpha mating type originate via unisexual mating. PLoS Pathog 5: e1000283.

Lin X, Jackson JC, Feretzaki M, Xue C, Heitman J. 2010. Transcription factors Mat2 and Znf2 operate cellular circuits orchestrating opposite- and same-sex mating in *Cryptococcus neoformans*. PLoS Genet 6: e1000953.

Litvintseva AP, Marra RE, Nielsen K, Heitman J, Vilgalys R, Mitchell TG. 2003. Evidence of sexual recombination among *Cryptococcus neoformans* serotype A isolates in sub-Saharan Africa. Eukaryot Cell 2: 1162–1168.

Litvintseva AP, Thakur R, Reller LB, Mitchell TG. 2005. Prevalence of clinical isolates of *Cryptococcus gattii* serotype C among patients with AIDS in sub-Saharan Africa. J Infect Dis 192: 888–892.

Litvintseva AP, Carbone I, Rossouw J, Thakur R, Govender NP, Mitchell TG. 2011. Evidence that the human pathogenic fungus *Cryptococcus neoformans* var. *grubii* may have evolved in Africa. PLoS One 6: e19688.

Lockhart SR, Pujol C, Daniels KJ, Miller MG, Johnson AD, Pfaller MA, Soll DR. 2002. In *Candida albicans*, white-opaque switchers are homozygous for mating type. Genetics 162: 737–745.

Lockhart SR, Zhao R, Daniels KJ, Soll DR. 2003. Alpha-pheromone-induced "shmooing" and gene regulation require white-opaque switching during *Candida albicans* mating. Eukaryot Cell 2: 847–855.

Lockhart SR, Messer SA, Pfaller MA, Diekema DJ. 2008. *Lodderomyces elongisporus* masquerading as *Candida parapsilosis* as a cause of bloodstream infections. J Clin Microbiol 46: 374–376.

Logue ME, Wong S, Wolfe KH, Butler G. 2005. A genome sequence survey shows that the pathogenic yeast *Candida parapsilosis* has a defective *MTLa1* allele at its mating type locus. Eukaryot Cell 4: 1009–1017.

Lohse MB, Johnson AD. 2008. Differential phagocytosis of white versus opaque *Candida albicans* by *Drosophila* and mouse phagocytes. PLoS One 3: e1473.

Lohse MB, Johnson AD. 2009. White-opaque switching in *Candida albicans*. Curr Opin Microbiol 12: 650–654.

Lohse MB, Zordan RE, Cain CW, Johnson AD. 2010. Distinct class of DNA-binding domains is exemplified by a master regulator of phenotypic switching in *Candida albicans*. Proc Natl Acad Sci U S A 107: 14105–14110.

Love GL, Boyd GD, Greer DL. 1985. Large *Cryptococcus neoformans* isolated from brain abscess. J Clin Microbiol 22: 1068–1070.

Magee BB, Magee PT. 2000. Induction of mating in *Candida albicans* by construction of *MTLa* and *MTLalpha* strains. Science 289: 310–313.

Mcclelland CM, Chang YC, Varma A, Kwon-Chung KJ. 2004. Uniqueness of the mating system in *Cryptococcus neoformans*. Trends Microbiol 12: 208–212.

Metin B, Findley K, Heitman J. 2010. The mating type locus (MAT) and sexual reproduction of *Cryptococcus heveanensis*: Insights into the evolution of sex and sex-determining chromosomal regions in fungi. PLoS Genet 6: e1000961.

Michielse CB, Van Wijk R, Reijnen L, Manders EM, Boas S, Olivain C, Alabouvette C, Rep M. 2009. The nuclear protein Sge1 of *Fusarium oxysporum* is required for parasitic growth. PLoS Pathog 5: e1000637.

Miller MG, Johnson AD. 2002. White-opaque switching in *Candida albicans* is controlled by mating-type locus homeodomain proteins and allows efficient mating. Cell 110: 293–302.

Morgan J, Wannemuehler KA, Marr KA, Hadley S, Kontoyiannis DP, Walsh TJ, Fridkin SK, Pappas PG, Warnock DW. 2005. Incidence of invasive aspergillosis following hematopoietic stem cell and solid organ transplantation: Interim results of a prospective multicenter surveillance program. Med Mycol 43(Suppl. 1): S49–S58.

Morschhauser J. 2010. Regulation of white-opaque switching in *Candida albicans*. Med Microbiol Immunol (Berl) 199: 165–172.

Nguyen VQ, Sil A. 2008. Temperature-induced switch to the pathogenic yeast form of *Histoplasma capsulatum* requires Ryp1, a conserved transcriptional regulator. Proc Natl Acad Sci U S A 105: 4880–4885.

Ni M, Feretzaki M, Sun S, Wang X, Heitman J. 2011. Sex in Fungi. Annu Rev Genet 45: 405–430.

Nielsen K, Heitman J. 2007. Sex and virulence of human pathogenic fungi. Adv Genet 57: 143–173.

Nielsen K, Cox GM, Wang P, Toffaletti DL, Perfect JR, Heitman J. 2003. Sexual cycle of *Cryptococcus neoformans* var. *grubii* and virulence of congenic a and alpha isolates. Infect Immun 71: 4831–4841.

Nielsen K, Cox GM, Litvintseva AP, Mylonakis E, Malliaris SD, Benjamin DK, Jr., Giles SS, Mitchell TG, Casadevall A, Perfect JR, Heitman J. 2005a. *Cryptococcus neoformans* {alpha} strains preferentially disseminate to the central nervous system during coinfection. Infect Immun 73: 4922–4933.

Nielsen K, Marra RE, Hagen F, Boekhout T, Mitchell TG, Cox GM, Heitman J. 2005b. Interaction between genetic background and the mating-type locus in *Cryptococcus neoformans* virulence potential. Genetics 171: 975–983.

Nierman WC, Pain A, Anderson MJ, Wortman JR, Kim HS, Arroyo J, Berriman M, Abe K, Archer DB, Bermejo C, Bennett J, Bowyer P, Chen D, Collins M, Coulsen R, Davies R, Dyer PS, Farman M, Fedorova N, Feldblyum TV, Fischer R, Fosker N, Fraser A,

Garcia JL, Garcia MJ, Goble A, Goldman GH, Gomi K, Griffith-Jones S, Gwilliam R, Haas B, Haas H, Harris D, Horiuchi H, Huang J, Humphray S, Jimenez J, Keller N, Khouri H, Kitamoto K, Kobayashi T, Konzack S, Kulkarni R, Kumagai T, Lafon A, Latge JP, Li W, Lord A, Lu C, Majoros WH, May GS, Miller BL, Mohamoud Y, Molina M, Monod M, Mouyna I, Mulligan S, Murphy L, O'Neil S, Paulsen I, Penalva MA, Pertea M, Price C, Pritchard BL, Quail MA, Rabbinowitsch E, Rawlins N, Rajandream MA, Reichard U, Renauld H, Robson GD, Rodriguez De Cordoba S, Rodriguez-Pena JM, Ronning CM, Rutter S, Salzberg SL, Sanchez M, Sanchez-Ferrero JC, Saunders D, Seeger K, Squares R, Squares S, Takeuchi M, Tekaia F, Turner G, Vazquez De Aldana CR, Weidman J, White O, Woodward J, Yu JH, Fraser C, Galagan JE, Asai K, Machida M, Hall N, Barrell B, Denning DW. 2005. Genomic sequence of the pathogenic and allergenic filamentous fungus *Aspergillus fumigatus*. Nature 438: 1151–1156.

Nobile CJ, Mitchell AP. 2006. Genetics and genomics of *Candida albicans* biofilm formation. Cell Microbiol 8: 1382–1391.

Odds FC, Bougnoux ME, Shaw DJ, Bain JM, Davidson AD, Diogo D, Jacobsen MD, Lecomte M, Li SY, Tavanti A, Maiden MC, Gow NA, D'Enfert C. 2007. Molecular phylogenetics of *Candida albicans*. Eukaryot Cell 6: 1041–1052.

O'Gorman CM, Fuller HT, Dyer PS. 2009. Discovery of a sexual cycle in the opportunistic fungal pathogen *Aspergillus fumigatus*. Nature 457: 471–474.

Okagaki LH, Strain AK, Nielsen JN, Charlier C, Baltes NJ, Chretien F, Heitman J, Dromer F, Nielsen K. 2010. Cryptococcal cell morphology affects host cell interactions and pathogenicity. PLoS Pathog 6: e1000953.

Okagaki LH, Wang Y, Ballou ER, O'Meara TR, Alspaugh JA, Xue C, Nielsen K. 2011. Cryptococcal titan cell formation is regulated by G-protein signaling. Eukaryot Cell 10: 1306–1316.

Panwar SL, Legrand M, Dignard D, Whiteway M, Magee PT. 2003. MFalpha1, the gene encoding the alpha mating pheromone of *Candida albicans*. Eukaryot Cell 2: 1350–1360.

Paoletti M, Rydholm C, Schwier EU, Anderson MJ, Szakacs G, Lutzoni F, Debeaupuis JP, Latge JP, Denning DW, Dyer PS. 2005. Evidence for sexuality in the opportunistic fungal pathogen *Aspergillus fumigatus*. Curr Biol 15: 1242–1248.

Paoletti M, Seymour FA, Alcocer MJ, Kaur N, Calvo AM, Archer DB, Dyer PS. 2007. Mating type and the genetic basis of self-fertility in the model fungus *Aspergillus nidulans*. Curr Biol 17: 1384–1389.

Park BJ, Wannemuehler KA, Marston BJ, Govender N, Pappas PG, Chiller TM. 2009. Estimation of the current global burden of cryptococcal meningitis among persons living with HIV/AIDS. AIDS 23: 525–530.

Park BJ, Lockhart SR, Brandt ME, Chiller TM. 2011. Public health importance of cryptococcal disease: Epidemiology, burden, and control. In: Heitman J, Kozel TR, Kwon-Chung KJ, Perfect JR, Casadevall A, eds. *Cryptococcus: From Human Pathogen to Model Yeast*. ASM Press, Washington, DC.

Pendrak ML, Yan SS, Roberts DD. 2004. Hemoglobin regulates expression of an activator of mating-type locus alpha genes in *Candida albicans*. Eukaryot Cell 3: 764–775.

Poggeler S. 2002. Genomic evidence for mating abilities in the asexual pathogen *Aspergillus fumigatus*. Curr Genet 42: 153–160.

Pontecorvo G, Roper JA, Hemmons LM, Macdonald KD, Bufton AW. 1953. The genetics of *Aspergillus nidulans*. Adv Genet 5: 141–238.

Pringle A, Baker DM, Platt JL, Wares JP, Latge JP, Taylor JW. 2005. Cryptic speciation in the cosmopolitan and clonal human pathogenic fungus *Aspergillus fumigatus*. Evolution 59: 1886–1899.

Pyrzak W, Miller KY, Miller BL. 2008. Mating type protein Mat1-2 from asexual *Aspergillus fumigatus* drives sexual reproduction in fertile *Aspergillus nidulans*. Eukaryot Cell 7: 1029–1040.

Ramirez-Prado JH, Moore GG, Horn BW, Carbone I. 2008. Characterization and population analysis of the mating-type genes in *Aspergillus flavus* and *Aspergillus parasiticus*. Fungal Genet Biol 45: 1292–1299.

Ramirez-Zavala B, Reuss O, Park YN, Ohlsen K, Morschhauser J. 2008. Environmental induction of white-opaque switching in *Candida albicans*. PLoS Pathog 4: e1000089.

Reedy JL, Heitman J. 2007. Evolution of *MAT* in the *Candida* species complex: Sex, ploidy, and complete sexual cycles in *C. lusitaniae*, *C. guilliermondii*, and *C. krusei*. In: Heitman J, ed. *Sex in Fungi*. ASM Press, Washington, DC.

Reedy JL, Floyd AM, Heitman J. 2009. Mechanistic plasticity of sexual reproduction and meiosis in the *Candida* pathogenic species complex. Curr Biol 19: 891–899.

Ruhnke M, Maschmeyer G. 2002. Management of mycoses in patients with hematologic disease and cancer—Review of the literature. Eur J Med Res 7: 227–235.

Rydholm C, Szakacs G, Lutzoni F. 2006. Low genetic variation and no detectable population structure in *Aspergillus fumigatus* compared to closely related *Neosartorya* species. Eukaryot Cell 5: 650–657.

Sahni N, Yi S, Daniels KJ, Huang G, Srikantha T, Soll DR. 2010. Tec1 mediates the pheromone response of the white phenotype of *Candida albicans*: Insights into the evolution of new signal transduction pathways. PLoS Biol 8: e1000363.

Sai S, Holland L, Mcgee CF, Lynch DB, Butler G. 2011. Evolution of mating within the *Candida parapsilosis* species group. Eukaryot Cell 10: 578–587.

Saul N, Krockenberger M, Carter D. 2008. Evidence of recombination in mixed-mating-type and alpha-only populations of *Cryptococcus gattii* sourced from single eucalyptus tree hollows. Eukaryot Cell 7: 727–734.

Scazzocchio C. 2006. *Aspergillus* genomes: Secret sex and the secrets of sex. Trends Genet 22: 521–525.

Schaefer D, Cote P, Whiteway M, Bennett RJ. 2007. Barrier activity in *Candida albicans* mediates pheromone degradation and promotes mating. Eukaryot Cell 6: 907–918.

Schoustra SE, Debets AJ, Slakhorst M, Hoekstra RF. 2007. Mitotic recombination accelerates adaptation in the fungus *Aspergillus nidulans*. PLoS Genet 3: e68.

Schurko AM, Logsdon JM, Jr. 2008. Using a meiosis detection toolkit to investigate ancient asexual "scandals" and the evolution of sex. Bioessays 30: 579–589.

Selmecki A, Forche A, Berman J. 2006. Aneuploidy and isochromosome formation in drug-resistant *Candida albicans*. Science 313: 367–370.

Selmecki A, Gerami-Nejad M, Paulson C, Forche A, Berman J. 2008. An isochromosome confers drug resistance in vivo by amplification of two genes, *ERG11* and *TAC1*. Mol Microbiol 68: 624–641.

Simwami SP, Khayhan K, Henk DA, Aanensen DM, Boekhout T, Hagen F, Brouwer AE, Harrison TS, Donnelly CA, Fisher MC. 2011. Low diversity *Cryptococcus neoformans* variety *grubii* multilocus sequence types from Thailand are consistent with an ancestral African origin. PLoS Pathog 7: e1001343.

Slutsky B, Staebell M, Anderson J, Risen L, Pfaller M, Soll DR. 1987. "White-opaque transition": A second high-frequency switching system in *Candida albicans*. J Bacteriol 169: 189–197.

Soll DR. 2009. Why does *Candida albicans* switch? FEMS Yeast Res 9: 973–989.

Srikantha T, Borneman AR, Daniels KJ, Pujol C, Wu W, Seringhaus MR, Gerstein M, Yi S, Snyder M, Soll DR. 2006. *TOS9* regulates white-opaque switching in *Candida albicans*. Eukaryot Cell 5: 1674–1687.

Staib F, Mishra SK, Rajendran C, Voigt R, Steffen J, Neumann KH, Hartmann CA, Heins G. 1980. A notable *Aspergillus* from a mortal aspergilloma of the lung. New aspects of the epidemiology, serodiagnosis and taxonomy of *Aspergillus fumigatus*. Zentralbl Bakteriol A 247: 530–536.

Stanton BC, Giles SS, Staudt MW, Kruzel EK, Hull CM. 2010. Allelic exchange of pheromones and their receptors reprograms sexual identity in *Cryptococcus neoformans*. PLoS Genet 6: e1000860.

Sukroongreung S, Kitiniyom K, Nilakul C, Tantimavanich S. 1998. Pathogenicity of basidiospores of *Filobasidiella neoformans* var. *neoformans*. Med Mycol 36: 419–424.

Szewczyk E, Krappmann S. 2010. Conserved regulators of mating are essential for *Asper-*

gillus fumigatus cleistothecium formation. Eukaryot Cell 9: 774–783.

Tang C, Toomajian C, Sherman-Broyles S, Plagnol V, Guo YL, Hu TT, Clark RM, Nasrallah JB, Weigel D, Nordborg M. 2007. The evolution of selfing in *Arabidopsis thaliana*. Science 317: 1070–1072.

Tavanti A, Hensgens LA, Ghelardi E, Campa M, Senesi S. 2007. Genotyping of *Candida orthopsilosis* clinical isolates by amplification fragment length polymorphism reveals genetic diversity among independent isolates and strain maintenance within patients. J Clin Microbiol 45: 1455–1462.

Tibayrenc M. 1997. Are *Candida albicans* natural populations subdivided? Trends Microbiol 5: 253–254.

Tscharke RL, Lazera M, Chang YC, Wickes BL, Kwon-Chung KJ. 2003. Haploid fruiting in *Cryptococcus neoformans* is not mating type alpha-specific. Fungal Genet Biol 39: 230–237.

Tsong AE, Miller MG, Raisner RM, Johnson AD. 2003. Evolution of a combinatorial transcriptional circuit: A case study in yeasts. Cell 115: 389–399.

Tuch BB, Mitrovich QM, Homann OR, Hernday AD, Monighetti CK, De La Vega FM, Johnson AD. 2010. The transcriptomes of two heritable cell types illuminate the circuit governing their differentiation. PLoS Genet 6: e1001070.

Varga J, Toth B. 2003. Genetic variability and reproductive mode of *Aspergillus fumigatus*. Infect Genet Evol 3: 3–17.

Velagapudi R, Hsueh YP, Geunes-Boyer S, Wright JR, Heitman J. 2009. Spores as infectious propagules of *Cryptococcus neoformans*. Infect Immun 77: 4345–4355.

Wang X, Hsueh YP, Li W, Floyd A, Skalsky R, Heitman J. 2010. Sex-induced silencing defends the genome of *Cryptococcus neoformans* via RNAi. Genes Dev 24: 2566–2582.

Wickes BL, Mayorga ME, Edman U, Edman JC. 1996. Dimorphism and haploid fruiting in *Cryptococcus neoformans*: Association with the alpha-mating type. Proc Natl Acad Sci U S A 93: 7327–7331.

Wisplinghoff H, Bischoff T, Tallent SM, Seifert H, Wenzel RP, Edmond MB. 2004. Nosocomial bloodstream infections in US hospitals: Analysis of 24,179 cases from a prospective nationwide surveillance study. Clin Infect Dis 39: 309–317.

Wrobel L, Whittington JK, Pujol C, Oh SH, Ruiz MO, Pfaller MA, Diekema DJ, Soll DR, Hoyer LL. 2008. Molecular phylogenetic analysis of a geographically and temporally matched set of *Candida albicans* isolates from humans and nonmigratory wildlife in central Illinois. Eukaryot Cell 7: 1475–1486.

Wu W, Pujol C, Lockhart SR, Soll DR. 2005. Chromosome loss followed by duplication is the major mechanism of spontaneous mating-type locus homozygosis in *Candida albicans*. Genetics 169: 1311–1327.

Wu W, Lockhart SR, Pujol C, Srikantha T, Soll DR. 2007. Heterozygosity of genes on the sex chromosome regulates *Candida albicans* virulence. Mol Microbiol 64: 1587–1604.

Zaragoza O, Garcia-Rodas R, Nosanchuk JD, Cuenca-Estrella M, Rodriguez-Tudela JL, Casadevall A. 2010. Fungal cell gigantism during mammalian infection. PLoS Pathog 6: e1000945.

Zordan RE, Galgoczy DJ, Johnson AD. 2006. Epigenetic properties of white-opaque switching in *Candida albicans* are based on a self-sustaining transcriptional feedback loop. Proc Natl Acad Sci U S A 103: 12807–12812.

Zordan RE, Miller MG, Galgoczy DJ, Tuch BB, Johnson AD. 2007. Interlocking transcriptional feedback loops control white-opaque switching in *Candida albicans*. PLoS Biol 5: e256.

CHAPTER 10

WORLDWIDE MIGRATIONS, HOST SHIFTS, AND REEMERGENCE OF *PHYTOPHTHORA INFESTANS*, THE PLANT DESTROYER

JEAN BEAGLE RISTAINO

INTRODUCTION

Phytophthora species are oomycete plant pathogens that are responsible for devastating diseases on a wide range of food and ornamental crops, natural vegetation, and forest trees worldwide (Erwin and Ribeiro, 1996). Indeed the name *Phytophthora* is derived from the Greek phytó, "plant," and phthorá, "destruction"; thus, members of this genus are often referred to as "the plant destroyer." They represent an emerging food security threat, in large part due to increases in plant movement via international trade (Brasier, 2008). *Phytophthora infestans*, the causal agent of late blight, exemplifies this threat; it was the first species in the genus described and left a path of devastation on potato in its wake in the United States, Ireland, and Europe in the 19th century (Berkeley, 1846; Bourke, 1964; deBary, 1876). Movement of infected potato tubers led to the potato famine epidemics of the 19th century that resulted in widespread human hunger, disease, and, ultimately, death of two million people in Ireland. The pathogen causes a destructive foliar blight of potato and also infects potato tubers and tomato fruit and is still wreaking havoc more than 160 years after the Irish famine (Fry, 2008; Hu et al., 2012).

Late blight is the most important biotic constraint to potato production worldwide and is major threat to food security, particularly in the developing world where use of fungicides is often uneconomical (Anderson et al., 2004; Pennisi, 2010). Worldwide losses due to late blight on potato and tomato exceed $7 billion annually (Haverkort et al., 2008). Evolution of strains varying in sensitivity to fungicides and novel pathotypes of *P. infestans* continue to challenge the sustainable production of potatoes. The pathogen has also reemerged as a significant disease threat to the organic tomato industry in the United States where synthetic chemicals are not used (Stone, 2009). In 2009, late blight epidemics of potato and tomato in the eastern United States were the worst in recent history due to widespread inoculum distribution and weather conducive for disease (Hu et al., 2012; Moskin, 2009).

This chapter provides an overview of the emergence of potato late blight in the 19th century and the evolutionary position, life

Evolution of Virulence in Eukaryotic Microbes, First Edition. Edited by L. David Sibley, Barbara J. Howlett, and Joseph Heitman.

history, and the population biology of *P. infestans*. Worldwide migrations of the pathogen and the role of host shifts in the evolution of species in the genus are also discussed.

EMERGENCE OF LATE BLIGHT IN THE UNITED STATES AND EUROPE

The potato blight struck with a vengeance in the United States in 1843 before it appeared on the European continent and in the British Isles. The potato crop in the United States was a good one in 1842, but the situation changed in 1843. One account documented in the *Annual Report of the Commissioner of Patents in the United States* stated that the "potato yield was nearly 50 percent less owing to a rot which seized them before the time for taking them out of the ground" (Ellsworth, 1843). A second report in the same document stated "The potato crop has been attacked. The cause is generally attributed to the peculiarity of the weather" (Ellsworth, 1843). The late blight epidemic began in 1843 in the United States in a five-state area starting from the ports of New York and Philadelphia (Bourke, 1964; Ellsworth, 1843). The first appearance of the disease near the two port cities of Philadelphia and New York suggested an introduction via imported tubers. Potatoes and bat guano, which were used as fertilizer, were being shipped at that time from Peru in South America into many ports in both the United States and Europe to improve the stock of seed potatoes that had declined from *Fusarium* dry rot. These potatoes probably provided the source of inoculum for the first disease outbreaks.

In Europe, late blight was first noticed on the coast of Belgium in the Courtrai area in June of 1845 (Bourke, 1964). By mid-July, late blight had moved southward into Flanders and parts of the Netherlands and France. By August, it had spread into the lower Rhineland, Switzerland, and to southern England. In 1845, on the 16th of August, it was seen in the Isle of Wight and on the 23rd of August in the South of England. The potato blight made its way into Ireland by the 7th of September, 1845, and later in the year to Scotland (Bourke, 1964).

The farmers and learned men and women of the time immediately started to speculate as to the cause of the "evil" that was now widely affecting potatoes grown everywhere. James Teschemacher, a member of Boston's Natural History Society, associated a minute fungus with the lesions on the potato plant (Teschemacher, 1844). There was by no means a consensus on the subject of the cause of the potato blight, which was attributed to the weather, insects, the wrath of God, and the minute "fungus." Speculation began about the cause of the disease and lively discussions ensued in the *Gardeners Chronicle* in the United Kingdom: "That minute and rapidly propagating fungi have been observed on the affected potato plants, there is no reason to doubt. But when we know, from other facts, that such appear on plants after they are dead or diseased, and not when they are alive and healthy, we are justified in affirming that we are yet ignorant of the cause of the malady in question" (MacKenzie, 1845). Others wrote, "It is certain that a fungus appears in the leaves, stems, and tubers of the plants which have been attacked, but it is uncertain how far the fungus is the cause or the consequence of the disease - how far it is to be considered as a parasite upon the living potato, or as a mere devourer of its dying parts" (Johnson, 1845).

The pathogen was described and named by C. Montagne in France and in the United Kingdom by the Reverend M. J. Berkeley in 1846 as *Botrytis infestans* (Berkeley, 1846; Montagne, 1845). Morren had observed the disease in Belgium in 1844 and called it *Botrytis devastatrix* and

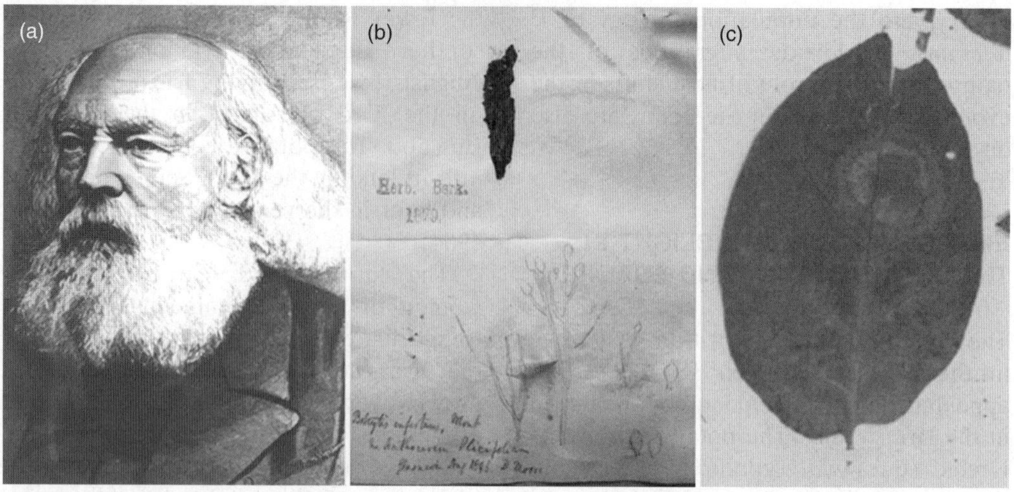

Figure 10.1 (a) M. J. Berkeley, the British mycologist who described the pathogen *Botrytis infestans* (later named as *Phytophthora infestans*) on potato in 1846; (b) sporangia and sporangiophores of *P. infestans* drawn on a famine-era archival specimen label by David Moore, Glasnevin, Ireland, that was sent to Berkeley for confirmation; (c) herbarium specimen with leaf lesion caused by *P. infestans* collected by Krieger in Germany in 1888. See color insert.

presented a paper to the Royal Society of Lille (Morren, 1844). Libert had also called the pathogen *B. devastatrix* (alternate spelling *vastatrix* or *devastrix*) Lib. 1845, but Berkeley chose to use the species name *B. infestans*. In 1847 and 1852, Unger and Caspary, respectively, considered it in the genus *Peronospora* (Unger, 1847). deBary initially accepted this opinion (Cline et al., 2008; deBary, 1863).

Miles J. Berkeley (Fig. 10.1a) and later deBary resolved the debate by documenting that the "fungus-like" organism *P. infestans* was the cause and not the consequence of the disease. David Moore of Ireland drew pictures of the sporangiophores and sporangia of the pathogen and sent specimens to Berkeley, who named the pathogen *B. infestans* (Fig. 10.1b). The work on late blight by Berkeley predated work by Louis Pasteur on the germ theory and clearly documented that a microbe could cause disease (Berkeley, 1846) and was a major contribution to both the development of the science of plant pathology and the germ theory. In 1876, deBary elucidated the life cycle of the pathogen and based on sporangial development and sporangiophore characteristics and changed its name from *Peronospora* to *P. infestans* (deBary, 1876).

EVOLUTIONARY POSITION AND PHYLOGENETIC RELATIONSHIPS OF *PHYTOPHTHORA* SPP.

The genus *Phytophthora* now encompasses more than 100 species that are classified within the diploid, algae-like Oomycetes in the Kingdom Stramenopila (Adl et al., 2005; Bauldauf, 2003; Gunderson et al., 1987). The number of *Phytophthora* species described has increased rapidly due to intensive monitoring of plants and waterways in the past 10 years for *Phytophthora ramorum*, the cause of sudden oak death (Cline et al., 2008; Rizzo et al., 2002). Oomycetes were once grouped with true fungi (ascomycetes and basidiomycetes) since they share a filamentous habit of growth by mycelium and obtain nutrition by absorption. However, oomycetes differ from true fungi in many ways and are now placed in a separate kingdom in a distinct branch in the eukaryotic tree of life and are more closely related to brown algae and diatoms than true fungi (Adl et al., 2005).

Based upon sequences of ribosomal genes and the introns associated with them (internal transcribed spacer [ITS] regions), 10 clades in the genus *Phytophthora* were described (Cooke et al., 2000). Subsequently, multilocus sequencing more clearly elucidated phylogenetic relationships among a larger group of *Phytophthora* species (Blair et al., 2008; Kroon et al., 2004; Martin and Tooley, 2003). *P. infestans* is a member of the Ic clade of *Phytophthora* (Blair et al., 2008; Cooke et al., 2000). Other species in this clade include the more distantly related *Phytophthora phaseoli* and *Phytophthora mirabilis*, *Phytophthora ipomoeae*, and *Phytophthora andina* (Flier et al., 2002; Galindo and Hohl, 1985; Oliva et al., 2010).

LIFE HISTORY OF THE PATHOGEN

P. infestans is a hemibiotrophic pathogen that infects host tissue and after a few days produces symptoms. One reason why late blight is such a destructive disease is due to the explosive, polycyclic nature of asexual sporulation that can occur on plant tissue (Fig. 10.2a).

The pathogen produces black water-soaked lesions on potato leaf tissue that produce asexual sporangia (Fig. 10.2a). Tuber infections lead to a purplish brown discoloration of the tissue and eventually the tuber rots (Fig. 10.2b). The pathogen causes brown lesions on the stems, petioles, and leaves of tomatoes (Fig. 10.2c). Brown zonate lesions occur on the fruit (Fig. 10.2d).

Sporangia (Fig. 10.2e) can be dispersed by wind and rain at local and national scales (hundreds of meters) and entire fields can be destroyed in a few days. Late blight is truly a community disease, and

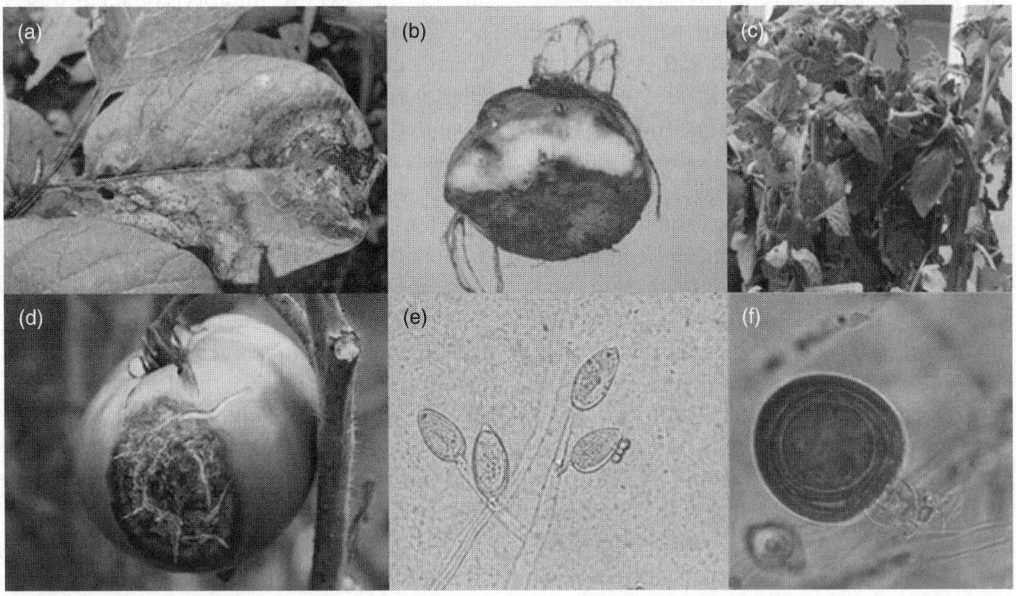

Figure 10.2 Symptoms of late blight caused by *P. infestans* on (a) potato leaf, (b) tuber, (c) tomato stem, (d) tomato fruit, (e) plants in a field, and (f) a tomato transplant. (e) Asexual sporangia borne on compound sporangiophores; (f) sexual oospore produced by fusion of anisogamous gametangia (male antheridium and female oogonia) of opposite mating types (A1 and A2). See color insert.

inoculum in fields left untreated can spread to neighboring fields due to aerial dispersal of sporangia. The pathogen has caused major losses mostly by the unintentional spread on infected plant materials such as potato tubers or tomato transplants that are transported over large areas, as is illustrated by the widespread epidemics of 2009 on tomato in the United States (Hu et al., 2012). The pathogen typically survives from season to season as mycelium in infected potato tubers, volunteer potato plants, or infected culled potatoes that contribute to epidemic development on subsequent crops.

P. infestans is heterothallic and requires two opposite mating types, A1 and A2, for sexual reproduction and formation of the nonmotile sexual oospore (Fig. 10.2f) (Gallegly and Galindo, 1958). Oogonia develop when A1 and A2 mating type isolates are paired. Hormonal heterothallism is involved in the formation of gametangia with one thallus producing a hormone that stimulates gametangia formation of the opposite mating type (Ko, 1988). Each thallus is bisexual and can act as either male or female and induces either oogonia or antheridia formation in the opposite mating type (Judelson, 1997). Studies of the inheritance of mating type show that the A1 types have a heterozygous locus (Aa), while A2 types are homozygous recessive (aa) (Fabritius and Judelson, 1997; Judelson et al., 1995). The position of the mating type locus on a genetic linkage map (Van der Lee et al., 2004) and in genomic clones (Randall et al., 2003) has been identified and in several strains has been linked to genetic abnormalities such as balanced lethality and translocations (Judelson, 1996a,b). The A1 mating-type hormone MH-1 α has been purified and reported to be a diterpene (Harutyunyan et al., 2008; Qi et al., 2005). A second structurally similar mating hormone, MH-2 α, has recently been purified, and both hormones are proposed to be derived from the chlorophyll-related plant hormone phytol (Ojika et al., 2011). This exciting discovery may lead to novel methods to stop sexual reproduction in the pathogen.

Oospores were reported first in Mexico, but the pathogen now also reproduces sexually in the Netherlands, Scandinavia, and Canada (Andersson et al., 2009; Drenth et al., 1994; Gavino et al., 2000). Oospores are not common in the United States and sexual reproduction is rare. The oospore is a thick-walled survival structure and can persist in soils at low temperatures for several years but does not survive well at higher temperatures (Fay and Fry, 1997; Maytoun et al., 2000).

POPULATION BIOLOGY: MIGRATION THEORIES OF *P. INFESTANS* IN THE 19TH CENTURY

P. infestans has four mitochondrial haplotypes (Ia, Ib, IIa, and IIb) (Avila-Adame et al., 2006). The mitochondrial genomes of all the extant haplotypes have been sequenced (Avila-Adame et al., 2006; Carter et al., 1990; Lang and Forget, 1993). Phylogenetic and coalescent analysis revealed that although the type I and II haplotypes share a common ancestor, they form two distinct lineages that evolve independently. Type II haplotypes diverge earlier than type I haplotypes. The type II lineages did not evolve from type I lineages, as previously suggested (Gavino and Fry, 2002). Our data support the hypothesis that all the extant mitochondrial lineages of *P. infestans* evolved from a common ancestor in South America since the most ancient mutations in the four lineages occur there (Gomez et al., 2007).

The center of origin and diversity of the late blight pathogen is proposed to be in Mexico (Fry and Goodwin, 1997; Goodwin et al., 1994b; Reddick, 1939). This hypothesis is based on the fact that in Mexico, (i) both mating types occur;

(ii) host resistance genes are present in wild *Solanum* populations; and (iii) pathogen populations are highly diverse for neutral DNA markers and pathotypes. Isolates of *P. infestans* of the A2 mating type and oospores in infected plant material were first discovered in central Mexico in 1956 (Fry et al., 1992; Gallegly and Galindo, 1958; Goodwin et al., 1994a). Prior to 1980, both the A1 and A2 mating types were reported only in Mexico, while the A1 mating type was reported elsewhere in the world (Fry, 2008; Fry and Goodwin, 1997; Gallegly and Galindo, 1958). This situation changed in the 1980s when the A2 mating type was observed in Switzerland (Hohl and Iselin, 1984).

Mexico has also been proposed to have provided the source inoculum for the late blight epidemics of the 1840s (Fry and Goodwin, 1997; Goodwin et al., 1994b). Genetic analysis of modern worldwide populations of *P. infestans* demonstrated that they were dominated by a single clonal lineage known as the US-1 "old" genotype (mtDNA haplotype Ib) in the mid 20th century (Fry, 2008; Goodwin et al., 1994b). Mexican populations are highly diverse for genotypic and phenotypic markers and thus, Mexico clearly represents a present-day center of diversity of the pathogen. However, since Mexican populations are highly diverse, the pathogen population is proposed to have undergone a genetic bottleneck in the 1840s during the first migration event, thus greatly reducing genetic diversity, in order to explain the dominance of a single "old" clonal genotype (US-1 genotype) in modern worldwide populations prior to the 1980s (Fry and Goodwin, 1997; Goodwin et al., 1994b). Domesticated potatoes were not grown for export in Mexico in the 1840s and tuber blight was not common. It was suggested that the pathogen may have been accidentally introduced by a plant collector (Goodwin et al., 1994b). These authors have provided well-documented evidence for migrations of the pathogen from Mexico after the 1970s (Goodwin et al., 1994a), but no definitive evidence for migrations of *P. infestans* during the interval between 1840 and 1970. The fact that the putative "old" US-1 lineage (Ib mtDNA) has not been widely reported in Mexico, does not support the theory that Mexico provided the source inoculum for the late blight epidemics of the 1840s.

The Mexican "bottleneck" theory of Goodwin and colleagues was challenged by several authors (Abad and Abad, 1997; Andrivon, 1996; Ristaino, 1998; Tooley et al., 1989). A second "hybrid" migration theory was proposed, suggesting that Mexico may represent the center of origin of the disease but that inoculum that caused the 19th century potato famine epidemics originated in Peru (Andrivon, 1996; Bourke, 1964). This theory was based on both historical data and an evaluation of the published population genetic data at the time. A third theory suggests that the Peruvian Andes represents both the center of origin of the disease and the source of inoculum for 19th century epidemics. Evidence to support this theory has been provided by both historical data about disease occurrence (Abad and Abad, 1997; Berkeley, 1846) and population genetic data (Gomez et al., 2007).

WHAT mtDNA HAPLOTYPE CAUSED THE FAMINE AND WHERE DID IT COME FROM?

Nineteenth and early twentieth century scientists collected and preserved potato and tomato leaves infected with *P. infestans*, and specimens exist from the Irish potato famine (Fig. 10.1b,c). Historical specimens have been analyzed to answer questions about the evolution and population biology of the pathogen (Ristaino, 1998, 2002, 2006; Ristaino et al., 2001; May and Ristaino, 2004). The polymerase

TABLE 10.1 Identity of the Mitochondrial DNA Lineage of *P. infestans* in Archival Herbarium Specimens Collected Worldwide

Geographic Region	Year Collected	Number of Specimens	mtDNA Haplotype[a]
Central and South America	1889–1969	18	Ia
			Ib—Bolivia (1944), Ecuador (1967)
			IIb—Nicaragua (1956)
North America	1855–1958	88	Ia
			Ib (1931 and later—potato, tomato, *Solanum sarrachoides*)
Northern Europe	1866–1905	11	Ia
Western Europe	1845–1982	52	Ia
Eastern Europe	1892–1978	7	Ia
United Kingdom	1845–1974	52	Ia
China	1935–1982	10	Ia
			Ib (1952—potato, 1956—tomato)
			Ib (1982—*Solanum lyratum*)
Southeast Asia and Australia (Japan, Philippines India, Nepal, Peninsular Malaysia, Thailand)	1901–1987	13	All Ia except:
			Ib India (1968, 1974—potato)
			Ib Thailand, (1981—tomato)

[a] Based on PCR methods of Griffith and Shaw (1998).

chain reaction (PCR) has been used to amplify minute amounts of DNA from leaves infected with *P. infestans* from historic epidemics.

The mtDNA haplotypes present in specimens collected during the Irish potato famine and later in the 19th and early 20th century were identified (May and Ristaino, 2004; Ristaino et al., 2001). First, a 100-bp fragment of DNA from ITS region 2 specific for *P. infestans* was amplified from 90% of the leaves tested, confirming infection by *P. infestans* (Trout et al., 1997; Ristaino, 1998, 2001). Primers were then designed that amplify short segments of mtDNA around variable restriction sites that separate the four mtDNA haplotypes (Griffith and Shaw, 1998) and the DNA was sequenced. Surprisingly, 86% of lesions from leaves collected during historic epidemics were caused by the mtDNA haplotype Ia (May and Ristaino, 2004), not Ib (Goodwin et al., 1994b), as previously believed.

Interestingly, both the Ia and IIb haplotypes were found in potato leaves from specimens collected in 1954 and 1956 in Nicaragua (Table 10.1). These data challenge the hypothesis that a single clonal lineage of *P. infestans* existed outside of Mexico since two mtDNA haplotypes were present in samples from Nicaragua. Thus, late blight populations in the mid 20th century in Central America did not consist of a single clonal lineage and pathogen diversity was greater than previously believed.

WHEN DID THE mtDNA HAPLOTYPE Ib MIGRATE FROM SOUTH AMERICA?

The Ib haplotype of *P. infestans* was found in several plant samples from herbaria including a sample from Bolivia (1944) and Ecuador (1967) in the mid-20th century (Table 10.1) (May and Ristaino, 2004). We recently examined mtDNA haplotypes of several hundred samples of *P. infestans* from infected potato and tomato leaves from herbaria collected from the United States between 1855 and 1953. In all cases, haplotype Ia was predominant and found before the haplotype Ib was detected (May and Ristaino, 2004; Ristaino and Lassiter, unpublished data). The Ia haplotype of *P. infestans* was also found earlier in China than other mtDNA haplotypes (Ristaino and Hu, 2009). In contrast, the earliest record of haplotype Ib in China was in 1952 on potato in 1954 and in 1956 on tomato (Beijing). The haplotype Ib still occurs in the Beijing area on tomato (Guo et al., 2010). Modern Chinese populations contain all four haplotypes, suggesting more recent introductions of IIa and IIb haplotypes (Guo et al., 2010). In other areas in Asia, the earliest documentation of haplotype Ia was in Japan in 1901. The haplotype Ia was also found in India (1913), the Philippines (1910), and Russia and Australia (1917). In contrast, haplotype Ib was only found in three samples in India on potato (1968 and 1974) and in Thailand on tomato (1981). All the rest of the herbarium plant samples from other Southeast Asian countries and regions were infected with haplotype Ia (Table 10.1). Thus, our data suggest that the haplotype Ib was dispersed in early to mid-20th century migrations into the United States, China, and Southeast Asia, whereas earliest introductions during the famine-era epidemic in the United States and Europe, no doubt on imported potato tubers, were the Ia haplotype (May and Ristaino, 2004; Ristaino and Hu, 2009).

HISTORIC MIGRATIONS: "OUT-OF-SOUTH AMERICA" MIGRATION HYPOTHESIS

We have assessed the genealogical history of *P. infestans* using multilocus sequences from portions of two nuclear genes (β-*tubulin* and *ras*) and several mitochondrial loci P3 (*rpl14*, *rpl5*, tRNA) and P4 (*Cox1*) from 94 "modern" isolates from South America, Central America, North America, and Ireland in order to test an out-of-South America migration hypothesis (Gomez et al., 2007). Summary statistics, migration analyses, and the genealogy of populations of *P. infestans* for both nuclear (*ras* gene) and mitochondrial loci are consistent with an out-of-South America origin for *P. infestans* rather than an "out-of-Mexico" origin. Mexican populations of *P. infestans* from the putative center of origin in Toluca Valley of Mexico harbored less nucleotide and haplotype diversities than Andean populations. Coalescent-based genealogies of mitochondrial and nuclear loci were congruent and demonstrate the existence of two lineages leading to present-day haplotypes of *P. infestans* on potatoes. Mitochondrial haplotypes found in Toluca, Mexico, were from the type I haplotype lineage, whereas those from Peru and Ecuador were derived from both type I and type II lineages (Fig. 10.3). Mitochondrial and nuclear haplotypes in populations from both the United States and Ireland were derived from both ancestral lineages that occur in South America, suggesting a common ancestry among these populations. The oldest mitochondrial lineage was associated with isolates from the botanical section Anarrhichomenum including *Solanum tetrapetalum* from Ecuador (Gomez et al., 2007). This lineage (EC-2 Ic) shares a common ancestor with *P. infestans* and has been named *P. andina* (Gomez et al., 2008; Kroon et al., 2004; Olivia et al., 2010). The geographic distribution of mutations on the rooted coalescent tree for both nuclear and mitochondrial loci demonstrates that the oldest mutations in *P. infestans* originated in South America; this is consistent with an Andean origin (Gomez et al., 2007).

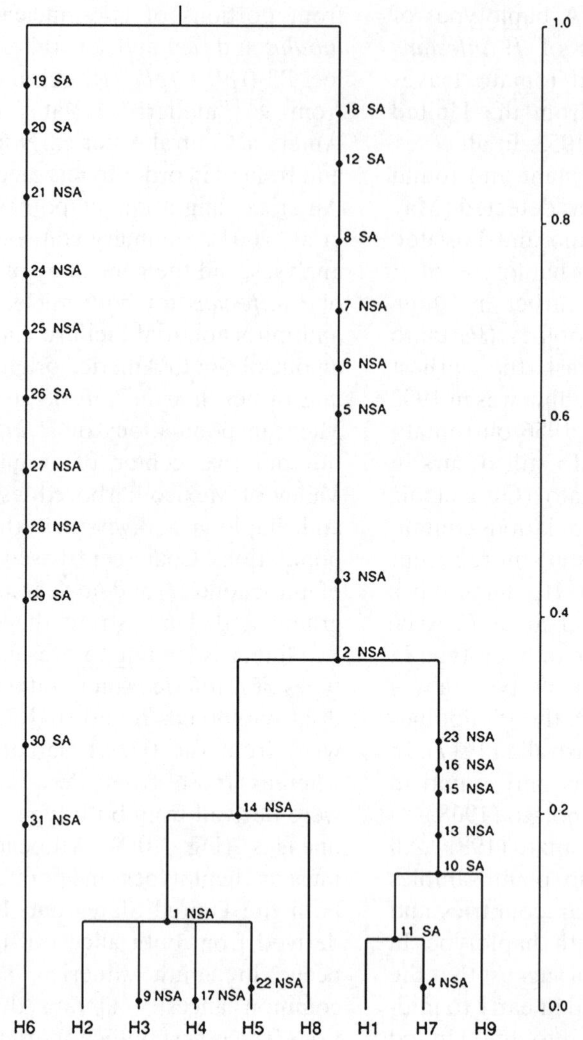

H6	H2	H3	H4	H5	H8	H1	H7	H9	
3	24	1	1	3	2	23	5	1	Count
3	7	.	.	.	.	20	.	1	South American
.	17	1	1	3	2	3	5	.	Non-South American
.	.	.	.	.	.	3	.	.	Bolivia
.	.	.	.	.	.	5	.	.	Brazil
.	6	.	.	.	.	5	.	1	Peru
3	1	.	.	.	.	7	.	.	Ecuador
.	7	.	.	.	.	.	.	.	Costa Rica
.	5	1	1	3	.	.	.	.	Mexico
.	3	.	.	.	2	.	5	.	United States
.	2	.	.	.	.	3	.	.	Ireland
A2	A1/A2	A1	A2	A1/A2	A1/A2	A1/A2	A1/A2	A1	Mating type
Ic	Ia/1b	Ia	Ia	Ia	Ia	IIa	IIb	IIa	Carter's mtDNA haplotype
Type I						Type II			

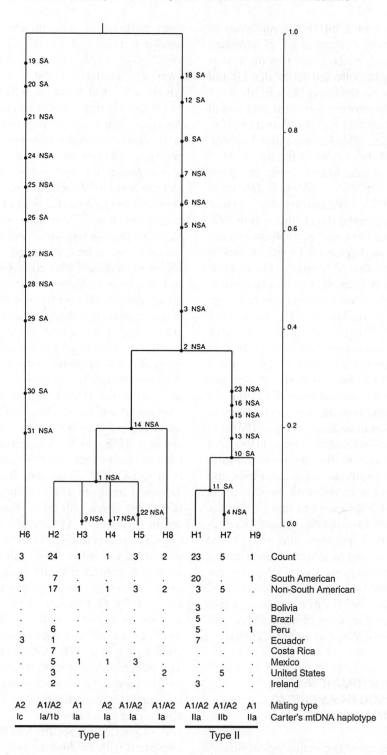

Figure 10.3 The rooted coalescent-based gene genealogy showing the distribution of mutations for South American (SA: Peru, Ecuador, Bolivia, Brazil) and non-South American (NSA: Costa Rica, Mexico, United States, Ireland) populations for the mitochondrial (P3 + P4) loci of *Phytophthora infestans* generated using GENETREE. The timescale is in coalescent units of effective population size and the direction of divergence is from the top (past–oldest) to the bottom (present–youngest). Numbers below the tree from top to bottom designate each distinct haplotype (H) and its count (i.e., the number of occurrences of the haplotype in the sample, where. = 0), the count of each haplotype in each population, the mating type of the isolates, and the mtDNA haplotype according to Carter et al. (1990). Image reproduced with permission from the *Proceedings of the National Academy of Sciences of the United States of America* from Gomez et al. (2007).

HOST SHIFTS AND JUMPS

P. infestans can infect other *Solanum* species and has shifted hosts on numerous occasions to exploit new niches. After its first introduction into Ireland, it was reported by Moore in Dublin in 1846 on *Anthocercis illicfolia*, a solanaceous shrub that had been imported from Australia into the Botanic Gardens at Glasnevin (Fig. 10.1b). The pathogen was also reported in petunia in the United Kingdom by M. C. Cooke in 1856. The haplotype Ib of *P. infestans* was identified on *Solanum lyratum*, a weed host that occurs commonly alongside potato fields in China in 1982 (Table 10.1). The occurrence of the Ib haplotype on a weed host in China, coincident with the time of first use of the fungicide metalaxyl in the country, suggests that this host shift may have enabled this fungicide-sensitive strain to survive by avoiding the fungicide that had been sprayed in fields. Weeds such as *Solanum nigrum*, *Solanum dulcamara*, and *Solanum sisymbriifolium* are also hosts of *P. infestans* and may act as a refuge and overwintering host for the pathogen in the absence of a potato or tomato crop in the field (Flier et al., 2003).

P. infestans occurs sympatrically in Mexico with *P. mirabilis* a pathogen that infects *Mirabilis jalapa* and *P. ipomoeae* a pathogen that infects *Ipomoea longipedunculata* (morning glory). The evolution in the Ic clade has involved host jumps to these unrelated plant species (Raffaele et al., 2010). *P. infestans* occurs sympatrically in Ecuador with *P. andina*. *P. andina* can infect other non-tuber-bearing species of *Solanum*, *Solanum muricatum*, and *Solanum betaceum*, indicating that *P. andina* has an expanded host range compared to *P. infestans*. *P. mirabilis* was first considered the closest known relative of *P. infestans* (Galindo and Hohl, 1985). However, phylogenetic analysis of *P. andina* from Ecuador suggests that *P. andina* (EC-2 Ic lineage) is also closely related to *P. infestans* (Adler et al., 2004; Blair et al., 2008; Gomez et al., 2008; Kroon et al., 2004; Oliva et al., 2010), and this lineage shares a common ancestor with *P. infestans* in the Andean region (Gomez et al., 2007).

Whether *P. andina* evolved as a hybrid of *P. infestans* and *P. mirabilis* (Gomez et al., 2008) or some other very close relative of *P. mirabilis* is still uncertain (Goss et al., 2011; Lassiter et al., 2010). The *ras* intron 1 sequence suggests that *P. andina* may have arisen from a hybridization possibly between *P. infestans* and *P. mirabilis* (Gomez et al., 2008). However, *P. mirabilis* has not been found in Ecuador where *P. andina* occurs. *P. mirabilis* was first reported in Mexico (Galindo and Hohl, 1985), so further exploration for *P. mirabilis* in the Andean region is warranted. *P. infestans* and *P. mirabilis* can hybridize in the laboratory and produce viable progeny (Goodwin and Fry, 1994). *M. jalapa* has not been widely surveyed for *P. mirabilis* in South America, although the host evolved in Peru and its common name is the flower of Peru. We are currently using nuclear gene genealogies and Bayesian statistics to clarify the evolutionary relationships of the Ic clade species (Lassiter et al., 2010). One interpretation of the data is that the Andean region is the center of evolutionary origin for all the species in the Ic clade since *P. andina* and *P. infestans* coexist there and the oldest mutations in the *P. andina* EC-2 Ic lineage are of South American origin. Further surveys are needed to test the possibility of the occurrence of *P. mirabilis* and *P. ipomoeae* in the Andean region.

The patterns of sequence variation involved in host jumps in several Ic clades species of *Phytophthora* including *P. phaseoli*, *P. ipomoeae*, and *P. mirabilis* have been compared to *P. infestans* by resequencing genomes of the sister species in the Ic clade (Raffaele et al., 2010). Patterns of gene-sparse, repeat-rich regions in the genomes of the sister species were found, as reported in the *P. infestans* genome (Haas et al., 2009; Raffaele et al., 2010). These repeat-rich regions contained many effector genes that showed evidence of positive selection, uneven rates of evolution, and enrichment of genes induced *in planta* during preinfection and infection, suggesting host adaptation has led to the evolution of species in the clade. Unfortunately, these authors (Raffaele et al., 2010) did not analyze the genome of *P. andina* due to difficulties in reassembly of the genome, which contained many loci with heterozygous sites, a situation suggestive of *P. andina* being a hybrid (Brasier et al., 1999; Gomez et al., 2008; Goss et al., 2011; Kroon et al., 2004; Lassiter et al., 2010).

WHY IS LATE BLIGHT A REEMERGING DISEASE?

Late blight has reemerged as a significant threat to production for both tomato and potato. The varied dispersal mechanisms of the pathogen including the ability to move both as airborne inoculum (sporangia) and in plant material (tubers or transplants) have made local and long-distance spread common. The polycyclic nature of the life cycle enables the pathogen to produce copious amounts of spores in a short time period on aerial parts of plants, and these spores can move rapidly (within days) through untreated fields. Pathogen populations became resistant to the phenylamide fungicides shortly after their introduction. The widespread monoculture of highly susceptible potato cultivars in the United States has made them vulnerable to disease. *P. infestans* also can infect a wide range of other solanaceous plants. The plasticity of the genome as revealed by genome sequencing and the diversity of pathogen effectors that can overcome host R (resistance) genes make host resistance a moving target at best and make the search for more durable host resistance a more viable goal then single-gene resistance (Haas et al., 2009; Kamoun and Smart, 2005).

Populations of *P. infestans* in the United States have consisted of a series of mostly asexual clonal genotypes and 19 genotypes have been reported previously in regional populations (Fraser et al., 1999; Fry et al., 1992; Gavino et al., 2000; Goodwin et al., 1994a, 1998; Wangsomboondee et al., 2002). We recently identified five new multilocus genotypes in the eastern United States (US-20 to US-24) (Hu et al., 2012). The presence of sexually reproducing populations in several European countries has been documented by restriction fragment length polymorphism (RFLP) markers and later by microsatellite markers (Cooke et al., 2007; Drenth et al., 1994; Knapova and Gisi, 2002; Lees et al., 2006; Zwankhuizen et al., 1998, 2000).

In 2009, a widespread strain, US-22 (Hu et al., 2012), identified by multilocus genotyping, spread in the northeast of the United States on tomato transplants and rapidly infected potatoes. There was a complete loss of the tomato crop in many states and an increase in fungicide use on potatoes to curb epidemic spread. A combination of weather conducive for disease and widespread inoculum distribution exacerbated the epidemic. Strain US-22 was very widespread (12 states) on tomato and potato, and this genotype accounted for about 60% of all the isolates genotyped in 2009 in the United States. Organic tomato growers were devastated by the disease since few options for control besides copper-based sprays are available in these production systems. At least 400 farms were affected by the disease in New England (Moskin, 2009). The lack of deployment of resistant varieties of both potato and tomato in the United States has exacerbated the situation. On the basis of sequence analysis of the *ras* gene, US-22, US-23, and US-24 appear to have been derived from a common ancestor (Hu et al., 2012).

The US-8 genotype, which is still common on potatoes in the United States, is mefenoxam resistant. Isolates of the widespread US-22 genotype and several other clonal lineages (US-23 and US-24) are largely sensitive to mefenoxam. However, this information was not known until after the widespread epidemics in the United States in 2009 since fungicide sensitivity assays are not rapid enough to deliver information to growers in a timely manner. Microsatellite markers can be used to quickly separate some of the genotypes. The phenylamide fungicides such as mefenoxam target RNA polymerase I (Davidse, 1988; Gisi and Cohen, 1996). Markers associated with mefenoxam resistance mapped to a single locus (Fabritius et al., 1997; Lee et al., 1999). A rapid molecular assay to identify mefenoxam insensitivity in the pathogen is needed.

Hundreds of effector genes are present in the genome (Haas et al., 2009). Avirulence alleles can now be tracked in the field to study the evolution of these genes in pathogen populations in response to deployed host resistance genes. This should help document the pathogen–host arms race and lead to strategies to slow the development of new pathotypes of *P. infestans*.

CONCLUSIONS

Further field and laboratory studies are under way exploiting the *P. infestans* genome sequence to develop rapid genotyping methods and to monitor further changes in the structure of both selectively neutral and evolving traits of the populations of the pathogen in North America. A national portal (http://www.USAblight.org) has been developed to track disease occurrences, send disease alerts, provide a venue for submitting samples for diagnosis and genotyping, publish pathogen multilocus genotypes, and provide management information to growers developed by the extension and research community on late blight. Over 160 years after the great famine, late blight continues to challenge the sustainable production of potato worldwide.

REFERENCES

Abad ZG, Abad JA. 1997. Another look at the origins of late blight of potatoes, tomatoes, and pear melon in the Andes of South America. Plant Dis 81: 682–688.

Adl SM, et al. 2005. The new higher level of classification of eukaryotes with emphasis on the taxonomy of protists. J Eukaryot Microbiol 52: 399–451.

Adler NE, Erselius LJ, Chacon MG, Flier WG, Ordoñez ME, Kroon LPNM, Forbes GA. 2004. Genetic diversity of *Phytophthora infestans* sensu lato in Ecuador provides new insight into the origin of this important plant pathogen. Phytopathology 94: 154–162.

Anderson PK, Cunningham AA, Patel NG, Morales FJ, Epstein PR, Daszak P. 2004. Emerging infectious diseases of plants: Pathogen pollution, climate change and agrotechnology drivers. Trends Ecol Evol 19: 535–544.

Andersson B, Widmark AK, Yuen JE, Nielsen B, Ravnskov S, Kessel GJT, Evenhuis A, Turkensteen LJ, Hansen JG, Lehtinen A, Hermansen A, Brurberg MB, Nordskog B. 2009. The role of oospores in the epidemiology of potato late blight. Acta Hortic 834: 61–68.

Andrivon D. 1996. The origin of *Phytophthora infestans* populations present in Europe in the 1840's: A critical review of historical and scientific evidence. Plant Pathol 45: 1027–1035.

Avila-Adame C, Gómez L, Buell CR, Ristaino JB. 2006. Mitochondrial genome sequencing of the haplotypes of the Irish potato famine pathogen, *Phytophthora infestans*. Curr Genet 49: 39–46.

Bauldauf SL. 2003. The deep roots of eukaryotes. Science 300: 1703–1706.

Berkeley MJ. 1846. Observations, botanical and physiological on the potato murain. J Hortic Soc Lond 1: 9–34.

Blair JE, Coffey MD, Park SY, Geiser DM, Kang S. 2008. A multi-locus phylogeny for *Phytophthora* utilizing markers derived from complete genome sequences. Fungal Genet Biol 45: 266–277.

Bourke PM. 1964. Emergence of potato blight. Nature 203: 805–808.

Brasier CM. 2008. The biosecurity threat to the UK and the global environment from international trade in plants. Plant Pathol 57: 792–808.

Brasier CM, Cooke DEL, Duncan JM. 1999. Origin of a new *Phytophthora pathogen* through interspecific hybridization. Proc Natl Acad Sci U S A 96: 5878–5883.

Carter DA, Archer SA, Buck KW. 1990. Restriction fragment length polymorphism of mitochondrial DNA of *Phyotophthora infestans*. Mycol Res 8: 1123–1128.

Cline ET, Farr DF, Rossman AY. 2008. A Synopsis of *Phytophthora* with accurate scientific names, host range, and geographic distribution. Plant Health Prog doi: 10.1094/PHP-2008-0318-01-RS.

Cooke DEL, Drenth A, Duncan JM, Wagels G, Brasier CM. 2000. A molecular phylogeny of *Phytophthora* and related oomycetes. Fungal Genet Biol 30: 17–32.

Cooke DEL, Schena L, Cacciola SO. 2007. Tools to detect, identify and monitor *Phytophthora* species in natural ecosystems. J Plant Pathol 89: 13–28.

Davidse LC. 1988. Phenylamide fungicides: Mechanism of action and resistance. In: Delp CJ, ed. *Fungicide Resistance in North America*. APS Press, St. Paul, MN, pp. 63–65.

deBary A. 1863. Du developpement de quelques champignons parasites. Annales des Sciences Naturelles, 4 Serie. T 20: 1–143.

deBary A. 1876. Researches into the nature of the potato-fungus - *Phytophthora infestans*. J R Agric Soc 12: 239–268.

Drenth A, Tas I, Govers F. 1994. DNA fingerprinting uncovers a new sexually reproducing population of *Phytophthora infestans* in the Netherlands. Eur J Plant Pathol 100: 97–107.

Ellsworth H. 1843. Annual Report of Commissioner of Patents, 28th Congress, No. 150.

Erwin DC, Ribeiro OK. 1996. *Phytophthora Diseases Worldwide*. Amercan Phytopathological Society Press, St. Paul, MN.

Fabritius AL, Judelson HS. 1997. Mating type loci segregate aberrantly in *Phytophthora infestans* but normally in *Phytophthora parasitica*: Implications for models of mating type determination. Curr Genet 32: 60–65.

Fabritius AL, Shattock RC, Judelson HS. 1997. Genetic analysis of metalaxyl insensitivity loci in *Phytophthora infestans* using linked DNA markers. Phytopathology 87: 1034–1040.

Fay JC, Fry WE. 1997. Effect of hot and cold temperatures on the survival of oospores produced by United States strains of *Phytophthora infestans*. Am Potato J 74: 315–323.

Flier WG, GruNwald NK, Kroon LPNM, Sturbaum AK, van den Bosch TBM, Garay-Serrano E, Lozoya-Saldana H, Bonants PJM, Turkensteen LJ. 2002. *Phytophthora ipomoeae* sp. nov., a new homothallic species causing leaf blight on *Ipomoea longipedunculata* in the Toluca Valley of central Mexico. Mycol Res 106: 848–856.

Flier WG, van den Bosch GBM, Turkensteen LS. 2003. Epidemiological impacts of *Solanum nigrum*, *Solanum dulcamara* and *Solanum sisymbrifolium* as alternative hosts for *Phytophthora infestans*. Plant Pathol 52: 595–603.

Fraser DE, Shoemaker PB, Ristaino JB. 1999. Characterization of isolates of *Phytophthora infestans* from tomato and potato in North Carolina from 1993-1995. Plant Dis 83: 633–638.

Fry W, Goodwin SB. 1997. Resurgence of the Irish potato famine fungus. Bioscience 47: 363–371.

Fry WE. 2008. *Phytophthora infestans*: The plant (and R gene) destroyer. Mol Plant Pathol 9: 385–402.

Fry WE, Goodwin SB, Matuszak J, Spielman L, Milgroom M. 1992. Population genetics and intercontinental migrations of *Phytophthora infestans*. Annu Rev Phytopathol 30: 107–129.

Galindo AJ, Hohl HR. 1985. *Phytophthora mirabilis*, a new species of *Phytophthora*. Sydowia Ann Mycol Ser II 38: 87–96.

Gallegly ME, Galindo AJ. 1958. Mating types and oospore of *Phytophthora infestans* in nature in Mexico. Phytopathology 48: 274–277.

Gavino PD, Fry WE. 2002. Diversity in and evidence for selection of the mitochondrial genome of *Phytophthora infestans*. Mycologia 94: 781–793.

Gavino PD, Smart CD, Sandrock RW, Miller JS, Hamm PB, Yun Lee T, Davis RM, Fry WF. 2000. Implications of sexual reproduction for *Phytophthora infestans* in the United States: Generation of an aggressive lineage. Plant Dis 84: 731–735.

Gisi U, Cohen Y. 1996. Resistace to phenylamide fungicides: A case study with *Phytophthora infestans* involving mating type and race structure. Annu Rev Phytopathol 34: 549–572.

Gomez L, Carbone I, Ristaino JB. 2007. An Andean origin for *Phytophthora infestans* inferred from nuclear and mitochondrial DNA sequences. Proc Natl Acad Sci U S A 104: 3306–3311.

Gomez L, Hu C, Olivia R, Forbes G, Ristaino JB. 2008. Phylogenetic relationships of a new species, *Phytophthora andina*, from the highlands of Ecuador that is closely related to the Irish potato famine pathogen *Phytophthora infestans*. Mycologia 100: 590–602.

Goodwin SB, Fry WE. 1994. Genetic analysis of interspecific hybrids between *Phytophthora infestans* and *Phytophthora mirabilis*. Exp Mycol 18: 20–32.

Goodwin SB, Cohen BA, Deahl KL, Fry WE. 1994a. Migration from Northern Mexico as the probable cause of recent genetic changes in populations of *Phytophthora infestans* in the United States and Canada. Phytopathology 84: 553–558.

Goodwin SB, Cohen BA, Fry WE. 1994b. Panglobal distribution of a single clonal lineage of the Irish potato famine fungus. Proc Natl Acad Sci U S A 91: 11591–11595.

Goodwin SB, Smart CD, Sandrock RW, Deahl KL, Punja ZK, Fry WE. 1998. Genetic change within populations of *Phytophthora infestans* in the United States and Canada during 1994-1996: Role of migration and recombination. Phytopathology 88: 939–949.

Goss EM, Cardenas ME, Myers K, Forbes GA, Fry WE, Restrepo S, Grünwald NJ. 2011. The plant pathogen *Phytophthora andina* emerged via hybridization of an unknown *phytophthora* species and the Irish potato famine pathogen, *P. infestans*. (S. Allodi, ed.). PLoS One 6: e24543.

Griffith GW, Shaw DS. 1998. Polymorphisms in *Phytophthora infestans*: Four mitochondrial haplotypes are detected after PCR amplification of DNA from pure cultures or from host lesions. Appl Environ Microbiol 64: 4007–4114.

Gunderson JH, Elwood H, Ingold H, Kindle A, Sogin ML. 1987. Phylogenetic relationships between chlororphytes, chrysophytes, and oomycetes. Proc Natl Acad Sci U S A 84: 5823–5827.

Guo L, Zhu X, Hu C, Ristaino JB. 2010. Genetic structure of *Phytophthora infestans* populations in China indicates multiple migration events. Phytopathology 100: 997–1006.

Haas BJ, et al. 2009. Genome sequence and comparative analysis of the Irish potato famine pathogen *Phytophthora infestans*. Nature 461: 393–398.

Harutyunyan SR, Zhao Z, den Hartog T, Bouwmeester K, Minnaard AJ, Feringa BL, Govers F. 2008. Biologically active Phytophthora mating hormone prepared by catalytic asymmetric total synthesis. Proc Natl Acad Sci U S A 105: 8507–8512.

Haverkort AJ, Boonekamp PM, Hutten R, Jacobsen E, Lotz LAP, Kessel GJT, Visser RGF, van der Vossen EAG. 2008. Societal costs of late blight in potato and prospects of durable resistance through cisgenic modification. Potato Res 51: 1871–4528.

Hohl H, Iselin K. 1984. Strains of *Phytophthora infestans* from Switzerland with A2 mating type behavior. Trans Br Mycol Soc 83: 529–530.

Hu C-H, Perez FG, Donahoo R, McLeod A, Myers K, Ivors K, Roberts PD, Fry WE, Deahl KL, Ristaino JB. 2012. Recent genotypes of *Phytophthora infestans* in eastern USA reveal clonal populations and reappearance of mefenoxam sensitivity. Plant Dis 96: in press.

Johnson JFW. 1845. *The Potato Disease in Scotland: Being Results of Investigations into its Nature and Origins*. William Blackwood and Sons, Edinburgh, Scotland.

Judelson HS. 1996a. Chromosomal heteromorphism linked to the mating type locus of the oomycete *Phytophthora infestans*. Mol Gen Genet 252: 155–161.

Judelson HS. 1996b. Genetic and physical variability at the mating type locus of the oomycete, *Phytophthora infestans*. Genetics 144: 1005–1013.

Judelson HS. 1997. Expression and inheritance of sexual preference and selfing propensity in *Phytophthora infestans*. Fungal Genet Biol 21: 188–197.

Judelson HS, Spielman LJ, Shattock RC. 1995. Genetic mapping and non-mendelian segregation of mating type loci in the oomycete. Genetics 141: 503–512.

Kamoun S, Smart CD. 2005. Late blight of potato and tomato in the genomics era. Plant Dis 89: 692–699.

Knapova G, Gisi U. 2002. Phenotypic and genotypic structure of *Phytophthora infestans* populations on potato and tomato in France and Switzerland. Plant Pathol 51: 641–553.

Ko WH. 1988. Hormonal heterothallism and homothallism in Phytophthora. Annu Rev Phytopathol 26: 57–73.

Kroon LPNM, Bakker FT, van den Bosch GBM, Bonants PJM, Flier WG. 2004. Phylogenetic analysis of *Phytophthora* species based on mitochondrial and nuclear DNA sequences. Fungal Genet Biol 41: 766–782.

Lang BF, Forget L. 1993. The mitochondrial genome of *Phytophthora infestans*. In: O'Brien SJ, ed. *Genetic Maps. Locus Maps of Complex Genomes*. Cold Springs Harbor, New York, pp. 3.133–3.135.

Lassiter E, Russ C, Nusbaum C, Zheng Q, Hu C, Thorne J, Ristaino JB. 2010. Inferring evolutionary relationships of species in the *Phytophthora* Ic clade using nuclear and mitochondrial genes. Phytopathology 100: S68.

Lee TY, Mizubuti E, Fry WE. 1999. Genetics of metalaxyl resistance in *Phytophthora infestans*. Fungal Genet Biol 26: 118–130.

Lees AK, Wattier R, Sullivan L, Williams NA, Cooke DEL. 2006. Novel microsatellite markers for the analysis of *Phytophthora infestans* populations. Plant Pathol 5: 311–319.

MacKenzie GS. 1845. The potato murrain. *Gardener's Chronicle*, September 27.

Martin FN, Tooley PW. 2003. Phylogenetic relationships among *Phytophthora* species inferred from sequence analysis of mitochondrially encoded cytochrome oxidase I and II genes. Mycologia 95: 269–284.

May K, Ristaino JB. 2004. Identify of the Mitochondrial DNA haplotype(s) of *Phytophthora infestans* in historical specimens from the Irish potato famine. Mycol Res 108: 171–179.

Maytoun H, Smart CD, Marovec BC, Mizubuti ESG, Muldoon AW, Fry WE. 2000. Oospore survival and pathogenicity of a single oospore recombinant progeny from a cross involving the US-8 and US-17 lineages of *Phytophthora infestans*. Plant Dis 84: 1190–1196.

Montagne JFC. 1845. Note sur la maladie qui ravage les pommes de terre et caractères du *Botrytis infestans* (Note on the disease that ravages potatoes and is characterized as *Botrytis infestans*). Mem Inst Fr 609: 989–101 (in French).

Morren C. 1844. Notice sur le *Botyrtis devastateur* ou le champignon des pommes de terre. Annales de la Société Royale d'Agriculture et de Botanique de Gand I: 287–292.

Moskin J. 2009. Outbreak of fungus threatens tomato crop. *New York Times*, July 17, 2009.

Ojika M, Molli SD, Kanazawa H, Yajima A, Toda K, Nukada T, Mao H, Murata R, Asano T, Qi J, Sakagami Y. 2011. The second *Phytophthora* mating hormone defines interspecies biosynthetic crosstalk. Nat Chem Biol 7: 591–593.

Oliva RF, Flier W, Kroon L, Ristaino JB, Forbes GA. 2010. *Phytophthora andina* sp. Nov., a newly identified heterothallic pathogen of solanaceous hosts in the Andean highlands. Plant Pathol 59: 613–625.

Pennisi E. 2010. Armed and dangerous. Science 32: 804–805.

Qi J, Asano T, Jinno M, Matsui K, Atsumi K, Sakagami Y, Ojika M. 2005. Characterization of a *Phytophthora* mating type hormone. Science 309: 1828.

Raffaele S, Farrer RA, Cano LM, Studholme DJ, MacLean DT, Hines M, Jiang RHY, Zody M, Kunjeti S, Donofrio N, Meyers BC, Nusbaum C, Komoun S. 2010. Genome evolution following host jumps in the Irish potato famine pathogen lineage. Science 330: 1540–1543.

Randall TA, Ah Fong A, Judelson HS. 2003. Chromosomal heteromorphism and an apparent translocation detected using a BAC contig spanning the mating type locus of *Phytophthora infestans*. Fungal Genet Biol 38: 75–84.

Reddick D. 1939. Whence came *Phytophthora infestans*? Chron Bot 5: 410–412.

Ristaino JB. 1998. The importance of archival and herbarium materials in understanding the role of oospores in late blight epidemics of the past. Phytopathology 88: 1120–1130.

Ristaino JB. 2002. Tracking historic migrations of the Irish potato famine pathogen *Phytophthora infestans*. Microbes Infect 4: 1369–1377.

Ristaino JB. 2006. Tracking the evolutionary history of the potato blight pathogen with historical collections. Outlooks Pest Manag 17: 228–231.

Ristaino JB, Hu C. 2009. DNA sequence analysis of the late-blight pathogen gives clues to the world-wide migration. Acta Horticulturae 834: 27–40.

Ristaino JB, Groves CT, Parra G. 2001. PCR amplification of the Irish potato famine pathogen from historic specimens. Nature 41: 695–697.

Rizzo DM, Garbelotto M, Davisdson JM, Slaughter GW, Koike ST. 2002. *Phytophthora ramorum* as the cause of extensive mortality of *Quercus* spp. and *Lithocarpus densiflorus* in California. Plant Dis 86: 205–214.

Stone A. 2009. Organic management of late blight of potato and tomato. Available from http://www.extension.org/article/18361.

Teschemacher JE. 1844. The disease in potatoes. N Engl Farmer Hortic Regist 23: 125.

Tooley PW, Therrien CD, Ritch DL. 1989. Mating type, race composition, nuclear DNA content, and isozyme analysis of Peruvian isolates of *Phytophthora infestans*. Phytopathology 79: 478–481.

Trout CL, Ristaino JB, Madritch M, Wangsomboondee T. 1997. Rapid detection of *Phytophthora infestans* in late blight infected tissue of potato and tomato using PCR. Plant Dis 81: 1042–1048.

Unger F. 1847. Beitrag zur Kenntnis der in der Kartoffelkrankheit vorkommenden Pilze und der Ursache ihres Entstehens (Contributions to the knowledge of the fungi occurring in the potato disease and the cause of their emergence). Bot Zeitschr 5: 314.

van der Lee T, Testa A, Robold A, Van't Klooster J, Govers F. 2004. High-density genetic linkage maps of *Phytophthora infestans* reveal trisomic progeny and chromosomal rearrangements. Genetics 167: 1643–1661.

Wangsomboondee T, Groves CT, Shoemaker PB, Cubeta MA, Ristaino JB. 2002. *Phytophthora infestans* populations from Tomato and Potato in North Carolina differ in the genetic diversity and structure. Phytopathology 92: 1189–1195.

Zwankhuizen MJ, Govers F, Zadoks JC. 1998. Development of potato late blight epidemics: Disease foci, disease gradients, and infection sources. Phytopathology 88: 754–763.

Zwankhuizen MJ, Govers F, Zadoks JC. 2000. Inoculum sources and genotypic diversity of *Phytophthora infestans* in Southern Flevoland the Netherlands. Eur J Plant Pathol 106: 667–680.

CHAPTER 11

EXPERIMENTAL AND NATURAL EVOLUTION OF THE *CRYPTOCOCCUS NEOFORMANS* AND *CRYPTOCOCCUS GATTII* SPECIES COMPLEX

ALEXANDER IDNURM and JIANPING XU

INTRODUCTION TO THE *CRYPTOCOCCUS NEOFORMANS* SPECIES COMPLEX (CNSC)

The CNSC contains two closely related species, *C. neoformans* and *Cryptococcus gattii*, that are currently recognized by the medical mycology community. The CNSC is unusual among pathogenic fungi by being a set of related, yet distinct, lineages, all of which are capable of causing disease in humans and in other animals. The biology of this species complex has been reviewed on numerous occasions, and two books provide comprehensive background information on this species (Casadevall and Perfect, 1998; Heitman et al., 2011). The emphasis of research on these species has been from the context of how the organisms cause disease, so there is a vast body of literature directly or indirectly pertinent to the evolution of pathogenicity. This chapter uses examples from a subset of these studies.

Based on cell surface antigenic properties, the CNSC was traditionally classified into four haploid serotypes corresponding to various varieties/species: serotype A (*C. neoformans* var. *grubii*), serotypes B and C (*C. gattii*), serotype D (*C. neoformans* var. *neoformans*), and diploid/aneuploid serotype AD that are recent hybrids of strains of serotypes A and D (Casadevall and Perfect, 1998; Xu and Mitchell, 2003; Xu et al., 2002). The species/varieties/serotypes have different geographic distributions, differential tolerance or susceptibility to antifungal drugs and environmental stresses, and different virulence properties (Litvintseva et al., 2011b; Nakyak and Xu, 2010; Xu et al., 2001, 2011). Therefore, serological and species identification is of practical importance in determining treatment regimes, particularly in individuals with underlying immune-compromising disorders. However, because commercial monoclonal serotyping kits are no longer available, serotype identification is no longer a common practice. Instead, molecular markers are used almost exclusively for species and/or variety identifications.

Over the past 20 years, a series of molecular markers have been developed for

Evolution of Virulence in Eukaryotic Microbes, First Edition. Edited by L. David Sibley, Barbara J. Howlett, and Joseph Heitman.

strain identifications in the CNSC, including randomly amplified polymorphic DNA (RAPD), polymerase chain reaction (PCR) fingerprinting, polymerase chain reaction combined with restriction fragment length polymorphism (PCR-RFLP), Southern hybridization using repetitive DNA fragments, and DNA sequencing of one or multiple gene fragments. These molecular markers identified that the CNSC could be divided into multiple distinct lineages within both *C. neoformans* and *C. gattii* and that these lineages likely represent different phylogenetic species (Fig. 11.1). Specifically, *C. neoformans* is composed of five recognized lineages: VNI, VNII, VNIII, VNIV, and VNB, and *C. gattii* is composed of at least four lineages: VGI, VGII, VGIII, and VGIV. The VNB lineage in *C. neoformans* var. *grubii* contains strains almost exclusively from Botswana (Litvintseva et al., 2011b). This lineage can be further divided into three sublineages (VNB-A, VNB-B, and VNB-C) with each receiving strong bootstrap support in phylogenetic analyses. In contrast to the geographically limited distribution of the VNB lineage, VNI and VNII are two major lineages within *C. neoformans* var. *grubii* (serotype A) and they are composed of strains from many geographic regions, spanning from tropical to subtropical, temperate, and arctic regions. The VNIV lineage corresponds to *C. neoformans* var. *neoformans*

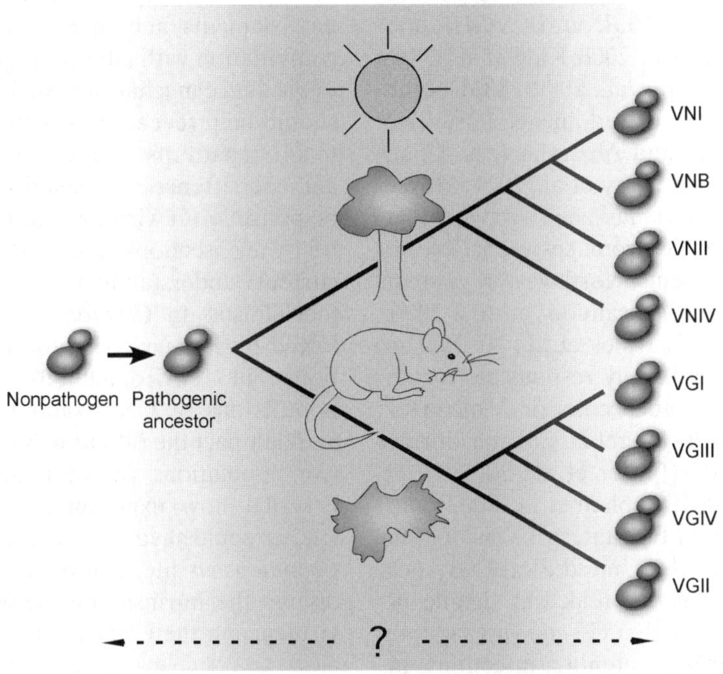

Figure 11.1 A summary of the questions related to the evolution of virulence in the *Cryptococcus neoformans* species complex. The complex comprises distinct lineages (VN for variety *neoformans* or VG for variety *gattii*) that likely derive from a progenitor pathogenic species, as indicated by the phylogeny. Environmental factors played a role in selection for the ability to cause disease in humans, although which factors are unknown. They are represented by the sun for climate conditions, tree for the association with plant species, mouse for links to mammalian pathogenesis, and amoeba for interactions with microbial predators. The gray overlay indicates the current lack of knowledge about when, where, and for how long the evolution of this complex has been influenced by the environment.

(serotype D). Compared to the large number of strains analyzed for VNI, the number of strains analyzed so far for the VNIV lineage is relatively limited. The VNIV lineage also has a global distribution but is more commonly found in Europe. Similar to those performed on VNI, VNII, and VNB, preliminary analyses suggest that multiple differentiated clones and clonal lineages exist in VNIV. In contrast to the haploid VNI, VNII, VNB, and VNIV lineages, strains in the VNIII genotype group (serotype AD) were recently derived from hybridizations between strains of VNI (or VNII) with those of VNIV and are mostly diploid or aneuploid.

While *C. neoformans* has a global distribution, its sibling species *C. gattii* was considered geographically more limited, until recently. Currently, four main lineages are found in *C. gattii*: VGI, VGII, VGIII, and VGIV (Bovers et al., 2008; Kidd et al., 2005; Ngamskulrungroj et al., 2009). VGI is the most commonly isolated lineage in several Asian countries and Australia (e.g., Chen et al., 2000; Chowdhary et al., 2011; Meyer et al., 2003). VGII has been recovered at greater frequencies from Australia, Colombia, and the Pacific Northwest region of North America (Escandón et al., 2006; Kidd et al., 2005; Trilles et al., 2003). The VGII lineage is largely responsible for an outbreak of cryptococcosis on Vancouver Island in British Columbia, Canada, during the last decade (Fraser et al., 2003, 2005; Kidd et al., 2004; Stephen et al., 2002) and more recently in the states of Washington and Oregon in the United States (Byrnes et al., 2010). This outbreak was significant because it showed that *C. gattii* was capable of causing life-threatening infections in apparently healthy individuals in a temperate climate: Why this disease emerged in the last 15 years where it did is an evolutionary mystery at present. VGIII and VGIV are relatively rare but have been recovered from both environmental and clinical samples (e.g., Bovers et al., 2008; Kidd et al., 2005). For example, VGIV has been implicated in causing secondary infections in AIDS patients in sub-Saharan Africa (Litvintseva et al., 2005b). However, more surveys and experimental analyses are required to confirm if the seemingly nonrandom geographic distribution of the four distinct lineages in *C. gattii* is due to incomplete sampling or represents true geographic or biological differences among the lineages.

The surveys conducted to date have shown that both *C. neoformans* and *C. gattii* can be found in a variety of environmental niches, from trees to soil to wild animals. It is believed that most human cryptococcal infections are acquired directly from environmental sources. Therefore, it is important to study the ecology and population biology of this organism in both its natural environments and in clinical settings. In combination with laboratory investigations of virulence mechanisms, such comparisons should help reveal the selection and transmission patterns and uncover the origin and persistence of genetic mechanisms responsible for virulence in CNSC. In the following sections, we first review our current understanding of the virulence mechanisms in *Cryptococcus*. This is followed by surveys of natural genetic variations in environmental and clinical populations of CNSC and how selection could impact the differences between these two populations. In the third section, we describe how experimental evolutionary studies could shed light on the evolution of *Cryptococcus* life history traits. Lastly, we discuss the intrinsic link between sex and virulence in the CNSC.

GENETIC REQUIREMENTS FOR VIRULENCE

The two *Cryptococcus* species have a number of established virulence traits. Like all human pathogens, the species must be able to grow at mammalian body temperature. Two more unique properties of *Cryptococcus* are the production of a polysaccharide capsule and the pigment melanin, which both aid in distinguishing *C. neoformans* and *C. gattii* from other pathogenic yeasts encountered in clinical settings and are also traits required for full virulence (Bose et al., 2003; Casadevall et al., 2000; Doering, 2009; Langfelder et al., 2003; Nosanchuk and Casadevall, 2003). Experimental approaches have led to the discovery of many *Cryptococcus* genes required for virulence, as assessed by creating mutants and testing their virulence in animal models, as part of a continuing effort to define the genetic basis of the species' ability to cause disease. These studies have mostly focused on individual genes (Perfect, 2005). One of the larger-scale approaches has been to screen pools of strains, each bar-coded with a signature tag, and to assess changes in relative abundance during disease progression (Idnurm et al., 2009; Liu et al., 2008; Nelson et al., 2001). The most comprehensive study analyzed 1201 defined gene deletion strains for growth in the lung (Liu et al., 2008). Both one-by-one and pooling strategies have revealed approximately 200 *C. neoformans* genes that are required for virulence. In summarizing the functions of those genes, they include factors required for the production of capsule and melanin and growth at 37°C, the three well-known virulence attributes of the CNSC. Other factors include those involved in environmental sensing, signal transduction, ion homeostasis, cell wall integrity, and protection against reactive oxygen or nitrogen species.

The majority of the CNSC genes required for virulence also play roles beyond causing disease in animals; that is, there are no clear-cut virulence factors, such as toxins, in the *Cryptococcus* spp. that account for their pathogenicity. The capsule material comes close to being a clear-cut virulence factor—it can directly cause damage to host immune cells. However, this is likely a reflection of their natural ecology, with their primary niche outside of mammalian hosts. Consequentially, the genes required for virulence in mammalian hosts all should have functions in addition to a role just for virulence, and they should contribute to reproductive success in the environment (Casadevall et al., 2003). This hypothesis is borne out by the identification of *in vitro* phenotypes of mutant strains with reduced virulence, suggesting that mutation of those genes would be a selective disadvantage to the organism.

One direction for future investigation is to search for signatures of positive selection in genes implicated in virulence versus those that are not. As discussed in the following sections, it is unclear how much contribution mammalian infection or exposure to environmental predators has on the evolution of the *Cryptococcus* species. Knowledge about those genes required for mammalian virulence, together with growing genome sequence resources, now provides this opportunity. For instance, testing for a correlation between the signature tag scores, reflecting fitness in mouse lung (Liu et al., 2008), and the K_a/K_s values for those genes would be informative.

Another bioinformatic approach is the use of comparative genomics to explain the pathogenicity of *Cryptococcus* spp. Comparisons could be between different varieties within species or between *Cryptococcus* and related species such as *Tremella mesenterica* whose genome has been completed by the U.S. Department of Energy. The genome sequences of five *C. neoformans* or *C. gattii* strains were generated by Sanger technology. Strains B3501 and JEC21 (var. *neoformans*, VNIV) are closely related, with B3501 being a parent of JEC21 (Heitman et al., 1999; Loftus et al., 2005). The genome of strain H99 (var. *grubii*, VNI) has been completed and is available at the Broad Institute but is not yet published. Sequences of two *C. gattii* strains,

WM276 and R265 (VGI and VGII, respectively), are available also from the University of British Columbia (UBC) Vancouver Genome Center and Broad Institute, and their genomes were recently published (D'Souza et al., 2011). Each of these strains offers different perspectives on the evolution of virulence because these model strains represent some of the best-studied members of the CNSC and exhibit differences in origin and virulence. Despite this knowledge, comparisons between the genomes have, as yet, not revealed major features associated with virulence. Nevertheless, this probably reflects the inability to know what to look for within the ~19 Mb of each genome. Furthermore, next-generation sequencing technology is a promising area to explore differences within the varieties or groups of *Cryptococcus* or for comparisons of genomes from clinical versus environmental sources.

The closest relatives of the CNSC are found in association with insects or as mycoparasites, while more distant Tremellales relatives are found in association with trees (Findley et al., 2009), suggesting that interaction with these organisms could have contributed to the evolution of a clade of species able to cause disease in diverse animals (Fig. 11.1). Further support for this hypothesis is the observation that *C. gattii* is commonly isolated from trees, and *C. neoformans* and *C. gattii* induce a disease response and respond to stimuli from *Arabidopsis thaliana* and *Eucalyptus camaldulensis* under laboratory conditions (Xue et al., 2007). Strains of *Cryptococcus* spp. have been isolated from insect frass and beehives (Ergin et al., 2004; Kidd et al., 2003). One hypothesis, discussed in subsequent sections, is that virulence in humans is an indirect consequence of selection by environmental predators such as amoeba. If so, then discovery and understanding of the function of genes required for virulence is essential to understand how this species complex has evolved and diverged from other related but nonpathogenic species.

The factors that distinguish *C. neoformans* from *C. gattii* related to pathogenesis are unknown at present. For instance, *C. neoformans* tends to infect patients with immunodeficiencies, whereas *C. gattii* tends to infect immunocompetent individuals. As mentioned above, the genome sequences of both species have generated no obvious leads from comparative analyses that could explain the differences in their geographic distributions and epidemiological patterns between the two species. One distinguishing feature of the most highly virulent *C. gattii* strains isolated from the recent outbreak in the Pacific Northwest is a tubular mitochondrial morphology during infection of host cells and the upregulation of transcription of mitochondria-encoded genes (Ma et al., 2009).

An additional challenge in understanding virulence is to relate how phenotypic variation and differences between strains of *C. neoformans* or *C. gattii* contibute to their epidemiology. The best-analyzed differences are between var. *grubii* and var. *neoformans* strains when examining the phenotypes of gene disruption strains. For example, there are two protein kinase A (PKA) catalytic subunits in *C. neoformans*, but each copy plays a different role in var. *grubii* and var. *neoformans* strains (Hicks et al., 2004). Tup1 also plays different roles between isolates: It is involved in sensing cell density in the var. *neoformans* strain but not in var. *grubii* (Lee et al., 2007, 2009). Other variety differences have been reported for p21-kinase and calcineurin functions (Cruz et al., 2000; Wang et al., 2002). The use of a single wild-type strain in genetic research is essential for community-based discovery and comparison of results; however, this may lead to loss of information about the variation found within a population. In terms of understanding pathogenesis, particularly for a species complex like the CNSC, the importance to analyze gene function in multiple representative isolates is becoming increasingly evident.

ANALYSES OF NATURAL POPULATIONS OF THE *C. NEOFORMANS* SPECIES COMPLEX

Ecological Niches

Ever since its discovery as an etiological agent of human diseases over 100 years ago, the ecology and population biology of CNSC in its natural environments has attracted significant attention among mycologists, medical microbiologists, clinical physicians, and population biologists. Surveys conducted so far have found strains of CNSC to be ubiquitous in many ecological niches across broad geographic areas. For example, strains of *C. neoformans* have been found in the soil, air, avian excreta, trees, fruits, water, and resting places of a variety of animals such as bats, birds, and rabbits. While recent studies from certain geographic regions suggest that decayed tree hollows could be a very significant source for *C. neoformans*, the most common ecological niche for this species is bird droppings, especially pigeon excreta. Indeed, pigeons and other birds have been proposed as the likely vectors for the dispersal of *C. neoformans* across geographic regions and continents (see population genetic analyses below).

Similarly, a variety of environmental sources have been identified for *C. gattii* from tropical to subtropical and temperate regions. These include soil, fruits, vegetables, freshwater and seawater reservoirs, insect frass, decayed wood of tree trunk hollows, plant debris, desiccated excreta of a few species of caged birds, and tissues of infected animal hosts such as pigs, goats, dolphins, horses, and humans (e.g., Casadevall and Perfect, 1998; Kidd et al., 2005; Xu et al., 2011). However, instead of bird droppings, trees and decayed wood seem the most frequent ecological niches harboring *C. gattii*. The focus on trees and decayed wood and not bird droppings as an ecological niche for *C. gattii* is not due to the lack of effort to investigate bird droppings for *C. gattii*. Rather, most investigations have identified few or no strains of *C. gattii* from bird droppings. In contrast, the emphasis on trees and decayed wood as sources of *C. gattii* was likely influenced by the earliest isolation of *C. gattii* in Australia from *Eucalyptus* trees (Ellis and Pfeiffer, 1990b). Following that initial success, many studies have subsequently found *C. gattii* from a variety of tree species (including *Eucalyptus*) in other geographic regions such as the United States, Mexico, India, Italy, Colombia, Argentina, Brazil, Jordan, Egypt, Paraguay, Africa, Papua New Guinea, Canada, and Southeast Asian countries (Chakrabarti et al., 1997; Davel et al., 2003; Granados and Castañeda, 2005, 2006; Kidd et al., 2005; Laurenson et al., 1997; Mahmoud, 1999; Montenegro and Paula, 2000; Pfeiffer and Ellis, 1992; Randhawa et al., 2003, 2008, 2006; Xu et al., 2011).

The host tree species identified so far for *C. gattii* are very broad and include both angiosperms and gymnosperms. As expected, several in the genus *Eucalyptus* have been found to harbor *C. gattii*, including *E. camaldulensis* (river red gum), *Eucalyptus tereticornis* (forest red gum), *Eucalyptus grandis* (flooded gum), *Eucalyptus blakelyi* (Blakely's red gum), *Eucalyptus rudis* (West Australian flooded gum), *Eucalyptus microcorys* (tallow wood), *Eucalyptus gomphocephala* (tuart), and *Eucalyptus globulus* (blue gum) (Ellis and Pfeiffer, 1990a,b; Granados and Castañeda, 2005; Vilcins et al., 2002). In South America, *C. gattii* has been recovered from the following tree species: *Terminalia catappa* (almond tree), *Cassia* sp., *Moquilea tomentosa* (pottery tree), *Ficus soatensis*,

Croton bogotanus, *Croton funckianus*, *Coussapoa* sp., *Pinus radiata* (Monterey pine), *Acacia decurrens* (black wattle), *Cupressus lusitanica* (Mexican cypress), *Cassia grandis* (pink shower tree) (Granados and Castañeda, 2005), and *Guettarda acreana* (quina quina) (Fortes et al., 2001). Following the outbreak of *C. gattii* in British Columbia, Kidd and colleagues (2007) surveyed the Pacific Northwest over a 5-year period (2001–2006) for *C. gattii*. The organism was recovered at high frequencies from *Pseudotsuga menziesii* var. *menziesii* (Douglas fir), *Alnus rubra* (alder), *Arbutus menziesii* var. *menziesii* (arbutus), *Thuja plicata* (red cedar), and *Quercus garryana* (Garry oak) and at variable frequencies from *Prunus emarginata* (bitter cherry), *Acer* sp. (maple), *Pinus* spp. (pine), *Picea* spp. (spruce), and the logs and stumps of some unidentified trees. In India, *Syzygium cumini* trees are among the most frequently colonized, although *C. gattii* was also found to be associated with decayed wood in tree hollows of *Polyalthia longifolia* (mast tree), *Azadirachta indica* (neem), *Tamarindus indica* (tamarind), *Acacia nilotica* (babul), *Mimusops elengi* (bullet wood), *Manilkara hexandra* (khirni), *Mangifera indica* (mango), and *Pithecolobium dulce* (Manila tamarind) (Randhawa et al., 2003, 2008, 2006; Xu et al., 2011). The interaction with *C. gattii* and *C. neoformans* with the model plant *A. thaliana* has also been investigated under laboratory conditions, revealing that plants and plant-derived molecules support the sexual cycle (Xue et al., 2007).

The initial discovery of the association of *C. gattii* with *Eucalyptus* trees in Australia led to the hypothesis that the global distribution of this organism might be related to the export of *Eucalyptus* seeds and saplings from Australia to other parts of the world (Ellis and Pfeiffer, 1990a,b). However, support for *Eucalyptus* seeds as vectors is weak as a histological investigation did not isolate *C. gattii* from eucalypt ovaries/anthers/seeds (Randhawa et al., 2003). In addition, recent population genetic analyses revealed significant diversity from other areas that are not found in Australia (see also below), inconsistent with the notion that Australia was the center of diversity and origin for other geographic populations of *C. gattii*.

Population Genetic Patterns

Three measures are typically used to describe genetic variation in microbial populations: the amount of genetic variation (e.g., nucleotide diversity, gene diversity, and genotype diversity), the ecological and geographic distributions of the observed genetic variation (e.g., population subdivisions and lineage diversification), and the potential mechanisms for generating the diversity (e.g., mutation, selection, and modes of reproduction). Here, we briefly review our current knowledge about the modes of reproduction and population subdivision in the CNSC.

The mode of reproduction can significantly impact the patterns of genetic variation in natural populations of CNSC, including the virulence potential of the organism. In laboratory environments, CNSC can reproduce through several pathways: asexual reproduction through mitotic division and budding, haploid fruiting involving meiosis and basidiospore generation (a form of same-sex mating), and the traditional opposite-sex mating that also involves meiosis and basidiospore generation. The opposite-sex mating system is controlled by a single locus with two alternative alleles: *MAT*α and *MAT***a** (Kwon-Chung, 1975, 1976b). However, as in the majority of microorganisms, asexual reproduction is expected to be common in natural populations of the CNSC. Asexual reproduction leads to biased association among alleles (i.e., linkage disequilibrium) and the propagation of clones and clonal lineages within a population, and through dispersal, to clonal expansion across ecological niches and geographic regions. On the other hand, sexual reproduction recombines genetic materials from different strains, results in linkage equilibrium and the formation of basidiospores, one cell type proposed as the infectious particle to initiate cryptococcal infections (Giles et al., 2009; Sukroongreung et al., 1998; Velagapudi et al., 2009). Because of their small size and low water content, basidiospores are more easily dispersed than vegetative cells.

Genetic analyses of natural populations of CNSC indicated variable modes of reproduction among geographic populations, from completely clonal to random mating (e.g., Brandt et al., 1996; Chowdhary et al., 2011; Hiremath et al., 2008; Kidd et al., 2005; Litvintseva et al., 2011b; Simwami et al., 2011). Signatures of clonality include overrepresentation of a single genotype and significant nonrandom association between alleles from pairs of loci (Xu, 2005). For example, using the highly discriminatory amplified fragment length polymorphism (AFLP) genotyping method, Litvintseva et al. (2005a) analyzed 826 clinical and environmental isolates of *C. neoformans* var. *grubii* from the United States and found only 12 genotypes. The predominance of *MAT*α strains in most environmental and clinical samples is also consistent with clonal population structures (e.g., Yan et al., 2002). However, evidence for mating and recombination has been found in several natural populations such as those from the hybrid AD population for the VNI, VNII, and VNIV lineages in the United States (Xu and Mitchell, 2003), the unisexual environmental population of the VNI lineage in India (Hiremath et al., 2008), the VGI lineage in Australia (Campbell et al., 2005a,b) and India (Chowdhary et al., 2011), the VGII lineage in the Americas (Xu et al., 2009), and the VNB lineage in Botswana. The VNB lineage in Botswana was especially interesting because of its high level genetic diversity and the presence of a high proportion of *MAT***a** strains (Litvintseva et al., 2003). These findings have led to the "out-of-Africa" hypothesis for the origin and subsequent spread of *C. neoformans* var. *grubii* around the world (Litvintseva et al., 2011a,b).

Similar to the highly variable nature of the modes of reproduction in natural populations of CNSC, recent results suggested that the patterns of relationships among geographic/ecological populations of the CNSC are also highly variable. Early findings of genetic differences between geographic/ecological populations were mainly due to the differential distributions of the various species (i.e., *C. neoformans* vs. *C. gattii*), varieties (i.e., *C. neoformans* var. *neoformans*, *C. neoformans* var. *grubii*, and *C. neoformans* var. *gattii*), serotypes (i.e., A, B, C, D, and AD), and phylogenetic lineages (i.e., among VNI, VNII, VNB, VNIV, VGI-VGIV), as discussed in the "Introduction" section. However, as the phylogenetic evidence showed, most such previous analyses dealt with issues at the above-species level. Indeed, despite the large number of studies, formal analysis of the relationships among geographic and ecological populations within individual phylogenetic lineages is still relatively limited. Two recent studies analyzed environmental populations of the VNI and VGI lineages from a variety of native and non-native trees located in different parts of India (Chowdhary et al., 2011; Hiremath et al., 2008). Their geographic locations and their host tree species were very similar and both studies employed the multiple gene genealogical approach based on an overlapping set of loci. However, the population relationships were quite different for the two lineages. The ecological and geographic populations of the VNI lineage were very similar to each other with little evidence of genetic differentiation and abundant evidence for gene flow. In contrast, those of the VGI lineages showed significant differentiation with little evidence for gene flow. At present, there is little systematic analysis of the relationships among clinical populations within individual lineages.

MICROEVOLUTION DURING INFECTION

The evolution of the CNSC is estimated to have occurred over 37 million years to lead to the currently recognized lineages (Xu and Mitchell, 2003). However, change can occur much more rapidly, including the important time frame during the course of disease development in a host. The steps in causing disease by *Cryptococcus* start with a pulmonary infection, from where the fungus can spread from the lungs to almost any organ of the body. The most serious secondary infections are within the central nervous system. Two hypotheses have been developed relating to the infection process in humans (reviewed in Dromer et al., 2011). In one, susceptible people are infected; the fungus establishes itself in the body and causes disease over a time frame of months. The other hypothesis is that people are infected early in life through exposure to the fungus; the initial infection is prevented from spreading by the host immune system, but the fungus is not eliminated from the body and rather remains in a dormant state. If a person becomes immunocompromised later in life, then the fungus can reactivate, grow, and cause fatal disease. This reactivation is similar to disease development of *Mycobacterium tuberculosis* from a latent infection, one of the correlations between this bacterial pathogen and the CNSC. Another factor is the high incidences of these two diseases in AIDS patients in sub-Saharan Africa.

The two models are not necessarily mutually exclusive and available evidence so far is mixed. Evidence for the first comes from studies suggesting recent acquisition of the fungus, including case studies in which patients infected with cryptococcosis had pet birds whose fecal material was positive for *C. neoformans* (Nosanchuk et al., 2000; Wegener and Staib, 1983). On the other hand, a study in New York showed that 70% of children over 5 years old had antibodies against *C. neoformans*, indicating an early exposure to the fungus (Goldman et al., 2001) and consistent with detection of antibodies in healthy adult populations (Dromer et al., 2011). Confounding these results are observations that isolates of the fungus from the environment do not cause disease in mouse models (Litvintseva and Mitchell, 2009), the lack of information on the inoculum doses needed to cause disease in humans, and evidence for multiple infections in up to 20% of cases (Desnos-Ollivier et al., 2010). Another factor necessary for consideration is growth conditions prior to infection as in experimental animal infections, the outcome is influenced by how the inocula were generated (Springer et al., 2010).

In either situation, the timescale for disseminated disease within the host postinfection is likely over time frames of months, providing ample time and cell division cycles for the fungus to change. Analyses of serial isolates from infected people suggest that they have undergone microevolution (Blasi et al., 2001; Fries and Casadevall, 1998; Jain et al., 2005; Sullivan et al., 1996). However, due to the treatment strategies, sampling may be months later or on a single colony isolated from a patient, making it difficult to distinguish whether this is microevolution or that the patient was at that time infected by multiple related strains or subsequently reinfected. Studies under controlled environments in animal models have confirmed that changes occur during the time course of disease development, such that better adapted subpopulations predominate (Desnos-Ollivier et al., 2010; Fries et al., 2001; Goldman et al., 1998).

The frequency of microevolution in *C. neoformans* has been investigated under *in vitro* conditions. This includes phenotypic switching between colony forms (wild-type smooth to mucoid, wrinkled, or pseudohyphal) that occurs at rates as high as one per 10,000 colonies (Fries et al., 1999; Goldman et al., 1998). Rates of switching are variable and can be influenced by the growth conditions like exposure to UV light, as cells age, or within biofilms (Fries et al., 1999; Jain et al., 2009; Martinez et al., 2008). The genetic basis for microevolution or phenotypic switching and their rate determination is unknown in *C. neoformans* or *C. gattii*. An alternative approach to assess rates of change has been to study changes in fertility and fitness during *in vitro* passage (Xu, 2002, 2004) and as described below.

Aside from causing disease in mammals, *C. neoformans* and *C. gattii* naturally cause disease in many animal species and also in other nonvertebrates under laboratory conditions. Each of these nonmammalian models has advantages and disadvantages as experimental systems and the conclusions that can be reached in their use can also differ (Mylonakis et al., 2007). Casadevall and colleagues developed a hypothesis that *Cryptococcus* spp. are preselected for virulence by exposure to environmental predatory microbes such as amoeba or nematodes (Steenbergen and Casadevall, 2003; Steenbergen et al., 2001). The evidence for this hypothesis is compelling. Genes required for virulence in mice are often also required for virulence in alternate hosts (Apidianakis et al., 2004; Mylonakis et al., 2002, 2005; Steenbergen et al., 2001). Screens of mutants in the nematode *Caenorhabditis elegans* have identified strains with mutations in the *KIN1* and *ROM2* genes that when tested in murine models are also avirulent (Mylonakis et al., 2004; Tang et al., 2005). Passage of the fungus through the slime mold *Dictyostelium discoideum* enhances virulence in subsequent mouse infections (Steenbergen et al., 2003). Other models for studying *C. neoformans* are the larva of the wax moth *Galleria mellonella* and the fruit fly *Drosophila melanogaster* (Apidianakis et al., 2004; Mylonakis et al., 2005). Similar trends in nonvertebrate hosts have been found for other environmentally acquired pathogenic fungi (Chamilos et al., 2010; Renwick et al., 2006; Slater et al., 2011; Steenbergen et al., 2004), suggesting that this selective pressure may be common to this class of pathogens.

The microbial predators of *Cryptococcus* spp. in nature are largely unknown and may vary and differ between groups. For instance, *C. gattii* strains respond differently than *C. neoformans* in response to amoeba, including being less frequently phagocytosed (Malliaris et al., 2004). Thus far, the closest associations for *C. neoformans* are with amoeba, with coisolation and interaction observed with three *Acanthamoeba* species (Castellani, 1930; Neilson et al., 1978, 1981). In the 1970s, coculturing experiments revealed that amoeba could engulf and kill *C. neoformans*, but surviving cells had developed a pseudohyphal morphology (Neilson et al., 1978). When tested in animals, the pseudohyphal strains were nonpathogenic. Thus, exposure to amoeba selects for nonpathogenic strains and remains one piece of evidence contrary to the model of selection in nonvertebrate hosts (Levitz, 2001). However, very little is known about the microbes sharing the environment with *Cryptococcus* species, and this is clearly another area in which rapid advances could be made toward understanding the ability of these fungi to cause disease.

A second direction for future research is to establish which genes are required for pathogenesis in mammalian species and which are required in nonmammalian models. On the other side of the host–pathogen interaction, specific host features should protect against *C. neoformans* disease. Recent screens, like the one using the *Drosophila* S2 cell line (Qin et al., 2011), have sought for host components to identify those that influence disease progression. This information would provide the benefit of understanding what role pathogenesis in alternative hosts might have played in the evolution of the *Cryptococcus* species.

EXPERIMENTAL EVOLUTION STUDY OF HIGH TEMPERATURE GROWTH AND IMPLICATIONS FOR LIFE HISTORY EVOLUTION IN HUMAN PATHOGENS

Despite the significant mortality on humans and other mammals, it is predicted that humans and other mammalian hosts are accidental hosts and are not an essential part of the life cycle of CNSC (Casadevall and Perfect, 1998). As mentioned in the preceding section, its pathogenicity to mammals is thought to have evolved as a by-product from interactions with soil protozoa and/or other soil organisms (Steenbergen and Casadevall, 2003; Steenbergen et al., 2001) (Fig. 11.1). The obvious question is how could a by-product be actively maintained if it does not convey a selective advantage in its primary and essential ecological niche? One pathogenicity trait found in all human pathogens, including members of the CNSC, is their ability to grow at the mammalian body temperature of 37°C. Because soil environments typically have temperatures below those of human hosts, the specific question is how could human pathogens with soil or nonmammalian hosts as the primary reservoir be able to maintain their ability to grow vigorously at 37°C in humans?

To help answer this question, an experimental evolutionary study was initiated to determine the relative change of vegetative fitness at two different temperatures, 25°C (a temperature more similar to ambient environment) and 37°C (mimicking the mammalian body temperature). In this study, eight asexual parallel mutation accumulation (MA) lines were established at each of the two temperatures and a single colony was subcultured from each of the 16 MA lines every 48 hours for a total of 30 transfers. Clones from transfers 5, 10, 20, and 30 were permanently stored in a –70°C freezer for each of the 16 MA lines. This protocol maximizes genetic drift by eliminating selection, thus allowing mutations to accumulate freely. Because most mutations are neutral or deleterious, and given that different MA lines will likely accumulate different mutations, the growth phenotypes among the MA lines were expected to decline and diverge. At the end of the MA phase, all stored clones were tested for their growth on four different environments: on the rich complex medium yeast extract–peptone–dextrose (YEPD) and on the minimum synthetic medium synthetic dextrose (SD), at 25°C, and 37°C.

Analyses identified that after 30 transfers (about 600 mitotic divisions), clones from MA lines maintained at both the 25 and 37°C environments showed significantly decreased vegetative fitness in all four testing environments. However, the amount of decrease varied: Those isolates that had accumulated mutations at the 25°C environment showed greater declines than those at the 37°C environment, especially at the 37°C fitness-testing environment (Table 11.1). For example, in the 37°C/YEPD fitness-testing environment, the mean vegetative fitness of the clones from the 37°C MA environment declined by about 20%, while those of the clones from the 25°C MA environment declined by over 40%.

These results suggest that periodic exposure to 37°C would be critical for strains of CNSC to maintain their abilities to grow at the 37°C environment, a requirement for them to cause diseases in humans and in other mammals. Where might such environments be? There are two candidate high

TABLE 11.1 Relative Vegetative Fitness of Mutation Accumulation (MA) Lines at Generation 600 in Four Environments

	Vegetative Fitness Testing Environments			
MA condition	25°C/SD	25°C/YEPD	37°C/SD	37°C/YEPD
25°C/YEPD	0.644 ± 0.338	0.857 ± 0.066	0.349 ± 0.293	0.588 ± 0.229
37°C/YEPD	0.715 ± 0.195	0.809 ± 0.132	0.768 ± 0.254	0.795 ± 0.052

Fitness of the original starting clone was scaled to 1 in each environment. Each data point represents the mean ± standard deviation of eight mutation accumulation lines. The media used was yeast extract–peptone–dextrose (YEPD) or synthetic dextrose (SD).

temperature environments for CNSC: (i) in warm-blooded animals and (ii) in microenvironmental niches in soil particles, bird droppings, trees, or other organic materials like a compost. These microenvironmental niches could periodically experience high temperatures due to heat generated by the degradation of organic compounds. However, while strains of the CNSC have been isolated from such microenvironmental niches, unlike many other fungi, no strain of *C. neoformans* has been found in large organic composts where temperatures could reach 50–60°C. A weeklong exposure to 40–41°C environments would kill most strains of the CNSC (Casadevall and Perfect, 1998; Heitman et al., 2011).

The alternative is that warm-blooded animals likely play a greater than expected role in the life history of pathogenic strains of CNSC. This hypothesis is also supported by data from population genetic studies of CNSC. Epidemiological surveys have identified that a few clones or clonal lineages dominate the global clinical strains of CNSC and that clonal dispersals are common over wide geographic areas (e.g., Brandt et al., 1996; Xu et al., 2000). If pathogenic strains of CNSC spent most of their time in soil, trees, and bird droppings and that warm-blooded animals were only accidental hosts, it would be expected that geographic populations should diverge from each other rather rapidly. Such divergences should lead to significant geographic structuring and the lack of clonal dispersal over long distances—phenomena rarely seen in clinical or environmental populations of CNSC (e.g., Brandt et al., 1996; Xu et al., 2000). Indeed, current population genetic surveys are consistent with the hypothesis that warm-blooded animals such as humans and/or other mammalian hosts might act as important reservoirs and selective agents for pathogenic strains of CNSC. Such a reservoir and the selective pressure for growth at high temperatures could purge nonpathogenic or attenuated strains from human and other mammalian populations, leading to limited genetic diversity, extensive clonality, and long-distance clonal dispersal (also by humans) of these pathogenic strains affecting humans. In addition, direct animal–human and animal–animal transmissions could also help maintain the growth ability at 37°C. At present, direct evidence for these two possibilities is lacking.

The importance of a high temperature environment for the maintenance of growth at 37°C of CNSC is also supported by vegetative fitness comparisons between clinical and environmental samples (Table 11.2). Overall, strains from clinical sources showed a greater vegetative fitness at 37°C and a lower fitness at 25°C than those from environmental sources, and significant interactions between strain source and temperature were observed (Table 11.2). However, it should be noted that while overall consistent, the magnitude and statistical significance of differences between

TABLE 11.2 Summary of Vegetative Fitness Comparisons between Clinical and Environmental Samples of *C. neoformans* at Two Different Temperatures, 25 and 37°C, Grown in a Yeast Extract–Peptone–Dextrose Medium

Population Sample	Number of Strains	Testing Temperature	
		25°C	37°C
Environmental	45	0.251 ± 0.091	0.516 ± 0.181
Clinical	35	0.192 ± 0.057	0.637 ± 0.096

Data are presented as mean ± standard deviation of colony sizes in millimeters.

laboratory populations were not identical to those between clinical and environmental populations (Tables 11.1 and 11.2). If the ancestors of the environmental strains had only lived in the environment (i.e., outside of human or other mammalian hosts), we should expect the environmental strains to lose the ability to grow at high temperature rather quickly. This expectation was not met. Their relatively vigorous growth at 37°C suggests that these environmental strains may have had significant exposure to high temperature environments such as humans or other mammalian hosts. Results from population genetic studies of environmental samples are also consistent with this hypothesis (see above). At present, aside from the sources prior to the isolation of each strain (patient vs. environment), their previous environments, selection pressure, and the lengths of time in their current environments are unknown for any strains of CNSC.

It should be noted that an extended exposure to a high temperature environment, for example, warm-blooded animals such as humans, does not suggest direct and frequent human-to-human transmission of infectious particles nor does it mean that environmental reservoirs are insignificant since many parts of the world that harbor the CNSC, particularly *C. gattii*, have daily maxima near 37°C. Indeed, natural environments with low temperatures (such as 25°C) and low nutrient levels (such as poor sources or low concentrations of nitrogen) are necessary for mating and sexual reproduction in CNSC (Kwon-Chung, 1976a). Laboratory studies have shown that high temperatures (such as 37°C) or a nutrient-rich medium (such as YEPD) is not conducive for mating and sexual recombination in CNSC. However, a brief exposure to 37°C and a subsequent return to permissive temperature has been reported to enhance the monokaryotic fruiting sexual process (Fu et al., 2011). Sexual recombination in nature could generate pools of genetically diverse offspring, some of which may have equivalent or greater fitness and pathogenicity than the parental strains. Furthermore, sexual recombination in natural environments could help avoid the mutational meltdown that can occur in asexual clones, as demonstrated in the experimental evolution populations here. Rare sexual recombination exists in environmental and clinical populations of the CNSC (Litvintseva et al., 2011b; Xu and Mitchell, 2003; Xu et al., 2000). Given that strains of the CNSC do not appear to mate at 37°C, the recombination events must have occurred in natural environments outside of warm-blooded hosts. One cannot rule out the possibility of mating on cooler parts of animals. For instance, *Candida albicans* was originally thought to be unable to mate in the host because the necessary opaque cells are unstable at 37°C. However, mating in *C. albicans* has been observed, that is, on the surface of the skin of mice that is lower than 37°C (Lachke et al., 2003).

RELATIONSHIP BETWEEN SEX AND VIRULENCE

Natural selection requires variation and diversity for evolution to occur. One natural source of this variation is produced during sexual reproduction in which different allelic combinations are generated. For many of the human pathogenic fungi, sex appears to be a rare occurrence (Heitman, 2006). This generalization once included *Cryptococcus* species, with over 70 years separating the first reports of the species with the discoveries of the sexual cycles for both *C. neoformans* and *C. gattii* (Kwon-Chung, 1975, 1976a). The population genetic studies discussed above highlight that recombination does occur in the wild, but probably in most locations at low frequencies and mediated by the monokaryotic fruiting/same-sex mating pathway. Further, there is variation in strain fertility for both *C. neoformans* and *C. gattii*. That said, strains of *C. neoformans* and *C. gattii* undergo a full sexual cycle with alteration of generations under laboratory conditions, a trait that has, to date, eluded researchers working on other common human pathogenic fungi.

Why maintain sex at all in pathogenic fungi? For some species, it makes sense to provide diversity in allelic combinations to survive in variable environments. Other species appear now as obligate pathogens, but even then, there is compelling evidence for a sexual cycle in some species such as in the Microsporidia or *Pneumocystis* species (Hazard and Brookbank, 1984; Kutty et al., 2010; Lee et al., 2008; Matsumoto and Yoshida, 1984). As proposed previously, occasional sexual reproduction may be beneficial, even in nonchanging environments (Heitman, 2006). Indeed, one hypothesis based on diversity analysis is that the *C. gattii* outbreak strains originated from a rare α–α mating event to generate a new and more aggressive pathogen (Fraser et al., 2005).

Fungi are experts at making spores, and sex can generate a unique spore type. The sexual basidiospores produced by the CNSC have been proposed as an infectious particle, originally because of their reduced size (Ellis and Pfeiffer, 1990a). An experimental analysis of the virulence properties of spores demonstrates that they can initiate infect in animal models before germinating into the yeast cell type that causes the disease symptoms (Giles et al., 2009; Sukroongreung et al., 1998; Velagapudi et al., 2009). Thus, one link between sex and virulence is the potential role the sexual process has in generating the inocula to cause disease in animals.

The other possible selection toward maintaining sex, as related to virulence is the overlap in proteins required for both mating and virulence. For instance, numerous genes in *C. neoformans* have been characterized that when mutated cause either a reduction or a complete loss of fertility and at the same time impair virulence. This includes small GTPase signaling pathways (e.g., Alspaugh et al., 1997, 2000; Wang et al., 2000), the cyclic AMP (cAMP)–PKA pathway (e.g., Alspaugh et al., 2002; D'Souza et al., 2001; Shen et al., 2008; Xue et al., 2008), cell morphogenesis and stress sensing (e.g., Bahn et al., 2005; Wang et al., 2002), and metal ion homeostasis (e.g., Cruz et al., 2001; Jung and Kronstad, 2011; Jung et al., 2008; Lin et al., 2006; Odom et al., 1997). Within the *MAT* locus itself, mutation of the alpha pheromones, the **a**–allele pheromone receptor, and *STE20* also reduce virulence (Chang et al., 2003; Shen et al., 2002; Wang et al., 2002), although these phenotypes are strain or serotype dependent because mutation of the same genes in other strain backgrounds can have no effect on virulence (Okagaki et al., 2010; Wang et al., 2002). However, not all genes required for mating are also required for virulence as mutation of other genes (e.g., *SXI1*α, *STE50*, or the Cpk1 MAP kinase pathway; Davidson et al., 2003; Fu et al.,

2011; Hull et al., 2004; Lin et al., 2010; Liu et al., 2008) reduce fertility but have no effect on virulence.

A curious link between sex and virulence in *C. neoformans* is that in some genetic backgrounds, mating-type alleles are linked to differences in virulence, with the α mating-type strains being more prevalent (>98%) than the **a** mating type in both wild and clinical isolates and the α mating type being more virulent than **a** mating types. These experiments have been performed in strains that are isogenic apart from the *MAT* locus, thus ensuring that the effect is specific to the *MAT* locus itself (Kwon-Chung et al., 1992; Nielsen et al., 2003, 2005b). The observation that the underlying strain background influences this effect illustrates the complexity of genetic interactions that occur in *C. neoformans*, and other microbes, to influence the host–pathogen interaction.

The global prevalence of the α mating type over the **a** mating type is not fully understood. *C. neoformans* var. *neoformans* (VNIV) undergoes the process of monokaryotic fruiting, in which basidiospores are produced by a sexual process with just one mating type (Lin et al., 2005; Wickes et al., 1996). Strains with the α mating type are better at fruiting (Lin et al., 2006; Tscharke et al., 2003; Wickes et al., 1996). The underlying genetic basis for the benefit of the α over the **a** mating type is unknown, in part because few investigators examine the mutational effects of the same genes on both mating-type backgrounds, although this in part reflects the relatively recent development of the tools in var. *grubii* for such analyses (Nielsen et al., 2003). However, there are known mating-type differences in response to certain mutations, such as *tup1* (Lee et al., 2005). An explanation for the skewed ratios of mating-type alleles is the interaction between the two mating partners. While inoculations with single strains cause a similar virulence, inoculation with mixes of **a** and α cells resulted in a skew of more α cells from the brains of mice (Nielsen et al., 2003, 2005a). One hypothesis that has been explored is that pheromone signaling between the strains leads to changes in cell morphology or other properties such that the **a** cells are less fit in the animal and are unable to cross the blood–brain barrier (BBB). Recently, it was found that coculturing the two mating types resulted in preferential development of giant cells of the **a** mating type, suggesting that the size increase may be a physical limitation for crossing the BBB (Okagaki et al., 2010). This is an interesting suggestion when considering the global prevalence of the α mating types and the role of environment–animal–environment selection for the species complex.

CONCLUDING REMARKS

C. neoformans and *C. gattii* represent globally distributed species capable of causing disease in a wide range of mammals. How these species evolved to be virulent remains a mystery, but given that members from all contingents of this species complex can cause disease, it is most parsimonious that the clade was founded by a pathogenic ancestor (Fig. 11.1). Clearly, there is a need for more experiments in *Cryptococcus* spp. to understand why they cause disease and how the environment has influenced their evolution. The pathogenic properties of the CNSC are not unique to this one lineage. There is a suite of other fungi that are commonly found in the environment that can also cause disease in susceptible people. Furthermore, disease outbreaks in animals caused by diverse fungi continue to occur on a regular basis. Examples include *Fusarium* spp. in contact lens users, *Geomyces destructans* in bats, *Nosema apis* in bees, and *Batrochochytrium dendrobatidis* in amphibians, in addition to *C. gattii* in the Pacific Northwest. Whether these newly emerged diseases are a consequence of global climate change, population expansion, environmental disturbance, or other factors remains unknown. Elucidating the evolutionary trajectory that has given rise to the pathogenic *Cryptococcus* clade may provide insights to aid in preventing or combating diseases caused by this important group of pathogens and at the same time other pathogenic fungi.

REFERENCES

Alspaugh JA, Perfect JR, Heitman J. 1997. *Cryptococcus neoformans* mating and virulence are regulated by the G-protein α subunit GPA1 and cAMP. Genes Dev 11: 3206–3217.

Alspaugh JA, Cavallo LM, Perfect JR, Heitman J. 2000. *RAS1* regulates filamentation, mating and growth at high temperature of *Cryptococcus neoformans*. Mol Microbiol 36: 352–365.

Alspaugh JA, Pukkila-Worley R, Harashima T, Cavallo LM, Funnell D, Cox GM, Perfect JR, Kronstad JW, Heitman J. 2002. Adenylyl cyclase functions downstream of the Gα protein Gpa1 and controls mating and pathogenicity of *Cryptococcus neoformans*. Eukaryot Cell 1: 75–84.

Apidianakis Y, Rahme LG, Heitman J, Ausubel FM, Calderwood SB, Mylonakis E. 2004. Challenge of *Drosophila melanogaster* with *Cryptococcus neoformans* and role of the innate immune response. Eukaryot Cell 3: 413–419.

Bahn Y-S, Kojima K, Cox GM, Heitman J. 2005. Specialization of the HOG pathway and its impact on differentiation and virulence of *Cryptococcus neoformans*. Mol Biol Cell 16: 2285–2300.

Blasi E, Brozzetti A, Francisci D, Neglia R, Cardinali G, Bistoni F, Vidotto V, Baldelli F. 2001. Evidence of microevolution in a clinical case of recurrent *Cryptococcus neoformans* meningoencephalitis. Eur J Clin Microbiol Infect Dis 20: 535–543.

Bose I, Reese AJ, Ory JJ, Janbon G, Doering TL. 2003. A yeast under cover: The capsule of *Cryptococcus neoformans*. Eukaryot Cell 2: 655–663.

Bovers M, Hagen F, Kuramae EE, Boekhout T. 2008. Six monophyletic lineages identified within *Cryptococcus neoformans* and *Cryptococcus gattii* by multi-locus sequence typing. Fungal Genet Biol 45: 400–421.

Brandt ME, Hutwagner LC, Klug LA, Baughman WS, Rimland D, Graviss EA, Hamill RJ, Thomas C, Pappas PG, Reingold AL, Pinner RW. 1996. Molecular subtype distribution of *Cryptococcus neoformans* in four areas of the United States. J Clin Microbiol 34: 912–917.

Byrnes EJ, 3rd, Li W, Lewit Y, Ma H, Voelz K, Ren P, Carter DA, Chaturvedi V, Bildfell RJ, May RC, Heitman J. 2010. Emergence and pathogenicity of highly virulent *Cryptococcus gattii* genotypes in the northwest United States. PLoS Pathog 6: e1000850.

Campbell LT, Currie BJ, Krockenberger M, Malik R, Meyer W, Heitman J, Carter D. 2005a. Clonality and recombination in genetically differentiated subgroups of *Cryptococcus gattii*. Eukaryot Cell 4: 1403–1409.

Campbell LT, Fraser JA, Nichols CB, Dietrich FS, Carter D, Heitman J. 2005b. Clinical and environmental isolates of *Cryptococcus gattii* from Australia that retain sexual fecundity. Eukaryot Cell 4: 1410–1419.

Casadevall A, Perfect J. 1998. *Cryptococcus neoformans*. American Society for Microbiology Press, Washington, DC.

Casadevall A, Rosas AL, Nosanchuk JD. 2000. Melanin and virulence in *Cryptococcus neoformans*. Curr Opin Microbiol 3: 354–358.

Casadevall A, Steenbergen JN, Nosanchuk JD. 2003. "Ready made" virulence and "dual use" virulence factors in pathogenic environmental fungi—The *Cryptococcus neoformans* paradigm. Curr Opin Microbiol 6: 332–337.

Castellani A. 1930. An amoeba growing in cultures of a yeast: Preliminary note. J Trop Med Hyg 33: 160.

Chakrabarti A, Jatana M, Kumar P, Chatha L, Kaushal A, Padhye AA. 1997. Isolation of *Cryptococcus neoformans* var. *gattii* from *Eucalyptus camaldulensis* in India. J Clin Microbiol 35: 3340–3342.

Chamilos G, Bignell EM, Schrettl M, Lewis RE, Leventakos K, May GS, Haas H, Kontoyiannis DP. 2010. Exploring the concordance of *Aspergillus fumigatus* pathogenicity in mice and Toll-deficient flies. Med Mycol 48: 506–510.

Chang YC, Miller GF, Kwon-Chung KJ. 2003. Importance of a developmentally regulated pheromone receptor of *Cryptococcus neoformans* for virulence. Infect Immun 71: 4953–4960.

Chen S, Sorrell T, Nimmo G, Speed B, Currie B, Ellis D, Marriott D, Pfeiffer T, Parr D, Byth K. 2000. Epidemiology and host- and variety-dependent characteristics of infection due to *Cryptococcus neoformans* in Australia and New Zealand. Australasian Cryptococcal Study Group. Clin Infect Dis 31: 499–508.

Chowdhary A, Hiremath SS, Sun S, Kowshik T, Randhawa HS, Xu J. 2011. Genetic differentiation, recombination and clonal expansion in environmental populations of *Cryptococcus gattii* in India. Environ Microbiol 13: 1875–1888.

Cruz MC, Fox DS, Heitman J. 2001. Calcineurin is required for hyphal elongation during mating and haploid fruiting in *Cryptococcus neoformans*. EMBO J 20: 1020–1032.

Cruz MC, Sia RA, Olson M, Cox GM, Heitman J. 2000. Comparison of the roles of calcineurin in physiology and virulence in serotype D and serotype A strains of *Cryptococcus neoformans*. Infect Immun 68: 982–985.

Davel G, Abrantes R, Brudny M, Córdoba S, Rodero L, Canteros CE, Perrotta D. 2003. 1st environmental isolation of *Cryptococcus neoformans* var. *gattii* in Argentina. Rev Argent Microbiol 35: 110–112.

Davidson RC, Nichols CB, Cox GM, Perfect JR, Heitman J. 2003. A MAP kinase cascade composed of cell type specific and non-specific elements controls mating and differentiation of the fungal pathogen *Cryptococcus neoformans*. Mol Microbiol 49: 469–485.

Desnos-Ollivier M, Patel S, Spaulding AR, Charlier C, Garcia-Hermoso D, Nielsen K, Dromer F. 2010. Mixed infections and in vivo evolution in the human fungal pathogen *Cryptococcus neoformans*. MBio 1: e00091–e00010.

Doering TL. 2009. How sweet it is! Cell wall biogenesis and polysaccharide capsule formation in *Cryptococcus neoformans*. Annu Rev Microbiol 63: 223–247.

Dromer F, Casadevall A, Perfect J, Sorrell T. 2011. *Cryptococcus neoformans*: Latency and disease. In: Heitman J, Kozel TR, Kwon-Chung KJ, Perfect JR, Casadevall A, eds. *Cryptococcus: From Human Pathogen to Model Yeast*. ASM Press, Washington, DC. pp. 431–439.

D'Souza CA, Alspaugh JA, Yue C, Harashima T, Cox GM, Perfect JR, Heitman J. 2001. Cyclic AMP-dependent protein kinase controls virulence of the fungal pathogen *Cryptococcus neoformans*. Mol Cell Biol 21: 3179–3191.

D'Souza CA, Kronstad JW, Taylor G, Warren R, Yuen M, Hu G, Jung WH, Sham A, Kidd SE, Tangen K, Lee N, Zeilmaker T, Sawkins J, McVicker G, Shah S, Gnerre S, Griggs A, Zeng Q, Bartlett K, Li W, Wang X, Heitman J, Stajich JE, Fraser JA, Meyer W, Carter D, Schein J, Krzywinski M, Kwon-Chung KJ, Varma A, Wang J, Brunham R, Fyfe M, Ouellette BFF, Siddiqui A, Marra M, Jones S, Holt R, Birren BW, Galagan JE, Cuomo CA. 2011. Genome variation in *Cryptococcus gattii*, an emerging pathogen of immunocompetent hosts. MBio 2: e00342–e00310.

Ellis DH, Pfeiffer TJ. 1990a. Ecology, life-cycle, and infectious propagule of *Cryptococcus neoformans*. Lancet 336: 923–925.

Ellis DH, Pfeiffer TJ. 1990b. Natural habitat of *Cryptococcus neoformans* var. *gattii*. J Clin Microbiol 28: 1642–1644.

Ergin C, Ilkit M, Kaftanoglu O. 2004. Detection of *Cryptococcus neoformans* var. *grubii* in honeybee (*Apis mellifera*) colonies. Mycoses 47: 431–434.

Escandón P, Sánchez A, Martínez M, Meyer W, Castañeda E. 2006. Molecular epidemiology of clinical and environmental isolates of the *Cryptococcus neoformans* species complex reveals a high genetic diversity and the presence of the molecular type VGII mating type a in Colombia. FEMS Yeast Res 6: 625–635.

Findley K, Rodriguez-Carres M, Metin B, Kroiss J, Fonseca Á, Vilgalys R, Heitman J. 2009. Phylogeny and phenotypic characterization of pathogenic *Cryptococcus* species and closely related saprobic taxa in the Tremellales. Eukaryot Cell 8: 353–361.

Fortes ST, Lazéra MS, Nishikawa MM, Macedo RCL, Wanke B. 2001. First isolation of *Cryptococcus neoformans* var. *gattii* from a native jungle tree in the Brazilian Amazon rainforest. Mycoses 44: 137–140.

Fraser JA, Subaran RL, Nichols CB, Heitman J. 2003. Recapitulation of the sexual cycle of the primary fungal pathogen *Cryptococcus neoformans* var. *gattii*: Implications for an outbreak on Vancouver Island, Canada. Eukaryot Cell 2: 1036–1045.

Fraser JA, Giles SS, Wenink EC, Geunes-Boyer SG, Wright JR, Diezmann S, Allen A, Stajich JE, Dietrich FS, Perfect JR, Heitman J. 2005. Same-sex mating and the origin of the Vancouver Island *Cryptococcus gattii* outbreak. Nature 437: 1360–1364.

Fries BC, Casadevall A. 1998. Serial isolates of *Cryptococcus neoformans* from patients with AIDS differ in virulence for mice. J Infect Dis 178: 1761–1766.

Fries BC, Goldman DL, Cherniak R, Ju R, Casadevall A. 1999. Phenotypic switching in *Cryptococcus neoformans* results in changes in cellular morphology and glucuronoxylomannan structure. Infect Immun 67: 6076–6083.

Fries BC, Taborda CP, Serfass E, Casadevall A. 2001. Phenotypic switching of *Cryptococcus neoformans* occurs in vivo and influences the outcome of infection. J Clin Invest 108: 1639–1648.

Fu J, Mares C, Lizcano A, Liu Y, Wickes BL. 2011. Insertional mutagenesis combined with an inducible filamentation phenotype reveals a conserved *STE50* homologue in *Cryptococcus neoformans* that is required for monokaryotic fruiting and sexual reproduction. Mol Microbiol 79: 990–1007.

Giles SS, Dagenais TR, Botts MR, Keller NP, Hull CM. 2009. Elucidating the pathogenesis of spores from the human fungal pathogen *Cryptococcus neoformans*. Infect Immun 77: 3491–3500.

Goldman DL, Fries BC, Franzot SP, Montella L, Casadevall A. 1998. Phenotypic switching in the human pathogenic fungus *Cryptococcus neoformans* is associated with changes in virulence and pulmonary inflammatory response in rodents. Proc Natl Acad Sci U S A 95: 14967–14972.

Goldman DL, Khine H, Abadi J, Lindenberg DJ, Pirofski L-A, Niang R, Casadevall A. 2001. Serologic evidence for *Cryptococcus neoformans* infection in early childhood. Pediatrics 107: E66.

Granados DP, Castañeda E. 2005. Isolation and characterization of *Cryptococcus neoformans* varieties recovered from natural sources in Bogotá, Colombia, and study of ecological conditions in the area. Microb Ecol 49: 282–290.

Granados DP, Castañeda E. 2006. Influence of climatic conditions on the isolation of members of the *Cryptococcus neoformans* species complex from trees in Colombia from 1992–2004. FEMS Yeast Res 6: 636–644.

Hazard EI, Brookbank JW. 1984. Karyogamy and meiosis in an *Amblyospora* sp. (Microsporidia) in the mosquito *Culex salinarius*. J Invertebr Pathol 44: 3–11.

Heitman J. 2006. Sexual reproduction and the evolution of microbial pathogens. Curr Biol 16: R711–R725.

Heitman J, Allen B, Alspaugh JA, Kwon-Chung KJ. 1999. On the origins of congenic *MAT*α and *MATa* strains of the pathogenic yeast *Cryptococcus neoformans*. Fungal Genet Biol 28: 1–5.

Heitman J, Kozel TR, Kwon-Chung KJ, Perfect JR, Casadevall A. 2011. *Cryptococcus: From Human Pathogen to Model Yeast.* ASM Press, Washington, DC.

Hicks JK, D'Souza CA, Cox GM, Heitman J. 2004. Cyclic AMP-dependent protein kinase catalytic subunits have divergent roles in virulence factor production in two varieties of the fungal pathogen *Cryptococcus neoformans*. Eukaryot Cell 3: 14–26.

Hiremath SS, Chowdhary A, Kowshik T, Randhawa HS, Sun S, Xu J. 2008. Long-distance dispersal and recombination in environmental populations of *Cryptococcus neoformans* var. *grubii* from India. Microbiology 154: 1513–1524.

Hull CM, Cox GM, Heitman J. 2004. The α-specific cell identity factor Sxi1α is not required for virulence of *Cryptococcus neoformans*. Infect Immun 72: 3643–3645.

Idnurm A, Walton FJ, Floyd A, Reedy JL, Heitman J. 2009. Identification of ENA1 as a virulence gene of the human pathogenic fungus *Cryptococcus neoformans* through signature-tagged insertional mutagenesis. Eukaryot Cell 8: 315–326.

Jain N, Wickes BL, Keller SM, Fu J, Casadevall A, Jain P, Ragan MA, Banerjee U, Fries BC. 2005. Molecular epidemiology of clinical *Cryptococcus neoformans* strains from India. J Clin Microbiol 43: 5733–5742.

Jain N, Cook E, Xess I, Hasan F, Fries D, Fries BC. 2009. Isolation and characterization of senescent *Cryptococcus neoformans* and implications for phenotypic switching and pathogenesis in chronic cryptococcosis. Eukaryot Cell 8: 858–866.

Jung WH, Kronstad JW. 2011. The iron-responsive, GATA-type transcription factor Cir1 influences mating in *Cryptococcus neoformans*. Mol Cells 31: 73–77.

Jung WH, Sham A, Lian T, Singh A, Kosman DJ, Kronstad JW. 2008. Iron source preference and regulation of iron uptake in *Cryptococcus neoformans*. PLoS Pathog 4: e45.

Kidd SE, Sorrell TC, Meyer W. 2003. Isolation of two molecular types of *Cryptococcus neoformans* var. *gattii* from insect frass. Med Mycol 41: 171–176.

Kidd SE, Hagen F, Tscharke RL, Huynh M, Bartlett KH, Fyfe M, Macdougall L, Boekhout T, Kwon-Chung KJ, Meyer W. 2004. A rare genotype of *Cryptococcus gattii* caused the cryptococcosis outbreak on Vancouver Island (British Columbia, Canada). Proc Natl Acad Sci U S A 104(49): 17258–17263.

Kidd SE, Guo H, Bartlett KH, Xu J, Kronstad JW. 2005. Comparative gene genealogies indicate that two clonal lineages of *Cryptococcus gattii* in British Columbia resemble strains from other geographical areas. Eukaryot Cell 4: 1629–1638.

Kidd SE, Chow Y, Mak S, Bach PJ, Chen H, Hingston AO, Kronstad JW, Bartlett KH. 2007. Characterization of environmental sources of the human and animal pathogen *Cryptococcus gattii* in British Columbia, Canada, and the Pacific Northwest of the United States. Appl Environ Microbiol 73: 1433–1443.

Kutty G, Achaz G, Maldarelli F, Varma A, Shroff R, Becker S, Fantoni G, Kovacs JA. 2010. Characterization of the meiosis-specific recombinase Dmc1 of *Pneumocystis*. J Infect Dis 202: 1920–1299.

Kwon-Chung KJ. 1975. A new genus, *Filobasidiella*, the perfect state of *Cryptococcus neoformans*. Mycologia 67: 1197–1200.

Kwon-Chung KJ. 1976a. A new species of *Filobasidiella*, the sexual state of *Cryptococcus neoformans* B and C serotypes. Mycologia 68: 943–946.

Kwon-Chung KJ. 1976b. Morphogenesis of *Filobasidiella neoformans*, the sexual state of *Cryptococcus neoformans*. Mycologia 68: 821–833.

Kwon-Chung KJ, Edman JC, Wickes BL. 1992. Genetic association of mating types and virulence in *Cryptococcus neoformans*. Infect Immun 60: 602–605.

Lachke SA, Lockhart SR, Daniels KJ, Soll DR. 2003. Skin facilitates *Candida albicans* mating. Infect Immun 71: 4970–4976.

Langfelder K, Streibel M, Jahn B, Haase G, Brakhage AA. 2003. Biosynthesis of fungal melanins and their importance for human pathogenic fungi. Fungal Genet Biol 38: 143–158.

Laurenson IF, Lalloo DG, Naraqi S, Seaton RA, Trevett AJ, Matuka A, Kevau IH. 1997. *Cryptococcus neoformans* in Papua New Guinea: A common pathogen but an elusive source. J Med Vet Mycol 35: 437–440.

Lee H, Chang YC, Kwon-Chung KJ. 2005. TUP1 disruption reveals biological differences between *MATa* and *MAT*α strains of *Cryptococcus neoformans*. Mol Microbiol 55: 1222–1232.

Lee H, Chang YC, Nardone G, Kwon-Chung KJ. 2007. *TUP1* disruption in *Cryptococcus neoformans* uncovers a peptide-mediated density-dependent growth phenomenon that mimics quorum sensing. Mol Microbiol 64: 591–601.

Lee H, Chang YC, Varma A, Kwon-Chung KJ. 2009. Regulatory diversity of TUP1 in *Cryptococcus neoformans*. Eukaryot Cell 8: 1901–1908.

Lee SC, Corradi N, Byrnes EJ, 3rd, Torres-Martinez S, Dietrich FS, Keeling PJ, Heitman

J. 2008. Microsporidia evolved from ancestral sexual fungi. Curr Biol 18: 1675–1679.

Levitz SM. 2001. Does amoeboid reasoning explain the evolution and maintenance of virulence factors in *Cryptococcus neoformans*? Proc Natl Acad Sci U S A 98: 14760–14762.

Lin X, Hull CM, Heitman J. 2005. Sexual reproduction between partners of the same mating type in *Cryptococcus neoformans*. Nature 434: 1017–1021.

Lin X, Huang JC, Mitchell TG, Heitman J. 2006. Virulence attributes and hyphal growth of *C. neoformans* are quantitative traits and the *MAT*α allele enhances filamentation. PLoS Genet 2: e187.

Lin X, Jackson JC, Feretzaki M, Xue C, Heitman J. 2010. Transcription factors Mat2 and Znf2 operate cellular circuits orchestrating opposite- and same-sex mating in *Cryptococcus neoformans*. PLoS Genet 6: e1000953.

Litvintseva AP, Mitchell TG. 2009. Most environmental isolates of *Cryptococcus neoformans* var. *grubii* (serotype A) are not lethal for mice. Infect Immun 77: 3188–3195.

Litvintseva AP, Marra RE, Nielsen K, Heitman J, Vilgalys R, Mitchell TG. 2003. Evidence of sexual recombination among *Cryptococcus neoformans* serotype A isolates in sub-Saharan Africa. Eukaryot Cell 2: 1162–1168.

Litvintseva AP, Kestenbaum L, Vilgalys R, Mitchell TG. 2005a. Comparative analysis of environmental and clinical populations of *Cryptococcus neoformans*. J Clin Microbiol 43: 556–564.

Litvintseva AP, Thakur R, Reller LB, Mitchell TG. 2005b. Prevalence of clinical isolates of *Cryptococcus gattii* serotype C among patients with AIDS in sub-Saharan Africa. J Infect Dis 192: 888–892.

Litvintseva AP, Carbone I, Rossouw J, Thakur R, Govender NP, Mitchell TG. 2011a. Evidence that the human pathogenic fungus *Cryptococcus neoformans* var. *grubii* may have evolved in Africa. PLoS One 6: e19688.

Litvintseva AP, Xu J, Mitchell TG. 2011b. Population structure and ecology of *Cryptococcus neoformans* and *Cryptococcus gattii*. In: Heitman J, Kozel TR, Kwon-Chung KJ, Perfect JR, Casadevall A, eds. *Cryptococcus: From Human Pathogen to Model Yeast*. ASM Press, Washington, DC. pp. 97–111.

Liu OW, Chun CD, Chow ED, Chen C, Madhani HD, Noble SM. 2008. Systematic genetic analysis of virulence in the human fungal pathogen *Cryptococcus neoformans*. Cell 135: 174–188.

Loftus BJ, Fung E, Roncaglia P, Rowley D, Amedeo P, Bruno D, Vamathevan J, Miranda M, Anderson IJ, Fraser JA, Allen JE, Bosdet IE, Brent MR, Chiu R, Doering TL, Donlin MJ, D'Souza CA, Fox DS, Grinberg V, Fu J, Fukushima M, Haas BJ, Huang JC, Janbon G, Jones SJM, Koo HL, Krzywinski MI, Kwon-Chung JK, Lengeler KB, Maiti R, Marra MA, Marra RE, Mathewson CA, Mitchell TG, Pertea M, Riggs FR, Salzberg SL, Schein JE, Shvartsbeyn A, Shin H, Shumway M, Specht CA, Suh BB, Tenney A, Utterback TR, Wickes BL, Wortman JR, Wye NH, Kronstad JW, Lodge JK, Heitman J, Davis RW, Fraser CM, Hyman RW. 2005. The genome of the basidiomycetous yeast and human pathogen *Cryptococcus neoformans*. Science 307: 1321–1324.

Ma H, Hagen F, Stekel DJ, Johnston SA, Sionov E, Falk R, Polacheck I, Boekhout T, May RC. 2009. The fatal fungal outbreak on Vancouver Island is characterized by enhanced intracellular parasitism driven by mitochondrial regulation. Proc Natl Acad Sci U S A 106: 12980–12985.

Mahmoud YA-G. 1999. First environmental isolation of *Cryptococcus neoformans* var. *neoformans* and var. *gatti* from the Gharbia Governorate, Egypt. Mycopathologia 148: 83–86.

Malliaris SD, Steenbergen JN, Casadevall A. 2004. *Cryptococcus neoformans* var. *gattii* can exploit *Acanthamoeba castellanii* for growth. Med Mycol 42: 149–158.

Martinez LR, Ibom DC, Casadevall A, Fries BC. 2008. Characterization of phenotypic switching in *Cryptococcus neoformans* biofilms. Mycopathologia 166: 175–180.

Matsumoto Y, Yoshida Y. 1984. Sporogony in *Pneumocystis carinii*: Synaptonemal complexes and meiotic nuclear divisions observed in precysts. J Protozool 31: 420–428.

Meyer W, Castañeda A, Jackson S, Huynh M, Castañeda E. 2003. Molecular typing of

IberoAmerican *Cryptococcus neoformans* isolates. Emerg Infect Dis 9: 189–195.

Montenegro H, Paula CR. 2000. Environmental isolation of *Cryptococcus neoformans* var. *gattii* and *C. neoformans* var. *neoformans* in the city of São Paulo, Brazil. Med Mycol 38: 385–390.

Mylonakis E, Ausubel FM, Perfect JR, Heitman J, Calderwood SB. 2002. Killing of *Caenorhabditis elegans* by *Cryptococcus neoformans* as a model of yeast pathogenesis. Proc Natl Acad Sci U S A 99: 15675–15680.

Mylonakis E, Idnurm A, Moreno R, El Khoury J, Rottman JB, Ausubel FM, Heitman J, Calderwood SB. 2004. *Cryptococcus neoformans* Kin1 protein kinase homologue, identified through a *Caenorhabditis elegans* screen, promotes virulence in mammals. Mol Microbiol 54: 407–419.

Mylonakis E, Moreno R, El Khoury JB, Idnurm A, Heitman J, Calderwood SB, Ausubel FM, Diener A. 2005. *Galleria mellonella* as a model system to study *Cryptococcus neoformans* pathogenesis. Infect Immun 73: 3842–3850.

Mylonakis E, Casadevall A, Ausubel FM. 2007. Exploiting amoeboid and non-vertebrate animal model systems to study the virulence of human pathogenic fungi. PLoS Pathog 3: e101.

Nakyak R, Xu J. 2010. Effects of sertraline hydrochloride and fluconazole combinations on *Cryptococcus neoformans* and *Cryptococcus gattii*. Mycology 1: 99–105.

Neilson JB, Ivey MH, Bulmer GS. 1978. *Cryptococcus neoformans*: Pseudohyphal forms surviving culture with *Acanthamoeba polyphaga*. Infect Immun 20: 262–266.

Neilson JB, Fromtling RA, Bulmer GS. 1981. Pseudohyphal forms of *Cryptococcus neoformans*: Decreased survival in vivo. Mycopathologia 73: 57–59.

Nelson RT, Hua J, Pryor B, Lodge JK. 2001. Identification of virulence mutants of the fungal pathogen *Cryptococcus neoformans* using signature-tagged mutagenesis. Genetics 157: 935–947.

Ngamskulrungroj P, Gilgado F, Faganello J, Litvintseva AP, Leal AL, Tsui KM, Mitchell TG, Vainstein MH, Meyer W. 2009. Genetic diversity of the *Cryptococcus* species complex suggests that *Cryptococcus gattii* deserves to have varieties. PLoS One 4: e5862.

Nielsen K, Cox GM, Wang P, Toffaletti DL, Perfect JR, Heitman J. 2003. Sexual cycle of *Cryptococcus neoformans* var. *grubii* and virulence of congenic a and α isolates. Infect Immun 71: 4831–4841.

Nielsen K, Cox GM, Litvintseva AP, Mylonakis E, Malliaris SD, Benjamin DK Jr, Giles SS, Mitchell TG, Casadevall A, Perfect JR, Heitman J. 2005a. *Cryptococcus neoformans* α strains preferentially disseminate to the central nervous system during coinfection. Infect Immun 73: 4922–4933.

Nielsen K, Marra RE, Hagen F, Boekhout T, Mitchell TG, Cox GM, Heitman J. 2005b. Interaction between genetic background and the mating-type locus in *Cryptococcus neoformans* virulence potential. Genetics 171: 975–983.

Nosanchuk JD, Casadevall A. 2003. The contribution of melanin to microbial pathogenesis. Cell Microbiol 5: 203–223.

Nosanchuk JD, Shoham S, Fries BC, Shapiro DS, Levitz SM, Casadevall A. 2000. Evidence of zoonotic transmission of *Cryptococcus neoformans* from a pet cockatoo to an immunocompromised patient. Ann Intern Med 132: 205–208.

Odom A, Muir S, Lim E, Toffaletti DL, Perfect J, Heitman J. 1997. Calcineurin is required for virulence of *Cryptococcus neoformans*. EMBO J 16: 2576–2589.

Okagaki LH, Strain AK, Nielsen JN, Charlier C, Baltes NJ, Chretien F, Heitman J, Dromer F, Nielsen K. 2010. Cryptococcal cell morphology affects host cell interactions and pathogenicity. PLoS Pathog 6: e1000953.

Perfect JR. 2005. *Cryptococcus neoformans*: A sugar-coated killer with designer genes. FEMS Immunol Med Microbiol 45: 395–404.

Pfeiffer TJ, Ellis DH. 1992. Environmental isolation of *Cryptococcus neoformans* var. *gattii* from *Eucalyptus tereticornis*. J Med Vet Mycol 30: 407–408.

Qin Q-M, Luo J, Lin X, Pei J, Li L, Ficht TA, de Figueiredo P. 2011. Functional analysis of host factors that mediate the intracellular lifestyle of *Cryptococcus neoformans*. PLoS Pathog 7: e1002078.

Randhawa HS, Kowshik T, Khan ZU. 2003. Decayed wood of *Syzygium cumini* and *Ficus religiosa* living trees in Delhi/New Delhi metropolitan area as natural habitat of *Cryptococcus neoformans*. Med Mycol 41: 199–209.

Randhawa HS, Kowshik T, Preeti Sinha K, Chowdhary A, Khan ZU, Yan Z, Xu J, Kumar A. 2006. Distribution of *Cryptococcus gattii* and *Cryptococcus neoformans* in decayed trunk wood of *Syzygium cumini* trees in northwestern India. Med Mycol 44: 623–630.

Randhawa HS, Kowshik T, Chowdhary A, Preeti Sinha K, Khan ZU, Sun S, Xu J. 2008. The expanding host tree species spectrum of *Cryptococcus gattii* and *Cryptococcus neoformans* and their isolations from surrounding soil in India. Med Mycol 46: 823–833.

Renwick J, Daly P, Reeves EP, Kavanagh K. 2006. Susceptibility of larvae of *Galleria mellonella* to infection by *Aspergillus fumigatus* is dependent upon stage of conidial germination. Mycopathologia 161: 377–384.

Shen G, Wang YL, Whittington A, Li L, Wang P. 2008. The RGS protein Crg2 regulates pheromone and cyclic AMP signaling in *Cryptococcus neoformans*. Eukaryot Cell 7: 1540–1548.

Shen W-C, Davidson RC, Cox GM, Heitman J. 2002. Pheromones stimulate mating and differentiation via paracrine and autocrine signaling in *Cryptococcus neoformans*. Eukaryot Cell 1: 366–377.

Simwami SP, Khayhan K, Henk DA, Aanensen DM, Boekhout T, Hagen F, Brouwer AE, Harrison TS, Donnelly CA, Fisher MC. 2011. Low diversity *Cryptococcus neoformans* variety *grubii* multilocus sequence types from Thailand are consistent with an ancestral African origin. PLoS Pathog 7: e1001343.

Slater JL, Gregson L, Denning DW, Warn PA. 2011. Pathogenicity of *Aspergillus fumigatus* mutants assessed in *Galleria mellonella* matches that in mice. Med Mycol 49S1: S107–S113.

Springer DJ, Ren P, Raina R, Dong Y, Behr MJ, McEwen BF, Bowser SS, Samsonoff WA, Chaturvedi S, Chaturvedi V. 2010. Extracellular fibrils of pathogenic yeast *Cryptococcus gattii* are important for ecological niche, murine virulence and human neutrophil interactions. PLoS One 5: e10978.

Steenbergen JN, Casadevall A. 2003. The origin and maintenance of virulence for the human pathogenic fungus *Cryptococcus neoformans*. Microbes Infect 5: 667–675.

Steenbergen JN, Shuman HA, Casadevall A. 2001. *Cryptococcus neoformans* interactions with amoebae suggest an explanation for its virulence and intracellular pathogenic strategy in macrophages. Proc Natl Acad Sci U S A 98: 15245–15250.

Steenbergen JN, Nosanchuk JD, Malliaris SD, Casadevall A. 2003. *Cryptococcus neoformans* virulence is enhanced after growth in the genetically malleable host *Dictyostelium discoideum*. Infect Immun 71: 4862–4872.

Steenbergen JN, Nosanchuk JD, Malliaris SD, Casadevall A. 2004. Interaction of *Blastomyces dermatitidis*, *Sporothrix schenckii*, and *Histoplasma capsulatum* with *Acanthamoeba castellanii*. Infect Immun 72: 3478–3488.

Stephen C, Lester S, Black W, Fyfe M, Raverty S. 2002. Multispecies outbreak of cryptococcosis on southern Vancouver Island, British Columbia. Can Vet J 43: 792–794.

Sukroongreung S, Kitiniyom K, Nilakul C, Tantimavanich S. 1998. Pathogenicity of basidiospores of *Filobasidiella neoformans* var. *neoformans*. Med Mycol 36: 419–424.

Sullivan D, Haynes K, Moran G, Shanley D, Coleman D. 1996. Persistence, replacement, and microevolution of *Cryptococcus neoformans* strains in recurrent meningitis in AIDS patients. J Clin Microbiol 34: 1739–1744.

Tang RJ, Breger J, Idnurm A, Gerik KJ, Lodge JK, Heitman J, Calderwood SB, Mylonakis E. 2005. *Cryptococcus neoformans* gene involved in mammalian pathogenesis identified by a *Caenorhabditis elegans* progeny-based approach. Infect Immun 73: 8219–8225.

Trilles L, Lazera M, Wanke B, Theelen B, Boekhout T. 2003. Genetic characterization of environmental isolates of the *Cryptococcus neoformans* species complex from Brazil. Med Mycol 41: 383–390.

Tscharke RL, Lazera M, Chang YC, Wickes BL, Kwon-Chung KJ. 2003. Haploid fruiting in *Cryptococcus neoformans* is not mating type α-specific. Fungal Genet Biol 39: 230–237.

Velagapudi R, Hsueh Y-P, Geunes-Boyer S, Wright JR, Heitman J. 2009. Spores as infectious propagules of *Cryptococcus neoformans*. Infect Immun 77: 4345–4355.

Vilcins I, Krockenberger M, Agus H, Carter D. 2002. Environmental sampling for *Cryptococcus neoformans* var. *gattii* from the Blue Mountains National Park, Sydney, Australia. Med Mycol 40: 53–60.

Wang P, Perfect JR, Heitman J. 2000. The G-protein β subunit GPB1 is required for mating and haploid fruiting in *Cryptococcus neoformans*. Mol Cell Biol 20: 352–362.

Wang P, Nichols CB, Lengeler KB, Cardenas ME, Cox GM, Perfect JR, Heitman J. 2002. Mating-type-specific and nonspecific PAK kinases play shared and divergent roles in *Cryptococcus neoformans*. Eukaryot Cell 1: 257–272.

Wegener HH, Staib F. 1983. Fatal cryptococcosis in a bird fancier. A clinical case report on pathology, diagnosis and epidemiology of cryptococcosis. Zentralbl Bakteriol Mikrobiol Hyg [A] 256: 231–238.

Wickes BL, Mayorga ME, Edman U, Edman JC. 1996. Dimorphism and haploid fruiting in *Cryptococcus neoformans*: Association with the α-mating type. Proc Natl Acad Sci U S A 93: 7327–7331.

Xu J. 2002. Estimating the spontaneous mutation rate of loss of sex in the human pathogenic fungus *Cryptococcus neoformans*. Genetics 162: 1157–1167.

Xu J. 2004. Genotype-environment interactions of spontaneous mutations for vegetative fitness in the human pathogenic fungus *Cryptococcus neoformans*. Genetics 168: 1177–1188.

Xu J. 2005. Fundamentals of fungal molecular population genetic analyses. In: Xu J, ed. *Evolutionary Genetics of Fungi*. Horizon Scientific Press, Norfolk, UK, pp. 87–116.

Xu J, Mitchell TG. 2003. Comparative gene genealogical analyses of strains of serotype AD identify recombination in populations of serotypes A and D in the human pathogenic yeast *Cryptococcus neoformans*. Microbiology 149: 2147–2154.

Xu J, Vilgalys R, Mitchell TG. 2000. Multiple gene genealogies reveal recent dispersion and hybridization in the human pathogenic fungus *Cryptococcus neoformans*. Mol Ecol 9: 1471–1481.

Xu J, Onyewu C, Yoell HJ, Ali RY, Vilgalys RJ, Mitchell TG. 2001. Dynamic and heterogeneous mutations to fluconazole resistance in *Cryptococcus neoformans*. Antimicrob Agents Chemother 45: 420–427.

Xu J, Luo G, Vilgalys RJ, Brandt ME, Mitchell TG. 2002. Multiple origins of hybrid strains of *Cryptococcus neoformans* with serotype AD. Microbiology 148: 203–212.

Xu J, Yan Z, Guo H. 2009. Divergence, hybridization, and recombination in the mitochondrial genome of the human pathogenic yeast *Cryptococcus gattii*. Mol. Ecol. 18: 2628–2642.

Xu J, Manosuthi W, Banerjee U, Zhu L-P, Chen J, Kohno S, Izumikawa K, Chen Y, Sungkanuparph S, Harrison TS, Fisher M. 2011. Cryptococcosis in Asia. In: Heitman J, Kozel TR, Kwon-Chung KJ, Perfect JR, Casadevall A, eds. *Cryptococcus: From Human Pathogen to Model Yeast*. ASM Press, Washington, DC. pp. 287–297.

Xue C, Tada Y, Dong X, Heitman J. 2007. The human fungal pathogen *Cryptococcus* can complete its sexual cycle during a pathogenic association with plants. Cell Host Microbe 1: 263–273.

Xue C, Hsueh YP, Chen L, Heitman J. 2008. The RGS protein Crg2 regulates both pheromone and cAMP signalling in *Cryptococcus neoformans*. Mol Microbiol 70: 379–395.

Yan Z, Li X, Xu J. 2002. Geographic distribution of mating type alleles of *Cryptococcus neoformans* in four areas of the United States. J Clin Microbiol 40: 965–972.

CHAPTER 12

POPULATION GENETICS, DIVERSITY, AND SPREAD OF VIRULENCE IN *TOXOPLASMA GONDII*

BENJAMIN M. ROSENTHAL and JAMES W. AJIOKA

IMPORTANCE

Globally, an estimated third of the human population harbors infection with *Toxoplasma gondii*, a single-celled eukaryotic parasite belonging to the phylum Apicomplexa (Dubey, 2010). Most infected persons are unaware of, and evidently unharmed by, the parasite cysts established in their muscles and/or neurological tissues. Nonetheless, this parasite has recently been ranked second in importance among all foodborne pathogens in the United States, causing over 300 deaths, 4000 hospitalizations, and costing nearly $3 billion annually in the United States alone (Batz et al., 2011), and although available evidence argues that most instances of human infection induce little, if any, harm, the full clinical spectrum of chronic infection remains incompletely understood (Henriquez et al., 2009; Weiss and Dubey, 2009).

Infections acquired during pregnancy, and in persons with compromised cell-mediated immunity, pose marked health risks (Dubey, 2010). Concern for the health of pregnant mothers and their babies, and for the health of persons with HIV/AIDS, motivates efforts to minimize the consequences of infection with this ubiquitous parasite. These efforts have included successful establishment of *in vitro* and animal models of infection that have uncovered marked phenotypic differences among isolates and evolutionary lineages of *T. gondii* (Boothroyd, 2009; Sibley, 2009). Outbred mice, for example, survive experimental exposure to various isolates of *T. gondii* to very different extents, enabling fruitful research into the heritable basis of differences in "virulence" (as discussed below). Such work gains relevance in the light of epidemiological evidence that recognizably atypical parasite genotypes are responsible for outbreaks of particularly severe disease (Khan et al., 2006b).

VARIABLE CLINICAL OUTCOMES

The degree and type of pathology resulting from infection with *T. gondii* therefore vary among hosts and parasites, and we wish to emphasize at the outset that although real

Evolution of Virulence in Eukaryotic Microbes, First Edition. Edited by L. David Sibley, Barbara J. Howlett, and Joseph Heitman.

progress has been made in seeking mechanistic explanations for such variable outcomes, the complex interactions mediating the clinical consequences of infection remain substantially open questions. Below we offer our view of the present state of knowledge, but before doing so, it seems appropriate to admit several sources of uncertainty: (i) *T. gondii* infects an extraordinarily wide range of host species, rendering observations made in any given model of infection difficult to generalize ; (ii) the most tractable experimental model for studying virulence (mortality in mice) provides only imperfect information concerning the clinical outcomes of human infection; (iii) the risk of placental infection and congenital transmission varies according to gestational age; (iv) available models of infection fail to adequately incorporate the obvious importance of the contribution of host immune status to the clinical outcome of infection. Furthermore, it is clear that the variable clinical outcome of infection with *T. gondii* arises from interactions between parasite and host. In spite of these real-world complexities, our discussion of virulence will emphasize the most successful empirical approach, which emphasizes differences among parasite isolates in outbred mice.

PHYLOGENETIC CONTEXT AND TYPICAL TRANSMISSION MODES

Virulence in *T. gondii* might best be understood by means of comparison to its closest relatives. *T. gondii* is a member of a broad array of parasites within the family Coccidia, phylum Apicomplexa (Morrison, 2009); a brief consideration of these diverse and important eukaryotic microbes provides useful context for understanding the biological and epidemiological features that lend interest and importance to *T. gondii*.

Like other members of the Apicomplexa, *T. gondii* is a single-celled intracellular parasite that bears an "apical complex" at its anterior end composed of secretory organelles (rhoptries) and a ringlike structure (the conoid) that facilitates host cell recognition and invasion, and that demarcates apicomplexans from cililates and dinoflagellates with which they form the alveolates (Bachvaroff et al., 2011; Dubremetz and Ferguson, 2009; Fast et al., 2002; Lukes et al., 2009). As discussed below, products secreted from the rhoptries play important roles in effecting host cell invasion and in modifying host cell physiology. With increasing precision, the molecular motors enabling parasite motility have been dissected, and their coordinated interaction (all necessary for virulence, discussed below) enables parasites to "pull themselves into" host cells that have been primed to accept them (Carruthers and Boothroyd, 2007).

Apicomplexans exploit various means to disseminate to new hosts, often exploiting feeding relationships among alternative hosts. For example, the requirement that female mosquitoes feed on mammalian blood in order to embryonate their eggs serves as an opportunity for malaria parasites (genus *Plasmodium*, but consider also *Hepatocystis* and *Leucocytozoon*). In these parasites, mammals serve as resource-rich environment for metabolism and mitotic growth, whereas the arthropod vector provides a means for wider dissemination (and serves as the host in which sexual reproduction is completed). Piroplasms (i.e., *Theileria* and *Babesia* spp.) manifest similar life cycles in which ticks serve as the definitive arthropod vector.

The tissue cyst-forming members of the family Coccidia (also known as the Sarcocystidae, of which *T. gondii* is but one member) likewise exploit feeding relationships (Fayer, 2004). For these, however, the definitive hosts (in whom sexual reproduction occurs) are predatory carnivores rather than blood-feeding arthropods. In the case of *T. gondii*, cats serve in this role. Related parasites complete their life cycles in carnivorous birds, snakes, and mammals (variously assigned to genera such as *Frenkelia*, *Sarcocystis*, *Hammondia*, *Neospora*). Although such "two-host" (heteroxenous) life cycles have occasionally been characterized in marine environments (involving fish and tubefaciid worms), this biome has never been given the attention that it undoubtedly deserves as the likely venue in which Apicomplexan parasitism originated and diversified (Fournie et al., 2000).

To survive and persist in the environment, infectious agents may adapt to new modes of transmission, new hosts, and new reproductive strategies. *T. gondii* exemplifies an extraordinarily successful generalist that has retained the means of propagation typical of its closest relatives but that has also evolved new means of transmission (see Fig. 12.1).

All coccidian species complete their life cycle by means of sexual reproduction in the gut of an animal termed its definitive host (Tenter et al., 2002). Some of these parasites (e.g., species of *Isospora*, *Eimeria*, and *Cyclospora*) complete their life cycles in just this one definitive host. Such parasites establish such infections when their oocysts (containing highly infectious sporozoites) are ingested. After the haploid sporozoites differentiate into an asexual, vegetatively replicating form, some of these further differentiate into micro- and macrogametocytes. Fusion of pairs of such gametocytes forms diploid zygotes. Ultimately, these are shed in feces, where they undergo sporulation and meiosis, resulting in oocysts containing haploid sporozoites (Pfefferkorn et al., 1977). A single infected animal may produce millions of sporozoites during an infection cycle (Dubey and Frenkel, 1976), so meiotic recombination could, in principle, produce a myriad of new variants. However, simultaneous infection by genetically different strains is rare, so most infections probably become the equivalent to a self-cross, resulting in an effectively clonal propagation (Wendte et al., 2010).

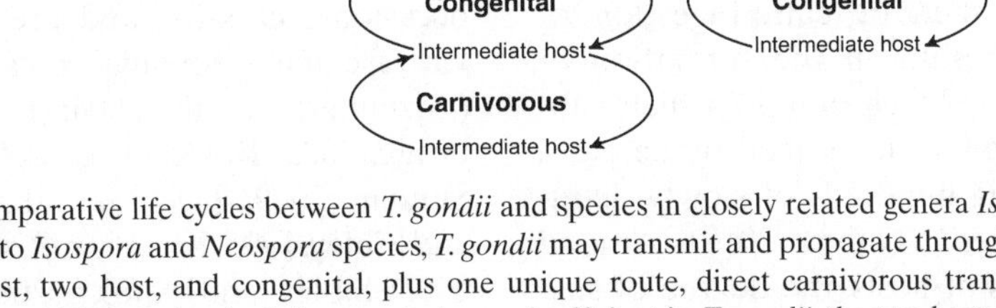

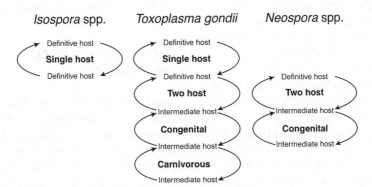

Figure 12.1 Comparative life cycles between *T. gondii* and species in closely related genera *Isospora* and *Neospora*. Compared to *Isospora* and *Neospora* species, *T. gondii* may transmit and propagate through three conserved routes: single host, two host, and congenital, plus one unique route, direct carnivorous transmission between intermediate hosts. Although the single-host route is very inefficient in *T. gondii*, the two-host route is extremely efficient by comparison. *T. gondii* oocyst production in cats is orders of magnitude greater than what is observed in *Neospora* species such that *T. gondii* can infect marine mammals that never come into direct contact with cats. The direct carnivorous transmission between intermediate hosts probably accounts for the global distribution of *T. gondii* and the high frequency of chronic infection seen in humans.

By contrast, *T. gondii* belongs to a second monophyletic assemblage whose members are typically acquired by a particular carnivorous definitive host when they consume prey in whose tissues the parasite has encysted. Like other tissue cyst-forming coccidia, *T. gondii* may be transmitted to its definitive hosts (cats) when they consume infected prey (Dubey, 1980, 1985; Dubey and Frenkel, 1998; Dubey and Thulliez, 1993; Dubey et al., 1984). The oocysts excreted by the second major group of coccidian parasites, including *Neospora* spp. and *Sarcocystis* spp., are not themselves typically infectious for those hosts. Indeed, the sporozoites of *T. gondii* shed by one cat pose relatively little risk for other cats (Freyre et al., 1989). Autologous transmission among cats therefore likely contributes little, if at all, to the persistence and dissemination of *T. gondii*.

Instead, intermediate hosts become infected by ingesting these oocysts, either by grazing contaminated vegetation or by imbibing contaminated water. Asexual parasite development within the intermediate hosts culminates in the maturation of tissue cysts. Definitive hosts acquire infection by eating meat infected with such tissue cysts, triggering parasite sexual maturation and culminating in the excretion of oocysts. This two-host (definitive and intermediate) life cycle occurs for *T. gondii*, with cats serving as the definitive host. Millions of sporozoites, infectious for a wide variety of intermediate hosts (including humans), are shed by cats. Ingestion of such oocysts results in the formation of actively dividing "tachyzoites," which subsequently engender long-lived tissue cysts, each containing hundreds of slowly dividing "bradyzoites" (Dubey, 2010). Among different kinds of hosts, the degree of susceptibility to *T. gondii* infection by ingestion of oocysts ranges widely (Innes, 1997). Merely a single oocyst is sufficient to establish infection in mice (Dubey et al., 1999; Fujii et al., 1983).

REMARKABLE TRANSMISSION MODALITIES FOR *T. GONDII*

Two extraordinary features set *T. gondii* apart from other members of the tissue cyst-forming coccidia studied to date. First, hosts other than cats may become infected when consuming tissues in which the parasite has encysted (Weinman and Chandler, 1954). Although such nonfeline intermediate hosts are incapable of supporting the development of oocysts, they do become chronically infected with tissue cysts of their own. Thus, chains of transmission among omnivorous intermediate hosts can take place in the absence of cats. Such a transmission route is unavailable for other types of tissue cyst-forming coccidia, which require their definitive hosts (Frenkel, 1977).

Second, *T. gondii* establishes infection in a remarkably broad range of intermediate hosts, which may potentially include all warm-blooded vertebrates. The selection and adaptation for the combination of direct intermediate host transmission and the immense potential host range has allowed *T. gondii* to become one of the world's most prevalent parasites.

CLONAL AND SEXUAL PROPAGATION

Some lineages of *T. gondii* are especially widespread, having diversified through occasional crossing and are disseminated via asexual propagation and/or selfing (Ajzenberg et al., 2004; Boothroyd and Grigg, 2002; Boyle et al., 2006; Grigg and Sundar, 2009; Grigg et al., 2001; Howe and Sibley, 1995; Khan et al., 2011; Morrison, 2005; Sibley and Ajioka, 2008). Although the observed *T. gondii* population structure is dominated by clonality, natural selection for recombinants produced by rare crossing events appears to be a major driver for subsequent clonal expansion.

Three clonal lineages/haplotypes have long been recognized to account for the vast majority of isolates derived from either human or other animal hosts in Europe (E) and in North America (NA; types I, II, and III). A fourth lineage has recently been found to be prevalent in certain wildlife hosts in North America (Khan et al., 2011). The pattern of single-nucleotide polymorphisms (SNPs) across the NA and E lineages is biallelic; that is, each polymorphic SNP site has only two nucleotides (reviewed in Sibley and Ajioka, 2008; see Fig. 12.2). The scarcity of such polymorphic positions suggests the recent common ancestry of all such parasite lineages where one allele is due to mutation and the other allele is the ancestral sequence. Recombination between offspring, followed by clonal expansion, has resulted in extensive haplotype blocks defined by a suite of biallelic SNPs; each lineage is identifiable by its unique combination of these relatively infrequent SNPs.

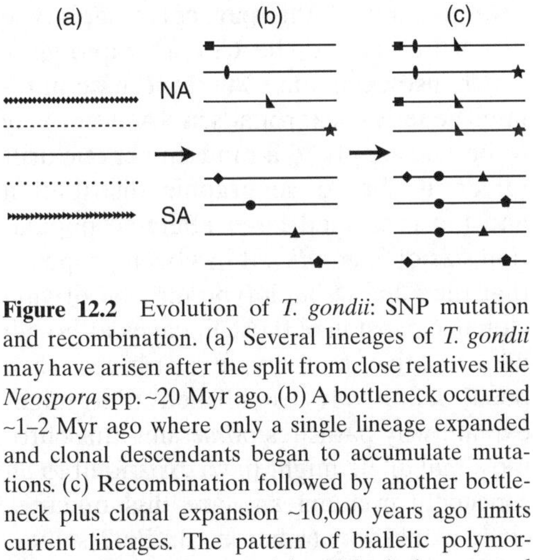

Figure 12.2 Evolution of *T. gondii*: SNP mutation and recombination. (a) Several lineages of *T. gondii* may have arisen after the split from close relatives like *Neospora* spp. ~20 Myr ago. (b) A bottleneck occurred ~1–2 Myr ago where only a single lineage expanded and clonal descendants began to accumulate mutations. (c) Recombination followed by another bottleneck plus clonal expansion ~10,000 years ago limits current lineages. The pattern of biallelic polymorphism arises because one of the alleles is the ancestral nucleotide sequence (solid line) and that multiple changes at one site are extremely rare. Geographic separation and limited gene flow may produce different biallelic patterns, for example, North America (NA) and South America (SA). Lineage-specific polymorphisms as defined by the NA types I, II, and III are generated by a simple consensus between three lineages. By comparison, relatively ancient isolates that survived but did not clonally expand such as COUG may be represented by the dashed or other nonsolid lines in (a).

The genetic composition of *T. gondii* populations elsewhere remains to be adequately defined, but work from certain intensively studied locales, notably in South America (SA; French Guyana and tropical Brazil) have revealed strikingly different and more genetically heterogeneous parasite populations (Ajzenberg et al., 2002, 2004; Lehmann et al., 2006). The nature of this diversity has been assessed by various means, including simple restriction fragment length polymorphism (RFLP) typing at loci known to discriminate among types I, II, and III. However, the truly distinct composition of these populations was not fully appreciated until they were characterized by unbiased means, for example, by sequencing an array of neutral genetic loci. Phylogenetic analysis based on SNPs in selectively neutral regions of the genome (introns) revealed a more diverse population in SA compared with NA and E, such that a total of 12 haplotypes are now recognized (Khan et al., 2007, 2011). Like NA and E, the common haplotypes in SA appear clonal, but comparison of the SA haplotype SNPs defines a different set of bialleles than those observed in NA and E. The lineages in NA/E, and those in SA, may each derive from common ancestors that lived 10^6 years ago (Khan et al., 2007). Because such estimates are highly dependent on unknown variables such as the mutation rate, it cannot be ruled out that their ancestors lived a few million years ago.

Interestingly, the vast majority of bases that are variable among strains in NA and E are monomorphic in SA; the reciprocal is also true. This suggests that regionally specific SNPs accumulated *in situ* since separation from a common ancestor (see Fig. 12.2). Geographic separation probably accounts for this difference in founder populations, and a Bayesian analysis of genotypic data suggests that the current lineages are derived from about four original lineages (Khan et al., 2007). Limited parentage, and common ancestry around 10^6 years ago, coincides with the reconnection of the Panamanian land bridge and migration/speciation of the cat family (Felidae; Johnson et al., 2006). Comigration of the host and parasite may explain the correlation between the greater diversity of cats in SA and greater *T. gondii* diversity.

The notion of a limited genetic origin for the majority of current strains is supported by the observation that an ancestral type II lineage appears to have been a parent of the other common NA and E lineages; in the case of the type III lineage, the other parent appears to be an ancestral haplotype 9 (Boyle et al., 2006; Khan et al., 2011; Sibley and Ajioka, 2008). Given that the highly represented NA and E lineages share a portion of the type II lineage genome, recombination and the inheritance of a common genetic material may have been the basis for a significant selective advantage in NA and E. The predominance of a very few *T. gondii* clonal lineages in NA and E, in light of the ubiquity of the parasite, suggests that strong forces must have shaped the current population structure.

Dating the period over which clonal lineages have expanded would provide insight into its causes. One means to do so entails enumerating mutational substitutions that have accumulated within each of the clonal lineages since their establishment. Each clonal lineage can be treated as an independent radiation accumulating neutral mutations as a "star phylogeny." Applying a molecular clock, the age of each such phylogeny can be estimated (malaria reference). Such lineage-specific SNPs were found to be extremely rare in introns of types I, II, and III, consistent with origins of each within the last 10,000 years (Su et al., 2003). Combining these data elucidates the origin of all three lineages. Rapid clonal expansion of all three may have been driven by an advantageous selective sweep (Berry et al., 1991; Maynard-Smith and Haigh, 1974). The vestiges of older diversity may be represented by a small minority of isolates, such as COUG, that diverged on the order of 10^7 years ago (Khan et al., 2007). The timing of the clonal expansion coincides with the development of human agriculture (expanded rodent populations and the domestication of the cat), which in combination would be a "perfect storm" for selecting parasites that could exploit this increasingly concentrated definitive and intermediate host interaction (Rosenthal, 2009; Sibley and Ajioka, 2008; Su et al., 2003).

A notable dearth of polymorphism in chromosome 1a (Chr1a) was identified when sequencing isolates representing types I, II, and III (Khan et al., 2006a). The near fixation of this particular haplotype of Chr1a (termed the Chr1a* haplotype), which also occurs in some otherwise unrelated genetic backgrounds in SA, is unlikely to be due simply to a random genetic drift (given its broad geographic distribution and the presumed large effective population size of *T. gondii*). It has been proposed that the Chr1a* haplotype may be driving clonal expansion of those lineages, although no causative phenotypic attributes have been definitively established that might explain why parasites possessing this chromosomal allele might have expanded to an especially great extent or through an especially clonal way (Khan et al., 2007).

Selection for parasites bearing this special chromosomal variant could conceivably operate through a variety of mechanisms, including (i) differential production or viability of diploid oocysts, (ii) differential sporogenesis/meiosis (which would constitute a form of meiotic drive), or (iii) differential generation or viability of haploid merozoites and/or tissue cysts (which would constitute a form of gametic selection). Differential production or viability of oocysts driven by Chr1a* could drive strain expansion; clonality would result in an overrepresentation of sporozoites carrying Chr1a*, regardless of coinfection with another strain. Similarly, differential generation or viability of haploid stages, even in the face of equal oocyst production in a mixed infection, would bias toward sporozoites carrying Chr1a*.

Bridging these two possible mechanisms is differential sporogenesis/meiosis as a form of meiotic drive (Novitski and Sandler, 1957). It is possible that *T. gondii* Chr1* has evolved a meiotic drive-like mechanism such as those observed in a wide variety of other species, including the *Drosophila* segregation distorter (SD) system (Lyttle, 1991). In *Drosophila melanogaster*, sperm carrying the wild-type chromosome 2 are disrupted in heterozygote males also bearing an SD, resulting in the predominance of sperm carrying SDs. This distortion is counterbalanced by modifiers and deleterious mutations/effects, including lower sperm production of the SD bearing males, resulting in a notional worldwide frequency of 1–5% for the SD allele. However, a recent study has revealed a new SD chromosome that has caused a large-scale selective sweep in Africa. There, 10 of 12 SD chromosomes bore no polymorphism whatsoever over >8 kb in the SD region and an extremely low polymorphism over >14 Mb (representing 39% of chromosome 2 euchromatin) (Presgraves et al., 2009). The SD region of this newly discovered chromosome has multiple inversions, and its proximity to the centromere also severely suppresses recombination. Coupled with strong selection and epistatic interactions in the SD region, this has driven a strong selective sweep that might have become complete but for the fact that accumulated mutations appear to be deleterious or lethal when expressed at other life cycle stages.

If an analogous situation occurs in *T. gondii*, it would span meiotic drive and gametic selection. Recombination would be suppressed by epistatic interactions and/or inversions, but the accumulation of deleterious mutations would be deterred by immediate expression in the long-lived haploid phase. Moreover, an SD-like system in *T. gondii* would not require direct interaction with the non-SD chromosome-bearing cells, as occurs in the disruption of spermatogenesis in *Drosophila*, but could instead entail something as simple as greater environmental persistence of sporozoites carrying the Chr1a* haplotype.

It had been proposed that types I, II, and III might owe their especially great abundance to a unique capacity to circumvent cats owing to tissue cysts capable of infecting other *intermediate* hosts (Su et al., 2003). If this property (termed "direct oral infectivity") had originated in a common ancestor to these abundant strains, a single explanation might be available for understanding their success (a new means of dispersal) and for their genotypic stability (asexual propagation). However, subsequent study showed that evidently unrelated isolates are capable of direct oral infectivity (Khan et al., 2006a). It remains to be determined what renders tissue cysts of *T. gondii* infectious for nonfeline hosts and when this property may first have evolved. Other potential properties, such as massive oocyst shedding, are also possible explanations for clonal expansion as self-crosses result in clonal propagation generating on the order of 10^7 infective oocysts capable of infecting a very large number of intermediate hosts (Dubey et al., 1970). The magnitude of *T. gondii* oocyst production has only been tested in a limited number of strains, and natural infection of closely related species such as *Neospora caninum* sheds very low numbers of oocysts (Basso et al., 2001), so variation for this trait may still exist in the *T. gondii* population. Testing some of these properties will be key to understanding the relationship between monomorphic Chr1a and clonal success.

ACUTE INFECTION AND VIRULENCE IN THE MOUSE MODEL

An account of virulence, in its broadest sense, would include the spectrum of pathogenesis resulting from infection with *T. gondii* in a wide range of intermediate hosts. The mouse is an important natural intermediate host. Because it is relatively sensitive to toxoplasma infection, it also provides a useful model for assessing morbidity and mortality associated with specific parasite strains. The infection of outbred Swiss mice is the standard method for measuring mortality and is referred to as "mouse virulence" (Sibley and Boothroyd, 1992). Only one infectious oocyst is sufficient to kill 100% of mice in highly virulent strains such as the type I RH strain (LD100 ~1); low virulence strains, such as the type II ME49 strain, require several orders of magnitude higher inoculum to guarantee mortality (LD100 ~105). Strains also differ in their capacity to induce disease and in the rate at which they induce death.

An early study of *T. gondii* isolates revealed that a particular genotype was especially virulent (Sibley and Boothroyd, 1992). Multilocus RFLP analysis showed isolates characterized by especially great virulence, including the lab standard RH strain, all shared a distinctive genotype. A similar analysis of an expanded set of isolates defined three common genotypes/clonal lineages: the type I lineage composed of high virulence strains, and type II and III lineages composed of low virulence strains (Howe and Sibley, 1995).

The extreme differences in pathogenicity among such clonal lineages raised the possibility that genetic crosses between parasite strains might enable discovery of genes accounting for such differences. Crosses established for genetic mapping (Sibley et al., 1992, 1993) also facilitated analysis of virulence phenotypes among the resulting F1 progeny. The avirulent type III lineage was crossed both with a lineage of intermediate virulence (type II) and the highly virulent type I lineage. The F1 recombinant progeny from the type II × type III cross displayed a wide range of mouse virulence, including some progeny that were at least three orders of magnitude more virulent than either parental strain (although none was as virulent as is typical for type I strains) (Grigg et al., 2001). The large variation in virulence among the F1 progeny showed that increased virulence could result from recombination between type II and type III parental alleles, suggesting a role for epistasis in determining the extent of pathogenesis.

By contrast, the analysis of F1 progeny from the type I × type III cross addressed a specific hypothesis: that most of the virulence of type I strains could be attributed to a single major locus (Su et al., 2002). Analyses of linkage and quantitative trait loci (QTLs) in the F1 progeny showed that virulence partitioned very strongly among the parasite progeny, where most of the variance in mortality was associated with a single locus on chromosome VII. Together, these two studies provided compelling evidence that virulence to mice is a multigenic trait driven by a few major loci.

Since allele combinations at the loci identified in the QTL analyses contribute to virulence in mice, the genetic relationships among the major lineages is key to understanding how particular sets of alleles produce a given mouse virulence phenotype. Genomic analysis of expressed sequence tag polymorphisms for type I, II, and III strains indicated that a type II strain was likely a parent of both the type I and III lineages (Boyle et al., 2006). This close relationship between the three common clonal lineages reduces the possible number of distinct allelic combinations and enables a more robust positional cloning of loci that contribute to mouse virulence because a great deal of the genetic background is shared.

QTL mapping, and subsequent positional cloning of candidate genes, established that rhoptry proteins substantially drive virulence in *T. gondii* infections (Khan et al., 2005; Ong et al., 2010; Reese et al., 2011; Saeij et al., 2006; Taylor et al., 2006; see Chapter 15 for a detailed discussion of ROPs and virulence). ROP proteins are secreted and "injected" into the host cell cytoplasm upon the initiation of invasion (Hakansson et al., 2001). The virulence-associated loci *ROP16* and *ROP18* are single genes that encode serine/threonine kinases which target STAT3/6 and immunity-related GTPases (IRGs), respectively (Fentress et al., 2010; Ong et al., 2010; Saeij et al., 2007; Yamamoto et al., 2009).

By contrast, the *ROP5* gene cluster, a family of 4–10 tandem, highly divergent isoform genes, encodes pseudokinases with unknown functions (Reese et al., 2011). Nucleotide and amino acid alignments of alleles at the *ROP5* loci correlate sequence identity with virulence. Molecular genetic analyses of the *ROP5* genes/proteins demonstrate that transforming otherwise avirulent parasites with a high virulence allele (at any of the *ROP5* loci) can increase mouse virulence by orders of magnitude (see Table 12.1). The three loci seem to act independently, and the combination of alleles appears to determine a substantial portion of the observed difference in lineage-specific mouse virulence. Since a type II strain appears to be one parent of type I and III strains, the recombinant mixture of alleles may have provided a selective advantage for type I and III strains (e.g., to exploit a different range of intermediate hosts). The evolution of the *ROP5* gene cluster, and specific amplification of genes within the tandem array, echoes the overall ROP diversification and has been proposed as a "Swiss army knife" mechanism for selection and adaptation across a variety of hosts, rather than as a precisely evolved, all-purpose "single blade" (Reese et al., 2011; see Table 12.1).

Combining forward genetics, QTL mapping, and positional cloning has proved invaluable in identifying ROP loci as the genetic basis for differences in the virulence, to mice, of the three most common *T. gondii* clonal lineages in North America and in Europe. It will be of great interest to see if similar analyses in less related strains (derived from other locales) will implicate these same loci as principally responsible for differences in virulence.

Rare recombination, interspersed by long intervals of asexual propagation, appears to have played a significant role in the emergence of the common-type strains in NA and Europe, resulting in the persistence and broad dissemination of particular combinations of alleles, especially if

TABLE 12.1 Genetic Contributions to Virulence Differences

Locus	Allele: High Virulence	Allele: Low Virulence	Log Difference
ROP18/VIR4	*ROP18I, ROP18II*	*ROP18III*	4
ROP16/VIR3	*ROP16I, ROP16III*	*ROP16II*	1
ROP5 cluster/*VIR1*	*ROP5AI, ROP5AIII* *ROP5BI, ROP5BIII* *ROP5CI, ROP5CIII*	*ROP5AII* *ROP5BII* *ROP5CII*	10[a]

[a] Compared to deletion of the *ROP5* cluster in a type I strain.

these allelic combinations provide selective advantages. Where recombination occurs more frequently, such "coadapted" allelic combinations would be continually broken up, precluding the formation and persistence of discrete clonal lineages. Since selection may occur in a wide variety of hosts, differing in their immunological characteristics, a trade-off may occur between exploiting a specific host and adapting to a wide variety of hosts. It may be that the variants of *T. gondii* represent alternative solutions to a patchy and fluctuating immunological environment. Selection for diversity at specific loci, such as *ROP5*, may provide the means to adapt to particular host environments (via the proverbial Swiss army knife approach) while preserving an essential set of allele combinations at other loci.

CHRONIC INFECTION, BEHAVIOR, AND MORBIDITY

Chronic infections probably occur in most hosts susceptible to *T. gondii*, but the consequences of chronic infection have only been studied experimentally in rodent models. Chronic infection in rats appears to alter their behavior toward cats, presumably making them more susceptible to predation (see, e.g., Webster, 2001). General behavioral abnormalities in chronically infected mice were first observed in conventionally housed mice kept for an extended period postinfection (Hutchinson et al., 1980; Hutchison et al., 1980). A systematic study of chronic infection comparing susceptible Swiss Webster mice infected with the low virulence *T. gondii* ME49 strain to controls (either uninfected Swiss Webster or infected resistant BALB/c) showed specific behavioral and neurological abnormalities, including poor grooming, lack of balance and motor coordination, and tremors (Hermes et al., 2008). These behavioral and neurological phenotypes correlated with lower brain weight, perivascular inflammatory cells contiguous to the hippocampus, and activated microglia. Microarray analysis of gene expression showed significant transcriptional upregulation of inflammatory and immune modulators, including immunoglobulin genes, *Gfap* (neuronal damage), *Complement 1q* (synapse remodeling), *PD-1L* (persistent viral infection), and *CD36* (proinflammation in Alzheimer's disease). Cell and molecular processes known to drive neuronal inflammation and damage appear to be associated with *T. gondii* chronic infection in mice, and the resulting in behavioral abnormalities would predispose the chronically infected mouse to predation. Although mice as a species may be generally more susceptible to neurodegeneration due to *T. gondii* infection, extrapolating from these observations to other species, including humans, suggests that there may be a sector of any population that is genetically susceptible to a similar disease process (see, e.g., Derouin et al., 2002). Given the frequency of chronic infection among human groups, even an extremely small proportion of susceptible individuals within the population potentially represent a very large number of individuals on a global scale.

FUTURE PROSPECTS

The desire to reduce the public health impact of toxoplasmosis impels us to investigate the biological basis and epidemiological consequences of virulence in toxoplasmosis, as do broader interests concerning the biology of intracellular parasitism. We require better understanding concerning the relationship between infection and disease in this especially widespread, occasionally dangerous, but more typically asymptomatic form of parasitism. Ingesting either oocysts (excreted by cats) or tissue cysts (contaminating any of various types of meat) poses risks of human infection. The relative contribution of each route of infection has been inferred from epidemiological data (Cvetkovic et al., 2010; Jones et al., 2009; Petersen et al., 2010; Sroka et al., 2010; Villena et al., 2010). More direct estimates of their relative importance may now be possible because stage-specific expression of parasite antigens have been exploited to differentiate among each type of exposure (Munoz-Zanzi et al., 2010). Human infections undoubtedly derive from the consumption of lamb, mutton, pork, beef, poultry, and any of a myriad of game animals, or from the ingestion of food or water contaminated with cat feces. Worldwide estimates suggest that upward of one-third of all adults harbor tissue cysts of *T. gondii*, a figure that varies regionally and according to dietary custom. It is likely that a similar proportion of warm-blooded vertebrates are also chronically infected. Given the vast number of infected people, and the virtually limitless numbers who remain at risk of becoming infected, protecting public health demands a better understanding of the basis and nature of pathogenesis.

In addition to its public health significance, *T. gondii* is also an important model for studying the evolutionary genetics, cell biology, and pathobiology of apicomplexan parasites (including the causative agents of malaria, a leading killer) owing to forward and reverse genetics, *in vitro* culture, *in vivo* murine models of infection, and so on (Boothroyd, 2009; Rosenthal, 2011; Sibley, 2009). Owing to its tractability as an experimental model, *T. gondii* provides valuable means by which to understand intracellular parasitism and pathogenesis in a broader sense.

As overviewed here, very significant advances have been made in understanding what processes contribute to the development of disease, most especially in the case of severe disease in mice, and plausible historical scenarios (not testable by direct experimentation but constrained by extant data) have been elaborated concerning forces that may have promoted or constrained virulence in this nearly ubiquitous parasite.

In the future, we anticipate that a few key developments will contribute most markedly to our understanding of pathogenesis in toxoplasmosis. These include the development of far more comparative genetic data (by which to assess the contribution of particular alleles and allelic combinations to the course of infection) and the development of additional models through which to assess such phenotypes. Of late, attention has turned to subclinical sequelae of chronic human infection; as carefully controlled, prospective epidemiological studies mature, we may learn much more about the true social costs of toxoplasmosis. Finally, as discussed above, there is reason to believe that we will soon have a firmer grasp on how most people acquire their infections. In addition to aiding in the most effective means of preventing future infections, such data may allow us to understand whether and how the route of exposure may influence the probability or severity of ensuing disease.

REFERENCES

Ajzenberg D, Banuls AL, Tibayrenc M, Dardé ML. 2002. Microsatellite analysis of *Toxoplasma gondii* shows considerable polymorphism structured into two main clonal groups. International Journal for Parasitology 32: 27–38.

Ajzenberg D, Bañuls AL, Su C, Dumètre A, Demar M, Carme B, Dardé ML. 2004. Genetic diversity, clonality and sexuality in *Toxoplasma gondii*. International Journal for Parasitology 34: 1185–1196.

Bachvaroff TR, Handy SM, Place AR, Delwiche CF. 2011. Alveolate phylogeny inferred using concatenated ribosomal proteins. The Journal of Eukaryotic Microbiology 58(3): 223–233.

Basso W, Venturini L, Venturini MC, Hill DE, Kwok OC, Shen SK, Dubey JP. 2001. First isolation of *Neospora caninum* from the feces of a naturally infected dog. The Journal of Parasitology 87: 612–618.

Batz MB, Hoffmann S, Morris JG. 2011. Ranking the risks: The 10 pathogen-food combinations with the greatest burden on public health. In: *Emerging Pathogens Institute UOF*. University of Florida, Gainesville, FL.

Berry AJ, Ajioka JW, Kreitman M. 1991. Lack of polymorphism on the *Drosophila* fourth chromosome resulting from selection. Genetics 129: 1111–1117.

Boothroyd JC. 2009. *Toxoplasma gondii*: 25 years and 25 major advances for the field. International Journal for Parasitology 39: 935–946.

Boothroyd JC, Grigg ME. 2002. Population biology of *Toxoplasma gondii* and its relevance to human infection: Do different strains cause different disease? Current Opinion in Microbiology 5: 438–442.

Boyle JP, Rajasekar B, Saeij JP, Ajioka JW, Berriman M, Paulsen I, Roos DS, Sibley LD, White MW, Boothroyd JC. 2006. Just one cross appears capable of dramatically altering the population biology of a eukaryotic pathogen like *Toxoplasma gondii*. Proceedings of the National Academy of Sciences of the United States of America 103: 10514–10519.

Carruthers V, Boothroyd JC. 2007. Pulling together: An integrated model of *Toxoplasma* cell invasion. Current Opinion in Microbiology 10: 83–89.

Cvetkovic D, Bobic B, Jankovska G, Klun I, Panovski N, Djurkovic-Djakovic O. 2010. Risk factors for *Toxoplasma* infection in pregnant women in FYR of Macedonia. Parasite (Paris, France) 17: 183–186.

Derouin F, Thulliez P, Romand S. 2002. Schizophrenia and serological methods for diagnosis of toxoplasmosis. Clinical Infectious Diseases 34: 127–129.

Dubey JP. 1980. Persistence of encysted *Toxoplasma gondii* in caprine livers and public health significance of toxoplasmosis in goats. Journal of the American Veterinary Medical Association 177: 1203–1207.

Dubey JP. 1985. Persistence of encysted *Toxoplasma gondii* in tissues of equids fed oocysts. American Journal of Veterinary Research 46: 1753–1754.

Dubey JP. 2010. *Toxoplasmosis of Animals and Humans*. CRC Press, Boca Raton, FL.

Dubey JP, Frenkel JK. 1976. Feline toxoplasmosis from acutely infected mice and the development of *Toxoplasma* cysts. The Journal of Protozoology 23: 537–546.

Dubey JP, Frenkel JK. 1998. Toxoplasmosis of rats: A review, with considerations of their value as an animal model and their possible role in epidemiology. Veterinary Parasitology 77: 1–32.

Dubey JP, Thulliez P. 1993. Persistence of tissue cysts in edible tissues of cattle fed *Toxoplasma gondii* oocysts. American Journal of Veterinary Research 54: 270–273.

Dubey JP, Miller NL, Frenkel JK. 1970. The *Toxoplasma gondii* oocyst from cat feces. The Journal of Experimental Medicine 132: 636–662.

Dubey JP, Murrell KD, Fayer R. 1984. Persistence of encysted *Toxoplasma gondii* in tissues of pigs fed oocysts. American Journal of Veterinary Research 45: 1941–1943.

Dubey JP, Shen SK, Kwok OC, Frenkel JK. 1999. Infection and immunity with the RH strain of *Toxoplasma gondii* in rats and mice. The Journal of Parasitology 85: 657–662.

Dubremetz JF, Ferguson DJ. 2009. The role played by electron microscopy in advancing our understanding of *Toxoplasma gondii* and other apicomplexans. International Journal for Parasitology 39: 883–893.

Fast NM, Xue L, Bingham S, Keeling PJ. 2002. Re-examining alveolate evolution using multiple protein molecular phylogenies. The Journal of Eukaryotic Microbiology 49: 30–37.

Fayer R. 2004. *Sarcocystis* spp. in human infections. Clinical Microbiology Reviews 17: 894–902, Table of Contents.

Fentress SJ, Behnke MS, Dunay IR, Mashayekhi M, Rommereim LM, Fox BA, Bzik DJ, Taylor GA, Turk BE, Lichti CF, Townsend RR, Qiu W, Hui R, Beatty WL, Sibley LD. 2010. Phosphorylation of immunity-related GTPases by a *Toxoplasma gondii*-secreted kinase promotes macrophage survival and virulence. Cell Host & Microbe 8: 484–495.

Fournie JW, Vogelbein WK, Overstreet RM, Hawkins WE. 2000. Life cycle of *Calyptospora funduli* (Apicomplexa: Calyptosporidae). The Journal of Parasitology 86: 501–505.

Frenkel JK. 1977. Besnoitia wallacei of cats and rodents: With a reclassification of other cyst-forming isosporoid coccidia. The Journal of Parasitology 63: 611–628.

Freyre A, Dubey JP, Smith DD, Frenkel JK. 1989. Oocyst-induced *Toxoplasma gondii* infections in cats. The Journal of Parasitology 75: 750–755.

Fujii H, Kamiyama T, Hagiwara T. 1983. Species and strain differences in sensitivity to *Toxoplasma* infection among laboratory rodents. Japanese Journal of Medical Science and Biology 36: 343–346.

Grigg ME, Sundar N. 2009. Sexual recombination punctuated by outbreaks and clonal expansions predicts *Toxoplasma gondii* population genetics. International Journal for Parasitology 39(8): 925–933.

Grigg ME, Bonnefoy S, Hehl AB, Suzuki Y, Boothroyd JC. 2001. Success and virulence in Toxoplasma as the result of sexual recombination between two distinct ancestries. Science 294: 161–165.

Hakansson S, Charron AJ, Sibley LD. 2001. Toxoplasma evacuoles: A two-step process of secretion and fusion forms the parasitophorous vacuole. The EMBO journal 20: 3132–3144.

Henriquez SA, Brett R, Alexander J, Pratt J, Roberts CW. 2009. Neuropsychiatric disease and *Toxoplasma gondii* infection. Neuroimmunomodulation 16: 122–133.

Hermes G, Ajioka JW, Kelly KA, Mui E, Roberts F, Kasza K, Mayr T, Kirisits MJ, Wollmann R, Ferguson DJ, Roberts CW, Hwang JH, Trendler T, Kennan RP, Suzuki Y, Reardon C, Hickey WF, Chen L, Mcleod R. 2008. Neurological and behavioral abnormalities, ventricular dilatation, altered cellular functions, inflammation, and neuronal injury in brains of mice due to common, persistent, parasitic infection. Journal of Neuroinflammation 5: 48.

Howe DK, Sibley LD. 1995. *Toxoplasma gondii* comprises three clonal lineages: Correlation of parasite genotype with human disease. The Journal of Infectious Diseases 172: 1561–1566.

Hutchinson WM, Bradley M, Cheyne WM, Wells BW, Hay J. 1980. Behavioural abnormalities in Toxoplasma-infected mice. Annals of Tropical Medicine and Parasitology 74: 337–345.

Hutchison WM, Aitken PP, Wells BW. 1980. Chronic *Toxoplasma* infections and motor performance in the mouse. Annals of Tropical Medicine and Parasitology 74: 507–510.

Innes EA. 1997. Toxoplasmosis: Comparative species susceptibility and host immune response. Comparative Immunology, Microbiology and Infectious Diseases 20: 131–138.

Johnson WE, Eizirik E, Pecon-Slattery J, Murphy WJ, Antunes A, Teeling E, O'brien SJ. 2006. The late Miocene radiation of modern Felidae: A genetic assessment. Science 311: 73–77.

Jones JL, Dargelas V, Roberts J, Press C, Remington JS, Montoya JG. 2009. Risk factors for *Toxoplasma gondii* infection in the United States. Clinical Infectious Diseases 49: 878–884.

Khan A, Taylor S, Su C, Mackey AJ, Boyle J, Cole R, Glover D, Tang K, Paulsen IT, Berriman M, Boothroyd JC, Pfefferkorn ER, Dubey JP, Ajioka JW, Roos DS, Wootton JC, Sibley LD. 2005. Composite genome map and recombination parameters derived from three archetypal lineages of *Toxoplasma gondii.* Nucleic Acids Research 33: 2980–2992.

Khan A, Bohme U, Kelly KA, Adlem E, Brooks K, Simmonds M, Mungall K, Quail MA, Arrowsmith C, Chillingworth T, Churcher C, Harris D, Collins M, Fosker N, Fraser A, Hance Z, Jagels K, Moule S, Murphy L, O'neil S, Rajandream MA, Saunders D, Seeger K, Whitehead S, Mayr T, Xuan X, Watanabe J, Suzuki Y, Wakaguri H, Sugano S, Sugimoto C, Paulsen I, Mackey AJ, Roos DS, Hall N, Berriman M, Barrell B, Sibley LD, Ajioka JW. 2006a. Common inheritance of chromosome Ia associated with clonal expansion of *Toxoplasma gondii.* Genome research 16: 1119–1125.

Khan A, Jordan C, Muccioli C, Vallochi AL, Rizzo LV, Belfort R, Jr, Vitor RW, Silveira C,

Sibley LD. 2006b. Genetic divergence of *Toxoplasma gondii* strains associated with ocular toxoplasmosis, Brazil. Emerging Infectious Diseases 12: 942–949.

Khan A, Fux B, Su C, Dubey JP, Darde ML, Ajioka JW, Rosenthal BM, Sibley LD. 2007. Recent transcontinental sweep of *Toxoplasma gondii* driven by a single monomorphic chromosome. Proceedings of the National Academy of Sciences of the United States of America 104(37): 14872–14877.

Khan A, Dubey JP, Su C, Ajioka JW, Rosenthal BM, Sibley LD. 2011. Genetic analyses of atypical *Toxoplasma gondii* strains reveal a fourth clonal lineage in North America. International Journal for Parasitology 41: 645–655.

Lehmann T, Marcet PL, Graham DH, Dahl ER, Dubey JP. 2006. Globalization and the population structure of *Toxoplasma gondii*. Proceedings of the National Academy of Sciences of the United States of America 103: 11423.

Lukes J, Leander BS, Keeling PJ. 2009. Cascades of convergent evolution: The corresponding evolutionary histories of euglenozoans and dinoflagellates. Proceedings of the National Academy of Sciences of the United States of America 106(Suppl. 1): 9963–9970.

Lyttle TW. 1991. Segregation distorters. Annual Review of Genetics 25: 511–557.

Maynard-Smith J, Haigh J. 1974. The hitchhiking effect of a favourable gene. Genetical Research 23: 23–35.

Morrison DA. 2005. How old are the extant lineages of *Toxoplasma gondii*? Parassitologia 47: 205–214.

Morrison DA. 2009. Evolution of the Apicomplexa: Where are we now? Trends in Parasitology 25: 375–382.

Munoz-Zanzi CA, Fry P, Lesina B, Hill D. 2010. *Toxoplasma gondii* oocyst-specific antibodies and source of infection. Emerging Infectious Diseases 16: 1591–1593.

Novitski E, Sandler I. 1957. Are All products of spermatogenesis regularly functional? Proceedings of the National Academy of Sciences of the United States of America 43: 318–324.

Ong YC, Reese ML, Boothroyd JC. 2010. Toxoplasma rhoptry protein 16 (ROP16) subverts host function by direct tyrosine phosphorylation of STAT6. The Journal of Biological Chemistry 285: 28731–28740.

Petersen E, Vesco G, Villari S, Buffolano W. 2010. What do we know about risk factors for infection in humans with *Toxoplasma gondii* and how can we prevent infections? Zoonoses Public Health 57: 8–17.

Pfefferkorn ER, Pfefferkorn LC, Colby ED. 1977. Development of gametes and oocysts in cats fed cysts derived from cloned trophozoites of *Toxoplasma gondii*. The Journal of Parasitology 63: 158–159.

Presgraves DC, Gerard PR, Cherukuri A, Lyttle TW. 2009. Large-scale selective sweep among segregation distorter chromosomes in African populations of *Drosophila melanogaster*. PLoS Genetics 5: e1000463.

Reese ML, Zeiner GM, Saeij JP, Boothroyd JC, Boyle JP. 2011. Polymorphic family of injected pseudokinases is paramount in *Toxoplasma* virulence. Proceedings of the National Academy of Sciences of the United States of America 108(23): 9625–9630.

Rosenthal BM. 2009. How has agriculture influenced the geography and genetics of animal parasites?. Trends in Parasitology 25(2): 67–70.

Rosenthal BM. 2011. Impact of the *Toxoplasma gondii* genome project. In: Fratamico PYL, Kathariou S, eds. *Genomes of Foodborne and Waterborne Pathogens*. ASM Press, Washington, DC.

Saeij JP, Boyle JP, Coller S, Taylor S, Sibley LD, Brooke-Powell ET, Ajioka JW, Boothroyd JC. 2006. Polymorphic secreted kinases are key virulence factors in toxoplasmosis. Science 314: 1780–1783.

Saeij JPJ, Coller S, Boyle JP, Jerome ME, White MW, Boothroyd JC. 2007. Toxoplasma coopts host gene expression by injection of a polymorphic kinase homologue. Nature 445: 324–327.

Sibley LD. 2009. Development of forward genetics in *Toxoplasma gondii*. International Journal for Parasitology 39: 915–924.

Sibley LD, Ajioka JW. 2008. Population structure of *Toxoplasma gondii*: Clonal expansion driven by infrequent recombination and selective sweeps. Annual Review of Microbiology 62: 329–351.

Sibley LD, Boothroyd JC. 1992. Virulent strains of *Toxoplasma gondii* comprise a single clonal lineage. Nature 359: 82–85.

Sibley LD, Leblanc AJ, Pfefferkorn ER, Boothroyd JC. 1992. Generation of a restriction fragment length polymorphism linkage map for *Toxoplasma gondii*. Genetics 132: 1003–1015.

Sibley LD, Pfefferkorn ER, Boothroyd JC. 1993. Development of genetic systems for *Toxoplasma gondii*. Parasitology Today (Personal ed.) 9: 392–395.

Sroka S, Bartelheimer N, Winter A, Heukelbach J, Ariza L, Ribeiro H, Oliveira FA, Queiroz AJ, Alencar C, Jr, Liesenfeld O. 2010. Prevalence and risk factors of toxoplasmosis among pregnant women in Fortaleza, Northeastern Brazil. The American Journal of Tropical Medicine and Hygiene 83: 528–533.

Su C, Howe DK, Dubey JP, Ajioka JW, Sibley LD. 2002. Identification of quantitative trait loci controlling acute virulence in *Toxoplasma gondii*. Proceedings of the National Academy of Sciences of the United States of America 99: 10753–10758.

Su C, Evans D, Cole RH, Kissinger JC, Ajioka JW, Sibley LD. 2003. Recent expansion of *Toxoplasma* through enhanced oral transmission. Science 299: 414–416.

Taylor S, Barragan A, Su C, Fux B, Fentress SJ, Tang K, Beatty WL, El Hajj H, Jerome M, Behnke MS, White M, Wootton JC, Sibley LD. 2006. A secreted serine-threonine kinase determines virulence in the eukaryotic pathogen *Toxoplasma gondii*. Science 314: 1776–1780.

Tenter AM, Barta JR, Beveridge I, Duszynski DW, Mehlhorn H, Morrison DA, Thompson RC, Conrad PA. 2002. The conceptual basis for a new classification of the coccidia. International Journal for Parasitology 32: 595–616.

Villena I, Ancelle T, Delmas C, Garcia P, Brezin AP, Thulliez P, Wallon M, King L, Goulet V. 2010. Congenital toxoplasmosis in France in 2007: First results from a national surveillance system. Euro Surveillance 15(25): pii=19600.

Webster JP. 2001. Rats, cats, people and parasites: The impact of latent toxoplasmosis on behaviour. Microbes and Infection/Institut Pasteur 3: 1037–1045.

Weinman D, Chandler AH. 1954. Toxoplasmosis in swine and rodents; reciprocal oral infection and potential human hazard. Proceedings of the Society for Experimental Biology and Medicine 87: 211–216.

Weiss LM, Dubey JP. 2009. Toxoplasmosis: A history of clinical observations. International Journal for Parasitology 39: 895–901.

Wendte JM, Miller MA, Lambourn DM, Magargal SL, Jessup DA, Grigg ME. 2010. Self-mating in the definitive host potentiates clonal outbreaks of the apicomplexan parasites *Sarcocystis neurona* and *Toxoplasma gondii*. PLoS Genetics 6: e1001261.

Yamamoto M, Standley DM, Takashima S, Saiga H, Okuyama M, Kayama H, Kubo E, Ito H, Takaura M, Matsuda T, Soldati-Favre D, Takeda K. 2009. A single polymorphic amino acid on *Toxoplasma gondii* kinase ROP16 determines the direct and strain-specific activation of Stat3. The Journal of Experimental Medicine 206: 2747–2760.

PART III

FORWARD AND REVERSE GENETIC SYSTEMS FOR DEFINING VIRULENCE

CHAPTER 13

GENETIC CROSSES IN *PLASMODIUM FALCIPARUM*: ANALYSIS OF DRUG RESISTANCE

JOHN C. TAN and MICHAEL T. FERDIG

INTRODUCTION

A primary goal of biologists is to characterize the molecular mechanisms that link genotypes to phenotypes. The torrent of emerging technologies might suggest that traditional methods like genetic linkage mapping are losing their place in the discovery process. We propose, to the contrary, that modern genomics empowers classical genetics, which in turn offers unique perspectives to emerging systems biology approaches. Segregating among a set of sibling progeny parasite clones is the full set of information that connects genotypes to phenotypes. This includes the major drug resistance genes, but perhaps as important is the "genetic background," that is, the broader whole-genome context in which drug resistance evolves and persists. As science gets better at measuring and parsing interacting components of cellular systems, linkage mapping can reveal the architecture of the genetic effects controlling complex traits, whether or not a major gene is involved. Notably, there are many and sometimes prohibitive challenges to identifying these fine-scale features of the genome; consequently, genetic linkage mapping in controlled crosses is most effectively used in concert with complimentary methods ranging from gene-targeting methods in isolated genetic contexts to whole-genome and systems biology approaches.

In the 1980s, the National Institutes of Health hosted bold efforts to generate the first crosses of this deadly human malaria parasite (Walliker et al., 1987; Wellems et al., 1990). At the time, the completion of the *P. falciparum* genome sequencing project was still 15 years away, yet this first genome-wide exploration demonstrated that the basic principles of Mendelian inheritance could be powerfully employed in malaria parasites, shifting the discovery landscape, putting new questions within the grasp of researchers, and launching a generation of studies. The quest to map chloroquine (CQ) resistance and the ensuing hunt through the locus identified on chromosome (chr) 7 that ultimately led to the discovery of the *P. falciparum* CQ resistance transporter (pfcrt) gene, has had a

Evolution of Virulence in Eukaryotic Microbes, First Edition. Edited by L. David Sibley, Barbara J. Howlett, and Joseph Heitman.

TABLE 13.1 Key Advances in P. falciparum Genetic Crosses

Conceptual and Technological Advances Accompanying the Mapping of Drug Resistance Genes	Literature[a]
Making genetic crosses	[1]
Developing markers and linkage maps	[2]
Mapping the CQR locus	[3]
Large-scale genome sequencing and the physical genome	[4]
Positional cloning of *pfcrt*	[5]
Selective sweeps, drug resistance origins	[6]
Quantitative drug responses (QTL) and drug response relationships	[7]
Genetic complexity, background effects and secondary loci, physiological phenotypes	[8]
Systems genetics and expression level (e) QTL	[9]
Large-scale data integration: chemical genomics, GWAS	[10]
High resolution genome structure, copy number, recombination rates, and physical mapping	[11]
Inheritance of metabolites, fitness, rewired expression networks, epigenetic profiles	[12]

[a] Does not include many related and follow-up publications.
[1] (Hayton et al., 2008; Walliker et al., 1987; Wellems et al., 1990).
[2] (Alkhalil et al., 2009; Hayton et al., 2008; Jiang et al., 2011; Su et al., 1999; Walker-Jonah et al., 1992).
[3] (Su et al., 1997; Wellems et al., 1990, 1991).
[4] (Gardner et al., 2002; Jeffares et al., 2007; Mu et al., 2005, 2007; Volkman et al., 2007).
[5] (Cooper et al., 2002; Fidock et al., 2000a,b; Sidhu et al., 2002).
[6] (Cooper et al., 2002; Mu et al., 2003; Wootton et al., 2002).
[7] (Ferdig et al., 2004; Sa et al., 2009).
[8] (Beez et al., 2011; Bennett et al., 2007; Patel et al., 2010; Sa et al., 2009; Sanchez et al., in press).
[9] (Gonzales et al., 2008; Patel et al., 2010).
[10] (Yuan et al., 2009, 2011).
[11] (Jiang et al., 2011; Samarakoon et al., 2011a,b).
[12] Work in progress.

far-reaching impact on the malaria research field and has entrenched genetic linkage as a discovery tool for this species. This work includes nearly three decades of generating crosses, developing markers, mapping traits, locating and sequencing genome segments, and discovering genes and their causal mutations. Efforts to tie these mutations to mechanisms and to understand their broader biological and evolutionary context continue to inspire technological and conceptual advances in the field (Table 13.1).

The continued role for genetic linkage mapping in malaria research reflects both the urgency and challenges of studying a highly plastic and deadly pathogen, the study of which does not conform readily to standard molecular and biochemical approaches. Throughout this chapter, we point the reader to specialized reviews on the foundations of linkage mapping in *P. falciparum*, including the essential groundwork for developing experimental crosses such as advances in parasite culture, *in vitro* exposure of mosquitoes to gametocytes, and transmission to splenectomized chimpanzees, but will not attempt to recapitulate those excellent efforts (Hayton and Su, 2008; Sen and Ferdig, 2004; Su and Wootton, 2004; Su et al., 2007). Rather than a comprehensive cataloging of prior drug susceptibility studies, we describe the strengths of the genetic approach and highlight particular conceptual advances and novel applications of genetic mapping in *P. falciparum* over the last 25 years.

THE CLASSICAL GENETICS APPROACH

Controlled genetic crosses are uniquely good at assaying genome-wide variation and its influence on key phenotypes. Because genetic linkage mapping is phenotype driven, no a priori knowledge or expectation for involvement of specific genes or pathways is required. Furthermore, if a candidate gene is known, genetic mapping can effectively test for an effect from a candidate gene's locus. Major single-gene effects (e.g., a point mutation in a drug target or drug resistance gene) follow classical Mendelian inheritance. Such traits are bimodally distributed (e.g., as present–absent) with no intermediate values. In the haploid *P. falciparum* blood stages, each progeny clone has inherited an allele from one of two parents, allowing straightforward mapping and positional cloning; however, a confounding factor in genetic mapping is that multiple genes usually contribute to the phenotypic variation of most traits. Even the phenotypic effect of a major gene resides in a whole-genome context of additional gene actions and interactions that may play crucial roles in generating phenotypes. Such quantitative drug responses and complex genetic backgrounds can also be dissected using quantitative trait locus (QTL) mapping. The success of mapping a complex trait improves with larger numbers of progeny; however, the haploid inheritance, genetically divergent parent clones, and the ability to generate highly replicated phenotype measurements from cultured parasites render these small *P. falciparum* mapping populations highly informative.

In genetic mapping studies, the combination of alleles in each parent that is parsed out into the progeny clones represents a genetic signature of unique evolutionary history (Fig. 13.1). A genetic cross contrasts these distinct parental genome-wide signatures, and the trait-conferring differences in genes throughout the genome emerge from genetic linkage analysis as significant "peaks" in QTL mapping. For each of the three *P. falciparum* genetic crosses, the parent lines were carefully chosen to set up a dichotomy for a particular phenotype; however, countless additional phenotypes segregate in each cross, ensuring that the value of each cross is not exhausted after the initially targeted trait has been successfully mapped (Hayton and Su, 2008; Walliker et al., 1987; Wellems et al., 1990).

QTLs are simply positions in the genome carrying polymorphisms that affect a phenotype's expression. Mapping such quantitative traits in inbred crosses employs statistical and computational tools to characterize the "genetic architecture" (the number of loci, the magnitudes and directions of their effects, and the nature of their interactions), the detailed methods for which are described elsewhere (Sen and Ferdig, 2004). Using this approach, it is possible to follow the trait's variation to a locus that can subsequently be investigated for causal molecular variants. It is both powerful and problematic that the molecular determinants underpinning a phenotype can be anything, from a point mutation that alters structure and function in a drug target to a wide range of indirect factors such as signaling or regulatory cascades that can be difficult to map in the small numbers of independent progeny available from the *P. falciparum* crosses. To enhance the opportunity to find genes, phenotypes must be precisely and comprehensively measured; noisy or inaccurate phenotype measurements will reduce the resolution or may introduce spurious positives (i.e., false QTL peaks). Generally, the larger the variance around the measurement, the more divergent the phenotypic means must be for detection. Information gained from crosses is often specific to the parents of that particular cross and different genetic crosses can identify distinct

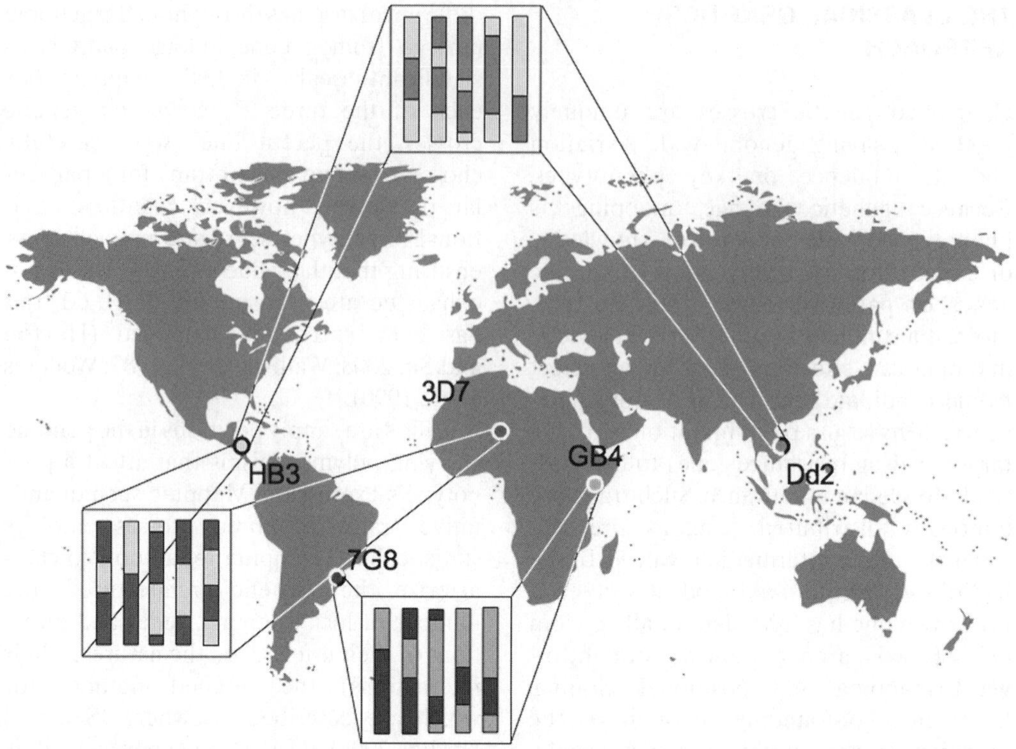

Figure 13.1 Genetic crosses capture the distinct evolutionary histories of the parent clones. The three *P. falciparum* crosses (HB3 × 3D7, HB3 × Dd2, and 7G8 × GB4) have captured a wide range of parasite biology in parent clones from various geographic origins with divergent drug selection histories. Many measurable aspects of parasite biology unique to these geographies can be dissected using QTL mapping. HB3, a clone of isolate H1 from Honduras (Bhasin and Trager, 1984; Nguyen-Dinh and Payne, 1980), is generally sensitive to drugs but has low level resistance to pyrimethamine (PYR). HB3 has been used in two different genetic crosses (green chromosomal segments). Dd2 was cloned from W2-Mef (Oduola et al., 1988a; Wellems et al., 1988), which was selected by mefloquine pressure from W2, a clone of Indochina III isolated from a Laotian patient who failed CQ therapy (Campbell et al., 1982; Oduola et al., 1988b); Dd2 is resistant to CQ, PYR, and mefloquine and also exhibits low level quinine resistance and decreased levels of susceptibilities to many other compounds. 3D7 is a drug-sensitive parasite cloned from isolate NF54, originating from a patient in The Netherlands who had never travelled overseas, and is believed to represent a case of airport malaria (Delemarre and van der Kaay, 1979; Ponnudurai et al., 1981); although the geographic origin of this parasite is unknown, high resolution genotyping clusters it with a Sudanese isolate, consistent with the general agreement that this parasite is from Africa (Mu et al., 2005). 7G8, a low level chloroquine-resistant (CQR) parasite, was cloned from Brazilian isolate IMTM22 (Burkot et al., 1984). GB4 is a CQR and multidrug-resistant clone of isolate Ghana III/CDC (Sullivan et al., 2003).

genetic components affecting the same phenotype. A trait of interest is often an "end-stage" phenotype that results from a cascade of discrete measurable steps, the control of which is inherited and can be handled individually as phenotypes (Fig. 13.2).

MAPPING MAJOR GENE EFFECTS

CQ Resistance: Mapping the Locus

The discovery of *pfcrt* stands as a foundational advance in malaria research and demonstrates the power of classical genetics to find novel genes and mechanisms governing critical phenotypes. Indeed, a classical genetic mapping approach performs especially well when a single gene is the major determinant of a trait because the locus can be mapped unambiguously to a genomic segment that can be surveyed for polymorphisms in candidate genes. The series of studies outlined in Table 13.1 marks the progression of the malaria field and its continued trajectory leading to and beyond the mapping of CQ resistance.

Building on the success of the first cross between HB3 × 3D7 (Walliker et al., 1987), Wellems and colleagues (1990) designed a new cross to map the determinant of CQ resistance between HB3 and Dd2 (Fig. 13.1). With an initial mapping population of 16 distinct recombinant progeny clones, the authors reported the absence of genetic linkage to *pfmdr1* polymorphisms on chr 5 earlier proposed to play a role in CQ resistance (Foote et al., 1990), effectively ruling out this gene as the primary determinant of CQ resistance in the context of this cross. Subsequent advances in *P. falciparum* genetic crosses and methodologies would later revive a complex role for *pfmdr1* in influencing quantitative CQ responses (described below).

The development of markers and the initial genetic linkage map was the essential first step. The initial mapping of the CQ resistance locus relied on 85 restriction fragment length polymorphism (RFLP) markers, where 1–14 markers were assigned to each of the 14 *P. falciparum* chromosomes using pulsed-field gradient electrophoresis separation of the chromosomes and Southern blots (Wellems et al., 1991). CQ response phenotypes were determined by microscopic analysis to classify parasites as "sensitive" or "resistant." This phenotype segregated perfectly only with a region on chr 7, which was resolved by long-range restriction mapping to a 400-kb locus.

With the locus clearly identified, a long, intensive project to isolate the gene began with the cloning of additional progeny from the original chimp blood. The goal of this work was to identify additional progeny with recombination events in this mapping interval to improve the resolution of the CQ resistance locus (Kirkman et al., 1996). To accompany this gain in resolution, more markers were needed to improve on the RFLP map (Walker-Jonah et al., 1992). Microsatellite (MS) markers were targeted to this genome segment (Su and Wellems, 1996), eventually narrowing the locus to a 36-kb segment and leading to the first major contiguous sequencing project, spanning 47 kb, in *P. falciparum* (Su et al., 1997). The methods of tiling cloned inserts and gap closure developed for the extraordinarily AT-rich genome, combined with the various physical and recombination mapping projects, would serve as an important guide for the malaria parasite genome sequencing project (Gardner et al., 2002). Those accomplishments also laid the groundwork for the current high resolution linkage maps consisting of hundreds or thousands of markers, completely sequenced genomes of many malaria parasite isolates, and catalogs of polymorphisms. Together, these databases now condense the timeline required to sift a genetic locus for candidate genes and polymorphisms from years to hours.

CQ Resistance: Candidate Gene Prioritization and Validation

The strength of the genetic mapping approach is that by using only genetic inheritance patterns, it is possible to find what is *not* already known or expected; however, once a locus is identified, the process of finding the causal variant is not trivial. The initial description of the chloroquine-resistant (CQR) locus included nine predicted genes, of which complex mutations in candidate gene 2 (*cg2*) were best able to distinguish a panel of CQR versus chloroquine-sensitive (CQS) parasite isolates from around the world, with the exception of one parasite, 106-1 (Su et al., 1997). As predicted by the incomplete association of these mutations with CQ resistance, subsequent allelic replacement studies using *cg2* failed to confer resistance (Fidock et al., 2000a), spurring a deeper investigation of the locus. Sequencing of cDNA libraries eventually pointed to a gene centered in the locus that had escaped recognition by gene-finding software due to its unusual 13-exon structure (Fidock et al., 2000b). This gene was dubbed *pfcrt* based on its transporter-like features. An amino acid position 76 lysine (K) to threonine (T) resistance-conferring mutation in PfCRT was shared among all CQR parasites, including 106-1, and a suite of various additional mutations identified independent origins of CQ resistance (Cooper et al., 2002; Djimde et al., 2001; Fidock et al., 2000b; Picot et al., 2009; Sidhu et al., 2002; Volkman et al., 2007; Wootton et al., 2002). PfCRT resides in the digestive vacuole membrane and is associated with reduced accumulation of CQ at this site of action (Cooper et al., 2002; Martin et al., 2009). This key finding inspired many follow-up studies including the demonstration of a selective sweep of the genome region surrounding *pfcrt* (Wootton et al., 2002) that provided a foundation for a new generation of genetic association studies in natural parasite populations (reviewed in Chapter 7, "Selective Sweeps in Human Malaria Parasites").

Even with this extraordinary success in identifying the gene and point mutation that causes CQ resistance, many questions remain, including the precise mechanism by which altered transport by mutant PfCRT causes reduced digestive vacuolar CQ concentrations and CQ resistance. Other questions include the endogenous function of *pfcrt* and how it is stabilized by compensatory mechanisms, the specific steps that led to CQ resistance evolution in surprisingly few independent origins, and the stability of resistance in some regions of the world after drug pressure is removed. Without question, there is more to CQ resistance than *pfcrt*, and this gene has other functions than simply mediating CQ resistance (Cooper et al., 2005); the answers to these questions reside in the genetic complexity that underlies the major effect of *pfcrt*.

MULTIGENIC TRAITS AND QTL MAPPING

Biological Complexity

Initial CQ resistance mapping efforts relied on bimodal categorization of CQR and CQS parasites. This approach works well when a major gene controls the trait of interest because it artificially eliminates the variation from the phenotype due to other genes and nongenetic sources. However, this strategy is blind to the much more common situation of multiple loci contributing quantitatively to a phenotype. If phenotypes are not precisely quantified, the system is effectively reduced to an unrealistically simple state negating the strength of genetic mapping to discover multigene pathways and cellular processes. To paraphrase Einstein's modification of Occam's razor: Seek the simplest possible explanation but not one that is simpler. From a drug development perspective, the silver bullet concept of simple drug targets and single-gene resistance mechanisms has largely failed. A drug's effect on parasite growth inhibition and/or survival often is indirect, not involving the target per se; furthermore, off-target and secondary response mechanisms are prevalent. To understand and harness these factors, it is necessary to devise methods geared to find this information.

The so-called missing heritability problem underscores the significance of biological complexity and the problem of overlooking it (Eichler et al., 2010). For example, it was expected that knowledge of the entire human genome, along with the capacity to genotype many thousands of disease cases and controls in human populations using massive genome-wide association studies (GWASs), would quickly reveal the molecular causes of diseases and accurately predict genetic risks. Surprisingly, these studies account for only a small fraction of the heritable phenotypic variation in populations (Musani et al., 2007); this is attributed to nonlinear, nonadditive contributions of gene × gene (i.e., epistatic) interactions and gene × environment interactions that are not accounted for by standard models for mapping genetic associations. The vast network of biomolecular interactions that regulate gene expression, drive cellular metabolism, and maintain homeostasis are highly redundant, ensuring the robustness of biological systems by buffering against genetic and environmental perturbations (Moore and Williams, 2009). This network organization also ensures that epistasis is ubiquitous in biological systems and can occur in the absence of single-gene effects.

CQ Resistance Revisited: Secondary Genes and the Role of Genetic Background

Examples of genetically simple traits are rare and, upon closer inspection, even simple inheritance masks subtle but crucial aspects of a trait's biology. This was recently demonstrated in the case of human cystic fibrosis (CF), a textbook example of simple Mendelian disease. Mutations in one gene encoding a membrane protein, *CFTR*, cause CF. However, the clinical progression and severity of pulmonary disease vary widely even among individuals with identical *CFTR*, and recent studies focused on disease variation within populations carrying mutant *CFTR* point to several disease modifiers that include regulatory functions (Wright et al., 2011). Moreover, different study populations exhibit different secondary factors that act in concert to influence CF outcome (Witt, 2011). Recent studies of CQ resistance demonstrate an analogous scenario of complexity around *pfcrt*. By leveraging the quantitative variation in CQ susceptibilities (e.g., IC_{50} levels) and comparing the genetic architectures in different genetic crosses, QTL mapping provides a way to look behind the main effect to understand the more nuanced biology and evolution of drug resistance. Given the range of observed *levels* of CQ resistance in natural isolates, additional modifiers are expected (Babiker et al., 2001; Chen et al., 2002; Mu et al., 2003). Also, it is reasonable to expect that intense CQ pressure selected a cohort of alleles that modulate levels of resistance and/or stabilize the basic biological function of the parasite. Compensatory mutations, often embedded in the same pathways and cellular processes as the gene conferring resistance, typically occur under drug selection to minimize the biological cost of acquiring the resistance mutation (Levin et al., 2000), highlighting the value of understanding the endogenous function of the resistance gene.

Using precise phenotypes and statistical scans for residual phenotypic variation, *pfmdr1* is indeed linked to CQ quantitative response levels in the 7G8 × GB4 cross (Sa et al., 2009). Furthermore, *pfmdr1* more modestly influences the degree of resistance in CQR progeny of the HB3 × Dd2 cross (Patel et al., 2010); in this cross, the combination of Dd2-type PfMDR1 (86Y) and PfCRT 76T confers low level CQR, while HB3-type PfMDR1 86N combined with PfCRT 76T confers the highest CQR levels. This observation is consistent with the coadapted combination of Dd2 alleles at these two loci reflecting a trade-off between the level of CQR and fitness.

Interestingly, while the progeny carrying the HB3-type *pfmdr1* allele carry a single copy, a nonparental copy number (CN) has been observed for the Dd2 allele, ranging from 1 to 4 (Patel et al., 2010; Wellems et al., 1990). Susceptibility to CQ could not be ascribed to CN differences. However, inherited combinations of *pfcrt* and *pfmdr1* suggest a fitness effect of CN because high CN is maintained only in the context of a Dd2 allele at both loci, and the wild type *pfcrt* is never paired with three or more copies of *pfmdr1*. Together, these studies highlight that combinations of alleles may play different roles in distinct genetic backgrounds as captured in the different *P. falciparum* crosses.

Using a qualitative mapping approach (resistant vs. sensitive), *pfmdr1* could not be identified in the HB3 × Dd2 cross because it was not a primary determinant; however, comparative quantitative mapping showed that *pfmdr1* robustly influences levels of CQ susceptibility in CQR parasites. The different contribution of *pfmdr1* in the two studies might also reflect different drug selection histories on different continents; Sa et al. (2009) recognized a complex interplay with another drug, amodiaquine, and proposed that drug pressure in different geographic regions resulted in differentially tuned allele combinations, such that *pfmdr1* mediates a more stable, persistent form of CQR in South America. The authors raised the concern that this advantaged *pfmdr1* × *pfcrt* allelic combination could be inadvertently induced in Africa by the increased use of amodiaquine.

The evolutionary steps to CQR remain unknown and the involvement of genes other than *pfcrt* may hold clues to these steps. In spite of massive drug pressure, it took more than a decade for the first CQR parasites to emerge. Laboratory-based drug selection of CQ resistance from CQS parasites has not been possible (Lim and Cowman, 1996), with the exception of 106-1 (Cooper et al., 2002), an unusual parasite that probably is a laboratory-generated back-mutant from a CQR clone, differing only at the single PfCRT K76T position. Extensive evidence indicates that genes in addition to *pfmdr1* can support CQR, underscored by the observation that parasites with identical combinations of *pfcrt* and *pfmdr1* alleles exhibit a wide range of CQ resistance levels (Chen et al., 2002; Ferdig et al., 2004; Sa et al., 2009). One candidate is an epistatic partner of *pfcrt* on chr 6 that regulates low level quinine (QN) resistance; the allele combinations of these two loci inherited in the HB3 × Dd2 progeny also suggest a fitness component reflected in slow *in vitro* expansion rates of progeny inheriting the HB3-type chr 6 and Dd2-type *pfcrt* loci (Ferdig et al., 2004). Another suggestive locus was identified on chr 7 (Patel et al., 2010) and was recently linked to the nested CQ accumulation trait (Sanchez et al., in press). A transporter encoding gene, *pfmrp*, was identified in another study (Mu et al., 2003).

Combined with the observation that some MDR strains also are primed to rapidly develop new resistances (Rathod et al., 1997), the shaping of the CQR genome around *pfcrt* has significant implications for malaria long after CQ has been discontinued. QTL mapping alone will not be adequate to find these interacting components of CQ resistance, but integration of QTL with other approaches holds great promise. Especially powerful tests of the role of genetic background in CQ resistance will come from using genetically manipulated parasites to insert mutant forms of drug resistance genes in various controlled genetic contexts. Only the Dd2 (Southeast Asia CQR origin) and 7G8 (South America CQR origin) mutant forms of *pfcrt* have been studied in identical genetic backgrounds to date (Sidhu et al., 2002). Notably, the 7G8 allele in three different genetic backgrounds produced differing levels of CQ resistance (Valderramos et al., 2010).

QN Susceptibility and Drug Response Relationships

Although the HB3 × Dd2 cross was developed to study CQ response, it is also ideal for studying response to other drugs. The first quantitative trait analysis in *P. falciparum* investigated genes contributing to low level QN resistance (Ferdig et al., 2004). The analysis revealed multiple loci on chrs 5, 7, and 13 corresponding to *pfmdr1*, *pfcrt*, and a locus containing a *P. falciparum* Na^+/H^+ exchanger gene (*pfnhe1*), acting in an additive fashion with interactive effects between two loci on chrs 9 and 6 and the two QTLs on chrs 13 and 7. Low level QN resistance in the HB3 × Dd2 cross exhibited a positive correlation with CQ IC_{90} values in CQR parasites, implying that secondary loci selected by QN pressure, including *pfmdr1*, could incrementally influence CQ and other drug susceptibilities through a complex adaptive role among various polymorphisms; that is, *pfmdr1* sequence polymorphisms and gene duplications associated with QN susceptibility (Cowman et al., 1994; Reed et al., 2000) could in turn influence the stability of CQ resistance in Southeast Asia and Africa through coadapted allele combinations. These findings are consistent with the observations of this gene's role in the CQR background, and also align with observations of Sa et al. that an interplay between different drug pressures can fine-tune the gene interactions (Sa et al., 2009). The subtle and combinatorial role of *pfcrt* in QN resistance is much different from its role in CQ resistance and may provide an avenue to better understand *pfcrt*'s relationship to other genes. Independent analysis supports a role for *pfcrt* and *pfmdr1* in influencing QN response (Cooper et al., 2002; Reed et al., 2000; Sidhu et al., 2002). Association of *pfcrt* with a QTL of QN resistance supports the argument for an evolutionary relationship between CQR and reduced QN sensitivity and corroborates a recent investigation in parasite populations (Mu et al., 2003).

CLASSICAL GENETICS IN THE ERA OF DATA-DRIVEN SCIENCE

Systems Genetics

The tools of genomics and systems biology offer great promise for dissecting the genetic complexity of drug responses and for illuminating new opportunities for intervention. The three existing *P. falciparum* genetic crosses have only begun to be tapped for the mass of information they can provide about subtle divergences in strain-specific regulation of cellular pathways and processes. "Systems genetics" is a functional genomics approach being impressively applied to model organisms to connect genotypes to phenotypes through a cascade of intervening traits ranging from transcript to protein to metabolite levels (Fig. 13.2). Among the kinds of high dimensional data, such as transcripts measured across developmental stages or time points, is the dimension of natural genetic variation, that is, measurements taken across a range of progeny genotypes. QTL mapping, by superimposing gene expression, proteins, and metabolites on important biological phenotypes, can provide an alternative, unbiased view of the network of gene actions that build traits.

Regulatory variation is emerging as a key component of phenotypic plasticity; gene expression is "tunable" by minor genetic changes; that is, structural gene changes that could have adverse fitness implications are not needed to generate sweeping phenotypic effects. Transcription is the first key step linking genotype to phenotype. Expression quantitative trait locus (eQTL) mapping identifies the regulatory architecture of the genome including both *cis* and *trans* regulatory factors (Gonzales et al., 2008). Inherited transcript levels in

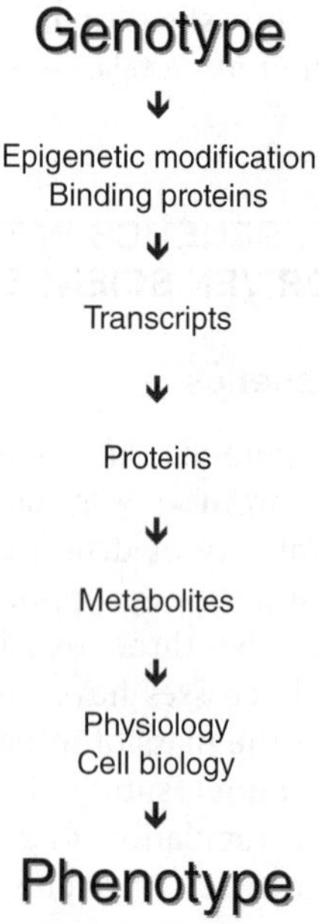

Figure 13.2 A cascade of measurable traits links genotype to phenotype. End-stage phenotypes such as drug dose–response IC_{50}s often are the target of genetic mapping studies. Such traits are produced from a cascade of underlying nested phenotypes. The output from each of these steps between DNA and phenotype can be measured and mapped. In theory, this approach can deconstruct the overall complexity of the end-stage trait into its constituent parts. For example, by taking a whole-systems approach first coined as "genetical genomics" (Jansen and Nap, 2001), the genetic regulation of inherited transcript levels for all the genes in the genome were mapped in *P. falciparum* (Gonzales et al., 2008). Other examples include the inheritance of an intermediate physiological/cell biological readouts including compartmental pH and drug accumulation (Bennett et al., 2007; Sanchez et al., in press).

the progeny point to an underlying genetic variation that regulates divergent strains such as the shifts in pathways associated with drug susceptibilities. Progenies inherit allelic variation of regulators and promoters that can directly alter transcript levels; however, eQTL "regulators" also include any DNA polymorphism that ultimately leads to varied transcript pools; these could include upstream signaling events, epigenetic modifications, and noncoding RNAs that can all contribute to strain-specific phenotype differences. In the *P. falciparum* progeny of the HB3 × Dd2 cross, nearly 20% of the transcripts are under divergent regulation at 18 hours post-red blood cell invasion, and the identified loci (eQTL) point to regulatory divergences in pathways that function differently between the multidrug-resistant Dd2 and the generally sensitive HB3. Several *trans* regulatory hot spots were identified including one associated with the amplification event on chr 5 harboring *pfmdr1* and 13 other genes (inherited from Dd2) that influenced the levels of 269 transcripts around the genome. This observation raises the interesting point that phenotypic associations (e.g., fitness effects described above) with *pfmdr1* could actually be reflecting the broader regulatory impact of the entire amplicon. The 874 genes regulated by eQTL encode proteins involved in numerous biochemical pathways and regulatory processes (Gonzales et al., 2008; Huang et al., 2009; Rider et al., 2010) that contribute to phenotypic divergence between HB3 and Dd2.

In addition to standard QTL approaches using systems biology readouts such as whole-genome transcription as described above, the profile of coinheritance of transcripts can be studied in a network framework, irrespective of whether QTLs are generated. In a systems genetics framework, the perturbation is the natural allelic variation introduced into a segregating population (e.g., a genetic cross). The dimensionality is provided by the series of random genetic profiles rather than a developmental time course, for example. Transcript levels and other nested traits can be connected to each other and end-stage phenotypes of interest by the strength of their coinheritance. These coinherited transcript levels provide a wiring diagram, and these transcripts can also be correlated by their inheritance across the full set of progeny with other whole-genome readouts such as metabolite levels and with classical traits like drug responses (Ayroles et al., 2009; Jumbo-Lucioni et al., 2010).

Ultraresolution Linkage Maps and the Structural Genome

Genetic crosses are a valuable prism through which to view the structural genome, including both Mendelian and non-Mendelian (e.g., gene conversion) recombinations, copy number variations (CNVs), and various genome rearrangements and mutations. With each iteration of technology development, the resolution of genome organization increases and the simple concept of linear order and linkage becomes more complex. Linkage maps remain an essential input parameter for linkage mapping, even given an assembled reference genome and various physical maps; notably, genomes are not completely colinear, various rearrangements distinguish parent lines, and assemblies are not perfect. Early studies were limited by the availability of genetic markers and the cost associated with developing high quality markers. However, this is changing with technological advancements and linkage maps are now moving toward purely physical maps.

The earliest work to relate genetic markers to the physical genome in *P. falciparum* crosses relied on pulsed-field gradient gel electrophoresis to isolate individual chromosomes to assist with linkage assignments of RFLPs (Peterson et al., 1988; Walker-Jonah et al., 1992; Wellems et al., 1990, 1991). MS-based maps increased resolution to include hundreds of informative markers (Alkhalil et al., 2009; Hayton et al., 2008; Su et al., 1999). Genomics technologies developed within the past 5 years allow uncharacterized *Plasmodium* genomes to be rapidly and comprehensively investigated with microarrays typing thousands of genetic markers (Dharia et al., 2009; Jiang et al., 2008; Mu et al., 2010; Neafsey et al., 2008; Tan et al., 2011), and ultraresolution is now available by next-generation sequencing (Kozarewa et al., 2009; Manske et al., in press; Samarakoon et al., 2011b).

Ultradense marker coverage brings into focus the relationship between genetic linkage (i.e., recombinational distance, in centimorgans [cM]) and physical distances. As recently as 1999, with the first detailed study of *P. falciparum* recombination parameters, 900 MS markers were considered extraordinarily high resolution (Su et al., 1999). These markers distinguished 326 crossover-defined genome segments in the HB3 × Dd2 cross and a recombination rate of 14.8 kb/cM. The 7G8 × GB4 cross was estimated to be 36 kb/cM when using 285 MS markers (Hayton et al., 2008); however, a subsequent microarray analysis significantly increased the predicted recombination rate to 9.6 kb/cM over a 2514 cM genome (Jiang et al., 2011). These various estimations underscore the challenges of comparing and integrating genetic and physical maps. In comparison to the MS-based analysis, the microarray analysis yielded a much greater marker density and coverage that included subtelomeric regions; many recombination events in the subtelomeric regions led to a greatly expanded recombination rate. Exclusion of these regions reduced the map length to 1654.7 cM and the recombination rate to be 12.8 kb/cM. Genome regions can reside in cold spots of limited recombination, as is typical of centromeres (Jiang et al., 2011; Kelly et al., 2006), or in hot spots rich with recombinations over short physical distances. Having sequence-level resolution enables searches for overrepresented motif sequences that could promote recombination; the 7G8 × GB4 cross progeny exhibit GC-rich motifs associated with recombination hot spots (Jiang et al., 2011).

The challenge posed by ultraresolution information is that genetic exchanges are detected that may not be due to standard Mendelian recombination, necessarily confounding the ability to detect genotype–phenotype relationships by coinheritance of markers and traits. An important feature of the physical genome that is difficult to ascertain with genetic markers is CNV. Based on observations from genetic crosses, laboratory-selected parasites, and field samples, CNVs arise through both meiotic and mitotic recombination (Samarakoon et al., 2011a). Gene amplification/deletion is present across the genomes of the parents of the various crosses (Hinterberg et al., 1994; Kidgell et al., 2006; Ribacke et al., 2007; Wellems et al., 1990). An analysis of Dd2, HB3, and the progeny of their genetic cross showed that some progeny inherited nonparental *pfmdr1* CNs (Patel et al., 2010), and mechanisms by which these events occur in meiosis and/or mitosis have been proposed based on whole-genome CN analysis of the progeny from this cross. A striking observation from the sequencing of two progeny genomes and comparative genome hybridization (CGH) of all progeny is the preponderance of small, de novo exchanges that could indicate widespread structural mutations and/or gene conversions (Samarakoon et al., 2011a,b). Given the assembly challenges and hybridization anomalies of this AT-rich genome, the predicted high level of non-Mendelian structural variation must be comprehensively validated using direct physical methods.

With microarrays and next-generation sequencing, probes are designed from and sequence reads are mapped to a reference genome. Some of the probes or genetic markers may appear to be in a different linear order or on a different chromosome with respect to the reference genome. It can be powerful to use dense sets of genetic markers without preconceived physical locations and in comparison with reference genome locations where genetic linkage helps to determine the physical location of rearrangement or translocation events. Genetic linkage mapping using the structural features of the genome, for example, CN at a particular locus, or hybridization signal of microarray probes can be used as a phenotype for QTL mapping and can reveal expected and unexpected linkage relationships.

Chemical Genomics

Recent publications of impressive scope and scale have illustrated a role for genetic crosses in a chemical genomic approach (Yuan et al., 2009, 2011). In such a data-driven approach, dimensionality and cross filtering are the basis for the power of this approach. Yuan et al. (2011) leveraged the independent strengths of high throughput compound library screens and classical genetics to develop response profiles across a diverse set of parasite strains from around the world. They further identified genome-wide associations in these diverse parasites and performed comparative linkage mapping in different genetic crosses. Each layer of information informed and refined the others, leading to expansive conclusions with major implications for malaria drug development. Rather than testing a handful of standard lab lines, as is typically the basis for live–dead compound screens, parasite isolates representing a comprehensive range of geographies and drug selection histories were assayed for quantitative dose responses to 2816 compounds registered for use in humans. The resulting matrix of drug susceptibilities across this spectrum of natural parasite variation generated a landscape of response signatures that were used to cluster responses on two axes, compounds and parasite response.

With this high dimensional data, correlated response patterns can point to shared drug targets and/or common parasite response pathways that can suggest related mechanisms of drug actions, drug synergies, and high priority drug candidates that are most effective against resistant parasite populations. Response classes identified in this study numbered much fewer than compound classes, indicating a convergence of drug action/killing mechanisms and/or response mechanisms across a range of chemotypes.

Using this global approach, the authors identified highly active compounds against the existing array of drug-resistant parasites and also identified categorical response signatures termed differential chemical phenotypes (DCPs). These DCPs were used to conduct GWAS to identify candidate genes controlling these responses. One hundred sixty-one DCPs distinguished the parents of the genetic crosses and were further screened in progeny clones of the appropriate cross to identify/verify genetic loci. GWAS alone in this small diverse parasite population lacks statistical power; however, when combined with targeted linkage mapping in the crosses, clear patterns emerged of three major loci (*pfcrt*, *pfmdr1*, and *dhfr*) that contribute significantly to the vast majority of DCPs.

This analysis demonstrates how drug selection has profoundly shaped parasite genomes throughout the world and that the path to reduced susceptibility to a wide range of chemical types passes through a small set of genes. Furthermore, this work sets a conceptual foundation by using high dimensionality (many compounds × many parasite strains) and layers of independent but complimentary information (library screens, genetic crosses, GWASs). Future advances to this approach will include more precise phenotype measurements and genome scans that focus specifically on the residual phenotypic variation, after the removal of the main effects, to see other mechanisms operating behind the well-known major drug resistance genes. Moreover, the inclusion of the genome structure (CN), transcript, and metabolite profiles from various populations, and the use of compounds chosen because of their potential to perturb the comprehensive metabolic space of the malaria parasite will each contribute to an even more powerful integrated chemical genomics approach.

Integration of Multiple Layers of Information: Data Mining

Historically, genetic studies have focused on the influence of single-gene variants; however, with genome projects and massive GWASs, the significant challenge and opportunity of genetic complexity have come to the forefront. The lesson from the early genomics era is that traits are rarely defined by single genes. Rather, they are properties generated by dynamic networks. To fully embrace biological complexity in our models of drug targets and resistance mechanisms, we must be able to find and interpret biomolecular interactions in the context of dynamic whole-genome networks. With the ever-increasing array of cost-effective whole-genome technologies, a wealth of data is at our fingertips representing the physical genome, variation across individual parasites, transcript/protein/metabolite levels, as well as epigenetic information and trait associations. In addition to these layers of the biological system, the data also exist across dimensions within these layers, for example, developmental stages, experimental treatments, and in genetic strains representing various geographies or selection histories. It is certain that these data must ultimately fit together to explain how phenotypes are made; however, as impressive as science's remarkable capacity to make data is, the severe bottleneck is in analyzing and interpreting these data.

Certainly, the classical genetic approach alone can find only the largest interactions due to being underpowered, but it gives a crucial view of the system. Information resources are rapidly accumulating to drive the systematic identification of parasite mechanisms that can be exploited for controlling malaria. There is no question that a community model of resource development and data integration holds great promise (Threadgill et al., 2002). In the emerging era of data-driven biomedical science, methods to integrate and mine large and diverse data sources are emerging, for example, by drawing on expression networks and incorporating functional enrichments and biochemical pathways into standard GWAS (Askland et al., 2009). New network-based computational models accurately predict the states generated by complex genetic networks. Systems genetics accesses and explores the strengths of classical genetics and modern genomics, becoming a useful tool for unraveling metabolic, regulatory, and developmental pathways.

Data quality and new analyses make it possible to explore subtle inheritance variation using classical genetics to enhance our opportunity to discover the "needle in the haystack," even without knowing what we are looking for. New analytical tools for finding molecular and genetic interactions (Huang et al., 2009; Wuchty et al., 2011) and for reconstructing transcriptional and metabolic networks can be effectively combined with the wealth of existing data and annotations (http://plasmodb.org/plasmo/) for end users to interface with available data resources in a hypothesis testing framework.

The use of molecular transgenic methodologies remains uniquely powerful for proving the role of specific genes (and mutations) in conferring phenotypes (Reed et al., 2000; Sidhu et al., 2002). Such studies are greatly influenced by genetic complexity for two reasons: Single genes corresponding to QTL may have very subtle phenotype effects, and the presence and magnitude of phenotypic effects will largely depend on the genetic context in which these mutations are introduced. Therefore, the choice of strains for these molecular studies will profoundly influence both the chance for success and the biological interpretation of the results. For this reason, genetic mapping and molecular genetics should be partnered to understand, for example, why gene mutations that appear to act alone in generating resistance may not be viable or may not lead to phenotype shifts in particular genetic backgrounds.

Each level of genomic data has particular strengths and weaknesses: QTLs tie phenotypes to genome regions but do not identify genes; primary sequence analysis provides the basic structure and content of genes but does not tie directly to traits; comparative sequencing explores the polymorphisms between genomes but does not identify which SNPs are functional; global transcriptional analysis generates a picture of coregulated genes but is subject to interpretation difficulties; gene knockouts can demonstrate phenotype shifts, but these are dependent on their genetic background. Discovery of new molecular processes depends on the integration of these distinct but interrelated data sets.

SUMMARY

The flurry of new age technologies might leave one with the impression that old-school methods such as linkage mapping are less relevant. In fact, modern genomics tools only empower the classical approaches. Inherited genetic factors include all of the information to link genotype to phenotype, even when this involves fine-tuned gene × gene and gene × environment interactions. Recombination gives us the means to unbiased dissection of the components of a complex trait (and all traits have multiple influencing factors) in ways that genetic manipulation cannot.

Much of our lack of understanding about the malaria parasite stems from our limited capacity to study how networks of genes interact to influence phenotypes. Selections by drugs on parasite genomes generate genetic signatures, that is, combinations of alleles that work well together to combat the drug while retaining normal physiological function. Knowledge of the biological processes that can incur mutations that render a parasite less susceptible to drugs will ultimately inform our understanding about how multiple drug resistances evolve. QTL mapping in small malaria crosses is limited in its capacity to find these genes; however, this approach is useful for finding larger gene effects and epistatic interactions, for generating testable hypotheses, and, in this era of high throughput genomics, for providing a bridge between real traits and the sometimes overwhelming quantity of data.

Genetic crosses will continue to provide a path for the discovery of basic biological mechanisms in *P. falciparum*. The success of this approach requires that we maximize the value of the three existing crosses by using them comparatively with precise phenotypes for as many progeny clones as possible. For some unexplored traits of interest, major genes will be identified; however, in most instances, the crosses will reveal the architecture of complex traits and relationships between allelic combinations of different genetic backgrounds. Integrating multiple layers of information will help discern gene interactions and will be a component of broader data-driven experimental designs.

REFERENCES

Alkhalil A, Pillai AD, Bokhari AA, Vaidya AB, Desai SA. 2009. Complex inheritance of the plasmodial surface anion channel in a *Plasmodium falciparum* genetic cross. Molecular Microbiology 72: 459–469.

Askland K, Read C, Moore J. 2009. Pathways-based analyses of whole-genome association study data in bipolar disorder reveal genes mediating ion channel activity and synaptic neurotransmission. Human Genetics 125: 63–79.

Ayroles JF, Carbone MA, Stone EA, Jordan KW, Lyman RF, Magwire MM, Rollmann SM, Duncan LH, Lawrence F, Anholt RR, Mackay TF. 2009. Systems genetics of complex traits in *Drosophila melanogaster*. Nature Genetics 41: 299–307.

Babiker HA, Pringle SJ, Abdel-Muhsin A, Mackinnon M, Hunt P, Walliker D. 2001. High-level chloroquine resistance in Sudanese isolates of *Plasmodium falciparum* is associated with mutations in the chloroquine resistance transporter gene pfcrt and the multidrug resistance gene pfmdr1. The Journal of Infectious Diseases 183: 1535–1538.

Beez D, Sanchez CP, Stein WD, Lanzer M. 2011. Genetic predisposition favors the acquisition of stable artemisinin resistance in malaria parasites. Antimicrobial Agents and Chemotherapy 55: 50–55.

Bennett TN, Patel J, Ferdig MT, Roepe PD. 2007. *Plasmodium falciparum* Na^+/H^+ exchanger activity and quinine resistance. Molecular and Biochemical Parasitology 153: 48–58.

Bhasin VK, Trager W. 1984. Gametocyte-forming and non-gametocyte-forming clones of *Plasmodium falciparum*. The American Journal of Tropical Medicine and Hygiene 33: 534–537.

Burkot TR, Williams JL, Schneider I. 1984. Infectivity to mosquitoes of *Plasmodium falciparum* clones grown in vitro from the same isolate. Transactions of the Royal Society of Tropical Medicine and Hygiene 78: 339–341.

Campbell CC, Collins WE, Nguyen-Dinh P, Barber A, Broderson JR. 1982. *Plasmodium falciparum* gametocytes from culture in vitro develop to sporozoites that are infectious to primates. Science 217: 1048–1050.

Chen N, Russell B, Fowler E, Peters J, Cheng Q. 2002. Levels of chloroquine resistance in *Plasmodium falciparum* are determined by loci other than pfcrt and pfmdr1. The Journal of Infectious Diseases 185: 405–407.

Cooper RA, Ferdig MT, Su XZ, Ursos LM, Mu J, Nomura T, Fujioka H, Fidock DA, Roepe PD, Wellems TE. 2002. Alternative mutations at position 76 of the vacuolar transmembrane protein PfCRT are associated with chloroquine resistance and unique stereospecific quinine and quinidine responses in

Plasmodium falciparum. Molecular Pharmacology 61: 35–42.

Cooper RA, Hartwig CL, Ferdig MT. 2005. pfcrt is more than the *Plasmodium falciparum* chloroquine resistance gene: A functional and evolutionary perspective. Acta Tropica 94: 170–180.

Cowman AF, Galatis D, Thompson JK. 1994. Selection for mefloquine resistance in *Plasmodium falciparum* is linked to amplification of the pfmdr1 gene and cross-resistance to halofantrine and quinine. Proceedings of the National Academy of Sciences of the United States of America 91: 1143–1147.

Delemarre BJ, Van Der Kaay HJ. 1979. Tropical malaria contracted the natural way in the Netherlands. Nederlands Tijdschrift Voor Geneeskunde 123: 1981–1982.

Dharia NV, Sidhu AB, Cassera MB, Westenberger SJ, Bopp SE, Eastman RT, Plouffe D, Batalov S, Park DJ, Volkman SK, Wirth DF, Zhou Y, Fidock DA, Winzeler EA. 2009. Use of high-density tiling microarrays to identify mutations globally and elucidate mechanisms of drug resistance in *Plasmodium falciparum*. Genome Biology 10: R21.

Djimde A, Doumbo OK, Cortese JF, Kayentao K, Doumbo S, Diourte Y, Dicko A, Su XZ, Nomura T, Fidock DA, Wellems TE, Plowe CV. 2001. A molecular marker for chloroquine-resistant falciparum malaria. The New England Journal of Medicine 344: 257–263.

Eichler EE, Flint J, Gibson G, Kong A, Leal SM, Moore JH, Nadeau JH. 2010. Missing heritability and strategies for finding the underlying causes of complex disease. Nature Reviews. Genetics 11: 446–450.

Ferdig MT, Cooper RA, Mu J, Deng B, Joy DA, Su XZ, Wellems TE. 2004. Dissecting the loci of low-level quinine resistance in malaria parasites. Molecular Microbiology 52: 985–997.

Fidock DA, Nomura T, Cooper RA, Su X, Talley AK, Wellems TE. 2000a. Allelic modifications of the cg2 and cg1 genes do not alter the chloroquine response of drug-resistant *Plasmodium falciparum*. Molecular and Biochemical Parasitology 110: 1–10.

Fidock DA, Nomura T, Talley AK, Cooper RA, Dzekunov SM, Ferdig MT, Ursos LM, Sidhu AB, Naude B, Deitsch KW, Su XZ, Wootton JC, Roepe PD, Wellems TE. 2000b. Mutations in the *P. falciparum* digestive vacuole transmembrane protein PfCRT and evidence for their role in chloroquine resistance. Molecules and Cells 6: 861–871.

Foote SJ, Kyle DE, Martin RK, Oduola AM, Forsyth K, Kemp DJ, Cowman AF. 1990. Several alleles of the multidrug-resistance gene are closely linked to chloroquine resistance in *Plasmodium falciparum*. Nature 345: 255–258.

Gardner MJ, Hall N, Fung E, White O, Berriman M, Hyman RW, Carlton JM, Pain A, Nelson KE, Bowman S, Paulsen IT, James K, Eisen JA, Rutherford K, Salzberg SL, Craig A, Kyes S, Chan MS, Nene V, Shallom SJ, Suh B, Peterson J, Angiuoli S, Pertea M, Allen J, Selengut J, Haft D, Mather MW, Vaidya AB, Martin DM, Fairlamb AH, Fraunholz MJ, Roos DS, Ralph SA, Mcfadden GI, Cummings LM, Subramanian GM, Mungall C, Venter JC, Carucci DJ, Hoffman SL, Newbold C, Davis RW, Fraser CM, Barrell B. 2002. Genome sequence of the human malaria parasite *Plasmodium falciparum*. Nature 419: 498–511.

Gonzales JM, Patel JJ, Ponmee N, Jiang L, Tan A, Maher SP, Wuchty S, Rathod PK, Ferdig MT. 2008. Regulatory hotspots in the malaria parasite genome dictate transcriptional variation. PLoS Biology 6: e238.

Hayton K, Su XZ. 2008. Drug resistance and genetic mapping in *Plasmodium falciparum*. Current Genetics 54: 223–239.

Hayton K, Gaur D, Liu A, Takahashi J, Henschen B, Singh S, Lambert L, Furuya T, Bouttenot R, Doll M, Nawaz F, Mu J, Jiang L, Miller LH, Wellems TE. 2008. Erythrocyte binding protein PfRH5 polymorphisms determine species-specific pathways of *Plasmodium falciparum* invasion. Cell Host & Microbe 4: 40–51.

Hinterberg K, Mattei D, Wellems TE, Scherf A. 1994. Interchromosomal exchange of a large subtelomeric segment in a *Plasmodium falciparum* cross. The EMBO Journal 13: 4174–4180.

Huang Y, Wuchty S, Ferdig MT, Przytycka TM. 2009. Graph theoretical approach to study eQTL: A case study of *Plasmodium falci-*

parum. Bioinformatics (Oxford, England) 25: i15–i20.

Jansen RC, Nap JP. 2001. Genetical genomics: The added value from segregation. Trends in Genetics: TIG 17: 388–391.

Jeffares DC, Pain A, Berry A, Cox AV, Stalker J, Ingle CE, Thomas A, Quail MA, Siebenthall K, Uhlemann AC, Kyes S, Krishna S, Newbold C, Dermitzakis ET, Berriman M. 2007. Genome variation and evolution of the malaria parasite *Plasmodium falciparum*. Nature Genetics 39: 120–125.

Jiang H, Yi M, Mu J, Zhang L, Ivens A, Klimczak LJ, Huyen Y, Stephens RM, Su XZ. 2008. Detection of genome-wide polymorphisms in the AT-rich *Plasmodium falciparum* genome using a high-density microarray. BMC Genomics 9: 398.

Jiang H, Li N, Gopalan V, Zilversmit MM, Varma S, Nagarajan V, Li J, Mu J, Hayton K, Henschen B, Yi M, Stephens R, Mcvean G, Awadalla P, Wellems TE, Su XZ. 2011. High recombination rates and hotspots in a *Plasmodium falciparum* genetic cross. Genome Biology 12: R33.

Jumbo-Lucioni P, Ayroles JF, Chambers MM, Jordan KW, Leips J, Mackay TF, De Luca M. 2010. Systems genetics analysis of body weight and energy metabolism traits in *Drosophila melanogaster*. BMC Genomics 11: 297.

Kelly JM, Mcrobert L, Baker DA. 2006. Evidence on the chromosomal location of centromeric DNA in *Plasmodium falciparum* from etoposide-mediated topoisomerase-II cleavage. Proceedings of the National Academy of Sciences of the United States of America 103: 6706–6711.

Kidgell C, Volkman SK, Daily J, Borevitz JO, Plouffe D, Zhou Y, Johnson JR, Le Roch K, Sarr O, Ndir O, Mboup S, Batalov S, Wirth DF, Winzeler EA. 2006. A systematic map of genetic variation in *Plasmodium falciparum*. PLoS Pathogens 2: e57.

Kirkman LA, Su XZ, Wellems TE. 1996. *Plasmodium falciparum*: Isolation of large numbers of parasite clones from infected blood samples. Experimental Parasitology 83: 147–149.

Kozarewa I, Ning Z, Quail MA, Sanders MJ, Berriman M, Turner DJ. 2009. Amplification-free Illumina sequencing-library preparation facilitates improved mapping and assembly of (G+C)-biased genomes. Nature Methods 6: 291–295.

Levin BR, Perrot V, Walker N. 2000. Compensatory mutations, antibiotic resistance and the population genetics of adaptive evolution in bacteria. Genetics 154: 985–997.

Lim AS, Cowman AF. 1996. *Plasmodium falciparum*: Chloroquine selection of a cloned line and DNA rearrangements. Experimental Parasitology 83: 283–294.

Manske M, et al. (in press) Analysis of *Plasmodium falciparum* diversity in natural infections by deep sequencing. Nature.

Martin RE, Marchetti RV, Cowan AI, Howitt SM, Broer S, Kirk K. 2009. Chloroquine transport via the malaria parasite's chloroquine resistance transporter. Science 325: 1680–1682.

Moore JH, Williams SM. 2009. Epistasis and its implications for personal genetics. American Journal of Human Genetics 85: 309–320.

Mu J, Ferdig MT, Feng X, Joy DA, Duan J, Furuya T, Subramanian G, Aravind L, Cooper RA, Wootton JC, Xiong M, Su XZ. 2003. Multiple transporters associated with malaria parasite responses to chloroquine and quinine. Molecular Microbiology 49: 977–989.

Mu J, Awadalla P, Duan J, Mcgee KM, Joy DA, Mcvean GA, Su XZ. 2005. Recombination hotspots and population structure in *Plasmodium falciparum*. PLoS Biology 3: e335.

Mu J, Awadalla P, Duan J, Mcgee KM, Keebler J, Seydel K, Mcvean GA, Su XZ. 2007. Genome-wide variation and identification of vaccine targets in the *Plasmodium falciparum* genome. Nature Genetics 39: 126–130.

Mu J, Myers RA, Jiang H, Liu S, Ricklefs S, Waisberg M, Chotivanich K, Wilairatana P, Krudsood S, White NJ, Udomsangpetch R, Cui L, Ho M, Ou F, Li H, Song J, Li G, Wang X, Seila S, Sokunthea S, Socheat D, Sturdevant DE, Porcella SF, Fairhurst RM, Wellems TE, Awadalla P, Su XZ. 2010. *Plasmodium falciparum* genome-wide scans for positive selection, recombination hot spots and resistance to antimalarial drugs. Nature Genetics 42: 268–271.

Musani SK, Shriner D, Liu N, Feng R, Coffey CS, Yi N, Tiwari HK, Allison DB. 2007. Detection of gene × gene interactions in genome-wide association studies of human population data. Human Heredity 63: 67–84.

Neafsey DE, Schaffner SF, Volkman SK, Park D, Montgomery P, Milner DA, Jr., Lukens A, Rosen D, Daniels R, Houde N, Cortese JF, Tyndall E, Gates C, Stange-Thomann N, Sarr O, Ndiaye D, Ndir O, Mboup S, Ferreira MU, Moraes Sdo L, Dash AP, Chitnis CE, Wiegand RC, Hartl DL, Birren BW, Lander ES, Sabeti PC, Wirth DF. 2008. Genome-wide SNP genotyping highlights the role of natural selection in *Plasmodium falciparum* population divergence. Genome Biology 9: R171.

Nguyen-Dinh P, Payne D. 1980. Pyrimethamine sensitivity in *Plasmodium falciparum*: Determination in vitro by a modified 48-hour test. Bulletin of the World Health Organization 58: 909–912.

Oduola AM, Milhous WK, Weatherly NF, Bowdre JH, Desjardins RE. 1988a. *Plasmodium falciparum*: Induction of resistance to mefloquine in cloned strains by continuous drug exposure in vitro. Experimental Parasitology 67: 354–360.

Oduola AM, Weatherly NF, Bowdre JH, Desjardins RE. 1988b. *Plasmodium falciparum*: Cloning by single-erythrocyte micromanipulation and heterogeneity in vitro. Experimental Parasitology 66: 86–95.

Patel JJ, Thacker D, Tan JC, Pleeter P, Checkley L, Gonzales JM, Deng B, Roepe PD, Cooper RA, Ferdig MT. 2010. Chloroquine susceptibility and reversibility in a *Plasmodium falciparum* genetic cross. Molecular Microbiology 78: 770–787.

Peterson DS, Walliker D, Wellems TE. 1988. Evidence that a point mutation in dihydrofolate reductase-thymidylate synthase confers resistance to pyrimethamine in *falciparum* malaria. Proceedings of the National Academy of Sciences of the United States of America 85: 9114–9118.

Picot S, Olliaro P, De Monbrison F, Bienvenu AL, Price RN, Ringwald P. 2009. A systematic review and meta-analysis of evidence for correlation between molecular markers of parasite resistance and treatment outcome in *falciparum* malaria. Malaria Journal 8: 89.

Ponnudurai T, Leeuwenberg AD, Meuwissen JH. 1981. Chloroquine sensitivity of isolates of *Plasmodium falciparum* adapted to in vitro culture. Tropical and Geographical Medicine 33: 50–54.

Rathod PK, Mcerlean T, Lee PC. 1997. Variations in frequencies of drug resistance in *Plasmodium falciparum*. Proceedings of the National Academy of Sciences of the United States of America 94: 9389–9393.

Reed MB, Saliba KJ, Caruana SR, Kirk K, Cowman AF. 2000. Pgh1 modulates sensitivity and resistance to multiple antimalarials in *Plasmodium falciparum*. Nature 403: 906–909.

Ribacke U, Mok BW, Wirta V, Normark J, Lundeberg J, Kironde F, Egwang TG, Nilsson P, Wahlgren M. 2007. Genome wide gene amplifications and deletions in *Plasmodium falciparum*. Molecular and Biochemical Parasitology 155: 33–44.

Rider AK, Siwo G, Chawla NV, Ferdig M, Emrich SJ. 2010. A statistical approach to finding overlooked genetic associations. BMC Bioinformatics 11: 526.

Sa JM, Twu O, Hayton K, Reyes S, Fay MP, Ringwald P, Wellems TE. 2009. Geographic patterns of *Plasmodium falciparum* drug resistance distinguished by differential responses to amodiaquine and chloroquine. Proceedings of the National Academy of Sciences of the United States of America 106: 18883–18889.

Samarakoon U, Gonzales JM, Patel JJ, Tan A, Checkley L, Ferdig MT. 2011a. The landscape of inherited and de novo copy number variants in a *Plasmodium falciparum* genetic cross. BMC Genomics 12: 457.

Samarakoon U, Regier A, Tan A, Desany BA, Collins B, Tan JC, Emrich SJ, Ferdig MT. 2011b. High-throughput 454 resequencing for allele discovery and recombination mapping in *Plasmodium falciparum*. BMC Genomics 12: 116.

Sanchez CP, Mayer S, Nurhasanah A, Stein WD, Lanzer M. 2011. Genetic linkage analyses redefine the roles of PfCRT and PfMDR1 in drug accumulation and susceptibility in *Plasmodium falciparum*. Molecular Microbiology 82: 865–878.

Sen S, Ferdig M. 2004. QTL analysis for discovery of genes involved in drug responses. Current Drug Targets. Infectious Disorders 4: 53–63.

Sidhu AB, Verdier-Pinard D, Fidock DA. 2002. Chloroquine resistance in *Plasmodium falciparum* malaria parasites conferred by pfcrt mutations. Science 298: 210–213.

Su X, Wellems TE. 1996. Toward a high-resolution *Plasmodium falciparum* linkage map: Polymorphic markers from hundreds of simple sequence repeats. Genomics 33: 430–444.

Su X, Kirkman LA, Fujioka H, Wellems TE. 1997. Complex polymorphisms in an approximately 330 kDa protein are linked to chloroquine-resistant *P. falciparum* in Southeast Asia and Africa. Cell 91: 593–603.

Su X, Ferdig MT, Huang Y, Huynh CQ, Liu A, You J, Wootton JC, Wellems TE. 1999. A genetic map and recombination parameters of the human malaria parasite *Plasmodium falciparum*. Science 286: 1351–1353.

Su X, Hayton K, Wellems TE. 2007. Genetic linkage and association analyses for trait mapping in *Plasmodium falciparum*. Nature reviews. Genetics 8: 497–506.

Su XZ, Wootton JC. 2004. Genetic mapping in the human malaria parasite *Plasmodium falciparum*. Molecular Microbiology 53: 1573–1582.

Sullivan JS, Sullivan JJ, Williams A, Grady KK, Bounngaseng A, Huber CS, Nace D, Williams T, Galland GG, Barnwell JW, Collins WE. 2003. Adaptation of a strain of *Plasmodium falciparum* from Ghana to *Aotus lemurinus griseimembra*, *A. nancymai*, and *A. vociferans* monkeys. The American Journal of Tropical Medicine and Hygiene 69: 593–600.

Tan JC, Miller BA, Tan A, Patel JJ, Cheeseman IH, Anderson TJ, Manske M, Maslen G, Kwiatkowski DP, Ferdig MT. 2011. An optimized microarray platform for assaying genomic variation in *Plasmodium falciparum* field populations. Genome Biology 12: R35.

Threadgill DW, Hunter KW, Williams RW. 2002. Genetic dissection of complex and quantitative traits: From fantasy to reality via a community effort. Mammalian Genome 13: 175–178.

Valderramos SG, Valderramos JC, Musset L, Purcell LA, Mercereau-Puijalon O, Legrand E, Fidock DA. 2010. Identification of a mutant PfCRT-mediated chloroquine tolerance phenotype in *Plasmodium falciparum*. PLoS Pathogens 6: e1000887.

Volkman SK, Sabeti PC, Decaprio D, Neafsey DE, Schaffner SF, Milner DA, Jr., Daily JP, Sarr O, Ndiaye D, Ndir O, Mboup S, Duraisingh MT, Lukens A, Derr A, Stange-Thomann N, Waggoner S, Onofrio R, Ziaugra L, Mauceli E, Gnerre S, Jaffe DB, Zainoun J, Wiegand RC, Birren BW, Hartl DL, Galagan JE, Lander ES, Wirth DF. 2007. A genome-wide map of diversity in *Plasmodium falciparum*. Nature Genetics 39: 113–119.

Walker-Jonah A, Dolan SA, Gwadz RW, Panton LJ, Wellems TE. 1992. An RFLP map of the *Plasmodium falciparum* genome, recombination rates and favored linkage groups in a genetic cross. Molecular and Biochemical Parasitology 51: 313–320.

Walliker D, Quakyi IA, Wellems TE, Mccutchan TF, Szarfman A, London WT, Corcoran LM, Burkot TR, Carter R. 1987. Genetic analysis of the human malaria parasite *Plasmodium falciparum*. Science 236: 1661–1666.

Wellems TE, Oduola AMJ, Fenton B, Desjardins R, Panton LJ, Dorosario VE. 1988. Chromosome size variation occurs in cloned *Plasmodium falciparum* on in vitro cultivation. Revista Brasileira de Genética 11: 813–825.

Wellems TE, Panton LJ, Gluzman IY, Do Rosario VE, Gwadz RW, Walker-Jonah A, Krogstad DJ. 1990. Chloroquine resistance not linked to mdr-like genes in a *Plasmodium falciparum* cross. Nature 345: 253–255.

Wellems TE, Walker-Jonah A, Panton LJ. 1991. Genetic mapping of the chloroquine-resistance locus on *Plasmodium falciparum* chromosome 7. Proceedings of the National Academy of Sciences of the United States of America 88: 3382–3386.

Witt H. 2011. New modifier loci in cystic fibrosis. Nature Genetics 43: 508–509.

Wootton JC, Feng X, Ferdig MT, Cooper RA, Mu J, Baruch DI, Magill AJ, Su XZ. 2002. Genetic diversity and chloroquine selective sweeps in *Plasmodium falciparum*. Nature 418: 320–323.

Wright FA, Strug LJ, Doshi VK, Commander CW, Blackman SM, Sun L, Berthiaume Y, Cutler D, Cojocaru A, Collaco JM, Corey M, Dorfman R, Goddard K, Green D, Kent JW, Jr., Lange EM, Lee S, Li W, Luo J, Mayhew GM, Naughton KM, Pace RG, Pare P, Rommens JM, Sandford A, Stonebraker JR, Sun W, Taylor C, Vanscoy LL, Zou F, Blangero J, Zielenski J, O'neal WK, Drumm ML, Durie PR, Knowles MR, Cutting GR. 2011. Genome-wide association and linkage identify modifier loci of lung disease severity in cystic fibrosis at 11p13 and 20q13.2. Nature Genetics 43: 539–546.

Wuchty S, Siwo GH, Ferdig MT. 2011. Shared molecular strategies of the malaria parasite *P. falciparum* and the human virus HIV-1. Molecular and Cellular Proteomics. 10: M111.009035.

Yuan J, Johnson RL, Huang R, Wichterman J, Jiang H, Hayton K, Fidock DA, Wellems TE, Inglese J, Austin CP, Su XZ. 2009. Genetic mapping of targets mediating differential chemical phenotypes in *Plasmodium falciparum*. Nature Chemical Biology 5: 765–771.

Yuan J, Cheng KC, Johnson RL, Huang R, Pattaradilokrat S, Liu A, Guha R, Fidock DA, Inglese J, Wellems TE, Austin CP, Su XZ. 2011. Chemical genomic profiling for antimalarial therapies, response signatures, and molecular targets. Science 333: 724–729.

CHAPTER 14

GENETIC MAPPING OF VIRULENCE IN RODENT MALARIAS

RICHARD CARTER and RICHARD CULLETON

THE MURINE RODENT MALARIA PARASITES

Malaria parasites of murine rodents from Africa have, over the course of the last 60 years, taken up a significant part of the laboratory research efforts devoted to malaria. The value of this effort, in contrast to that devoted to investigations directly upon the malaria parasites that infect humans, is often questioned. It is a debate we will not attempt to enter here, except by presenting our own knowledge and understanding of rodent malaria research within the present remit.

Unusually, if not uniquely, the malaria parasites of African rodents were first found not in their vertebrate hosts but in a mosquito, *Anopheles dureni*, among specimens collected in the course of a holiday fishing expedition out of pure curiosity by Ignace Vincke, a Belgian malariologist and colonial medical officer for Katanga in the then Belgian Congo. This was in 1943. Five years later following a detective's trail of clues, he discovered the vertebrate hosts—species of murine rodents, mainly *Grammomys surdaster*, or the African thicket rat—that inhabited the tree-lined verges of rivers and streams in the Katanga Highlands. He named the parasites *Plasmodium berghei* (Garnham, 1966).

In the course of the next almost 20 years, malaria parasites were discovered in the blood of murine rodents from locations across central and western Africa (Killick-Kendrick and Peters, 1978). The locations included others within Katanga itself, locations along the course of the River Congo, from Brazzaville toward its mouth and from Bangui in the Central African Republic on the northern banks of its middle reaches, and in Nigeria near its capital, Lagos. By 1967, what were considered to be three different species were described in the literature. These comprised *P. berghei* itself, mainly a blood reticulocyte inhabiting a malaria parasite that is transmitted through mosquitoes only at temperatures below 21°C (its natural environment in highland Katanga rarely exceeding this temperature), a morphologically indistinguishable parasite from the more torrid zones of Central Africa on the banks of the Congo and in Nigeria, which was named a subspecies of *P. berghei*, as *Plasmodium berghei yoelii*, and which transmitted optimally to mosquitoes at 26°C, a second full

Evolution of Virulence in Eukaryotic Microbes, First Edition. Edited by L. David Sibley, Barbara J. Howlett, and Joseph Heitman.

species, *Plasmodium vinckei*, a normocyte-inhabiting parasite found in all locations from which the murine rodent malaria parasites had been found and a second normocyte-inhabiting parasite found only along the banks of the Congo named *Plasmodium chabaudi*, the third species of rodent malaria parasite as they were classified at this time.

There remained, however, much uncertainty about the characteristics of, and genetic and phylogenetic relationships among, these parasites. These difficulties continued until around 1974 when detailed and precise descriptions of numerous cloned lines of the parasites had been undertaken. The characteristics that were measured were biological, including in mosquito and rodent hosts, and, for the first time, molecular genetic, using electrophoretic mobilities of enzyme proteins as genetic markers of the parasites (Carter, 1978; Killick-Kendrick and Peters, 1978). The outcome of these investigations was the unequivocal distinction not of three but of four species of the parasites *P. berghei*, *Plasmodium yoelii*, *P. vinckei*, and *P. chabaudi*, each represented by genetically distinct regional subspecies. These relationships have been upheld ever since including in postgenomic phylogentic analyses based on gene sequences (Perkins et al., 2007). The locations from which these parasites have been isolated are shown in Figure 14.1.

The murine rodent malaria parasites have been investigated in numerous ways, covering almost all aspects of the biology, molecular and otherwise, of malaria. Of particular note in the present context are genetic analyses of phenotypes of the parasites related to their "virulence."

THE DEFINITION OF VIRULENCE IN MALARIA

Virulence in malaria has attracted the attention of researchers primarily from two different biological perspectives. One is from that of the clinical observer asking how much harm does a parasite cause to its host during an infection. The other is from that of the evolutionary biologist asking how does a parasite's virulence affect its fitness to survive. This discussion will examine both perspectives. First, however, we need to look at what is, and has, been understood by the term virulence in relation to studies in malaria.

Virulence has sometimes carried two, actually independent, meanings in past literature on malaria. One meaning has referred to malaria parasites that multiply more rapidly than normal in the blood of a host. This meaning is associated especially with the appearance in the laboratory of lines of rodent malaria parasites that grew much more rapidly than did parasites of the stock from which they appeared to have come (Yoeli et al., 1975). The rapidly multiplying "virulent" parasites were much more clinically dangerous to their rodent hosts than were parasites of the "mild," slow-growing lines from which they had arisen. If allowed to run their course, infections with parasites of the "virulent" lines were almost invariably lethal to their hosts within a few days. Infections with the "mild" progenitors were almost invariably nonlethal within the same period of time.

The ideas of virulence, in the sense of a rapidly multiplying parasite, and virulence in the sense of a more clinically dangerous infection, were not usually distinguished or remarked upon. It is, nevertheless, essential to do so.

One plausible basis, but only one, for the virulence of a particular parasite stock, could, indeed, be the speed with which it multiplies in a host's tissues and a corresponding amount of damage to those tissues. This rapid parasite multiplication in host tissues, however, is by no means the only pathophysiological basis for damage to a host. Properties of the parasites that

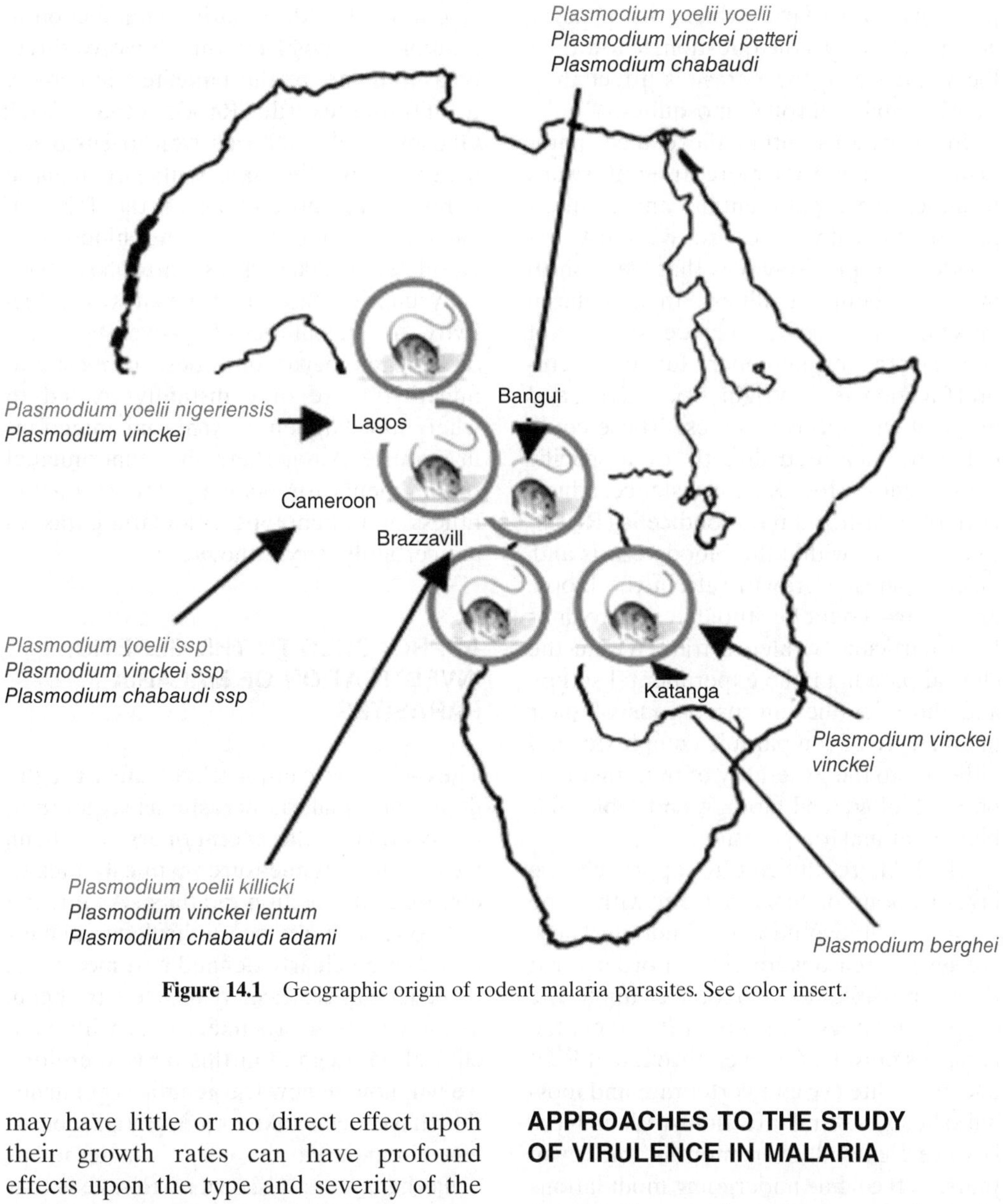

Figure 14.1 Geographic origin of rodent malaria parasites. See color insert.

may have little or no direct effect upon their growth rates can have profound effects upon the type and severity of the pathology that they cause. There are also situations where, within limits, growth rate differences have negligible effects upon the severity of the disease. For these reasons, therefore, parasite growth rate, as such, is not even implied in our definition of virulence in malaria. The virulence of a specific stock of malaria parasites is defined here solely by the harm they cause to their host.

APPROACHES TO THE STUDY OF VIRULENCE IN MALARIA

As already stated, studies on virulence in malaria have been undertaken from within the two paradigms of clinical medicine and evolutionary biology. A clinical investigator is interested directly in whether properties of the parasites affect the type and severity of associated illness. This is quite distinct from the evolutionary perspective, where an investigator tries to understand how a virulence trait affects the fitness of the parasites to survive, which, in malaria, equates closely to the question, "How does the virulence of the parasites affect their transmissibility through mosquitoes?"

Investigations within the clinical paradigm are certainly the more straightforward to undertake experimentally and to interpret scientifically. To do so, we must first decide what precisely it is that we wish to measure about the illness. In a malarial infection, this could include degree of anemia, renal or pulmonary function, cerebral function, body weight, risk of dying and many other aspects of illness. These could either be measured directly or a specific physiological process—for instance, adherence of parasitized red blood cells (RBCs) to other cells within the blood vessels and, indeed, parasite growth rate in the blood, and so on—could be studied as surrogates for a particular virulence trait. Within the clinical paradigm, the experimental setups, and the scientific interpretations of their findings, are of comparable complexity and difficulty to the generality of investigations on the biology, cell biology, and molecular biology of malaria parasites.

Much more difficult to approach are investigations of malaria from within the paradigm of evolutionary biology. There are several reasons for this. In order for it to be possible to pronounce upon the degree of "fitness" of a parasite stock, the parasites must be followed through at least one entire life cycle in vertebrate and mosquito hosts. Natural infections with malaria, however, extend over months and even years, all the while undergoing modulations in infectivity to mosquitoes. Cycles of transmission would, therefore, have to be tested throughout the natural duration of an infection to even begin to approach a realistic measure of parasite fitness. To do so presents almost insuperable practical difficulties for an investigation.

It is, therefore, nearly always the case that data collection is limited to a snapshot of an accessible part of the parasite's life, typically a few days during an infection in a laboratory host. This may involve direct measurements of the parasites' infectivity to mosquitoes (de Roode et al., 2005; Gadsby et al., 2009) as well as surrogate measures of the infectivity of malaria parasites to mosquitoes in the form of gametocyte densities in the blood (De Roode et al., 2003). These latter have typically unpredictable relationships to infectivity to mosquitoes. However, all such data—gametocyte densities or mosquito infectivity—are only distantly related to what may happen in malarial infections in nature. Altogether, the experimental requirements for testing a theory on the fitness of a phenotype of malaria parasites are certainly very difficult.

APPROACHES TO THE GENETIC INVESTIGATION OF MALARIA PARASITES

Thus, while we cannot reliably measure the fitness of a malaria parasite as required in the context of the evolutionary paradigm, we can directly measure manifestations of disease and infection in a host. As it is the only one within which a virulence phenotype can be clearly defined and measured, all genetic analyses of virulence have been, and are likely always to be, made within the clinical paradigm. On this basis, therefore, we will now review the genetic approaches that have been taken to investigate virulence in malaria.

Perhaps a first point of all to note is that genetic studies of virulence, even within the clinical paradigm, are virtually prohibitive in a context of human malaria parasites themselves. Only within a practically manageable and ethically acceptable animal host can the complex procedures and observations required for the measurements of clinical outcome and genetic manipulation of the parasites be undertaken. In this respect, the rodent malarias offer the only feasible option for such investigations on mammalian species of malaria parasites. The specific genetic analyses that have been undertaken to study virulence in rodent malaria will be discussed in a later section. First, we will review the principles and practice of undertaking genetic analysis of a rodent malaria parasite.

The Classical Genetic Approach

The earliest such studies were conducted in the late 1960s when a large collection of isolates of rodent malaria parasites was established at the University of Edinburgh (Beale et al., 1978). This collection was instrumental in the clarification of the characteristics and identities of the species and subspecies of these parasites as indicated above. Their primary purpose, however, had been as a source of suitable material for genetic analyses upon these parasites. A number of phenotypes or traits were identified and used in genetic analysis. These included drug resistance and sensitivity (Rosario, 1976), growth rate (virulence) differences (Walliker et al., 1976), and, much later, differences in immunogenicity (strain-specific protective immunity) (Martinelli et al., 2005) as well as genotype-specific differences in electrophoretic mobility of enzymes (Carter, 1978). In all the classical genetic studies on rodent malaria parasites, the experimental process is essentially the same.

Two cloned (genotypically pure), and genetically and phenotypically distinct for the trait(s) of interest, parental lines are crossed by coinfecting mice with both lines. Anopheles mosquitoes are allowed to feed upon the mixed-line infection at a stage in the infection when the gametocytes are most likely to infect the mosquitoes. This is something that is different for different species of rodent malaria parasite and can also be for different parasite phenotypes within a species. Within the mosquitoes, male and female gametocytes of both parental lines emerge from their host RBCs as male and female gametes. These cross-fertilize, leading to both hybrid as well as parental selfed genotypes. The resulting zygotes develop within the mosquitoes, eventually forming sporozoites in the salivary glands, each zygote having undergone meiosis with recombination between parental genomes and the production of haploid recombinant progeny as represented in the sporozoites.

At maturation of the mosquito infections, when the gland sporozoites are infectious, these are introduced into mice by mosquito bite or inoculation of the salivary glands. The resulting infections, when they reach the blood, contain numerous haploid, recombinant, as well as parental genotype parasites. These are cloned by dilution and each progeny clone is tested for the phenotypes of interest as well being characterized by other neutral genetic markers such as electrophoretic mobility differences in enzyme proteins, by which the parental parasites were distinguished.

The findings from such analyses indicate whether a phenotype, be it growth rate, drug resistance, or other, is inherited among the haploid recombinant parasites as a single trait with its properties always the same as in the parental lines, or whether more complex outcomes are found. The former outcome could be interpreted to indicate that a trait is determined at a single genetic locus; the latter that its genetic basis is more complex. With further genetic analysis using larger numbers of molecular markers and extensive cloning of cross progeny, this approach combined with quantitative trait locus (QTL) analyses can locate genes for parasite phenotypes to specific chromosomes and to approximate locations within them.

Postgenomic Approaches

Today, in the postgenomic era, classical genetic analyses of malaria parasites can be very powerful and, with genome saturation densities of genetic markers combined with QTL analysis, can identify a specific gene locus for involvement in a specific parasite trait or phenotype. All such investigations carry, however, the necessity to generate large numbers of cloned recombinant lines, running into the many hundreds, from each genetic cross between specific parental lines. This aspect of the process remains very expensive, time-consuming, and costly to animal use in the case of studies with rodent malaria parasites. This difficulty has now been largely overcome by a streamlined postgenomic approach called linkage group selection (LGS) (Carter et al., 2007; Culleton et al., 2005; Hunt et al., 2007; Martinelli et al., 2005; Pattaradilokrat et al., 2007).

The goal of the LGS approach to genetic analysis is to locate as near as possible the position of the gene or genes determining a specific parasite phenotype and to do this with the minimum number of experimental manipulations. As just noted, the traditional approach to the genetic analysis of malaria parasites involves the usually laborious preparation of numerous clones from the recombinant progeny. Each of these must be independently characterized for all genetic markers and traits involved in the particular analysis. In LGS analysis, on the other hand, no cloning is necessary as the crucial operation is performed in a single step upon the entire uncloned progeny of the cross. The details are, briefly, as follows.

Following the introduction of the recombinant sporozoite progeny of a genetic cross between two parental lines of malaria parasite, the parasites are allowed to grow in the vertebrate host (a mouse in the present context) under the influence of the phenotype of interest. If this is simply a growth rate difference, those recombinant parasites that have inherited the alleles for fast growth rate at the controlling loci will outgrow those that have inherited the alleles for slow growth. After an appropriate period of growth in the blood, the parasites are harvested and prepared for a molecular genetic analysis. This analysis involves screening for markers distributed across the entire parasite genome, the markers being in sufficient number and distributed as evenly as possible through the genome to allow for any region that is genetically linked to genes that determine the faster growth rate to be detected. Markers of the slower-growing parent that are linked to loci controlling the phenotype will be represented less and less in the progeny the more closely they are genetically linked to these loci. At the controlling loci, the slow growth-determining alleles will tend to become entirely replaced by the alleles for rapid growth, thus allowing their identification within the genome.

THE GENETIC ANALYSIS OF VIRULENCE IN RODENT MALARIA PARASITES

For reasons already discussed, the rodent malarias are perhaps the only practical system in which it is possible to investigate the genetic basis of virulence in malaria. Virulence differences among malaria parasites have been described in several different contexts.

Virulence is defined here in terms of pathogenicity to the host; some parasites cause a greater degree of clinical harm than other parasites, and this may be independent of growth rates. This is exemplified, in the rodent malarias, by the case of *P. berghei*, in which one strain, ANKA, causes cerebral malaria-like disease in some strains of laboratory mice resulting in death 6–8 days postinoculation; other strains of the same species, while having comparable growth rates, do not cause this particular pathology (Neill and Hunt, 1992). The phenotype has no association with the parasites' growth rate in the blood and its genetic basis has not been investigated.

Differences in parasite virulence (anemia and body weight) during infections of *P. chabaudi* have been investigated. These effects are directly associated with parasite growth rate in the blood. The phenotypes have been studied especially within the evolutionary paradigm and in relation to the interaction between virulence/growth rate and fitness/transmissibility through mosquitoes (De Roode et al., 2003, 2005; Gadsby et al., 2009). There are no published studies on the genetic bases of these virulence/growth rate differences. However, using the LGS approach, N. Gadsby (2008) has shown that the growth rate differences between two lines of *Plasmodium chabaudi adami*—differences that appear to have been present in these two lines from the time of their isolation from a wild-caught thicket rat (Carter and Walliker, 1976) and cannot, therefore, be considered to be a laboratory artifact or mutation—are determined at loci across a large section of *P. c. adami* chromosome 7 and also on chromosome 9. The growth rate/virulence phenotype, therefore, is controlled at several different loci in the parasite's genome in this example.

The most fully studied and understood virulence/growth rate phenotype in the rodent malarias is that of a line, or "stock," of *Plasmodium yoelii yoelii* that appears to have arisen spontaneously after its isolation into the laboratory. As in *P. berghei*, the blood-stage parasites of *P. y. yoelii* are classically restricted to very young RBCs (reticulocytes), being unable, apparently, to invade older erythrocytes (normocytes). This restricted tropism leads to the *P. y. yoelii* infection course illustrated in Figure 14.2. The parasitemia slowly increases from the day of inoculation, never reaching more than 10% by day 9, after which it climbs steadily in tandem with increasing reticulocytemia and can reach up to 50% by about the 20th day of infection. Thereafter, the parasitemia drops, presumably in response to host immune factors, is cleared by around days 30–35, with the animal subsequently refractory to challenge with the same strain, although they are generally *not* protected against other genetically distinct strains.

A dramatic deviation from this phenotype arose, apparently independently, at least three times in the laboratory from a single isolate of *P. y. yoelii* designated 17X (Pattaradilokrat et al., 2008). This stock of the parasites, derived by subinoculation of blood from a single wild-caught *Thamnomys rutilans* captured near Bangui in the Central African Republic in 1965 (Killick-Kendrick and Peters, 1978) was maintained by blood passage in laboratory mice and intermittent storage in liquid nitrogen for several years. During this time, it maintained the characteristic slow-growing reticulocyte-inhabiting "mild" phenotype, before the first appearance of the fast-growing, normocyte-inhabiting, and rapidly lethal virulent parasites. The best documented of these occurred in the laboratory of Meir Yoeli in New York University in 1971(Yoeli et al., 1975). A stabilate of the uncloned *P. y. yoelii* line "17X," previously phenotyped as a typical *P. y. yoelii* reticulocyte-restricted parasite, was removed from the deep freeze and inoculated into laboratory mice. Following a number of days of replication in the blood of the animal, it became apparent that this parasite line was no longer restricted to reticulocytes but could invade and survive in a much larger repertoire of RBCs. This new invasion phenotype manifested itself typically only on and after day 4 postinoculation. Up until this time, the growth resembled that of the typical *P. y. yoelii* parasite, being restricted to development within reticulocytes. At day 4 postinoculation, however, at the time when availability of noninvaded reticulocytes becomes scarce, the parasite was able to invade normocytes. Within the subsequent 4 days, the infection would achieve parasitemias in excess of 80%, killing the host by day 8 postinoculation.

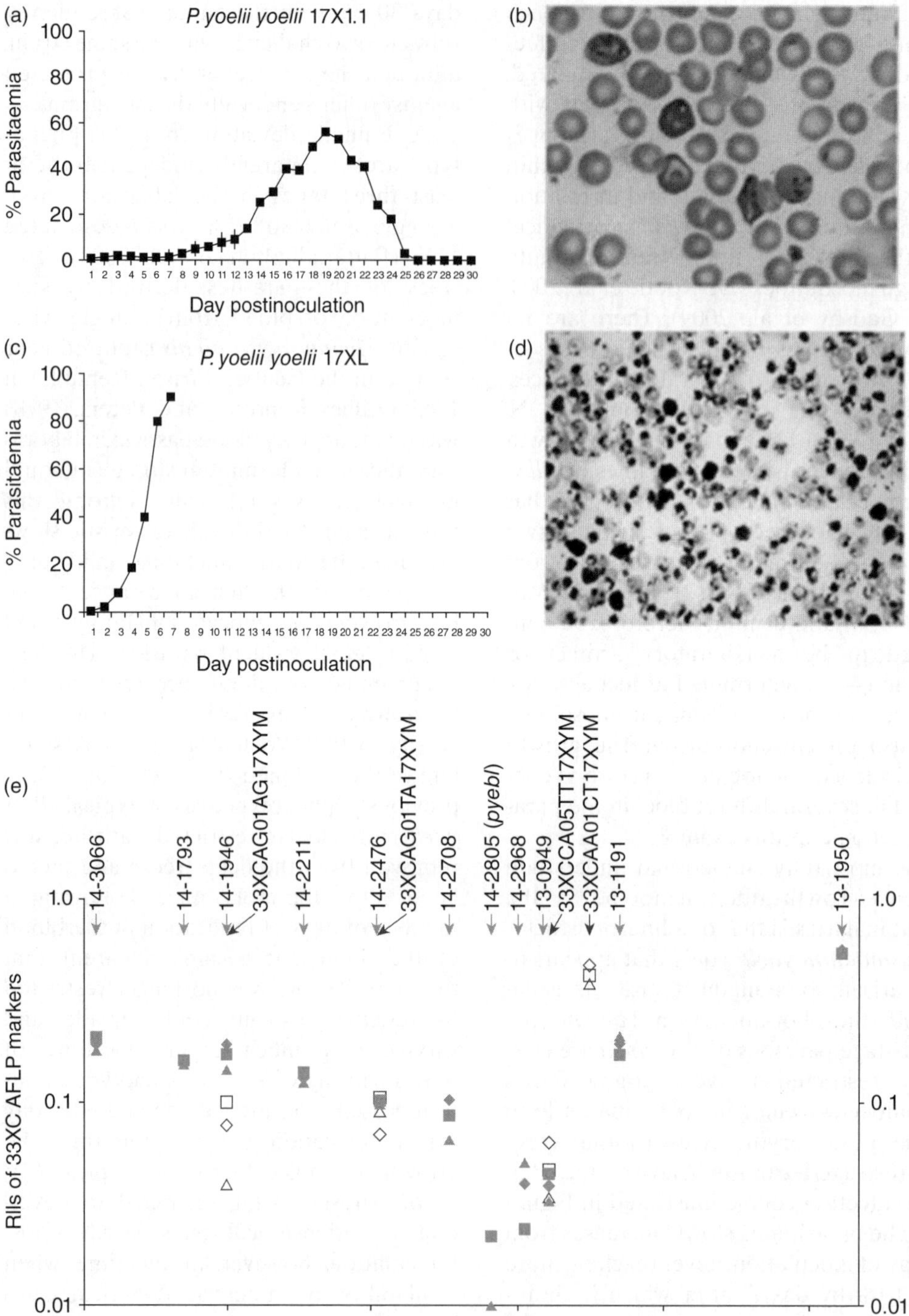
(a)
P. yoelii yoelii 17X1.1
% Parasitaemia
100
80
60
40
20
0
1 2 3 4 5 6 7 8 9 10 11 12 13 14 15 16 17 18 19 20 21 22 23 24 25 26 27 28 29 30
Day postinoculation
(b)
(c)
P. yoelii yoelii 17XL
% Parasitaemia
100
80
60
40
20
0
1 2 3 4 5 6 7 8 9 10 11 12 13 14 15 16 17 18 19 20 21 22 23 24 25 26 27 28 29 30
Day postinoculation
(d)
(e)
14-3066
14-1793
14-1946
33XCAG01AG17XYM
14-2211
14-2476
33XCAG01TA17XYM
14-2708
14-2805 (pyebl)
14-2968
13-0949
33XCCA05TT17XYM
33XCAA01CT17XYM
13-1191
13-1950
RIIs of 33XC AFLP markers
1.0
0.1
0.01
Proportion of parasites carrying the 33XC allele by Pyrosequencing™
0
500
1000
1500
2000
2500
Estimated physical distance along P. yoelii yoelii chromosome 13 (kb)

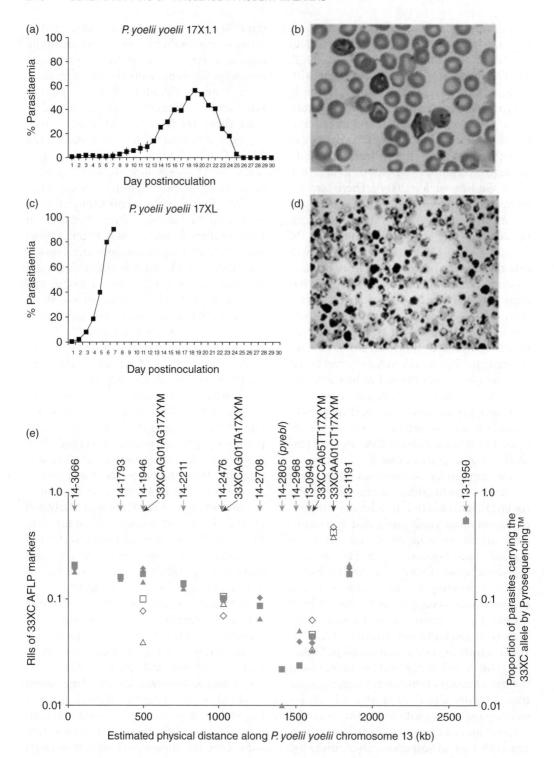

Figure 14.2 Parasitemia of *Plasmodium yoelii yoelii* 17X1.1 (a) and 17XL (c) in female CBA mice, and thin blood smears showing host cell preference at day 5 postinoculation, (b) and (d), respectively. Initial inocula were 1×10^6 parasites per mouse. Error bars show standard error of the mean for five mice. Data are representative of five repeat experiments. (e) Proportions of parasites carrying alleles on chromosome 13 of *P. y. yoelii* 33XC in the growth selected progeny of a genetic cross between 17XYM (phenotypically identical to 17XL) and 33XC. The relative intensity indices of four 33XC AFLP markers are represented by open green symbols. The proportions of parasites with the 33XC allele at 11 defined Pyrosequencing loci are represented by filled red symbols. Data obtained from the cross progeny after the first, second, and third rounds of multiplication rate selection in mice are represented by squares, diamonds, and triangles, respectively. Arrows indicate predicted locations of AFLP and Pyrosequencing markers along *P. y. yoelii* chromosome 13. Extrapolation to *P. y. yoelii* chromosome 13 is from synteny between the human malaria parasite *Plasmodium falciparum* and the rodent malaria parasite *P. y. yoelii*. *pyebl* stands for the gene encoding the erythrocyte-binding ligand. From Pattaradilokrat et al. (2009). See color insert.

The fact that this dramatic change in phenotype occurred in a spontaneous manner and remained stable following multiple passages through mice and through mosquitoes suggested that it was controlled by, and had arisen as a consequence of, a genetic mutation. Other possible explanations for the increased virulence were quickly discounted; contamination with viruses and bacteria was tested for and ruled out, and enzyme marker analysis following cloning confirmed that the line was, indeed, *P. y. yoelii*.

This parasite was subsequently cloned, named YM (an initialism of Yoeli's mouse), and subjected to classical genetic linkage analysis in which it was crossed with a genetically unrelated and characteristically slow-growing (avirulent) clone of *P. y. yoelii*, and the progeny was cloned and phenotyped. It was found that the fast-growing/virulent phenotype segregated independently from various enzyme markers and a marker for resistance to the antimalarial drug pyrimethamine, with which the parental parasites could be distinguished. Apparently inherited in a simple Mendelian fashion, it was postulated that the growth rate phenotype was genetically controlled, and very probably by a single gene, as all except one of 34 recombinant progeny clones characterized were phenotypically indistinguishable from either the avirulent or virulent parents (Walliker et al., 1976). Nevertheless, the existence of even one recombinant clone with an intermediate growth rate/virulence phenotype raised a question as to the possibility of the involvement of more than one genetic locus.

No further genetic characterization of the virulent line of *P. y. yoelii* 17X was carried out until very recently, when two complimentary studies, one using LGS (Pattaradilokrat et al., 2009) and the other replacement transfection technology (Otsuki et al., 2009), identified a single-nucleotide polymorphism (SNP) in the gene encoding the *P. yoelii* erythrocyte-binding ligand (PyEBL) protein as the major determinant of the virulence difference between YM and the avirulent *P. y. yoelii* strains.

LGS, as already discussed, involves the production of a genetic cross between two parasites that differ in some selectable aspect of their phenotypes, in this case their growth rates. If two parasite clones with differing growth rates are allowed to grow simultaneously in the blood of one host animal, then, after a certain number of cycles of schizogony, dependent on the strength of the growth differential between them, the faster-growing clone will come to dominate the infection. If the growth rate differences are extreme, as with YM and the avirulent *P. y. yoelii* parasites, then complete domination by the faster-growing parasite and the exclusion from the infection of the slower grower may be achieved following relatively few replication cycles.

If a single gene controls the growth rate difference between the slow- and the fast-growing parasites, then half of the recombinant progeny produced following crossing of the parental clones will inherit the slow-growing allele and half the fast-growing allele. If this uncloned recombinant progeny is then allowed to complete the necessary number of blood-stage schizogonic cycles required to exclude the slower-growing parasites from the infection, then the resulting infection should be composed of parasites that all carry the fast-growing parental allele of the gene that controls growth rate. Beyond the region in genetic linkage with the selected locus, the genome should be represented by equal proportions of alleles from both parents. This growth rate-selected parasite population can then be genotyped with a large number of genome-wide quantitative genetic markers that differentiate between the two parental clones. The genomic region that contains the gene that encodes the relevant protein is identified by the reduced proportion of markers from the slow-growing parent.

Such an experiment was conducted by Pattaradilokrat and colleagues at Edinburgh University in 2009 (Pattaradilokrat et al., 2009). They used the cloned parental lines YM and 33X (a genetically distinct, slow-growing *P. y. yoelii* clone), and genome-wide amplified fragment length polymorphism (AFLP) markers to analyze the uncloned recombinant progeny population at loci throughout the genome. Analysis of the growth rate-selected progeny revealed strong selection against four AFLP markers of the slow-growing parent that mapped to chromosome 13. Further analysis of this locus using 11 specifically designed quantitative Pyrosequencing markers (Cheesman et al., 2007) revealed the presence of a selection valley, the bottom of which located within a region of about 500 kb on chromosome 13 (Fig. 14.2).

Independent of this work, Otsuki and colleagues at Ehime University in Japan had sequenced the *P. y. yoelii* ortholog of the *Plasmodium* spp. Duffy binding-like (DBL)/erythrocyte binding protein-like (EBL) gene. This gene, while existing as a multigene family in *P. falciparum* (members of which include EBL, EBA-175, BAEBL, MAEBL, and JSEBL among others), appears to be single copy in the rodent malaria parasites including in *P. y. yoelii* clones YM, 17XL (an apparently YM isogenic line of 17X, which is also virulent; see Fig. 14.2), and 17X (the nonvirulent isogenic line (Pattaradilokrat et al., 2008).

Plasmodium *EBL* genes have been shown to control erythrocyte invasion in other malaria parasite species, most notably in the human parasite *Plasmodium vivax*, in which the EBL member *P. vivax* Duffy binding protein (PvDBP) recognizes and binds to the Duffy antigen receptor for chemokines (DARC) on the surface of human RBCs. Individuals who lack the expression of this receptor on the surface of their RBCs (the "Duffy negative" phenotype, found almost exclusively and at very high frequency among human populations native to West and Central Africa) are refractory to infection with *P. vivax*. Given this, Otsuki and colleagues speculated that there might be a role for this protein in controlling the differences in erythrocyte tropism between the virulent and avirulent *P. y. yoelii* parasites.

They found a SNP in region 6 of the *pyebl* gene between 17X, the avirulent parasite, and 17XL and YM, the isogenic virulent parasites. Region 6 of *Plasmodium* EBL proteins is thought to be involved in the cellular trafficking of the protein to the micronemes (Gilberger et al., 2003; Treeck et al., 2006); its tertiary structure is maintained by four disulfide bonds between pairs of eight conserved cysteine residues (Fig. 14.3). The

SNP identified between the virulent and avirulent parasites causes a replacement in the amino acid sequence of the protein so that one of the conserved cysteine residues is replaced by an arginine, presumably disrupting one of the disulfide bonds and altering the tertiary structure of the region.

Immunoelectron micrography revealed that, in the virulent 17XL parasites, the protein is not trafficked to the micronemes as it is in the avirulent parasites but is located in the dense granuoles. These observations support the theory that region 6 of EBL is required for its trafficking to the micronemes. This being the case, it follows that the removal of the EBL protein from the micronemes actually *increases* the repertoire of erythrocytes that the parasite can invade.

Upon communication of this data to Pattaradilokrat and colleagues, it was found that the *pyebl* gene marked the bottom of their selection valley on chromosome 13, providing strong evidence for the involvement of this gene in controlling the growth rate/virulence phenotype.

Replacement transfection experiments were subsequently carried out by Otsuki and colleagues in which the EBL allele of the virulent clone was replaced with that of the avirulent clone, following which the phenotype of the virulent parasites reverted to that of the avirulent type. However, when the EBL of the avirulent parasite was replaced with that of the virulent parasite, it changed not to a fast growth rate/virulent but to an "intermediate" growth rate phenotype. This intermediate growth rate was characterized by an ability to invade a larger repertoire of erythrocytes than just reticulocytes. This resulted in parasitemias reaching around 10% by day 5 postinoculation (in contrast to the 2–3% parasitemia of the avirulent parasite at the same time point) and in the survival of the host. It was shown that in these parasites, the PyEBL protein was not trafficked to the micronemes as in the virulent clones, suggesting that another, as yet unidentified, gene is involved in controlling the growth rate/virulence differences between lines of *P. y. yoelii*. This is consistent with the LGS analysis of Pattaradilokrat et al. (2009), which indicated another locus, possibly on chromosome 5 or 6, under growth rate selection, and with the findings of the original classical genetic analysis of these parasites (Walliker et al., 1976).

How the trafficking of a normally microneme-targeted protein to the dense granules can increase the repertoire of erythrocyte types susceptible to invasion by the parasites is not currently understood. One possible explanation is that, in the absence of a functioning *pyebl* gene, another less specific invasion ligand takes on its role. However, attempts to knock out the *pyebl* gene were unsuccessful (Otsuki et al., 2009), suggesting that the protein is essential for parasite growth; it is, of course, possible that the *pyebl* gene also plays a role other than that of an invasion ligand during the parasite's blood-stage growth.

An alternative explanation is that the PyEBL protein in the dense granules remains functional and that its removal from the micronemes creates free space at the tight junction formed between the invading merozoite and the host erythrocyte. This space may be filled by another ligand, possibly one of the py235 reticulocyte binding ligand (RBL) genes (Iyer et al., 2007a; see below), in a manner similar to that proposed for the erythrocyte receptor invasion switch observed in *P. falciparum* upon disruption of PfEBA-175 (a *P. falciparum* ortholog of *pyebl*) (Duraisingh et al., 2003). The exact molecular mechanism accounting for the virulence differences between the *P. y. yoelii* parasites, while almost certainly involving the alternative trafficking of the PyEBL protein from the micronemes to the dense granules, is currently unresolved.

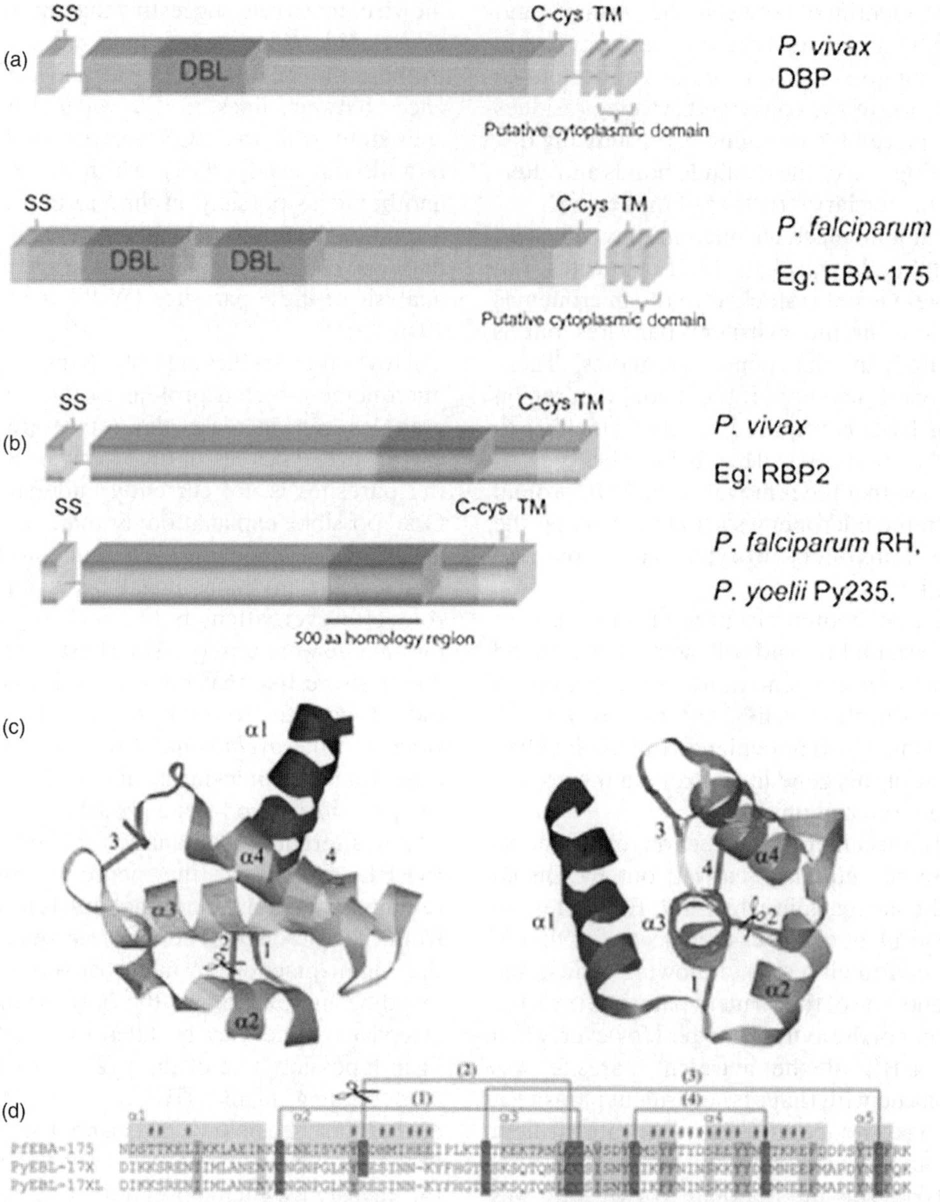

(a)
SS
DBL
C-cys TM
Putative cytoplasmic domain
P. vivax
DBP
SS
DBL
DBL
C-cys TM
Putative cytoplasmic domain
P. falciparum
Eg: EBA-175
(b)
SS
C-cys TM
P. vivax
Eg: RBP2
SS
C-cys TM
P. falciparum RH,
P. yoelii Py235.
500 aa homology region
(c)
α1
α2
α3
α4
1
2
3
4
(d)
(1)
(2)
(3)
(4)
α1
α2
α3
α4
α5
PfEBA-175
PyEBL-17X
PyEBL-17XL
PfJESEBL
PfBAEBL
PvDBP
PcyEBP
PkDBP-α
PkDBP-β
PkDBP-γ
PbEBL
PchEBL
PviEBL
PfMAEBL

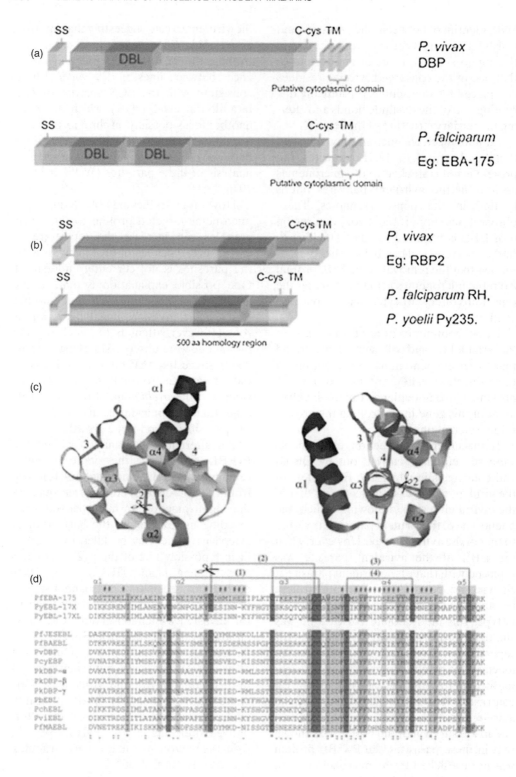

Figure 14.3 Schematic diagram of the EBLs and RBLs. Sequence comparisons between members showed high overall conservation denoted by DBL domains between the EBLs (a) and orange regions in the RBLs (b). Regions of variability are interspersed throughout the coding region. Conserved cysteine-rich (Cys-c) regions and the putative transmembrane (TM) regions are indicated. (c) Predicted structure of PyEBL region 6 (R6) from Iyer et al. (2007b). Four predicted disulfide bridges are shown by red bars; central α3 helices are surrounded by other helices (α1, α2, and α4). Note that α1 may not be related to other elements. (d) Amino acid sequence comparison of *Plasmodium* EBL R6. PfEBA-175 PfJESEBL, PfBAEBL, *Plasmodium falciparum* EBLs; PvDBP, *P. vivax* Duffy binding protein; PcyEBP, *Plasmodium cynomolgi* EBP; PkDBP, *Plasmodium knowlesi* Duffy binding proteins; PyEBL-17X and -17XL, *Plasmodium yoelii* EBL; PbEBL, *P. berghei* EBL (ANKA line) PchEBL, *P. chabaudi* EBL; PvEBL, *P. vinckei* EBL. In the PfEBA-175 sequence, helices are highlighted in gray; residues involved in protein-fold stabilization are highlighted in cyan and yellow for nonpolar and polar residues, respectively; all Cys residues are in red; and residues involved in the dimeric interaction are indicated by hash marks. Asterisks indicate the positions where amino acids are identical, and similar amino acids are indicated with colons or periods under the alignment. Disulfide bridges (red lines) are represented above the alignment. Scissors indicate the disulfide bridge that is abolished by substitution from Cys to Arg in the *P. y. yoelii* 17XL line. From Culleton and Kaneko (2010). See color insert.

THE RETICULOCYTE BINDING-LIKE PROTEINS (RBLs) OF *P. Y. YOELII*

The EBL proteins of *Plasmodia*, being type 1 integral transmembrane proteins, are characterized by two interspecies conserved cysteine-rich domains and a C-terminal transmembrane region. The cysteine-rich domain nearest the N-terminus, termed region 2, contains one (or sometimes two) DBL domains, which are thought to recognize receptors on the host erythrocyte surface (Adams et al., 1992; Chitnis and Miller, 1994; Sim et al., 1994) (Fig. 14.3). The other cysteine-rich domain, located just upstream of the transmembrane region and termed region 6, is apparently responsible for the trafficking of the protein to the micronemes, as described above.

Another major class of malaria parasite invasion proteins is the reticulocyte binding-like protein (RBL) superfamily family. These proteins, which were first described in *P. y. yoelii* in which they are referred to as the Py235 proteins, lack a DBL domain but also contain a cysteine-rich domain just downstream of a C-terminal transmembrane region (Fig. 14.3). RBL proteins are found in many species of plasmodia, including *P. falciparum* (the rhoptry homolog [PfRH] proteins), *P. vivax* (RBP 1 and 2), *Plasmodium cynomolgi* (also called RBP 1 and 2), *Plasmodium reichenowi*, *P. chabaudi*, and *P. berghei* (reviewed in Iyer et al., 2007a).

There are around 14 copies of the *Py235* genes in the *P. y. yoelii* genome (Carlton et al., 2002), and it has been proposed that differences in expression of these genes may be associated with differences in multiplication rates between lines of *P. y. yoelii* (Gruner et al., 2004; Iyer et al., 2007b). Furthermore, it was shown that either immunization with Py235 proteins or passive transfer of monoclonal antibodies specific to the protein was able to restrict the invasion of the normally lethal YM parasite to reticulocytes (Freeman et al., 1980; Holder and Freeman, 1981). Unlike the RH genes of *P. falciparum*, which are under epigenic control, all the members of the *Py235* gene family appear to be expressed simultaneously, and disruption of one gene does not alter the expression of the other members or lead to a change in invasion phenotype (Ogun et al., 2011). This suggests some functional redundancy in this family of proteins, which may have evolved as a way for the parasite to circumvent the action of inhibitory antibodies targeting these proteins.

The original classical genetic analysis of Walliker et al. (1976), as well as the LGS analysis of Pattaradilokrat and colleagues

(2009) and the replacement transfection studies of Otsuki and colleagues (2009), hinted at the role of a secondary genetic mediator of the growth rate differences between the virulent and avirulent lines of *P. y. yoelli*. In the study by Pattaradilokrat and colleagues, this was indicated by the presence of selection on loci other than on chromosome 13, and in that of Otsuki and colleagues by the fact that an intermediate phenotype was observed following the replacement of the EBL gene of the avirulent parasite with that of the virulent parasite. It was speculated by both groups that the second gene might be one of the members of the *Py235* family.

HOST FACTORS THAT MODULATE THE VIRULENCE OF *P. Y. YOELII* PARASITES

As well as parasite genetic factors, it appears that there is a role for the phenotype of the host animal in determining whether an infection with a given line of *P. y. yoelii* will be virulent or avirulent in nature. The growth of a 17X line that achieved intermediate virulence in BALB/c mice was shown to have a significantly altered course of infection in genetically modified mice that do not express DARC on the surface of their erythrocytes (Swardson-Olver et al., 2002). While apparently able to invade both reticulocytes and normocytes in wild-type mice, the parasites were unable to invade the normocytes of Duffy negative mice, suggesting that different receptors are used by *P. y. yoelii* to invade the two cell types. Furthermore, it appears that the immune status of the host may also affect the growth rate of *P. y. yoelii*. Two reports show that, in immunocompromised animals, the normally avirulent and reticulocyte-restricted lines of 17X are able to invade normocytes, achieve parasitemia levels akin to those of the virulent lines, and become lethal to the host (Fahey and Spitalny, 1984; Hisaeda et al., 2004). These observations complicate the simple invasion ligand-mediated growth rate story, and further experimentation will be necessary to understand how the two observations may be connected, if at all.

SUMMARY

We have discussed the nature of virulence in malaria under the definition that it is the amount of harm that a particular genetic stock of the parasites causes its host. We have noted that virulence defined in this way is amenable to measurement in the context of a clinical paradigm and also to investigation by experimental means including genetic analysis. We have also pointed out that there are at least several forms of clinical manifestation of virulence, so defined, some of which may be associated with growth rate and others not. The best-studied example of virulence in a malaria parasite is the case of laboratory mutant lines of the rodent malaria parasite *P. y. yoelii*. These parasites grow rapidly and are rapidly lethal to their rodent hosts. Genetic and molecular studies have shown that a single point mutation in the gene for the PyEBL protein is largely, but not entirely, responsible for the dramatic change in growth rate/virulence in these parasites.

REFERENCES

Adams JH, Sim BK, Dolan SA, Fang X, Kaslow DC, Miller LH. 1992. A family of erythrocyte binding proteins of malaria parasites. Proc Natl Acad Sci U S A 89: 7085–7089.

Beale GH, Carter R, Walliker D. 1978. Genetics. Chapter 5 in: Killick-Kendrick R, Peters W, eds. *Rodent Malaria*. Academic Press, London, New York, San Francisco.

Carlton JM, Angiuoli SV, Suh BB, Kooij TW, Pertea M, Silva JC, Ermolaeva MD, Allen JE, Selengut JD, Koo HL, Peterson JD, Pop M,

Kosack DS, Shumway MF, Bidwell SL, Shallom SJ, Van Aken SE, Riedmuller SB, Feldblyum TV, Cho JK, Quackenbush J, Sedegah M, Shoaibi A, Cummings LM, Florens L, Yates JR, Raine JD, Sinden RE, Harris MA, Cunningham DA, Preiser PR, Bergman LW, Vaidya AB, Van Lin LH, Janse CJ, Waters AP, Smith HO, White OR, Salzberg SL, Venter JC, Fraser CM, Hoffman SL, Gardner MJ, Carucci DJ. 2002. Genome sequence and comparative analysis of the model rodent malaria parasite *Plasmodium yoelii yoelii*. Nature 419: 512–519.

Carter R. 1978. Studies on enzyme variation in the murine malaria parasites *Plasmodium berghei*, *P. yoelii*, *P. vinckei* and *P. chabaudi* by starch gel electrophoresis. Parasitology 76: 241–267.

Carter R, Walliker D. 1976. Malaria parasites of rodents of the Congo (Brazzaville): *Plasmodium chabaudi adami* subsp. nov. and *Plasmodium vinckei lentum* Landau, Michel, Adam and Boulard, 1970. Ann Parasitol Hum Comp 51: 637–646.

Carter R, Hunt P, Cheesman S. 2007. Linkage group selection—A fast approach to the genetic analysis of malaria parasites. Int J Parasitol 37: 285–293.

Cheesman S, Creasey A, Degnan K, Kooij T, Afonso A, Cravo P, Carter R, Hunt P. 2007. Validation of Pyrosequencing for accurate and high throughput estimation of allele frequencies in malaria parasites. Mol Biochem Parasitol 152: 213–219.

Chitnis CE, Miller LH. 1994. Identification of the erythrocyte binding domains of *Plasmodium vivax* and *Plasmodium knowlesi* proteins involved in erythrocyte invasion. J Exp Med 180: 497–506.

Culleton R, Kaneko O. 2010. Erythrocyte binding ligands in malaria parasites: Intracellular trafficking and parasite virulence. Acta Trop 114(3): 131–137.

Culleton R, Martinelli A, Hunt P, Carter R. 2005. Linkage group selection: Rapid gene discovery in malaria parasites. Genome Res 15: 92–97.

De Roode JC, Read AF, Chan BH, Mackinnon MJ. 2003. Rodent malaria parasites suffer from the presence of conspecific clones in three-clone *Plasmodium chabaudi* infections. Parasitology 127: 411–418.

De Roode JC, Pansini R, Cheesman SJ, Helinski ME, Huijben S, Wargo AR, Bell AS, Chan BH, Walliker D, Read AF. 2005. Virulence and competitive ability in genetically diverse malaria infections. Proc Natl Acad Sci U S A 102: 7624–7628.

Duraisingh MT, Maier AG, Triglia T, Cowman AF. 2003. Erythrocyte-binding antigen 175 mediates invasion in *Plasmodium falciparum* utilizing sialic acid-dependent and -independent pathways. Proc Natl Acad Sci U S A 100: 4796–4801.

Fahey JR, Spitalny GL. 1984. Virulent and nonvirulent forms of *Plasmodium yoelii* are not restricted to growth within a single erythrocyte type. Infect Immun 44: 151–156.

Freeman RR, Trejdosiewicz AJ, Cross GA. 1980. Protective monoclonal antibodies recognising stage-specific merozoite antigens of a rodent malaria parasite. Nature 284: 366–368.

Gadsby, JN. 2008. A genetic analysis of two strains of *Plasmodium chabaudi adami* that differ in growth and pathogenicity. PhD thesis, University of Edinburgh.

Gadsby N, Lawrence R, Carter R. 2009. A study on pathogenicity and mosquito transmission success in the rodent malaria parasite *Plasmodium chabaudi adami*. Int J Parasitol 39: 347–354.

Garnham PCC. 1966. *Malaria Parasites and Other Haemosporidia*. Blackwell Scientific Publications, Oxford.

Gilberger TW, Thompson JK, Reed MB, Good RT, Cowman AF. 2003. The cytoplasmic domain of the *Plasmodium falciparum* ligand EBA-175 is essential for invasion but not protein trafficking. J Cell Biol 162: 317–327.

Gruner AC, Snounou G, Fuller K, Jarra W, Renia L, Preiser PR. 2004. The Py235 proteins: Glimpses into the versatility of a malaria multigene family. Microbes Infect 6: 864–873.

Hisaeda H, Maekawa Y, Iwakawa D, Okada H, Himeno K, Kishihara K, Tsukumo S, Yasutomo K. 2004. Escape of malaria parasites from host immunity requires CD4+ CD25+ regulatory T cells. Nat Med 10: 29–30.

Holder AA, Freeman RR. 1981. Immunization against blood-stage rodent malaria using purified parasite antigens. Nature 294: 361–364.

Hunt P, Afonso A, Creasey A, Culleton R, Sidhu AB, Logan J, Valderramos SG, Mcnae I, Cheesman S, Do Rosario V, Carter R, Fidock DA, Cravo P. 2007. Gene encoding a deubiquitinating enzyme is mutated in artesunate- and chloroquine-resistant rodent malaria parasites. Mol Microbiol 65: 27–40.

Iyer J, Gruner AC, Renia L, Snounou G, Preiser PR. 2007a. Invasion of host cells by malaria parasites: A tale of two protein families. Mol Microbiol 65: 231–249.

Iyer JK, Amaladoss A, Genesan S, Preiser PR. 2007b. Variable expression of the 235 kDa rhoptry protein of *Plasmodium yoelii* mediate host cell adaptation and immune evasion. Mol Microbiol 65: 333–346.

Killick-Kendrick R, Peters W, eds. 1978. *Rodent Malaria*. Academic Press, London.

Martinelli A, Cheesman S, Hunt P, Culleton R, Raza A, Mackinnon M, Carter R. 2005. A genetic approach to the de novo identification of targets of strain-specific immunity in malaria parasites. Proc Natl Acad Sci U S A 102: 814–819.

Neill AL, Hunt NH. 1992. Pathology of fatal and resolving *Plasmodium berghei* cerebral malaria in mice. Parasitology 105(Pt 2): 165–175.

Ogun SA, Tewari R, Otto TD, Howell SA, Knuepfer E, Cunningham DA, Xu Z, Pain A, Holder AA. 2011. Targeted disruption of py235ebp-1: Invasion of erythrocytes by *Plasmodium yoelii* using an alternative Py235 erythrocyte binding protein. PLoS Pathog 7: e1001288.

Otsuki H, Kaneko O, Thongkukiatkul A, Tachibana M, Iriko H, Takeo S, Tsuboi T, Torii M. 2009. Single amino acid substitution in *Plasmodium yoelii* erythrocyte ligand determines its localization and controls parasite virulence. Proc Natl Acad Sci U S A 106: 7167–7172.

Pattaradilokrat S, Cheesman SJ, Carter R. 2007. Linkage group selection: Towards identifying genes controlling strain specific protective immunity in malaria. PLoS One 2: e857.

Pattaradilokrat S, Cheesman SJ, Carter R. 2008. Congenicity and genetic polymorphism in cloned lines derived from a single isolate of a rodent malaria parasite. Mol Biochem Parasitol 157: 244–247.

Pattaradilokrat S, Culleton R, Cheesman S, Carter R. 2009. Gene encoding erythrocyte binding ligand linked to blood stage multiplication rate phenotype in *Plasmodium yoelii yoelii*. Proc Natl Acad Sci U S A 106: 7161–7166.

Perkins SL, Sarkar IN, Carter R. 2007. The phylogeny of rodent malaria parasites: Simultaneous analysis across three genomes. Infect Genet Evol 7: 74–83.

Rosario VE. 1976. Genetics of chloroquine resistance in malaria parasites. Nature 261: 585–586.

Sim BK, Chitnis CE, Wasniowska K, Hadley TJ, Miller LH. 1994. Receptor and ligand domains for invasion of erythrocytes by *Plasmodium falciparum*. Science 264: 1941–1944.

Swardson-Olver CJ, Dawson TC, Burnett RC, Peiper SC, Maeda N, Avery AC. 2002. *Plasmodium yoelii* uses the murine Duffy antigen receptor for chemokines as a receptor for normocyte invasion and an alternative receptor for reticulocyte invasion. Blood 99: 2677–2684.

Treeck M, Struck NS, Haase S, Langer C, Herrmann S, Healer J, Cowman AF, Gilberger TW. 2006. A conserved region in the EBL proteins is implicated in microneme targeting of the malaria parasite *Plasmodium falciparum*. J Biol Chem 281: 31995–32003.

Walliker D, Sanderson A, Yoeli M, Hargreaves BJ. 1976. A genetic investigation of virulence in a rodent malaria parasite. Parasitology 72: 183–194.

Yoeli M, Hargreaves B, Carter R, Walliker D. 1975. Sudden increase in virulence in a strain of *Plasmodium berghei yoelii*. Ann Trop Med Parasitol 69: 173–178.

CHAPTER 15

GENETIC MAPPING OF ACUTE VIRULENCE IN *TOXOPLASMA GONDII*

L. DAVID SIBLEY and JOHN C. BOOTHROYD

THE ORGANISM, LIFE CYCLE, AND TRANSMISSION

Toxoplasma gondii is one of the world's most abundant parasites; as a generalist, it infects all types of nucleated cells in a wide variety of warm-blooded mammals and birds (Dubey, 2010). Infection rates of wild and domestic animals vary widely, but seropositivity rates of 10–30% are common (Dubey, 2010), reflecting a high burden of chronic infection. First recognized as an intracellular parasite in the tissues of wild rodents in Tunisia, Africa, by Charles Nicole (Nicolle and Manceaux, 1908) and rabbits in Brazil by Alfonso Splendore (Splendore, 1908) in the early 1900s, the complete life cycle was not delineated until the mid-1970s (Dubey and Frenkel, 1972). We now appreciate that transmission to a wide variety of vertebrate hosts occurs by asexual propagation of tissue stages, which may be ingested by carnivorous feeding and scavenging or by ingestion of oocysts shed in feces following a sexual cycle that occurs exclusively in the intestinal epithelium of cats (Dubey and Frenkel, 1972).

During the acute phase of infection, the rapid replication of haploid stages called tachyzoites is associated with expansion and dissemination within the host. Following strong innate and adaptive immune responses, the parasite differentiates into an intracellular tissue cyst containing slowly replicating bradyzoites, which are also haploid. Tissue cysts form in a variety of organs, skeletal muscle, and the central nervous system (CNS) (Dubey, 2010). Successive rounds of bradyzoite replication, cyst growth, and rupture give rise to long-lived chronic infections. Humans are an accidental host and play no role in natural transmission, other than the risk of congenital transmission when infection occurs during pregnancy (Pfaff et al., 2007). Although human infections are widespread, with seropositivity levels of >50% in many countries, most infections are characterized by a brief acute phase that progresses to a subclinical chronic phase (Joynson and Wreghitt, 2001). Toxoplasmosis poses a significant risk in the immunocompromised, largely due to reactivation of chronic infections that are no longer adequately

Evolution of Virulence in Eukaryotic Microbes, First Edition. Edited by L. David Sibley, Barbara J. Howlett, and Joseph Heitman.

controlled in the absence of effective immune responses (Montoya and Liesenfeld, 2004). The pathogenicity of *T. gondii* is enhanced by its ability to widely disseminate to deep tissues including sites where the immune system operates in a more restricted fashion, such as the CNS and placenta.

A number of key developmental switches occur during the life cycle of *T. gondii*; for example, the transition from acute to chronic infection is associated with a switch from fast-growing tachyzoites to slow-growing bradyzoites, a process that reverses during reactivation. When orally ingested, bradyzoites are infectious to a wide variety of intermediate hosts, a trait distinguishing *T. gondii* from closely related parasites that typically have obligatory life cycles that alternate between intermediate (asexual replication) and definitive (sexual replication) hosts (Dubey, 2010). Such direct oral means of transmission may be responsible for spread through the food chain, hence contributing to the success of *T. gondii* (Su et al., 2003). Bradyzoites are able to survive the acidic pH of the stomach, enter through the intestinal epithelium of the small intestine, and cross this barrier to infect deeper tissues (Dubey, 1997). The ingestion of bradyzoites by members of the cat family triggers sexual development within enterocytes of the small intestine. Fusion of male and female gametes results in a diploid zygote, which undergoes meiosis following shedding of oocysts in the environment. Oocysts shed by infected cats are highly resistant to environmental stress, and long-term survival in the environment enables the infection of a variety of intermediate hosts via oral ingestion. Human infections can also occur as a consequence of oocyst ingestion from contaminated food or water (Jones et al., 2007; Mead et al., 1999; Robben and Sibley, 2004).

The outcome of infection is highly variable in humans and in laboratory mice, although not necessarily for the same reasons. In humans, most infections result in mild symptoms; however, occasionally the outcome is dramatic with severe blindness or even death occurring, even in those who are immunocompetent. Mounting evidence suggests that one of the variables producing these different disease outcomes is the strain of the parasite responsible for infection (Grigg et al., 2001b; Khan et al., 2005a, 2006; Silveira et al., 2001; Vasconcelos-Santos et al., 2009). As discussed in detail below, the situation is more clear in mice that are experimentally infected; different strains of the parasite produce dramatically different diseases, ranging from mild to uniformly fatal (Derouin and Garin, 1991; Gavrilescu and Denkers, 2001; Mordue et al., 2001; Suzuki et al., 1989, 1995).

T. gondii is highly unusual in that it belongs to a genus composed of a single species that infects a wide range of hosts (Dubey, 2010). Population genetic studies have revealed that three major clonal types (referred to as I, II, III or 1, 2, 3) define the majority of strains isolated from animal and human infections in North America and Europe (Sibley and Ajioka, 2008). These clonal lineages arose quite recently as the result of only a few natural meiotic recombinations, which involved unique parental types crossing with an ancestor closely related to the type II lineage (Boyle et al., 2006; Su et al., 2003). As such, the lineages are closely related and share a biallelic pattern at most loci. Crucially, the few genetic differences between the lineages often control important phenotypic differences such as those mentioned above and discussed further below. This highly clonal population structure in Europe and North America is in stark contrast to that found in South America, where strains are more diverse and undergo sexual recombination between strains at a much higher frequency (Khan et al., 2007; Lehmann et al., 2006; Pena et al., 2008). The origins and dynamics of these population patterns are considered more fully in Chapter 12 in this volume. For the purposes of this chapter, we will summarize the marked phenotypic differences between the three lineages and a series of genetic crosses aimed at identifying the genes that contribute to natural differences in virulence in rodents.

MOUSE MODEL: STRENGTHS AND LIMITATIONS

Rodents are natural hosts of *T. gondii*, and consequently, the laboratory mouse and rat have been developed as models to study pathogenesis. Laboratory rats are relatively resistant to infection, although host strain differences have been described and mapped to a single dominant locus in the genome (Cavailles et al., 2006). Differences in parasite strains have not been extensively evaluated in the rat model, perhaps because rats easily survive challenge by all three common clonal types. In comparison, laboratory mice are highly susceptible to many strains of *T. gondii*, and various models have been developed to explore aspects of pathogenesis (Munoz et al., 2011). The extreme susceptibility of laboratory mice to toxoplasmosis may be a result of narrow initial genetic diversity as most current laboratory strains were derived from just a few founder lines of so-called fancy mice, which were first bred as pets and were later adopted as laboratory strains (Gubter and Dhand, 2002). In contrast, wild rodents such as *Peromyscus* spp. in North America are relatively resistant to parasite strains that cause acute mortality in laboratory mice (Frenkel, 1953). House mice (*Mus musculus*), which are globally distributed and also show considerable genetic diversity (Boursot et al., 1993), have not been extensively evaluated for susceptibility to infection but may also express a wider range of susceptibility–resistance phenotypes. Although no longer fully reflective of natural hosts, the enhanced susceptibility of laboratory mice provides a magnifying lens to accentuate differences between *T. gondii* lineages.

The most dramatic strain-dependent phenotype is the marked difference in acute virulence exhibited in outbred mice challenged with representatives of the three clonal genotypes of *T. gondii*. Type I strains are characterized by having extremely high acute virulence with the lethal dose being approximately equal to the infectious dose ($LD_{100} = 1$) (Sibley and Boothroyd, 1992). Consistent with this, challenge with even with low numbers of parasites leads to death in ~10 days, regardless of the mouse strain type. A corollary of this is that any animals surviving low challenge inoculum (i.e., by limiting dilution) remain seronegative, indicating they were never infected. Type I strains also show enhanced migratory potential *in vitro*, leading to enhanced dissemination *in vivo* (Barragan and Sibley, 2002). In contrast, type II strains show intermediate virulence, causing nonlethal infections in outbred mice (Sibley and Boothroyd, 1992), while inbred mice (i.e., BALB/c, C57BL/6) show greater susceptibility than outbred mice with LD_{50}s ranging from 10^2 to 10^3 (Saeij et al., 2006). Infection by type II strains has also been shown to upregulate dendritic cell (DC) migration, thus enhancing entry of parasites into the CNS (Lambert et al., 2006), reflecting a different means of dissemination compared to type I strains. Finally, type III strains are largely avirulent and do not cause lethal infections in mice unless administered at very high doses (Saeij et al., 2006; Taylor et al., 2006). Differences in the mouse model are not the result of phenotypic changes due to laboratory passage, as many type I strains were reported to be acutely virulent upon initial isolation (Dubey, 1980; Dubey et al., 2007; Ferreira et al., 2001; Sibley and Boothroyd, 1992), including the widely used RH strain (Sabin, 1941). Although the acute challenge model is based on intraperitoneal inoculation with tachyzoites and is not a natural route of infection, it is highly reproducible and therefore useful for identifying genes that control acute virulence in different *T. gondii* strains, as described below.

Additional models for comparing pathogenicity include oral challenge with bradyzoites, which in C57BL/6 mice leads to acute ileitis and rapid death of animals challenged with the type II strain ME49 (Liesenfeld, 2002). Tissue damage results from an overly aggressive immune response and down-modulation of inducible nitric oxide synthase (iNOS), interferon gamma (IFN-γ), and tumor necrosis alpha (TNF-α) reduces pathology (Liesenfeld et al., 1999). Pathology in this model is also dependent on the microbial flora in the gut (Heimesaat et al., 2006), and it is likely that following the initial breach of the intestinal barrier due to parasite infection, the ensuing reaction to luminal bacterial products results in sepsis that leads to death. Susceptible mouse strains, including C57BL/6, also develop encephalitis when challenged with some type II strains, while type III strains are largely benign in this model (Suzuki and Joh, 1994). Genetic differences that underlie these latter two phenotypes have not been explored, although they should be amenable to the genetic mapping as described below for differences in acute virulence.

THE IMMUNE RESPONSE AND CONTROL OF INFECTION

The laboratory mouse has also served as an excellent model for resistance to infection by *T. gondii* and for illuminating general features of innate and adaptive immunity. Most studies have used the type II strain ME49 due to the fact that it readily produces a nonlethal infection in mice and therefore can be used to examine the enhanced susceptibility of mouse strains lacking particular immune effectors or cell populations. Infection with *T. gondii* is among the most potent inducers of IL-12, which in turn drives production of IFN-γ production from natural killer (NK), natural killer T (NKT), and T lymphocytes (Yap and Sher, 1999a). Together with TNF-α, which is produced primarily by monocytes, these cytokines are responsible for control of acute infection (Yap and Sher, 1999a). IFN-γ induces a variety of effector mechanisms to control and eliminate parasites, and adoptive transfer studies have shown that IFN-γ signaling is important in both hematopoietic and nonhematopoietic cells (Yap and Sher, 1999b). DCs (Scanga et al., 2002), macrophages (Robben et al., 2004), and neutrophils (Bliss et al., 1999) have been shown to produce IL-12 in response to infection *in vitro*. Studies using DTR-CD11c mice treated with diptheria toxin have shown that DC cells are also of major importance *in vivo* (Liu et al., 2006). Intravenous injection of soluble tachyzoite antigen (STAg) revealed that CD8α$^+$ DC cells in the spleen are important for the early production of IL-12 (Sousa et al., 1997), and recent studies based on selective ablation of the transcription factor Batf3 confirm the critical importance of these cells as a source of IL-12 (Mashayekhi et al., 2011). There are marked strain differences in the induction of IL-12 with infection of macrophages *in vitro* by type II strains being strongly stimulatory, while type I and III strains induce much lower levels (Robben et al., 2004). Induction of IL-12 relies on both MyD88-dependent and MyD88-independent pathways, and the relative contribution of these varies with parasite strain type (Kim et al., 2006). Infection of mice with either type I or type II strains induces a strong type I cytokine profile dominated by IL-12, TNF-α, and IFN-γ. This response is important in controlling infection by type II strains but is unable to control proliferation and dissemination by type I strains, leading to very high levels of cytokines that contribute to pathology and the demise of animals (Gavrilescu and Denkers, 2001; Mordue et al., 2001).

HOST CELL INVASION AND PROTEIN SECRETION

The process of host cell invasion by apicomplexan parasites occurs by active penetration that is governed by an actin–myosin-based motility system in the parasite (Sibley, 2010). During invasion, the parasite induces invagination of the host cell plasma membrane, sliding through a tight constriction known as the moving junction, an interface that may be important in sorting proteins within the membrane of the nascent vacuole. Regulated protein secretion is tightly coordinated with cell entry by the parasite and key to intracellular survival. Three distinct secretory compartments release their contents in precise waves during motility and cell invasion (Carruthers and Sibley, 1997). Adhesive proteins are released onto the surface of the invading parasite from apical microneme (MIC) organelles and are vital for motility, host cell recognition, and attachment (Carruthers and Tomley, 2008). Following intimate attachment, apical bulb-like organelles called rhoptries inject their contents into the host cell cytoplasm. Proteins found in the rhoptry necks (RONs) contribute to the host cell side of the moving junction, while those in the bulb of the rhoptry (ROPs) serve a variety of functions within the infected cell (Boothroyd and Dubremetz, 2008). The initial suggestion that ROP secretion might provide a mechanism to alter host cell signaling (Håkansson et al., 2001) has now been abundantly validated, as discussed below. Finally, dense granule (GRA) proteins are released into the lumen of the vacuole, where they contribute to the formation of an intricate membranous network (Mercier et al., 2007). Interestingly, some GRA proteins are also found on bead-like structures within the host cell early during invasion (Dunn et al., 2008; Jacobs et al., 1998), thus raising the possibility, also recently confirmed for at least one example discussed below, that they too could affect host signaling pathways.

SUBVERSION OF HOST CELL SIGNALING

Infection of human cells with *T. gondii* alters the expression of a large number of genes controlling pathways involved in immune responses, apoptosis, cell growth, and metabolism (Blader et al., 2001). Early response genes include those controlling glycolytic metabolism and transferrin uptake, pathways that are regulated by the transcription factor HIF1, which is important for optimal *in vitro* growth of the parasite (Spear et al., 2006). Infection by *T. gondii* also modulates other host transcription factors, including epidermal growth factor receptor (EGR) and activator protein 1 (AP-1) pathways that modulate cell growth and differentiation (Phelps et al., 2008), although their roles in parasite survival have not been determined.

Infection with *T. gondii* has been described to block a number of cellular pathways, suggesting that the parasite actively modulates host cell signaling pathways during infection, producing, for example, the following effects: blocking development of DCs (McKee et al., 2004), prevention of apoptosis (Carmen et al., 2006; Luder and Gross, 2005), modulation of nuclear factor kappa beta (NFκβ) (Leng et al., 2009a), and disruption of the cell cycle (Molestina et al., 2008). Infection of macrophages with type I strains has been shown to block the subsequent activation of IL-12 release by lipopolysaccharide (LPS) (Butcher et al., 2001). Additionally, the induction of TNF-α production by LPS stimulation, which acts through toll-like receptor 4 (TLR4), is blocked after infection by type I or II strain (Lee et al., 2006). Signaling of other TLRs is also repressed in *T. gondii*-infected cells, which appears to be mediated by altered chromatin remodeling that decreases transcription (Leng et al., 2009b).

Resistance to acute toxoplasmosis in the mouse model requires IFN-γ and hence, not surprisingly, it relies on the transcription factor STAT1 (Gavrilescu et al., 2004). However, infection by the parasite down-regulates STAT1-dependent signaling and decreases the induction of MHCII, leading to decreased antigen presentation (Leng et al., 2009a). Down-modulation of STAT1-dependent responses is also responsible for decreased iNOS expression in infected cells (Leng et al., 2009a), which likely limits the ability of nitric oxide to inhibit parasite growth. A more global analysis revealed that more than 100 genes induced by IFN-γ are misregulated in *T. gondii*-infected cells, suggesting a general defect in STAT1 signaling (Kim et al., 2007). Although debate exists about the mechanism of this inhibition, recent results suggest that STAT1 is normally phosphorylated and translocated to the nucleus but is less active there, perhaps due to an altered phosphorylated state (Kim et al., 2007). This blockage of STAT1 signaling is seen in all three strains types of *T. gondii* but is not shared by the closely related parasite *Neospora caninum* (Kim et al., 2007).

Collectively, it is evident that infection by *T. gondii* induces a wide range of altered responses in the host cell. Whether these pathways are differentially regulated by different parasite strain types has not been tested in many cases. However, in a few, as discussed below, there are strain-specific differences allowing genetic approaches to be used to define the genes involved.

EXPERIMENTAL GENETICS

Genetic approaches, be it classical "forward" or "reverse," provide one of the most powerful tools available to a biologist. One of the attractions of working with *T. gondii* is that its natural life cycle is highly amenable to classical genetic approaches; as discussed above, *T. gondii* has a well-described sexual cycle that culminates in the generation of oocysts, each of which has eight haploid progeny stemming from meiosis. The potential for forward genetics was first developed into an experimental tool by Elmer Pfefferkorn and colleagues, who showed that although a single strain can "self," simultaneous infection with two strains could be used to produce experimental crosses, yielding the expected ratios of recombinant F1 progeny (Pfefferkorn and Pfefferkorn, 1980; Pfefferkorn et al., 1977). With the advent of molecular biology tools, this potential was fully developed to generate F1 progeny from crosses between members of the three canonical lines described above (initially, types II and III, but later, other pairwise combinations). This resulted in the generation of the linkage maps needed to determine the parameters of genetic mapping and to perform segregation analyses (Khan et al., 2005b; Sibley et al., 1992; Su et al., 2002).

Reverse genetic techniques are also well developed in this parasite. The means now exist for the efficient introduction of transgenes (Kim and Boothroyd, 1995; Soldati and Boothroyd, 1993), including those under the control of regulated promoters or expressing fusions with regulatable "degradation domains" (Herm-Gotz et al., 2007; Meissner et al., 2001, 2002). Combined with the ability to efficiently target genes for deletion through homologous recombination (Donald and Roos, 1998; Donald et al., 1996), especially in strains lacking the repair enzyme Ku80 (Fox et al., 2009; Huynh and Carruthers, 2009), validation of genes initially identified as possibly important through classical genetics becomes relatively straightforward.

NATURE HAS PROVIDED US WITH HIGHLY INFORMATIVE "VARIANTS"

By serendipity, the first genetic cross between two different strains involved parents belonging to two of the three predominant lines (types II and III; Sibley et al., 1992). Approximately 20 F1 progenies from this cross, combined with a similar number of progeny selected from a second II × III cross, were genotyped at about ~135 loci, yielding a medium resolution genetic map (Khan et al., 2005b). More importantly, the progeny could be cryopreserved and used repeatedly for analyzing interesting phenotypes. All that was needed was a set of interesting phenotypes worth the effort of characterizing in each of the F1 progeny and the subsequent mapping of the loci involved. Conveniently, the three canonical strains differ in many important phenotypes. Such differences are likely not due to chance but rather are a result of natural selection; the things many of us who study pathogenesis are most interested in are the proteins involved in the subtle, but crucial, dialogue occurring between the intracellular parasite and its host cell. These proteins are likely to be exquisitely adapted to their intermediate host species, and given the wide range of hosts *T. gondii* infects, this suggests that host range will drive adaptation of proteins involved in virulence and transmission. When a given strain finds itself in a new intermediate host species, these "negotiator" or "effector" proteins may not be optimally tuned to that host, with serious and potentially disastrous consequences to the host–parasite interaction. Hence, when the different strains are used to infect just one host species, for example, laboratory mice, major differences are observed, as detailed above. Crosses between such strains, therefore, can be extremely informative about some parasite genes that are most crucial in the interaction with the host. It is important to remember, however, that at some important loci, any two parental strains being compared may carry identical alleles and such genes will therefore not be identifiable through genetic crosses involving those parents. Hence, while forward genetics is exquisitely tuned to find differences, it is also blind to shared traits and their underlying genes.

KEY PARAMETERS AND USES OF THE GENETIC MAP

From the II × III crosses described above, a genetic map was derived that has proven highly informative. To produce this, restriction fragment length polymorphisms (RFLPs) were identified at ~135 loci across the parasite's ~65 Mbp genome. By genotyping the ~40 F1 progeny at these loci, cosegregation analyses allowed the generation of a linkage map consisting of 14 chromosomes ranging from 2 to >10 mbp in size (Khan et al., 2005b). These same analyses also revealed that 1 centimorgan (cM) in this parasite corresponds to about 100 kbp (i.e., there will be an average of one recombination per 100 kbp per meioses) (Khan et al., 2005b). Genome sequencing of these same parental lines has yielded a saturating set of genetic markers (e.g., single-nucleotide polymorphisms [SNPs]) that can be used to refine the mapping even further (http://toxodb.org/toxo/). Microarrays containing probes that correspond to alleles defined by single nucleotide polymorphisms (SNPs) that were discovered by expressed sequence tag (EST) sequencing have also provided the potential for relatively high throughput and high resolution genetic mapping of recombinants (Bahl et al., 2010).

Centimorgan size is key for genealogical studies. It allows an estimation of the number of meioses separating any two strains with a shared pedigree. This approach has been applied to the three canonical strains and revealed that surprisingly, they are not independently evolved lineages. Instead, they represent one abundant line (type II), and progeny from crosses between the type II line and another lineage (Boyle et al., 2006). Remarkably, it appears that the type III line is the F1 progeny from such a cross (i.e., involving just one cat), and type I is likely to be the F2 progeny from a pair of related crosses (Boyle et al., 2006). The predominance of these three recently emerged recombinants (types I, II, and III) indicates that they have extremely well-suited genotypes for an important environmental niche, perhaps a particularly abundant host species.

MAPPING BIOLOGICALLY IMPORTANT PHENOTYPIC DIFFERENCES

Crosses between all three pairwise combinations of the three canonical types have been performed and have enabled the mapping of genes responsible for many phenotypic differences, including virulence, growth, and effects on host signaling pathways. As might be expected for complex biological phenotypes, often multiple genes contribute to the outcome of infection. Although classical genetic mapping is straightforward for simple Mendelian traits, analysis of more complex traits is facilitated by quantitative trait locus (QTL) mapping (Lander and Botstein, 1989). QTL mapping provides a powerful statistical means of evaluating the contribution of individual loci to complex traits and for analyzing potential interactions between them (Churchill and Doerge, 1994).

TABLE 15.1 Summary of Pairwise Genetic Crosses, Phenotypes Analyzed, and Genes Involved in Acute Virulence of *T. gondii* in the Mouse Model

Cross	Assay Systems: *In Vitro*	Assay Systems: *In Vivo*	Phenotypes	QTLs[a]	Genes Implicated
II × III (Reese et al., 2011; Rosowski et al., 2011; Saeij et al., 2006)		Inbred mice (BALB/c, CBA/J)	Low dose (n = 100) time until death[b]	*XII* left, *VIIb*, *XII* right	*ROP5*, *ROP16*, -
			Mortality at any dose	*XII* left, *VIIa*	*ROP5*, *ROP18*
			High dose ($n = 10^5$) time until death[b]	*XII* left, VIIa	*ROP5*, *ROP18*
	Microarray		Host gene expression	*VIIb*, *X*	*ROP16*, *GRA15*
I × III (Taylor et al., 2006)	Intracellular growth		Replication rate	*VIIa*, *XI*, *XII*, *Ia*	Multigenic
	Soft agar		Migration	*VIIa*	–
	MDCK monolayers		Transmigration	*VIIa*	–
		Outbred mice (CD1)	Cumulative mortality[c]	*VIIa*, *Ia*	*ROP18*
			Serology in surviving mice	*VIIa*, *Ia*	*ROP18*
I × II (Behnke et al., 2011)		Outbred mice (CD1)	Acute mortality at low dose (n = 100) challenge[b]	*XII*	*ROP5*

[a] Chromosome locations of major peaks ($P < 0.05$)
[b] Challenge i.p. with tachyzoites.
[c] Challenge i.p. with 10, 100, and 1000 tachyzoites.

Virulence is a complex quantitative trait as can be seen by the range of phenotypes in the progeny of a cross, for example, in their relative lethality in mice, with respect to the parental types. For example, comparing the F1 progeny of a cross between types II and III revealed clones that differed in their LD_{50} by more than five orders of magnitude in inbred mice (although the difference is much less in outbred mice), including some that were much less and some much more virulent than either parent (Grigg et al., 2001a). Genotyping these progenies as well as progenies from I × III and I × II crosses allowed five QTLs that are involved in virulence to be mapped (*VIR1-5*) (Saeij et al., 2006) (Table 15.1). Three of these loci have now been precisely determined and, remarkably, all three encode members of an extended and highly polymorphic family of ROPs (Behnke et al., 2011; Reese et al., 2011; Saeij et al., 2006, 2007; Taylor et al., 2006) (Table 15.1, Fig. 15.1). The ROP family was defined by the first member to

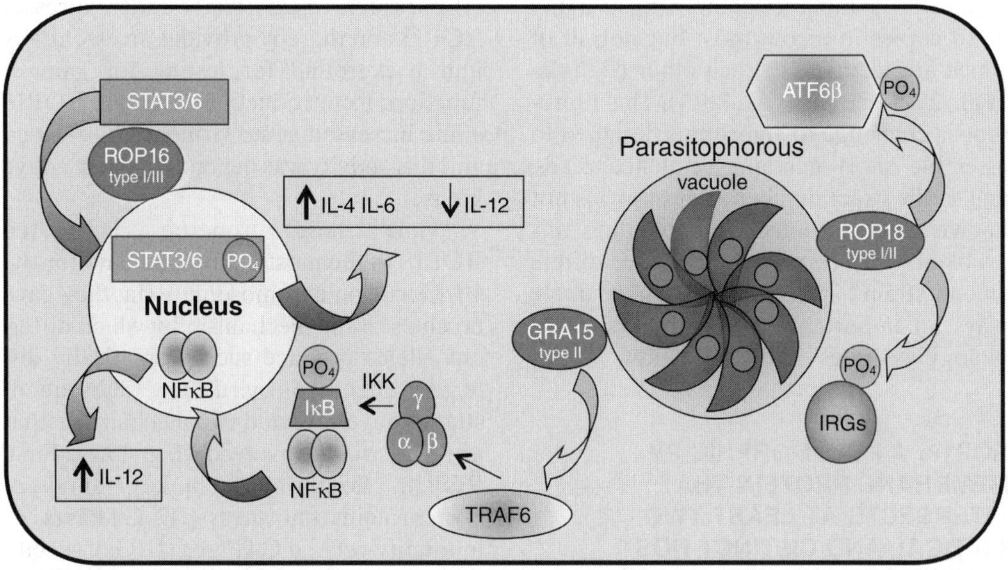

Figure 15.1 Model for the cellular roles of *T. gondii* secretory proteins in modulating host cell signaling and controlling acute virulence. Diagram of a host cell infected by *T. gondii* showing the distribution of key secretory proteins and functional pathways altered within the cell. ROP16 is secreted into the host cell and accumulates in the nucleus (Saeij et al., 2007). ROP16 contains a nuclear localization signal, although this is not essential for its activity on STATs. Recent evidence suggests ROP16 affects host gene transcription by directly phosphorylating STAT3 (Yamamoto et al., 2009) and STAT6 (Ong et al., 2010), host cell transcription factors that upregulate IL-4- and IL-6-dependent responses while impairing IL-12 production. GRA15 is also released upon or after invasion and this activates the NFκB pathway, resulting in induction of IL-12 by type II strains (Rosowski et al., 2011). Activation requires the host signaling proteins TRAF6 and the IKK complex that phosphorylates IκB, resulting in nuclear translocation of NFκB and activation of gene transcription. GRA15 has been localized to the parasitophorous vacuole (PV) membrane and to tubule extensions that emanate from it. ROP18 is secreted into the host cell where it remains tethered to the PV membrane through an N-terminal low complexity region (LCR) (Labesse et al., 2009; Reese and Boothroyd, 2009). ROP18 directly phosphorylates immunity-related GTPases (IRGs), thus blocking their accumulation on the vacuole and avoiding clearance of the parasite by this innate immunity mechanism (Fentress et al., 2010; Steinfeldt et al., 2010). ROP18 has also been described to interact with and phosphorylate ATF6β, an ER stress response transcription factor that is important for controlling dendritic cell function in response to *T. gondii* infection (Yamamoto et al., 2011). See color insert.

be identified, ROP2 (Beckers et al., 1994), which was originally thought to be an integral membrane protein in the PV membrane but is now known to contain a protein kinase fold (El Hajj et al., 2006b; Labesse et al., 2009; Wernimont et al., 2010). ROP2 family members associate to the PV membrane via a low complexity N-terminal extension (Reese and Boothroyd, 2009). Some ROP2 family members are active kinases, while others are predicted to be pseudokinases (El Hajj et al., 2006a; Peixoto et al., 2010). The exact size of this gene family is difficult to state categorically because many of the genes are tandemly duplicated, sometimes comprising as many as 10 copies in a row, most but not all of which are identical to each other (Behnke et al., 2011; Reese et al., 2011). This represents a challenge to algorithms designed to assemble short genomic sequence reads, and so the exact number at most loci is not known with certainty. Compounding this problem, the copy number also differs among strains. This extreme fluidity likely plays an important role in the parasite's biology, as discussed further below.

ROP18: A POLYMORPHIC PV MEMBRANE PROTEIN THAT INTERSECTS AT LEAST TWO CRITICAL AND DISTINCT HOST PATHWAYS INVOLVED IN THE IMMUNE RESPONSE

The first such *VIR* QTL to be identified was *ROP18* (Saeij et al., 2006; Taylor et al., 2006), which emerged as a major virulence locus in both type II × III and type I × III crosses (Table 15.1). The initial genetic mapping localized the *VIR* locus to within several hundred kilobase pairs on chromosome VIIa. Within this region, *ROP18* stood out as the most likely candidate for several reasons. First, it encodes an active kinase (ROP18) that decorates the cytosolic face of the PV membrane in infected cells (Bradley et al., 2005; El Hajj et al., 2007); this endowed it with an attractive enzymatic activity and placed it in a position where it could easily interact with host proteins. Second, ROP18 is highly polymorphic between strains with a very high ratio of nonsynomymous to synonymous substitutions (Saeij et al., 2006; Taylor et al., 2006). Third, it is one of the genes with expression levels that differ the most between different strains (Boyle et al., 2008). These inferences were confirmed by generating transgenic parasites expressing the type I (Taylor et al., 2006) or type II (Saeij et al., 2006) alleles of ROP18 in type III parasites, which greatly underexpresses ROP18 and thereby provides an essentially null background for testing for gain of function. Reintroduction of either ROP18 allele increased acute virulence by >4 logs, and this activity was dependent on an active kinase.

While these properties implicated ROP18 as the most likely candidate for the *VIR* locus on chromosome VIIa, they gave no clues to the mechanism by which different alleles produce such dramatically different effects on virulence. Subsequent studies have revealed two mechanisms that are potentially involved (Fig. 15.1). First, ROP18 phosphorylates a key family of host proteins known as p47 GTPases, or immunity-related GTPases (IRGs), specifically Irga6, Irgb6, and Irgb10 (Fentress et al., 2010; Steinfeldt et al., 2010). IRG proteins are specifically targeted to pathogen-containing vacuoles in infected cells, where they somehow mediate the disruption of the vacuole membrane and the subsequent destruction of the pathogen (Taylor et al., 2004). Clues to the target of ROP18 were provided by the fact that type I strains routinely avoid decoration by IRGs and hence resist clearance in activated cells, while types II and III both recruit IRGs and are cleared rapidly from the cell (Khaminets et al., 2010; Zhao et al., 2009). The interaction between ROP18 and

IRGs leading to phosphorylation of several family members was detected in infected host cells and *in vitro*, and mass spectrometry was used to map the phosphorylation sites (Fentress et al., 2010; Steinfeldt et al., 2010). Interestingly, phosphorylation occurs at key conserved Thr residues in a critical region of the switch region 1, thereby inhibiting GTPase activity, preventing oligomerization, and blocking accumulation on the vacuole (Fentress et al., 2010; Steinfeldt et al., 2010). ROP18 was shown to be both necessary and sufficient for this activity, which led to increased survival in IFN-γ activated cells, including macrophages, that are normally key to infection control (Fentress et al., 2010). The type III allele of ROP18 is a hypomorph with a promoter region that contains a large insertion relative to the alleles of types I and II, resulting in very low expression. Comparison to the outgroup *N. caninum* reveals that this "insertion" is in fact the ancestral state, suggesting that it was lost in types I and II, leading to upregulation of expression (Khan et al., 2009). The promoter and resulting expression differences in *ROP18* readily explain the phenotypic differences of progeny of type IIII and II × III crosses. However, *ROP18* was not a significant VIR QTL in the subsequently performed cross between high virulence type I and intermediate virulence type II strains (Table 15.1). Interestingly, the types I and II alleles of ROP18 encode many polymorphisms and their resulting proteins differ in their relative ability to efficiently phosphorylate Irga6 in infected cells (Steinfeldt et al., 2010). Apparently, this difference is not enough for a *VIR* QTL to be detectable at the ROP18 locus in I × II crosses. The reasons for this are not clear, but one explanation is that other loci may have epistatic interactions with ROP18 that affect activity and hence virulence.

More recently, ROP18 has been shown to have a second completely unrelated target in the host cell: the transcription factor ATF6β (Yamamoto et al., 2011). This host protein has been reported to be part of the endoplasmic reticulum (ER) stress response associated with the unfolded protein response (Wang et al., 2000); however, other reports indicate that while ATF6α is essential for this pathway, ATF6β is not (Yoshida et al., 2001). The interaction of ROP18 with ATF6β was discovered using yeast 2 hybrid assays, and this interaction is mediated by binding between the N-terminus of ROP18 and the C-terminus of ATF6β (Yamamoto et al., 2011). Although the effect on the endogenous levels of the host protein have not been evaluated, infection with *T. gondii* induces increased proteasome-dependent turnover of a transgenic version of ATF6β (Yamamoto et al., 2011). *In vitro* data suggest that ROP18 can phosphorylate ATF6β, and indeed, a kinase-dead version of ROP18 loses its capacity to induce turnover of ATF6β in cells expressing a transgenic, tagged copy. Direct phosphorylation of ATF6β in infected cells has not yet been demonstrated, and the location of the C-terminus of this protein in the lumen of the ER poses a topological challenge for the kinase, which is on the cytoplasmic surface of the PV membrane. Regardless, mice that are deficient in ATF6β are more susceptible than wild-type mice to infection with parasites lacking ROP18 (normally attenuated), supporting a key role for this host protein in defense against *Toxoplasma*. This phenotype appears to be due to a DC-specific defect in inducing IFN-γ production by T cells (Yamamoto et al., 2011), although the precise reasons why targeting ATF6β in these cells leads to such a defect remain unclear. Although additional studies are necessary to validate this interaction in infected cells and reveal the precise mechanism by which it compromises immunity, this finding illustrates that ROP kinases may affect multiple targets within host cells to thwart host cell signaling.

ROP16: A TYROSINE KINASE TARGETED TO THE HOST NUCLEUS THAT DRAMATICALLY ALTERS THE CHARACTER OF THE HOST IMMUNE RESPONSE

The second *VIR* QTL to be identified was *ROP16* (Table 15.1), encoding a protein that was first detected in proteomic studies of the rhoptries (Bradley et al., 2005) and predicted to be an active protein kinase (Peixoto et al., 2010). As with ROP18, ROP16 emerged as the most likely candidate for the gene responsible for one of the *VIR* loci, in this case on chromosome VIIb; it is highly polymorphic, injected into the host cell, and is an active kinase (Saeij et al., 2007). However, unlike ROP18, ROP16 does not localize to the PV membrane but instead concentrates in the host nucleus by a typical eukaryotic nuclear localization signal (NLS) (Saeij et al., 2007) (Fig. 15.1).

Clues to how ROP16 functions emerged from complementary genetic studies to those used for mapping ROP18. The hypothesis was that virulence differences might be detected by analyzing the different impact of the F1 progeny on host gene expression using microarrays. Preliminary experiments showed striking differences between the infection of human foreskin fibroblasts with each of the three canonical lines of *T. gondii* (Saeij et al., 2007). Subsequent analysis of the F1 progeny of the same type II × III cross used for the virulence mapping showed many of these host expression differences mapped to unique loci in the *T. gondii* genome. The parasite locus with the greatest impact in terms of the number of host genes that were affected mapped to the same region of chromosome VIIb as one of the *VIR* QTLs. For the same reasons that *ROP16* was an attractive candidate for the *VIR* locus on chromosome VIIb, it also stood out as the most likely gene on this chromosome responsible for the differences in human fibroblast gene expression. Reverse genetic techniques demonstrated that ROP16 is indeed the VIIb locus responsible for the strain-specific differences in host gene expression and, likely, virulence (Saeij et al., 2007).

The role of ROP16 in the host–parasite interaction has been evaluated by analyzing host genes whose expression differed depending on the ROP16 allele of the infecting strain (Table 15.1). These responses showed a strong enrichment for genes regulated by signal transducers and activators of transcription (STAT) proteins, especially STAT3 and STAT6. These proteins are key intermediaries in the process by which an immune signal is transduced from outside a host cell to the nucleus. STATs are regulated by phosphorylation; therefore, it was tempting to speculate that ROP16 might directly phosphorylate these proteins. Arguing against this was the fact that ROP16 was predicted to be a serine/threonine kinase rather than a tyrosine kinase that is normally required for STAT activation. Additionally, commercial inhibitors that were purported to be highly specific for the host kinases (i.e., Janus kinases [JAKs]) responsible for STAT phosphorylation blocked the effects of ROP16 during host cell infection. Both of these phenomena turned out to be misleading; ROP16 is indeed a tyrosine kinase that can directly phosphorylate STAT3 (Yamamoto et al., 2009) and STAT6 (Ong et al., 2010) (Fig. 15.1). As it happens, the commercial inhibitor is even more active on ROP16 than it is on JAKs!

All three allelic forms of ROP16 are expressed at similar levels (contrary to the situation with ROP18), predicting that the key phenotypic differences are due to differences in the amino acids encoded by different alleles. Indeed, Yamamoto et al. (2009) identified position 503 as the key difference between alleles based on comparisons of sequences from different strains and their effect on STAT phosphorylation. The biochemical basis for how these differences translate into different effects on STATs has yet to be determined.

Interestingly, the type I, II, and III alleles of ROP16 all activate (tyrosine-phosphorylate) STATs in the first few minutes after invasion (*in vitro*, at least). The alleles differ, however, in the length of time that the activation persists; infection with strains carrying the type I and III alleles results in sustained STAT activation that lasts at least 20 hours, whereas the type II allele gives only a transient activation with no detectable activation after a few hours. Again, the biochemical basis for this difference is not known, but the biological outcome is profound. As STAT3 is a negative regulator of IL-12 expression, its sustained activation in type I and III strains leads to a much lower level of this key inflammatory cytokine than infection with type II strains (Saeij et al., 2007) (Fig. 15.1). This explains, in part, the previously noted difference in IL-12 levels resulting from infection with the different strains (Robben et al., 2004).

STAT6 is an important driver of so-called alternative (M2) activation of macrophages, a noninflammatory phenotype characterized by high levels of arginase expression. Hence, infection of mouse macrophages with type I and III strains leads to this alternative activation phenotype. In contrast, infection with type II strains results in a classically activated (M1) phenotype characterized by inflammatory markers, including inducible nitric oxide synthase (Jensen et al., 2011).

STAT phosphorylation by JAKs normally occurs in the cytoplasm, but there is no a priori reason why ROP16 could not also carry out this function in the nucleus. Importantly, the NLS on ROP16 is necessary for nuclear localization but not for the sustained STAT activation phenotype (Saeij et al., 2007), suggesting that it is performing some other role in this compartment. Some JAKs are known to work in both the cytoplasm and nucleus where JAK2, for example, phosphorylates histones. As of yet, there are no data to support such an activity for ROP16, but it seems likely that there is an additional, non-STAT target of ROP16 in the nucleus.

ROP5: A CRUCIALLY IMPORTANT BUT ENIGMATIC PSEUDOKINASE

The most recent ROP2 family member to be identified as having a role in pathogenesis is ROP5, the major virulence locus that was mapped in both the type II × III cross and a recently completed I × II cross (Behnke et al., 2011; Reese et al., 2011) (Table 15.1). Interestingly, unlike the single-copy genes encoding ROP16 and ROP18, ROP5 is encoded by a series of tandemly repeated genes consisting of about 4–10 copies depending on the parasite strain. While many copies are identical to each other, some show differences even within a given strain and these differences appear to be important for the virulence differences observed in mice. Deletion of all *ROP5* copies renders even the highly virulent type I strain completely avirulent in mouse models; that is, the LD_{50} changes from "<1" to $>10^6$. The number of repeats likely also plays a role in modulating virulence, as restoration of the knockout with a single allele only partially restores virulence (Reese et al., 2011), while complementation with the entire locus completely restores virulence (Behnke et al., 2011).

What most sets ROP5 apart from the active kinases ROP16 and ROP18 is the fact that it is a pseudokinase, that is, a protein with a kinase fold but no ability to transfer phosphates onto other proteins. It is nonintuitive that such a catalytically inactive protein should have such an important role, but like ROP18, ROP5 is present on the PV membrane, and it is tempting to speculate that it may facilitate the action of other parasite proteins that similarly localize, perhaps even ROP18, as suggested recently (Behnke et al., 2011). Direct demonstration of such an interaction, however, has yet to be reported, and so the precise role of this unusual set of proteins remains enigmatic.

GRA15: POLYMORPHIC DENSE GRA PROTEINS CAN ALSO IMPACT THE HOST IN A STRAIN-SPECIFIC MANNER

One strain-specific effect observed in fibroblasts is the activation of NF-κB and translocation of p50/REL-A heterodimers into the host nucleus (Rosowski et al., 2011). Type II strains do this much more efficiently than type I or III strains, and this difference was used to map the locus involved by the genetic analysis of the F1 progeny from the same type II × III cross described above (Table 15.1). The locus was found to be on chromosome X, but the gene involved was not so obvious as for the ROP examples described above. Using similar forward and reverse genetic strategies, however, a secreted protein dubbed GRA15 was determined to be responsible for this effect (Rosowski et al., 2011) (Table 15.1, Fig. 15.1).

GRA15 induces phosphorylation of IκBα, leading to release and translocation of NF-κB to the nucleus (Rosowski et al., 2011). Although this pathway relies on TNF-receptor associated factor 6 (TRAF6) and I kappa B kinase (IKK), it does not require MyD88 or TIR domain containing adaptor inducing interferon beta (TRIF) (Fig. 15.1) (Rosowski et al., 2011). *In vitro*, infection with a type II strain with a deletion of *GRA15* results in substantially lower levels of this key cytokine produced by infected bone marrow macrophages (Rosowski et al., 2011). Type II strains possess a *GRA15* allele that is much more active in this process, and this may further explain (in addition to the differences discussed above for ROP16) the much higher production of IL-12 in animals infected with type II strains versus types I and III.

RELEVANCE TO OTHER HOSTS, TRANSMISSION, AND POPULATION

The remarkable, strain-specific differences described above beg the question of what forces have led to the current situation. It seems unlikely that the major, strain-specific differences in ROP5/16/18 discussed above emerged by chance. Instead, they are presumed to each offer a selective advantage, singly or in combination, to a strain that carries them for one or other ecological niche. As already mentioned, one hypothesis is that these differences arose as a result of natural selection within different intermediate hosts. As yet, this hypothesis has not been explored as almost all studies have been on laboratory rodents for their experimental ease of analysis or on humans because of the clinical relevance. However, in the case of *ROP18*, there is evidence that the type III hypomorphic allele has a very long ancestry during which it has evolved under neutral selection, while both types I and II show strong evidence of selective pressure (Khan et al., 2009). The specific environmental niches (hosts) that drove this process remain undefined but might relate to host differences. Additionally, it is also important to consider that "virulence" adaptations for a particular host may also not participate in infection of hosts that lack the target pathway. For example, although ROP18 targets IRGs in rodent cells, this family of innate defense molecules is highly expanded in rodents while rare or absent in other vertebrate lineages, including cats (Bekpen et al., 2005). Humans express only one truncated and one tissue-specific member of this family, and so this interaction is unlikely to participate in human resistance to toxoplasmosis. In contrast, a related family of GTPases called guanylate binding proteins (or p67 GTPases) is represented by numerous paralogs in both mouse and human and may participate more broadly in resistance to infection (Shenoy et al., 2008). Elucidating these relationships will require assessment of potential resistance mechanism and virulence factors across multiple species.

The data in humans suggest that strain-specific differences may also influence disease outcome, but the numbers analyzed have been necessarily small and/or the possibility of confounding variables large; thus, drawing definitive conclusions on the role of strains in these disease differences has been difficult (Gilbert et al., 2008; Vasconcelos-Santos et al., 2009). One complication of human studies is that it is often difficult to determine the genotype of the parasite that caused a specific infection because isolation of the strains is not straightforward, and serological differences have limited resolution for genotyping (Kong et al., 2003). More importantly for the topic of this review, while human studies are obviously of great clinical importance, they likely have little bearing on the forces driving the emergence of different strains because humans are a "dead-end" host for *T. gondii* (transmission from an infected human to any other animal has likely been a very rare event in the parasite's evolution). This, of course, is not to say that the parasite has not had an impact on human evolution; *T. gondii* infection could have played a major role in the differences seen in immune response genes both within the human species and between our species and other mammalian species. Studies that look at different hosts that are likely important in the transmission of the parasite, perhaps birds, wild rodents, and other common hosts, are needed to further elucidate the parasite's evolutionary past.

Determining the relevant strain/host combination behind the evolution of the different *T. gondii* strains will be an exceedingly difficult challenge; even natural infection of hosts in the environment does not necessarily reveal the key relationship because what strain a given host picks up does not necessarily represent what strain it efficiently transmits. For example, a given rodent species might be found to be frequently infected with a strain that does not produce large numbers of tissue cysts that can be readily transmitted, and this might be because an alternative host (e.g., an avian species) is the normal reservoir for that strain type. Ultimately, studies of the incidence and prevalence of infection within a host and transmissibility between hosts will be needed to unravel the ecology of *T. gondii* in the wild.

SUMMARY AND FUTURE DIRECTIONS

Overall, the recent work summarized here provides a glimpse into the interaction of an extremely successful infectious agent and its myriad hosts. The combination of classical and reverse genetic analyses has revealed that the parasite has evolved a diverse set of proteins that it uses to interface with the infected host cell, enabling it to actively manage the interaction that ensues. Suggestions for the forces driving this evolution are, as always, speculative, but the available data make a compelling case for differences in different host species being an important variable. Further studies of more parasite strains and a wider range of hosts will help reveal the forces that may be operating. It is especially important to remember that the type of genetic studies done here will not reveal important, nonpolymorphic genes common to the success of all strains; rather, additional laboratory-generated variants (mutants) will be needed for their discovery. Similarly, focusing on *in vitro* studies or even *in vivo* infections within the unnatural setting of a laboratory is likely to result in many key genes being missed that are crucial for subtle aspects of transmission, be they the intriguing behavioral changes in the host that sometimes occur or other factors that could affect transmission (e.g., the load and distribution of tissue cysts).

ACKNOWLEDGMENTS

We are grateful to the many individuals who have contributed to the development of genetics in *T. gondii*, most notably Elmer Pfefferkorn and members of his laboratory who pioneered the development of genetic crosses. Recent work in the authors' laboratories discussed here was supported by grants from the National Institutes of Health.

REFERENCES

Bahl A, Davis PH, Behnke M, Dzierszinski F, Jagalur M, Chen F, Shanmugam D, White MW, Kulp D, Roos DS. 2010. A novel multifunctional oligonucleotide microarray for *Toxoplasma gondii*. BMC Genomics 11: 603.

Barragan A, Sibley LD. 2002. Transepithelial migration of *Toxoplasma gondii* is linked to parasite motility and virulence. The Journal of Experimental Medicine 195: 1625–1633.

Beckers CJM, Dubremetz JF, Mercereau-Puijalon O, Joiner KA. 1994. The *Toxoplasma gondii* rhoptry protein ROP2 is inserted into the parasitophorous vacuole membrane, surrounding the intracellular parasite, and is exposed to the host cell cytoplasm. The Journal of Cell Biology 127: 947–961.

Behnke MS, Khan A, Wootton JC, Dubey JP, Tang K, Sibley LD. 2011. Virulence differences in *Toxoplasma* mediated by amplification of a family of polymorphic pseudokinases. Proceedings of the National Academy of Sciences of the United States of America 108: 9631–9636.

Bekpen C, Hunn JP, Rohde C, Parvanova I, Guethlein L, Dunn DM, Glowalla E, Leptin M, Howard JC. 2005. The interferon-inducible p47 (IRG) GTPases in vertebrates: Loss of the cell autonomous resistance mechanism in the human lineage. Genome Biology 6: R92.

Blader I, Manger ID, Boothroyd JC. 2001. Microarray analysis reveals previously unknown changes in *Toxoplasma gondii* infected human cells. The Journal of Biological Chemistry 276: 24223–24231.

Bliss SK, Zhang Y, Denkers EY. 1999. Murine neutrophil stimulation by *Toxoplasma gondii* antigen drives high level production of IFN-γ-independent IL-12. Journal of Immunology 163: 2081–2088.

Boothroyd JC, Dubremetz JF. 2008. Kiss and spit: The dual roles of *Toxoplasma* rhoptries. Nature Reviews. Microbiology 6: 79–88.

Boursot P, Auffray JC, Britton-Davidian J, Bonhomme F. 1993. The evolution of the house mouse. Annual Review of Ecology and Systematics 24: 119–152.

Boyle JP, Rajasekar B, Saeij JPJ, Ajioka JW, Berriman M, Paulsen I, Sibley LD, White M, Boothroyd JC. 2006. Just one cross appears capable of dramatically altering the population biology of a eukaryotic pathogen like *Toxoplasma gondii*. Proceedings of the National Academy of Sciences of the United States of America 103: 10514–10519.

Boyle JP, Saeij JP, Hrada SY, Ajioka JW, Boothroyd JC. 2008. Expression QTL mapping of *Toxoplasma* genes reveals multiple mechanisms for strain-specific differences in gene expression. Eukaryotic Cell 7: 1403–1414.

Bradley PJ, Ward C, Cheng SJ, Alexander DL, Coller S, Coombs GH, Dunn JD, Ferguson DJ, Sanderson SJ, Wastling JM, Boothroyd JC. 2005. Proteomic analysis of rhoptry organelles reveals many novel constituents for host-parasite interactions in *T. gondii*. The Journal of Biological Chemistry 280: 34245–34258.

Butcher BA, Kim L, Johnson PF, Denkers EY. 2001. *Toxoplasma gondii* tachyzoites inhibit proinflammatory cytokine induction in infected macrophages by preventing nuclear translocation of the transcription factor NF-κB. Journal of Immunology 167: 2193–2201.

Carmen JC, Hardi L, Sinai AP. 2006. *Toxoplasma gondii* inhibits ultraviolet light-induced apoptosis through multiple interactions with the mitochondrion-dependent programmed cell death pathway. Cellular Microbiology 8: 301–315.

Carruthers VB, Sibley LD. 1997. Sequential protein secretion from three distinct organelles of *Toxoplasma gondii* accompanies invasion of human fibroblasts. European Journal of Cell Biology 73: 114–123.

Carruthers VB, Tomley FM. 2008. Microneme proteins in apicomplexans. Sub-cellular Biochemistry 47: 33–45.

Cavailles P, Sergent V, Bisanz C, Papapietro O, Colacios C, Mas M, Subra JF, Lagrange D, Calise M, Appolinaire S, Faraut T, Druet P, Saoudi A, Bessieres MH, Pipy B, Cesbron-Delauw MF, Fournie GJ. 2006. The rat *Toxo1* locus directs toxoplasmosis outcome and controls parasite proliferation and spreading by macrophage-dependent mechanisms. Proceedings of the National Academy of Sciences of the United States of America 103: 744–749.

Churchill GA, Doerge RW. 1994. Empirical threshold values for quantitative trait mapping. Genetics 138: 963–971.

Derouin F, Garin YJF. 1991. *Toxoplasma gondii*: Blood and tissue kinetics during acute and chronic infections in mice. Experimental Parasitology 73: 460–468.

Donald RGK, Roos DS. 1998. Gene knock-outs and allelic replacements in *Toxoplasma gondii*: HXGPRT as a selectable marker for hit-and-run mutagenesis. Molecular and Biochemical Parasitology 91: 295–305.

Donald RGK, Carter D, Ullman B, Roos DS. 1996. Insertional tagging, cloning, and expression of the *Toxoplasma gondii* hypoxanthine-xanthine-guanine phosphoribosyltransferase gene. The Journal of Biological Chemistry 271: 14010–14019.

Dubey J. 1980. Mouse pathogenicity of *Toxoplasma gondii* isolated from a goat. American Journal of Veterinary Research 41: 427–429.

Dubey JP. 1997. Bradyzoite-induced murine toxoplasmosis: Stage conversion pathogenesis, and tissue cyst formation in mice fed bradyzoites of different strains of *Toxoplasma gondii*. Journal of Eukaryotic Microbiology 44: 592–602.

Dubey JP. 2010. *Toxoplasmosis of Animals and Humans*. CRC Press, Boca Raton, FL.

Dubey JP, Frenkel JF. 1972. Cyst-induced toxoplasmosis in cats. Journal of Protozoology 19: 155–177.

Dubey JP, Lopez-Torres HY, Sundar N, Velmurugan GV, Azjzenerg D, Kwok OC, Hill R, Darde ML, Su C. 2007. Mouse-virulent *Toxoplasma gondii* isolated from feral cats on Mona Island, Puerto Rico. The Journal of Parasitology 93: 1365–1359.

Dunn JD, Ravindran S, Kim SK, Boothroyd JC. 2008. The *Toxoplasma gondii* dense granule protein GRA7 is phosphorylated upon invasion and forms an unexpected association with the rhoptry proteins ROP2 and ROP4. Infection and Immunity 76: 5853–5861.

El Hajj H, Demey E, Poncet J, Lebrun M, Wu B, Galeotti N, Fourmaux MN, Mercereau-Puijalon O, Vial H, Dubremetz JF. 2006a. The ROP2 family of *Toxoplasma gondii* rhoptry proteins: Proteomic and genomic characterization and molecular modeling. Proteomics 6: 5773–5784.

El Hajj H, Lebrun M, Fourmaux MN, Vial H, Dubremetz JF. 2006b. Inverted topology of the *Toxoplasma gondii* ROP5 rhoptry protein provides new insights into the association with the parasitophorous vacuole membrane. Cellular Microbiology 9: 54–64.

El Hajj H, Lebrun M, Arold ST, Vial H, Labesse G, Dubremetz JF. 2007. ROP18 is a rhoptry kinase controlling the intracellular proliferation of *Toxoplasma gondii*. PLoS Pathogens 3: e14.

Fentress SJ, Behnke MS, Dunay IR, Moashayekhi M, Rommereim LM, Fox BA, Bzik DJ, Tayor GA, Turk BE, Lichti CF, Townsend RR, Qiu W, Hui R, Beatty WL, Sibley LD. 2010. Phosphorylation of immunity-related GTPases by a parasite secretory kinase promotes macrophage survival and virulence. Cell Host & Microbe 16: 484–495.

Ferreira AM, Martins MS, Vitor RW. 2001. Virulence for BALB/c mice and antigenic diversity of eight *Toxoplasma gondii* strains isolated from animals and humans in Brazil. Parasite 8: 99–105.

Fox BA, Ristuccia JG, Gigley JP, Bzik DJ. 2009. Efficient gene replacements in *Toxoplasma gondii* strains deficient for nonhomologous end joining. Eukaryotic Cell 8: 520–529.

Frenkel JK. 1953. Host, strain and treatment variation as factors in the pathogenesis of toxoplasmosis. American Journal of Tropical Medicine and Hygiene 2: 390–415.

Gavrilescu LC, Denkers EY. 2001. IFN-γ overproduction and high level apoptosis are associated with high but not low virulence *Toxoplasma gondii* infection. Journal of Immunology 167: 902–909.

Gavrilescu LC, Butcher BA, Del Rio L, Taylor GA, Denkers EY. 2004. STAT1 is essential for antimicrobial effector function but dispensable for gamma interferon production during *Toxoplasma gondii* infection. Infection and Immunity 72: 1257–1264.

Gilbert RE, Freeman K, Lago EG, Bahia-Oliveira LM, Tan HK, Wallon M, Buffolano W, Stanford MR, Petersen E. 2008. Ocular sequelae of congenital toxoplasmosis in Brazil compared with Europe. PLoS Neglected Tropical Diseases 2: e277.

Grigg ME, Bonnefoy S, Hehl AB, Suzuki Y, Boothroyd JC. 2001a. Success and virulence in Toxoplasma as the result of sexual recombination between two distinct ancestries. Science 294: 161–165.

Grigg ME, Ganatra J, Boothroyd JC, Margolis TP. 2001b. Unusual abundance of atypical strains associated with human ocular toxoplasmosis. The Journal of Infectious Diseases 184: 633–639.

Gubter C, Dhand R. 2002. Human biology by proxy: The mouse genome timeline. Nature doi: 10.1038/420509a.

Håkansson S, Charron AJ, Sibley LD. 2001. *Toxoplasma* evacuoles: A two-step process of secretion and fusion forms the parasitophorous vacuole. The EMBO Journal 20: 3132–3144.

Heimesaat MM, Berewill S, Fischer D, Fuchs D, Struck D, Niebergall J, Jahn HK, Dunay IR, Moter A, Gescher DM, Schumann RR, Gobel UB, Liesenfeld O. 2006. Gram-negative bacteria aggravate murine small intestinal Th1-type immunopathology following oral infection with *Toxoplasma gondii*. Journal of Immunology 177: 8785–8795.

Herm-Gotz A, Agop-Nersesian C, Munter S, Grimley JS, Wandless TJ, Frischknecht F, Meissner M. 2007. Rapid control of protein level in the apicomplexan *Toxoplasma gondii*. Nature Methods 4: 1003–1005.

Huynh MH, Carruthers VB. 2009. Tagging of endogenous genes in a *Toxoplasma gondii* strain lacking Ku80. Eukaryotic Cell 8: 530–539.

Jacobs D, Dubremetz J, Loyens A, Bosman F, Saman E. 1998. Identification and heterologous expression of a new dense granule protein (GRA7) from *Toxoplasma gondii*. Molecular and Biochemical Parasitology 91: 237–249.

Jensen KD, Wang Y, Wojno ED, Shastri AJ, Hu K, Cornel L, Boedec E, Ong YC, Chien YH, Hunter CA, Boothroyd JC, Saeij JP. 2011. Toxoplasma polymorphic effectors determine macrophage polarization and intestinal inflammation. Cell Host & Microbe 9: 472–483.

Jones JL, Kruszon-Moran D, Sanders-Lewis K, Wilson M. 2007. *Toxoplasma gondii* infection in the United States, 1999–2004, decline from the prior decade. The American Journal of Tropical Medicine and Hygiene 77: 405–410.

Joynson DH, Wreghitt TJ. 2001. *Toxoplasmosis: A Comprehensive Clinical Guide*. Cambridge University Press, Cambridge, UK.

Khaminets A, Hunn JP, Konen-Waisman S, Zhao YO, Preukschat D, Coers J, Boyle JP, Ong YC, Boothroyd JC, Reichmann G, Howard JC. 2010. Coordinated loading of IRG resistance GTPases on to the *Toxoplasma gondii* parasitophorous vacuole. Cellular Microbiology 12: 939–961.

Khan A, Su C, German M, Storch GA, Clifford D, Sibley LD. 2005a. Genotyping of *Toxoplasma gondii* strains from immunocompromised patients reveals high prevalence of type I strains. Journal of Clinical Microbiology 43: 5881–5887.

Khan A, Taylor S, Su C, Mackey AJ, Boyle J, Cole RH, Glover D, Tang K, Paulsen I, Berriman M, Boothroyd JC, Pfefferkorn ER, Dubey JP, Roos DS, Ajioka JW, Wootton JC, Sibley LD. 2005b. Composite genome map and recombination parameters derived from three archetypal lineages of *Toxoplasma gondii*. Nucleic Acids Research 33: 2980–2992.

Khan A, Jordan C, Muccioli C, Vallochi AL, Rizzo LV, Belfort R Jr, Vitor RW, Silveira C, Sibley LD. 2006. Genetic divergence of *Toxoplasma gondii* strains associated with ocular toxoplasmosis Brazil. Emerging Infectious Diseases 12: 942–949.

Khan A, Fux B, Su C, Dubey JP, Darde ML, Ajioka JW, Rosenthal BM, Sibley LD. 2007. Recent transcontinental sweep of *Toxoplasma gondii* driven by a single monomorphic chromosome. Proceedings of the National Academy of Sciences of the United States of America 104: 14872–14877.

Khan A, Taylor S, Ajioka JW, Rosenthal BM, Sibley LD. 2009. Selection at a single locus leads to widespread expansion of *Toxoplasma gondii* lineages that are virulence in mice. PLoS Genetics 5: e1000404.

Kim K, Boothroyd JC. 1995. *Toxoplasma gondii*: Stable complementation of *sag1* (p30) mutants using *SAG1* transfection and fluorescence-activated cell sorting. Experimental Parasitology 80: 46–53.

Kim L, Butcher BA, Lee CW, Uematsu S, Akira S, Denkers EY. 2006. *Toxoplasma gondii* genotype determines MyD88-dependent signaling in infected macrophages. Journal of Immunology 177: 2584–2591.

Kim SK, Fouts AE, Boothroyd JC. 2007. *Toxoplasma gondii* dysregulates IFN-γ inducible gene expression in human fiboblasts: Insights from a genome-wide transcriptional profiling. Journal of Immunology 178: 5154–5165.

Kong JT, Grigg ME, Uyetake L, Parmley SF, Boothroyd JC. 2003. Serotyping of *Toxoplasma gondii* infections in humans using synthetic peptides. The Journal of Infectious Diseases 187: 1484–1495.

Labesse G, Gelin M, Bessin Y, Lebrun M, Papoin J, Cerdan R, Arold ST, Dubremetz JF. 2009. ROP2 from *Toxoplasma gondii*: A virulence factor with a protein-kinase fold and no enzymatic activity. Structure 17: 139–146.

Lambert H, Hitziger N, Dellacasa I, Svensson M, Barragan A. 2006. Induction of dendritic cell migration upon *Toxoplasma gondii* infection potentiates parasite dissemination. Cellular Microbiology 8: 1611–1623.

Lander ES, Botstein D. 1989. Mapping mendelian factors underlying quantitative traits using RFLP linkage maps. Genetics 121: 185–199.

Lee CW, Bennouna S, Denkers EY. 2006. Screening for *Toxoplasma gondii*-regulated transcriptional responses in lipopolysaccharide-activated macrophages. Infection and Immunity 74: 1916–1923.

Lehmann T, Marcet PL, Graham DH, Dahl ER, Dubey JP. 2006. Globalization and the population structure of *Toxoplasma gondii*. Proceedings of the National Academy of Sciences of the United States of America 103: 11423–11428.

Leng J, Butcher BA, Denkers EY. 2009a. Dysregulation of macrophage signal transduction by *Toxoplasma gondii*: Past progress and recent advances. Parasite Immunology 31: 717–728.

Leng J, Butcher BA, Egan CE, Abdallah DS, Denkers EY. 2009b. *Toxoplasma gondii* prevents chromatin remodeling initiated by TLR-triggered macrophage activation. Journal of Immunology 182: 489–497.

Liesenfeld O. 2002. Oral infection of C57BL/6 mice with *Toxoplasma gondii*: A new model of inflammatory bowel disease? The Journal of Infectious Diseases 185: S96–101.

Liesenfeld O, Kang H, Park D, Nguyen TA, Parkhe CV, Watanabe H, Abo T, Sher A, Remington JS, Suzuki Y. 1999. TNFα, nitric oxide and IFN-γ are all critical for development of necrosis in the small intestine and early mortality in genetically susceptible mice infected perorally with *Toxoplasma gondii*. Parasite Immunology 21: 365–376.

Liu CH, Fan YT, Dias A, Esper L, Corn RA, Bafica A, Machado FS, Aliberti J. 2006. Cutting edge: Dendritic cells are essential for *in vivo* IL-12 production and development of resistance against *Toxoplasma gondii* infection in mice. Journal of Immunology 177: 31–35.

Luder CG, Gross U. 2005. Apoptosis and its modulation during infection with *Toxoplasma gondii*: Molecular mechanisms and role in pathogenesis. Current Topics in Microbiology and Immunology 289: 219–237.

Mashayekhi M, Sandau MM, Dunay IR, Frickel EM, Khan A, Goldszmid RS, Sher A, Ploegh HL, Murphy TL, Sibley LD, Murphy KM. 2011. CD8 alpha dendritic cells are the critical source of IL-12 controlling acute infection by *Toxoplasma gondii* tachyzoites. Immunity 35: 249–259.

Mckee AS, Dzierszinski F, Boes M, Roos DS, Pearce EJ. 2004. Functional inactivation of immature dendritic cells by the intracellular parasite *Toxoplasma gondii*. Journal of Immunology 173: 2632–2640.

Mead PS, Slutsker L, Dietz V, Mccaig LF, Bresee JS, Shapiro C, Griffin PM, Tauxe RV. 1999. Food-related illness and death in the United States. Emerging Infectious Diseases 5: 607–625.

Meissner M, Brecht S, Bujard H, Soldati D. 2001. Modulation of myosin A expression by a newly established tetracycline repressor based inducible system in *Toxoplasma gondii*. Nucleic Acids Research 29: E115.

Meissner M, Schluter D, Soldati D. 2002. Role of *Toxoplasma gondii* myosin A in powering parasite gliding and host cell invasion. Science 298: 837–840.

Mercier C, Cesbron-Delauw MF, Ferguson DJP. 2007. Dense granules of the infectious stages of *Toxoplasma gondii*: Their central role in the host-parasite relationship. In: Ajioka JW, Soldati D, eds. *Toxoplasma, Molecular and Cellular Biology*. Horizon, Norfolk, UK, pp. 475–492.

Molestina RE, El-Guendy N, Sinai AP. 2008. Infection with *Toxoplasma gondii* results in dysregulation of the host cell cycle. Cellular Microbiology 10: 1153–1165.

Montoya JG, Liesenfeld O. 2004. Toxoplasmosis. Lancet 363: 1965–1976.

Mordue DG, Monroy F, La Regina M, Dinarello CA, Sibley LD. 2001. Acute toxoplasmosis leads to lethal overproduction of Th1 cytokines. Journal of Immunology 167: 4574–4584.

Munoz M, Liesenfeld O, Heimesaat MM. 2011. Immunology of *Toxoplasma gondii*. Immunological Reviews 240: 269–285.

Nicolle C, Manceaux LH. 1908. Sur une infection a corp de Leishman (ou organismes voisins) du gondi. Comptes Rendus de l Academie des Sciences. Serie III, Sciences de la Vie 147: 763–766.

Ong YC, Reese ML, Boothroyd JC. 2010. Toxoplasma rhoptry protein 16 (ROP16) subverts host function by direct tyrosine phosphorylation of STAT6. The Journal of Biological Chemistry 285: 28731–28740.

Peixoto L, Chen F, Harb OS, Davis PH, Beiting DP, Brownback CS, Ouluguem D, Roos DS. 2010. Integrative genomics approaches highlight a family of parasite-specific kinases that regulate host responses. Cell Host & Microbe 8: 208–218.

Pena HF, Gennari SM, Dubey JP, Su C. 2008. Population structure and mouse-virulence of *Toxoplasma gondii* in Brazil. International Journal for Parasitology 38: 561–569.

Pfaff AW, Liesenfeld O, Candolfi E. 2007. Congenital toxoplasmosis. In: Ajioka JW, Soldati D, eds. *Toxoplasma: Molecular and Cellular Biology*. Horizon Bioscience, Norfolk, UK, pp. 93–110.

Pfefferkorn ER, Pfefferkorn LC, Colby ED. 1977. Development of gametes and oocysts in cats fed cysts derived from cloned trophozoites of *Toxoplasma gondii*. Journal of Parasitology 63: 158–159.

Pfefferkorn LC, Pfefferkorn ER. 1980. *Toxoplasma gondii*: Genetic recombination between drug resistant mutants. Experimental Parasitology 50: 305–316.

Phelps ED, Sweeney KR, Blader IJ. 2008. *Toxoplasma gondii* rhoptry discharge correlates with activation of the early growth response 2 host cell transcription factor. Infection and Immunity 76: 4703–4712.

Reese ML, Boothroyd JC. 2009. A helical membrane-binding domain targets the *Toxoplasma* ROP2 family to the parasitophorous vacuole. Traffic 10: 1458–1470.

Reese ML, Zeiner GM, Saeij JP, Boothroyd JC, Boyle JP. 2011. Polymorphic family of injected pseudokinases is paramount in *Toxoplasma* virulence. Proceedings of the National Academy of Sciences of the United States of America 108: 962509630.

Robben PM, Sibley LD. 2004. Food- and waterborne pathgens: You are (infected by) what you eat! Microbes and Infection 6: 406–413.

Robben PM, Mordue DG, Truscott SM, Takeda K, Akira S, Sibley LD. 2004. Production of IL-12 by macrophages infected with *Toxoplasma gondii* depends on the parasite genotype. Journal of Immunology 172: 3686–3694.

Rosowski EE, Lu D, Julien L, Rodda L, Gaiser RA, Jensen KD, Saeij JP. 2011. Strain-specific activation of the NF-kappaB pathway by GRA15, a novel *Toxoplasma gondii* dense granule protein. The Journal of Experimental Medicine 208: 195–212.

Sabin AB. 1941. Toxoplasmic encephalitis in children. Journal American Medical Association 116: 801–807.

Saeij JPJ, Boyle JP, Coller S, Taylor S, Sibley LD, Brooke-Powell ET, Ajioka JW, Boothroyd JC. 2006. Polymorphic secreted kinases are key virulence factors in toxoplasmosis. Science 314: 1780–1783.

Saeij JPJ, Coller S, Boyle JP, Jerome ME, White ME, Boothroyd JC. 2007. *Toxoplasma* co-opts host gene expression by injection of a polymorphic kinase homologue. Nature 445: 324–327.

Scanga CA, Aliberti J, Jankovic D, Tilloy F, Bennouna S, Denkers EY, Medzhitov R, Sher A. 2002. Cutting edge: MyD88 is required for resistance to *Toxoplasma gondii* infection and regulates parasite-induced IL-12 production by dendritic cells. Journal of Immunology 168: 5997–6001.

Shenoy AR, Kim BH, Choi HP, Matsuzawa T, Tiwari S, Macmicking JD. 2008. Emerging themes in IFN-gamma-induced macropahge immunity by the p47 and p65 GTPase families. Immunobiology 212: 771–784.

Sibley LD. 2010. How apicomplexan parasites move in and out of cells. Current Opinion in Biotechnology 21: 592–598.

Sibley LD, Ajioka JW. 2008. Population structure of *Toxoplasma gondii*: Clonal expansion driven by infrequent recombination and selective sweeps. Annual Review of Microbiology 62: 329–351.

Sibley LD, Boothroyd JC. 1992. Virulent strains of *Toxoplasma gondii* comprise a single clonal lineage. Nature 359: 82–85.

Sibley LD, Leblanc AJ, Pfefferkorn ER, Boothroyd JC. 1992. Generation of a restriction fragment length polymorphism linkage map for *Toxoplasma gondii*. Genetics 132: 1003–1015.

Silveira C, Belfort R Jr, Muccioli C, Abreu MT, Martins MC, Victora C, Nussenblatt RB, Holland GN. 2001. A follow-up study of *Toxoplasma gondii* infection in southern Brazil. American Journal of Ophthalmology 131: 351–354.

Soldati D, Boothroyd JC. 1993. Transient transfection and expression in the obligate intracellular parasite *Toxoplasma gondii*. Science 260: 349–352.

Sousa CR, Hieny S, Scharton-Kersten T, Jankovic D, Charest H, Germain RN, Sher A. 1997. In vivo microbial stimulation induces rapid CD40 ligand-independent production of interleukin 12 by dendritic cells and their redistribution to T cell areas. The Journal of Experimental Medicine 186: 1819–1829.

Spear W, Chan D, Coppens I, Johnson RS, Giaccia A, Blader IJ. 2006. The host cell transcription factor hypoxia-inducible factor 1 is required for *Toxoplasma gondii* growth and survival at physiological oxygen levels. Cellular Microbiology 8: 339–352.

Splendore A. 1908. Un nuovo protozoa parassita de' conigli. Incontrato nelle lesioni anatomiche d'une malattia che ricorda in molti punti il Kala-azar dell'uomo. Nota preliminare pel. Revista de Sociedade Scientifica de São Paulo 3: 109–112.

Steinfeldt T, Konen-Waisman S, Tong L, Pawlowski N, Lamkemeyer T, Sibley LD, Hunn JP, Howard JC. 2010. Phosphorylation of mouse immunity-related GTPase (IRG) resistance proteins is an evasion strategy for virulent *Toxoplasma gondii*. PLoS Biology 8: e1000576.

Su C, Howe DK, Dubey JP, Ajioka JW, Sibley LD. 2002. Identification of quantitative trait loci controlling acute virulence in *Toxoplasma gondii*. Proceedings of the National Academy of Sciences of the United States of America 99: 10753–10758.

Su C, Evans D, Cole RH, Kissinger JC, Ajioka JW, Sibley LD. 2003. Recent expansion of Toxoplasma through enhanced oral transmission. Science 299: 414–416.

Suzuki Y, Joh K. 1994. Effect of the strain of *Toxoplasma gondii* on the development of toxoplasmic encephalitis in mice treated with antibody to interferon-gamma. Parasitology Research 80: 125–130.

Suzuki Y, Conley FK, Remington JS. 1989. Differences in virulence and development of encephalitis during chronic infection vary with the strain of *Toxoplasma gondii*. The Journal of Infectious Diseases 159: 790–794.

Suzuki Y, Yang Q, Remington JS. 1995. Genetic resistance against acute toxoplasmosis depends on the strain of *Toxoplasma gondii*. Journal of Parasitology 81: 1032–1034.

Taylor GA, Feng CG, Sher A. 2004. p47 GTPases: Regulators of immunity to intracellular pathogens. Nature Reviews. Immunology 4: 100–109.

Taylor S, Barragan A, Su C, Fux B, Fentress SJ, Tang K, Beatty WL, Haijj EL, Jerome M, Behnke MS, White M, Wootton JC, Sibley LD. 2006. A secreted serine-threonine kinase determines virulence in the eukaryotic pathogen *Toxoplasma gondii*. Science 314: 1776–1780.

Vasconcelos-Santos DV, Machado Azevedo DO, Campos WR, Orefice F, Queiroz-Andrade GM, Carellos EV, Castro Romanelli RM, Januario JN, Resende LM, Martins-Filho OA, De Aguiar Vasconcelos Carneiro AC, Almeida Vitor RW, Caiaffa WT. 2009. Congenital toxoplasmosis in southeastern Brazil: Results of early ophthalmologic examination of a large cohort of neonates. Ophthalmology 116: 2199-205 e1.

Wang Y, Shen J, Arenzana N, Tirasophon W, Kaufman RJ, Prywes R. 2000. Activation of ATF6 and an ATF6 DNA binding site by the endoplasmic reticulum stress response. The Journal of Biological Chemistry 275: 27013–27020.

Wernimont AK, Artz JD, Finerty P, Lin Y, Amani M, Allali-Hassani A, Senisterra G, Vedadi M, Tempel W, Mackenzie F, Chau I, Lourido S, Sibley LD, Hui R. 2010. Structures of apicomplexan calcium-dependent protein kinases reveal mechanism of activation by calcium. Nature Structural & Molecular Biology 17: 596–601.

Yamamoto M, Standley DM, Takashima S, Saiga H, Okuyama M, Kayama H, Kubo E, Ito H, Takaura M, Matsuda T, Soldati-Favre D, Takeda K. 2009. A single polymorphic amino acid on *Toxoplasma gondii* kinase ROP16 determines the direct and strain-specific activation of Stat3. The Journal of Experimental Medicine 206: 2747–2760.

Yamamoto M, Ma JS, Mueller C, Kamiyama N, Saiga H, Kubo E, Kimura T, Okamoto T, Okuyama M, Kayama H, Nagamune K, Takashima S, Matsuura Y, Soldati-Favre D, Takeda K. 2011. ATF6-beta is a host cellular target of the *Toxoplasma gondii* virulence factor ROP18. The Journal of Experimental Medicine 208: 1533–1546.

Yap GS, Sher A. 1999a. Cell-mediated immunity to *Toxoplasma gondii*: Initiation, regulation and effector function. Immunobiology 201: 240–247.

Yap GS, Sher A. 1999b. Effector cells of both nonhemopoietic and hemopoietic origin are required for interferon (IFN)-gamma- and tumor necrosis factor (TNF)-alpha- dependent host resistance to the intracellular pathogen, *Toxoplasma gondii*. The Journal of Experimental Medicine 189: 1083–1091.

Yoshida H, Okada T, Haze K, Yanagi H, Yura T, Negishi M, Mori K. 2001. Endoplasmic reticulum stress-induced formation of transcription factor complex ERSF including NF-Y (CBF) and activating transcription factors 6alpha and 6beta that activates the mammalian unfolded protein response. Molecular and Cellular Biology 21: 1239–1248.

Zhao Y, Ferguson DJ, Wilson DC, Howard JC, Sibley LD, Yap GS. 2009. Virulent *Toxoplasma gondii* evade immunity-related GTPase-mediated parasite vacuole disruption within primed macrophages. Journal of Immunology 182: 3775–3781.

CHAPTER 16

VIRULENCE IN AFRICAN TRYPANOSOMES: GENETIC AND MOLECULAR APPROACHES

ANNETTE MACLEOD, LIAM J. MORRISON, and ANDY TAIT

INTRODUCTION

The study of the key genetic determinants of virulence in African trypanosomes is still at an early stage, although there have been recent advances of our understanding both at the level of the parasite and the host. In this chapter, we will discuss these developments, as well as consider the future opportunities provided by both the availability of a system for laboratory genetic analysis and the rapid advances in genome sequencing technology, which allow new approaches. In considering virulence, it is important to clarify the context within which it is defined. A priori, many genes can have an effect on virulence when defined as the growth of the parasite *in vivo* or *in vitro* or the level of parasitemia. The availability of efficient reverse genetic and RNAi tools (particularly in one subspecies of trypanosome, *Trypanosoma brucei brucei*) means that genes can be knocked out or their expression knocked down and their function essentially removed. In many cases, the resultant genetically manipulated parasites will have reduced growth or rates of replication and so will be less virulent. However, such drastic events involving essentially gene deletion are likely to be very rare or nonexistent in natural populations of the parasite. This is particularly pertinent with respect to quantitative phenotypes such as growth and virulence, where allelic variation will be the main determinant of phenotypic alterations on which selection and evolution will act. So, if we are to understand the range of virulence that is evident in the natural host, it is necessary to identify the genes, and particularly the alleles of those genes, that are associated with variation in virulence.

African Trypanosomes

In considering virulence in this important group of parasitic protozoa, it is important to recognize that there are a large number of species of African trypanosomes. However, three species are particularly important in terms of disease; *Trypanosoma congolense*, *Trypanosoma vivax*, and *Trypanosoma brucei*. They infect a broad range of mammals across a wide area of sub-Saharan Africa and are predominantly transmitted from host to host by the insect

Evolution of Virulence in Eukaryotic Microbes, First Edition. Edited by L. David Sibley, Barbara J. Howlett, and Joseph Heitman.

vector—the tsetse fly. Infection of livestock results in significant losses, with some 60 million cattle at risk (as well as significant numbers of small ruminants and equines) and an estimated financial burden of around $1300 million per annum (Shaw, 2004). In addition, the *T. brucei* subspecies *T. b. gambiense* and *T. b. rhodesiense* cause approximately 70,000 cases of human African trypanosomiasis per year (WHO, 2006), although this number is probably a substantial underestimate (Fevre et al., 2008). Using molecular markers, it has been shown that *T. vivax* can be divided into two distinct clades, one found in West Africa and the other in East Africa (Gardiner, 1989). Similarly, *T. congolense* has been subdivided into three distinct clades (Savanah, Forest, and Kilifi), although these are not restricted to specific geographic regions and may constitute separate subspecies (Gashumba et al., 1988; Majiwa et al., 1986). Different species, subspecies, and clades are all transmitted by the tsetse fly vector and so are distributed throughout sub-Saharan Africa where this fly is prevalent. In contrast to the other species mentioned, *T. vivax*, together with *Trypanosoma evansi* (closely related to *T. brucei*), can also be efficiently transmitted by several groups of biting flies (mechanical transmission), and this has led to these species having a broader geographic range, including South America and many regions of the Middle East and Asia (Hoare, 1972; Jones and Davila, 2001).

The human infective trypanosomes, *T. b. rhodesiense* and *T. b. gambiense*, infect humans and animals and are therefore zoonotic, while *T. b. brucei* only infects domestic and wild animals throughout sub-Saharan Africa. In addition to the classically described *T. b. gambiense* (group 1), a distinct clade of *T. b. gambiense* (group 2) has been identified by marker analysis (Balmer et al., 2011; Gibson, 1986). The human infective subspecies exist as geographically discrete, long established foci across Africa with *T. b. rhodesiense* restricted to East and Southern Africa and *T. b. gambiense* to West and Central Africa (Hoare, 1972). The genetic relatedness between these subspecies has been a matter of debate, but based on a large body of work from the 1980s, it was concluded that *T. b. rhodesiense* was a host range variant of *T. b. brucei*, while *T. b. gambiense* was considered a distinct subspecies (Gibson, 1986; Gibson et al., 1980; Godfrey et al., 1987; Mehlitz et al., 1982; Tait et al., 1984). More recently, a set of isolates of all three subspecies from across Africa was analyzed phylogeographically using sequence data from the mitochondrial cytochrome oxidase subunit 1 gene (CO1) and microsatellite polymorphisms, and showed that the interpretation of the earlier work was correct in terms of the genetic relatedness of the subspecies (Balmer et al., 2011). Furthermore, the *T. b. gambiense* group 2 isolates clustered as a distinct group that was more closely related to East African *T. b. brucei* and *T. b. rhodesiense* than to *T. b. gambiense* group 1. Therefore, it can be seen that the situation as regards speciation is complex in trypanosomes, and this complexity will clearly impact upon genetic diversity and phenotype expression.

The Genetic System

How these parasites generate genetic diversity is a key question, and their ability or otherwise to undergo sexual recombination will obviously play a significant role in our understanding of phenotypic variation. Trypanosomes are unicellular protozoa that have relatively small genomes of ~35 Mb (Berriman et al., 2005) and are diploid (Turner et al., 1990). During their life cycle in the mammalian host and tsetse vector, the parasite undergoes a series of morphological and biochemically distinct stages associated with the adaptation (or preadaptation) to the different environments (mammalian bloodstream, tsetse gut, and salivary glands/mouthparts) that they encounter. It has been known since 1986 that genetic crosses can be undertaken in the laboratory between different strains of *T. brucei* (Jenni et al., 1986), and it is now clear that mating takes place in the salivary gland stages of the tsetse fly (Gibson et al., 2008; Peacock et al., 2011; Tait et al., 2007). Available evidence supports a single round of mating between parental strains when they coinfect a tsetse fly, with the products of mating being the equivalent of F1 progeny in a diploid Mendelian genetic system (Gibson and Stevens, 1999; MacLeod et al., 2007). Analysis of the segregation of parental markers in such F1 progeny is consistent with the occurrence of meiosis (MacLeod et al., 2005a; Turner et al., 1990), and this has been more directly shown by the detection of the expression of several meiosis-specific genes in morphologically distinct salivary gland stages of the parasite (Peacock et al., 2011). Importantly, unlike most other protozoan parasites such as *Plasmodium*, mating is not an obligatory part of the life cycle (Gibson et al., 2008; Schweizer et al., 1988). The products of mating (both cross-fertilization between different parasite strains and self-fertilization; Peacock et al., 2009; Tait et al., 1996) occur together with parasites that have not undergone sexual recombination in the same tsetse transmission. To date, some 12 crosses between different strains and subspecies have been undertaken, with genetic exchange being demonstrated between different strains of *T. b. brucei*, *T. b. brucei* and *T. b. gambiense* group 2; *T. b. rhodesiense* and *T. b. brucei* and *T. b. rhodesiense* and *T. b. gambiense* group 2 (Gibson and Stevens, 1999; MacLeod et al., 2007). Thus, there appear to be limited barriers to mating even between subspecies under laboratory conditions. The one notable exception is that no crosses with *T. b. gambiense* group 1 have been reported. Genetic maps of both *T. b. brucei* and *T. b. gambiense* group 2 have been constructed using >120 microsatellite markers, and the resultant linkage groups align with the physical map of the genome provided by the *T. b. brucei* TREU927 genome sequence (Berriman et al., 2005; Cooper et al., 2008; MacLeod et al., 2005a). The construction and analysis of the genetic maps has illustrated two further conventional properties of the trypanosome genetic system, namely, crossing over between pairs of homologous chromosomes and regions of high and low recombination (hot and cold spots) along a chromosome. Thus, overall *T. brucei* has a conventional diploid, Mendelian genetic system in common with many other eukaryotes. Although the system has a number of unique features, it does provide the potential for the identification of the genetic determinants of important phenotypes by classical genetic analysis (Tait et al., 2002).

Laboratory genetic analysis has focused on *T. brucei* and no reports have been published on crosses between the other species of trypanosome. Recent population genetic analysis of *T. vivax* (Duffy et al., 2009) and *T. congolense* (Morrison et al., 2009b) have shown that there is no evidence for mating in the former species but good evidence for genetic exchange in *T. congolense* (Savannah). While the data for *T. vivax* do not mean that genetic exchange does not occur in some epidemiological situations, the data for *T. congolense* provide strong evidence for its occurrence. The recent report of the ability to replicate the whole life cycle of *T. congolense in vitro* (Coustou et al., 2010) offers the opportunity of testing whether crosses with this species could be undertaken in the laboratory.

Genetic Diversity

If genetic or population genomic analysis is to be used as a tool to identify the genes determining phenotypes such as virulence, parasite strains need to show a significant level of both genotypic and phenotypic diversity. Studies over many years have shown that there are high levels of genotypic strain diversity within each species of African trypanosome (Agbo et al., 2002; Duffy et al., 2009; MacLeod et al., 2000; Morrison et al., 2009b) even though the population structures can vary from clonal to panmictic in relation to the role of genetic exchange in generating this diversity (Koffi et al., 2009; MacLeod et al., 2000). Similarly, albeit on a smaller sample set, variation in the phenotype of different strains has been reported, for example, in drug resistance (Kibona et al., 2006), tsetse transmission (Masumu et al., 2006b; Welburn et al., 1995), and in virulence (Morrison, 2011). There have been significant advances in our understanding of drug resistance using molecular and biochemical approaches (reviewed by Delespaux and de Koning 2007) but limited or no research on the basis of variation in tsetse transmission. However, there have been recent advances in our understanding of virulence from both a host and parasite perspective using a range of approaches.

THE VIRULENCE PHENOTYPE

Virulence has been defined and measured in a number of different ways in a range of pathogens, and this is also the case with African trypanosomes. At one extreme, strains are defined as virulent on the basis of causing mortality and at the other of having a higher growth rate but with no obvious impact on their host (although the latter scenario may possibly affect transmission). In the absence of a full understanding of the parasite genes that are involved, it is difficult to determine which measured phenotypes are the outcomes of a single determinant and which are due to multiple or separate determinants. In considering what genes are likely to be responsible for virulence of relevance to the field, a direct approach is to use the variation between recently field-isolated strains as a tool. The alternative approach of using reverse genetics or RNAi in screens for altered growth or replication is likely to identify many nonessential genes, which could include those responsible for virulence in the field. Equally, as the field phenotypes are likely to be determined by allelic variation in genes rather than deletions or a knockdown of expression, the reverse genetic approach might not detect these genes. The definition of virulence is further complicated by the fact that different host genetic backgrounds can influence the virulence phenotype of a parasite strain. Virulence is even more difficult to define when considering clinical data from the natural host and in the natural setting, where an additional layer of complexity can be added, for example, in livestock where it is common to find several species of trypanosome infecting a single host. These considerations have meant than in order to define and measure virulence, host diversity has been minimized by laboratory studies using inbred strains of mice, but of course, such studies raise questions about the relevance to the field situation.

Species and Subspecies Variation

Despite these caveats, differences in human disease severity between infections with the different subspecies of *T. brucei* are well recognized, with *T. b. rhodesiense* classically causing acute disease and *T. b. gambiense* group 1 causing a more chronic infection (Hoare, 1972). However, there is also a range of clinical outcomes within these two subspecies, for example, both "mild" and "severe" *T. b. rhodesiense* diseases are observed in sleeping sickness foci in Malawi and Uganda (MacLean et al., 2004), and this distinction can occur even between geographically different foci within Uganda (Maclean et al., 2007; Sternberg and Maclean, 2010), where it was suggested that the different parasite genotypes could have been responsible for the differing levels of the severity of human African trypanosomiasis. Although these studies provided strong circumstantial evidence for a spectrum of virulence influenced by parasite variation, the divergence of clinical signs could potentially be due to the host or other, as yet undetermined, factors. In *T. b. gambiense* group 1 infections, there is a range of clinical features of the human disease. The two main parameters that have been monitored are the progression from early stage (infection of the tissues and bloodstream—stage 1) to late stage (infection of the brain—stage 2). Recently, asymptomatic individuals have also been identified with detectable parasites and high antibody titers, yet no clinical progression. Although there is suggestive evidence that variation in the virulence of the parasites may be the cause of this heterogeneity in disease phenotype (Jamonneau et al., 2000, 2004), again it is also possible that the different symptoms are due to host variation (Bucheton et al., 2011; Garcia et al., 2006; Kabore et al., 2011).

In livestock trypanosomiasis, *T. congolense* is considered to be more virulent than *T. vivax*, largely based on the level of anemia (measured as the packed cell volume [PCV] of red blood cells) induced by infection (Stephen, 1970). However, there are reports of hemorrhagic variants of *T. vivax* (Magona et al., 2008; Wellde et al., 1983), which are highly pathogenic (virulent). Because livestock in the field can be coinfected with several trypanosome species, direct clinical monitoring is potentially problematic for comparative analysis between species. However, experimental infection of both bovine and murine hosts has shown differences in virulence/pathogenicity between the three different clades (Savannah, Kilifi, and Forest) of *T. congolense* (Bengaly et al., 2002a,b). In these studies, there was good concordance in the clinical parameters measured (survival, prepatent period, and drop in PCV) between mice and cattle.

Strain-Specific Diversity

Although identifying differences in virulence between different trypanosome species, subspecies, and clades supplies important information in terms of understanding the disease profile, such studies do not provide an obvious route to defining the basis for these differences at the level of the gene and the molecule. At the present time, there are potentially three routes toward such a goal: first, biochemical/molecular analysis of candidate genes *in vitro* and *in vivo* models of the disease process; second, the inheritance of the phenotype in crosses of different strains; and third, population genomics coupled with association analysis. In addition, while not a separate approach, it would be important to screen for differences in gene expression as recent studies in other organisms have suggested that variation in *cis*-acting elements are responsible for a large component of phenotypic variation (Goring et al., 2007; Stranger et al., 2007). Such an analysis could be incorporated as a component of genetic analysis to analyze so-called expression quantitative trait loci (e-QTLs). These approaches are discussed in more detail in the next section. For the second and third approaches, it is necessary to identify strain variation in the virulence phenotype within a species or subspecies as both rely on identifying genes that determine strain-specific variation in the phenotype of interest. While the clinical studies in humans imply that such variation occurs, the potential for other factors and determinants being responsible for the differences in clinical disease is a real possibility. So what is the available evidence for strain variation in virulence?

TABLE 16.1 Measures of Trypanosome Virulence

Phenotype	Parasite Species	Host	References
Survival	*T. b. gambiense* and *T. congolense*	Mouse	1, 2
Prepatent period	*T. b. gambiense* and *T. congolense*	Mouse	1, 2
Maximum parasitemia	*T. b. gambiense* and *T. congolense*	Mouse	1, 2
Anemia (PCV)[a]	*T. congolense* and *T. b. brucei*	Mouse	2, 3
Organomegaly	*T. b. brucei*	Mouse	3
Reticulocytosis	*T. b. brucei*	Mouse	3
Macrophage activation	*T. b. brucei* and *T. b. gambiense*	Mouse and human	1, 4
Blood–brain barrier	*T. b. brucei* and *T. b. gambiense*	*In vitro*	5
Stage1/2 progression	*T. b. rhodesiense*	Human	6
Assymptomatic	*T. b. gambiense*	Human	7

[a] Packed cell volume of red blood cells.

References:

1. Holzmuller et al., 2008.
2. Masumu et al., 2006a.
3. Morrison et al., 2009a.
4. Morrison et al., 2010.
5. Grab et al., 2004.
6. MacLean et al., 2004.
7. Jamonneau et al., 2004.

The majority of studies have used the mouse as a host and it should be stressed that caution needs to be taken in translating any findings into another host. However, once studies reach the point of defining the genes involved, appropriate field studies could be used to validate any experimental findings from the mouse model. The measures that have been used to define virulence are summarized in Table 16.1 and this illustrates the multiplicity of phenotypes that have been used to define virulence in both model and natural hosts. Studies in *T. brucei* have demonstrated that there are significant differences between strains in terms of various measures of virulence. With *T. b. gambiense* group 1, 10 strains from Cote d'Ivoire were used to infect mice and differences were observed in infectivity, mortality, prepatent period, and the level of parasitemia between strains. Two strains with the most different features were studied further and statistically significant differences were demonstrated in the prepatent period, maximum parasitemia, and survival time (Holzmuller et al., 2008). Additionally, differences in the ability to stimulate arginase activity in isolated macrophages were also demonstrated between the two strains (Holzmuller et al., 2008). Strain-specific differences have also been demonstrated in *T. b. brucei* infections of mice, but, in contrast to the *T. b. gambiense* group 1 study, no differences in mortality or maximum parasitemia were reported, but significant differences in organomegaly, reticulocytosis, and anemia were found (Morrison et al., 2010). In a further study, 31 strains of *T. congolense* (Savannah) from several locations in Zambia were classified into three groups: virulent, moderately virulent, and low virulence (Masumu et al., 2006a). The parameters that differed between the groups were survival time, prepatent period, and PCV. Thus, overall, there is strong evidence for variation in virulence in both *T. brucei* and *T. congolense* as measured by survival time, prepatent period, anemia (PCV), maximum parasitemia, organomegaly, and reticulocytosis. It seems likely that these have a genetic basis and are unlikely to be determined by a single gene.

GENETIC BASIS OF VIRULENCE

One approach is to investigate candidate genes, but, given the number of potentially independent phenotypes, this could be problematic. Therefore, it is important to use methods that exploit the extensive natural phenotypic variation. Additionally, the genes determining this natural variation are obviously more relevant to virulence in the field and so important for understanding the evolution of virulence. The observed diversity in the virulence phenotype opens up genetic approaches on a genome-wide scale for the identification of the parasite genes that determine these phenotypes. In this context, two approaches can be considered: first, crosses between strains that differ in virulence phenotypes and, second, the as yet untried approach with trypanosomes, population genomics using association analysis between phenotype and genotype.

Quantitative Trait Locus (QTL) Analysis

Given the number of different phenotypes that can be defined as measures of virulence (Table 16.1), it is likely that more than one gene is involved. Genetic analysis can be used to accommodate this by treating the segregation of the phenotype in crosses as a quantitative trait and assigning the variance in the phenotype to several loci (QTL analysis). The genetic approach taken is illustrated in Figure 16.1 and is based on the analysis of the phenotype in infections of the F1 progeny, followed by linkage analysis to the markers on the existing genetic map (MacLeod et al., 2005b).

In the cross used for the analysis (strain TREU927 × strain STIB247), the availability of the F1 progeny allows the linkage analysis of loci that are heterozygous in one of the parents (formally the equivalent of F2 progeny in this context) but does not allow the analysis of loci that are homozygous but different between the parental lines. There has been one study undertaken (from our laboratory) which mapped the loci that determine strain-specific differences in organomegaly, reticulocytosis, and anemia (Morrison et al., 2009a). The progeny from a cross between two strains that differed in these phenotypes was scored (in mice) for each of the parameters and these were shown to segregate in a semiquantitative manner. This segregation pattern implied that these phenotypes are determined by allelic variation of several parasite genes, and so a genetic linkage analysis was undertaken assuming these were QTLs as illustrated diagrammatically in Figure 16.1. Splenomegaly and hepatomegaly showed evidence for a highly significant QTL (LOD scores >7) on chromosome 3 accounting for 66% and 64%, respectively, of the phenotypic variance, thus demonstrating a parasite gene or genes determine a major component of these phenotypes. The region on chromosome 3 is large with over 300 genes, but this analysis does show that the phenotypes are genetically determined and offers an approach to identify the individual genes involved. Further analysis would require fine-scale mapping with more markers, analysis of more progeny clones to identify crossovers within the designated region, analysis of stage-specific expression of the genes as one could assume that the genes involved must be expressed in the bloodstream stage of the parasite (see Morrison et al., 2009a for a discussion). Once the number of genes is reduced by these analyses, a reverse genetic approach would be feasible to test which alleles/genes are responsible for the phenotype. In addition

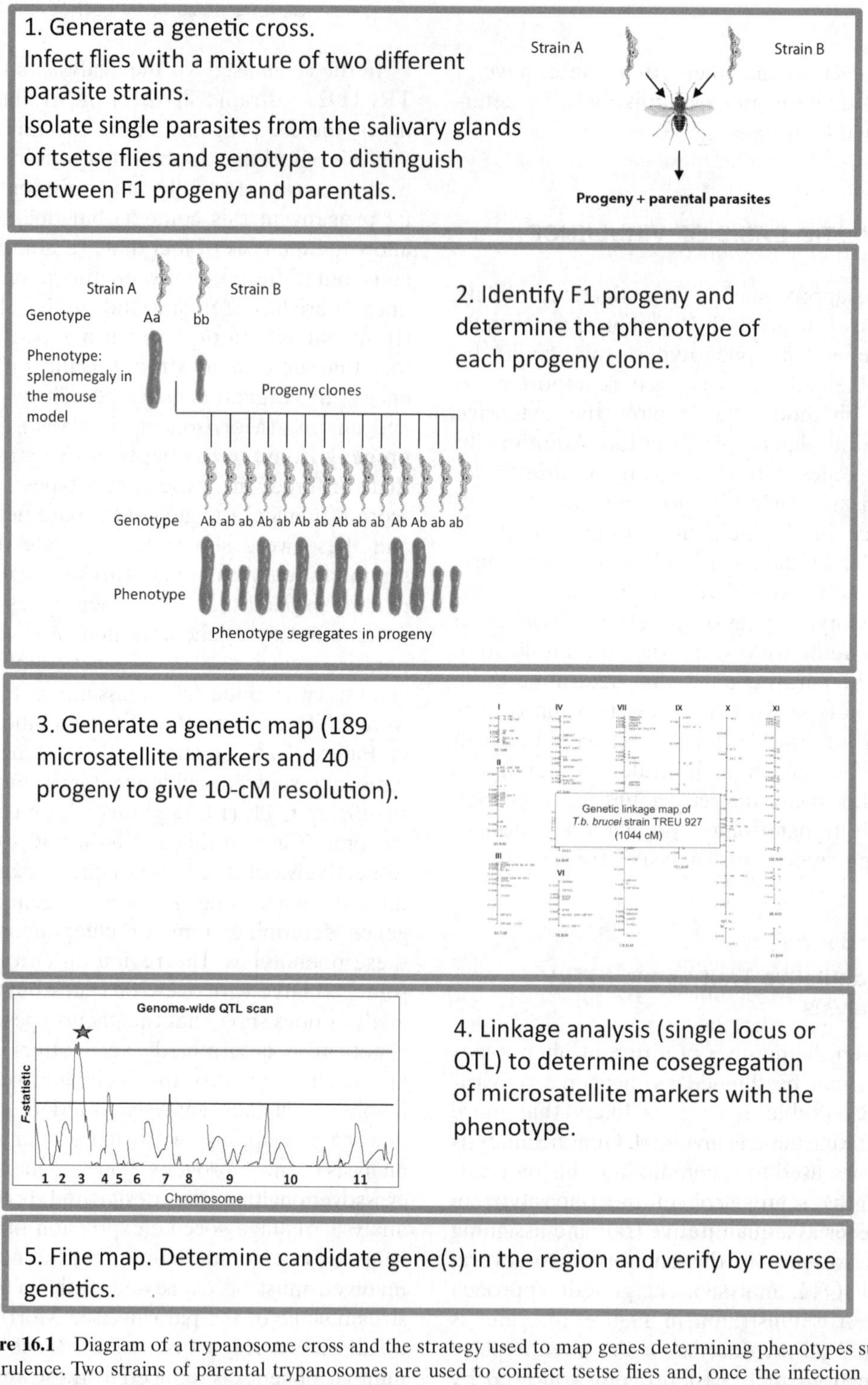

1. Generate a genetic cross.
Infect flies with a mixture of two different parasite strains.
Isolate single parasites from the salivary glands of tsetse flies and genotype to distinguish between F1 progeny and parentals.
Strain A
Strain B
Progeny + parental parasites
Strain A
Strain B
Genotype
Aa
bb
Phenotype: splenomegaly in the mouse model
Progeny clones
2. Identify F1 progeny and determine the phenotype of each progeny clone.
Genotype
Ab ab ab Ab ab Ab ab Ab ab ab Ab Ab ab ab
Phenotype
Phenotype segregates in progeny
3. Generate a genetic map (189 microsatellite markers and 40 progeny to give 10-cM resolution).
I
II
III
IV
VI
VII
IX
X
XI
Genetic linkage map of
T.b. brucei strain TREU 927
(1044 cM)
Genome-wide QTL scan
F-statistic
1 2 3 4 5 6 7 8 9 10 11
Chromosome
4. Linkage analysis (single locus or QTL) to determine cosegregation of microsatellite markers with the phenotype.
5. Fine map. Determine candidate gene(s) in the region and verify by reverse genetics.

Figure 16.1 Diagram of a trypanosome cross and the strategy used to map genes determining phenotypes such as virulence. Two strains of parental trypanosomes are used to coinfect tsetse flies and, once the infection has developed to the salivary gland stages and after mating, individual trypanosomes are expanded vegetatively. These "clones" are then genotyped with microsatellite markers to define independent F1 progeny and subsequently phenotyped by infection of mice. In this example, splenomegaly is illustrated as a phenotype that differs between parental strains A and B and segregates in the progeny. The phenotype is treated as being determined by more than one locus and the segregation of the quantitative trait used to map the loci by linkage analysis to the genetic map. See color insert.

to this locus, significant QTLs were identified for reticulocytosis, anemia and organomegaly on other chromosomes (Morrison et al., 2009a). The importance of this work is that it shows that genetic analysis can be used to map the loci determining natural variation in virulence and provides a route to identifying the genes involved. To investigate other phenotypes that did not differ between the two parental lines, it would be necessary to undertake crosses between other parasite strains or to phenotype strains that have already been crossed (MacLeod et al., 2007). The discovery of mating in *T. congolense* (Morrison et al., 2009b) would potentially also allow the genetic analysis of virulence phenotypes in this species, particularly as all stages of the *T. congolense* life cycle have now been replicated *in vitro* (Coustou et al., 2010). However, this would require considerable investment in setting up a system for laboratory crosses.

Potential Association Studies

The recent development of more rapid and much cheaper genomic sequencing technologies opens up the possibility of another genetic approach in identifying the genes determining virulence phenotypes using population genomics. This approach has been used for the analysis of the genes determining drug resistance in *Plasmodium falciparum* (Mu et al., 2010) and the analysis of a wide range of phenotypes in the yeast *Saccharomyces cerevisiae* (Liti et al., 2009). In principle (with trypanosomes), a large collection of strains from one or a series of populations need to be phenotyped for virulence in mice and then the genome sequence of each strain determined. By grouping the strains into different phenotypic classes and comparative analysis of the genome sequences of the strains in each class, it would be possible to define single-nucleotide polymorphisms (SNPs), copy number variants, or haplotypes that uniquely associate with the phenotype and so define genes that are candidates for determining virulence. This approach is illustrated in Figure 16.2, where it can be seen that a particular haplotype associates with the virulent phenotype, thus identifying a region of the genome where a virulence gene lies. Additionally, strain variation in the level of RNA transcripts could also be analyzed (transcriptome analysis) and any difference used to associate with virulence and thus identify genes determining virulence.

The size of the haplotype will depend on whether recombination occurs at a high or low rate. Identifying a specific gene within such a region could be undertaken in a similar way to genetic analysis once a locus is identified. Thus, genes not expressed in the bloodstream stages and those not encoding a secreted protein could be eliminated, narrowing down the number of candidate genes. These could be subsequently tested by functional analysis using reverse genetics. With this approach, there are a number of unknowns such as the size of the haplotypes, the level of mating, the level of required polymorphism and, therefore, the number of strains required to give the analysis sufficient power. However, these could readily be addressed by a large-scale strain sequencing project. The major advantage of this approach is that it could be applied to any phenotype for which there is data on diversity.

Biochemical/Cell Biological Analysis

An additional approach, not dependent on strain variation, is to use *in vitro* systems such as macrophages or cultured brain cells. If protein extracts can be shown to produce a phenotypic effect, then fractionation of the proteins and use of their amino acid sequence can be used to identify the gene(s) involved. Alternatively, candidate genes could be tested either by expressing their products as recombinant proteins or

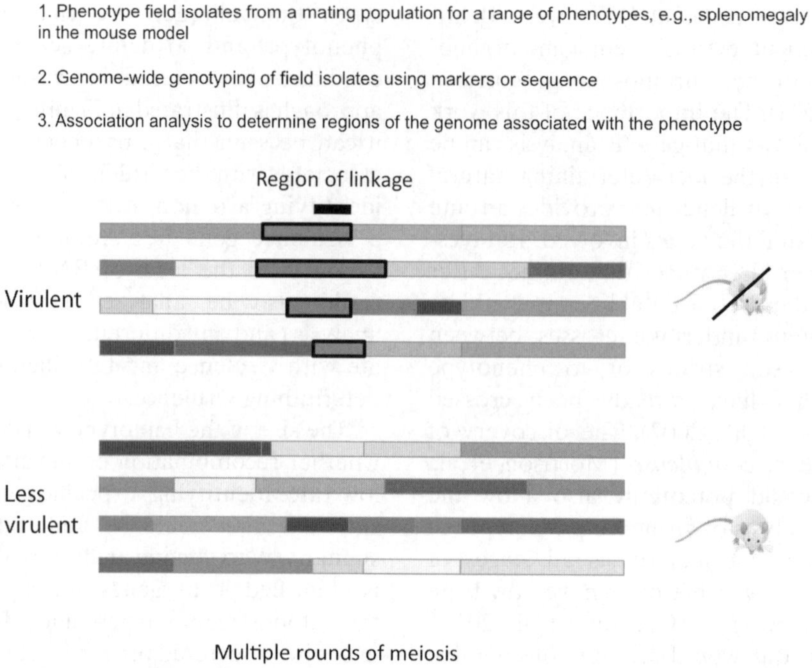

Figure 16.2 Population genomic association analysis. The diagram illustrates the division of strains into two phenotypic classes (virulent and less virulent) using one or more measures of virulence The isolates are then genotyped using markers distributed over the whole genome or by whole-genome sequencing. In this example, the haplotype structure of each strain is illustrated just using a single chromosome (for simplicity). The red haplotype appears to be associated with high virulence and is not present in the avirulent class. Other haplotypes are randomly distributed across the virulence classes and so show no association. Thus, the genes determining the particular virulence phenotype are located within the region of the red haplotype common to all strains. In a population with a high frequency of mating, many rounds of meiosis will have occurred (rather than a single round in the laboratory cross) and so the common haplotype could be physically quite small. See color insert.

by knocking down expression in the trypanosome to mimic the phenotypic effect. For example, the penetration of brain microvascular endothelial cells by *T. brucei* can be measured *in vitro* and provides a model system for invasion of the blood–brain barrier by the parasite (Grab et al., 2004). It has been shown that this process is blocked when inhibitors of a parasite-secreted/excreted cysteine proteinase (brucipain) are introduced and so suggests that this proteinase mediates penetration. From the perspective of parasite diversity, experiments have been undertaken with *T. b. rhodesiense* (previously described in error as *T. b. gambiense*) and two strains of *T. b. brucei*. The results show that the human infective subspecies crosses the endothelial cells six times as effectively as the *T. b. brucei* strains and has an eightfold higher level of proteinase activity, suggesting that this is likely to be responsible for the difference in the ability to penetrate the endothelial monolayer (Nikolskaia et al., 2006). These studies suggest that this cysteine proteinase is a virulence factor, although to date, the findings made *in vitro* have not been examined *in vivo*. A further example of a system that has the potential to identify virulence genes is the activation of macrophage arginase by *T. b. gambiense* (Holzmuller et al., 2008), where it was shown that two strains with different *in vivo* virulence phenotypes induced different levels of arginase expression in macrophages isolated from the infected animals. Interestingly, coincubation of macrophages from uninfected mice *in vitro* with the two parasite strains lead to the same differential activation of arginase. It was found that the secreted/excreted trypanosome proteins could reproduce this effect on macrophages, and these findings open up the possibility of biochemical fractionation of the secretome/excretome to identify the active protein(s). Furthermore, as extracts from the two strains differ in their ability to induce arginase, this adds a further criterion for identifying the active protein(s). In a different study, the secretome of two strains of *T. congolense*, which differ in their virulence phenotype (Bengaly et al., 2002a,b), has been characterized and differences in the level of expression of 21 proteins identified (Grebaut et al., 2009), although their effect upon macrophage arginase expression was not examined. These results would potentially allow the genes encoding these protein differences to be tested as determinants of virulence.

CONCLUSIONS AND FUTURE PROSPECTS

Research on the genetic basis of trypanosome virulence is at a relatively early stage, although it is clear that there is both inter- and intraspecies variation in a number of measures of virulence. Molecular/biochemical analysis has identified a specific cysteine proteinase (brucipain) that mediates the transfer of parasites across the cell monolayers in an *in vitro* model of the blood–brain barrier (Grab et al., 2004). RNAi knockdown of the expression of brucipain *in vivo* results in increased survival and a reduction in the level of splenomegaly (Abdulla et al., 2008), suggesting that this gene affects multiple virulence phenotypes, although the *in vivo* effect on penetration of the blood–brain barrier was not tested in these experiments. Different strains of *T. b. gambiense* that differ in virulence and in their secreted/excreted proteins have been shown to differentially activate host macrophage arginase, suggesting that secreted proteins mediate this effect (Holzmuller et al., 2008). The identity of the protein(s) is not known but could potentially include brucipain. However, studies of the differences in virulence between two strains of *T. b. brucei* have shown that the differences are associated with the differential activation of the innate immune response in the spleen, which also at least in part involves arginase (Morrison et al., 2010). The major locus determining splenomegaly has been mapped (Morrison et al., 2009a), and this locus does not include the gene encoding brucipain, thus suggesting different gene(s) are involved and implicated as virulence factors. The genetic analysis has also identified a further locus on chromosome 2 (determining splenomegaly/reticulocytosis), which again does not include the gene for brucipain. Thus, overall, at least three loci/genes have been identified as virulence determinants of variant phenotypes, although this is obviously a very minimum estimate. The genetic analysis suggests that further loci are involved, but the linkage is based on single markers so further analysis is required to determine whether these are significant. With a minimum of three different genes and their allelic variants identified, virulence is clearly a complex phenotype, and this is consistent with the range of different phenotypes observed during infection (Table 16.1). None of the studies to date have investigated the basis for the variation in prepatent period and level of parasitemia observed between strains (Table 16.1). Whether these phenotypes are also determined by the genes/loci already identified or whether further loci are involved is an open question.

All the studies discussed rely on either *in vitro* or mouse models of virulence, and so an important question is whether the genes/loci identified are relevant to clinical infections in the field. The analysis of such clinical infections is complicated both by practical issues such as the difficulties of making a very detailed and longitudinal analysis in the field and by the fact that there is significant genetic diversity in the host compared to that in inbred lines of mice. However, any parasite genes/alleles identified in model systems can be examined in natural populations to test for their association with virulence in the "natural" host. With livestock, this could be undertaken experimentally by infecting animals with different parasite variants with defined genes/alleles known to encode determinants of virulence. In the case of humans, one would, of course, have to rely on natural infections. It should be restated, however, that the studies described above are utilizing and examining natural, preexisting, nonselected virulence variation, albeit in experimental hosts. This aspect makes it more likely that any genes identified may well have relevance to virulence in the field.

There is a considerable body of data on the variation in the host response to infection and the genetic basis of tolerance or "resistance" to infection both in human and livestock disease (Courtin et al., 2008; Kemp et al., 1997). In terms of human disease, a number of association studies have been undertaken and evidence for specific host genetic variants obtained, which can affect the response to infection (reviewed by Courtin et al., 2008). In the case of animal trypanosomiasis (using *T. congolense*), genetic analysis in mice and, importantly, cattle has been undertaken to identify loci that are major determinants of disease outcome (Hanotte et al., 2003; Hill et al., 2005; Iraqi et al., 2000; Nganga et al., 2010; Noyes et al., 2011)—commonly referred to as "trypanotolerence" loci. Thus, the available evidence suggests that there is a significant host genetic component that determines how virulent an infection can be. What we do not know is how the host and parasite components of virulence interact. Put simply, what is the dominance relationship between a virulent parasite and a tolerant host? The tools and technologies are now available to define the host and parasite loci that are involved and to identify the alleles that are responsible for different phenotypes. Once these have been identified, it would be possible to begin to define the multiple routes to symptomatic and asymptomatic infection and identify the essential interactions between parasite molecules and host pathways. While identifying these would be an important advance in our understanding of virulence and pathogenesis, there would also be the opportunity to develop interventions that could prevent or minimize the pathological consequences of infection.

ACKNOWLEDGMENTS

A.M.L. would like to thank the Wellcome Trust for financial support through a Wellcome Trust Career Development Fellowship and L.J. M. would like to thank the Royal Society (London) for financial support through a University Research Fellowship.

REFERENCES

Abdulla MH, O'Brien T, Mackey ZB, Sajid M, Grab DJ, McKerrow JH. 2008. RNA interference of *Trypanosoma brucei* cathepsin B and L affects disease progression in a mouse model. PLoS Neglected Tropical Diseases 2: e298.

Agbo EE, Majiwa PA, Claassen HJ, te Pas MF. 2002. Molecular variation of *Trypanosoma brucei* subspecies as revealed by AFLP fingerprinting. Parasitology 124: 349–358.

Balmer O, Beadell JS, Gibson W, Caccone A. 2011. Phylogeography and taxonomy of *Try-*

panosoma brucei. PLoS Neglected Tropical Diseases 5: e961.

Bengaly Z, Sidibe I, Boly H, Sawadogo L, Desquesnes M. 2002a. Comparative pathogenicity of three genetically distinct *Trypanosoma congolense*-types in inbred Balb/c mice. Veterinary Parasitology 105: 111–118.

Bengaly Z, Sidibe I, Ganaba R, Desquesnes M, Boly H, Sawadogo L. 2002b. Comparative pathogenicity of three genetically distinct types of *Trypanosoma congolense* in cattle: Clinical observations and haematological changes. Veterinary Parasitology 108: 1–19.

Berriman M, Ghedin E, Hertz-Fowler C, Blandin G, Renauld H, Bartholomeu DC, Lennard NJ, Caler E, Hamlin NE, Haas B, Bohme U, Hannick L, Aslett MA, Shallom J, Marcello L, Hou L, Wickstead B, Alsmark UC, Arrowsmith C, Atkin RJ, Barron AJ, Bringaud F, Brooks K, Carrington M, Cherevach I, Chillingworth TJ, Churcher C, Clark LN, Corton CH, Cronin A, Davies RM, Doggett J, Djikeng A, Feldblyum T, Field MC, Fraser A, Goodhead I, Hance Z, Harper D, Harris BR, Hauser H, Hostetler J, Ivens A, Jagels K, Johnson D, Johnson J, Jones K, Kerhornou AX, Koo H, Larke N, Landfear S, Larkin C, Leech V, Line A, Lord A, Macleod A, Mooney PJ, Moule S, Martin DM, Morgan GW, Mungall K, Norbertczak H, Ormond D, Pai G, Peacock CS, Peterson J, Quail MA, Rabbinowitsch E, Rajandream MA, Reitter C, Salzberg SL, Sanders M, Schobel S, Sharp S, Simmonds M, Simpson AJ, Tallon L, Turner CM, Tait A, Tivey AR, Van Aken S, Walker D, Wanless D, Wang S, White B, White O, Whitehead S, Woodward J, Wortman J, Adams MD, Embley TM, Gull K, Ullu E, Barry JD, Fairlamb AH, Opperdoes F, Barrell BG, Donelson JE, Hall N, Fraser CM, Melville SE, El-Sayed NM. 2005. The genome of the African trypanosome *Trypanosoma brucei*. Science 309: 416–422.

Bucheton B, Macleod A, Jamonneau V. 2011. Human host determinants influencing the outcome of *T. b. gambiense* infections. Parasite Immunology 33: 438–447.

Cooper A, Tait A, Sweeney L, Tweedie A, Morrison L, Turner CM, MacLeod A. 2008. Genetic analysis of the human infective trypanosome *Trypanosoma brucei gambiense*: Chromosomal segregation, crossing over, and the construction of a genetic map. Genome Biology 9: R103.

Courtin D, Berthier D, Thevenon S, Dayo GK, Garcia A, Bucheton B. 2008. Host genetics in African trypanosomiasis. Infection, Genetics and Evolution 8: 229–238.

Coustou V, Guegan F, Plazolles N, Baltz T. 2010. Complete in vitro life cycle of *Trypanosoma congolense*: Development of genetic tools. PLoS Neglected Tropical Diseases 4: e618.

Delespaux V, de Koning HP. 2007. Drugs and drug resistance in African trypanosomiasis. Drug Resistance Updates 10: 30–50.

Duffy CW, Morrison LJ, Black A, Pinchbeck GL, Christley RM, Schoenefeld A, Tait A, Turner CM, MacLeod A. 2009. *Trypanosoma vivax* displays a clonal population structure. International Journal for Parasitology 39: 1475–1483.

Fevre EM, Wissmann BV, Welburn SC, Lutumba P. 2008. The burden of human African trypanosomiasis. PLoS Neglected Tropical Diseases 2: e333.

Garcia A, Courtin D, Solano P, Koffi M, Jamonneau V. 2006. Human African trypanosomiasis: Connecting parasite and host genetics. Trends in Parasitology 22: 405–409.

Gardiner PR. 1989. Recent studies of the biology of *Trypanosoma vivax*. Advances in Parasitology 28: 229–317.

Gashumba JK, Baker RD, Godfrey DG. 1988. *Trypanosoma congolense*: The distribution of enzymic variants in east and West Africa. Parasitology 96(Pt 3): 475–486.

Gibson W, Stevens J. 1999. Genetic Exchange in the Trypanosomatidae. Advances in Parasitology 43: 1–45.

Gibson W, Peacock L, Ferris V, Williams K, Bailey M. 2008. The use of yellow fluorescent hybrids to indicate mating in *Trypanosoma brucei*. Parasites & Vectors 1: 4.

Gibson WC. 1986. Will the real *Trypanosoma b. gambiense* please stand up. Parasitology Today 2: 255–257.

Gibson WC, Marshall Tfdc, Godfrey DG. 1980. Numerical analysis of enzyme polymorphism: A new approach to the epidemiology and taxonomy of trypanosomes of the subgenus *Trypanozoon*. Advances in Parasitology 18: 175–246.

Godfrey DG, Scott CM, Gibson WC, Mehlitz D, Zillmann U. 1987. Enzyme polymorphism and the identity of *Trypanosoma brucei gambiense*. Parasitology 94: 337–347.

Goring HH, Curran JE, Johnson MP, Dyer TD, Charlesworth J, Cole SA, Jowett JB, Abraham LJ, Rainwater DL, Comuzzie AG, Mahaney MC, Almasy L, MacCluer JW, Kissebah AH, Collier GR, Moses EK, Blangero J. 2007. Discovery of expression QTLs using large-scale transcriptional profiling in human lymphocytes. Nature Genetics 39: 1208–1216.

Grab DJ, Nikolskaia O, Kim YV, Lonsdale-Eccles JD, Ito S, Hara T, Fukuma T, Nyarko E, Kim KJ, Stins MF, Delannoy MJ, Rodgers J, Kim KS. 2004. African trypanosome interactions with an in vitro model of the human blood-brain barrier. The Journal of Parasitology 90: 970–979.

Grebaut P, Chuchana P, Brizard JP, Demettre E, Seveno M, Bossard G, Jouin P, Vincendeau P, Bengaly Z, Boulange A, Cuny G, Holzmuller P. 2009. Identification of total and differentially expressed excreted-secreted proteins from *Trypanosoma congolense* strains exhibiting different virulence and pathogenicity. International Journal for Parasitology 39: 1137–1150.

Hanotte O, Ronin Y, Agaba M, Nilsson P, Gelhaus A, Horstmann R, Sugimoto Y, Kemp S, Gibson J, Korol A, Soller M, Teale A. 2003. Mapping of quantitative trait loci controlling trypanotolerance in a cross of tolerant West African N'Dama and susceptible East African Boran cattle. Proceedings of the National Academy of Sciences of the United States of America 100: 7443–7448.

Hill EW, O'Gorman GM, Agaba M, Gibson JP, Hanotte O, Kemp SJ, Naessens J, Coussens PM, MacHugh DE. 2005. Understanding bovine trypanosomiasis and trypanotolerance: The promise of functional genomics. Veterinary Immunology and Immunopathology 105: 247–258.

Hoare CA. 1972. *The Trypanosomes of Mammals*. Blackwell Scientific Publications, Oxford.

Holzmuller P, Biron DG, Courtois P, Koffi M, Bras-Goncalves R, Daulouede S, Solano P, Cuny G, Vincendeau P, Jamonneau V. 2008. Virulence and pathogenicity patterns of *Trypanosoma brucei gambiense* field isolates in experimentally infected mouse: Differences in host immune response modulation by secretome and proteomics. Microbes and Infection 10: 79–86.

Iraqi F, Clapcott SJ, Kumari P, Haley CS, Kemp SJ, Teale AJ. 2000. Fine mapping of trypanosomiasis resistance loci in murine advanced intercross lines. Mammalian Genome 11: 645–648.

Jamonneau V, Garcia A, Frezil JL, N'Guessan P, N'Dri L, Sanon R, Laveissiere C, Truc P. 2000. Clinical and biological evolution of human trypanosomiasis in Cote d'Ivoire. Annals of Tropical Medicine and Parasitology 94: 831–835.

Jamonneau V, Ravel S, Garcia A, Koffi M, Truc P, Laveissiere C, Herder S, Grebaut P, Cuny G, Solano P. 2004. Characterization of *Trypanosoma brucei* s.l. infecting asymptomatic sleeping-sickness patients in Cote d'Ivoire: A new genetic group? Annals of Tropical Medicine and Parasitology 98: 329–337.

Jenni L, Marti S, Schweizer J, Betschart B, Le Page RW, Wells JM, Tait A, Paindavoine P, Pays E, Steinert M. 1986. Hybrid formation between African trypanosomes during cyclical transmission. Nature 322: 173–175.

Jones TW, Davila AM. 2001. *Trypanosoma vivax*—Out of Africa. Trends in Parasitology 17: 99–101.

Kabore J, Koffi M, Bucheton B, Macleod A, Duffy C, Ilboudo H, Camara M, De Meeus T, Belem AM, Jamonneau V. 2011. First evidence that parasite infecting apparent aparasitemic serological suspects in human African trypanosomiasis are *Trypanosoma brucei gambiense* and are similar to those found in patients. Infection, Genetics and Evolution 11: 1250–1255.

Kemp SJ, Iraqi F, Darvasi A, Soller M, Teale AJ. 1997. Localization of genes controlling resistance to trypanosomiasis in mice. Nature Genetics 16: 194–196.

Kibona SN, Matemba L, Kaboya JS, Lubega GW. 2006. Drug-resistance of *Trypanosoma b. rhodesiense* isolates from Tanzania. Tropical Medicine & International Health 11: 144–155.

Koffi M, De Meeus T, Bucheton B, Solano P, Camara M, Kaba D, Cuny G, Ayala FJ, Jamonneau V. 2009. Population genetics of *Trypanosoma brucei gambiense*, the agent of sleeping sickness in Western Africa. Proceedings of the National Academy of Sciences of the United States of America 106: 209–214.

Liti G, Carter DM, Moses AM, Warringer J, Parts L, James SA, Davey RP, Roberts IN, Burt A, Koufopanou V, Tsai IJ, Bergman CM, Bensasson D, O'Kelly MJ, van Oudenaarden A, Barton DB, Bailes E, Nguyen AN, Jones M, Quail MA, Goodhead I, Sims S, Smith F, Blomberg A, Durbin R, Louis EJ. 2009. Population genomics of domestic and wild yeasts. Nature 458: 337–341.

MacLean L, Chisi JE, Odiit M, Gibson WC, Ferris V, Picozzi K, Sternberg JM. 2004. Severity of human african trypanosomiasis in East Africa is associated with geographic location, parasite genotype, and host inflammatory cytokine response profile. Infection and Immunity 72: 7040–7044.

Maclean L, Odiit M, Macleod A, Morrison L, Sweeney L, Cooper A, Kennedy PG, Sternberg JM. 2007. Spatially and genetically distinct African trypanosome virulence variants defined by host interferon-gamma response. The Journal of Infectious Diseases 196: 1620–1628.

MacLeod A, Tweedie A, Welburn SC, Maudlin I, Turner CM, Tait A. 2000. Minisatellite marker analysis of *Trypanosoma brucei*: Reconciliation of clonal, panmictic, and epidemic population genetic structures. Proceedings of the National Academy of Sciences of the United States of America 97: 13442–13447.

MacLeod A, Tweedie A, McLellan S, Hope M, Taylor S, Cooper A, Sweeney L, Turner CM, Tait A. 2005a. Allelic segregation and independent assortment in *T. brucei* crosses: Proof that the genetic system is Mendelian and involves meiosis. Molecular and Biochemical Parasitology 143: 12–19.

MacLeod A, Tweedie A, McLellan S, Taylor S, Hall N, Berriman M, El-Sayed NM, Hope M, Turner CM, Tait A. 2005b. The genetic map and comparative analysis with the physical map of *Trypanosoma brucei*. Nucleic Acids Research 33: 6688–6693.

MacLeod A, Turner CM, Tait A. 2007. The system of genetic exchange in *Trypanosoma brucei* and other trypanosomatids. In JD Barry, R McCulloch, JC Mottram, A Acosta-Serrano, eds. *Trypanosomes: After the Genome*. Horizon Bioscience, Wymondham, UK.

Magona JW, Walubengo J, Odimin JT. 2008. Acute haemorrhagic syndrome of bovine trypanosomosis in Uganda. Acta Tropica 107: 186–191.

Majiwa PA, Hamers R, Van Meirvenne N, Matthyssens G. 1986. Evidence for genetic diversity in *Trypanosoma (Nannomonas) congolense*. Parasitology 93(Pt 2): 291–304.

Masumu J, Marcotty T, Geysen D, Geerts S, Vercruysse J, Dorny P, den Bossche PV. 2006a. Comparison of the virulence of *Trypanosoma congolense* strains isolated from cattle in a trypanosomiasis endemic area of eastern Zambia. International Journal for Parasitology 36: 497–501.

Masumu J, Marcotty T, Ndeledje N, Kubi C, Geerts S, Vercruysse J, Dorny P, van den Bossche P. 2006b. Comparison of the transmissibility of *Trypanosoma congolense* strains, isolated in a trypanosomiasis endemic area of eastern Zambia, by *Glossina morsitans morsitans*. Parasitology 133: 331–334.

Mehlitz D, Zilmann U, Scott CM, Godfrey DG. 1982. Epidemiological studies on the animal reservoir of *Gambiense* sleeping sickness. Part III. Characterization of *Trypanozoon* stocks by iso-enzymes and sensitivity to human serum. Tropenmedizin und Parasitologie 33: 113–118.

Morrison LJ. 2011. Parasite-driven pathogenesis in *Trypanosoma brucei* infections. Parasite Immunology 33: 448–455.

Morrison LJ, Tait A, McLellan S, Sweeney L, Turner CM, MacLeod A. 2009a. A major genetic locus in *Trypanosoma brucei* is a determinant of host pathology. PLoS Neglected Tropical Diseases 3: e557.

Morrison LJ, Tweedie A, Black A, Pinchbeck GL, Christley RM, Schoenefeld A, Hertz-Fowler C, MacLeod A, Turner CM, Tait A. 2009b. Discovery of mating in the major African livestock pathogen *Trypanosoma congolense*. PLoS One 4: e5564.

Morrison LJ, McLellan S, Sweeney L, Chan CN, Macleod A, Tait A, Turner CM. 2010. Role for parasite genetic diversity in differential host responses to *Trypanosoma brucei* infection. Infection and Immunity 78: 1096–1108.

Mu J, Myers RA, Jiang H, Liu S, Ricklefs S, Waisberg M, Chotivanich K, Wilairatana P, Krudsood S, White NJ, Udomsangpetch R, Cui L, Ho M, Ou F, Li H, Song J, Li G, Wang X, Seila S, Sokunthea S, Socheat D, Sturdevant DE, Porcella SF, Fairhurst RM, Wellems TE, Awadalla P, Su XZ. 2010. *Plasmodium falciparum* genome-wide scans for positive selection, recombination hot spots and resistance to antimalarial drugs. Nature Genetics 42: 268–271.

Nganga JK, Soller M, Iraqi FA. 2010. High resolution mapping of trypanosomosis resistance loci Tir2 and Tir3 using F12 advanced intercross lines with major locus Tir1 fixed for the susceptible allele. BMC Genomics 11: 394.

Nikolskaia OV, De ALAP, Kim YV, Lonsdale-Eccles JD, Fukuma T, Scharfstein J, Grab DJ. 2006. Blood-brain barrier traversal by African trypanosomes requires calcium signaling induced by parasite cysteine protease. The Journal of Clinical Investigation 116: 2739–2747.

Noyes H, Brass A, Obara I, Anderson S, Archibald AL, Bradley DG, Fisher P, Freeman A, Gibson J, Gicheru M, Hall L, Hanotte O, Hulme H, McKeever D, Murray C, Oh SJ, Tate C, Smith K, Tapio M, Wambugu J, Williams DJ, Agaba M, Kemp SJ. 2011. Genetic and expression analysis of cattle identifies candidate genes in pathways responding to *Trypanosoma congolense* infection. Proceedings of the National Academy of Sciences of the United States of America 108: 9304–9309.

Peacock L, Ferris V, Bailey M, Gibson W. 2009. Intraclonal mating occurs during tsetse transmission of *Trypanosoma brucei*. Parasites & Vectors 2: 43.

Peacock L, Ferris V, Sharma R, Sunter J, Bailey M, Carrington M, Gibson W. 2011. Identification of the meiotic life cycle stage of *Trypanosoma brucei* in the tsetse fly. Proceedings of the National Academy of Sciences of the United States of America 108: 3671–3676.

Schweizer J, Tait A, Jenni L. 1988. The timing and frequency of hybrid formation in African trypanosomes during cyclical transmission. Parasitology Research 75: 98–101.

Shaw APM. 2004. Economics of African trypanosomiasis. In: Maudlin I, Holmes PH, Miles MA, eds. *The Trypanosomiases*. CABI Publishing, Wallingford, UK, pp. 369–402.

Stephen LE. 1970. Clinical manifestations of the Trypanosomiasis in livestock and other domestic animals. In: Mulligan HW, ed. *The African Trypanosomiases*. Allen and Unwin, London, pp. 774–794.

Sternberg JM, Maclean L. 2010. A spectrum of disease in human African trypanosomiasis: The host and parasite genetics of virulence. Parasitology 137: 2007–2015.

Stranger BE, Nica AC, Forrest MS, Dimas A, Bird CP, Beazley C, Ingle CE, Dunning M, Flicek P, Koller D, Montgomery S, Tavare S, Deloukas P, Dermitzakis ET. 2007. Population genomics of human gene expression. Nature Genetics 39: 1217–1224.

Tait A, Babiker EA, LeRay D. 1984. Enzyme variation in *T. brucei spp*. I. Evidence for the sub-speciation of *T. b. gambiense*. Parasitology 89: 311–326.

Tait A, Buchanan N, Hide G, Turner CMR. 1996. Self-fertilisation in *Trypanosoma brucei*. Molecular and Biochemical Parasitology 76: 31–42.

Tait A, Masiga D, Ouma J, MacLeod A, Sasse J, Melville S, Lindegard G, McIntosh A, Turner M. 2002. Genetic analysis of phenotype in *Trypanosoma brucei*: A classical approach to potentially complex traits. Philosophical Transactions of the Royal Society of London. Series B, Biological Sciences 357: 89–99.

Tait A, Macleod A, Tweedie A, Masiga D, Turner CM. 2007. Genetic exchange in *Trypanosoma brucei*: Evidence for mating prior to metacyclic stage development. Molecular and Biochemical Parasitology 151: 133–136.

Turner CMR, Sternberg J, Buchanan N, Smith E, Hide G, Tait A. 1990. Evidence that the mechanism of gene exchange in *Trypanosoma brucei* involves meiosis and syngamy. Parasitology 101: 377–386.

Welburn SC, Maudlin I, Milligan PJ. 1995. *Trypanozoon*: Infectivity to humans is linked to reduced transmissibility in tsetse. I. Comparison of human serum-resistant and human serum-sensitive field isolates. Experimental Parasitology 81: 404–408.

Wellde BT, Chumo DA, Adoyo M, Kovatch RM, Mwongela GN, Opiyo EA. 1983. Haemorrhagic syndrome in cattle associated with *Trypanosoma vivax* infection. Tropical Animal Health and Production 15: 95–102.

WHO. 2006. Human African trypanosomiasis (sleeping sickness): Epidemiological update. Weekly Epidemiological Record 81: 69–80.

CHAPTER 17

THE EVOLUTION OF ANTIGENIC VARIATION IN AFRICAN TRYPANOSOMES

ANDREW P. JACKSON and J. DAVID BARRY

INTRODUCTION

Virulence can be defined as the cost a parasite levies on its host's fitness. It is a key trait that is thought to arise as a consequence of the parasite population maximizing its abundance and longevity in the host. Although the parasite gains in transmission capacity from such maximization, the trade-off is host death or pathogenesis, leading to a decrease in parasite transmissibility. The trade-off hypothesis proposes that the parasite will evolve to optimum virulence, balancing host damage against transmissibility (Mackinnon and Read, 2004). In the complex relationships between parasite and host, it might be expected therefore that some mechanisms adapt to contribute to optimal virulence. Antigenic variation is a parasite survival mechanism that appears to achieve such balance, working in concert with other factors such as growth rate and density-dependent effects. Here, we outline trypanosome population phenotypes associated with antigenic variation and the evolution of major trypanosome traits at the cellular and molecular levels. Using comparative genomics, we track the evolution of trypanosome antigenic variation to its current dominant role in balancing virulence and transmission, and we reveal interspecies differences in the generation of variation.

Trypanosoma brucei, a single-cell eukaryote (i.e., protozoan), displays a classical type of antigenic variation that occurs also in other pathogens inhabiting the blood and facing the enormous scale, efficiency, and flexibility of adaptive immunity (Barry and McCulloch, 2001; Morrison et al., 2009). The parasite is covered completely by a dense coat composed of variant surface glycoprotein (VSG). The coat thwarts various innate immune mechanisms and, critically, shields necessarily conserved macromolecules from the immune system. As lethal antibodies arise against VSG, trypanosomes that have switched to expression of a distinct VSG coat survive and can proliferate. The switching process, which occurs in about every 100 dividing cells, is spontaneous and is preemptive of immunity, so that a diverse population is always present. In addition, variants emerge in a hierarchy, which is thought to result in chronic, rather than acute, overwhelming infection. All of these features are shared with many bacterial

Evolution of Virulence in Eukaryotic Microbes, First Edition. Edited by L. David Sibley, Barbara J. Howlett, and Joseph Heitman.

and protozoal pathogens; this form of antigenic variation is highly successful, having evolved convergently and independently many times (Deitsch et al., 2009).

The VSG is highly diverse, especially in its N-terminal domain (~350 amino acids), which is exposed to the host (Carrington et al., 1991; Marcello and Barry, 2007). There are two families of N-domain, termed a and b, which, respectively, have modal peptide identities of 10% and 15%. That these values are so low reflects three important features of the N-domain: It apparently has no function other than to form a dense coat; the protein is composed mostly of alpha-helix, which can be formed by a significant proportion of common amino acids; and it diversifies evenly along its entire length rather than by selection for the few epitopes exposed to immune pressure. The C-terminal domain (~90–160 amino acids) is disposed toward the plasma membrane and is also diverse, although to a lesser extent.

Trypanosome antigenic variation is based on an archive of nearly 2000 distinct, silent *VSG* genes and pseudogenes (Berriman et al., 2005; Marcello and Barry, 2007). Each cell expresses a single *VSG*, and new variants arise by transcription of another *VSG* instead. Although ~70% of *T. brucei VSG* are pseudogenes, it is emerging that, after the first few waves of parasitemia, novel, temporary *VSGs* become assembled from segments of intact genes and pseudogenes with high homology to them (Barbet and Kamper, 1993; Marcello and Barry, 2007). The scale of trypanosome antigenic variation is potentially enormous due to this combinatorial route to variant production, a level of hyperdiversity that is thought to confer the capacity to reinfect previously immune hosts (Barry et al., 2005; Futse et al., 2008). In this review, we investigate mechanisms for generating novel *VSG* by inference from the diversity of *VSG* repertoires.

It is commonly thought that solely the interplay of antigenic variation and adaptive immunity can account for the relapsing nature of parasitemia. Even the simplest mathematical modeling shows this cannot be the case. Instead, there is a complex relationship between switching, immunity, and the self-limitation of trypanosome growth (Gjini et al., 2010; Lythgoe et al., 2007; MacGregor et al., 2011). This limitation arises from density-dependent differentiation of trypanosomes to the nondividing, short-lived transmission stage that awaits uptake by the tsetse vector (MacGregor et al., 2011). Even the switching process is highly complex, with the *VSG* gene archive being substructured into groups (Morrison et al., 2005), each of which helps specify how many variants coexist in a growth peak and the rate of transition to the next set of variants. Major outcomes of the intrinsic interplaying of these layers of complexity are amelioration of virulence due to parasite persistence at sublethal levels, persistently high level of the transmission stage and consequent high transmissibility, and perhaps the ability of this parasite to adapt its growth dynamically to the widely different physiology of its very broad host range. A key element, the substructuring of the archive into locus types and sequence subfamilies that drive the switching hierarchy, has been reviewed recently (Morrison et al., 2009) and is currently the subject of intensive study.

The trypanosome bloodstream stages are unusual at the cellular level. They are adapted to high production and processing of VSG, which comprises ~10% of soluble protein, and they have remarkable surface dynamics that see VSG turning over very rapidly, especially when antibodies are bound, providing a means of disposing of low levels of antibody (Engstler et al., 2007); high antibody concentrations are lethal rather than being cleared. Whether this turnover has any role *in vivo* has not been investigated but conceivably could delay the effective onset of antibody-mediated lethality. It seems likely that aspects of this unusual cell biology have arisen as a consequence of antigenic variation.

There is further evidence of unusual adaptations to antigenic variation at the genome level. A large set of minichromosomes has evolved, apparently the sole function of which is to harbor a subset of the silent *VSG* archive (Wickstead et al., 2003). The positioning of this subset at the telomeres of the minichromosomes gives them a high probability for being activated during switches, which contributes substantially to the hierarchical nature of expression, and therefore prolonged growth at sublethal levels. At the molecular level, a number of nuclear adaptations are becoming apparent, including, among others, the unusual *VSG* transcription factory and the dynamics of chromosomal segregation; these features have been reviewed recently (Daniels et al., 2010). All in all, the large molecular and physiological burden placed on trypanosomes by antigenic variation appears to have led to marked cellular and molecular adaptations. These adaptations appear to be tailored to provide the efficient and organized *VSG* switching system and the exclusive *VSG* expression that, when extended to the trypanosome population dynamics level, make antigenic variation such an effective mechanism for prolonged infection with ameliorated virulence.

For the remainder of this review, we shall focus on several key evolutionary questions about the *VSG*, its family, and its functions that can only now be addressed with the advent of comparative genomics: (i) What was the origin of the *VSG*? (ii) How did the *VSG* family become so large and diverse and how is it evolving at present? (iii) Are the features of *T. brucei* VSG antigenic variation applicable to all trypanosomes with antigenic variation?

To address the first question, we begin by defining the *VSG* repertoires in three related species, *T. brucei*, *Trypanosoma congolense*, and *Trypanosoma vivax*, and then we examine the phylogenetic relationships of *VSG* subfamilies within and between species, culminating in a model of *VSG* macroevolution. Phylogenetic analysis within *VSG* subfamilies allows us to address the second question, through description of the molecular evolution of contemporary *VSG*, and species differences in the mechanisms of sequence variation. We then address the third question, how applicable the current model for antigenic variation in *T. brucei* is for all African trypanosomes.

WHAT WAS THE ORIGIN OF THE VSG?

Antigenic variation is known in three species of African trypanosome: *T. brucei* sspp., *T. congolense* and *T. vivax*. These species are relatively divergent and classified in distinct subgenera. We have analyzed the assembled genomes for *T. brucei* TREU927, *T. congolense* IL3000, and *T. vivax* Y486 (available from the TritrypDB and GeneDB Web sites) and have found that, despite major differences in life cycles, the only immediately obvious differences are in the predicted surface proteomes. This is not wholly surprising as plasma membrane or secreted proteins are at the most direct host–parasite molecular interface. Here, we focus on the *VSG* system, including not just the silent archive of canonical *VSG*, that is, those that function as variant antigens, but the whole repertoire of sequences resembling *VSG*. We identified all *VSG* and *VSG*-like genes in *T. congolense* and *T. vivax* and have analyzed them phylogenetically, individually, and in combination. These results are summarized in Table 17.1 and Figure 17.1 and form part of a cell surface "phylome" consisting of phylogenetic trees for every gene family in African trypanosomes predicted to have a cell surface role. The trees can be accessed

TABLE 17.1 A Summary of a- and b-Type *VSG*-Like Gene Families Present in the Genomes of *T. brucei* TREU927, *T. congolense* IL3000, and *T. vivax* Y486

	T. brucei	*n*	*T. congolense*	*n*	*T. vivax*	*n*
a-Type	a-Type variant surface glycoprotein (Fam0a)	429[a]	–	–	a-*VSG*-like genes (Fam23)	540
	Transferrin receptor (*ESAG6/7*, Fam15)	~40[b]	*ESAG6/7*-like (Fam15)	43	–	–
	Procyclin-associated genes (*PAG1-2,4-5*; Fam14)	9	*PAG*-like (Fam14)	22	–	–
b-Type	b-Type variant surface glycoprotein (Fam0b)	350[a]	(See below)		b-*VSG*-like genes (Fam24)	279
	VSG-related genes (Fam9)	39	*VR*-like genes (Fam16)	512	–	–
	Expression site-associated gene 2 (Fam13)	~50[b]	*ESAG2*-like genes (Fam13)	302	–	–
	VSG-related genes (Fam1)	5	–	–	–	–
Other	–	–	–	–	*VSG*-like genes (Fam25)	227
	–	–	–	–	*VSG*-like genes (Fam26)	87

Gene families known to include functional variant antigens are underlined. It is not known whether *T. vivax* families 24–26 (dashed underline) function as variant antigens.

[a] The number of full-length *VSG* genes and pseudogenes in *T. brucei* 927 is given; an indeterminate number of gene fragments are also present.

[b] For expression site-associated genes, copy number within the *T. brucei* genome depends on the number of telomeric expression sites and their completeness, which varies between strains. The numbers given are an approximate maximum assuming single representation in expression sites on every telomere, combined with related genes on core chromosomes ("GRESAGs").

via the *Trypanosoma* pages of the GeneDB Web site. The outcomes reveal more complexity than imagined.

As described above, canonical *VSG* in *T. brucei* have two types of N-domain (a and b), which are similar in length and tertiary structure, share a homologous C-domain, but are only distantly related and differentiated by distinct patterns of conserved cysteine residues (Carrington et al., 1991; Marcello and Barry, 2007). In addition to these canonical *VSGs*, the *T. brucei* genome contains related genes without this function. Transferrin receptor proteins and their homologs (collectively referred to here as "TFR family") are encoded by genes homologous to the a-*VSG*, in both vertebrate (*ESAG6* and *7*; Salmon et al., 1997) and insect (*PAG1,2,4* and *5*; Koenig-Martin et al., 1992) life stages. The structures and mechanism of TFR proteins are not homologous to transferrin-binding proteins in other eukaryotes (Salmon et al., 1997), prompting some to suggest that this transferrin receptor evolved uniquely in *T. brucei* from the a-*VSG* (Borst and Fairlamb, 1998; Carrington and Boothroyd, 1996). Similarly, *T. brucei* has various *VSG*-related (*VR*) genes, which have sequence similarity to b-*VSG*, but expression patterns, structural features (Marcello and Barry, 2007), and genomic conservation (Jackson et al., 2010) inconsistent with antigenic variation. *VR* also have been suggested to have originated from functional *VSG* in *T. brucei* (Marcello and Barry, 2007).

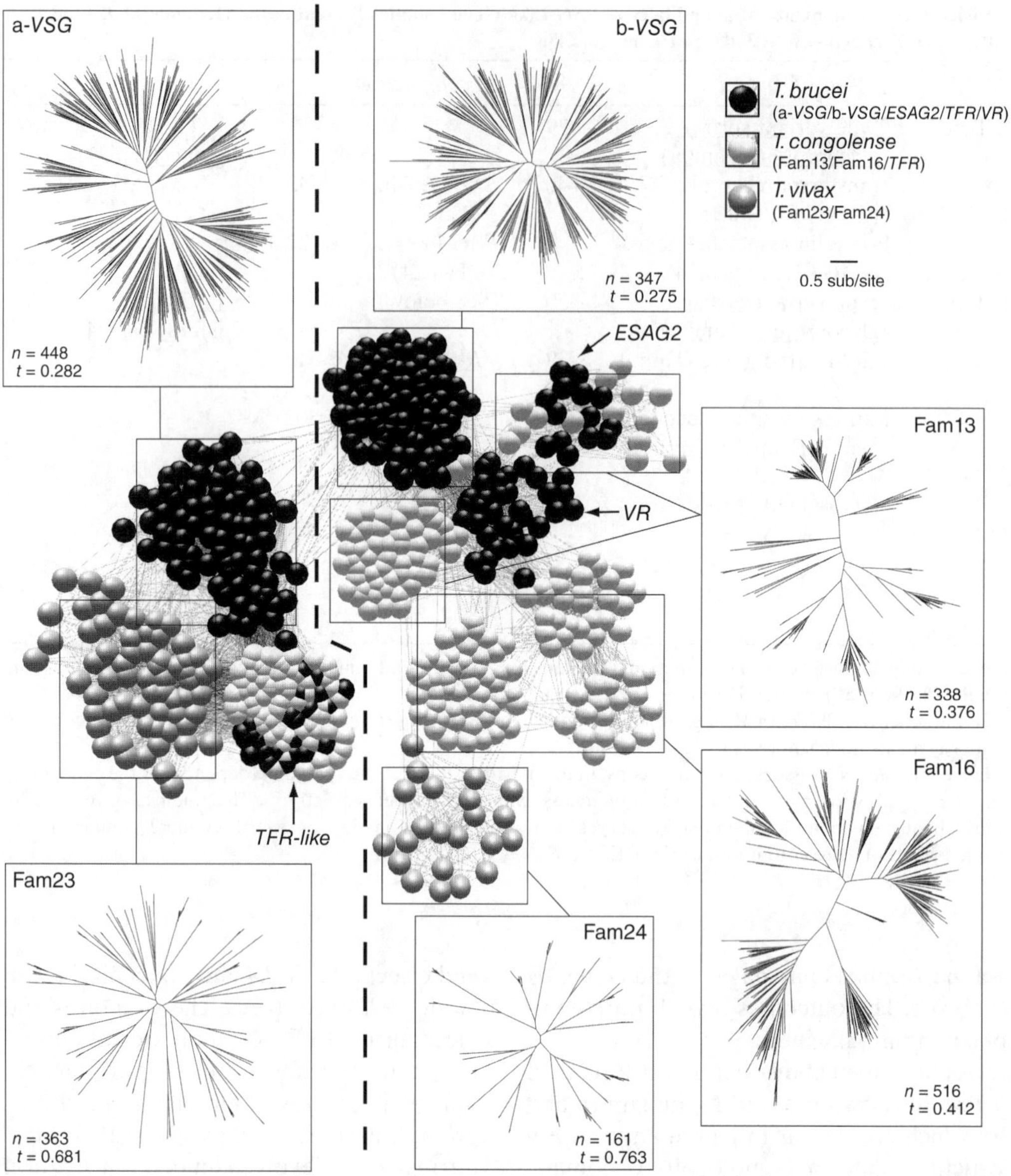

Figure 17.1 Evolutionary relationships and phylogenetic patterns among major *VSG* subfamilies in African trypanosomes. At the center, a sequence similarity network represents the evolutionary relationships of principal *VSG* subfamilies from *T. brucei* (black), *T. congolense* (white), and *T. vivax* (gray). Individual sequences are shown as spheres, with connections to other sequences with scores above an empirically derived threshold. The network was estimated with BioLayout Express 3D using pairwise, global alignment (i.e., FASTA) scores estimated from alignments of a- and b-*VSG*-like protein sequences. The program arranges the spheres to minimize distance, such that related elements cluster together. A vertical dashed line separates a- (left) and b-*VSG*-like (right) subfamilies. Subfamilies identified in the text are marked with arrows. Placed around the network are Bayesian phylogenies for six subfamilies representing known or predicted variant antigens in each species; the number of sequences included (*n*) is given, as well as the "treeness" value (*t*), which defines the proportion of tree length taken up by internal branches. All phylogenies are drawn to the same scale.

The *VSG* repertoires of the single sequenced strains of *T. congolense* and *T. vivax* demonstrate that the a- and b-type distinctions are conserved throughout, but otherwise there are substantial species differences. There is no a-type *VSG* gene family in *T. congolense*, although this species does have a-type genes corresponding to both *ESAG6/7* and *PAG* members of the *TFR* family in *T. brucei*. It remains to be seen whether these *T. congolense* genes encode functional transferrin receptors, but some residues crucial to transferrin binding in *T. brucei* (Salmon et al., 1997) are among those conserved across species. Canonical *VSG* in *T. congolense* belong to two b-*VSG*-like subfamilies (Fam13 and Fam16 in the cell surface phylome), which encode proteins with predicted signal peptides and glycosylphosphatidylinositol (GPI) anchors. Both families can be split further into several clades, each with a characteristic C-domain, such that the whole repertoire may consist of 15–20 distinct lineages. The familiar *T. brucei* C-domain is entirely absent from *T. congolense* and *T. vivax*. Fam13 and 16 are divided because they are not monophyletic; that is, the closest relatives of Fam13 and 16 are not each other but b-*VSG*-like genes in *T. brucei* including, respectively, *ESAG2* and *VR* genes.

Whereas the putative *VSGs* in *T. congolense* are easy to identify due to their homology with *T. brucei* b-*VSG* and with published VSG sequences from *T. congolense* (Eshita et al., 1992; Rausch et al., 1994; Strickler et al., 1987), assignation in *T. vivax* is more difficult because there is only one published *T. vivax* VSG sequence (Gardiner et al., 1996), and there are many *T. vivax*-specific genes. These complications are understandable, given that *T. vivax* is the most divergent of the three species (Hamilton et al., 2007). Certainly, there are large gene families encoding proteins of the expected size with predicted signal peptides, which have modest basic local alignment search tool (BLAST) matches to *T. brucei VSG*. However, sequence variation within these families is distributed very differently in *T. vivax*; genes encoding VSG-like proteins form ~330 clusters of near-identical sequences, the clusters ranging in size from 4 to 50. While sequence variation within clusters is very low, the protein distance between clusters is very high; indeed, it is often impossible to align sequences from different clusters. By careful comparison of conserved cysteine and tryptophan residues, however, the clusters can be sorted into four subfamilies (Fam23–26, Table 17.1). Fam23 has the pattern of conserved cysteine residues found in a-type VSG and TFR family proteins in *T. brucei* and *T. congolense*; we refer to Fam23 as a-*VSG*-like. Fam24 has the conserved cysteine pattern of b-type VSG in *T. brucei* and *T. congolense*, and so is referred to as b-*VSG*-like. Fam25 and 26 encode predicted proteins of length and composition similar to those of known VSG but have conserved cysteine residues that are unlike a- or b-type VSG. We consider these families to be *T. vivax*-specific *VSG*. Sequences and alignments for all families are available from the cell surface phylome page. As sequence similarity alone does not guarantee that function is conserved, we have directly determined the peptide sequence of a candidate expressed VSG in the *T. vivax* genome strain. We find this protein to be identical to gene TvY486_0027060, which is a-*VSG*-like, showing that Fam23 encodes at least one variant antigen. It remains to be seen what proportion of all these putative *VSGs* encode functional variant antigens in *T. vivax*.

Comparison of the structural properties of these different *VSG* families reveals significant differences. Table 17.2 shows that *T. brucei VSG* are longer than homologs elsewhere due largely to their shared C-domain. The shortest *VSG* belong to Fam16 in *T. congolense*, which encode proteins with smaller N-domains than other VSG. Base composition also shows

TABLE 17.2 Base Composition, Codon Usage, and Pseudogene Content of *VSG*-Like Gene Families

Species	Family	n^a	GC		Total Codons		Nc		CBI		Hydrophobicity		Aromaticity		Pseudogenes %	Pseudogenes Mutations[b]	
T. brucei	0a	429	0.474	*0.017*	489.2	*31.0*	52.9	*2.738*	0.030	*0.044*	−0.495	*0.208*	0.051	*0.010*	69.2	3.432	*2.498*
	0b	350	0.481	*0.021*	492.6	*47.6*	52.8	*2.794*	0.028	*0.055*	−0.566	*0.221*	0.054	*0.011*	72.2	2.988	*2.382*
T. congolense	13	302	0.467	*0.027*	415.4	*68.0*	56.5	*2.856*	0.002	*0.051*	−0.489	*0.142*	0.074	*0.013*	21.1	3.154	*2.522*
	16	512	0.461	*0.039*	366.8	*39.5*	56.7	*3.081*	−0.011	*0.052*	−0.624	*0.178*	0.078	*0.016*	29.7	5.557	*3.447*
T. vivax	23	540	0.592	*0.032*	429.5	*46.8*	43.3	*4.997*	0.078	*0.050*	−0.488	*0.313*	0.038	*0.011*	15.5	1.360	*0.775*
	24	279	0.598	*0.016*	420.9	*88.1*	42.4	*2.495*	0.101	*0.041*	−0.563	*0.215*	0.033	*0.007*	27.2	1.418	*0.794*
	25	227	0.598	*0.020*	419.5	*52.5*	42.5	*3.066*	0.096	*0.070*	−0.356	*0.140*	0.033	*0.007*	26.5	1.341	*0.883*
	26	87	0.595	*0.015*	398.8	*104.3*	42.6	*3.936*	0.078	*0.051*	−0.500	*0.259*	0.039	*0.011*	36.9	1.619	*0.804*

Note: Percentage GC content (GC); effective number of codons (Nc); codon bias index (CBI). Standard deviations are given in italics.

[a] The number of full-length *VSG* genes in *T. brucei* 927 is given; an indeterminate number of gene fragments are also present. Genuine gene fragments were negligible in other species.

[b] For each gene family, the average number of pseudogenic mutations (i.e., internal stop codons and/or frameshifts) is given.

consistent species differences; *T. vivax* *VSGs* are 59.8% GC on average, compared with 46.4% and 47.8% in *T. congolense* and *T. brucei*, respectively. This significant difference is specific to *VSG*, as base composition within chromosomal cores does not differ significantly, and results in biased codon usage. *T. vivax* VSGs are encoded using a more restricted range of codons, which coincides with certain amino acids being underrepresented, for instance, aromatic residues. In these respects, the effect is species specific, suggesting that conditions within individual trypanosome species are shaping *VSG* sequence composition rather than structural features evolved by, and inherent to, particular *VSG* lineages.

When we consider the relationships between these various *VSG* subfamilies in each species, the principal issue is whether *VSG* lineages are shared between species, suggesting that those lineages were present in their ancestor or, alternatively, whether *VSG* lineages are species specific, indicating that evolutionary turnover is so fast that nothing is shared between species beyond the basic structural signatures of a- and b-type proteins. Phylogenies for a- and b-*VSG* superfamilies, as well as individual *VSG* subfamilies from each species, are available from the Cell Surface Phylome at GeneDB. The basal nodes of these trees are not robust, reflecting a biological reality that the relationships of major clades rest on relatively few evolutionary changes. So, to provide a more tangible picture of these relationships, one that entertains the distances between all subfamilies and not just the bifurcations shown in a phylogeny, we have estimated sequence similarity networks for a- and b-*VSG* based on FASTA protein distances (see Fig. 17.1). Only the highest 25% FASTA scores were used; this threshold was determined empirically to produce a network that clearly displays the closest relationships.

There are five principal results to take from the similarity network. (i) The a- and b-*VSG* lineages form separate networks; the only connection between them occurs between *T. brucei* a- and b-*VSG*, which is entirely due to the shared C-domain, unique to this species. Hence, the a- and b-*VSG* types are ancient lineages that were likely present in the common ancestor of all African trypanosomes. (ii) The b-*VSG* repertoire in *T. congolense* (i.e., all *VSG*) forms several distinct clusters, indicating that it is derived from multiple evolutionary origins, unlike counterparts in *T. brucei* and *T. vivax*, which form tight clusters, indicative of a single ancestral lineage. (iii) The closest match to Fam13 is not another *T. congolense* family but *ESAG2*, demonstrating that *ESAG2* shared an ancestor with a *T. congolense* variant antigen in the common ancestor of *T. brucei* and *T. congolense*, and suggests that *ESAG2* was co-opted to a novel function (as yet unknown) from a *VSG* in *T. brucei* after speciation. (iv) Likewise, the closest relatives of Fam16 are Fam9 *VR* genes in *T. brucei*, as indicated by the position of *VR* intermediate between *T. brucei* and *T. congolense* b-*VSG* in Figure 17.1. This relationship indicates that the *VR* evolved in *T. brucei* from an ancestral lineage that previously functioned as variant antigens and are not derived from contemporary b-*VSG* in *T. brucei* through the secondary loss of functional domains. (v) Finally, the network demonstrates the very close relationship between *T. congolense* Fam14 and 15 and, respectively, the *PAG* and *TFR* family sequences in *T. brucei*. In fact, their phylogeny shows that Fam14/*PAG* and Fam15/*TFR* families are sibling lineages, suggesting that they are orthologs and are likely to perform similar roles in both species. More information regarding the orthology and functional conservation of these proteins is available from the cell surface phylome. *T. vivax* does not possess clear orthologs of these genes; if it did, those genes would otherwise cluster at the same point. Therefore, the transferrin receptor family evolved in the common ancestor of *T. brucei* and *T. congolense* and is retained in both daughter species. *T. vivax* lacks any member of this *TFR* family but presumably could bind transferrin in other ways.

Based on these observations, we can produce a model of *VSG* evolution. The a- and b-*VSGs* comprise two ancient lineages that were in the common ancestor. *T. vivax*-specific families may constitute further lineages of the same age that were not inherited by the *T. brucei–congolense* ancestor. This ancestral genome included a b-*VSG* multigene family, presumably encoding authentic variant antigens, as in both daughter species. Two lineages within this family were inherited by both *T. brucei* and *T. congolense*, one becoming *VR* + *VSG*/Fam16, respectively, and the other becoming *ESAG2*/Fam13, respectively. In *T. brucei*, *VR* and *ESAG2* lost variant antigen function, which instead is focused on a single lineage (contemporary b-*VSG*). In *T. congolense*, in contrast, both Fam13 and Fam16 include sequences loosely related to known variant antigens (data not shown). In effect, *T. brucei* b-*VSGs* pass through a bottleneck emerging structurally uniform, while *T. congolense* variant antigens continue to be drawn from a phylogenetically diverse group of b-*VSG* clades and are therefore structurally heterogeneous by comparison.

With respect to the a-*VSG* lineage, the ancestral African trypanosome possessed a diverse gene repertoire, part of which at least encoded variant antigens. The *TFR* family originated prior to the speciation of *T. brucei* and *T. congolense*, but after the separation of *T. vivax* when one type of a-*VSG* was co-opted to bind transferrin in the flagellar pocket and ceased to function as a variant antigen. To reconcile the absence of a-*VSG* variant antigens in *T. congolense* with the presence of *TFR* family genes, we must either posit a duplication in the *T. brucei–congolense* ancestor to create a-*VSG* and *TFR* lineages, followed by the loss of a-*VSG* in *T. congolense* exclusively, or suggest the secondary gain of a-*VSG* variant antigens from a *TFR* family member in *T. brucei*. The loss hypothesis is simple, but it is difficult to imagine the mechanism that could delete hundreds of a-*VSG* from the ancestral *T. congolense* genome. The gain hypothesis does not require an extra duplication or copious gene deletions, but this would leave a signature in the form of shared characters between a-*VSG* and *TFR* family genes in *T. brucei*, which are not apparent. Ultimately, we require greater sampling of these species' genomes to determine whether the a-*VSG* lineage originally encoded variant antigens or transferrin receptors or combined both functions in the same protein.

HOW DID THE VSG FAMILY BECOME SO LARGE AND DIVERSE AND HOW IS IT EVOLVING AT PRESENT?

In *T. brucei*, the *VSG* family evolves through the standard process of gene duplication followed by point mutation and recombination. Recombination clearly plays a prominent role in shuffling domain combinations, with the various types of N- and C-domains being interchangeable, apparently in any combination (Hutchinson et al., 2003, 2007; Marcello and Barry, 2007). Additionally, ectopic recombination moves whole genes between silent *VSG* arrays in the subtelomeres, often by duplication (Marcello and Barry, 2007). More importantly, for the creation of novel VSGs, intragenic shuffling of sequences between silent *VSG* also occurs (L. Plenderleith and J.D. Barry, unpublished data). Interestingly, N-domain diversity is distributed across the whole domain, with only a minimal sign of greater level in the region of presumed epitopes that are exposed to the host (Hutchinson et al., 2007). This spread of diversity implies that, because individual *VSG* are rarely expressed, they are under secondary selection pressure, such as via a domain-wide mutational system. Expressed *VSGs* show mutations, including some point mutations and the generation of mosaic sequences. If some of these expressed genes then recombine back into the archive, such mutation could contribute to overall diversity.

There are also striking differences in the molecular evolution of the *VSG* subfamilies in each species, as revealed by Bayesian phylogenies (Fig. 17.1). These differences are reflected in "treeness" indices (White et al., 2007), which range from $t = 0$ (completely starlike) to $t = 1$ (completely treelike). *T. brucei VSGs* produce more starlike trees ($t = 0.282$ and 0.275), while *T. congolense VSG* phylogenies contain distinct clades separated by long basal branches and a defined root ($t = 0.376$ and 0.412). *T. vivax* presents a third topology with clusters of almost identical *VSG* separated by very long internal branches and poorly resolved basal nodes ($t = 0.681$ and 0.763). Topologies that are more treelike retain information about the deep past, which can persist only in the absence of genetic exchange, so recombination may contribute differently to *VSG* evolution in each species. The incidence of pseudogenes also points toward mechanistic differences between the species. While 69.2% and 72.7% of full-length a- and b-*VSG* in *T. brucei* are predicted pseudogenes, only 21.1% and 29.7% of Fam13 and 16 (*T. congolense*), respectively, are so defined. Likewise, in *T. vivax*, the percentage of pseudogenes among predicted *VSG* is as low as 15.5%. Although *VSG* pseudogenes in *T. brucei* arise primarily through short indel events (L. Plenderleith and J.D. Barry, unpublished data), those mutations spread in the archive through recombination events, and specifically by conversions (E. Gjini, D. Haydon, J.D. Barry, and C. Cobbold, unpublished data). As *T. brucei VSGs* contain less phylogenetic information and encode intact reading frames less often than those of other species, we compared the evidence for recombination among *VSG* in all three species.

Phylogenetic incompatibility describes the presence of multiple, distinct phylogenetic signals within a single sequence alignment, and it is a signature of historical recombination (Weiller, 2008). We determined the rate of phylogenetic incompatibility in the alignments of the *VSG* subfamilies in Table 17.3. Given that exchange of the shared C-domain between *VSGs* is known to occur (Hutchinson et al., 2003; Marcello and Barry, 2007), we removed the C-domain from *T. brucei VSG* alignments. For each alignment, 100 bootstrap replicates were generated. For each replicate alignment, 100 samples were taken by selecting 10 sequences at random for each sample. Phylogenetic incompatibility was estimated for each sample using the pairwise homoplasy index, which returns a single probability value for the whole alignment and is robust in the presence of rate heterogeneity (Bruen et al., 2006). The proportion of 100 samples with $P < 0.05$ is here termed P_{pi}; the mean P_{pi} is reported for six *VSG* subfamilies in Table 17.3. Evidence for recombination is greatest in the Fam16 alignment ($P_{pi} = 43.3$), moderate in the *T. brucei* alignments (without the C-domain; $P_{pi} = 23.4$ and 15.2), and lowest in the *T. vivax* ($P_{pi} = 13.8$ and 12.6) and Fam13 ($P_{pi} = 12.5$) alignments. As the recombination rate usually is dependent on sequence homology between interacting sequences, the more structurally heterogeneous subfamilies contain sequences less likely to successfully recombine; this explains why Fam13, 23, and 24 have lower values for P_{pi}. It follows that the evidence for recombination will be greater if more closely related sequences are sampled. We repeated the analysis but only sampling sequences from crown clades, as defined by topologies in Figure 17.1. P_{pi} increased for all *T. brucei* and *T. congolense* subfamilies, suggesting that homology

TABLE 17.3 Proportion of Amino Acid Alignments of *VSG*-Like Subfamilies Displaying Significant Phylogenetic Incompatibility (P_{pi})

Species	Alignment		n	Length bp	P_{pi} μ	σ
T. brucei	a-*VSG*	A	459	677	39.3	6.9
		B	459	481	23.4	4.8
		C	112	677	68.1	8.7
	b-*VSG*	A	380	675	45.0	6.3
		B	380	512	15.2	5.3
		C	90	675	64.2	8.5
T. congolense	Fam13	A	305	649	12.5	4.2
		B	289	583	36.5	7.6
		C	190	649	46.6	10.9
	Fam16	A	517	517	43.3	8.9
		B	372	372	50.3	9.5
		C	257	517	82.4	7.5
T. vivax	Fam23	A	580	580	13.8	4.6
		C	109	580	7.49	4.3
	Fam24	A	552	552	12.6	4.3
		C	84	552	13.1	7.2

Note: Three types of alignment were analyzed. A, random samples of full-length sequences; B, random samples of sequences without C-terminal domains; and C, samples taken from single crown clades exclusively (i.e., from highly related sequences). The proportion of 100 sampled alignments showing significant (i.e., $P < 0.05$) phylogenetic incompatibility (P_{pi}) was calculated for 100 bootstrapped replicates of each alignment. The mean and standard deviation of P_{pi} are given in each case.

limits the prevalence of recombination within these subfamilies, but not in *T. vivax*, where evidence for recombination remained scant. We also reestimated P_{pi} using the entire *T. brucei* alignments, and this resulted in a 41% increase in P_{pi}, confirming the importance of the C-domain in generating diversity. However, this was not true for *T. congolense*: Removing the C-domain resulted in increased P_{pi}, suggesting that it did not contribute to phylogenetic incompatibility.

In summary, the shapes of *VSG* phylogenies differ substantially between species. In part, these differences can be explained simply due to historical differences in the number and relatedness of ancestral lineages inherited. However, because each evolutionary dynamics characterizes a genome and not a particular gene family (i.e., a- and b-*VSG* phylogenies have the same shape despite being better related to gene families in other species), compositional differences are not sufficient explanation. Instead, the phylogenetic data suggest species differences in the molecular evolution of *VSG* genome-wide. Thus, we find that the role of recombination is much greater in *T. brucei* and *T. congolense* than in *T. vivax*, and while the frequency of recombination in *T. brucei* and *T. congolense* may be comparable, differences in the mechanism nonetheless may exist.

ARE THE FEATURES OF *T. BRUCEI* VSG ANTIGENIC VARIATION APPLICABLE TO ALL TRYPANOSOMES WITH ANTIGENIC VARIATION?

Attempts to classify antigenic variation systems according to their molecular genetic mechanisms, as often occur, are flawed because antigenic variation is a phenotype. The overwhelming evidence, from all taxa with antigenic variation, is that those molecular mechanisms supplying the phenotype will become genetically fixed in each organism. The question we are addressing, then, concerns the phenotype. Little is known about *T. congolense*, but there have been several studies of the population phenotype of antigenic variation in several *T. brucei* sspp. (Capbern et al., 1977; Miller and Turner, 1981; Morrison et al., 2005; Robinson et al., 1999) and one large study in *T. vivax* (Barry, 1986). Besides a generally similar tempo of variant appearance, both species display the simultaneous presence of several variants and, importantly, hardwired hierarchical expression. At the cellular level, all these African trypanosomes first display their VSG surface coat as the infective, metacyclic population arises in the tsetse. In *T. brucei* and *T. congolense*, the metacyclic population expresses a specific set of variants. The gross population phenotype appears to be conserved—and in general terms also in widely diverse genera with antigenic variation. At the genome level, we have shown that the molecular basis to antigenic variation has diverged, as might have been expected, but that all three species possess large, diverse *VSG* repertoires. Indeed, the total sequence variation is approximately equal, although it should be remembered that the repertoires are not equally old.

Our data show, nevertheless, that the divergent trajectories taken by each trypanosome genome have produced compositional differences that affect the molecular evolution of contemporary *VSG*. Comparison of *T. brucei* with *T. congolense* suggests that sequence heterogeneity among *T. congolense VSG* places greater limitations on the prevalence of recombination than in *T. brucei*, where b-*VSG* are more homogeneous. The difference also appears to be mechanistic, given that the C-domain plays a role in recombination of *T. brucei VSG*, but not *T. congolense*. Coupled with the variation in the frequency of pseudogenes, these differences indicate fundamental differences in the processes creating *VSG* sequence variation and, therefore, in the genomic basis of antigenic variation. So, while the current population phenotypic model may be applicable to all African trypanosomes, the underlying mechanistic model in *T. brucei* may depart substantially from the situation in other species.

SUMMARY

VSG-based antigenic variation is a defining characteristic of the African trypanosomes and evolved in their common ancestor. Although the gross phenotype is conserved, the underlying genomic basis to antigenic variability has diverged in each species, which has produced compositional differences in contemporary *VSG* repertoires. This historical legacy has effects on molecular evolution and the generation of novel *VSG* that are best demonstrated by comparing *T. congolense*, in which multiple, ancient lineages combine in a relatively heterogeneous repertoire, with *T. brucei*, in which homogeneous *VSGs* are drawn exclusively from two recently derived lineages. The heterogeneity of the *T. congolense* repertoire places limits on the prevalence of recombination, while in *T. brucei*, recombination appears more pervasive, mediated by exchange of the ubiquitous C-domain. This difference could have important consequences for virulence; for instance, a high proportion of pseudogenes in *T. brucei* may contribute combinatorially to the creation of a potentially enormous set of expressed, mosaic genes and therefore could increase probabilities for reinfection of already immune hosts, perhaps resulting in greater host range than other species. Although it could be that *T. vivax* Y486 has only ~330 distinct *VSGs* to deploy (based on the untested assumption that near-identical *VSG* copies are serologically equivalent), compared with >2000 in *T. brucei* 927 and ~850 in *T. congolense* IL3000.

Thus, while the current phenotypic model for antigenic variation may be adequate for all species, comparative genomics shows that the mechanisms underlying phenotypic conservatism, in particular the contribution of recombination, may be quite distinct in different trypanosomes.

REFERENCES

Barbet AF, Kamper SM. 1993. The importance of mosaic genes to trypanosome survival. Parasitology Today 9: 63–66.

Barry JD. 1986. Antigenic variation during *Trypanosoma vivax* infections of different host species. Parasitology 92: 51–65.

Barry JD, Mcculloch R. 2001. Antigenic variation in trypanosomes: Enhanced phenotypic variation in a eukaryotic parasite. Advances in Parasitology 49: 1–70.

Barry JD, Marcello L, Morrison LJ, Read AF, Lythgoe K, Jones N, Carrington M, Blandin G, Bohme U, Caler E, Hertz-Fowler C, Renauld H, El-Sayed N, Berriman M. 2005. What the genome sequence is revealing about trypanosome antigenic variation. Biochemical Society Transactions 33: 986–989.

Berriman M, Ghedin E, Hertz-Fowler C, Blandin G, Renauld H, Bartholomeu DC, Lennard NJ, Caler E, Hamlin NE, Haas B, et al. 2005. The genome of the African trypanosome *Trypanosoma brucei*. Science 309: 416–422.

Borst P, Fairlamb AH. 1998. Surface receptors and transporters of *Trypanosoma brucei*. Annual Review Of Microbiology 52: 745–778.

Bruen TC, Philippe H, Bryant D. 2006. A simple and robust statistical test for detecting the presence of recombination. Genetics 172: 2665–2681.

Capbern A, Giroud C, Baltz T, Mattern P. 1977. *Trypanosoma equiperdum*: Etude des variations antigeniques au cours de la trypanosomose experimentale du lapin. Experimental Parasitology 42: 6–13.

Carrington M, Boothroyd J. 1996. Implications of conserved structural motifs in disparate trypanosome surface proteins. Molecular and Biochemical Parasitology 81: 119–126.

Carrington M, Miller N, Blum M, Roditi I, Wiley D, Turner M. 1991. Variant specific glycoprotein of *Trypanosoma brucei* consists of two domains each having an independently conserved pattern of cysteine residues. Journal of Molecular Biology 221: 823–835.

Daniels JP, Gull K, Wickstead B. 2010. Cell biology of the trypanosome genome. Microbiology and Molecular Biology Reviews 74: 552–569.

Deitsch KW, Lukehart SA, Stringer JR. 2009. Common strategies for antigenic variation by bacterial, fungal and protozoan pathogens. Nature Reviews. Microbiology 7: 493–503.

Engstler M, Pfohl T, Herminghaus S, Boshart M, Wiegertjes G, Heddergott N, Overath P. 2007. Hydrodynamic flow-mediated protein sorting on the cell surface of trypanosomes. Cell 131: 505–515.

Eshita Y, Urakawa T, Hirumi H, Fish WR, Majiwa PA. 1992. Metacyclic form-specific variable surface glycoprotein-encoding genes of *Trypanosoma* (*Nannomonas*) congolense. Gene 113: 139–148.

Futse JE, Brayton KA, Dark MJ, Knowles DP Jr, Palmer GH. 2008. Superinfection as a driver of genomic diversification in antigenically variant pathogens. Proceedings of the National Academy of Sciences of the United States of America 105: 2123–2127.

Gardiner PR, Nene V, Barry MM, Thatthi R, Burleigh B, Clarke MW. 1996. Characterization of a small variable surface glycoprotein from *Trypanosoma vivax*. Molecular and Biochemical Parasitology 82: 1–11.

Gjini E, Haydon DT, Barry JD, Cobbold CA. 2010. Critical interplay between parasite differentiation, host immunity, and antigenic variation in trypanosome infections. The American Naturalist 176: 424–439.

Hamilton PB, Gibson WC, Stevens JR. 2007. Patterns of co-evolution between trypanosomes and their hosts deduced from ribosomal RNA and protein-coding gene phylogenies. Molecular Phylogenetics and Evolution 44: 15–25.

Hutchinson OC, Smith W, Jones NG, Chattopadhyay A, Welburn SC, Carrington M. 2003. VSG structure: Similar N-terminal domains can form functional VSGs with different

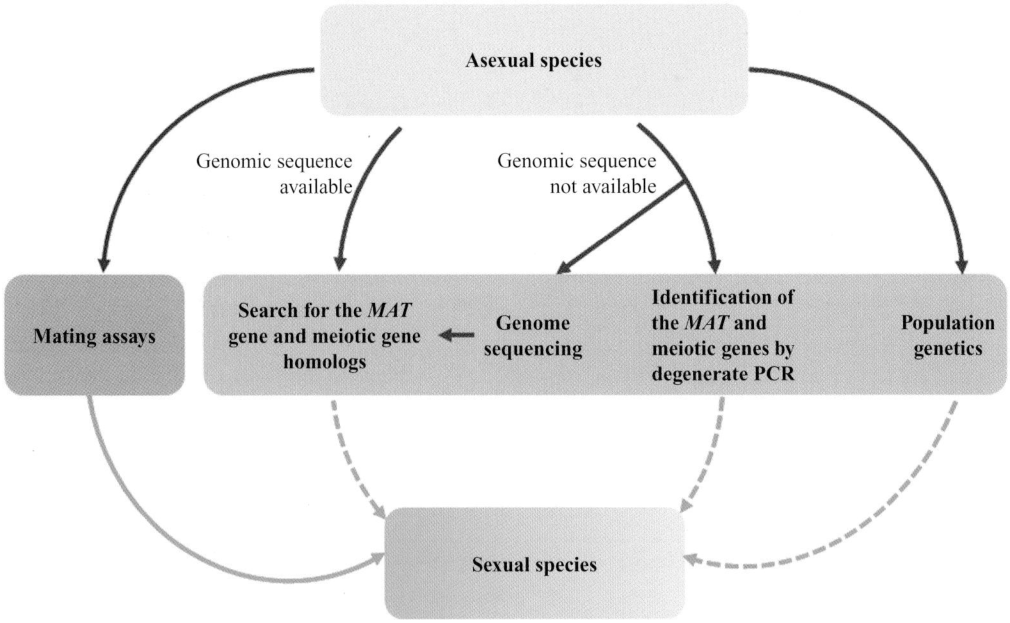

Figure 2.1 Identifying sexual cycles of microbial eukaryotes. (See text for full caption.)

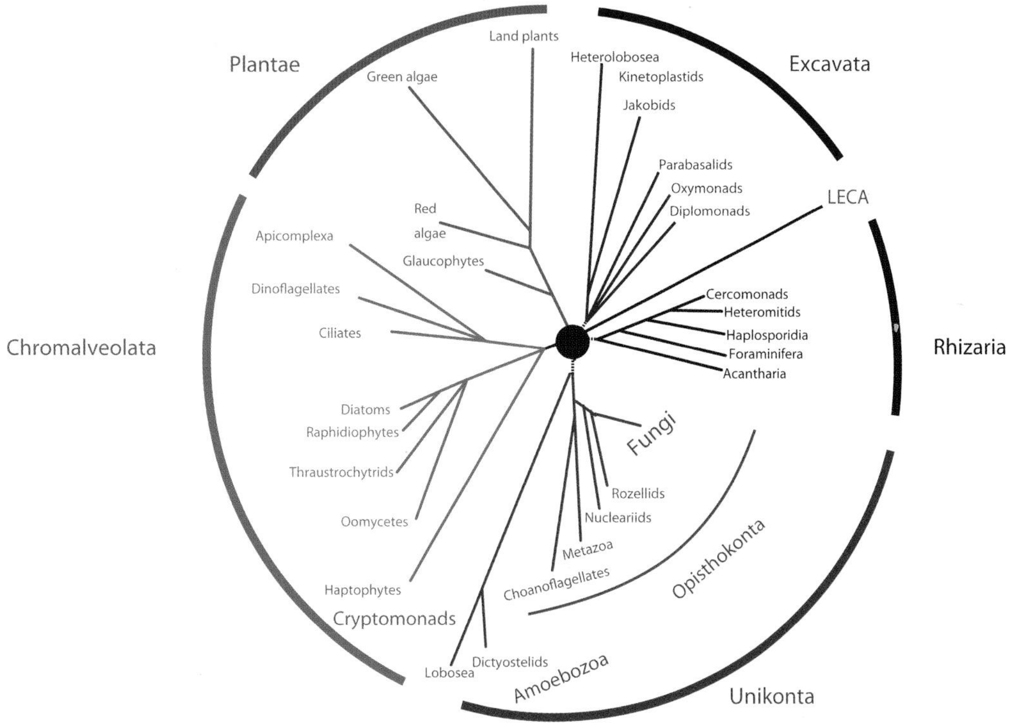

Figure 4.1 Eukaryote phylogenetic tree. (See text for full caption.)

Evolution of Virulence in Eukaryotic Microbes, First Edition. Edited by L. David Sibley, Barbara J. Howlett, and Joseph Heitman.

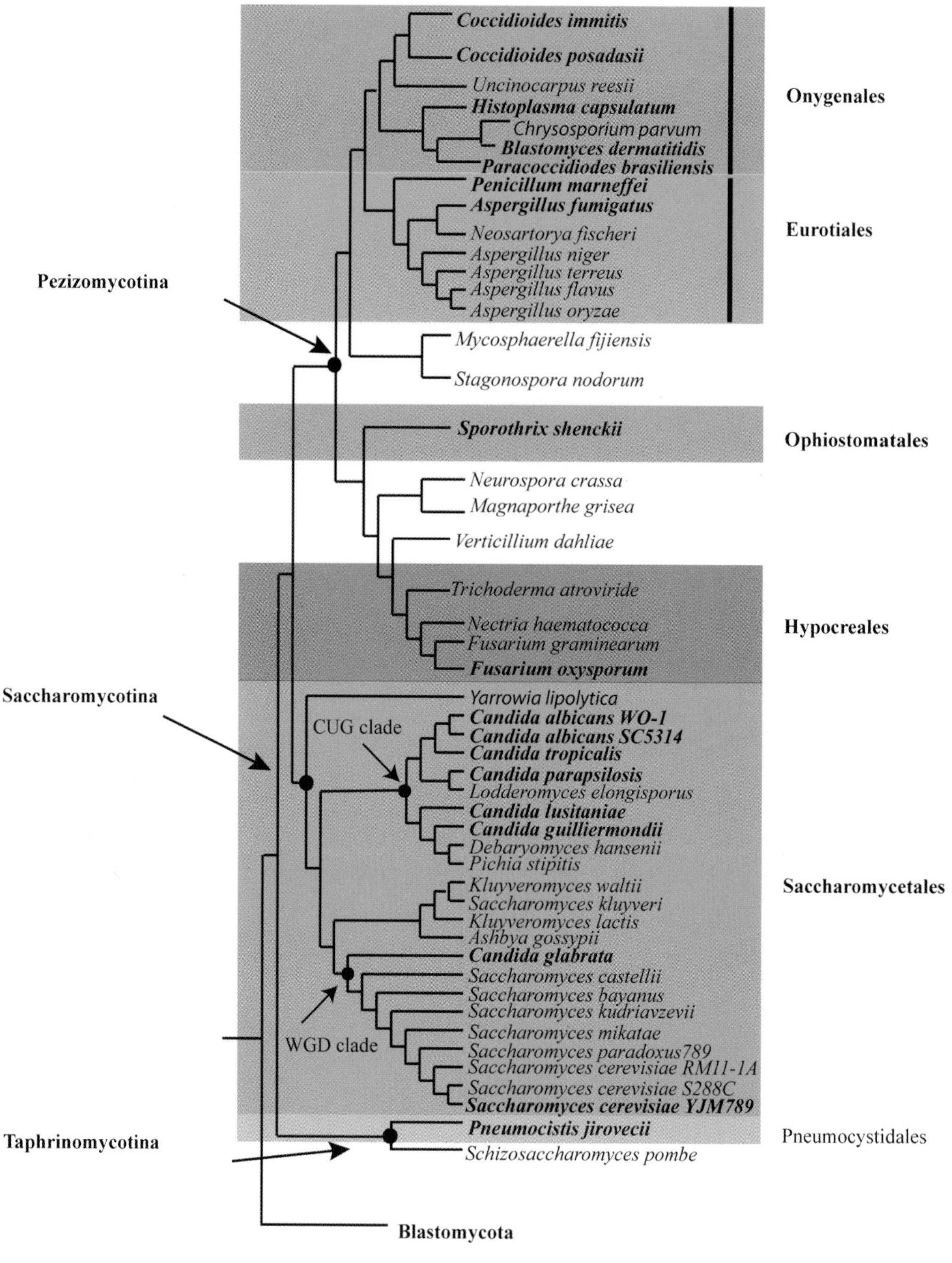

Figure 4.3 Phylogenetic tree highlighting the main human pathogenic Ascomycota. The majority of the pathogenic ascomycetes are concentrated in three orders: Onygenales, Eurotiales, and Saccharomycetales. Pathogenic species are shown in bold characters. The two major clades within the Saccharomycetales are shown; the CUG clade includes *Candida* species that translate the CUG codon into serine, and the other clade underwent a whole-genome duplication event and comprises *Saccharomyces*, which has emerged recently as an opportunistic pathogen. Modified from Wang et al. (2009).

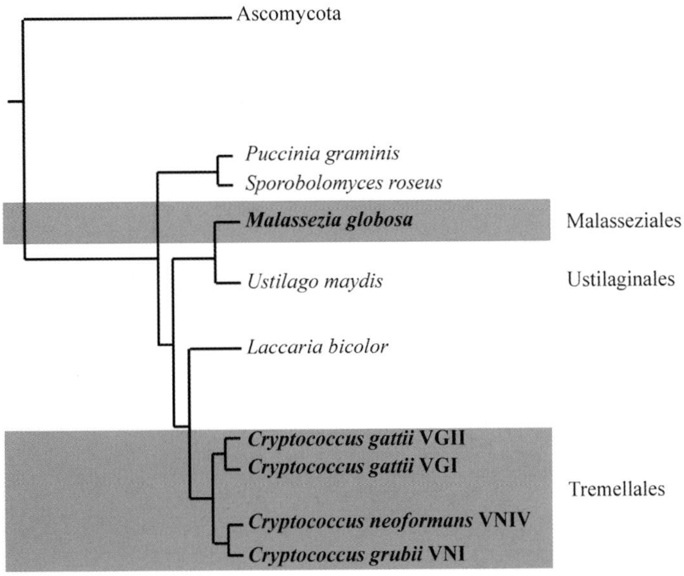

Figure 4.4 Phylogenetic tree of the human pathogenic Basidiomycota. Pathogenic species are shown in bold characters. Notice that the two human pathogens is this group, *Malassezia* and *Cryptococcus*, are not closely related. *Cryptococcus* comprises a species complex formed by at least three different species: *neoformans*, *grubii*, and *gattii*. Modified from Wang et al. (2009).

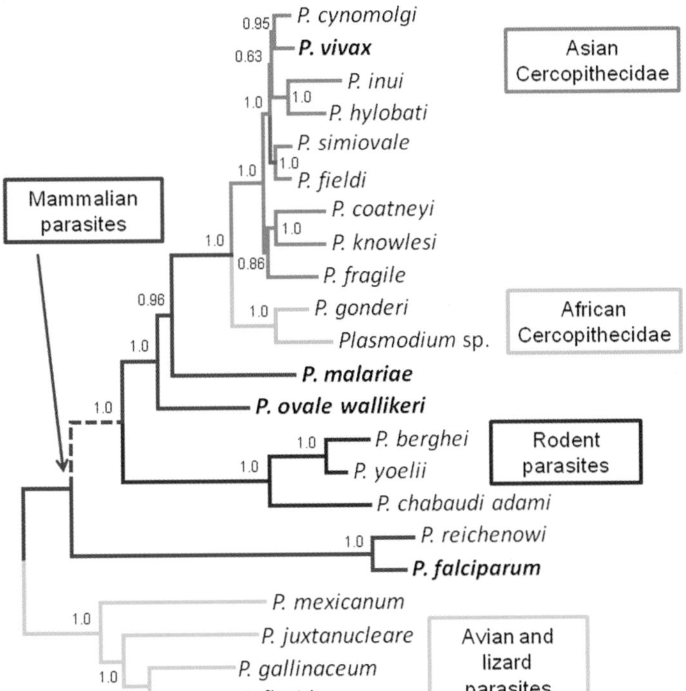

Figure 5.1 Bayesian phylogenetic tree of primate malarial parasites based on complete mitochondrial genomes. (See text for full caption.)

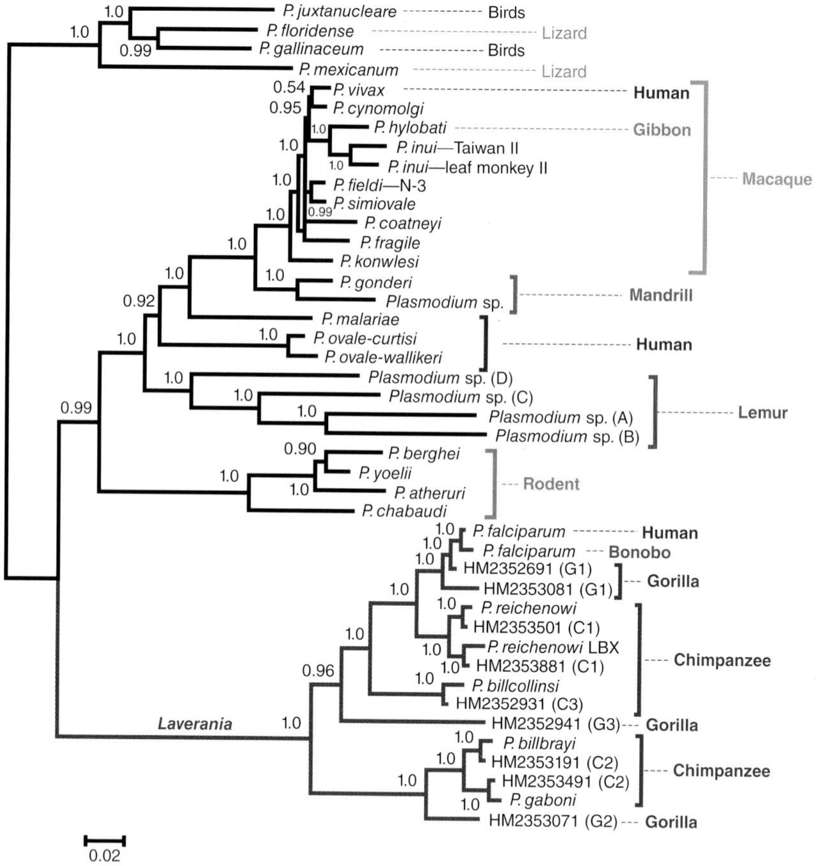

Figure 5.2 Bayesian phylogenetic tree using partial mitochondrial genomes, approximately 3400 bp in length. (See text for full caption.)

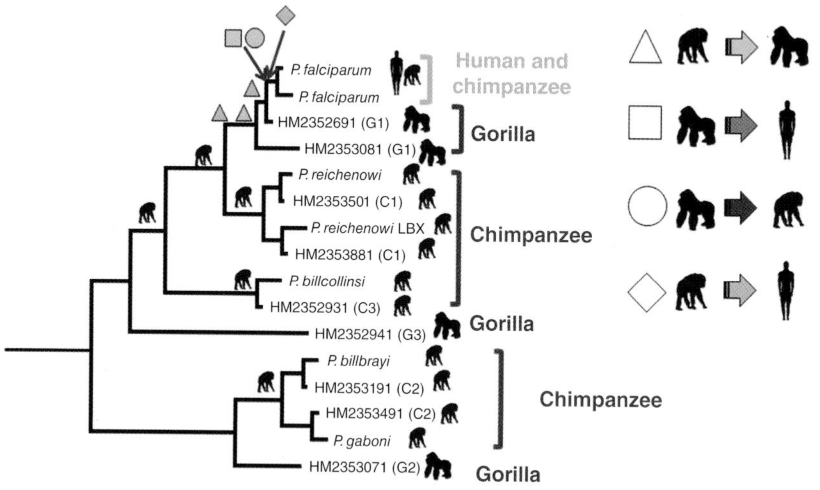

Figure 5.3 Phylogenetic tree of the malarial parasites found in African apes indicating whether lineages or species are found in *Pan* (chimpanzees), *Gorilla*, and/or humans. (See text for full caption.)

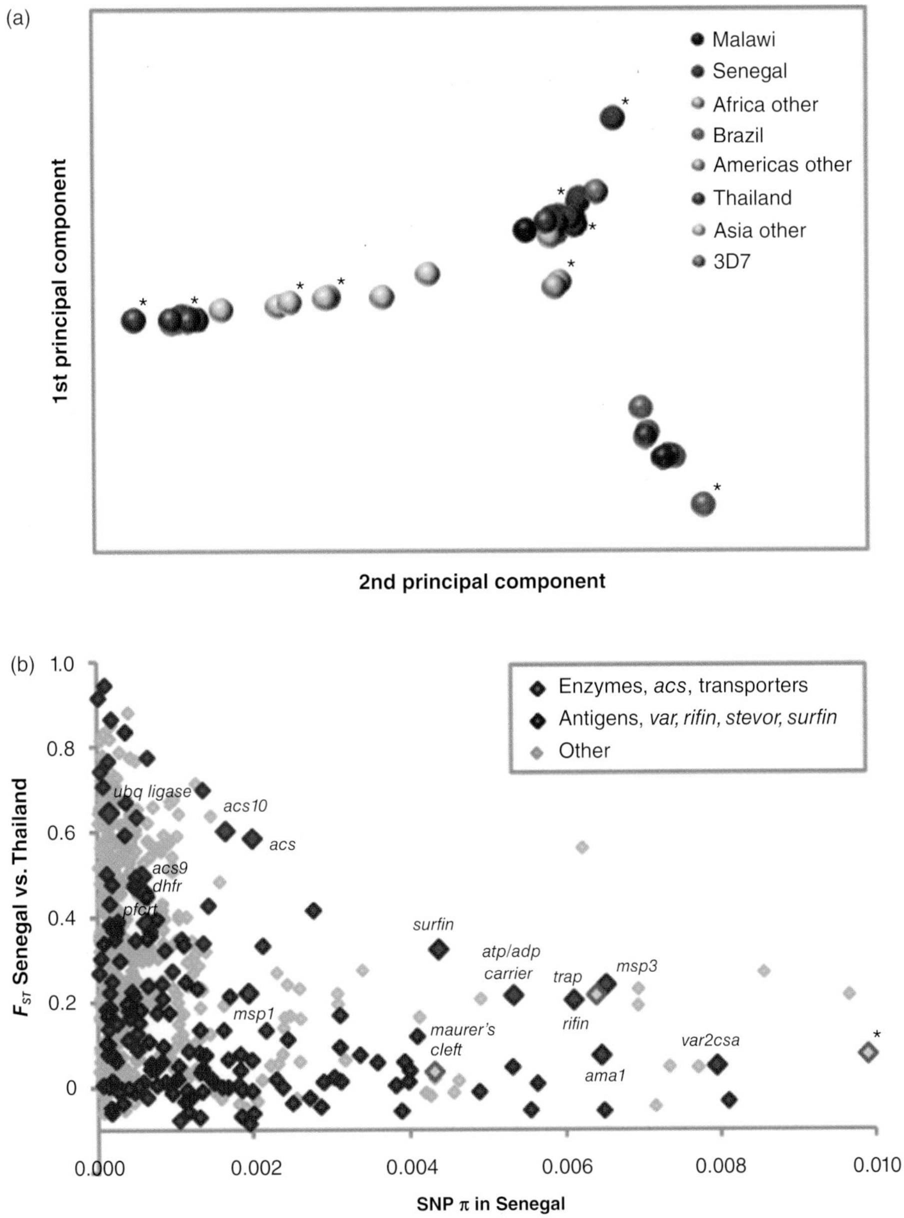

Figure 6.1 Example of global parasite population structure and genetic diversity versus divergence (Van Tyne et al., 2011). (a) Population structure is visualized using the first two principal components of genetic variation for 57 parasites. Solid circles represent individual parasites, with colors assigned by reported origin: Africa in red, America in blue, and Asia in green. The nine strains used for ascertainment sequencing are indicated with (*). (b) Genetic diversity (SNP π) in Senegal versus divergence (F_{ST}) between Senegal and Thailand is reported for 688 genes containing >3 successfully genotyped SNPs. Blue diamonds: enzymes, acyl-CoA synthetases (ACS), or transporters; red diamonds: antigens, vars, rifins, stevors, or surfins; gray diamonds: all other genes. Gene IDs (PlasmoDB.org) for highlighted genes are listed in table S7 from Van Tyne et al. (2011). A gene with an unknown function is flagged with an asterisk (*) to indicate that SNP π is off-scale (0.014).

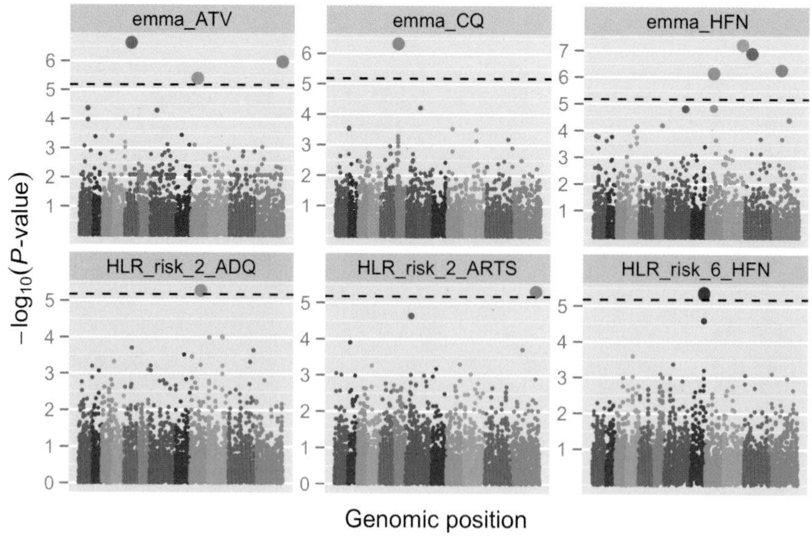

Figure 6.2 Genome-wide association study (GWAS) results (Van Tyne et al., 2011). (See text for full caption.)

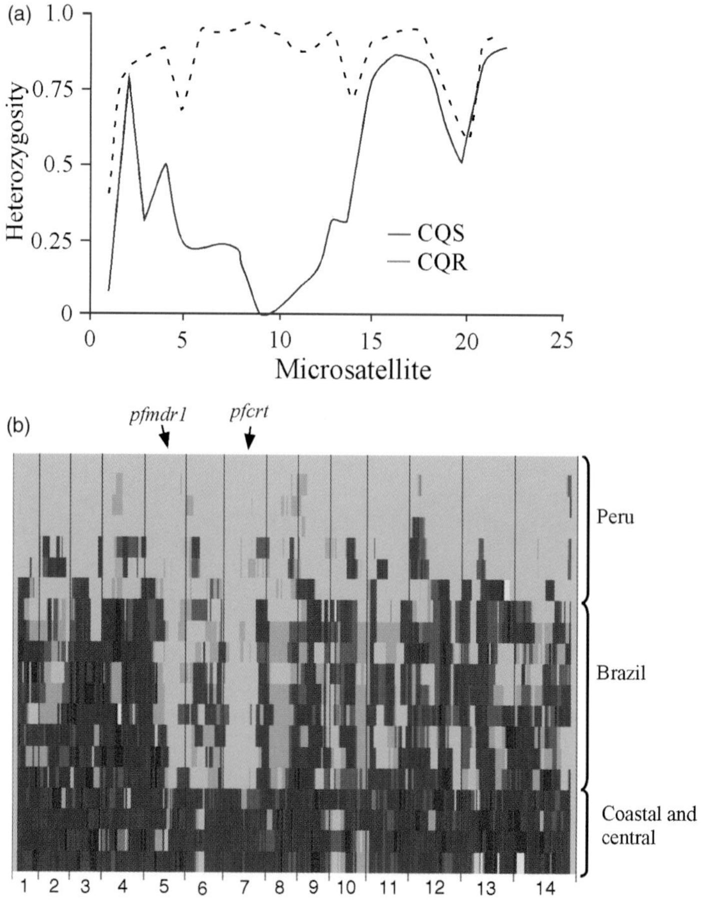

Figure 7.1 Detection of drug selective sweeps by searching for loci with reduced microsatellite (MS) diversity. (See text for full caption.)

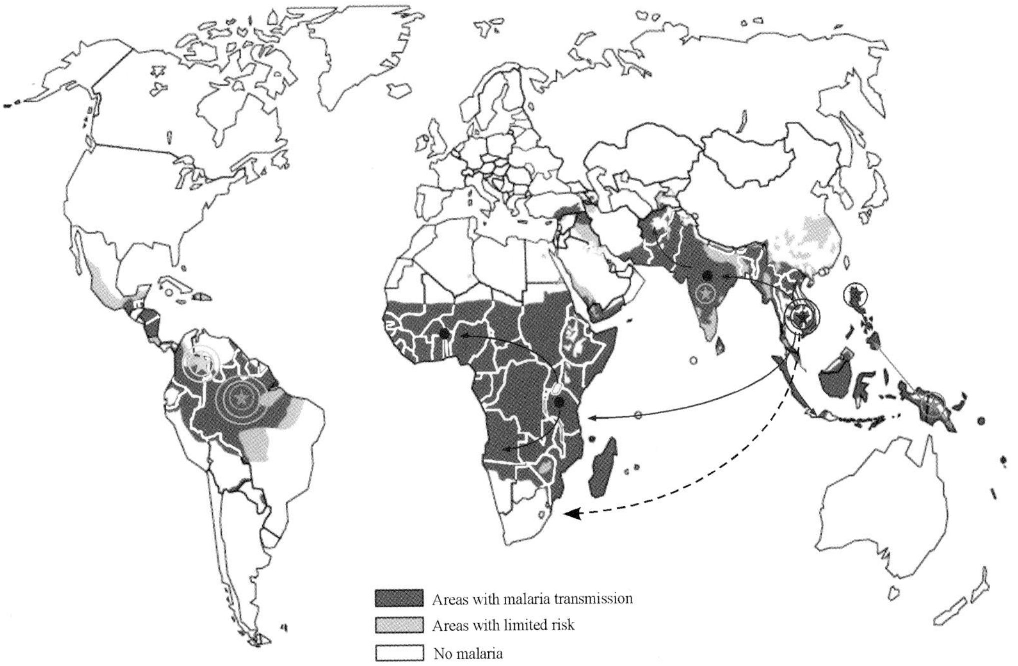

Figure 7.2 Chloroquine resistance (CQ) founder mutations and drug selective sweeps in *Plasmodium falciparum*. At least five independent CQ founder events, indicated by the stars of different colors, have been detected through haplotype analysis of *pfcrt* alleles and their flanking microsatellites (MSs). Each star represents one site where unique *pfcrt* and/or flanking MS haplotypes have been observed, although selective sweeps and the number of founder mutations in India are still being debated. The dashed line represents a sweep of high level resistant *pfdhfr* allele from Southeast Asia to Africa. The colored areas of the world map represent malaria-endemic regions with two levels of transmission risk as defined on the World Health Organization (WHO) Web site (http://www.who.int/malaria/en).

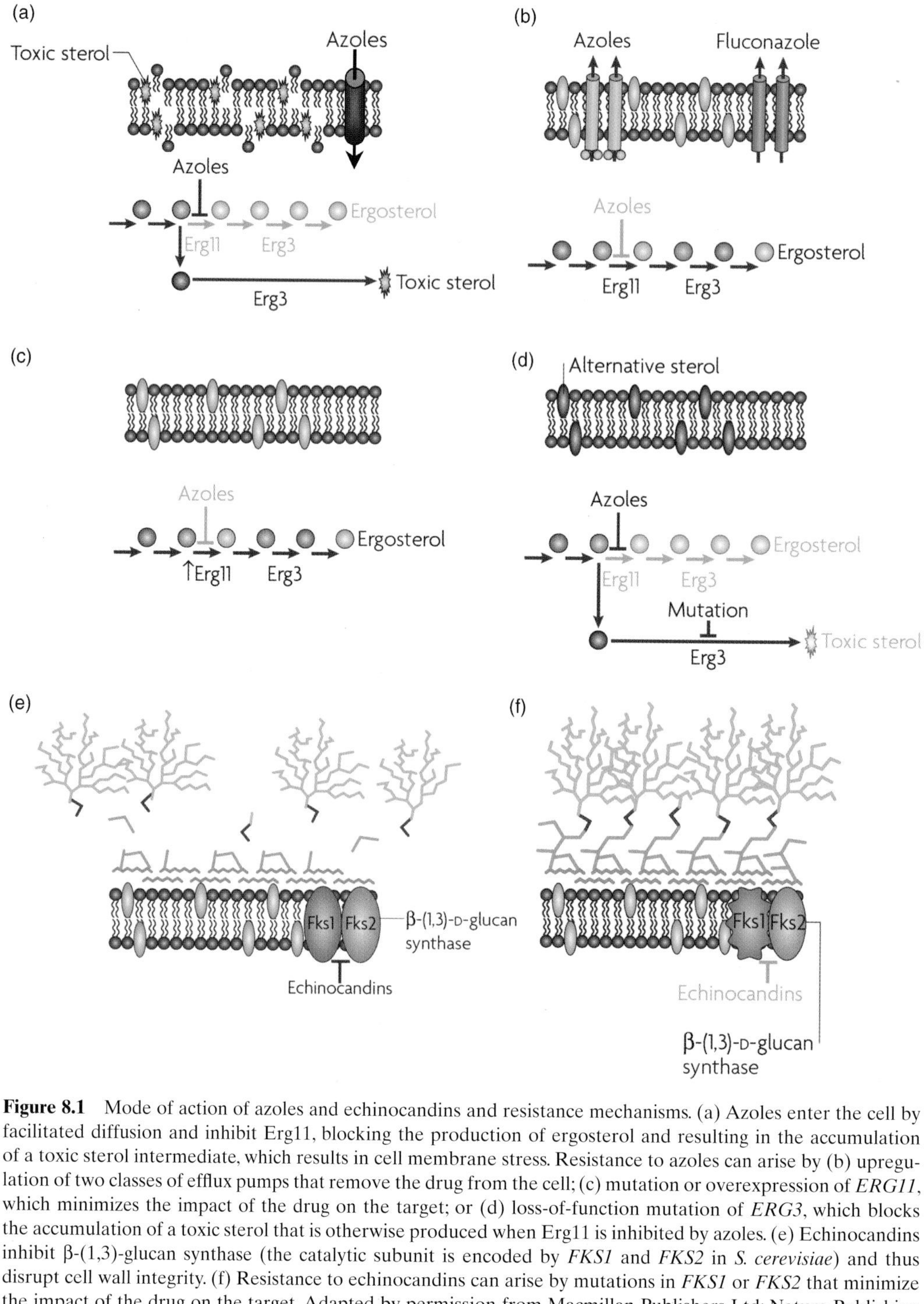

Figure 8.1 Mode of action of azoles and echinocandins and resistance mechanisms. (a) Azoles enter the cell by facilitated diffusion and inhibit Erg11, blocking the production of ergosterol and resulting in the accumulation of a toxic sterol intermediate, which results in cell membrane stress. Resistance to azoles can arise by (b) upregulation of two classes of efflux pumps that remove the drug from the cell; (c) mutation or overexpression of *ERG11*, which minimizes the impact of the drug on the target; or (d) loss-of-function mutation of *ERG3*, which blocks the accumulation of a toxic sterol that is otherwise produced when Erg11 is inhibited by azoles. (e) Echinocandins inhibit β-(1,3)-glucan synthase (the catalytic subunit is encoded by *FKS1* and *FKS2* in *S. cerevisiae*) and thus disrupt cell wall integrity. (f) Resistance to echinocandins can arise by mutations in *FKS1* or *FKS2* that minimize the impact of the drug on the target. Adapted by permission from Macmillan Publishers Ltd: Nature Publishing Group, Nature Reviews Microbiology (Cowen, 2008), © 2008.

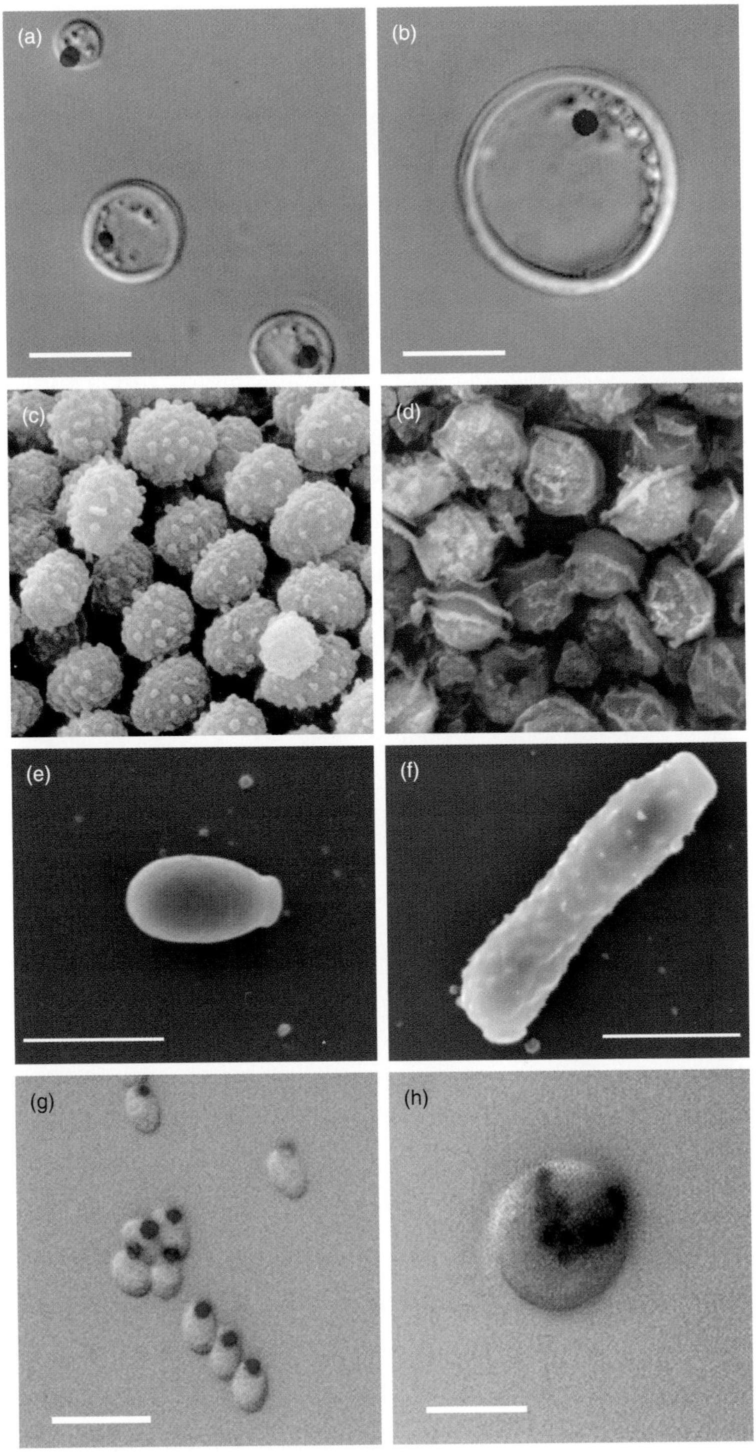

Figure 9.2 Sexual reproduction is associated with morphological diversity in human fungal pathogens. (See text for full caption.)

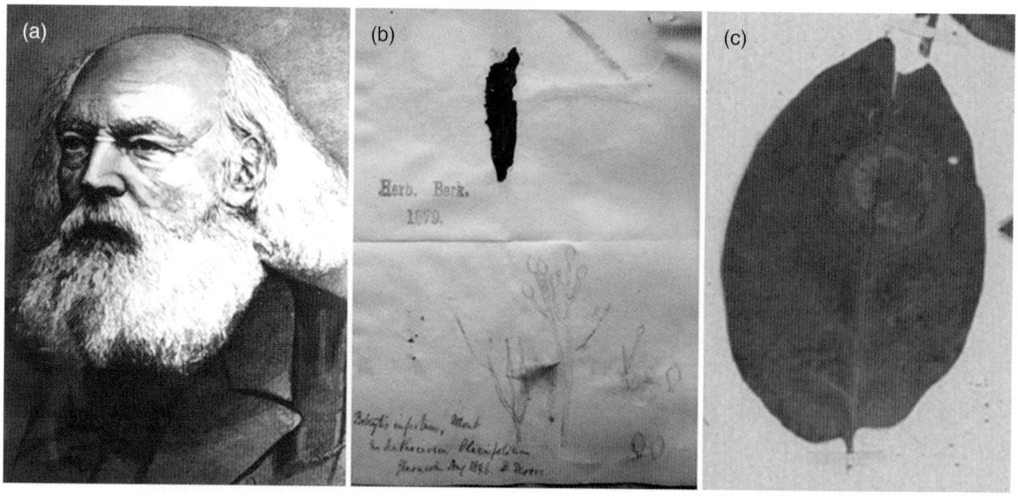

Figure 10.1 (a) M. J. Berkeley, the British mycologist who described the pathogen *Botrytis infestans* (later named as *Phytophthora infestans*) on potato in 1846; (b) sporangia and sporangiophores of *P. infestans* drawn on a famine-era archival specimen label by David Moore, Glasnevin, Ireland, that was sent to Berkeley for confirmation; (c) herbarium specimen with leaf lesion caused by *P. infestans* collected by Krieger in Germany in 1888.

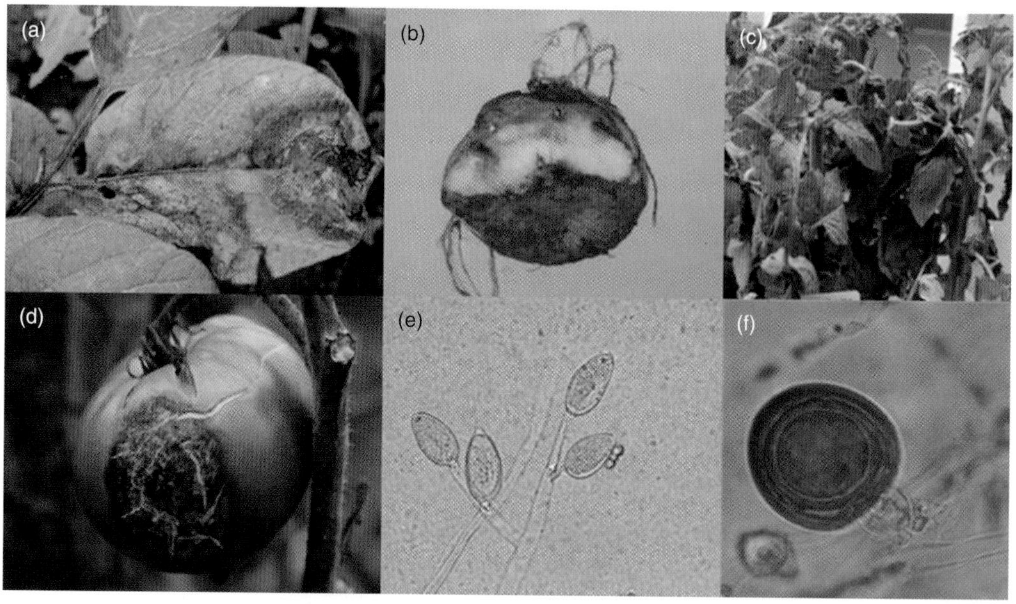

Figure 10.2 Symptoms of late blight caused by *P. infestans* on (a) potato leaf, (b) tuber, (c) tomato stem, (d) tomato fruit, (e) plants in a field, and (f) a tomato transplant. (e) Asexual sporangia borne on compound sporangiophores; (f) sexual oospore produced by fusion of anisogamous gametangia (male antheridium and female oogonia) of opposite mating types (A1 and A2).

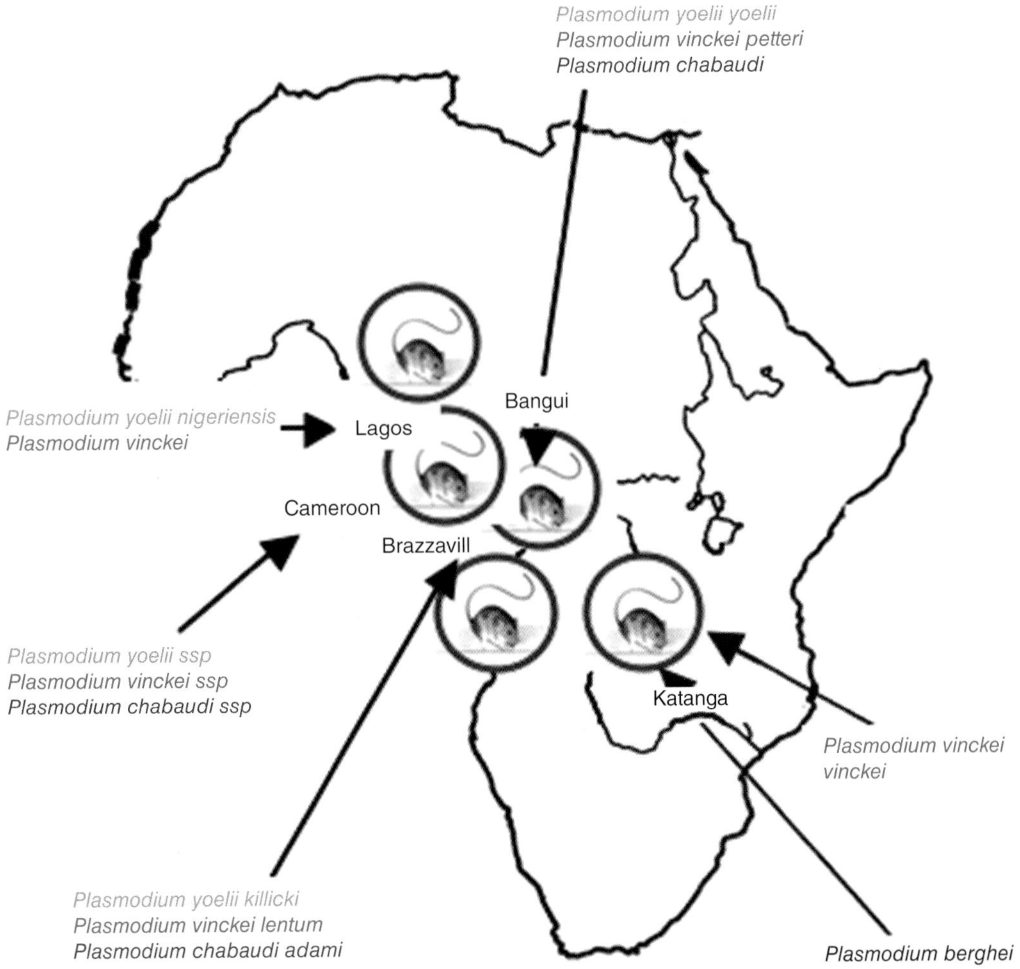

Figure 14.1 Geographic origin of rodent malaria parasites.

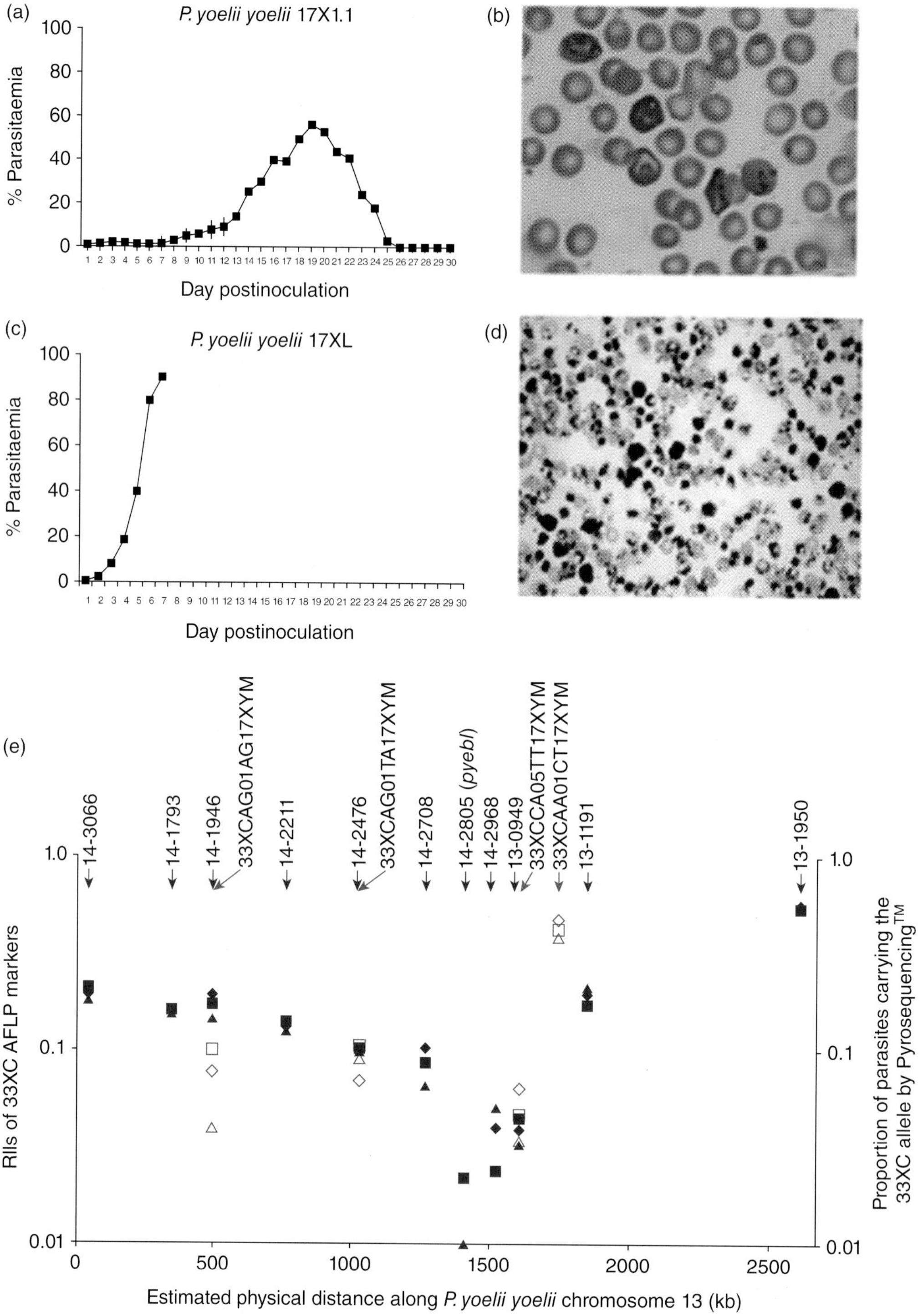

Figure 14.2 Parasitemia of *Plasmodium yoelii yoelii* 17X1.1 (a) and 17XL (c) in female CBA mice, and thin blood smears showing host cell preference at day 5 postinoculation, (b) and (d), respectively. (See text for full caption.)

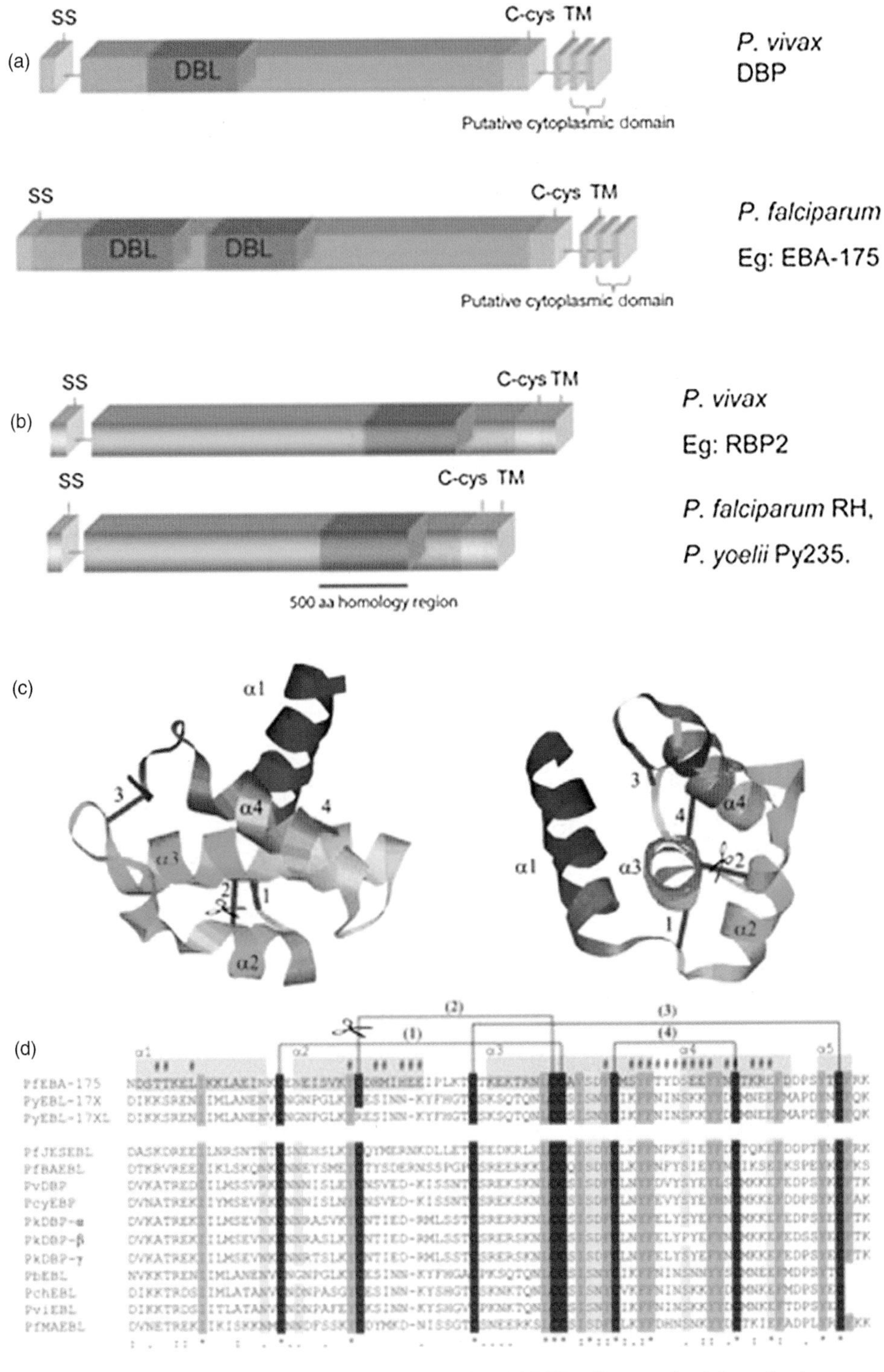

Figure 14.3 Schematic diagram of the EBLs and RBLs. (See text for full caption.)

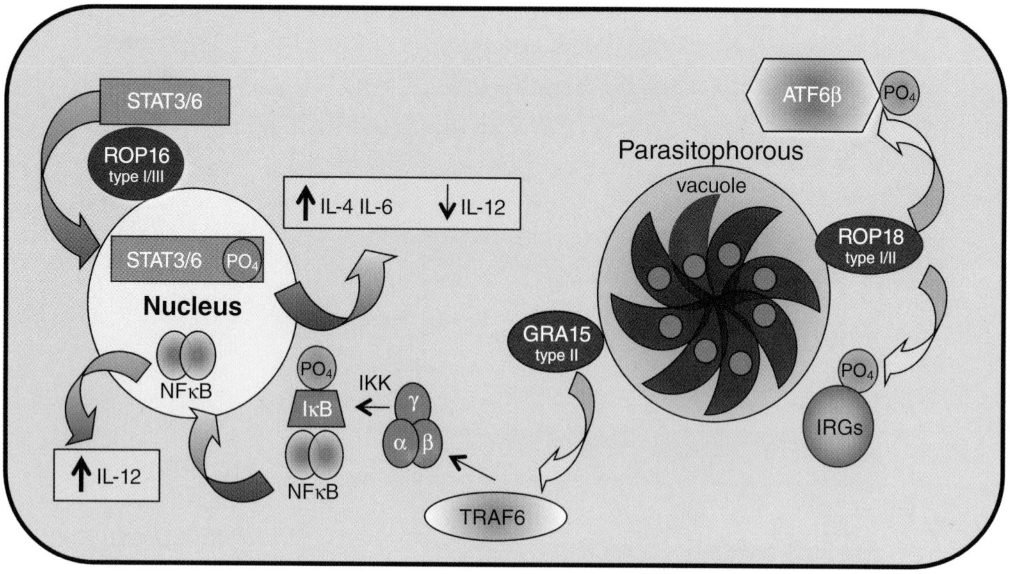

Figure 15.1 Model for the cellular roles of *T. gondii* secretory proteins in modulating host cell signaling and controlling acute virulence. Diagram of a host cell infected by *T. gondii* showing the distribution of key secretory proteins and functional pathways altered within the cell. ROP16 is secreted into the host cell and accumulates in the nucleus (Saeij et al., 2007). ROP16 contains a nuclear localization signal, although this is not essential for its activity on STATs. Recent evidence suggests ROP16 affects host gene transcription by directly phosphorylating STAT3 (Yamamoto et al., 2009) and STAT6 (Ong et al., 2010), host cell transcription factors that upregulate IL-4- and IL-6-dependent responses while impairing IL-12 production. GRA15 is also released upon or after invasion and this activates the NFκB pathway, resulting in induction of IL-12 by type II strains (Rosowski et al., 2011). Activation requires the host signaling proteins TRAF6 and the IKK complex that phosphorylates IκB, resulting in nuclear translocation of NFκB and activation of gene transcription. GRA15 has been localized to the PV membrane and to tubule extensions that emanate from it. ROP18 is secreted into the host cell where it remains tethered to the parasitophorous vacuole (PV) membrane through an N-terminal low complexity region (LCR) (Labesse et al., 2009; Reese and Boothroyd, 2009). ROP18 directly phosphorylates immunity-related GTPases (IRGs), thus blocking their accumulation on the vacuole and avoiding clearance of the parasite by this innate immunity mechanism (Fentress et al., 2010; Steinfeldt et al., 2010). ROP18 has also been described to interact with and phosphorylate ATF6β, an ER stress response transcription factor that is important for controlling dendritic cell function in response to *T. gondii* infection (Yamamoto et al., 2011).

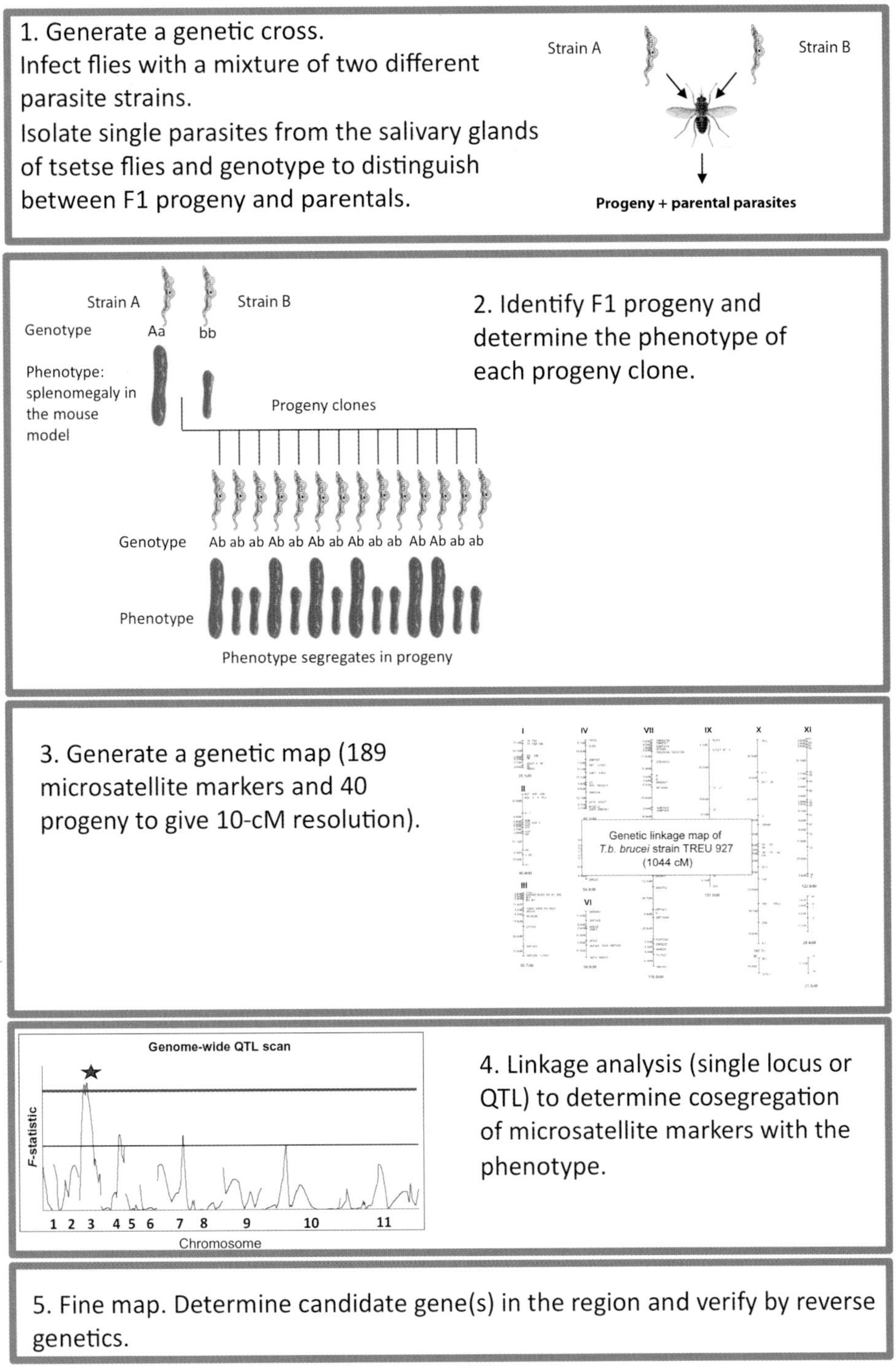
1. Generate a genetic cross.
Infect flies with a mixture of two different parasite strains.
Isolate single parasites from the salivary glands of tsetse flies and genotype to distinguish between F1 progeny and parentals.
Strain A
Strain B
Progeny + parental parasites
Strain A
Strain B
Genotype
Aa
bb
Phenotype: splenomegaly in the mouse model
Progeny clones
2. Identify F1 progeny and determine the phenotype of each progeny clone.
Genotype
Ab ab ab Ab ab Ab ab Ab ab ab Ab Ab ab ab
Phenotype
Phenotype segregates in progeny
3. Generate a genetic map (189 microsatellite markers and 40 progeny to give 10-cM resolution).
Genetic linkage map of T.b. brucei strain TREU 927 (1044 cM)
Genome-wide QTL scan
F-statistic
1 2 3 4 5 6 7 8 9 10 11
Chromosome
4. Linkage analysis (single locus or QTL) to determine cosegregation of microsatellite markers with the phenotype.
5. Fine map. Determine candidate gene(s) in the region and verify by reverse genetics.

Figure 16.1 Diagram of a trypanosome cross and the strategy used to map genes determining phenotypes such as virulence. Two strains of parental trypanosomes are used to coinfect tsetse flies and, once the infection has developed to the salivary gland stages and after mating, individual trypanosomes are expanded vegetatively. These "clones" are then genotyped with microsatellite markers to define independent F1 progeny and subsequently phenotyped by infection of mice. In this example, splenomegaly is illustrated as a phenotype that differs between parental strains A and B and segregates in the progeny. The phenotype is treated as being determined by more than one locus and the segregation of the quantitative trait used to map the loci by linkage analysis to the genetic map.

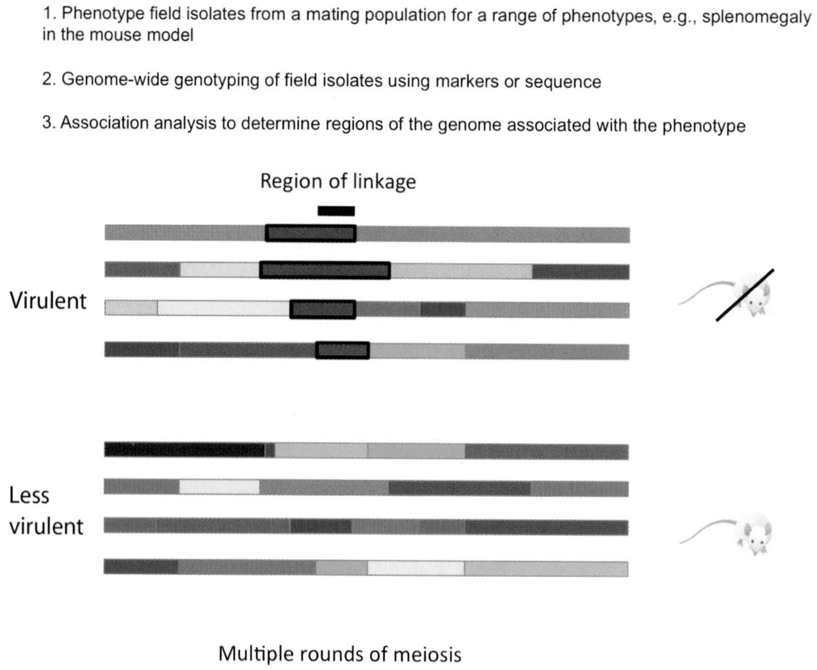

Figure 16.2 Population genomic association analysis. The diagram illustrates the division of strains into two phenotypic classes (virulent and less virulent) using one or more measures of virulence The isolates are then genotyped using markers distributed over the whole genome or by whole-genome sequencing. In this example, the haplotype structure of each strain is illustrated just using a single chromosome (for simplicity). The red haplotype appears to be associated with high virulence and is not present in the avirulent class. Other haplotypes are randomly distributed across the virulence classes and so show no association. Thus, the genes determining the particular virulence phenotype are located within the region of the red haplotype common to all strains. In a population with a high frequency of mating, many rounds of meiosis will have occurred (rather than a single round in the laboratory cross) and so the common haplotype could be physically quite small.

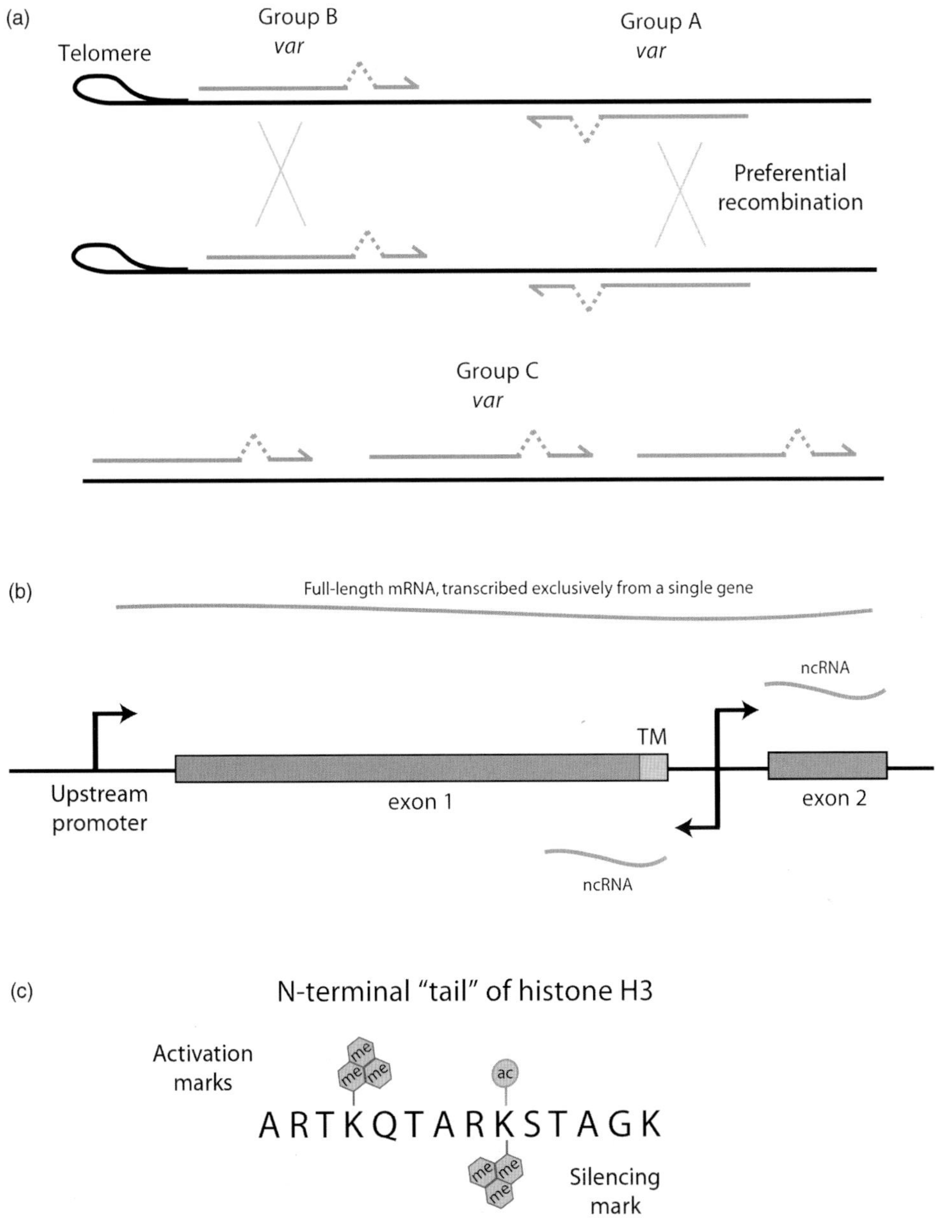

Figure 18.1 Gene structure and chromosomal organization of the *var* gene family. (a) Group B genes are typically located immediately adjacent to the telomere and transcribed toward the center of the chromosome, while group A genes are also within the subtelomeric regions but in the opposite orientation. Telomeres are found in "bouquet" structures that are theorized to align *var* genes for preferential recombination. Group C genes are arranged in tandem arrays in the central parts of chromosomes. (b) *var* genes have a two-exon structure, with an upstream promoter that mediates transcription of the full-length mRNA. In addition, a bidirectional promoter found within the intron results in transcription of noncoding RNAs from most or all members of the family. The transmembrane domain (TM) of the protein is encoded at the very 3 end of exon 1. (c) The amino acid sequence of the N-terminal "tail" of histone H3. The three histone modifications that are proposed to be most important for regulating *var* gene expression are drawn, including methylation of lysine 4 and both acetylation and methylation of lysine 9.

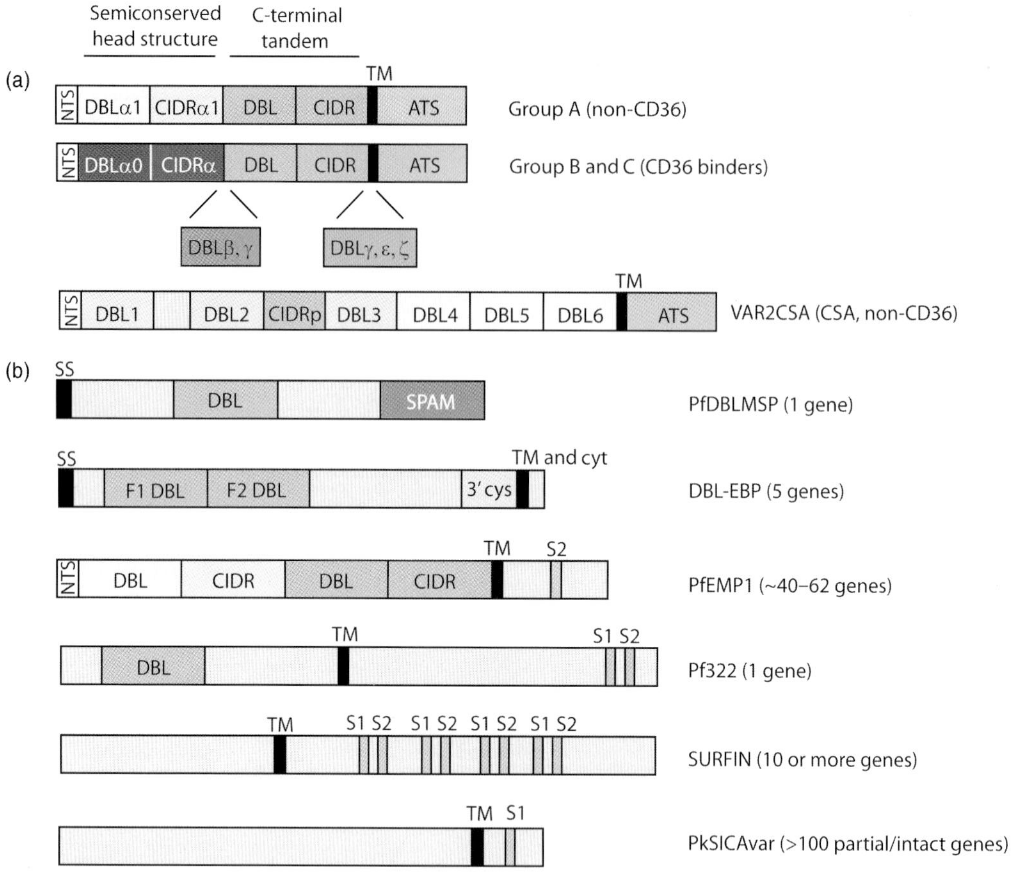

Figure 18.2 Protein architecture of PfEMP1 proteins and their evolutionary relationship to other *Plasmodium* proteins. (a) PfEMP1 proteins exhibit a high degree of compositional order. The extracellular region of small PfEMP1 proteins contains two DBL–CIDR tandem domains. The first tandem is known as the semiconserved head structure and is a different sequence type from the C-terminal tandem. Group A PfEMP1 proteins do not bind CD36 and have a different type of protein head structure from the group B and C PfEMP1 proteins. Large PfEMP1 proteins have additional DBL domains of specific sequence types that are located between or after the two DBL–CIDR tandems. VAR2CSA is an unusually conserved PfEMP1 family member with a distinct protein architecture. (b) PfEMP1 proteins are members of the DBL superfamily. This family participates in both parasite invasion of erythrocytes (DBL-EBP and PfDBLMSP) and modification of the infected erythrocyte (PfEMP1 and Pf322). The PfEMP1 cytoplasmic tail is encoded on a separate exon from the extracellular region. It contains an S2 motif that is homologous to the S1/S2 tryptophan-rich domain (WRD) in the C-terminus of the PfSURFIN family, Pf332, and the *P. knowlesi* variant antigen family (SICAvar).

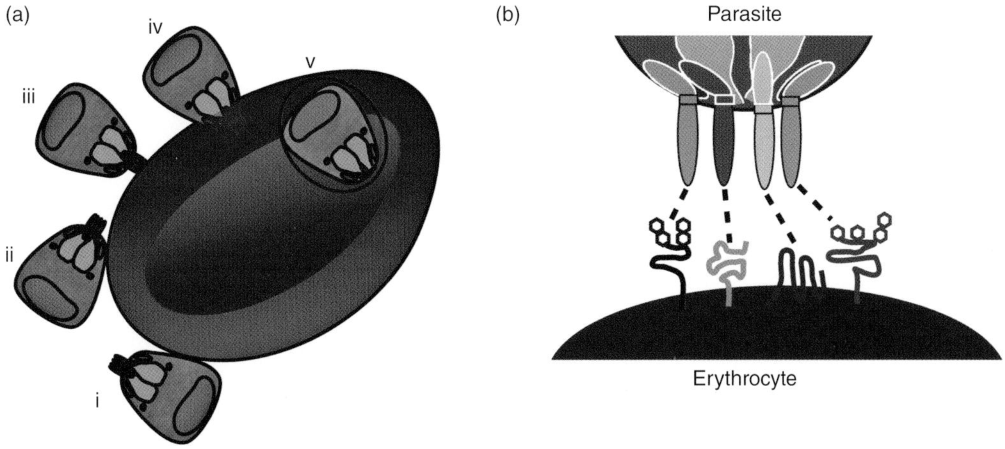

(c)

Parasite	DBL Family		RBL Family	
	Parasite Ligand	Erythrocyte Receptor	Parasite Ligand	Erythrocyte Receptor
P. falciparum	PfEBA175	Gly A	PfRH1	"Y"
	PfEBA140	Gly C	PfRH2a	?
	PfEBA181	"E"	PfRH2b	"Z"
	PfEBL1	Gly B	*PfRH3*	*pseudogene*
	PfEBA165	*pseudogene*	PfRH4	CR1
			PfRH5	?
			PfRH6	*pseudogene*
P. vivax	PvDBP	Duffy (DARC)	PvRBP1a, PvRBP1b, PvRBP2a, PvRBP2b, PvRBP2c, PvRBP2c, PvRBP3	?
P. knowlesi	PkDBP-α	Duffy (DARC)	PkRBP1, PkRBP2, PkRBP3a, PkRBP3b	?
	PkDBP-β, PkDBP-γ	?		
P. yoelii	PyEBP	Duffy	Py235 family	?
P. berghei	PbEBP	?	Pb235 family	?

Figure 19.1 (a) Schematic of *Plasmodium* spp. invasion of the erythrocyte. In the first step (i), the merozoite binds the surface of the erythrocyte. This binding is reversible and likely involves multiple members of the merozoite surface proteins. Following binding, the merozoite must undergo reorientation (ii) to position the parasite apical organelles in close proximity to the site of erythrocyte binding. Apical membrane antigen 1 (AMA-1) has been identified as a potential mediator of parasite reorientation (Mitchell et al., 2004). Following apical reorientation (iii), parasite ligands interact with erythrocyte surface proteins, forming a host–parasite junction and committing the parasite to invasion. The merozoite then actively invades the erythrocyte (iv) via actinomyosin-based motor proteins. Finally, the merozoite completes invasion (v) into the newly formed parasitophorous vacuole. (b) Schematic of the parasites ligands binding to host erythrocyte receptors. It is likely that multiple ligand–receptor interactions are necessary to form the tight junction and allow efficient parasite invasion. (c) Table of parasite ligands and erythrocyte receptors. *P. falciparum* has an expanded repertoire of DBLs. The erythrocyte receptors have been identified as the Duffy receptor (or Duffy antigen/receptor for chemokines) for PvDBP, PkDBPalpha, and PyEBP. The only other receptors known include PfEBA175/Gly A, PfEBA140/Gly C, PfEBL1/Gly B, and PfRh4/CR1. Pseudogenes are shown in italics.

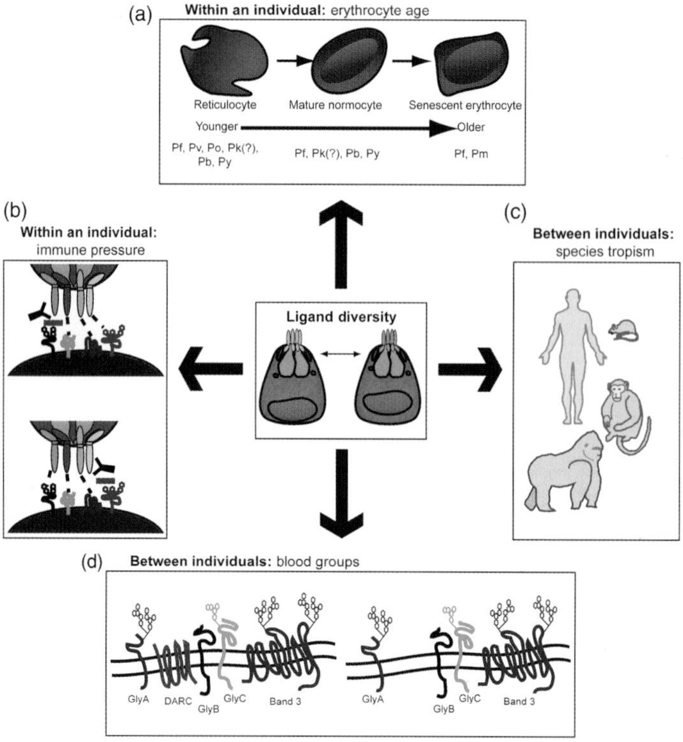

Figure 19.2 Parasite ligand diversity allows the *Plasmodium* merozoites to invade erythrocytes of different stages of maturity (a), in the presence of immune pressure (b), with different blood group polymorphisms (c), and from different species (d). (See text for full caption.)

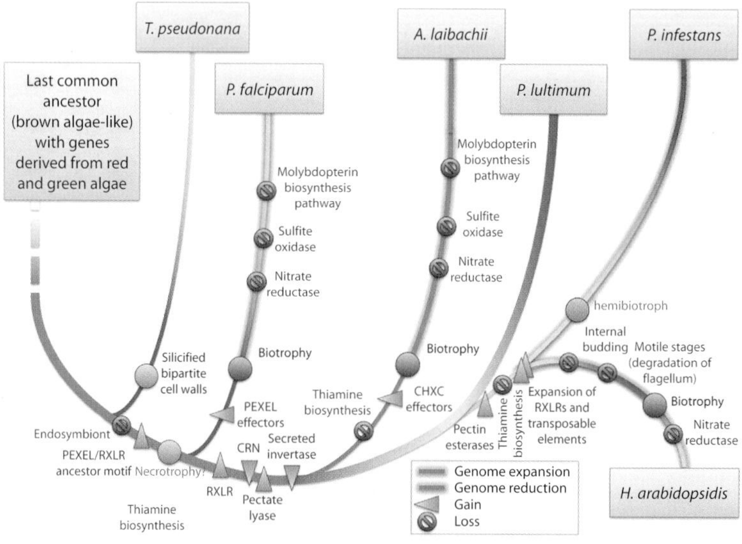

Figure 20.1 Model for evolution of gain and loss of genes and pathways for selected Chromalveolata (from Kemen et al., 2011). (See text for full caption.)

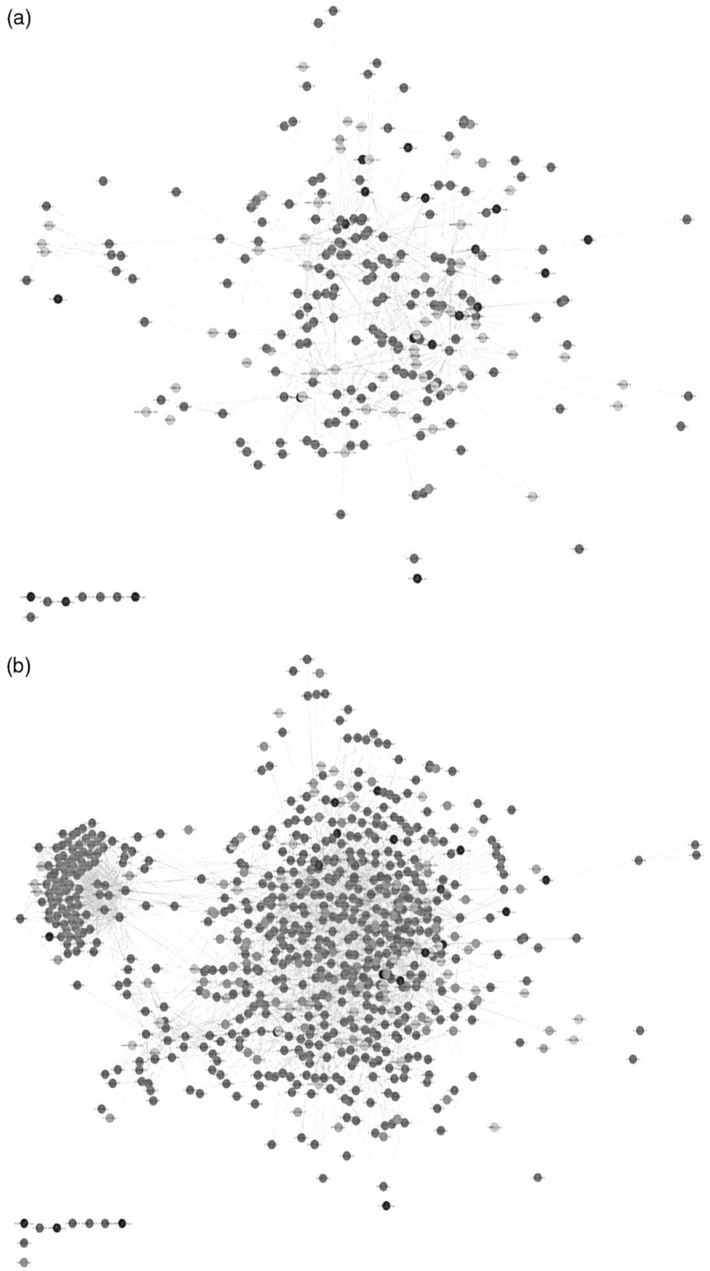

Figure 20.2 Protein–protein interactions between *Arabidopsis* host proteins and pathogen effector protein domains determined by yeast two-hybrid experiments and literature curation (Mukhtar et al., 2011). Pink: *Pseudomonas syringiae* effector; orange: *Hyaloperanospera arabidopsidis* effector; green: *Arabidopsis* immune system protein (NB-LRR, receptor-like kinase, or protein annotated as defense related); gray: other *Arabidopsis* protein. (a) Direct interactions between pathogen effector and *Arabidopsis* host proteins: Host proteins interacting with effectors form a densely connected subnetwork within the host protein–protein interactome. In a number of cases, independently evolved effectors from the two pathogens interact with the same host protein. (b) Direct pathogen effector–host protein interactions in the context of other protein interactions in which the host proteins participate. A number of the host proteins interacting with effectors are major interactome hubs. Interactome hubs as a whole are significantly overrepresented among immune interactors. Many host proteins annotated as immune related interact with proteins interacting with effectors. Host proteins with orthologs among land plants, but not in more distant taxa, are overrepresented in immune interactors.

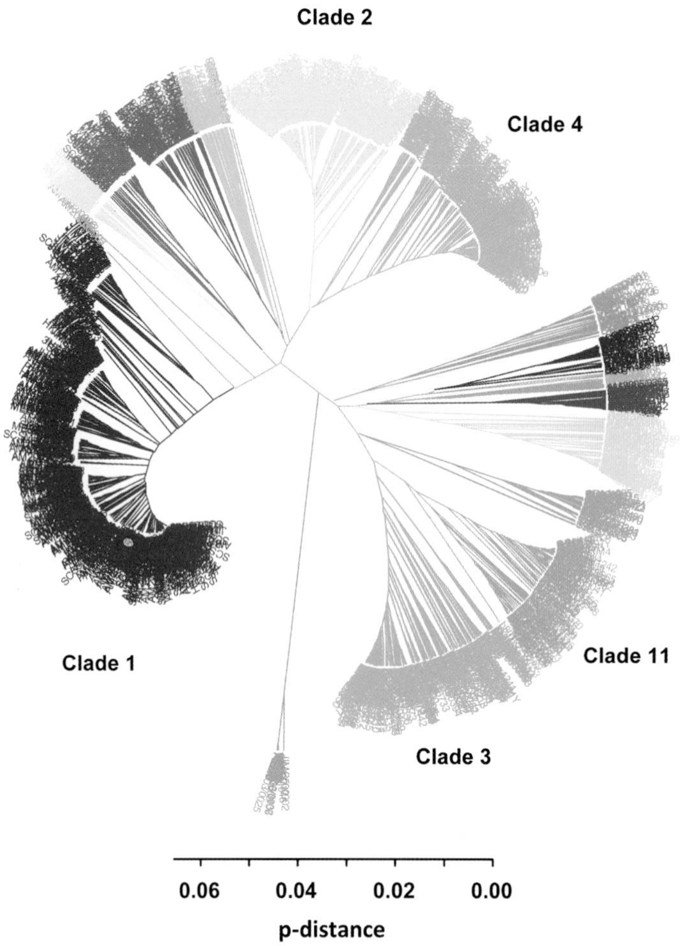

Figure 21.2 UPGMA dendrogram based on 1410 MLST diploid sequence types of *C. albicans* showing the five major clades (1, 2, 3, 4, and 11) and the more distantly related clade 13 (in brown at 6 o'clock).

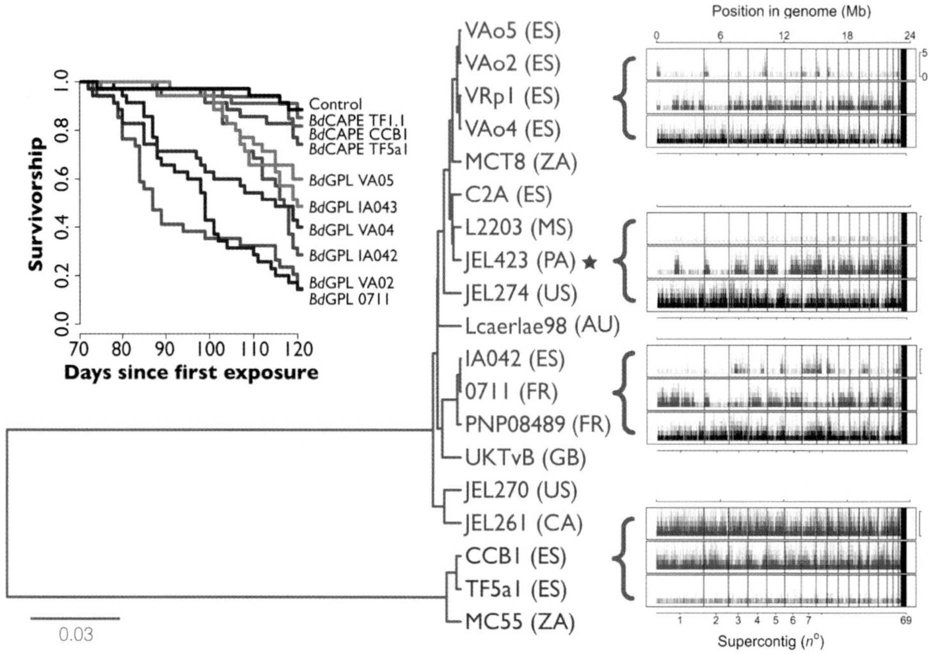

Figure 24.1 A tree of 19 *Bd* nuclear genomes made using the UPGMA algorithm in PAUP with Kaplan–Meier survival curves and representations of the polymorphic sites within the genome. (See text for full caption.)

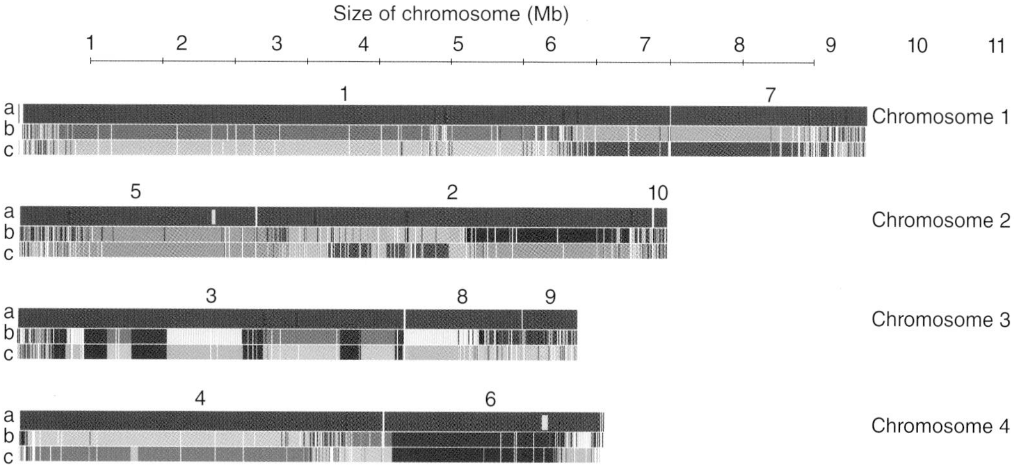

Figure 26.1 Global view of syntenic alignments of *F. graminearum* (*Fg*) to *F. verticillioides* (*Fv*) and *F. oxysporum* (*Fo*) using *Fg* chromosomes as reference. (See text for full caption.)

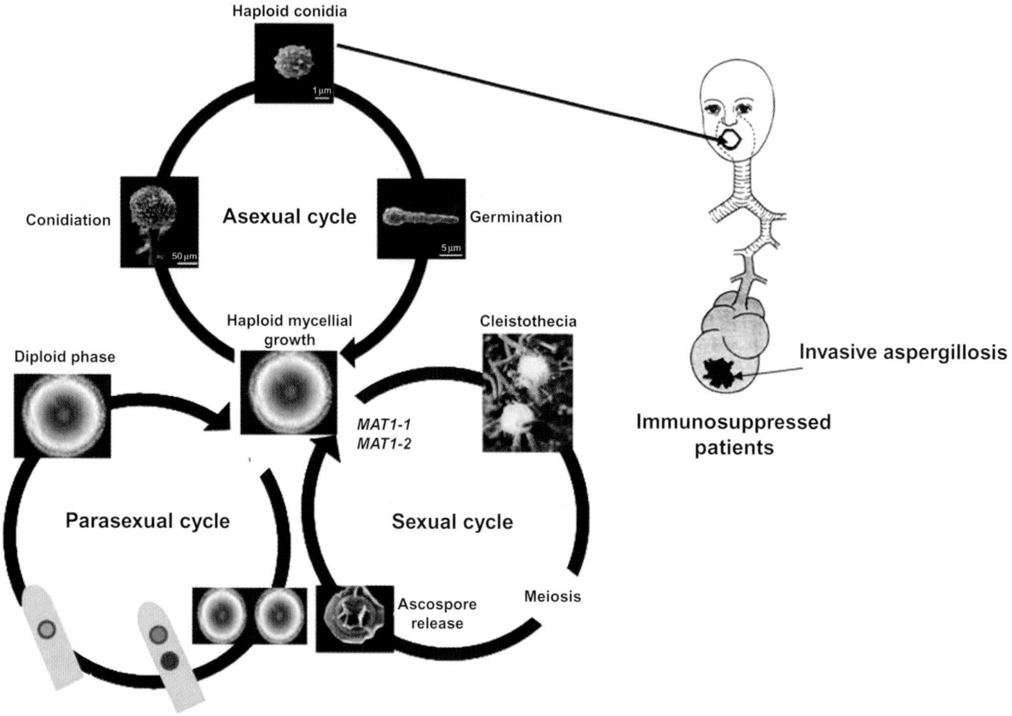

Figure 27.1 Life cycles of *Aspergillus fumigatus*. The asexual cycle is the common form of *A. fumigatus* growth. Fungi grow on decaying vegetal material until nutrient deprivation, which induces conidiation. Haploid conidia are dispersed to the air and can then colonize a new substrate. *A. fumigatus* possesses all genes required to complete a sexual cycle, and the formation of cleistothecia and ascospores are observed in laboratory conditions even though sexual reproduction has never been observed in natural conditions. *A. fumigatus* can be manipulated in a diploid form to undergo a parasexual cycle also not observed in nature. The return to haploid form can be induced with chemicals. Humans are continuously exposed to small-sized haploid conidia that can infiltrate pulmonary alveoli. In immunosuppressed patients, the immune system fails to eliminate conidia, which can then germinate, colonize lung tissue, and cause invasive aspergillosis.

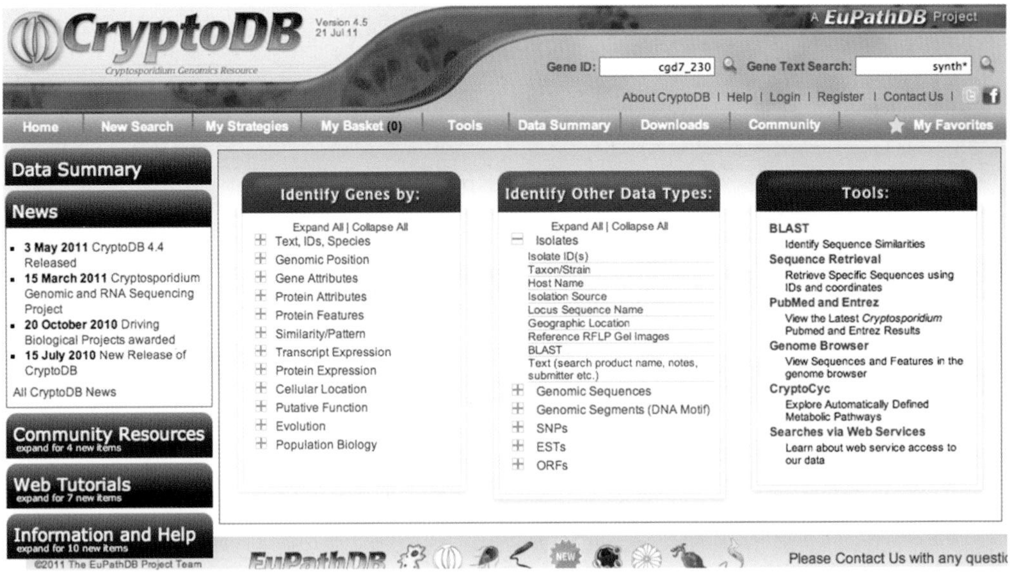

Figure 28.1 Screenshot of the *Cryptosporidium* Database CryptoDB. Most molecular data (genomic, transcriptomic, proteomic, etc.) are housed and available for comparative analysis and complex querying at CryptoDB. In this figure, the second panel, "Identify Other Data Types," is expanded to reveal a rich set of tools that can be used to query isolate data from epidemiological studies. Particular isolates can be identified by geographic region, species, molecular marker used for typing, host species, and environmental source, for example, drinking water, feces, and so on. As a service to the community, common RFLP profiles are also available to facilitate isolate identification. The database maintains a BLAST server of reference type sequences for each species to facilitate quick identification of and polymorphisms in common molecular sequence tags used by the community.

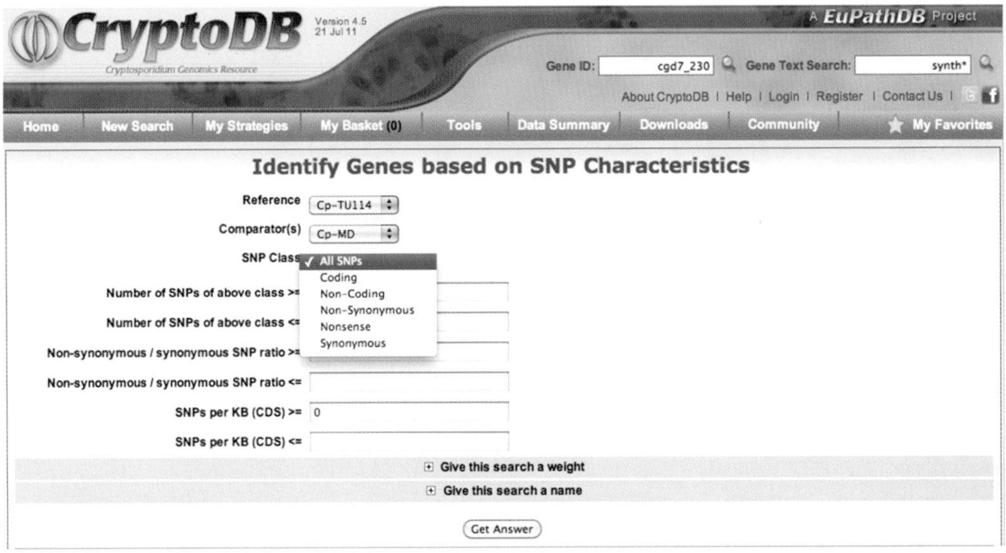

Figure 28.2 Screenshot of a CryptoDB single-nucleotide polymorphism (SNP) search between *Cryptosporidium* strains. SNP data are available for three *C. parvum* strains. Users of the database can identify SNPs between the strains based upon a number of criteria including the location of the SNP in a coding or noncoding region. If an SNP is located in a protein-encoding region, the SNP's potential effect (synonymous or nonsynonymous substitution) on the resulting amino acid can be queried. Searches by SNP density and dN/dS ratio are possible.

types of C-terminal domain. Molecular and Biochemical Parasitology 130: 127–131.

Hutchinson OC, Picozzi K, Jones NG, Mott H, Sharma R, Welburn SC, Carrington M. 2007. Variant surface glycoprotein gene repertoires in *Trypanosoma brucei* have diverged to become strain-specific. BMC Genomics 8: 234.

Jackson AP, Sanders M, Berry A, Mcquillan J, Aslett MA, Quail MA, Chukualim B, Capewell P, Macleod A, Melville SE, Gibson W, Barry JD, Berriman M, Hertz-Fowler C. 2010. The genome sequence of *Trypanosoma brucei gambiense*, causative agent of chronic human African trypanosomiasis. PLoS Neglected Tropical Diseases 4: e658.

Koenig-Martin E, Yamage M, Roditi I. 1992. A procyclin-associated gene in *Trypanosoma brucei* encodes a polypeptide related to ESAG 6 and 7 proteins. Molecular and Biochemical Parasitology 55: 135–145.

Lythgoe KA, Morrison LJ, Read AF, Barry JD. 2007. Parasite-intrinsic factors can explain ordered progression of trypanosome antigenic variation. Proceedings of the National Academy of Sciences of the United States of America 104: 8095–8100.

Macgregor P, Savill NJ, Hall D, Matthews KR. 2011. Transmission stages dominate trypanosome within-host dynamics during chronic infections. Cell Host and Microbe 9: 310–318.

Mackinnon MJ, Read AF. 2004. Virulence in malaria: An evolutionary viewpoint. Philosophical Transactions of the Royal Society of London B Biological Sciences 359: 965–986.

Marcello L, Barry JD. 2007. Analysis of the VSG gene silent archive in *Trypanosoma brucei* reveals that mosaic gene expression is prominent in antigenic variation and is favored by archive substructure. Genome Research 17: 1344–1352.

Miller EN, Turner MJ. 1981. Analysis of antigenic types appearing in first relapse populations of clones of *Trypanosoma brucei*. Parasitology 82: 63–80.

Morrison LJ, Majiwa P, Read AF, Barry JD. 2005. Probabilistic order in antigenic variation of *Trypanosoma brucei*. International Journal for Parasitology 35: 961–972.

Morrison LJ, Marcello L, Mcculloch R. 2009. Antigenic variation in the African trypanosome: Molecular mechanisms and phenotypic complexity. Cellular Microbiology 11: 1724–1734.

Rausch S, Shayan P, Salnikoff J, Reinwald E. 1994. Sequence determination of three variable surface glycoproteins from *Trypanosoma congolense*. Conserved sequence and structural motifs. European Journal of Biochemistry 223: 813–821.

Robinson NP, Burman N, Melville SE, Barry JD. 1999. Predominance of duplicative *VSG* gene conversion in antigenic variation in African trypanosomes. Molecular and Cellular Biology 19: 5839–5846.

Salmon D, Hanocq-Quertier J, Paturiaux-Hanocq F, Pays A, Tebabi P, Nolan DP, Michel A, Pays E. 1997. Characterization of the ligand-binding site of the transferrin receptor in *Trypanosoma brucei* demonstrates a structural relationship with the N-terminal domain of the variant surface glycoprotein. The EMBO Journal 16: 7272–7278.

Strickler JE, Binder DA, L'italien JJ, Shimamoto GT, Wait SW, Dalheim LJ, Novotny J, Radding JA, Konigsberg WH, Armstrong MY, et al. 1987. *Trypanosoma congolense*: Structure and molecular organization of the surface glycoproteins of two early bloodstream variants. Biochemistry 26: 796–805.

Weiller GF. 2008. Detecting genetic recombination. Methods in Molecular Biology 452: 471–483.

White WT, Hills SF, Gaddam R, Holland BR, Penny D. 2007. Treeness triangles: Visualizing the loss of phylogenetic signal. Molecular Biology and Evolution 24: 2029–2039.

Wickstead B, Ersfeld K, Gull K. 2003. The mitotic stability of the minichromosomes of *Trypanosoma brucei*. Molecular and Biochemical Parasitology 132: 97–100.

INTERNET RESOURCE

The African Trypanosome Cell Surface Phylome: http://www.genedb.org/Page/trypanosoma_surface_phylome

CHAPTER 18

ANTIGENIC VARIATION, ADHERENCE, AND VIRULENCE IN MALARIA

JOSEPH SMITH and KIRK W. DEITSCH

INTRODUCTION

In the study of the interactions of malaria parasites with their vertebrate hosts, of particular interest is the interface between the surface of the infected red blood cell (iRBC) and the circulatory system of the infected individual. It is at this surface that many of the key mechanisms underlying disease operate, including antigenic variation, cytoadhesion, and many of the determinants of virulence. All *Plasmodium* species studied to date have devoted large portions of their genomes specifically to processes that modify the red cell membrane and cytoskeleton, and have evolved large, multicopy gene families that encode variants of proteins that are displayed on the extracellular surface and directly interact with the host.

Of the five *Plasmodium* species that infect humans, *Plasmodium falciparum* is the most extensively studied and best understood. Several large families of genes have been identified that encode proteins that are exported to the red blood cell (RBC), including *var*, *rif*, *stevor*, and *Pfmc-2TM*. Of these, the largest amount of work and the clearest understanding of the function of the encoded protein have developed around the *var* gene family and the corresponding protein *Plasmodium falciparum* erythrocyte protein 1 (PfEMP1). This protein is widely considered the predominant virulence determinant of *P. falciparum* infections and remains a prominent vaccine candidate and target of intervention. This chapter will focus specifically on *var* genes and PfEMP1 and will attempt to provide a perspective on what is known regarding the role of the protein in virulence as well as the underlying mechanisms controlling *var* gene expression and antigenic variation.

PfEMP1, CYTOADHERENCE, AND *var* GENES

P. falciparum is the most virulent of the malaria species that infect humans, and much of this virulence is a result of the parasite's ability to alter the iRBC through modifications made to the cytoskeleton and surface of the RBC membrane (Miller et al., 2002; Scherf et al., 2008). In particular, iRBCs become cytoadherent, leading to their sequestration in the capillary beds of several tissues, frequently resulting in

Evolution of Virulence in Eukaryotic Microbes, First Edition. Edited by L. David Sibley, Barbara J. Howlett, and Joseph Heitman.

organ-specific damage, including the potentially lethal syndrome of cerebral malaria. Sequestration of the infected cells enables the parasites to avoid splenic clearance, and many pioneering studies identified a hypervariable, high molecular weight protein at the iRBC surface that was proposed to mediate this process, a protein termed PfEMP1 (Berendt et al., 1990; Howard et al., 1988). In 1995, the *var* gene family was definitively identified as encoding PfEMP1, with members of the family encoding unique variants of a very large protein (200–350 kDa), each with a single transmembrane domain and multiple putative extracellular adhesive motifs (Baruch et al., 1995; Smith et al., 1995; Su et al., 1995). The sequence variability was sufficient to result in antigenic variation while still maintaining a conserved basic structure for adhesion. The identification of the *var* gene family opened a window into our understanding of the virulence of *P. falciparum* infections, the processes underlying antigenic variation, the evolution of host–parasite interactions, and the prospects for targeting this process for disease intervention.

THE *var* GENE REPERTOIRE AND GENOMIC ORGANIZATION OF THE FAMILY

With the advent of large-scale genome sequencing projects and the resulting influx of data, an understanding of conserved aspects of *var* gene organization is now available. Genome comparisons have revealed a conserved *var* gene organization with the majority of the gene family located in clusters and tandem arrays within portions of the chromosomes that appear to be devoted to hypervariable, multicopy gene families (Gardner et al., 2002; Kraemer et al., 2007; Rask et al., 2010). For example, the subtelomeric regions of most chromosomes contain arrays of *var*, *rifin*, *stevor*, and *Pfmc-2TM* genes, with additional clusters of genes found in a few internal regions of chromosomes 4, 7, 8, and 12. The functional significance behind the partitioning of these gene families into specific regions of the genome is not entirely clear, although it likely involves the need for a specific chromatin environment for accelerated rates of recombination or transcriptional regulation. *var* gene repertoires have been compared from seven parasite strains that had between 40 and 61 intact *var* gene copies (Rask et al., 2010). The most conserved *var* genes are three strain-transcendent variants named *var1csa*, *var2csa*, and type 3 *var*. These three genes are found in most or all isolates and encode proteins with upward of 75% amino acid identity (Kraemer and Smith, 2003, 2007; Rowe et al., 2002; Trimnell et al., 2006). By comparison, most PfEMP1 proteins have less than 50% amino acid identity between individual domains (Flick and Chen, 2004; Kraemer et al., 2007). The PfEMP1 repertoire also contains several "domain cassettes," composed of two to four domains, which are commonly inherited as a unit in multiple parasite genotypes (Rask et al., 2010). Domain cassettes are less conserved than isolate-transcendent genes but may be under unusual selection pressures to maintain them at higher frequency in the parasite population.

Based on chromosome location, upstream sequence, and protein architecture features, *var* genes are classified into three main groups (A, B, and C) and two transitional groups (B/A and B/C) (Kraemer and Smith, 2003; Lavstsen et al., 2003). Group A, B, and C genes are present in similar proportions across *P. falciparum* strains (Kraemer et al., 2007; Rask et al., 2010). Furthermore, representatives of the UpsA, UpsB, and UpsC categories and partial *var1csa* and *var2csa* homologs are present in the chimpanzee parasite *Plasmodium reichenowi* (Rask et al., 2010; Trimnell et al., 2006), which may have diverged from *P. falciparum* ~5–7 million years ago (Escalante et al., 1995). The fact that this gene organization has been maintained in the parasite population over this long evolutionary history argues that it has an important role in *var* gene regulation and/or gene recombination.

ASSESSING THE *var* REPERTOIRE

Assessing the global *var* repertoire is challenging because genes are large (9–12 kB) and it is difficult to design unbiased primers that will recognize all family members. To estimate diversity, primers have been designed that recognize semiconserved blocks and amplify *var* sequence tags (~200–500 bp) (Blomqvist et al., 2010; Duffy et al., 2002; Kraemer and Smith, 2003; Kyes et al., 1997; Taylor et al., 2000a). An emerging conclusion is that *var* diversity differs between Africa and other regions of the world (Afonso et al., 2002; Albrecht et al., 2006, 2010; Barry et al., 2007; Chen et al., 2011). In one study, 2158 unique sequences were identified from 88 African isolates (Barry et al., 2007; Chen et al., 2011). In this set, there was a limited overlap of *var* tags between parasite strains, and it was estimated that each local African population may have as many as 5000–7000 types (Chen et al., 2011). By comparison, local parasite diversity is more limited in Papua New Guinea and particularly in South America with greater sharing of *var* types between strains (Albrecht et al., 2006, 2010; Barry et al., 2007; Chen et al., 2011). The factors influencing the size of local *var* repertoires are likely determined by parasite transmission intensity/outcrossing rates, acquired immunity in local human populations, and various parasite population bottlenecks that *P. falciparum* might have underwent as it accompanied humans out of Africa. Accordingly, *P. falciparum* is thought to have experienced a severe bottleneck when introduced to the Americas as a result of the African slave trade (Anderson et al., 2000). The vast antigenic diversity of PfEMP1 proteins in the parasite population may help explain why individuals are repeatedly susceptible to *P. falciparum* infections and may never develop sterile immunity (Mackinnon and Marsh, 2010). Over successive infections, African children acquire a broad repertoire of anti-PfEMP1 antibodies (Cham et al., 2009). Antibody breadth is acquired at an earlier age in children living in high transmission regions, and by adulthood, hyperimmune plasma displays considerable breadth (Cham et al., 2009). The predominant agglutinating antibody response is variant specific (Marsh and Howard, 1986; Newbold et al., 1992), although parasites from East or West Africa display antigenic overlap (Aguiar et al., 1992) and worldwide placental isolates have extensive antigenic overlap (Fried et al., 1998).

MECHANISMS OF VARIANT ANTIGEN DIVERSIFICATION

The extreme level of diversity observed when *var* genes are sampled from various geographic regions suggests that this family might be subject to unique mechanisms that accelerate rates of recombination or mutation. Although the relative contribution of new mutation versus gene recombination is not yet clear, there are several lines of evidence for frequent gene recombination within the *var* gene family and segmental gene recombination appears to be an important mechanism of gene diversification (Frank et al., 2008; Kraemer et al., 2007; Rask et al., 2010). Prior to the completion of various genome sequencing projects and thus before significant amounts of *var* gene sequence data were available, analysis of the progeny of an experimental genetic cross identified ectopic recombination between *var* genes that resulted in the generation of new chimeric PfEMP1 proteins (Freitas-Junior et al., 2000; Taylor et al., 2000b). Similarly, isolation of a sibling parasite to 3D7, the parasite clone used for the initial genome sequencing project, also identified several genes that appeared to be chimeric, suggesting that segmental gene conversions are common within the *var* gene family and might be frequent during meiotic division (Frank et al., 2008).

As more *var* gene sequences have become available through the analysis of several independent isolates, computational sequence alignments have provided additional support for frequent gene conversion events (Kraemer et al., 2007; Rask et al., 2010). Further, the fact that *var* genes have diverged into identifiable subfamilies (A, B, and C types) suggests that recombination between *var* genes might not be happening at random, but rather preferentially within subfamilies (Kraemer and Smith, 2003; Lavstsen et al., 2003). The location of *var* genes along the chromosome has been proposed to play a role in this process by aligning *var* genes in a way that could facilitate recombination (Figueiredo et al., 2002; Marty et al., 2006). Telomeres have been shown to cluster into "bouquets" at the nuclear periphery, thereby aligning genes that are in similar positions and orientations with respect to the telomeric repeats (Figueiredo et al., 2002; Freitas-Junior et al., 2000). Consistent with this model, group B genes are found immediately adjacent to the telomeric repeats and transcribed toward the centromere; group A genes are found more proximal within the subtelomeric domain and transcribed toward the chromosome ends; and group C genes are located in the interior of chromosomes 4, 7, 8, and 12, typically in tandem arrays (Gardner et al., 2002) (Fig. 18.1). Gene conversion events are initiated by the creation of a double-strand break in the DNA strand, either by random DNA damage or by a specific endonuclease. *Plasmodium* is thought to be uniquely dependent on gene conversions for repairing double-strand breaks because the alternative repair pathway (nonhomologous end joining) appears to be absent from the genome of all parasite strains sequenced (Gardner et al., 2002). Thus, gene conversion events might be a frequent occurrence in malaria parasites, and the positioning of large, hypervariable gene families within the subtelomeric regions could have evolved to take advantage of telomeric clustering to facilitate rapid and efficient gene diversification through standard DNA repair.

EVOLUTIONARY ORIGINS OF *var* GENES

All malaria species studied to date modify the surface of infected erythrocytes with clonally variant antigens, but unlike *P. falciparum*, most malaria species do not undergo massive sequestration of trophozoite-infected erythrocytes (Kriek et al., 2003). Remarkably, several distinct multigene families have independently evolved in different parasite species of rodents, primates, and humans that participate in antigenic variation of *Plasmodium*-infected erythrocytes (al Khedery et al., 1999; Carlton et al., 2002, 2008; Cheng et al., 1998; del Portillo et al., 2001; Kyes et al., 1999; Pain et al., 2008), but *var* homologs have not been identified in most *Plasmodium* spp. with the exception of the chimpanzee malaria *P. reichenowi* (Rask et al., 2010; Trimnell et al., 2006). Recently, several additional close relatives of *P. falciparum* were discovered in great apes (Krief et al., 2010; Ollomo et al., 2009; Prugnolle et al., 2010, 2011). Further investigation of this clade may permit a more detailed understanding of the origins and evolution of the *var* gene family.

Clues into *var* gene origins have been revealed by bioinformatic analysis of the gene structure and protein architecture. *var* genes have two exons. The first exon encodes the extracellular binding region and transmembrane domain, and the

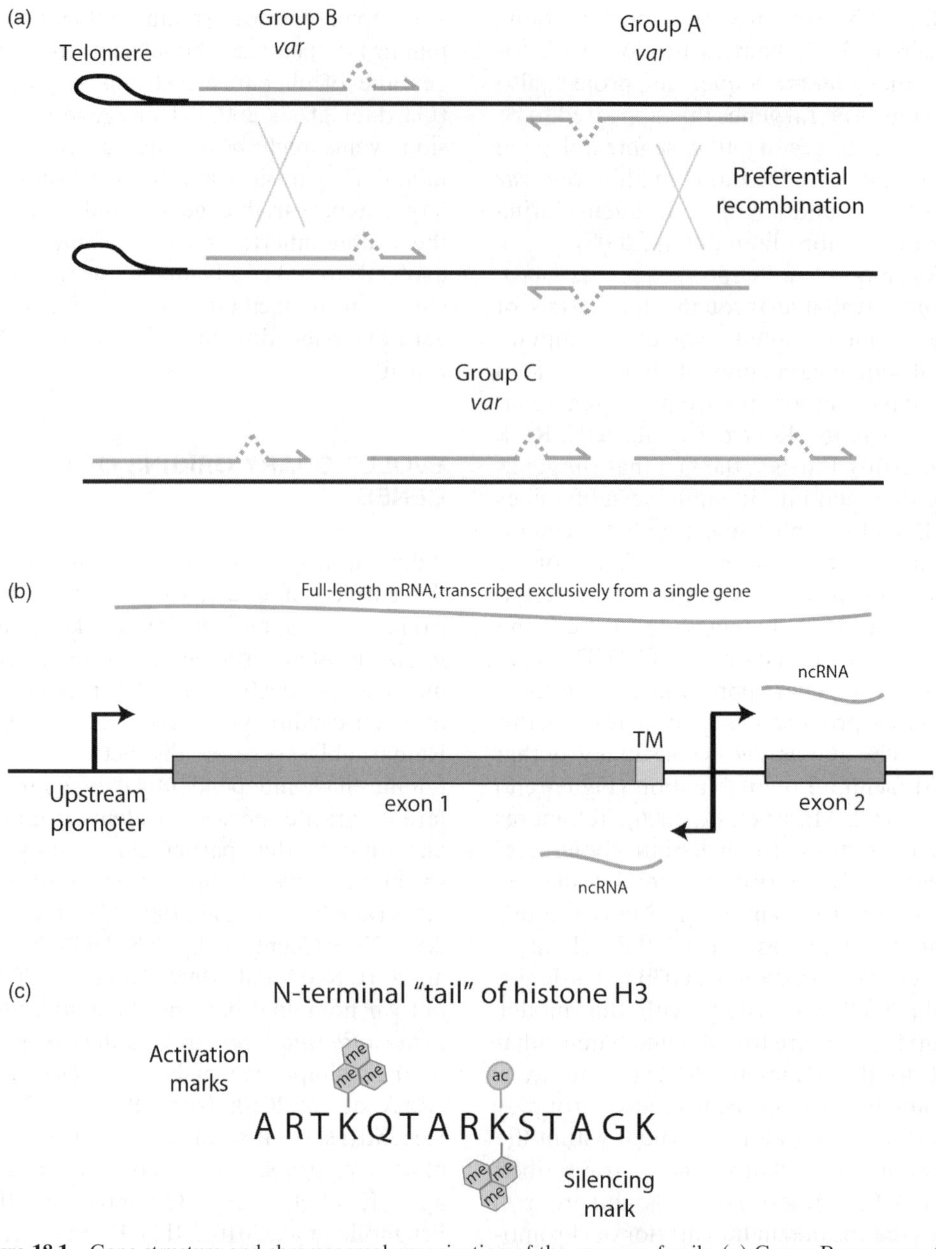

Figure 18.1 Gene structure and chromosomal organization of the *var* gene family. (a) Group B genes are typically located immediately adjacent to the telomere and transcribed toward the center of the chromosome, while group A genes are also within the subtelomeric regions but in the opposite orientation. Telomeres are found in "bouquet" structures that are theorized to align *var* genes for preferential recombination. Group C genes are arranged in tandem arrays in the central parts of chromosomes. (b) *var* genes have a two-exon structure, with an upstream promoter that mediates transcription of the full-length mRNA. In addition, a bidirectional promoter found within the intron results in transcription of noncoding RNAs from most or all members of the family. The transmembrane domain (TM) of the protein is encoded at the very 3′ end of exon 1. (c) The amino acid sequence of the N-terminal "tail" of histone H3. The three histone modifications that are proposed to be most important for regulating *var* gene expression are drawn, including methylation of lysine 4 and both acetylation and methylation of lysine 9. See color insert.

second exon encodes the cytoplasmic tail (Fig. 18.1). The extracellular binding region of PfEMP1 proteins is composed of multiple adhesion domains, classified as Duffy binding-like (DBL) and cysteine-rich interdomain regions (CIDR) (Baruch et al., 1995; Peterson et al., 1995; Smith et al., 1995; Su et al., 1995). The building blocks for this region of the protein appear to be a unique *Plasmodium* adaptation. The DBL domain has not been found in other organisms but is present in other *Plasmodium* proteins belonging to the erythrocyte binding protein family (DBL-EBP), where it functions as an adhesion module for parasite invasion of RBCs (Adams et al., 1990, 1992; Chitnis and Miller, 1994; Sim et al., 1994). The DBL superfamily in *P. falciparum* also contains at least two other proteins (Fig. 18.2). Pf322 is an extremely large protein that is exported to the red cell. Although not displayed at the iRBC surface, Pf322 appears to associate with the erythrocyte cytoskeleton and has a role in the mechanical deformability properties of infected erythrocytes (Glenister et al., 2009; Hodder et al., 2009; Moll et al., 2007). PfDBLMSP (PF10_0348) is a novel merozoite surface protein that encodes RBC binding activity (Wickramarachchi et al., 2009).

The specific evolutionary relationship between DBL superfamily members is not fully characterized, but since all *Plasmodium* species infect RBCs and few have *var* homologs, PfEMP1 proteins likely evolved from DBL domains involved in erythrocyte invasion. This expansion of PfEMP1 could have happened by partial gene duplication of a DBL domain and recombination with another genetic element that directed protein export to the erythrocyte. Transport of most parasite proteins across the parasitophorous vacuole membrane is enabled by a five-amino acid sequence called the *Plasmodium* export element (PEXEL) or the vacuolar transport signal (VTS) (Hiller et al., 2004; Marti et al., 2004). The N-terminus in PfEMP1 proteins contains an unconventional "PEXEL-like" sequence that is a strong candidate for the export signal (Marti et al., 2004; Melcher et al., 2010). Structural investigations of DBL and CIDR recombinant proteins show the CIDR fold is similar to the C-terminal region of the DBL domain (Klein et al., 2008; Singh et al., 2006; Su et al., 1995; Tolia et al., 2005). These findings suggest that DBL and CIDR domains have a related ancestry and that both extracellular adhesion modules in PfEMP1 proteins were probably inherited from *Plasmodium* invasion ligands that have a critical role in the hematopoietic lifestyle of the parasite.

The PfEMP1 cytoplasmic tail anchors the protein at knob-like protrusions at the iRBC surface by associating with the parasite encoded knob-associated protein (KAHRP) (Baruch et al., 1995; Crabb et al., 1997) and the RBC cytoskeleton proteins actin and spectrin (Kilejian et al., 1991; Oh et al., 2000; Waller et al., 1999, 2002). The PfEMP1 cytoplasmic tail contains an S2 motif (Fig. 18.2), which has homology to the S1/S2 tryptophan-rich domain (WRD) in the C-terminus of the *Plasmodium knowlesi* variant antigen (SICAvar), Pf332, and a different protein family in *P. falciparum* called SURFIN (Winter et al., 2005). Conservation of the WRD element between various parasite proteins that are exported to the iRBC suggests the possibility that it may be an interaction region with red cell cytoskeleton or other host erythrocyte proteins.

PfEMP1 PROTEIN STRUCTURE–FUNCTION

To date, at least 12 host receptors have been reported to mediate *P. falciparum* iRBC binding (Rowe et al., 2010), but CD36 and ICAM-1 binding are most common (Newbold et al., 1997; Rogerson et al., 1999). These two receptors synergize to mediate iRBC binding to microvasculature

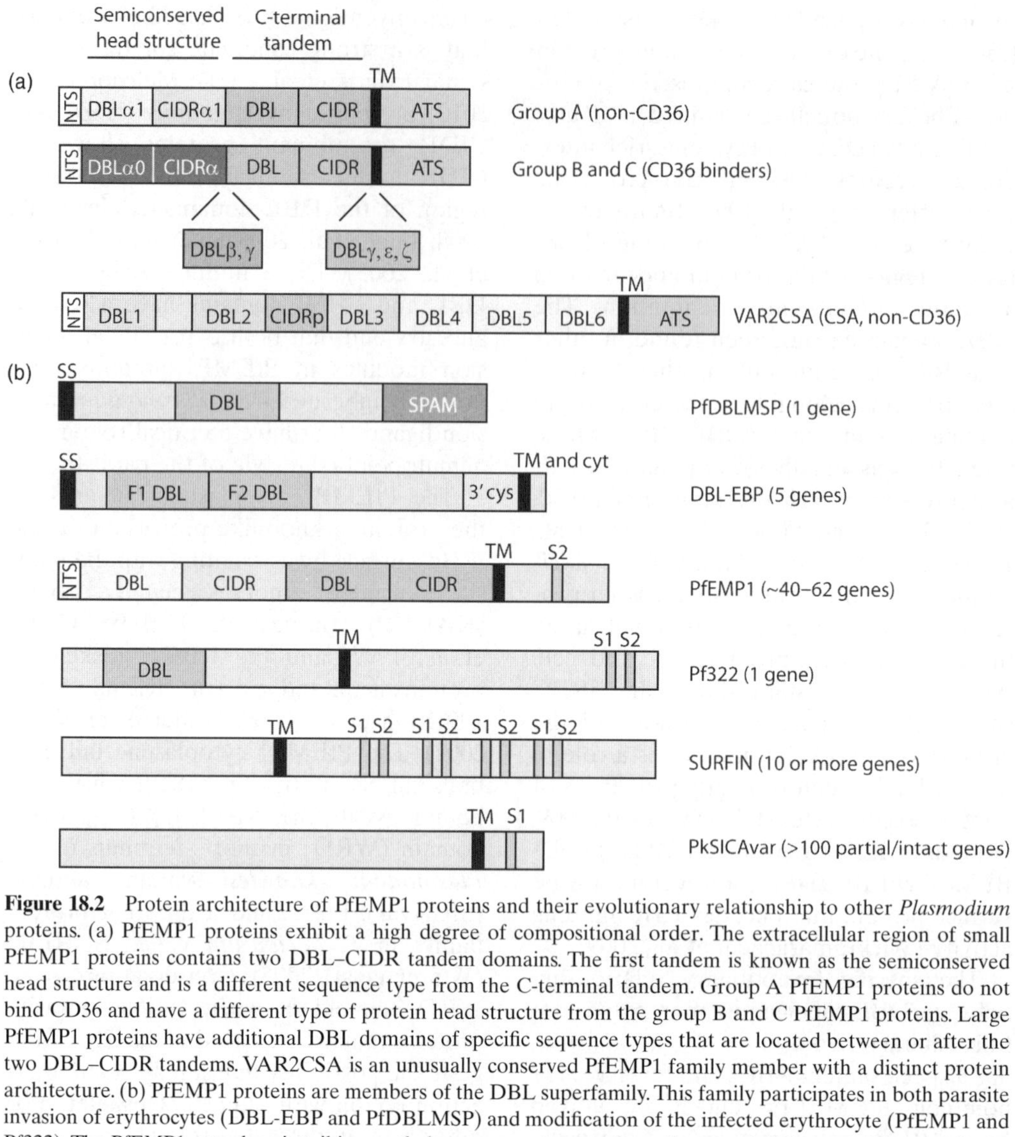

Figure 18.2 Protein architecture of PfEMP1 proteins and their evolutionary relationship to other *Plasmodium* proteins. (a) PfEMP1 proteins exhibit a high degree of compositional order. The extracellular region of small PfEMP1 proteins contains two DBL–CIDR tandem domains. The first tandem is known as the semiconserved head structure and is a different sequence type from the C-terminal tandem. Group A PfEMP1 proteins do not bind CD36 and have a different type of protein head structure from the group B and C PfEMP1 proteins. Large PfEMP1 proteins have additional DBL domains of specific sequence types that are located between or after the two DBL–CIDR tandems. VAR2CSA is an unusually conserved PfEMP1 family member with a distinct protein architecture. (b) PfEMP1 proteins are members of the DBL superfamily. This family participates in both parasite invasion of erythrocytes (DBL-EBP and PfDBLMSP) and modification of the infected erythrocyte (PfEMP1 and Pf322). The PfEMP1 cytoplasmic tail is encoded on a separate exon from the extracellular region. It contains an S2 motif that is homologous to the S1/S2 tryptophan-rich domain (WRD) in the C-terminus of the PfSURFIN family, Pf332, and the *P. knowlesi* variant antigen family (SICAvar). See color insert.

endothelium under flow conditions (Cooke et al., 1994; Gray et al., 2003; Ho et al., 2000) and potentially mediate other iRBC–host cell interactions (Serghides et al., 2003; Urban et al., 1999). DBL and CIDR domains in PfEMP1 proteins are classified by phylogeny into six major DBL classes (α, β, γ, δ, ε, and ζ) and five major CIDR classes (α, β, γ, δ, and pam) (Rask et al., 2010; Smith et al., 2000b). Major classes can be further subdivided into subtypes (α1, α2, etc.) (Rask et al., 2010). PfEMP1 protein architectures exhibit a high degree of compositional order (Rask et al., 2010; Smith et al., 2000b). Nearly all proteins have a DBLα–CIDR semiconserved "head structure" (Fig. 18.2).

In small proteins, the head structure is usually followed by a second C-terminal tandem that is a different sequence type from the head structure (DBLδ–CIDRnon-α). This arrangement could have arisen by duplication of the DBL–CIDR head structure in an ancestral 2-domain extracellular protein and subsequent divergence. In large proteins, additional DBL types (β, γ, ζ) are located between or after the two DBL–CIDR cassettes (Fig. 18.2).

The organization of PfEMP1 protein architectures probably evolved under selection for various host receptor interactions. For instance, CD36 binding has been mapped to the CIDRα domain in the PfEMP1 head structure (Baruch et al., 1997; Smith et al., 1998) and ICAM-1 binding to the DBLβ domain that immediately follows the protein head structure (Chattopadhyay et al., 2004; Smith et al., 2000a). This arrangement may allow the PfEMP1 head structure to interact with the more compact CD36 protein (Tao et al., 1996) while simultaneously positioning the distally located DBLβ domain to engage the N-terminal immunoglobulin domain in ICAM-1 (Berendt et al., 1992; Ockenhouse et al., 1992; Staunton et al., 1990). It could also explain why the second, membrane-proximal CIDR domain is not under strong selection for CD36 binding (Robinson et al., 2003).

PfEMP1 binding properties are partially predictable by their adhesion domain sequence types. Most PfEMP1 proteins encode CD36 binding activity (Robinson et al., 2003), but only a subset binds ICAM-1 (Howell et al., 2008; Oleinikov et al., 2009). Of the five major CIDR classes (α, β, γ, δ, and pam), only the CIDRα binds CD36 and the CIDRα1 subtype identifies domains that do not bind CD36 (Robinson et al., 2003). CIDRα1 domains are only found in group A PfEMP1 proteins and not in group B or C PfEMP1 proteins. In addition, the CIDRα–DBLβ5 combination identifies a category of PfEMP1 that is under dual selection for CD36 and ICAM-1 binding (Janes et al., 2011). A subset of DBLβ3 domains has also been shown to bind ICAM-1, but DBLβ3 domains are present in both CD36 binding and non-binding proteins (Howell et al., 2008; Janes et al., 2011; Oleinikov et al., 2009). Thus, CD36 and ICAM-1 binding has left strong signatures of selection on the PfEMP1 family that can be partially decoded by adhesion domain classification, despite the extensive sequence divergence in the family.

Sequence comparisons have further revealed gene recombination hierarchies that underlie the functional specialization of PfEMP1 proteins (Kraemer and Smith, 2003, 2007; Lavstsen et al., 2003; Rask et al., 2010). For instance, the group A proteins all tend to be large proteins with complex domain compositions except for the type 3 *var* gene. Group A proteins are more closely related to each other (Kraemer et al., 2007; Warimwe et al., 2009) and, as noted above, they possess a distinct type of PfEMP1 head structure that does not bind CD36 (Kraemer and Smith, 2003; Robinson et al., 2003). Consequently, by limiting recombination between CD36 binding (groups B and C) and non-CD36 binding (group A) proteins, the parasite may evolve specialized sets of genes. A prime example is VAR2CSA and placental malaria. VAR2CSA has a unique upstream sequence (UpsE) and distinct protein architecture (Salanti et al., 2003). This may make it more difficult to line up and recombine with other genes (Kraemer and Smith, 2003). Sequence comparisons show that *var2CSA* has little sequence overlap with other *var* genes and primarily recombines by self–self recombination (Bockhorst et al., 2007; Kraemer et al., 2007; Trimnell et al., 2006). Thus, gene recombination hierarchies may cause nonoverlapping sets of adhesion traits to evolve in different PfEMP1 groups and provide the parasite greater adaptability to exploit distinct host biological niches.

VAR2CSA AND PLACENTAL MALARIA

Studies of pregnant women have revealed how a single PfEMP1 protein can precipitate disease in women who have otherwise acquired significant immunity to malaria. First-time pregnant women are highly susceptible to severe placental infections (Brabin et al., 2004). Placental isolates bind to the uniquely low sulfated chondroitin sulfate A (CSA) in the placenta (Fried and Duffy, 1996; Muthusamy et al., 2004) and do not adhere to CD36 on peripheral microvasculature. Placental binding is mediated by selective upregulation of VAR2CSA at the surface of placental iRBCs (Duffy et al., 2006; Magistrado et al., 2011; Salanti et al., 2003, 2004; Viebig et al., 2005). Over successive pregnancies, women acquire antibodies that react broadly to the surface of diverse placental infected erythrocytes (Beeson et al., 1999; Fried et al., 1998) and may have a role in protection (Duffy and Fried, 2003; Fried et al., 1998; Staalsoe et al., 2004).

The identification of a specific PfEMP1 protein associated with disease provides a unique paradigm to investigate the structural and functional adaptations of PfEMP1 proteins as immunoevasive binding molecules. VAR2CSA has a distinct protein architecture. The extracellular region contains six extracellular DBL domains but lacks a typical protein head structure (Fig. 18.2) (Salanti et al., 2003). This domain structure may account for its inability to bind CD36. Atypically, VAR2CSA contains a unique CIDR-like domain between the second and third DBL domain, referred to as $CIDR_{PAM}$ (Andersen et al., 2008). It has been shown that the full-length VAR2CSA extracellular region binds with high affinity and specificity to CSA (Khunrae et al., 2010; Srivastava et al., 2010). A minimal high affinity binding region has been mapped to the DBL2–$CIDR_{PAM}$–DBL3 region (Dahlback et al., 2011), and CSA binding activity has been mapped to several individual domains in VAR2CSA (DBL2, DBL3, and DBL6) (Gamain et al., 2005; Singh et al., 2008). In contrast, a DBL4-6 recombinant protein had only low CSA binding activity (Srivastava et al., 2010). Thus, multiple N-terminal domains in VAR2CSA interact with CSA to confer specific and high affinity binding.

The comparison of different VAR2CSA sequences has suggested a novel mechanism by which the parasite may evade immunity while introducing polymorphism at PfEMP1 interactions sites. VAR2CSA polymorphism is highly structured. Different *var2csa* alleles display extensive segmental gene relationships and polymorphic sites group into segments of limited diversity (Bockhorst et al., 2007). Within these polymorphic segments, two or three basic types characterize a substantial majority of parasite isolates from globally dispersed parasite populations. Remarkably, related polymorphic segment types are found in *P. reichenowi*, indicating their ancient origins (Bockhorst et al., 2007).

Polymorphic segments are concentrated to exposed surfaces and flexible loops that connect the more conserved DBL scaffolding elements (Andersen et al., 2008; Bockhorst et al., 2007). Through a process of segmental gene recombination, the parasite is able to create extensive combinatorial diversity at known DBL interaction sites (Howell et al., 2006, 2008) by reshuffling nearby adjacent loops (Bockhorst et al., 2007). This clever strategy may allow the parasite to subtly alter and experiment with new host binding interactions while at the same time facilitating parasite escape from antibody-dependent protective mechanisms. It may also contribute to cross-reactive antibody responses between different CSA binding parasite lines (Avril et al., 2008; Hommel et al., 2010). A similar strategy likely exists for other *var* family members except that they participate in larger groups of recombining genes (Frank et al., 2008; Ward et al., 1999; Warimwe et al., 2009).

To overcome this problem for pregnancy malaria vaccine development, it may be necessary to identify functional conserved regions in VAR2CSA that are less able to vary under vaccine-induced antibody pressure. Relatively broad inhibitory antibody responses have been elicited to an FCR3-DBL4 recombinant protein (Magistrado et al., 2011; Nielsen et al., 2009). Inhibitory antibodies have also been generated to various full-extracellular and multidomain VAR2CSA immunogens (Dahlback et al., 2011; Khunrae et al., 2010). However, in one case where it was studied, antibodies to full-length immunogen were highly strain specific (Avril et al., 2011). Thus, overcoming VAR2CSA polymorphism remains a key hurdle for pregnancy malaria vaccine development.

PfEMP1 EXPRESSION IN THE YOUNG, NONIMMUNE, AND IN NONPLACENTAL SEVERE DISEASE

VAR2CSA provides a clear example for how expression of a particular *var* gene can result in a very specific disease syndrome through the resulting binding phenotype of the iRBC. Attempts to associate disease phenotypes with expression of other *var* genes has been complicated by the extensive variability of genes and therefore the difficulty in equating *var* genes from one parasite to those of another. Field studies of nonimmune children in endemic regions, however, have begun to shed light on the influence that expression of specific *var* genes has on disease severity. In a postmortem study of children in Malawi, the authors found that despite detecting transcripts from large numbers of *var* gene in any given child, parasites adhering within brain or cardiac tissue preferentially expressed only one or two *var* genes, consistent with the notion that disease severity is strongly influenced by which *var* gene is expressed and supporting the identification of PfEMP1 as a primary virulence factor (Montgomery et al., 2007).

The increase in *var* gene sequence data has made it possible to differentiate group A, B, and C type genes based on PCR products amplified from the circulating blood of infected children. By obtaining cDNA from parasites in children with differing degrees of disease severity or probing PfEMP1 protein arrays with malaria-endemic plasma, several groups have attempted to associate expression of particular *var* groups to immunity and disease phenotype (Jensen et al., 2004; Kaestli et al., 2004, 2006; Kalmbach et al., 2010; Kyriacou et al., 2006; Rottmann et al., 2006). In general, these studies have found that parasites obtained from young children with limited immunity and those suffering from severe disease, including cerebral malaria and severe anemia, tend to be expressing group A or B/A *var* genes, while parasites from older, asymptomatic children, who presumably have acquired some level of immunity, are often found to be expressing group B or C genes (Cham et al., 2009, 2010).

These studies have led to the hypothesis that expression of PfEMP1s encoded by group A and B/A *var* genes might provide parasites with a growth advantage, perhaps through particularly tight binding to host receptors, resulting in their dominance within populations of parasites growing in individuals lacking immunity. However, in older children who have experienced several previous infections, immunity to group A and B/A genes has been acquired, leading to the dominance of group B or C genes and less severe infections. Intriguingly, this analysis implies that there may also be a shift in parasite binding preference with early infections dominated by non-CD36 binding variants and subsequent infections dominated by CD36 binding variants. This pattern switches again to a non-CD36 binding variant in pregnant women. Of interest, studies using cultured parasites have found that genes from the different *var* groups have different switching rates, suggesting that parasites might have evolved varying switching frequencies that enable them to adjust to the immune status of their hosts (Frank et al., 2007; Recker et al., 2011).

CONTROL OF *var* GENE TRANSCRIPTION, SWITCHING, AND EPIGENETIC MEMORY

Antigenic variation is a phenomenon displayed by many infectious organisms, including bacteria, fungi, and protozoan pathogens (Deitsch et al., 1997, 2009). In many of these systems, the process involves the activation and silencing of individual genes from within a large hypervariable family, analogous to the *var* gene family. Expression of individual genes within these families is typically coordinated such that it is mutually exclusive; only a single gene is actively expressed and all the others are silent. This process of exclusive expression is thought to represent the most efficient way to extend the life of the repertoire by not prematurely exposing antigens to the host's immune system. Mutually exclusive expression was presumed to apply to *var* genes, and early experiments with parasites that were selected for specific binding phenotypes (e.g., binding to CSA) yielded populations expressing a single dominant gene, consistent with this model (Scherf et al., 1998). Subsequent experiments using transgenic lines of parasites in which the PfEMP1 coding region of a specific *var* gene was replaced with a drug selectable marker demonstrated that under selection, only the recombinant *var* gene was actively transcribed and all other genes were silent (Dzikowski et al., 2006; Voss et al., 2006). These findings validated the mutually exclusive expression paradigm and also demonstrated that PfEMP1 expression is not necessary for parasite viability in culture. Recently, however, two examples have been published of cultured parasites that express more than one *var* gene simultaneously (Brolin et al., 2009; Joergensen et al., 2010), suggesting that while mutually exclusive expression might be the typical situation, it appears not to be an absolute rule. Currently, the underlying mechanism that limits expression to one or a few *var* genes is only partially understood.

Using iRBC binding assays to select for parasites that express particular variants of PfEMP1, it has been possible to enrich for populations that express individual *var* genes. Such experimental manipulation has enabled investigators to compare several properties of *var* genes when they are in their active or silent state and, thus, to begin to decipher the molecular mechanisms responsible for their transcriptional control. It was first established that the DNA sequence does not change when genes switch from silent to active (or back), and therefore this is an epigenetic phenomenon (Scherf et al., 1998). Several studies have investigated changes in chromatin structure associated with transcriptional status, specifically focusing on histone modifications. In general, the active *var* gene is associated with histone marks (Fig. 18.1) that are typically found at transcriptionally active chromatin in other eukaryotes (H3K4me2, H3K4me3, and H3K9ac) (Duraisingh et al., 2005; Freitas-Junior et al., 2005; Lopez-Rubio et al., 2007), suggesting that this aspect of gene regulation is conserved even with organisms of very distant evolutionary lineages. Similarly, the silent mark H3K9me3 is found at silent *var* genes (Chookajorn et al., 2007; Lopez-Rubio et al., 2007); however, unlike most eukaryotes where this mark is found at inactive genes throughout the genome, in *P. falciparum*, it appears to be devoted specifically to the hypervariable, multicopy gene families including *var*, *rifin*, *stevor*, and *Pfmc-2TM* (Flueck et al., 2009; Lopez-Rubio et al., 2009; Ponts et al., 2010; Westenberger et al., 2009). The evolutionarily conserved protein HP1 also associates with silent *var* genes, presumably recruited by the H3K9me3 mark (Flueck et al., 2009; Perez-Toledo et al., 2009). Further, it has recently been shown that the variant histone H2AZ is incorporated into the chromatin surrounding the transcription start site of the actively transcribed *var* gene (Bartfai et al., 2010; Petter et al., 2011).

Interestingly, unlike most eukaryotes, the *Plasmodium* genome encodes two copies of the histone deacetylase SIR2, called PfSIR2a and PfSIR2b (Gardner et al., 2002; Tonkin et al., 2009). These proteins are associated with telomeres and help to maintain the heterochromatin found at the ends of chromosomes. In *P. falciparum*, PfSIR2 not only binds within the telomeric repeats but also is found within the adjacent subtelomeric regions that contain *var*, *rifin*, *stevor*, and *Pfmc-2TM* genes (Freitas-Junior et al., 2005). The functions of the two SIR2 orthologs appear to be at least partially redundant. Knockouts of either gene individually result in a partial desilencing of a portion of the *var* gene family, and a double knockout appears lethal (Duraisingh et al., 2005; Tonkin et al., 2009). Given the role that SIR2 plays in telomere biology, and the proximity of a large portion of the *var* gene family to the telomeres, the thought that *var* gene regulation might be related to the known phenomenon of telomere position effect has been considered (Figueiredo and Scherf, 2005). However, it is worth noting that approximately a third of the gene family is found at a significant distance from the telomeres (Gardner et al., 2002), and thus while *var* gene silencing might share some components with telomere position effect, the two mechanisms are likely to be quite different.

Several laboratories have used transgenic parasites and reporter gene constructs to identify the DNA elements that are involved in *var* gene regulation (Calderwood et al., 2003; Deitsch et al., 2001; Frank et al., 2006; Voss et al., 2000). Two essential DNA regions found at each individual gene have been identified as playing important roles, the upstream region surrounding the transcription start site and a second element found within the single, conserved intron of each *var* gene (Fig. 18.1) (Deitsch et al., 2001; Voss et al., 2006). Gel shift assays have identified protein binding sites within the upstream region (Voss et al., 2003); however, the transcription factors that bind there have not been identified. The significance of the intron for proper regulation was discovered through transfection studies (Calderwood et al., 2003; Deitsch et al., 2001). It was found that when isolated on a transfected episome, *var* upstream regions behave as typical promoters and are transcriptionally active, even though a chromosomal *var* gene is also actively transcribed (Dzikowski et al., 2006; Frank et al., 2006; Gannoun-Zaki et al., 2005). In other words, mutually exclusive expression has been lost and two *var* promoters are active simultaneously. However, when a *var* intron is also included on the plasmid, an episomal *var* promoter is silenced, and if forced to be active through drug selection, all other *var* genes in the parasite genome become silenced (Dzikowski et al., 2006). Thus, it is now recognized by the mechanism that limits expression to a single *var* gene. How the intron produces this effect is not clear, although it appears to depend on promoter activity found within the intron itself (Calderwood et al., 2003; Gannoun-Zaki et al., 2005). Late in the asexual cycle, during DNA replication, *var* introns produce long, noncoding RNAs in both the sense and antisense directions (Kyes et al., 2003; Su et al., 1995). These transcripts remain nuclear and are associated with chromatin (Epp et al., 2008); however, their function is not understood. Malaria parasites appear to lack the RNAi pathway (Gardner et al., 2002); however, significant work in higher eukaryotes has recently identified several instances in which long noncoding RNAs are important for epigenetic gene regulation (Morey and Avner, 2004), suggesting that *var* noncoding RNAs might play a similar role.

In addition to the DNA regulatory elements and the chromatin modifications associated with *var* gene regulation, subnuclear localization might also play a significant role. Florescent *in situ* hybridization (FISH) studies have shown that *var* genes are found clustered at the nuclear periphery (Freitas-Junior et al., 2000), a portion of the nucleus that is typically enriched in transcriptionally silent heterochromatin (Sutherland and Bickmore, 2009). Interestingly, the transcriptionally active *var* gene remains at the nuclear periphery; however, it appears to move to a unique position near the nuclear membrane, a region that is referred to as the *var* expression site (Duraisingh et al., 2005; Lopez-Rubio et al., 2009; Ralph et al., 2005). These initial observations led to the suggestion that this region might only be able to accommodate a single *var* gene at a time, thereby ensuring mutually exclusive expression. However, subsequent studies using episomal *var* promoters that had been rendered constitutively active by removing the influence of the intron showed that it was possible for parasites to actively express multiple *var* promoters from the same subnuclear position (Dzikowski et al., 2007; Dzikowski and Deitsch, 2008). In addition, it appears that active *rif* and *stevor* genes might also utilize the same expression site (Howitt et al., 2009). Therefore, while the exact significance of the *var* expression site is not yet clear, considering the general heterochromatic nature of the nuclear periphery, the movement of *var* genes to a specific subnuclear position to facilitate activation might be an important aspect of transcriptional regulation.

An additional aspect of *var* gene regulation that is important for antigenic variation is called epigenetic memory. This state refers to the fact that once a particular *var* gene has become activated, it will remain active for many cell divisions (Frank et al., 2007; Horrocks et al., 2004; Recker et al., 2011). Epigenetic memory ensures that the switching rate is low enough that during an infection, the *var* repertoire will not be exposed to the host's immune system too rapidly and therefore a long infection can be sustained. Since malaria parasites undergo schizogony, the epigenetic marks that are found at the active locus must be replicated repeatedly as the parasite undergoes multiple rounds of genome replication. Not much is known about the specifics of this process; however, by using a technique called promoter titration, it is possible to prevent the active *var* gene from being transcribed for several cell cycles (Dzikowski and Deitsch, 2008). This titration resulted in complete loss of epigenetic memory, suggesting that active transcription is required for the maintenance of memory. In higher eukaryotes, the movement of RNA polymerase II through a locus can recruit specific histone modifiers to the surrounding chromatin, thus influencing the epigenetic marks and reinforcing the active state of a gene (Carty and Greenleaf, 2002). A similar mechanism might be playing a role in maintaining *var* gene expression patterns through multiple rounds of schizogeny.

CONCLUSIONS

The *var* gene family has a central role in antigenic variation and adherence of *P. falciparum*-infected erythrocytes. This gene family evolved in a branch of *Plasmodium* that is specialized to infect great apes. The adhesion domains in PfEMP1 proteins have their origins in parasite invasion ligands used to invade RBCs and have subsequently been adapted for a variety of adhesive interactions between iRBCs and the host circulatory system. To evade antibody responses, the parasite generates combinatorial diversity at known DBL domain–host interaction sites by segmental gene recombination. Different *P. falciparum* strains have nonoverlapping *var* repertoires but share a common gene organization at subtelomeric and interior chromosome regions. The *var* gene organization underlies recombination hierarchies that have resulted in the functional specialization of different PfEMP1 subfamilies with nonoverlapping adhesion traits. A particular PfEMP1 protein is responsible for placental malaria, and emerging evidence suggest the parasite may preferentially employ different sets of *var* genes to parasitize young children with limited malaria immunity and older children who have acquired partial malaria immunity, which has implications for PfEMP1 types associated with severe malaria. Expression of *var* genes is tightly coordinated such that a single gene is typically expressed at a time. Control of gene expression is regulated at multiple levels including chromatin modifications, coordination between the *var* upstream promoter and *var* intron promoter, subnuclear localization, and epigenetic marks laid down during *var* transcription to ensure that the switching rate is low enough that the entire *var* repertoire is not rapidly exhausted.

REFERENCES

Adams JH, Hudson DE, Torii M, Ward GE, Wellems TE, Aikawa M, Miller LH. 1990. The Duffy receptor family of *Plasmodium knowlesi* is located within the micronemes of invasive malaria merozoites. Cell 63: 141–153.

Adams JH, Sim BK, Dolan SA, Fang X, Kaslow DC, Miller LH. 1992. A family of erythrocyte binding proteins in malaria parasites. Proc Natl Acad Sci U S A 89: 7085–7089.

Afonso NP, Wunderlich G, Shugiro TM, D'arc Neves CJ, Jose MM, Scherf A, Pereira-Da-Silva LH. 2002. *Plasmodium falciparum*: Analysis of transcribed var gene sequences in natural isolates from the Brazilian Amazon region. Exp Parasitol 101: 111–120.

Aguiar JC, Albrecht GR, Cegielski P, Greenwood BM, Jensen JB, Lallinger G, Martinez A, Mcgregor IA, Minjas JN, Neequaye J, Patarroyo ME, Sherwood JA, Howard RJ. 1992. Agglutination of *Plasmodium falciparum*-infected erythrocytes from East and West African isolates by human sera from distant geographic regions. Am J Trop Med Hyg 47: 621–632.

al Khedery B, Barnwell JW, Galinski MR. 1999. Antigenic variation in malaria: A 3′ genomic alteration associated with the expression of a *P. knowlesi* variant antigen. Mol Cell 3: 131–141.

Albrecht L, Merino EF, Hoffmann EH, Ferreira MU, Mattos Ferreira RG, Osakabe AL, Dalla Martha RC, Ramharter M, Durham AM, Ferreira JE, Del Portillo HA, Wunderlich G. 2006. Extense variant gene family repertoire overlap in Western Amazon *Plasmodium falciparum* isolates. Mol Biochem Parasitol 150: 157–165.

Albrecht L, Castineiras C, Carvalho BO, Ladeia-Andrade S, Santos Da Silva N, Hoffmann EH, Dalla Martha RC, Costa FT, Wunderlich G. 2010. The South American *Plasmodium falciparum* var gene repertoire is limited, highly shared and possibly lacks several antigenic types. Gene 453: 37–44.

Andersen P, Nielsen MA, Resende M, Rask TS, Dahlback M, Theander T, Lund O, Salanti A. 2008. Structural insight into epitopes in the pregnancy-associated malaria protein VAR2CSA. PLoS Pathog 4: e42.

Anderson TJ, Haubold B, Williams JT, Estrada-Franco JG, Richardson L, Mollinedo R, Bockarie M, Mokili J, Mharakurwa S, French N, Whitworth J, Velez ID, Brockman AH, Nosten F, Ferreira MU, Day KP. 2000. Microsatellite markers reveal a spectrum of population structures in the malaria parasite *Plasmodium falciparum*. Mol Biol Evol 17: 1467–1482.

Avril M, Kulasekara BR, Gose SO, Rowe C, Dahlback M, Duffy PE, Fried M, Salanti A, Misher L, Narum DL, Smith JD. 2008. Evidence for globally shared, cross-reacting polymorphic epitopes in the pregnancy-associated malaria vaccine candidate VAR2CSA. Infect Immun 76: 1791–1800.

Avril M, Hathaway MJ, Srivastava A, Dechavanne S, Hommel M, Beeson JG, Smith JD, Gamain B. 2011. Antibodies to a full-length VAR2CSA immunogen are broadly strain-transcendent but do not cross-inhibit different placental-type parasite isolates. PLoS One 6: e16622.

Barry AE, Leliwa-Sytek A, Tavul L, Imrie H, Migot-Nabias F, Brown SM, Mcvean GA, Day KP. 2007. Population genomics of the immune evasion (*var*) genes of *Plasmodium falciparum*. PLoS Pathog 3: e34.

Bartfai R, Hoeijmakers WA, Salcedo-Amaya AM, Smits AH, Janssen-Megens E, Kaan A, Treeck M, Gilberger TW, Francoijs KJ, Stunnenberg HG. 2010. H2A.Z demarcates intergenic regions of the *Plasmodium falciparum* epigenome that are dynamically marked by H3K9ac and H3K4me3. PLoS Pathog 6: e1001223.

Baruch DI, Pasloske BL, Singh HB, Bi X, Ma XC, Feldman M, Taraschi TF, Howard RJ. 1995. Cloning the *P. falciparum* gene encoding PfEMP1, a malarial variant antigen and adherence receptor on the surface of parasitized human erythrocytes. Cell 82: 77–87.

Baruch DI, Ma XC, Singh HB, Bi X, Pasloske BL, Howard RJ. 1997. Identification of a region of PfEMP1 that mediates adherence of *Plasmodium falciparum* infected erythrocytes to CD36: Conserved function with variant sequence. Blood 90: 3766–3775.

Beeson JG, Brown GV, Molyneux ME, Mhango C, Dzinjalamala F, Rogerson SJ. 1999. *Plasmodium falciparum* isolates from infected pregnant women and children are associated with distinct adhesive and antigenic properties. J Infect Dis 180: 464–472.

Berendt AR, Ferguson DJP, Newbold CI. 1990. Sequestration in *Plasmodium falciparum* malaria: Sticky cells and sticky problems. Parasitol Today 6: 247–254.

Berendt AR, Mcdowall A, Craig AG, Bates PA, Sternberg MJ, Marsh K, Newbold CI, Hogg N. 1992. The binding site on ICAM-1 for *Plasmodium falciparum*-infected erythrocytes overlaps, but is distinct from, the LFA-1-binding site. Cell 68: 71–81.

Blomqvist K, Normark J, Nilsson D, Ribacke U, Orikiriza J, Trillkott P, Byarugaba J, Egwang TG, Kironde F, Andersson B, Wahlgren M. 2010. var gene transcription dynamics in *Plasmodium falciparum* patient isolates. Mol Biochem Parasitol 170: 74–83.

Bockhorst J, Lu F, Janes JH, Keebler J, Gamain B, Awadalla P, Su XZ, Samudrala R, Jojic N, Smith JD. 2007. Structural polymorphism and diversifying selection on the pregnancy malaria vaccine candidate VAR2CSA. Mol Biochem Parasitol 155: 103–112.

Brabin BJ, Romagosa C, Abdelgalil S, Menendez C, Verhoeff FH, Mcgready R, Fletcher KA, Owens S, D'Alessandro U, Nosten F, Fischer PR, Ordi J. 2004. The sick placenta-the role of malaria. Placenta 25: 359–378.

Brolin KJ, Ribacke U, Nilsson S, Ankarklev J, Moll K, Wahlgren M, Chen Q. 2009. Simultaneous transcription of duplicated var2csa gene copies in individual Plasmodium falciparum parasites. Genome Biol 10: R117.

Calderwood MS, Gannoun-Zaki L, Wellems TE, Deitsch KW. 2003. *Plasmodium falciparum var* genes are regulated by two regions with separate promoters, one upstream of the coding region and a second within the intron. J Biol Chem 278: 34125–34132.

Carlton JM, Angiuoli SV, Suh BB, Kooij TW, Pertea M, Silva JC, Ermolaeva MD, Allen JE, Selengut JD, Koo HL, Peterson JD, Pop M, Kosack DS, Shumway MF, Bidwell SL, Shallom SJ, van Aken SE, Riedmuller SB, Feldblyum TV, Cho JK, Quackenbush J, Sedegah M, Shoaibi A, Cummings LM, Florens L, Yates JR, Raine JD, Sinden RE, Harris MA, Cunningham DA, Preiser PR, Bergman LW, Vaidya AB, Van Lin LH, Janse CJ, Waters AP, Smith HO, White OR, Salzberg SL, Venter JC, Fraser CM, Hoffman SL, Gardner MJ, Carucci DJ. 2002. Genome sequence and comparative analysis of the model rodent malaria parasite *Plasmodium yoelii yoelii*. Nature 419: 512–519.

Carlton JM, Adams JH, Silva JC, Bidwell SL, Lorenzi H, Caler E, Crabtree J, Angiuoli SV, Merino EF, Amedeo P, Cheng Q, Coulson RM, Crabb BS, Del Portillo HA, Essien K, Feldblyum TV, Fernandez-Becerra C, Gilson PR, Gueye AH, Guo X, Kang'a S, Kooij TW, Korsinczky M, Meyer EV, Nene V, Paulsen I,

White O, Ralph SA, Ren Q, Sargeant TJ, Salzberg SL, Stoeckert CJ, Sullivan SA, Yamamoto MM, Hoffman SL, Wortman JR, Gardner MJ, Galinski MR, Barnwell JW, Fraser-Liggett CM. 2008. Comparative genomics of the neglected human malaria parasite *Plasmodium vivax*. Nature 455: 757–763.

Carty SM, Greenleaf AL. 2002. Hyperphosphorylated C-terminal repeat domain-associating proteins in the nuclear proteome link transcription to DNA/chromatin modification and RNA processing. Mol Cell Proteomics 1: 598–610.

Cham GK, Turner L, Lusingu J, Vestergaard L, Mmbando BP, Kurtis JD, Jensen AT, Salanti A, Lavstsen T, Theander TG. 2009. Sequential, ordered acquisition of antibodies to *Plasmodium falciparum* erythrocyte membrane protein 1 domains. J Immunol 183: 3356–3363.

Cham GK, Turner L, Kurtis JD, Mutabingwa T, Fried M, Jensen AT, Lavstsen T, Hviid L, Duffy PE, Theander TG. 2010. Hierarchical, domain type-specific acquisition of antibodies to *Plasmodium falciparum* erythrocyte membrane protein 1 in Tanzanian children. Infect Immun 78: 4653–4659.

Chattopadhyay R, Taneja T, Chakrabarti K, Pillai CR, Chitnis CE. 2004. Molecular analysis of the cytoadherence phenotype of a *Plasmodium falciparum* field isolate that binds intercellular adhesion molecule-1. Mol Biochem Parasitol 133: 255–265.

Chen DS, Barry AE, Leliwa-Sytek A, Smith TA, Peterson I, Brown SM, Migot-Nabias F, Deloron P, Kortok MM, Marsh K, Daily JP, Ndiaye D, Sarr O, Mboup S, Day KP. 2011. A molecular epidemiological study of var gene diversity to characterize the reservoir of *Plasmodium falciparum* in humans in Africa. PLoS One 6: e16629.

Cheng Q, Cloonan N, Fischer K, Thompson J, Waine G, Lanzer M, Saul A. 1998. *stevor* and *rif* are *Plasmodium falciparum* multicopy gene families which potentially encode variant antigens. Mol Biochem Parasitol 97: 161–176.

Chitnis CE, Miller LH. 1994. Identification of the erythrocyte binding domains of *Plasmodium vivax* and *Plasmodium knowlesi* proteins involved in erythrocyte invasion. J Exp Med 180: 497–506.

Chookajorn T, Dzikowski R, Frank M, Li F, Jiwani AZ, Hartl DL, Deitsch KW. 2007. Epigenetic memory at malaria virulence genes. Proc Natl Acad Sci U S A 104: 899–902.

Cooke BM, Berendt AR, Craig AG, Macgregor J, Newbold CI, Nash GB. 1994. Rolling and stationary cytoadhesion of red blood cells parasitized by *Plasmodium falciparum*: Separate roles for ICAM-1, CD36 and thrombospondin. Br J Haematol 87: 162–170.

Crabb BS, Cooke BM, Reeder JC, Waller RF, Caruana SR, Davern KM, Wickham ME, Brown GV, Coppel RL, Cowman AF. 1997. Targeted gene disruption shows that knobs enable malaria-infected red cells to cytoadhere under physiological shear stress. Cell 89: 287–296.

Dahlback M, Joergensen LM, Nielsen MA, Clausen TM, Ditlev SB, Resende M, Pinto VV, Arnot DE, Theander TG, Salanti A. 2011. The chondroitin sulphate A-binding site of the VAR2CSA protein involves multiple N-terminal domains. J Biol Chem 286: 15908–15917.

Deitsch KW, Moxon ER, Wellems TE. 1997. Shared themes of antigenic variation and virulence in bacterial, protozoal, and fungal infections. Microbiol Mol Biol Rev 61: 281–293.

Deitsch KW, Calderwood MS, Wellems TE. 2001. Malaria. Cooperative silencing elements in *var* genes. Nature 412: 875–876.

Deitsch KW, Lukehart SA, Stringer JR. 2009. Common strategies for antigenic variation by bacterial, fungal and protozoan pathogens. Nat Rev Microbiol 7: 493–503.

del Portillo HA, Fernandez-Becerra C, Bowman S, Oliver K, Preuss M, Sanchez CP, Schneider NK, Villalobos JM, Rajandream MA, Harris D, da Silva LHP, Barrell B, Lanzer M. 2001. A superfamily of variant genes encoded in the subtelomeric region of *Plasmodium vivax*. Nature 410: 839–842.

Duffy MF, Brown GV, Basuki W, Krejany EO, Noviyanti R, Cowman AF, Reeder JC. 2002. Transcription of multiple *var* genes by individual, trophozoite-stage *Plasmodium falciparum* cells expressing a chondroitin sulphate A binding phenotype. Mol Microbiol 43: 1285–1293.

Duffy MF, Maier AG, Byrne TJ, Marty AJ, Elliott SR, O'neill MT, Payne PD, Rogerson SJ, Cowman AF, Crabb BS, Brown GV. 2006. VAR2CSA is the principal ligand for chondroitin sulfate A in two allogeneic isolates of *Plasmodium falciparum*. Mol Biochem Parasitol 148: 117–124.

Duffy PE, Fried M. 2003. Antibodies that inhibit *Plasmodium falciparum* adhesion to chondroitin sulfate A are associated with increased birth weight and the gestational age of newborns. Infect Immun 71: 6620–6623.

Duraisingh MT, Voss TS, Marty AJ, Duffy MF, Good RT, Thompson JK, Freitas-Junior LH, Scherf A, Crabb BS, Cowman AF. 2005. Heterochromatin silencing and locus repositioning linked to regulation of virulence genes in *Plasmodium falciparum*. Cell 121: 13–24.

Dzikowski R, Deitsch KW. 2008. Active transcription is required for maintenance of epigenetic memory in the malaria parasite *Plasmodium falciparum*. J Mol Biol 382: 288–297.

Dzikowski R, Frank M, Deitsch K. 2006. Mutually exclusive expression of virulence genes by malaria parasites is regulated independently of antigen production. PLoS Pathog 2: e22.

Dzikowski R, Li F, Amulic B, Eisberg A, Frank M, Patel S, Wellems TE, Deitsch KW. 2007. Mechanisms underlying mutually exclusive expression of virulence genes by malaria parasites. EMBO Rep 8: 959–965.

Epp C, Li F, Howitt CA, Chookajorn T, Deitsch KW. 2008. Chromatin associated sense and antisense noncoding RNAs are transcribed from the *var* gene family of virulence genes of the malaria parasite *Plasmodium falciparum*. RNA 15: 116–127.

Escalante AA, Barrio E, Ayala FJ. 1995. Evolutionary origin of human and primate malarias: Evidence from the circumsporozoite protein gene. Mol Biol Evol 12: 616–626.

Figueiredo L, Scherf A. 2005. *Plasmodium* telomeres and telomerase: The usual actors in an unusual scenario. Chromosome Res 13: 517–524.

Figueiredo LM, Freitas-Junior LH, Bottius E, Olivo-Marin JC, Scherf A. 2002. A central role for *Plasmodium falciparum* subtelomeric regions in spatial positioning and telomere length regulation. EMBO J 21: 815–824.

Flick K, Chen Q. 2004. *var* genes, PfEMP1 and the human host. Mol Biochem Parasitol 134: 3–9.

Flueck C, Bartfai R, Volz J, Niederwieser I, Salcedo-Amaya AM, Alako BT, Ehlgen F, Ralph SA, Cowman AF, Bozdech Z, Stunnenberg HG, Voss TS. 2009. *Plasmodium falciparum* heterochromatin protein 1 marks genomic loci linked to phenotypic variation of exported virulence factors. PLoS Pathog 5: e1000569.

Frank M, Dzikowski R, Constantini D, Amulic B, Burdougo E, Deitsch K. 2006. Strict pairing of var promoters and introns is required for *var* gene silencing in the malaria parasite *Plasmodium falciparum*. J Biol Chem 281: 9942–9952.

Frank M, Dzikowski R, Amulic B, Deitsch K. 2007. Variable switching rates of malaria virulence genes are associated with chromosomal position. Mol Microbiol 64: 1486–1498.

Frank M, Kirkman L, Costantini D, Sanyal S, Lavazec C, Templeton TJ, Deitsch KW. 2008. Frequent recombination events generate diversity within the multi-copy variant antigen gene families of *Plasmodium falciparum*. Int J Parasitol 38: 1099–1109.

Freitas-Junior LH, Bottius E, Pirrit LA, Deitsch KW, Scheidig C, Guinet F, Nehrbass U, Wellems TE, Scherf A. 2000. Frequent ectopic recombination of virulence factor genes in telomeric chromosome clusters of *P. falciparum*. Nature 407: 1018–1022.

Freitas-Junior LH, Hernandez-Rivas R, Ralph SA, Montiel-Condado D, Ruvalcaba-Salazar OK, Rojas-Meza AP, Mancio-Silva L, Leal-Silvestre RJ, Gontijo AM, Shorte S, Scherf A. 2005. Telomeric heterochromatin propagation and histone acetylation control mutually exclusive expression of antigenic variation genes in malaria parasites. Cell 121: 25–36.

Fried M, Duffy PE. 1996. *Plasmodium falciparum*-infected erythrocytes adhere to chondroitin sulfate A in the human placenta. Science 272: 1502–1504.

Fried M, Nosten F, Brockman A, Brabin BJ, Duffy PE. 1998. Maternal antibodies block malaria. Nature 395: 851–852.

Gamain B, Trimnell AR, Scheidig C, Scherf A, Miller LH, Smith JD. 2005. Identification of multiple chondroitin sulfate A (CSA)-binding domains in the var2CSA gene transcribed in CSA-binding parasites. J Infect Dis 191: 1010–1013.

Gannoun-Zaki L, Jost A, Mu JB, Deitsch KW, Wellems TE. 2005. A silenced *Plasmodium falciparum var* promoter can be activated in vivo through spontaneous deletion of a silencing element in the intron. Eukaryot Cell 4: 490–492.

Gardner MJ, Hall N, Fung E, White O, Berriman M, Hyman RW, Carlton JM, Pain A, Nelson KE, Bowman S, Paulsen IT, James K, Eisen JA, Rutherford K, Salzberg SL, Craig A, Kyes S, Chan MS, Nene V, Shallom SJ, Suh B, Peterson J, Angiuoli S, Pertea M, Allen J, Selengut J, Haft D, Mather MW, Vaidya AB, Martin DM, Fairlamb AH, Fraunholz MJ, Roos DS, Ralph SA, Mcfadden GI, Cummings LM, Subramanian GM, Mungall C, Venter JC, Carucci DJ, Hoffman SL, Newbold C, Davis RW, Fraser CM, Barrell B. 2002. Genome sequence of the human malaria parasite *Plasmodium falciparum*. Nature 419: 498–511.

Glenister FK, Fernandez KM, Kats LM, Hanssen E, Mohandas N, Coppel RL, Cooke BM. 2009. Functional alteration of red blood cells by a megadalton protein of *Plasmodium falciparum*. Blood 113: 919–928.

Gray C, Mccormick C, Turner G, Craig A. 2003. ICAM-1 can play a major role in mediating *P. falciparum* adhesion to endothelium under flow. Mol Biochem Parasitol 128: 187–193.

Hiller NL, Bhattacharjee S, van Ooij C, Liolios K, Harrison T, Lopez-Estrano C, Haldar K. 2004. A host-targeting signal in virulence proteins reveals a secretome in malarial infection. Science 306: 1934–1937.

Ho M, Hickey MJ, Murray AG, Andonegui G, Kubes P. 2000. Visualization of *Plasmodium falciparum*-endothelium interactions in human microvasculature: Mimicry of leukocyte recruitment. J Exp Med 192: 1205–1211.

Hodder AN, Maier AG, Rug M, Brown M, Hommel M, Pantic I, Puig-De-Morales-Marinkovic M, Smith B, Triglia T, Beeson J, Cowman AF. 2009. Analysis of structure and function of the giant protein Pf332 in *Plasmodium falciparum*. Mol Microbiol 71: 48–65.

Hommel M, Elliott SR, Soma V, Kelly G, Fowkes FJ, Chesson JM, Duffy MF, Bockhorst J, Avril M, Mueller I, Raiko A, Stanisic DI, Rogerson SJ, Smith JD, Beeson JG. 2010. Evaluation of the antigenic diversity of placenta-binding *Plasmodium falciparum* variants and the antibody repertoire among pregnant women. Infect Immun 78: 1963–1978.

Horrocks P, Pinches R, Christodoulou Z, Kyes SA, Newbold CI. 2004. Variable *var* transition rates underlie antigenic variation in malaria. Proc Natl Acad Sci U S A 101: 11129–11134.

Howard RJ, Barnwell JW, Rock EP, Janet N, Ofori-Adjei D, Maloy WL, Lyon JA, Saul A. 1988. Two approximately 300 kilodalton *Plasmodium falciparum* proteins at the surface membrane of infected erythrocytes. Mol Biochem Parasitol 27: 207–224.

Howell DP, Samudrala R, Smith JD. 2006. Disguising itself—Insights into *Plasmodium falciparum* binding and immune evasion from the DBL crystal structure. Mol Biochem Parasitol 148: 1–9.

Howell DP, Levin EA, Springer AL, Kraemer SM, Phippard DJ, Schief WR, Smith JD. 2008. Mapping a common interaction site used by *Plasmodium falciparum* Duffy binding-like domains to bind diverse host receptors. Mol Microbiol 67: 78–87.

Howitt CA, Wilinski D, Llinas M, Templeton TJ, Dzikowski R, Deitsch KW. 2009. Clonally variant gene families in *Plasmodium falciparum* share a common activation factor. Mol Microbiol 73: 1171–1185.

Janes JH, Wang CP, Levin-Edens E, Vigan-Womas I, Guillotte M, Melcher M, Mercereau-Puijalon O, Smith JD. 2011. Investigating the host binding signature on the *Plasmodium falciparum* PfEMP1 Protein Family. PLoS Pathog 7: e1002032.

Jensen AT, Magistrado P, Sharp S, Joergensen L, Lavstsen T, Chiucchiuini A, Salanti A, Vestergaard LS, Lusingu JP, Hermsen R, Sauerwein R, Christensen J, Nielsen MA, Hviid L, Sutherland C, Staalsoe T, Theander TG. 2004. *Plasmodium falciparum* associated with severe childhood malaria preferentially expresses PfEMP1 encoded by group A *var* genes. J Exp Med 199: 1179–1190.

Joergensen L, Bengtsson DC, Bengtsson A, Ronander E, Berger SS, Turner L, Dalgaard MB, Cham GK, Victor ME, Lavstsen T, Theander TG, Arnot DE, Jensen AT. 2010. Surface co-expression of two different PfEMP1 antigens on single *Plasmodium falciparum*-infected erythrocytes facilitates binding to ICAM1 and PECAM1. PLoS Pathog 6: e1001083.

Kaestli M, Cortes A, Lagog M, Ott M, Beck HP. 2004. Longitudinal assessment of *Plasmodium falciparum var* gene transcription in naturally infected asymptomatic children in Papua New Guinea. J Infect Dis 189: 1942–1951.

Kaestli M, Cockburn IA, Cortes A, Baea K, Rowe JA, Beck HP. 2006. Virulence of malaria is associated with differential expression of *Plasmodium falciparum var* gene subgroups in a case-control study. J Infect Dis 193: 1567–1574.

Kalmbach Y, Rottmann M, Kombila M, Kremsner PG, Beck HP, Kun JF. 2010. Differential var gene expression in children with malaria and antidromic effects on host gene expression. J Infect Dis 202: 313–317.

Khunrae P, Dahlback M, Nielsen MA, Andersen G, Ditlev SB, Resende M, Pinto VV, Theander TG, Higgins MK, Salanti A. 2010. Full-length recombinant *Plasmodium falciparum* VAR2CSA binds specifically to CSPG and induces potent parasite adhesion-blocking antibodies. J Mol Biol 397: 826–834.

Kilejian A, Rashid MA, Aikawa M, Aji T, Yang Y-F. 1991. Selective association of a fragment of the knob protein with spectrin, actin and the red cell membrane. Mol Biochem Parasitol 44: 175–182.

Klein MM, Gittis AG, Su HP, Makobongo MO, Moore JM, Singh S, Miller LH, Garboczi DN. 2008. The cysteine-rich interdomain region from the highly variable *Plasmodium falciparum* erythrocyte membrane protein-1 exhibits a conserved structure. PLoS Pathog 4: e1000147.

Kraemer SM, Smith JD. 2003. Evidence for the importance of genetic structuring to the structural and functional specialization of the *Plasmodium falciparum var* gene family. Mol Microbiol 50: 1527–1538.

Kraemer SM, Kyes SA, Aggarwal G, Springer AL, Nelson SO, Christodoulou Z, Smith LM, Wang W, Levin E, Newbold CI, Myler PJ, Smith JD. 2007. Patterns of gene recombination shape *var* gene repertoires in *Plasmodium falciparum*: Comparisons of geographically diverse isolates. BMC Genomics 8: 45.

Krief S, Escalante AA, Pacheco MA, Mugisha L, Andre C, Halbwax M, Fischer A, Krief JM, Kasenene JM, Crandfield M, Cornejo OE, Chavatte JM, Lin C, Letourneur F, Gruner AC, Mccutchan TF, Renia L, Snounou G. 2010. On the diversity of malaria parasites in African apes and the origin of *Plasmodium falciparum* from Bonobos. PLoS Pathog 6: e1000765.

Kriek N, Tilley L, Horrocks P, Pinches R, Elford BC, Ferguson DJ, Lingelbach K, Newbold CI. 2003. Characterization of the pathway for transport of the cytoadherence-mediating protein, PfEMP1, to the host cell surface in malaria parasite-infected erythrocytes. Mol Microbiol 50: 1215–1227.

Kyes S, Taylor H, Craig AG, Marsh K, Newbold CI. 1997. Genomic representation of *var* gene sequences in *Plasmodium falciparum* field isolates from different geographic regions. Mol Biochem Parasitol 87: 235–238.

Kyes SA, Rowe JA, Kriek N, Newbold CI. 1999. Rifins: A second family of clonally variant proteins expressed on the surface of red cells infected with *Plasmodium falciparum*. Proc Natl Acad Sci U S A 96: 9333–9338.

Kyes SA, Christodoulou Z, Raza A, Horrocks P, Pinches R, Rowe JA, Newbold CI. 2003. A well-conserved *Plasmodium falciparum var* gene shows an unusual stage-specific transcript pattern. Mol Microbiol 48: 1339–1348.

Kyriacou HM, Stone GN, Challis RJ, Raza A, Lyke KE, Thera MA, Kone AK, Doumbo OK, Plowe CV, Rowe JA. 2006. Differential var gene transcription in *Plasmodium falciparum* isolates from patients with cerebral malaria compared to hyperparasitaemia. Mol Biochem Parasitol 150: 211–218.

Lavstsen T, Salanti A, Jensen ATR, Arnot DE, Theander TG. 2003. Sub-grouping of *Plasmodium falciparum* 3D7 *var* genes based on sequence analysis of coding and non-coding regions. Malar J 2: 27.

Lopez-Rubio JJ, Gontijo AM, Nunes MC, Issar N, Hernandez RR, Scherf A. 2007. 5' Flanking region of *var* genes nucleate histone modification patterns linked to phenotypic inheritance of virulence traits in malaria parasites. Mol Microbiol 66: 1296–1305.

Lopez-Rubio JJ, Mancio-Silva L, Scherf A. 2009. Genome wide analysis of heterochromatin associates clonally variant gene regulation with perinuclear repressive centers in malaria parasites. Cell Host Microbe 5: 179–190.

Mackinnon MJ, Marsh K. 2010. The selection landscape of malaria parasites. Science 328: 866–871.

Magistrado PA, Minja D, Doritchamou J, Ndam NT, John D, Schmiegelow C, Massougbodji A, Dahlback M, Ditlev SB, Pinto VV, Resende M, Lusingu J, Theander TG, Salanti A, Nielsen MA. 2011. High efficacy of anti DBL4varepsilon-VAR2CSA antibodies in inhibition of CSA-binding Plasmodium falciparum-infected erythrocytes from pregnant women. Vaccine 29: 437–443.

Marsh K, Howard RJ. 1986. Antigens induced on erythrocytes by *P. falciparum*: Expression of diverse and conserved determinants. Science 231: 150–153.

Marti M, Good RT, Rug M, Knuepfer E, Cowman AF. 2004. Targeting malaria virulence and remodeling proteins to the host erythrocyte. Science 306: 1930–1933.

Marty AJ, Thompson JK, Duffy MF, Voss TS, Cowman AF, Crabb BS. 2006. Evidence that *Plasmodium falciparum* chromosome end clusters are cross-linked by protein and are the sites of both virulence gene silencing and activation. Mol Microbiol 62: 72–83.

Melcher M, Muhle RA, Henrich PP, Kraemer SM, Avril M, Vigan-Womas I, Mercereau-Puijalon O, Smith JD, Fidock DA. 2010. Identification of a role for the PfEMP1 semi-conserved head structure in protein trafficking to the surface of *Plasmodium falciparum* infected red blood cells. Cell Microbiol 12: 1446–1462.

Miller LH, Baruch DI, Marsh K, Doumbo OK. 2002. The pathogenic basis of malaria. Nature 415: 673–679.

Moll K, Chene A, Ribacke U, Kaneko O, Nilsson S, Winter G, Haeggstrom M, Pan W, Berzins K, Wahlgren M, Chen Q. 2007. A novel DBL-domain of the *P. falciparum* 332 molecule possibly involved in erythrocyte adhesion. PLoS One 2: e477.

Montgomery J, Mphande FA, Berriman M, Pain A, Rogerson SJ, Taylor TE, Molyneux ME, Craig A. 2007. Differential *var* gene expression in the organs of patients dying of *falciparum* malaria. Mol Microbiol 65: 959–967.

Morey C, Avner P. 2004. Employment opportunities for non-coding RNAs. FEBS Lett 567: 27–34.

Muthusamy A, Achur RN, Bhavanandan VP, Fouda GG, Taylor DW, Gowda DC. 2004. *Plasmodium falciparum*-infected erythrocytes adhere both in the intervillous space and on the villous surface of human placenta by binding to the low-sulfated chondroitin sulfate proteoglycan receptor. Am J Pathol 164: 2013–2025.

Newbold C, Warn P, Black G, Berendt A, Craig A, Snow B, Msobo M, Peshu N, Marsh K. 1997. Receptor-specific adhesion and clinical disease in *Plasmodium falciparum*. Am J Trop Med Hyg 57: 389–398.

Newbold CI, Pinches R, Roberts DJ, Marsh K. 1992. *Plasmodium falciparum*: The human agglutinating antibody response to the infected red cell surface is predominantly variant specific. Exp Parasitol 75: 281–292.

Nielsen MA, Pinto VV, Resende M, Dahlback M, Ditlev SB, Theander TG, Salanti A. 2009. Induction of adhesion-inhibitory antibodies against placental Plasmodium falciparum parasites by using single domains of VAR2CSA. Infect Immun 77: 2482–2487.

Ockenhouse CF, Betageri R, Springer TA, Staunton DE. 1992. *Plasmodium falciparum*-infected erythrocytes bind ICAM-1 at a site distinct from LFA-1, Mac-1, and human rhinovirus. Cell 68: 63–69.

Oh SS, Voigt S, Fisher D, Yi SJ, Leroy PJ, Derick LH, Liu S, Chishti AH. 2000. *Plasmodium falciparum* erythrocyte membrane protein 1 is anchored to the actin-spectrin junction and knob-associated histidine-rich protein in the erythrocyte skeleton. Mol Biochem Parasitol 108: 237–247.

Oleinikov AV, Amos E, Frye IT, Rossnagle E, Mutabingwa TK, Fried M, Duffy PE. 2009. High throughput functional assays of the variant antigen PfEMP1 reveal a single domain in the 3D7 *Plasmodium falciparum* genome that binds ICAM1 with high affinity and is targeted by naturally acquired neutralizing antibodies. PLoS Pathog 5: e1000386.

Ollomo B, Durand P, Prugnolle F, Douzery E, Arnathau C, Nkoghe D, Leroy E, Renaud F. 2009. A new malaria agent in African hominids. PLoS Pathog 5: e1000446.

Pain A, Bohme U, Berry AE, Mungall K, Finn RD, Jackson AP, Mourier T, Mistry J, Pasini EM, Aslett MA, Balasubrammaniam S, Borgwardt K, Brooks K, Carret C, Carver TJ, Cherevach I, Chillingworth T, Clark TG, Galinski MR, Hall N, Harper D, Harris D, Hauser H, Ivens A, Janssen CS, Keane T, Larke N, Lapp S, Marti M, Moule S, Meyer IM, Ormond D, Peters N, Sanders M, Sanders S, Sargeant TJ, Simmonds M, Smith F, Squares R, Thurston S, Tivey AR, Walker D, White B, Zuiderwijk E, Churcher C, Quail MA, Cowman AF, Turner CM, Rajandream MA, Kocken CH, Thomas AW, Newbold CI, Barrell BG, Berriman M. 2008. The genome of the simian and human malaria parasite *Plasmodium knowlesi*. Nature 455: 799–803.

Perez-Toledo K, Rojas-Meza AP, Mancio-Silva L, Hernandez-Cuevas NA, Delgadillo DM, Vargas M, Martinez-Calvillo S, Scherf A, Hernandez-Rivas R. 2009. *Plasmodium falciparum* heterochromatin protein 1 binds to tri-methylated histone 3 lysine 9 and is linked to mutually exclusive expression of var genes. Nucleic Acids Res 37: 2596–2606.

Peterson DS, Miller LH, Wellems TE. 1995. Isolation of multiple sequences from the *Plasmodium falciparum* genome that encode conserved domains homologous to those in erythrocyte-binding proteins. Proc Natl Acad Sci U S A 92: 7100–7104.

Petter M, Lee CC, Byrne TJ, Boysen KE, Volz J, Ralph SA, Cowman AF, Brown GV, Duffy MF. 2011. Expression of *P. falciparum* var genes involves exchange of the histone variant H2A.Z at the promoter. PLoS Pathog 7: e1001292.

Ponts N, Harris EY, Prudhomme J, Wick I, Eckhardt-Ludka C, Hicks GR, Hardiman G, Lonardi S, Le Roch KG. 2010. Nucleosome landscape and control of transcription in the human malaria parasite. Genome Res 20: 228–238.

Prugnolle F, Durand P, Neel C, Ollomo B, Ayala FJ, Arnathau C, Etienne L, Mpoudi-Ngole E, Nkoghe D, Leroy E, Delaporte E, Peeters M, Renaud F. 2010. African great apes are natural hosts of multiple related malaria species, including *Plasmodium falciparum*. Proc Natl Acad Sci U S A 107: 1458–1463.

Prugnolle F, Durand P, Ollomo B, Duval L, Ariey F, Arnathau C, Gonzalez JP, Leroy E, Renaud F. 2011. A fresh look at the origin of *Plasmodium falciparum*, the most malignant malaria agent. PLoS Pathog 7: e1001283.

Ralph SA, Scheidig-Benatar C, Scherf A. 2005. Antigenic variation in *Plasmodium falciparum* is associated with movement of *var* loci between subnuclear locations. Proc Natl Acad Sci U S A 102: 5414–5419.

Rask TS, Hansen DA, Theander TG, Gorm PA, Lavstsen T. 2010. Plasmodium falciparum erythrocyte membrane protein 1 diversity in seven genomes—Divide and conquer. PLoS Comput Biol 6: e1000933.

Recker M, Buckee CO, Serazin A, Kyes S, Pinches R, Christodoulou Z, Springer AL, Gupta S, Newbold CI. 2011. Antigenic variation in *Plasmodium falciparum* malaria involves a highly structured switching pattern. PLoS Pathog 7: e1001306.

Robinson BA, Welch TL, Smith JD. 2003. Widespread functional specialization of *Plasmodium falciparum* erythrocyte membrane protein 1 family members to bind CD36 analysed across a parasite genome. Mol Microbiol 47: 1265–1278.

Rogerson SJ, Tembenu R, Dobano C, Plitt S, Taylor TE, Molyneux ME. 1999. Cytoadherence characteristics of *Plasmodium falciparum*-infected erythrocytes from Malawian children with severe and uncomplicated malaria. Am J Trop Med Hyg 61: 467–472.

Rottmann M, Lavstsen T, Mugasa JP, Kaestli M, Jensen AT, Muller D, Theander T, Beck HP. 2006. Differential expression of *var* gene groups is associated with morbidity caused by *Plasmodium falciparum* infection in Tanzanian children. Infect Immun 74: 3904–3911.

Rowe JA, Kyes SA, Rogerson SJ, Babiker HA, Raza A. 2002. Identification of a conserved *Plasmodium falciparum var* gene implicated in malaria in pregnancy. J Infect Dis 185: 1207–1211.

Rowe JA, Claessens A, Corrigan RA, Arman M. 2010. Adhesion of *Plasmodium falciparum*-infected erythrocytes to human cells: Molecular mechanisms and therapeutic implications. Expert Rev Mol Med 11: e16.

Salanti A, Staalsoe T, Lavstsen T, Jensen ATR, Sowa MPK, Arnot DE, Hviid L, Theander TG. 2003. Selective upregulation of a single distinctly structured *var* gene in chondroitin sulphate A-adhering *Plasmodium falciparum* involved in pregnancy-associated malaria. Mol Microbiol 49: 179–191.

Salanti A, Dahlback M, Turner L, Nielsen MA, Barfod L, Magistrado P, Jensen AT, Lavstsen T, Ofori MF, Marsh K, Hviid L, Theander TG. 2004. Evidence for the involvement of VAR2CSA in pregnancy-associated malaria. J Exp Med 200: 1197–1203.

Scherf A, Hernandez-Rivas R, Buffet P, Bottius E, Benatar C, Pouvelle B, Gysin J, Lanzer M. 1998. Antigenic variation in malaria: *In situ* switching, relaxed and mutually exclusive transcription of *var* genes during intra-erythrocytic development in *Plasmodium falciparum*. EMBO J 17: 5418–5426.

Scherf A, Lopez-Rubio JJ, Riviere L. 2008. Antigenic variation in *Plasmodium falciparum*. Annu Rev Microbiol 62: 445–470.

Serghides L, Smith TG, Patel SN, Kain KC. 2003. CD36 and malaria: Friends or foes? Trends Parasitol 19: 461–469.

Sim BKL, Chitnis CE, Wasniowska K, Hadley TJ, Miller LH. 1994. Receptor and ligand domains for invasion of erythrocytes by *Plasmodium falciparum*. Science 264: 1941–1944.

Singh K, Gittis AG, Nguyen P, Gowda DC, Miller LH, Garboczi DN. 2008. Structure of the DBL3x domain of pregnancy-associated malaria protein VAR2CSA complexed with chondroitin sulfate A. Nat Struct Mol Biol 15: 932–938.

Singh SK, Hora R, Belrhali H, Chitnis CE, Sharma A. 2006. Structural basis for Duffy recognition by the malaria parasite Duffy-binding-like domain. Nature 439: 741–744.

Smith JD, Chitnis CE, Craig AG, Roberts DJ, Hudson-Taylor DE, Peterson DS, Pinches R, Newbold CI, Miller LH. 1995. Switches in expression of *Plasmodium falciparum var* genes correlate with changes in antigenic and cytoadherent phenotypes of infected erythrocytes. Cell 82: 101–110.

Smith JD, Kyes S, Craig AG, Fagan T, Hudson-Taylor D, Miller LH, Baruch DI, Newbold CI. 1998. Analysis of adhesive domains from the A4VAR *Plasmodium falciparum* erythrocyte membrane protein-1 identifies a CD36 binding domain. Mol Biochem Parasitol 97: 133–148.

Smith JD, Craig AG, Kriek N, Hudson-Taylor D, Kyes S, Fagan T, Pinches R, Baruch DI, Newbold CI, Miller LH. 2000a. Identification of a *Plasmodium falciparum* intercellular adhesion molecule-1 binding domain: A parasite adhesion trait implicated in cerebral malaria. Proc Natl Acad Sci U S A 97: 1766–1771.

Smith JD, Subramanian G, Gamain B, Baruch DI, Miller LH. 2000b. Classification of adhesive domains in the *Plasmodium falciparum* erythrocyte membrane protein 1 family. Mol Biochem Parasitol 110: 293–310.

Srivastava A, Gangnard S, Round A, Dechavanne S, Juillerat A, Raynal B, Faure G, Baron B, Ramboarina S, Singh SK, Belrhali H, England P, Lewit-Bentley A, Scherf A, Bentley GA, Gamain B. 2010. Full-length extracellular region of the var2CSA variant of PfEMP1 is required for specific, high-affinity binding to CSA. Proc Natl Acad Sci U S A 107: 4884–4889.

Staalsoe T, Shulman CE, Bulmer JN, Kawuondo K, Marsh K, Hviid L. 2004. Variant surface antigen-specific IgG and protection against clinical consequences of pregnancy-associated *Plasmodium falciparum* malaria. Lancet 363: 283–289.

Staunton DE, Dustin ML, Erickson HP, Springer TA. 1990. The arrangement of the immunoglobulin-like domains of ICAM-1 and the binding sites for LFA-1 and rhinovirus. Cell 61: 243–254.

Su X, Heatwole VM, Wertheimer SP, Guinet F, Herrfeldt JV, Peterson DS, Ravetch JV, Wellems TE. 1995. A large and diverse gene family (*var*) encodes 200–350 kD proteins implicated in the antigenic variation and cytoadherence of *Plasmodium falciparum*-infected erythrocytes. Cell 82: 89–100.

Sutherland H, Bickmore WA. 2009. Transcription factories: Gene expression in unions? Nat Rev Genet 10: 457–466.

Tao N, Wagner SJ, Lublin DM. 1996. CD36 is palmitoylated on both N- and C-terminal cytoplasmic tails. J Biol Chem 271: 22315–22320.

Taylor HM, Kyes SA, Harris D, Kriek N, Newbold CI. 2000a. A study of var gene transcription in vitro using universal var gene primers. Mol Biochem Parasitol 105: 13–23.

Taylor HM, Kyes SA, Newbold CI. 2000b. Var gene diversity in Plasmodium falciparum is generated by frequent recombination events. Mol Biochem Parasitol 110: 391–397.

Tolia NH, Enemark EJ, Sim BK, Joshua-Tor L. 2005. Structural basis for the EBA-175 erythrocyte invasion pathway of the malaria parasite *Plasmodium falciparum*. Cell 122: 183–193.

Tonkin CJ, Carret CK, Duraisingh MT, Voss TS, Ralph SA, Hommel M, Duffy MF, Silva LM, Scherf A, Ivens A, Speed TP, Beeson JG, Cowman AF. 2009. Sir2 paralogues cooperate to regulate virulence genes and antigenic variation in *Plasmodium falciparum*. PLoS Biol 7: e84.

Trimnell AR, Kraemer SM, Mukherjee S, Phippard DJ, Janes JH, Flamoe E, Su XZ, Awadalla P, Smith JD. 2006. Global genetic diversity and evolution of *var* genes associated with placental and severe childhood malaria. Mol Biochem Parasitol 148: 169–180.

Urban BC, Ferguson DJ, Pain A, Willcox N, Plebanski M, Austyn JM, Roberts DJ. 1999. *Plasmodium falciparum*-infected erythrocytes modulate the maturation of dendritic cells. Nature 400: 73–77.

Viebig NK, Gamain B, Scheidig C, Lepolard C, Przyborski J, Lanzer M, Gysin J, Scherf A. 2005. A single member of the *Plasmodium falciparum var* multigene family determines cytoadhesion to the placental receptor chondroitin sulphate A. EMBO Rep 6: 775–781.

Voss TS, Thompson JK, Waterkeyn J, Felger I, Weiss N, Cowman AF, Beck HP. 2000. Genomic distribution and functional characterisation of two distinct and conserved *Plasmodium falciparum var* gene 5′ flanking sequences. Mol Biochem Parasitol 107: 103–115.

Voss TS, Kaestli M, Vogel D, Bopp S, Beck HP. 2003. Identification of nuclear proteins that interact differentially with *Plasmodium falciparum var* gene promoters. Mol Microbiol 48: 1593–1607.

Voss TS, Healer J, Marty AJ, Duffy MF, Thompson JK, Beeson JG, Reeder JC, Crabb BS, Cowman AF. 2006. A *var* gene promoter controls allelic exclusion of virulence genes in *Plasmodium falciparum* malaria. Nature 439: 1004–1008.

Waller KL, Cooke BM, Nunomura W, Mohandas N, Coppel RL. 1999. Mapping the binding domains involved in the interaction between the *Plasmodium falciparum* knob-associated histidine-rich protein (KAHRP) and the cytoadherence ligand *P. falciparum* erythrocyte membrane protein 1 (PfEMP1). J Biol Chem 274: 23808–23813.

Waller KL, Nunomura W, Cooke BM, Mohandas N, Coppel RL. 2002. Mapping the domains of the cytoadherence ligand *Plasmodium falciparum* erythrocyte membrane protein 1 (PfEMP1) that bind to the knob-associated histidine-rich protein (KAHRP). Mol Biochem Parasitol 119: 125–129.

Ward CP, Clottey GT, Dorris M, Ji DD, Arnot DE. 1999. Analysis of *Plasmodium falciparum* PfEMP-1/*var* genes suggests that recombination rearranges constrained sequences. Mol Biochem Parasitol 102: 167–177.

Warimwe GM, Keane TM, Fegan G, Musyoki JN, Newton CR, Pain A, Berriman M, Marsh K, Bull PC. 2009. Plasmodium falciparum var gene expression is modified by host immunity. Proc Natl Acad Sci U S A 106: 21801–21806.

Westenberger SJ, Cui L, Dharia N, Winzeler E, Cui L. 2009. Genome-wide nucleosome mapping of *Plasmodium falciparum* reveals histone-rich coding and histone-poor intergenic regions and chromatin remodeling of core and subtelomeric genes. BMC Genomics 10: 610.

Wickramarachchi T, Cabrera AL, Sinha D, Dhawan S, Chandran T, Devi YS, Kono M, Spielmann T, Gilberger TW, Chauhan VS, Mohmmed A. 2009. A novel *Plasmodium falciparum* erythrocyte binding protein associated with the merozoite surface, PfDBLMSP. Int J Parasitol 39: 763–773.

Winter G, Kawai S, Haeggstrom M, Kaneko O, von Euler A, Kawazu S, Palm D, Fernandez V, Wahlgren M. 2005. SURFIN is a polymorphic antigen expressed on *Plasmodium falciparum* merozoites and infected erythrocytes. J Exp Med 201: 1853–1863.

CHAPTER 19

INVASION LIGAND DIVERSITY AND PATHOGENESIS IN BLOOD-STAGE MALARIA

MANOJ T. DURAISINGH, JEFFREY D. DVORIN, and PETER R. PREISER

INTRODUCTION

Plasmodium spp. are unicellular organisms that are obligatory intracellular parasites belonging to the phylum Apicomplexa. This phylum also includes other medically important parasites such as *Toxoplasma* and *Cryptosporidium*. All of these parasites are distinguished by the possession of specialized organelles at their apical ends, which play a central role in the attachment to and entry into host cells. For *Plasmodium* spp. erythrocyte invasion is an active process of penetration by the parasite into host cells, powered by an actinomyosin motor (Keeley and Soldati, 2004), leading to the internalization of the parasite within a parasitophorous vacuolar membrane (Fig. 19.1a). The parasite grows within this vacuole and replicates, until it forms mature invasive merozoites that egress from an erythrocyte to invade a new host cell.

Due to their requirement for specific receptors on the surface of the host erythrocyte, merozoites of *Plasmodium* spp. are restricted in their ability to invade different types of cells. During erythrocyte invasion, multiple ligand–receptor interactions are required for the many steps including attachment, recognition, and subsequent entry into the host cell. Many of the proteins that mediate the high affinity adhesive steps of invasion appear to be encoded by multigene families (Carlton et al., 2002, 2008; Gardner et al., 2002). Variant expression as well as sequence polymorphism is observed for these invasion genes; this diversity may be driven by the observed variation in host erythrocytes at the level of species, between individuals, or within individuals, and may be one of the causes for the differential ability of *Plasmodium* parasites to efficiently invade erythrocytes both within as well as across species. Furthermore, expression level polymorphism in critical invasion ligands may also provide an ideal mechanism for antigenic variation. Emerging evidence suggests that the parasite uses epigenetic mechanisms for regulating this variant expression. Indeed, many of the molecular strategies employed for erythrocyte invasion are also shared with those employed by the parasite for cytoadherence of the infected erythrocyte to host cells.

Here, we discuss the molecular and cell biology approaches that have been used to unravel the intricate molecular interactions

Evolution of Virulence in Eukaryotic Microbes, First Edition. Edited by L. David Sibley, Barbara J. Howlett, and Joseph Heitman.

(a)

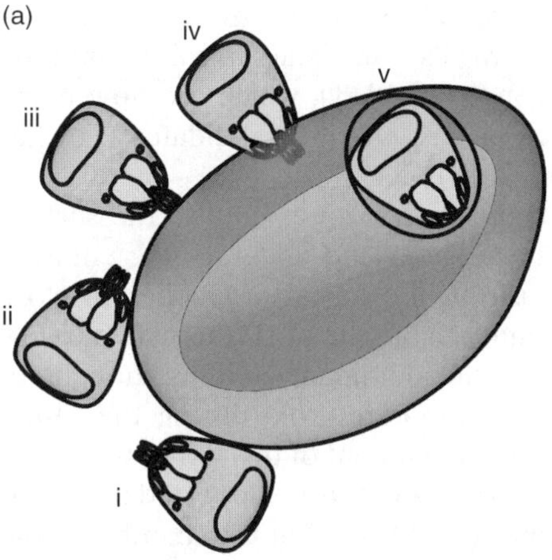

(b)

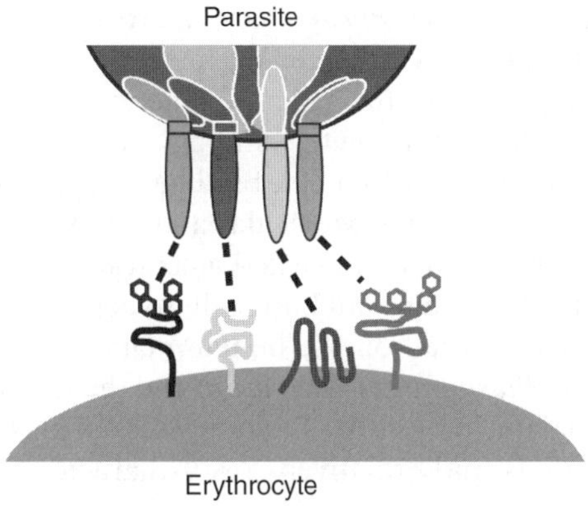

(c)

Parasite	DBL Family		RBL Family	
	Parasite Ligand	Erythrocyte Receptor	Parasite Ligand	Erythrocyte Receptor
P. falciparum	PfEBA175	Gly A	PfRH1	"Y"
	PfEBA140	Gly C	PfRH2a	?
	PfEBA181	"E"	PfRH2b	"Z"
	PfEBL1	Gly B	*PfRH3*	*pseudogene*
	PfEBA165	*pseudogene*	PfRH4	CR1
			PfRH5	?
			PfRH6	*pseudogene*
P. vivax	PvDBP	Duffy (DARC)	PvRBP1a, PvRBP1b, PvRBP2a, PvRBP2b, PvRBP2c, PvRBP2c, PvRBP3	?
P. knowlesi	PkDBP-α	Duffy (DARC)	PkRBP1, PkRBP2, PkRBP3a, PkRBP3b	?
	PkDBP-β, PkDBP-γ	?		
P. yoelii	PyEBP	Duffy	Py235 family	?
P. berghei	PbEBP	?	Pb235 family	?

Figure 19.1 (a) Schematic of *Plasmodium* spp. invasion of the erythrocyte. In the first step (i), the merozoite binds the surface of the erythrocyte. This binding is reversible and likely involves multiple members of the merozoite surface proteins. Following binding, the merozoite must undergo reorientation (ii) to position the parasite apical organelles in close proximity to the site of erythrocyte binding. Apical membrane antigen 1 (AMA-1) has been identified as a potential mediator of parasite reorientation (Mitchell et al., 2004). Following apical reorientation (iii), parasite ligands interact with erythrocyte surface proteins, forming a host–parasite junction and committing the parasite to invasion. The merozoite then actively invades the erythrocyte (iv) via actinomyosin-based motor proteins. Finally, the merozoite completes invasion (v) into the newly formed parasitophorous vacuole. (b) Schematic of the parasites ligands binding to host erythrocyte receptors. It is likely that multiple ligand–receptor interactions are necessary to form the tight junction and allow efficient parasite invasion. (c) Table of parasite ligands and erythrocyte receptors. *P. falciparum* has an expanded repertoire of DBLs. The erythrocyte receptors have been identified as the Duffy receptor (or Duffy antigen/receptor for chemokines) for PvDBP, PkDBPalpha, and PyEBP. The only other receptors known include PfEBA175/Gly A, PfEBA140/Gly C, PfEBL1/Gly B, and PfRH4/CR1. Pseudogenes are shown in italics. See color insert.

between the *Plasmodium* merozoite and the host erythrocyte. We place a specific emphasis on the reverse genetic analyses of the erythrocyte binding protein-like (EBL) and reticulocyte binding-like protein (RBL) families. Molecular and cell biological studies have revealed a central role for these proteins in host cell selection. We have focused on the human malaria parasite *Plasmodium falciparum* and the rodent malaria parasites as they possess sophisticated and relatively well-characterized genetic systems.

DISCOVERY OF HIGH AFFINITY PARASITE LIGANDS AND RECEPTORS FOR ERYTHROCYTE INVASION

Initial studies with *Plasmodium* spp. relied on the *in vivo* propagation of the parasite in experimental animal models, such as the macaque parasite *Plasmodium knowlesi* and the avian parasite *Plasmodium galinaceum*, for the generation of parasites. The continuous culture of the *P. falciparum* parasite was a landmark breakthrough (Trager and Jensen, 1976), which has now facilitated many molecular and genetic studies. However, there has been a renewed interest in the "other" malarias because of recent renewed efforts for the eradication of malaria and a growing interest in the comparative biology of *Plasmodium* spp.

The invasion process depends upon multiple receptor–ligand interactions between parasite and host cell proteins (Butcher et al., 1973; Cowman and Crabb, 2006) (Fig. 19.1b). Many studies over the last three decades have focused on the discovery of the specific parasite ligands used by *Plasmodium* merozoites for invasion into host erythrocytes. Parasite genomic expression libraries were created and screened with malaria-specific sera to identify antigenic proteins. This screening for candidates resulted in the identification of key proteins such as the merozoite surface proteins (MSPs) and the apical membrane antigen 1 (AMA-1), which are now being developed as vaccine candidates (Fowkes et al., 2010).

Invasive merozoites bind to the surface of erythrocytes and undergo reorientation to position the apical organelles in close proximity to the site of erythrocyte binding. MSP-1, the dominant MSP, is thought to play an important role during the initial random attachment of the merozoite to the erythrocyte, and this is supported by recent findings that MSP-1 binds to band 3 on the red blood cell (Goel et al., 2003). In contrast, AMA-1, a micronemal protein that is essential and conserved in all *Plasmodium* species, plays an important role in the reorientation of the merozoite after erythrocyte attachment (Mitchell et al., 2004).

Early seminal experiments showed that *Plasmodium* parasites secrete significant amounts of protein during the invasion process that binds to the erythrocyte surface with high affinity (Camus and Hadley, 1985; Galinski et al., 1992). Through these "erythrocyte binding assays," two superfamilies of *Plasmodium* proteins—the Duffy binding-like protein (DBL) proteins and the RBL proteins—have been identified. Following apical reorientation, these proteins interact with erythrocyte surface proteins, forming a high affinity or "tight" host–parasite junction and committing the parasite to cellular invasion.

DBL Family

The name for this family of erythrocyte binding proteins hails from the Duffy binding protein (DBP) identified in *Plasmodium vivax*. It was observed for a long time that infections with *P. vivax* were largely absent from West Africa, and experimental infections into humans from the same region were largely unsuccessful (Miller et al., 1976). As Duffy receptor negativity was found to be predominant in sub-Saharan Africa, it was thought that invasion into human erythrocytes by *P. vivax* was critically dependent on this receptor. Parasite-derived proteins that bound with high affinity to Duffy-positive but not to Duffy-negative erythrocytes were then identified. Following screening of a genomic expression library, the *P. vivax* DBP ortholog of *P. knowlesi*, PkDBPalpha, was identified (Adams et al., 1990). The availability of the sequence of PkDBPalpha then subsequently allowed the use of a degenerate PCR approach followed by sequencing to identify paralogous genes—PkDBPbeta and PkDBPgamma—in *P. knowlesi* as well as the *P. vivax* DBP gene (Adams et al., 1992; Fang et al., 1991). Analogous to the identification of the DBP protein in *P. vivax*, in *P. falciparum*, metabolic labeling experiments had also identified a 175-kDa parasite-derived erythrocyte binding protein, termed an erythrocyte binding antigen (EBA) (Camus and Hadley, 1985). The gene encoding for this EBA-175 was identified (Sim et al., 1990) and shown to be part of a larger family that also included PvDBP and the PkDBPs (Adams et al., 1992). When the *P. falciparum* genome was sequenced in 2002, the genes for the paralogous EBA proteins EBA-140, EBA-181, EBL-1, and the EBA-165 pseudogene were revealed (Adams et al., 2001; Gardner et al., 2002). With the gene sequences for a number of members of this protein family in hand, it was possible to build a canonical structure for a DBL family. Members of the DBL family are type I transmembrane proteins, where an N-terminal signal peptide is followed by a DBL "erythrocyte binding" domain together with a long C-terminal ectodomain region, which includes a cysteine-rich domain required for trafficking (Gilberger et al., 2003).

Domains of the different DBL proteins have been successfully expressed in and purified from bacterial and eukaryotic systems. Importantly, the relatively highly conserved DBL domain found in all members of this family possesses erythrocyte binding activity (Camus and Hadley, 1985; Gilberger et al., 2003; Lobo et al., 2003; Mayer et al., 2009; Orlandi et al., 1992; Thompson et al., 2001; Tran et al., 2005). These binding domains have also been expressed in mammalian systems as membrane fusion proteins in "rosetting" assays, where binding activity is determined by the formation of rosettes of erythrocytes binding via the erythrocyte binding domain fusions (Chitnis and Miller, 1994). Further, crystal structures of PfEBA-175, PkDBPalpha, and PvDBP, which are proving to be informative in structure–function studies and for aiding vaccine development, have been solved (Batchelor et al., 2011; Singh et al., 2006; Tolia et al., 2005). These adhesive domains have been used in immunological studies for measuring humoral responses in individuals from malaria-endemic regions (Fowkes et al., 2010).

RBL Family

A second family of adhesive proteins known as the RBL family has also been found. The first member of this family was identified in the rodent parasite *Plasmodium yoelii* as the target of a monoclonal antibody that restricts virulent infections of *P. yoelii* to reticulocytes, suggesting that this protein is involved in host cell selection (Holder and Freeman, 1984). The monoclonal antibody recognizes a 235-kDa rhopty protein (PY235) that was later shown to be part of a multigene family in this parasite (Keen et al., 1990). The subsequent identification of two paralogs of PY235 in *P. vivax*, PvRBP1 and PvRBP2, that preferentially bound to reticulocytes over normocytes, suggesting a role in the reticulocyte preference of *P. vivax*, gave this gene family its name (Galinski et al., 1992). Unlike for the EBL family members, the overall sequence conservation between the RBL genes is relatively low, making it difficult to identify paralogs for other *Plasmodium* spp. in the absence of genomic sequences. This lack of conservation meant that the discovery of the five RBL family members in *P. falciparum*, PfRH1, PfRH2a, PfRH2b, PfRH4, and PfRH5, as well as two pseudogenes, PfRH3 and PfRH6, did not happen until the *P. falciparum* genome sequencing project was in progress (Dvorin et al., 2010; Hayton et al., 2008; Kaneko et al., 2002; Rayner et al., 2000, 2001; Taylor et al., 2001; Triglia et al., 2001). The structure of the canonical RBL is similar to that of the DBL proteins as most are type 1 transmembrane proteins, with large ectodomains and small cytoplasmic tails, though clear domains as seen for the EBL have not yet been identified. The exception is PfRH5, which is much smaller and lacks a C-terminal transmembrane domain and a cytoplasmic tail (Hayton et al., 2008).

Antibodies raised against different members of the superfamily show that these proteins appear to be localized in secretory organelles at the apical ends of the invasive merozoite, primed and ready for interactions with the host erythrocyte (Duraisingh et al., 2003b; Galinski et al., 1992; Stubbs et al., 2005). Direct binding of different RBLs to erythrocytes has been demonstrated in erythrocyte binding studies (Gao et al., 2008; Gaur et al., 2007; Tham et al., 2009; Triglia et al., 2011). Interestingly, in *P. yoelii*, it has been shown that PY235, in addition to an erythrocyte binding domain, contains an ATP/ADP binding domain that may have a role in nucleotide sensing (Gruber et al., 2010; Ramalingam et al., 2008).

Expansion and Genomic Diversification of RBLs and DBLs in *Plasmodium* spp.

One interesting insight gained from the extensive sequence information of different *Plasmodium* spp. is the fact that despite significant variation in their overall copy number and diversity, there is at least one recognizable member of the DBL and RBL family in every species (Fig. 19.1c). In the case of the DBL family, *P. falciparum* appears to have the highest number (five) of paralogous genes (Gardner et al., 2002), while in *P. knowlesi*, three different DBLs are found (Pain et al., 2008). In contrast, *P. vivax* appears to only possess a single DBL (Carlton et al., 2008) similar to the rodent parasites *P. yoelii*, *Plasmodium chabaudi*, and *Plasmodium berghei* (Carlton et al., 2001; Hall et al., 2005). One interesting feature highlighted by the comparison between *P. falciparum* and the closely related parasite *Plasmodium reichenowi* is the fact that while EBA-165 is a pseudogene in the human parasite, it is maintained as a functional copy in the simian parasite, suggesting different evolutionary constraints on invasion for these two parasite species (Rayner et al., 2004a).

Similar to the EBL, the overall number of RBL paralogs varies significantly across different *Plasmodium* spp. In *P. yoelii*, the genome sequencing project has identified at least 14 paralogous genes (Carlton et al., 2002), while in *P. berghei* and in *P. chabaudi*, the approximate numbers are 8 and 4, respectively. All members of the *P. falciparum* RBL have a corresponding copy in the *P. reichenowi* genome (Rayner et al., 2004b). Similar to EBA-165, it is interesting to note that the ortholog of PfRH1 in *P. reichenowi* appears to be a pseudogene, with multiple frameshifts, while the ortholog of the *P. falciparum* PfRH3 pseudogene in *P. reichenowi* is functionally intact. In *P. vivax*, the overall number of RBL could be as high as 10, though due to frameshifts and incomplete reading frames, the total number of complete and functional genes is most likely much lower (Carlton et al., 2008). The genome sequence of *P. knowlesi* has so far only identified two complete RBL genes with additional partial sequences also being present (Pain et al., 2008).

One conclusion that can be drawn from the large sequence diversity observed for the RBL compared to the EBL proteins in the different species is that the RBL family is evolving more rapidly than the EBL family. Why these two different erythrocyte binding protein families might be under different evolutionary pressure remains to be established.

IDENTIFICATION OF MULTIPLE ERYTHROCYTE RECEPTORS FOR INVASION

Several approaches have been used to identify erythrocyte receptors for the high affinity DBL and RBL parasite ligands. These have included the use of erythrocyte binding assays, together with naturally occurring mutant erythrocytes and enzyme-treated erythrocytes, protein overlays, biochemical studies with parasite ligands, and parasite invasion assays.

P. vivax DBP Receptor

Historically, the only way to determine the specific effect of erythrocyte receptors on parasite invasion was to obtain erythrocytes from individuals with mutations or lacking putative receptors. This approach has been used to confirm the major role of the PvDBP–Duffy antigen interaction by demonstrating the inability of *P. knowlesi* to invade Duffy-negative human erythrocytes (Horuk et al., 1994). The absence of the Duffy receptor on host erythrocytes prevents invasion of *P. vivax*, identifying this protein as one of the first receptors important for invasion by malaria parasites. In further analyses, Duffy negativity was mapped to a single point mutation at nucleotide -46 of the Duffy gene (Tournamille et al., 1995). Following cloning and expression of the gene, it was evident that the Duffy receptor is also a receptor for chemokines, and hence is also called Duffy antigen receptor for chemokines (DARC) (Horuk et al., 1996). Several biochemical studies have now identified the N-terminal region of DARC as the binding site for PvDBP, with binding being sialic acid independent and dependent on tyrosine sulfation (Choe et al., 2005).

P. falciparum EBA Receptors

Erythrocyte mutants for most of the known erythrocyte receptors for *P. falciparum* have been identified. However, some of these are very rare and interpretation of results can be complicated as it is unknown whether there may be additional coinherited polymorphisms. Invasion of some, but not all, *P. falciparum* strains into glycophorin A null cells is reduced, providing the first evidence that different *P. falciparum* strains utilize alternative ligand–receptor interactions for invasion (Mitchell et al., 1986). In contrast, invasion of *P. falciparum* into glycophorin B null and glycophorin C null erythrocytes is not significantly reduced (Soubes et al., 1999). Recently, the ability to genetically manipulate RBCs *in vitro* has provided an additional tool to assess the contribution of individual receptors and allows for the production of erythrocyte receptor mutations that do not naturally occur in populations (Bei et al., 2010).

In contrast to single DBL erythrocyte binding domain of *P. vivax*, the members of the *P. falciparum* EBA proteins contain tandem DBL domains. Erythrocyte binding studies suggest that all of the members of the PfEBA family bind to erythrocytes in a sialic acid-dependent fashion (Gilberger et al., 2003; Peterson and Wellems, 2000; Sim et al., 1990; Thompson et al., 2001). Erythrocytes treated with neuraminidase to remove terminal sialic acid residues reduce *P. falciparum* invasion, as well as reduce binding of PfEBA proteins. It has been demonstrated using erythrocyte mutants that glycophorin A, glycophorin B, and glycophorin C are the cognate receptors for the parasite ligands EBA-175, EBL-1, and EBA-140 (Maier et al., 2003; Mayer et al., 2009; Sim et al., 1994). EBA-181 is the only remaining sialic acid-dependent PfEBA for which no receptor has been identified to date; however, it appears that it is a chymotrypsin-sensitive receptor. Recently, crystallographic structures have been obtained for the EBA-175, with sialyl-lactose as a receptor, revealing precise details of receptor binding (Tolia et al., 2005).

P. falciparum RBL Receptors

In contrast to the PfEBA proteins, only one receptor has been identified so far for the PfRBL proteins, which also bind erythrocytes. This is in large part due to the fact that the PfRBL proteins are larger and precise binding domains have not been defined. The complement receptor 1 (CR1) has been identified as a receptor for *P. falciparum* using erythrocytes expressing different amounts of CR1 (Spadafora et al., 2010), and binding of specific PfRH4 domains to CR1 has been demonstrated (Tham et al., 2010). Differing levels of CR1 on the surface of erythrocytes were shown to correlate with the level of erythrocyte invasion. Although receptors have not been identified for the other PfRBLs, evidence suggests that the PfRH1 protein binds in a sialic acid-dependent fashion, while PfRH2a/PfRH2b and RH5 (and PfRH4) bind in a sialic acid-independent fashion (Duraisingh et al., 2002; Gunalan et al., 2011; Hayton et al., 2008; Sahar et al., 2011; Triglia et al., 2011).

Rodent Malaria Receptors

Relatively little is known about the mouse erythrocyte receptors used by the rodent malaria parasites during the invasion process. Binding to both reticulocytes as well as normocytes has been shown for the Py235 protein family, and enzymatic treatment of the erythrocytes with either chymoptrypsin or trypsin significantly disrupts binding, while treatment with neuraminidase has no effect (Ogun and Holder, 1996; Ogun et al., 2000). The only mouse receptor that has so far been directly linked to merozoite invasion is the murine DARC, which appears to be important for normocyte invasion but not reticulocyte invasion (Swardson-Olver et al., 2002) and is required for the lethal effect seen during an infection with the *P. yoelii* 17XL strain (Akimitsu et al., 2004) (Fig. 19.2a). Studies using the DBL domain of the *P. yoelii* EBP shows that it binds to a receptor on the mouse erythrocyte that has the same sensitivity to neuraminidase, chymotrypsin, and trypsin as would be expected of the murine DARC (Prasad et al., 2003).

DIVERSITY IN SEQUENCE AND EXPRESSION OF *PLASMODIUM* INVASION LIGANDS DEFINES INVASION PATHWAYS

Parasite antigens, such as MSP-1 and AMA-1, are being developed as vaccine antigens, and both the DBL and RBL are also considered as putative vaccine candidates. It has become clear from both serological studies and through analyses of their DNA sequence that there are extensive sequence polymorphisms in MSP-1 and AMA-1 between parasite strains. Although less well studied, the genes encoding the DBL and RBL proteins are also polymorphic, particularly in the predicted erythrocyte binding domains. This diversity may not only impact on their immunogenicity but may also alter the receptor specificity as has been suggested for both EBA-140 and EBA-181 (Mayer et al., 2001, 2004).

An additional layer of complexity is added by the variation in expression levels of members of the DBL and RBL families in both laboratory and field isolates (Bei et al., 2007; Duraisingh et al., 2003b; Gomez-

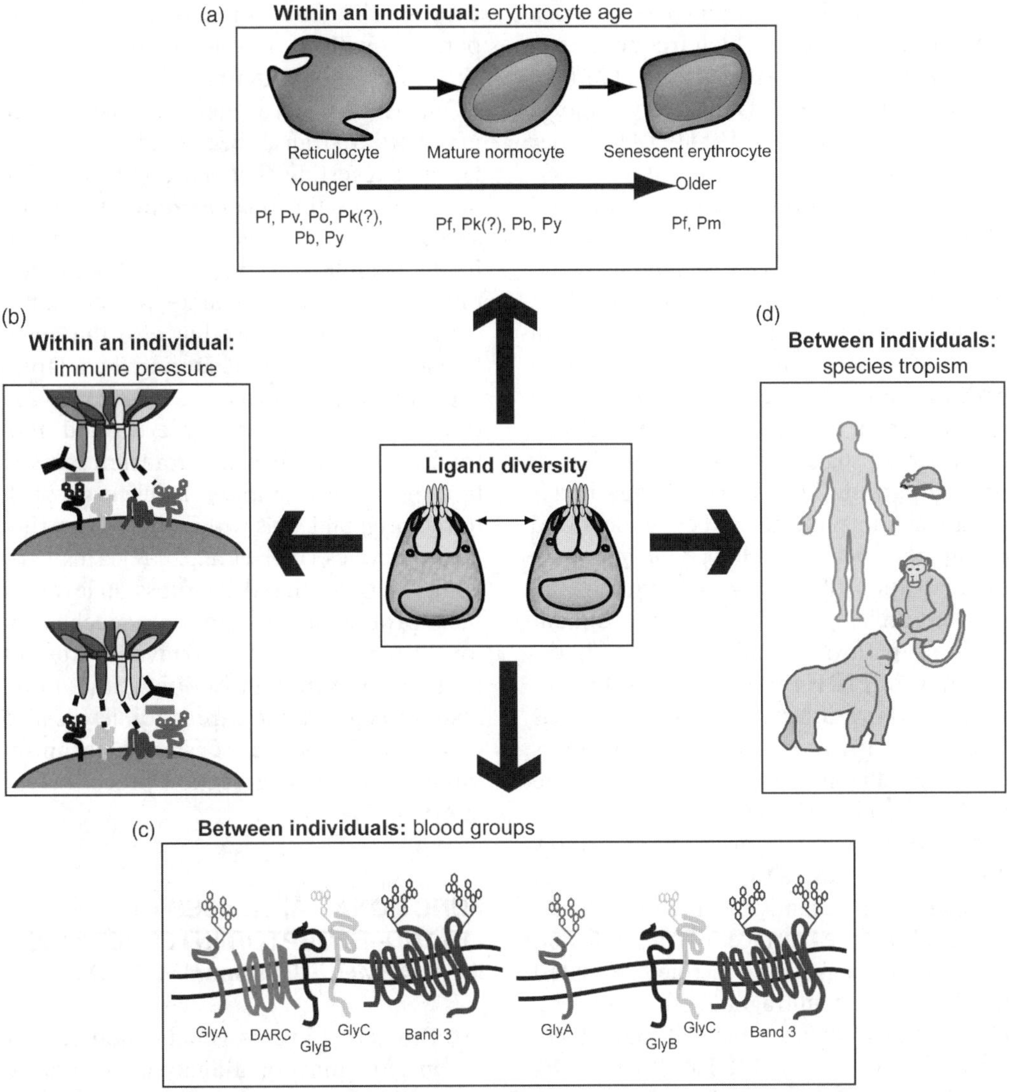

Figure 19.2 Parasite ligand diversity allows the *Plasmodium* merozoites to invade erythrocytes of different stages of maturity (a), in the presence of immune pressure (b), with different blood group polymorphisms (c), and from different species (d). Shown schematically at the center of the figure, the two merozoites display different sets of invasion ligands, thus allowing these parasites to invade under variant conditions. Human and animal drawings courtesy of Dr. Amy Bei. See color insert.

Escobar et al., 2010; Taylor et al., 2002). The expression of different sets of parasite ligands in isolates would result in the use of different erythrocyte receptors or sets of receptors, hence defining different "invasion pathways." Mirroring this variant expression of parasite ligands is the observed diversity of erythrocyte receptors. It is striking that all of the known *Plasmodium* erythrocyte receptors define polymorphic blood group proteins. Further, genetic mutations, including null mutants, exist for all of these proteins, some which have risen to relatively high frequencies in malaria-endemic areas, such as the Gerbich (glycophorin C) and Dantu (glycophorin B) polymorphisms (Field et al., 1994; Maier et al., 2003).

Consistent with the above-mentioned idea that RBLs and DBLs provide the parasite with the capacity to adapt to variations in host cell receptors or provide a mechanism of immune evasion, it has been known for some time that the invasion pathway used by certain *P. falciparum* parasites can be clonally variant (Dolan et al., 1990), suggesting that expression of specific parasite ligands can change under certain circumstances. The best understood example of such a switch in *P. falciparum* is the activation of PfRH4 that leads to the parasite utilizing a sialic acid-independent invasion pathway (Stubbs et al., 2005). Deletion of PfRH4 in these parasites prevents this invasion pathway switch. It has further been demonstrated that members of other schizont-stage gene families, the PfEBAs and PfCLAGs, also undergo variant expression (Cortes et al., 2007). The latter family, although expressed in schizonts, has recently been shown to be exported to the surface of the infected erythrocyte, suggesting a role unrelated to invasion (Nguitragool et al., 2011). Detailed analysis has revealed that silencing and activation of PfRH4, PfEBA-140, and the PfClag3.1 and PfClag3.2 genes are associated with specific histone marks that define regions of facultative chromatin, as well as changes in nucleosome occupancy (Comeaux et al., 2011; Cortes et al., 2007; Jiang et al., 2010). This association suggests that variant expression of these invasion ligands is regulated epigenetically.

The rodent malaria parasite *P. yoelii* has provided some insights on how multigene families are regulated *in vivo*. Studies of the *py235* gene family has shown that the merozoites derived from a single schizont express different members, providing the parasite with a novel immune evasion as well as adaptation mechanism (Preiser et al., 1999). Interestingly, *in vivo*, not all members are transcribed equally in a population, suggesting that there is a selection mechanism acting on merozoites with more successful merozoites becoming dominant in the population (Iyer et al., 2007). Not only the specific variant of *py235* that is expressed is important but also the absolute amount of protein. During the adaptation of *P. yoelii* from a mouse to a rat, *py235* transcription levels increase around 100-fold; if the parasite is then transferred back to a mouse, transcription levels drop back to the original levels within one to two replication cycles (Iyer et al., 2007). This variability suggests that the expression levels of these proteins are an important mechanism for the parasite to adjust to invade different host erythrocytes; in addition, the rapid ability to change the expression levels also indicates an active sensing mechanism rather than a slow selection process (Iyer et al., 2007).

FUNCTIONAL ANALYSES OF LIGAND–RECEPTOR INTERACTIONS IN *PLASMODIUM* spp. *IN VITRO*

Over the last 15 years, genetic methods for the *in vitro* functional analysis of parasite genes have been developed in *P. falciparum* and for the rodent malarias *in vivo*. This includes stable expression plasmids, as well as integration of targeting plasmids by homologous recombination, which allows for the epitope tagging and targeted gene disruptions. Several different strategies have been employed to determine the functional role of parasite ligands in *P. falciparum* including the generation of genetic knockouts of parasite ligands that reflect the natural variation in expression, the generation of parasites expressing essential gene chimaeras, and forward genetic studies to identify parasite genes that confer different invasion phenotypes.

Identification of Redundancy in Ligand–Receptor Interactions in *P. falciparum*

Once reverse genetic methodology was established, it was possible to study in more detail the role of individual DBL and RBL in defining different invasion pathways. The first invasion gene to be disrupted was PfEBA-175, and while no major growth phenotype was observed, a striking switch in invasion from sialic acid-dependent to independent invasion was observed (Duraisingh et al., 2003a; Reed et al., 2000). This alteration was later found to be associated with activation of the PfRH4 gene by microarray analysis of the switched and unswitched lines (Stubbs et al., 2005).

The subsequent disruption of the other EBA genes clearly demonstrated the redundancy of these proteins in *P. falciparum* for invasion (Gilberger et al., 2003; Maier et al., 2003). Furthermore, it also demonstrated a hierarchy in the importance of each of the ligand–receptor interactions (Baum et al., 2005), with the EBA-175/glycophorin A interaction being required for a major sialic acid-dependent invasion pathway. The disruption of the other EBA proteins did not have significant affects on either parasite multiplication rate or invasion pathway usage (Gilberger et al., 2003; Maier et al., 2003). Recently, it has been shown using a combination of genetic mutants and antibody inhibition experiments that loss of certain invasion ligands is accompanied by an increase in sensitivity to antibody inhibition of other invasion ligands, suggesting that the loss of function of one invasion ligand is compensated for by the increased utilization of another (Baum et al., 2005; Lopaticki et al., 2011).

A similar approach has been used to functionally analyze the importance of members of the PfRBL family. Knockouts of all the different members of this family have been obtained (Duraisingh et al., 2002, 2003b; Stubbs et al., 2005; Triglia et al., 2005), except for PfRH5, which appears to be essential (Hayton et al., 2008). As mentioned above, these studies showed that PfRH4 is essential for using sialic acid-independent receptors. Although the disruption of PfRH1 has been shown to correlate with loss of use of a sialic acid-dependent receptor (Triglia et al., 2005), the loss of PfRH2b is associated with the inability to use a major chymotrypsin-sensitive and sialic acid-independent receptor (Duraisingh et al., 2003b). Taken together, it appears that the PfRBL proteins play a central role in defining the erythrocyte receptors utilized during invasion.

Variant Expression for Immune Evasion

Knockout parasites have also provided insights on the role of variant expression of parasite ligands on the ability of parasites to evade humoral immune responses (Fig. 19.2b). As expected, knockout of PfEBA-175 was associated with an inability of specific sera raised in animals against PfEBA-175, or generated in individuals in endemic areas, to inhibit parasite invasion (Persson et al., 2008). Similar results were also obtained for PfRBL (Baum et al., 2005; Gao et al., 2008; Triglia et al., 2011). In a second approach, chimeric parasites have been generated where specific domains thought to be important targets of immunity have been replaced by domains that are distinct in sequence and hence immunogenicity, while maintaining their function. Such swaps have been done for both PfMSP1-19 and PfAMA-1 (Healer et al., 2004; O'Donnell et al., 2001).

These chimeric parasites demonstrate that sequence diversity within parasite ligands allows evasion of inhibitory antibodies.

Inhibition of Ligand–Receptor Interactions

New insights on specific parasite ligand–receptor interactions have also come from biochemical studies that either use recombinantly expressed receptor ectodomains or specific antibodies raised against a particular erythrocyte receptor to interrupt specific interactions. The ability of these reagents to inhibit parasite invasion *in vitro* has been determined for many DBL and RBL proteins (Fowkes et al., 2010). Due perhaps in part to the redundancy inherent in the ligand–receptor involved in erythrocyte invasion, the level of inhibition by individual reagents has not been very high.

Domains That Define Specificity

The well-conserved erythrocyte binding domains of the DBLs consist of multiple cysteine-rich domains that are well conserved between homologs within a species as well as between species, and it appears that each DBL domain of an invasion protein binds to a different erythrocyte receptor. Antibodies affinity purified against the DBL domains of PvDBP can inhibit invasion of *P. vivax* (Grimberg et al., 2007). In contrast, the erythrocyte binding domains of the RBL are not easily predictable based on the amino acid sequence of the protein and are therefore much less well studied, though binding domains are now slowly being defined, and antibodies against these regions can also inhibit invasion (Gao et al., 2008) (Triglia et al., 2011).

The current model of alternative invasion pathway utilization postulates that members of the DBL or RBL families bind different erythrocyte surface receptors via their diversified binding domains. Nevertheless, it is becoming clear that other parts of the molecule are important for function. For instance, it has been shown using a reverse genetic approach that the invasion pathway utilization of the PfRH2 proteins is dependent on the presence of either a PfRH2a or PfRH2b tail (Dvorin et al., 2010).

Functional Analyses of Essential Genes

The inability to disrupt specific PfEBA and PfRH proteins in certain genetic backgrounds (Duraisingh et al., 2003b; Gilberger et al., 2003) certainly suggests that the importance of an invasion ligand depends on the complete and specific set of ligand–receptor interactions in a given parasite strain. Interestingly, it has not been possible to disrupt the PfRH5 gene in any parasite line in which it was attempted, suggesting an essential role for this adhesin (Hayton et al., 2008). The refractoriness of PfAMA-1 and PfMSP-1 to genetic disruption suggests they are also essential. Due to the lack of a robust conditional expression system for *P. falciparum*, it has been difficult to study the functions of these proteins using a genetic approach.

To overcome this constraint, one alternative method that has been successfully used is the generation of chimeric parasites. In one study investigating the role of antibodies targeting the C-terminal domain of PfMSP-1 in protective immunity, this region was exchanged for that of *P. chabaudi* (O'Donnell et al., 2001). Even though the sequence of the two tails was very different, the role of the protein in erythrocyte invasion was conserved. Alternatively, in a very elegant study, a version of PfAMA-1 that conferred resistance to R1, a small invasion-blocking peptide (Li et al., 2002), was ectopically expressed in parasites that also expressed an endogenous R1-sensitive allele of PfAMA-1 (Treeck et al., 2009). In these chimeric parasites, utilization of the ectopically expressed PfAMA-1 could be forced via inhibition of the endogenous allele. This system thereby allowed the specific dissection of elements within PfAMA-1 that were required for function by site-directed mutagenesis.

Genetic Studies in *P. knowlesi*

The primate malaria parasite *P. knowlesi* demonstrates exquisite cell biology, close species conservation to *P. vivax*, a completely sequenced and annotated genome, and has been used in many of the early invasion studies. Uniquely, this parasite can be grown both *in vitro* and *in vivo*. *P. knowlesi* has also been developed into a powerful genetic system. Knockout of the PkDBPalpha gene has identified it as an essential parasite ligand for invasion into human erythrocytes (Singh et al., 2006). Interestingly, attempted invasion of PkDBPalpha-knockout parasites into Duffy-positive cells has demonstrated that these parasites can attach and apically orientate on these erythrocytes but that no tight junction is formed between the merozoite and the erythrocyte, implicating the DBL–receptor interaction in tight junction formation (Singh et al., 2005a).

Genetic Studies in *P. yoelii*

This parasite is an ideal model to investigate the relationship between invasion ligand expression and parasite virulence. Unlike in *P. falciparum*, this rodent parasite only has a single member of the DBL family, which, to date, has been impervious to genetic disruption, providing strong support to the notion that at least a single member of this protein family is essential for parasite survival.

BIOLOGICAL CONSEQUENCES OF DIVERSITY IN LIGAND–RECEPTOR INTERACTIONS

The identification of multiple and variable ligand interactions *in vitro*, however, immediately raises the question of their relevance *in vivo*. This question is particularly difficult to study in *Plasmodium* spp. due to the lack of an accessible and relevant animal model of the human malaria parasites. This limitation highlights the importance of both *ex vivo* human studies in endemic populations and *in vivo* rodent malaria studies.

Functional Correlates of Immunity

Clearly, one of the most important *in vivo* functions of variant expression and sequence polymorphisms in invasion is immune evasion. Seroepidemiological studies have demonstrated that specific domains are recognized in an allele-specific fashion by different sera. Antibodies to some antigens have been found to be acquired in an age-dependent fashion (Akpogheneta et al., 2008). However, the ultimate goal of such studies has been to identify antibodies that correlate with protection against severe malaria or clinical malaria infection. To identify functional correlates of immunity, it has been necessary to develop robust assays that can quantify the inhibition of invasion in the presence of specific antibodies, something that, to date has been difficult to successfully attain.

Receptor Selection in Populations

It has long been known that erythrocyte genetic polymorphisms can affect malaria infection. From the early observations that sickle cell trait can protect against *P. falciparum* malaria (Allison, 1954), it has become clear that many erythrocyte polymorphisms have most likely been selected for by *Plasmodium* spp. The most striking of these is Duffy negativity that protects against *P. vivax* infections. Many erythrocyte receptor mutants have been used to define the receptors themselves as they reduce the efficiency of invasion. Some mutations that are thought to have been selected for by *P. falciparum* in different populations are very rare, like glycophorin A nulls, while others are quite common, such as glycophorin B null and glycophorin C deletion mutants, which are thought to have been selected for by *P. falciparum* in different populations (Field et al., 1994) (Fig. 19.2c).

It has further been postulated that invasion ligand polymorphism is ancient and that it might play a role in speciation (Fig. 19.2d). The PfEBA and the PfRH genes have evolved dramatically between *P. falciparum* and *P. reichenowi*, the closely related chimpanzee malaria parasite. Comparison of sequences of invasion ligands has shown considerable sequence polymorphism, and as described above, in some cases, genes that are pseudogenes in one are uninterrupted genes in the other and vice versa (Rayner et al., 2004a,b). Interestingly, the CMAH gene active in chimpanzees is deleted in humans and is responsible for modifying sialic acid from Neu5GAC to Neu5GC (Chou et al., 2002). The EBA proteins of *P. falciparum* and the chimpanzee *P. reichenowi* preferentially recognize Neu5Ac and Neu5GC, which might be the molecular basis for the tropism of these parasites (Martin et al., 2005).

IN VIVO PATHOLOGICAL CONSEQUENCES OF DIVERSITY IN LIGAND–RECEPTOR INTERACTIONS

Little is known about the correlation between invasion properties and virulence in the human host. The more efficient a parasite is in invading erythrocytes, the more likely it is to proliferate rapidly and therefore to cause more severe diseases (Chotivanich et al., 2000). Similarly, a parasite that is restricted to only a subset of circulating erythrocytes, like *P. vivax*, automatically will reach a self-limiting level of infection, thereby reducing the risk of severe disease. How receptors and invasion pathways utilized during invasion relates to human pathology is not known. What is common in all the studies on the relationship between invasion and human pathology is that experimental validation is not possible, and importantly, a clear cause and effect at the molecular level is difficult to establish.

Rodent malarias, and in particular *P. yoelii*, provide powerful models to study the relationship between invasion properties and pathogenesis. Historically, there was a substantial body of evidence that indicated that the Py235 rhoptry protein multigene family played a major role in host cell selection and parasite virulence, although it was less clear on how this protein family was mediating this effect. In these studies, the question on whether actual differences in the amino acid sequence of different Py235 members found in virulent and avirulent strains of *P. yoelii* were the main factor or whether other mechanisms needed to be considered remained unresolved. There was some evidence that there may indeed be some differences in *py235* copy number and sequence, though it was not clear whether these could be directly linked to virulence (Iyer et al., 2006; Preiser and Jarra, 1998). Studies focusing on the expression of distinct *py235* genes in virulent and avirulent parasite clones provided new insights into the underlying mechanisms of how these proteins mediate virulence (Iyer et al., 2007). Further, using a genetic knockdown system, this work clearly showed that the overall expression level of *py235* directly impacts on the host cell repertoire that the parasite is able to invade, thereby directly impacting on virulence. In addition, direct genetic knockouts of a dominantly expressed *py235* demonstrated that other members of this gene family are able to compensate for the loss of this gene and that the impact on virulence appears to be influenced by which other *py235* is dominantly expressed in the new parasite population (Bapat et al., 2011; Ogun et al., 2011).

Recently, an elegant study focusing on the genetic cross of virulent and avirulent clones of *P. yoelii* identified a single chromosomal region containing *pyebl* as the virulence-containing locus (Pattaradilokrat et al., 2009). This finding is in line with initial genetic studies done more than several decades earlier that suggested that a single locus is responsible for the differences in virulence observed in *P. yoelii* (Walliker et al., 1976). The importance of PyEBL in virulence was then shown to be due to a single amino acid mutation in the C-terminal region of the protein, resulting in aberrant trafficking of PyEBL (Otsuki et al., 2009). Although the mislocation of PyEBL was essential for parasite virulence, it was not sufficient to completely convert an avirulent parasite into a virulent one, consistent with a role of Py235 as well.

OTHER INVASION LIGANDS INVOLVED IN ALTERNATIVE INVASION PATHWAY UTILIZATION?

The availability of the genome sequence of different *Plasmodium* spp., in combination with a range of functional genomics and proteomic approaches, has significantly increased the number of parasite proteins proposed to be involved in invasion. In terms of recognition and binding to specific receptors on the surface of the erythrocyte, these proteins can be broadly divided into MSPs, rhoptry proteins, and microneme proteins (Cowman and Crabb, 2006). Although the microneme protein family EBL as well as the rhoptry RBL family have already been discussed above, it is likely that other invasion proteins may be involved in binding to alternative receptors. Here we present an overview of other merozoite proteins where evidence indicates that they may play a direct role in receptor recognition/binding on the erythrocyte, and thus play a role in invasion.

The microneme protein AMA-1 appears to be processed and translocated to the circumference of the merozoite following schizont rupture (Narum and Thomas, 1994). Domain III of AMA-1 has recently been shown to bind to the erythrocyte membrane protein Kx, and erythrocytes lacking this receptor are less efficiently invaded by *P. falciparum* (Kato et al., 2005). The rhoptry protein apical sushi protein (ASP) contains a sushi-like domain potentially involved in erythrocyte binding (O'Keeffe et al., 2005). Another rhoptry protein, the rhoptry-associated membrane antigen (RAMA) in *P. falciparum*, is a GPI-anchored protein thought to play a dual role in rhoptry organelle biogenesis as well as merozoite invasion. An initial 170-kDa form of RAMA that is part of the budding rhoptry organelle is subsequently processed during schizont maturation producing a 60-kDa GPI anchored form that binds to the erythrocyte surface (Topolska et al., 2004).

MSP-1 is the most abundant protein found on the surface of the merozoite and is thought to play an important role during the initial random attachment of the merozoite to the erythrocyte. This hypothesis is supported by recent findings that MSP-1 binds to band 3 on the red blood cell (Goel et al., 2003), possibly via a complex including MSP9 (Li et al., 2004). It is encoded by a single-copy gene and, although essential, exhibits great sequence diversity between parasite strains. Functional conservation of the C-terminal domains despite divergence in primary amino acid sequences between different *Plasmodium* spp. has also been demonstrated (O'Donnell et al., 2000).

In contrast to the single-copy genes described above, expansions of MSP3 and MSP7 have resulted in significant variation in the total number of MSP3 and MSP7 related proteins (MSRP) found in a particular parasite species. Members of both families have been shown to be targets of invasion inhibitory antibodies (Oeuvray et al., 1994; Singh et al., 2005b). Unlike MSP3, the transcriptional profiles of the different MSP7 variants in the asexual cycle of the blood-stage differ from each other. In the case of MSRP2, it appears that the protein, although expressed in late-stage parasites is not found on the merozoite surface (Kadekoppala et al., 2010), suggesting that the expansion of MSP7 has been associated with a concurrent functional diversification. Interestingly, genetic disruption of MSP7 in the rodent malaria parasite *P. berghei* resulted in reduced parasite growth *in vivo* with a stronger relative preference to invade reticulocytes (Tewari et al., 2005), suggesting some role in host cell selection during invasion.

CONCLUSIONS AND FUTURE QUESTIONS

Despite the significant expansion of our understanding of erythrocyte invasion by the malaria parasite over the last decade, there are many questions that remain. For example, although much progress has made on the roles of the RBLs and DBLs in *P. falciparum*, their role in other *Plasmodium* spp. remains to be elucidated. In addition, much focus has been on the *qualitative* nature of phenotypic diversity; it will also be of importance to determine the *quantitative* aspects of diversity. The ability to identify and functionally analyze essential genes continues to be a problem due to the lack of a robust conditional system for gene expression. It will be of great interest to identify how the different ligands interact with each other and to determine whether and when transient complexes between them might form. If so, what is the combinatorial nature of these interactions and, most importantly, what would be the minimal number of interactions that would need to be interrupted to prevent the invasion of all *Plasmodium* strains? Finally, we clearly need to begin to look at these processes *in vivo*. Many studies showing successful disruption of invasion *in vitro* have failed when tested *in vivo*. This lack of direct correlation has significantly hampered the development of a successful malaria vaccine. It will be very important to assess the importance of phenotypic diversity *in vivo*, particularly in relation to immunity and pathology.

REFERENCES

Adams JH, Hudson DE, Torii M, Ward GE, Wellems TE, Aikawa M, Miller LH. 1990. The Duffy receptor family of *Plasmodium knowlesi* is located within the micronemes of invasive malaria merozoites. Cell 63: 141–153.

Adams JH, Sim BK, Dolan SA, Fang X, Kaslow DC, Miller LH. 1992. A family of erythrocyte binding proteins of malaria parasites. Proc Natl Acad Sci U S A 89: 7085–7089.

Adams JH, Blair PL, Kaneko O, Peterson DS. 2001. An expanding ebl family of *Plasmodium falciparum*. Trends Parasitol 17: 297–299.

Akimitsu N, Kim HS, Hamamoto H, Kamura K, Fukuma N, Arimitsu N, Ono K, Wataya Y, Torii M, Sekimizu K. 2004. Duffy antigen is important for the lethal effect of the lethal strain of *Plasmodium yoelii* 17XL. Parasitol Res 93: 499–503.

Akpogheneta OJ, Duah NO, Tetteh KK, Dunyo S, Lanar DE, Pinder M, Conway DJ. 2008. Duration of naturally acquired antibody responses to blood-stage *Plasmodium falciparum* is age dependent and antigen specific. Infect Immun 76: 1748–1755.

Allison AC. 1954. Protection afforded by sickle-cell trait against subtertian malareal infection. Br Med J 1: 290–294.

Bapat D, Huang X, Gunalan K, Preiser PR. 2011. Changes in parasite virulence induced by the disruption of a single member of the 235 kDa rhoptry protein multigene family of *Plasmodium yoelii*. PLoS One 6: e20170.

Batchelor JD, Zahm JA, Tolia NH. 2011. Dimerization of *Plasmodium vivax* DBP is induced upon receptor binding and drives recognition of DARC. Nat Struct Mol Biol 18: 908–914.

Baum J, Maier AG, Good RT, Simpson KM, Cowman AF. 2005. Invasion by *P. falciparum* merozoites suggests a hierarchy of molecular interactions. PLoS Pathog 1: e37.

Bei AK, Membi CD, Rayner JC, Mubi M, Ngasala B, Sultan AA, Premji Z, Duraisingh MT. 2007. Variant merozoite protein expression is associated with erythrocyte invasion phenotypes in *Plasmodium falciparum* isolates from Tanzania. Mol Biochem Parasitol 153: 66–71.

Bei AK, Brugnara C, Duraisingh MT. 2010. In vitro genetic analysis of an erythrocyte determinant of malaria infection. J Infect Dis 202: 1722–1727.

Butcher GA, Mitchell GH, Cohen S. 1973. Letter: Mechanism of host specificity in malarial infection. Nature 244: 40–41.

Camus D, Hadley TJ. 1985. A *Plasmodium falciparum* antigen that binds to host erythrocytes and merozoites. Science 230: 553–556.

Carlton JM, Muller R, Yowell CA, Fluegge MR, Sturrock KA, Pritt JR, Vargas-Serrato E, Galinski MR, Barnwell JW, Mulder N, Kanapin A, Cawley SE, Hide WA, Dame JB. 2001. Profiling the malaria genome: A gene survey of three species of malaria parasite with comparison to other apicomplexan species. Mol Biochem Parasitol 118: 201–210.

Carlton JM, Angiuoli SV, Suh BB, Kooij TW, Pertea M, Silva JC, Ermolaeva MD, Allen JE, Selengut JD, Koo HL, Peterson JD, Pop M, Kosack DS, Shumway MF, Bidwell SL, Shallom SJ, van Aken SE, Riedmuller SB, Feldblyum TV, Cho JK, Quackenbush J, Sedegah M, Shoaibi A, Cummings LM, Florens L, Yates JR, Raine JD, Sinden RE, Harris MA, Cunningham DA, Preiser PR, Bergman LW, Vaidya AB, van Lin LH, Janse CJ, Waters AP, Smith HO, White OR, Salzberg SL, Venter JC, Fraser CM, Hoffman SL, Gardner MJ, Carucci DJ. 2002. Genome sequence and comparative analysis of the model rodent malaria parasite *Plasmodium yoelii yoelii*. Nature 419: 512–519.

Carlton JM, Adams JH, Silva JC, Bidwell SL, Lorenzi H, Caler E, Crabtree J, Angiuoli SV, Merino EF, Amedeo P, Cheng Q, Coulson RM, Crabb BS, Del Portillo HA, Essien K, Feldblyum TV, Fernandez-Becerra C, Gilson PR, Gueye AH, Guo X, Kang'a S, Kooij TW, Korsinczky M, Meyer EV, Nene V, Paulsen I, White O, Ralph SA, Ren Q, Sargeant TJ, Salzberg SL, Stoeckert CJ, Sullivan SA, Yamamoto MM, Hoffman SL, Wortman JR, Gardner MJ, Galinski MR, Barnwell JW, Fraser-Liggett CM. 2008. Comparative genomics of the neglected human malaria parasite *Plasmodium vivax*. Nature 455: 757–763.

Chitnis CE, Miller LH. 1994. Identification of the erythrocyte binding domains of *Plasmodium vivax* and *Plasmodium knowlesi* proteins involved in erythrocyte invasion. J Exp Med 180: 497–506.

Choe H, Moore MJ, Owens CM, Wright PL, Vasilieva N, Li W, Singh AP, Shakri R, Chitnis CE, Farzan M. 2005. Sulphated tyrosines mediate association of chemokines and *Plasmodium vivax* Duffy binding protein with the Duffy antigen/receptor for chemokines (DARC). Mol Microbiol 55: 1413–1422.

Chotivanich K, Udomsangpetch R, Simpson JA, Newton P, Pukrittayakamee S, Looareesuwan S, White NJ. 2000. Parasite multiplication potential and the severity of *falciparum* malaria. J Infect Dis 181: 1206–1209.

Chou HH, Hayakawa T, Diaz S, Krings M, Indriati E, Leakey M, Paabo S, Satta Y, Takahata N, Varki A. 2002. Inactivation of CMP-N-acetylneuraminic acid hydroxylase occurred prior to brain expansion during human evolution. Proc Natl Acad Sci U S A 99: 11736–11741.

Comeaux CA, Coleman BI, Bei AK, Whitehurst N, Duraisingh MT. 2011. Functional analysis of epigenetic regulation of tandem RhopH1/clag genes reveals a role in *Plasmodium falciparum* growth. Mol Microbiol 80: 378–390.

Cortes A, Carret C, Kaneko O, Yim Lim BY, Ivens A, Holder AA. 2007. Epigenetic silencing of *Plasmodium falciparum* genes linked to erythrocyte invasion. PLoS Pathog 3: e107.

Cowman AF, Crabb BS. 2006. Invasion of red blood cells by malaria parasites. Cell 124: 755–766.

Dolan SA, Miller LH, Wellems TE. 1990. Evidence for a switching mechanism in the invasion of erythrocytes by *Plasmodium falciparum*. J Clin Invest 86: 618–624.

Duraisingh MT, Triglia T, Cowman AF. 2002. Negative selection of *Plasmodium falciparum* reveals targeted gene deletion by double crossover recombination. Int J Parasitol 32: 81–89.

Duraisingh MT, Maier AG, Triglia T, Cowman AF. 2003a. Erythrocyte-binding antigen 175 mediates invasion in *Plasmodium falciparum* utilizing sialic acid-dependent and -independent pathways. Proc Natl Acad Sci U S A 100: 4796–4801.

Duraisingh MT, Triglia T, Ralph SA, Rayner JC, Barnwell JW, McFadden GI, Cowman AF. 2003b. Phenotypic variation of *Plasmodium falciparum* merozoite proteins directs receptor targeting for invasion of human erythrocytes. EMBO J 22: 1047–1057.

Dvorin JD, Bei AK, Coleman BI, Duraisingh MT. 2010. Functional diversification between two related *Plasmodium falciparum* merozoite invasion ligands is determined by changes in the cytoplasmic domain. Mol Microbiol 75: 990–1006.

Fang XD, Kaslow DC, Adams JH, Miller LH. 1991. Cloning of the *Plasmodium vivax* Duffy receptor. Mol Biochem Parasitol 44: 125–132.

Field SP, Hempelmann E, Mendelow BV, Fleming AF. 1994. Glycophorin variants and *Plasmodium falciparum*: Protective effect of the Dantu phenotype in vitro. Hum Genet 93: 148–150.

Fowkes FJ, Richards JS, Simpson JA, Beeson JG. 2010. The relationship between anti-merozoite antibodies and incidence of *Plasmodium falciparum* malaria: A systematic review and meta-analysis. PLoS Med 7: e1000218.

Galinski MR, Medina CC, Ingravallo P, Barnwell JW. 1992. A reticulocyte-binding protein complex of *Plasmodium vivax* merozoites. Cell 69: 1213–1226.

Gao X, Yeo KP, Aw SS, Kuss C, Iyer JK, Genesan S, Rajamanonmani R, Lescar J, Bozdech Z, Preiser PR. 2008. Antibodies targeting the PfRH1 binding domain inhibit invasion of *Plasmodium falciparum* merozoites. PLoS Pathog 4: e1000104.

Gardner MJ, Hall N, Fung E, White O, Berriman M, Hyman RW, Carlton JM, Pain A, Nelson KE, Bowman S, Paulsen IT, James K, Eisen JA, Rutherford K, Salzberg SL, Craig A, Kyes S, Chan MS, Nene V, Shallom SJ, Suh B, Peterson J, Angiuoli S, Pertea M, Allen J, Selengut J, Haft D, Mather MW, Vaidya AB, Martin DM, Fairlamb AH, Fraunholz MJ, Roos DS, Ralph SA, McFadden GI, Cummings LM, Subramanian GM, Mungall C, Venter JC, Carucci DJ, Hoffman SL, Newbold C, Davis RW, Fraser CM, Barrell B. 2002. Genome sequence of the human malaria parasite *Plasmodium falciparum*. Nature 419: 498–511.

Gaur D, Singh S, Jiang L, Diouf A, Miller LH. 2007. Recombinant *Plasmodium falciparum* reticulocyte homology protein 4 binds to erythrocytes and blocks invasion. Proc Natl Acad Sci U S A 104: 17789–17794.

Gilberger TW, Thompson JK, Triglia T, Good RT, Duraisingh MT, Cowman AF. 2003. A novel erythrocyte binding antigen-175 paralogue from *Plasmodium falciparum* defines a new trypsin-resistant receptor on human erythrocytes. J Biol Chem 278: 14480–14486.

Goel VK, Li X, Chen H, Liu SC, Chishti AH, Oh SS. 2003. Band 3 is a host receptor binding merozoite surface protein 1 during the *Plasmodium falciparum* invasion of erythrocytes. Proc Natl Acad Sci U S A 100: 5164–5169.

Gomez-Escobar N, Amambua-Ngwa A, Walther M, Okebe J, Ebonyi A, Conway DJ. 2010. Erythrocyte invasion and merozoite ligand gene expression in severe and mild *Plasmodium falciparum* malaria. J Infect Dis 201: 444–452.

Grimberg BT, Udomsangpetch R, Xainli J, McHenry A, Panichakul T, Sattabongkot J, Cui L, Bockarie M, Chitnis C, Adams J, Zimmerman PA, King CL. 2007. *Plasmodium vivax* invasion of human erythrocytes inhibited by antibodies directed against the Duffy binding protein. PLoS Med 4: e337.

Gruber A, Manimekalai MS, Balakrishna AM, Hunke C, Jeyakanthan J, Preiser PR, Gruber G. 2010. Structural determination of functional units of the nucleotide binding domain (NBD94) of the reticulocyte binding protein

Py235 of *Plasmodium yoelii*. PLoS One 5: e9146.

Gunalan K, Gao X, Liew KJ, Preiser PR. 2011. Differences in erythrocyte receptor specificity of different parts of the *Plasmodium falciparum* reticulocyte binding protein homologue 2a. Infect Immun 79: 3421–3430.

Hall N, Karras M, Raine JD, Carlton JM, Kooij TW, Berriman M, Florens L, Janssen CS, Pain A, Christophides GK, James K, Rutherford K, Harris B, Harris D, Churcher C, Quail MA, Ormond D, Doggett J, Trueman HE, Mendoza J, Bidwell SL, Rajandream MA, Carucci DJ, Yates JR 3rd, Kafatos FC, Janse CJ, Barrell B, Turner CM, Waters AP, Sinden RE. 2005. A comprehensive survey of the *Plasmodium* life cycle by genomic, transcriptomic, and proteomic analyses. Science 307: 82–86.

Hayton K, Gaur D, Liu A, Takahashi J, Henschen B, Singh S, Lambert L, Furuya T, Bouttenot R, Doll M, Nawaz F, Mu J, Jiang L, Miller LH, Wellems TE. 2008. Erythrocyte binding protein PfRH5 polymorphisms determine species-specific pathways of *Plasmodium falciparum* invasion. Cell Host Microbe 4: 40–51.

Healer J, Murphy V, Hodder AN, Masciantonio R, Gemmill AW, Anders RF, Cowman AF, Batchelor A. 2004. Allelic polymorphisms in apical membrane antigen-1 are responsible for evasion of antibody-mediated inhibition in *Plasmodium falciparum*. Mol Microbiol 52: 159–168.

Holder AA, Freeman RR. 1984. Characterization of a high molecular weight protective antigen of *Plasmodium yoelii*. Parasitology 88: 211–219.

Horuk R, Wang ZX, Peiper SC, Hesselgesser J. 1994. Identification and characterization of a promiscuous chemokine-binding protein in a human erythroleukemic cell line. J Biol Chem 269: 17730–17733.

Horuk R, Martin A, Hesselgesser J, Hadley T, Lu ZH, Wang ZX, Peiper SC. 1996. The Duffy antigen receptor for chemokines: Structural analysis and expression in the brain. J Leukoc Biol 59: 29–38.

Iyer JK, Fuller K, Preiser PR. 2006. Differences in the copy number of the py235 gene family in virulent and avirulent lines of *Plasmodium yoelii*. Mol Biochem Parasitol 150: 186–191.

Iyer JK, Amaladoss A, Genesan S, Preiser PR. 2007. Variable expression of the 235 kDa rhoptry protein of *Plasmodium yoelii* mediate host cell adaptation and immune evasion. Mol Microbiol 65: 333–346.

Jiang L, Lopez-Barragan MJ, Jiang H, Mu J, Gaur D, Zhao K, Felsenfeld G, Miller LH. 2010. Epigenetic control of the variable expression of a *Plasmodium falciparum* receptor protein for erythrocyte invasion. Proc Natl Acad Sci U S A 107: 2224–2229.

Kadekoppala M, Ogun SA, Howell S, Gunaratne RS, Holder AA. 2010. Systematic genetic analysis of the *Plasmodium falciparum* MSP7-like family reveals differences in protein expression, location, and importance in asexual growth of the blood-stage parasite. Eukaryot Cell 9: 1064–1074.

Kaneko O, Mu J, Tsuboi T, Su X, Torii M. 2002. Gene structure and expression of a *Plasmodium falciparum* 220-kDa protein homologous to the *Plasmodium vivax* reticulocyte binding proteins. Mol Biochem Parasitol 121: 275–278.

Kato K, Mayer DC, Singh S, Reid M, Miller LH. 2005. Domain III of *Plasmodium falciparum* apical membrane antigen 1 binds to the erythrocyte membrane protein Kx. Proc Natl Acad Sci U S A 102: 5552–5557.

Keeley A, Soldati D. 2004. The glideosome: A molecular machine powering motility and host-cell invasion by Apicomplexa. Trends Cell Biol 14: 528–532.

Keen J, Holder A, Playfair J, Lockyer M, Lewis A. 1990. Identification of the gene for a *Plasmodium yoelii* rhoptry protein. Multiple copies in the parasite genome. Mol Biochem Parasitol 42: 241–246.

Li F, Dluzewski A, Coley AM, Thomas A, Tilley L, Anders RF, Foley M. 2002. Phage-displayed peptides bind to the malarial protein apical membrane antigen-1 and inhibit the merozoite invasion of host erythrocytes. J Biol Chem 277: 50303–50310.

Li X, Chen H, Oo TH, Daly TM, Bergman LW, Liu SC, Chishti AH, Oh SS. 2004. A co-ligand complex anchors *Plasmodium falciparum* merozoites to the erythrocyte invasion receptor band 3. J Biol Chem 279: 5765–5771.

Lobo CA, Rodriguez M, Reid M, Lustigman S. 2003. Glycophorin C is the receptor for the *Plasmodium falciparum* erythrocyte binding ligand PfEBP-2 (baebl). Blood 101: 4628–4631.

Lopaticki S, Maier AG, Thompson J, Wilson DW, Tham WH, Triglia T, Gout A, Speed TP, Beeson JG, Healer J, Cowman AF. 2011. Reticulocyte and erythrocyte binding-like proteins function cooperatively in invasion of human erythrocytes by malaria parasites. Infect Immun 79: 1107–1117.

Maier AG, Duraisingh MT, Reeder JC, Patel SS, Kazura JW, Zimmerman PA, Cowman AF. 2003. *Plasmodium falciparum* erythrocyte invasion through glycophorin C and selection for Gerbich negativity in human populations. Nat Med 9: 87–92.

Martin MJ, Rayner JC, Gagneux P, Barnwell JW, Varki A. 2005. Evolution of human-chimpanzee differences in malaria susceptibility: Relationship to human genetic loss of N-glycolylneuraminic acid. Proc Natl Acad Sci U S A 102: 12819–12824.

Mayer DC, Kaneko O, Hudson-Taylor DE, Reid ME, Miller LH. 2001. Characterization of a *Plasmodium falciparum* erythrocyte-binding protein paralogous to EBA-175. Proc Natl Acad Sci U S A 98: 5222–5227.

Mayer DC, Mu JB, Kaneko O, Duan J, Su XZ, Miller LH. 2004. Polymorphism in the *Plasmodium falciparum* erythrocyte-binding ligand JESEBL/EBA-181 alters its receptor specificity. Proc Natl Acad Sci U S A 101: 2518–2523.

Mayer DC, Cofie J, Jiang L, Hartl DL, Tracy E, Kabat J, Mendoza LH, Miller LH. 2009. Glycophorin B is the erythrocyte receptor of *Plasmodium falciparum* erythrocyte-binding ligand, EBL-1. Proc Natl Acad Sci U S A 106: 5348–5352.

Miller LH, Mason SJ, Clyde DF, McGinniss MH. 1976. The resistance factor to *Plasmodium vivax* in blacks. The Duffy-blood-group genotype, FyFy. N Engl J Med 295: 302–304.

Mitchell GH, Hadley TJ, McGinniss MH, Klotz FW, Miller LH. 1986. Invasion of erythrocytes by *Plasmodium falciparum* malaria parasites: Evidence for receptor heterogeneity and two receptors. Blood 67: 1519–1521.

Mitchell GH, Thomas AW, Margos G, Dluzewski AR, Bannister LH. 2004. Apical membrane antigen 1, a major malaria vaccine candidate, mediates the close attachment of invasive merozoites to host red blood cells. Infect Immun 72: 154–158.

Narum DL, Thomas AW. 1994. Differential localization of full-length and processed forms of PF83/AMA-1 an apical membrane antigen of *Plasmodium falciparum* merozoites. Mol Biochem Parasitol 67: 59–68.

Nguitragool W, Bokhari AA, Pillai AD, Rayavara K, Sharma P, Turpin B, Aravind L, Desai SA. 2011. Malaria parasite clag3 genes determine channel-mediated nutrient uptake by infected red blood cells. Cell 145: 665–677.

O'Donnell RA, Saul A, Cowman AF, Crabb BS. 2000. Functional conservation of the malaria vaccine antigen $MSP\text{-}1_{19}$ across distantly related *Plasmodium* species. Nat Med 6: 91–95.

O'Donnell RA, de Koning-Ward TF, Burt RA, Bockarie M, Reeder JC, Cowman AF, Crabb BS. 2001. Antibodies against merozoite surface protein (MSP)-1(19) are a major component of the invasion-inhibitory response in individuals immune to malaria. J Exp Med 193: 1403–1412.

Oeuvray C, Bouharoun-Tayoun H, Gras-Masse H, Bottius E, Kaidoh T, Aikawa M, Filgueira MC, Tartar A, Druilhe P. 1994. Merozoite surface protein-3: A malaria protein inducing antibodies that promote *Plasmodium falciparum* killing by cooperation with blood monocytes. Blood 84: 1594–1602.

Ogun SA, Holder AA. 1996. A high molecular mass *Plasmodium yoelii* rhoptry protein binds to erythrocytes. Mol Biochem Parasitol 76: 321–324.

Ogun SA, Scott-Finnigan TJ, Narum DL, Holder AA. 2000. *Plasmodium yoelii*: Effects of red blood cell modification and antibodies on the binding characteristics of the 235-kDa rhoptry protein. Exp Parasitol 95: 187–195.

Ogun SA, Tewari R, Otto TD, Howell SA, Knuepfer E, Cunningham DA, Xu Z, Pain A, Holder AA. 2011. Targeted disruption of py235ebp-1: Invasion of erythrocytes by *Plasmodium yoelii* using an alternative Py235 erythrocyte binding protein. PLoS Pathog 7: e1001288.

O'Keeffe AH, Green JL, Grainger M, Holder AA. 2005. A novel Sushi domain-containing protein of *Plasmodium falciparum*. Mol Biochem Parasitol 140: 61–68.

Orlandi PA, Klotz FW, Haynes JD. 1992. A malaria invasion receptor, the 175-kilodalton erythrocyte binding antigen of *Plasmodium falciparum* recognizes the terminal Neu5Ac (alpha2-3)Gal- sequences of glycophorin A. J Cell Biol 116: 901–909.

Otsuki H, Kaneko O, Thongkukiatkul A, Tachibana M, Iriko H, Takeo S, Tsuboi T, Torii M. 2009. Single amino acid substitution in *Plasmodium yoelii* erythrocyte ligand determines its localization and controls parasite virulence. Proc Natl Acad Sci U S A 106: 7167–7172.

Pain A, Bohme U, Berry AE, Mungall K, Finn RD, Jackson AP, Mourier T, Mistry J, Pasini EM, Aslett MA, Balasubrammaniam S, Borgwardt K, Brooks K, Carret C, Carver TJ, Cherevach I, Chillingworth T, Clark TG, Galinski MR, Hall N, Harper D, Harris D, Hauser H, Ivens A, Janssen CS, Keane T, Larke N, Lapp S, Marti M, Moule S, Meyer IM, Ormond D, Peters N, Sanders M, Sanders S, Sargeant TJ, Simmonds M, Smith F, Squares R, Thurston S, Tivey AR, Walker D, White B, Zuiderwijk E, Churcher C, Quail MA, Cowman AF, Turner CM, Rajandream MA, Kocken CH, Thomas AW, Newbold CI, Barrell BG, Berriman M. 2008. The genome of the simian and human malaria parasite *Plasmodium knowlesi*. Nature 455: 799–803.

Pattaradilokrat S, Culleton RL, Cheesman SJ, Carter R. 2009. Gene encoding erythrocyte binding ligand linked to blood stage multiplication rate phenotype in *Plasmodium yoelii yoelii*. Proc Natl Acad Sci U S A 106: 7161–7166.

Persson KE, McCallum FJ, Reiling L, Lister NA, Stubbs J, Cowman AF, Marsh K, Beeson JG. 2008. Variation in use of erythrocyte invasion pathways by *Plasmodium falciparum* mediates evasion of human inhibitory antibodies. J Clin Invest 118: 342–351.

Peterson DS, Wellems TE. 2000. EBL-1, a putative erythrocyte binding protein of *Plasmodium falciparum*, maps within a favored linkage group in two genetic crosses. Mol Biochem Parasitol 105: 105–113.

Prasad CD, Prasad Singh A, Chitnis CE, Sharma A. 2003. A *Plasmodium yoelii yoelii* erythrocyte binding protein that uses Duffy binding-like domain for invasion: A rodent model for studying erythrocyte invasion. Mol Biochem Parasitol 128: 101–105.

Preiser PR, Jarra W. 1998. *Plasmodium yoelii*: Differences in the transcription of the 235-kDa rhoptry protein multigene family in lethal and nonlethal lines. Exp Parasitol 89: 50–57.

Preiser PR, Jarra W, Capiod T, Snounou G. 1999. A rhoptry-protein-associated mechanism of clonal phenotypic variation in rodent malaria. Nature 398: 618–622.

Ramalingam JK, Hunke C, Gao X, Gruber G, Preiser PR. 2008. ATP/ADP binding to a novel nucleotide binding domain of the reticulocyte-binding protein Py235 of *Plasmodium yoelii*. J Biol Chem 283: 36386–36396.

Rayner JC, Galinski MR, Ingravallo P, Barnwell JW. 2000. Two *Plasmodium falciparum* genes express merozoite proteins that are related to *Plasmodium vivax* and *Plasmodium yoelii* adhesive proteins involved in host cell selection and invasion. Proc Natl Acad Sci U S A 97: 9648–9653.

Rayner JC, Vargas-Serrato E, Huber CS, Galinski MR, Barnwell JW. 2001. A *Plasmodium falciparum* homologue of *Plasmodium vivax* reticulocyte binding protein (PvRBP1) defines a trypsin-resistant erythrocyte invasion pathway. J Exp Med 194: 1571–1581.

Rayner JC, Huber CS, Barnwell JW. 2004a. Conservation and divergence in erythrocyte invasion ligands: *Plasmodium reichenowi* EBL genes. Mol Biochem Parasitol 138: 243–247.

Rayner JC, Huber CS, Galinski MR, Barnwell JW. 2004b. Rapid evolution of an erythrocyte invasion gene family: The *Plasmodium reichenowi* reticulocyte binding like (RBL) genes. Mol Biochem Parasitol 133: 287–296.

Reed MB, Caruana SR, Batchelor AH, Thompson JK, Crabb BS, Cowman AF. 2000. Targeted disruption of an erythrocyte binding antigen in *Plasmodium falciparum* is associated with a switch toward a sialic acid independent pathway of invasion. Proc Natl Acad Sci U S A 97: 7509–7514.

Sahar T, Reddy KS, Bharadwaj M, Pandey AK, Singh S, Chitnis CE, Gaur D. 2011. *Plasmodium falciparum* reticulocyte binding-like homologue protein 2 (PfRH2) is a key adhesive molecule involved in erythrocyte invasion. PLoS One 6: e17102.

Sim B, Orlandi PA, Haynes JD, Klotz FW, Carter JM, Camus D, Zegans ME, Chulay JD. 1990. Primary structure of the 175K *Plasmodium falciparum* erythrocyte binding antigen and identification of a peptide which elicits antibodies that inhibit malaria merozoite invasion. J Cell Biol 111: 1877–1884.

Sim BKL, Chitnis CE, Wasniowska K, Hadley TJ, Miller LH. 1994. Receptor and ligand domains for invasion of erythrocytes by *Plasmodium falciparum*. Science 264: 1941–1944.

Singh AP, Ozwara H, Kocken CH, Puri SK, Thomas AW, Chitnis CE. 2005a. Targeted deletion of *Plasmodium knowlesi* Duffy binding protein confirms its role in junction formation during invasion. Mol Microbiol 55: 1925–1934.

Singh S, Soe S, Roussilhon C, Corradin G, Druilhe P. 2005b. *Plasmodium falciparum* merozoite surface protein 6 displays multiple targets for naturally occurring antibodies that mediate monocyte-dependent parasite killing. Infect Immun 73: 1235–1238.

Singh SK, Hora R, Belrhali H, Chitnis CE, Sharma A. 2006. Structural basis for Duffy recognition by the malaria parasite Duffy-binding-like domain. Nature 439: 741–744.

Soubes SC, Reid ME, Kaneko O, Miller LH. 1999. Search for the sialic acid-independent receptor on red blood cells for invasion by *Plasmodium falciparum*. Vox Sang 76: 107–114.

Spadafora C, Awandare GA, Kopydlowski KM, Czege J, Moch JK, Finberg RW, Tsokos GC, Stoute JA. 2010. Complement receptor 1 is a sialic acid-independent erythrocyte receptor of *Plasmodium falciparum*. PLoS Pathog 6: e1000968.

Stubbs J, Simpson KM, Triglia T, Plouffe D, Tonkin CJ, Duraisingh MT, Maier AG, Winzeler EA, Cowman AF. 2005. Molecular mechanism for switching of *P. falciparum* invasion pathways into human erythrocytes. Science 309: 1384–1387.

Swardson-Olver CJ, Dawson TC, Burnett RC, Peiper SC, Maeda N, Avery AC. 2002. *Plasmodium yoelii* uses the murine Duffy antigen receptor for chemokines as a receptor for normocyte invasion and an alternative receptor for reticulocyte invasion. Blood 99: 2677–2684.

Taylor HM, Triglia T, Thompson J, Sajid M, Fowler R, Wickham ME, Cowman AF, Holder AA. 2001. *Plasmodium falciparum* homologue of the genes *for Plasmodium vivax* and *Plasmodium yoelii* adhesive proteins, which is transcribed but not translated. Infect Immun 69: 3635–3645.

Taylor HM, Grainger M, Holder AA. 2002. Variation in the expression of a *Plasmodium falciparum* protein family implicated in erythrocyte invasion. Infect Immun 70: 5779–5789.

Tewari R, Ogun SA, Gunaratne RS, Crisanti A, Holder AA. 2005. Disruption of *Plasmodium berghei* merozoite surface protein 7 gene modulates parasite growth in vivo. Blood 105: 394–396.

Tham WH, Wilson DW, Reiling L, Chen L, Beeson JG, Cowman AF. 2009. Antibodies to reticulocyte binding protein-like homologue 4 inhibit invasion of *Plasmodium falciparum* into human erythrocytes. Infect Immun 77: 2427–2435.

Tham WH, Wilson DW, Lopaticki S, Schmidt CQ, Tetteh-Quarcoo PB, Barlow PN, Richard D, Corbin JE, Beeson JG, Cowman AF. 2010. Complement receptor 1 is the host erythrocyte receptor for *Plasmodium falciparum* PfRH4 invasion ligand. Proc Natl Acad Sci U S A 107: 17327–17332.

Thompson JK, Triglia T, Reed MB, Cowman AF. 2001. A novel ligand from *Plasmodium falciparum* that binds to a sialic acid-containing receptor on the surface of human erythrocytes. Mol Microbiol 41: 47–58.

Tolia NH, Enemark EJ, Sim BK, Joshua-Tor L. 2005. Structural basis for the EBA-175 erythrocyte invasion pathway of the malaria parasite *Plasmodium falciparum*. Cell 122: 183–193.

Topolska AE, Lidgett A, Truman D, Fujioka H, Coppel RL. 2004. Characterization of a membrane-associated rhoptry protein of *Plasmodium falciparum*. J Biol Chem 279: 4648–4656.

Tournamille C, Colin Y, Cartron JP, Le Van Kim C. 1995. Disruption of a GATA motif in the Duffy gene promoter abolishes erythroid gene expression in Duffy-negative individuals. Nat Genet 10: 224–228.

Trager W, Jensen JB. 1976. Human malaria parasites in continuous culture. Science 193: 673–675.

Tran TM, Moreno A, Yazdani SS, Chitnis CE, Barnwell JW, Galinski MR. 2005. Detection of a *Plasmodium vivax* erythrocyte binding protein by flow cytometry. Cytometry A 63: 59–66.

Treeck M, Zacherl S, Herrmann S, Cabrera A, Kono M, Struck NS, Engelberg K, Haase S, Frischknecht F, Miura K, Spielmann T, Gilberger TW. 2009. Functional analysis of the leading malaria vaccine candidate AMA-1 reveals an essential role for the cytoplasmic domain in the invasion process. PLoS Pathog 5: e1000322.

Triglia T, Thompson J, Caruana SR, Delorenzi M, Speed T, Cowman AF. 2001. Identification of proteins from *Plasmodium falciparum* that are homologous to reticulocyte binding proteins in *Plasmodium vivax*. Infect Immun 69: 1084–1092.

Triglia T, Duraisingh MT, Good RT, Cowman AF. 2005. Reticulocyte-binding protein homologue 1 is required for sialic acid-dependent invasion into human erythrocytes by *Plasmodium falciparum*. Mol Microbiol 55: 162–174.

Triglia T, Chen L, Lopaticki S, Dekiwadia C, Riglar DT, Hodder AN, Ralph SA, Baum J, Cowman AF. 2011. *Plasmodium falciparum* merozoite invasion is inhibited by antibodies that target the PfRH2a and b binding domains. PLoS Pathog 7: e1002075.

Walliker D, Sanderson A, Yoeli M, Hargreaves BJ. 1976. A genetic investigation of virulence in a rodent malaria parasite. Parasitology 72: 183–194.

PART IV

COMPARATIVE "OMICS" APPROACHES TO DEFINING VIRULENCE

CHAPTER 20

EVOLUTION OF VIRULENCE IN OOMYCETE PLANT PATHOGENS

PAUL R. J. BIRCH, MARY E. COATES, and JIM L. BEYNON

INTRODUCTION

Oomycetes are a diverse group of organisms that often closely resemble fungi in their morphology and growth habit yet are members of the heterokont/chromist clade within the kingdom Stramenopila. As such, they are related to aquatic organisms such as diatoms, brown and golden-brown algae, and distantly related to fungi. They include a range of free-living water molds, as well as diverse pathogens of plants, insects, fish, crustaceans, mammals, and microbes, such as fungi (Kamoun, 2003). This chapter will focus on plant pathogens, which cause devastating diseases of ornamental, native, and crop species. The most notoriously damaging groups of oomycete plant pathogens comprise more than 80 characterized species of *Phytophthora* (Blair et al., 2008), the downy mildews, including genera such as *Peronospora*, *Plasmopora*, and *Bremia* (Goker et al., 2007), the white blister rusts, such as *Albugo*, and more than 110 species of *Pythium* (Levesque and de Cock, 2004).

Plant pathogenic oomycetes display a wide range of infection strategies. On one hand are the opportunistic or weakly pathogenic *Pythium* species, which are typically soilborne necrotrophs with wide host ranges. At the opposite extreme are the highly specialized (narrow or single host range), aerially disseminated obligate biotrophs. Obligate biotrophy evolved independently in the downy mildews, such as *Hyaloperonospora arabidopsidis*, and the white blister rusts, such as *Albugo laibachii* (Thines and Kamoun, 2010). *H. arabidopsidis* and *A. laibachii* are each pathogens of the model plant *Arabidopsis thaliana*, making them ideal models in which to investigate the requirements, and evolution, of obligate biotrophy (Coates and Beynon, 2010; Kemen et al., 2011). Spanning these extremes of infection modes are the hemibiotrophic *Phytophthora* species, which may be relatively narrow in host range, such as the infamous, airborne potato late blight pathogen, *Phytophthora infestans* (Birch and Whisson, 2001), and the soilborne pathogen of soybean, *Phytophthora sojae* (Kamoun, 2003), through to species with extremely wide host ranges, such as *Phytophthora cinnamomi* (Kamoun, 2003). The latter include emerging pathogens that threaten crop and natural plant

Evolution of Virulence in Eukaryotic Microbes, First Edition. Edited by L. David Sibley, Barbara J. Howlett, and Joseph Heitman.

species, such as *Phytophthora capsici* (Lamour et al., 2007) and *Phytophthora ramorum* (Tyler et al., 2006).

The infection cycles of oomycete plant pathogens may incorporate a diverse range of cell types and developmental stages. Sporangia, and their potential release of swimming zoospores, provide means of dispersal above and below ground. Appressoria and hyphae may be formed to penetrate and colonize host tissues, and haustoria can form an intimate, biotrophic interface between pathogen and host cells.

Each stage of colonization, encompassing biotrophic and necrotrophic modes of infection, involves the antagonistic exchange of molecules between host and pathogen, a battle of defense and attack that shapes their coevolution. The genomics era has provided the entire genetic blueprints of a number of oomycetes, including the necrotroph *Pythium ultimum* (Levesque et al., 2010); the obligate biotrophs *H. arabidopsidis* (Baxter et al., 2010) and *A. laibachii* (Kemen et al., 2011); and the hemibiotrophic *P. infestans* (Haas et al., 2009), *P. sojae*, and *Ph. ramorum* (Tyler et al., 2006). From each genome sequence is emerging a picture of the molecular armory, metabolic requirements, and developmental changes that characterize the diverse patterns of oomycete–plant coevolution.

THE BASIS AND BASICS OF PLANT–PATHOGEN COEVOLUTION

All plant cells possess plasma membrane-localized pattern-recognition receptors (PRRs) that perceive molecules that are conserved, secreted or surface exposed, and indispensable to many microbial species. Detection of these microbe-associated molecular patterns (MAMPs), also known as pathogen-associated molecular patterns (PAMPs), leads to the activation of pattern-triggered immunity (PTI), encompassing a range of defense responses that prevent further colonization by any microbial species. To counteract PTI, successful pathogens deploy effector proteins. Effectors act either outside or inside host cells to suppress or otherwise manipulate plant defenses, often through direct interaction with host proteins (Chisholm et al., 2006; Jones and Dangl, 2006). The range of biochemical activities associated with pathogen effectors, the interactions with their target proteins in the plant, and the roles that those targets play in host defense are the subject of intense research. At all stages of molecular interaction, from MAMPs to PRRs, to effectors and their targets in the host, evolutionary forces are predicted to constantly shape the outcomes of recognition in the host (activation of PTI in the plant or evasion of recognition/suppression of PTI by the pathogen). In addition to their suppression of plant defenses, some effectors can manipulate host metabolism in order to maximize the production of nutrients for the invading pathogen. Coevolution between effector and target proteins influences the outcomes of disease development or disease resistance, and indeed may determine which plant species a particular pathogen is able to infect.

Coevolution is further influenced by a second layer of plant defense, in the form of immune receptors, which are typically intracellular nucleotide-binding, leucine-rich repeat (NB-LRR) resistance (R) proteins that evolve to acquire and maintain recognition of pathogen effectors (which are then named avirulence [AVR] proteins). NB-LRR R proteins may detect AVR effectors through direct physical association or may indirectly detect their activities upon their target proteins in the plant cell (Jones and Dangl, 2006). When an AVR protein is detected, effector-triggered immunity (ETI) is activated, often involving the hypersensitive response (HR), a rapid, localized programmed cell death

(PCD). Patterns of mutation, indicating diversifying selection, presence/absence polymorphisms, and differential expression are all indications of the ways in which pathogens modify their effector repertoires to evade recognition by host R proteins.

OOMYCETE MAMPs: ACTIVATING THE FRONTLINE OF PLANT DEFENSE

The conserved oomycete molecules that trigger PTI, and the effectors that suppress these responses, have been recently reviewed (Hein et al., 2009), so the contents of that review will be briefly reiterated here. Nevertheless, this is a rapidly developing field and the generation of several oomycete genome sequences since 2009, including the late blight pathogen *P. infestans* (Haas et al., 2009), the obligate biotrophs *H. arabidopsidis* (Baxter et al., 2010) and *A. laibachii* (Kemen et al., 2011), and the necrotroph *Py. ultimum* (Levesque et al., 2010), have each added to our understanding of the mechanisms underpinning distinct oomycete infection strategies and have especially revealed more about the evolution and diverse repertoires of effectors (Fig. 20.1).

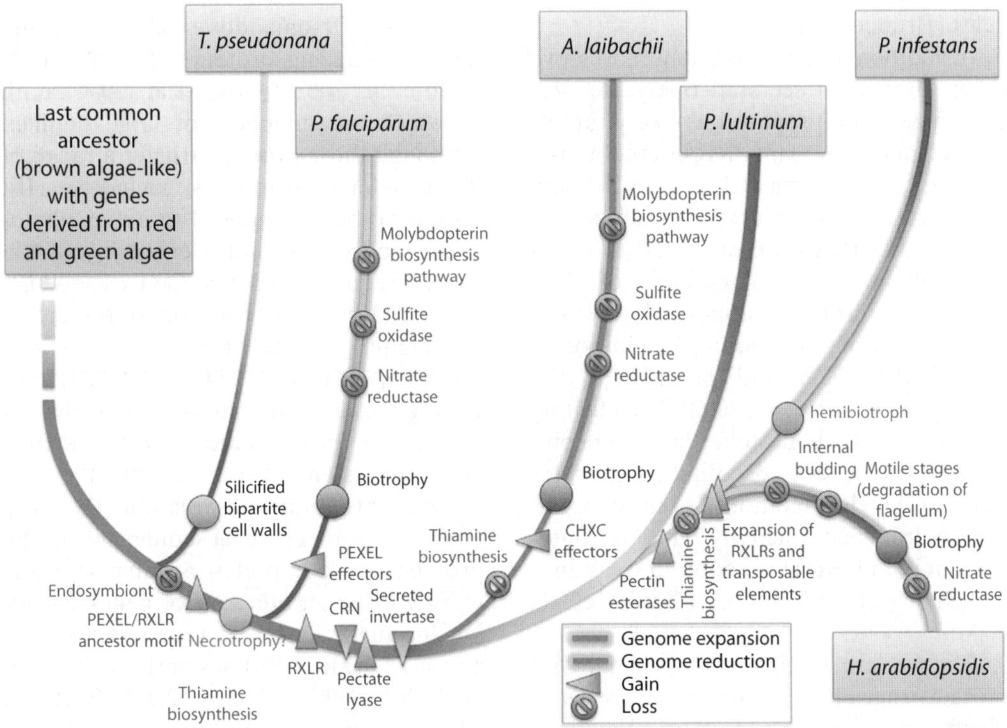

Figure 20.1 Model for evolution of gain and loss of genes and pathways for selected Chromalveolata (from Kemen et al., 2011). After splitting off from the diatom *Thalassiosira pseudonana* lineage, convergent evolution led to PEXEL (in *Plasmodium falciparum*) and RXLR effectors. In addition to RXLR effectors, oomycetes acquired a second class of effectors, CRNs and a secreted invertase that allows use of sucrose from plant hosts. Oomycetes that are biotrophs or hemibiotrophs lost their thiamine biosynthetic pathway and, in the case of *A. laibachii*, evolved a new CHXC effector class. The biotrophic lifestyles of *P. falciparum*, *A. laibachii*, and *H. arabidopsidis* led to the loss of key biosynthetic enzymes. *P. infestans* shows a strong genome expansion and an expansion in the number of RXLR effector genes. See color insert.

As has been described for fungi and bacteria (Zipfel, 2008), oomycete MAMPs that activate PTI include both nonproteinaceous molecules, such as polysaccharides (Umemoto et al., 1997), and defined, conserved peptide sequences of proteins, such as the pep-13 component of transglutaminases, which are found in all oomycetes (Brunner et al., 2002). Pep-13 is a 13-amino acid sequence within secreted transglutaminases that is necessary and sufficient to trigger PTI. Mutational analysis demonstrated that the key amino acid residues required to trigger host defenses are also essential for transglutaminase activity, supporting the concept that PRRs have evolved to detect motifs that are functionally critical to widely conserved pathogen molecules (Brunner et al., 2002).

In addition to the transglutaminases, two other families of secreted oomycete proteins have been proposed as likely targets for recognition by plant PRRs: the cellulose-binding elicitor lectin (CBEL) and elicitin families. Both are widespread across the oomycetes. CBELs are proposed to act in host cell adhesion; knockdown of their expression results in abnormal development, supporting the required definition of a MAMP as being indispensable to the pathogen (Gaulin et al., 2002). The eliciting portion of CBEL proteins has been pinpointed to two cellulose-binding domains (CBDs). Single mutations in either CBD can abolish elicitor activity, whereas mutations in both CBDs are required to attenuate plant cell wall binding (Gaulin et al., 2006).

Elicitins are small cysteine-rich (SCR) proteins that elicit a range of defenses in plants, including the HR in *Nicotiana* species (Kamoun, 2006). Perhaps one of the best-studied elicitins is INF1 from *P. infestans*. INF1 elicits HR in *Nicotiana benthamiana* that is dependent on heat shock proteins HSP70 and HSP90 (Kanzaki et al., 2003), the respiratory burst oxidase Nbrboh (Yoshioka et al., 2003), the ubiquitin ligase-associated protein NbSGT1 (Peart et al., 2002), a mitogen-activated protein kinase (MAPK) (Takahashi et al., 2008), and a receptor-like kinase (RLK) MbLRK1 (Kanzaki et al., 2008). The view that INF1 should be regarded as a MAMP is strongly supported by the observation that reactive oxygen species generation, following INF1 perception, is dependent also on the LRR RLK SERK3/BAK1, which integrates diverse MAMP perception events into a universal PTI response (Heese et al., 2007).

In addition to these proteinaceous MAMPs, other secreted oomycete proteins elicit defense responses in plants. These include SCRs, such as *Phytophthora cactorum-Fragaria* (PcF) from *Phytophthora cactorum* (Orsomando et al., 2001), and necrosis-inducing protein 1 (NPP1) (Fellbrich et al., 2002; Qutob et al., 2006), both of which are members of large families. Although these proteins trigger a range of plant defense responses, including HR, they have been proposed to act as phytotoxins and do not yet meet many of the criteria required to define them as MAMPs (reviewed in Hein et al., 2009). Nevertheless, members of the *SCR* gene family in *P. infestans*, such as *SCR74* and *SCR91*, have undergone diversifying selection that is reminiscent of coevolution with host surveillance systems (Liu et al., 2005).

One intriguing observation that has emerged from genome sequencing is the presence in oomycetes as diverse as *P. infestans* and *A. laibachii* of genes encoding the major enzymes required for lipopolysaccharide (LPS) biosynthesis (Kemen et al., 2011; Whitaker et al., 2009). Bacterial LPS has long been regarded as a MAMP that is detected by both plant and animal host cells (Livaja et al., 2008). It will be interesting to see whether LPS is synthesized by oomycetes and whether it has been targeted for detection by plant PRRs.

THE EXPANDING OOMYCETE EFFECTOR COLLECTION: SUPPRESSORS OF PTI AND MANIPULATORS OF HOST METABOLISM

The past few years have witnessed considerable strides in the identification of candidate effectors in oomycete plant pathogens and in gaining an understanding of how they may be involved in suppressing PTI. This progress has been accelerated by oomycete genome sequencing. As a starting point, all effector candidates are secreted and thus possess an N-terminal signal peptide for export from the pathogen cell. Broadly speaking, effectors may be "apoplastic," acting outside the plant cell, or "cytoplasmic," being delivered, or translocated, inside a living host cell (Hein et al., 2009; Schornack et al., 2009).

Apoplastic Effectors

As part of the basal defense mechanism (PTI) to microbial attack, plants secrete a range of hydrolytic enzymes. Many of these directly degrade pathogen cell walls. Some, such as glucanases, also release β 1,3-glucan oligosaccharide elicitors from the host cell wall, activating further defense responses (van Loon et al., 2006). Oomycete plant pathogens have evolved effective mechanisms to combat hydrolytic enzymes, secreting a range of inhibitors that target glucanases and proteases (Hein et al., 2009). Glucanase inhibitor proteins (GIPs) from *P. sojae* (Rose et al., 2002) and *P. infestans* (Damasceno et al., 2008) directly complex with, and inhibit, host endo-β 1,3-glucanases (EGases) in the apoplast. Evidence of diversifying selection at key amino acid residues in both GIPs and EGases, at the sites where they are predicted to interact, indicate tight coevolution between these pathogen and host proteins (Damasceno et al., 2008).

Protease inhibitors are also secreted by oomycete plant pathogens. *P. infestans* secretes at least 14 kazal-like serine extracellular protease inhibitors (EPIs) and 4 cystatin-like cysteine protease inhibitors (EPICs). EPI1 inhibits pathogenesis-related (PR) protein P69B, a host secreted subtilisin-like protease (Tian et al., 2004, 2005, 2007). P69B targets and degrades two of the cystatin-like inhibitors from *P. infestans*, EPIC1 and EPIC2B, thus providing the first evidence that this PR protein targets secreted pathogen effectors. EPI1 thus effectively protects these proteins by preventing their degradation (Tian et al., 2007). EPIC1 and EPIC2B target and inhibit two host cysteine proteases, *Phytophthora*-inhibited protease 1 (PIP1) and RCR3. Like GIPs, PIP1 is under diversifying selection, suggesting it is also coevolving with pathogen effectors, such as EPIC1 and EPIC2B (Shabab et al., 2008). Both PIP1 and RCR3 are also targeted by Avr2, a cystatin-like protease inhibitor from the fungal pathogen *Cladosporium fulvum*, which attacks tomato. Thus, pathogens from different kingdoms have evolved effectors with similar properties that target the same host defense proteins (Song et al., 2009). Analyses of oomycete plant pathogen genomes reveal that secreted protease inhibitors are likely to be a universal class of apoplastic effectors.

CYTOPLASMIC EFFECTORS

Bacterial effectors delivered inside the plant cell by the type III secretion system are well-characterized and have a range of different biochemical functions and host targets, including nucleic acids and proteins (Block et al., 2008). Cytoplasmic effectors from filamentous pathogens, such as fungi and oomycetes, are relatively poorly understood. Nevertheless, a wide diversity of candidate cytoplasmic effectors from oomycetes, often with characteristic, conserved peptide signatures, has been identified from studies of genome sequences (Fig. 20.1).

The most widely studied oomycete effectors are RXLR proteins (Birch et al., 2006, 2008, 2009; Kamoun, 2006; Schornack et al., 2009). They contain the amino acids Arg-X-Leu-Arg, often followed by an acidic region ending with the amino acids Glu-Glu-Arg (EER), approximately 30–40 residues downstream of a canonical signal peptide. The oomycete RXLR motif is similar in sequence and position to the host cell targeting signal (*Plasmodium* export element [PEXEL] motif) required for translocation of proteins from malaria parasites (*Plasmodium* spp.) into host red blood cells (Hiller et al., 2004; Marti et al., 2004). Indeed, the oomycete RXLR signal was demonstrated to function for effector delivery in the malarial system (Bhattacharjee et al., 2006) and has thus been proposed to share a common evolutionary origin (Fig. 20.1). The RXLR motif has since been shown to be required for the delivery of oomycete effectors inside living plant cells from haustoria (Whisson et al., 2007). Delivery can occur in the absence of the pathogen, suggesting that it involves endocytosis at the host cell surface (Dou et al., 2008a).

Recently, Kale et al. (2010) reported that RXLR-dependent effector delivery into plant or animal cells involves binding to phospholipids, a finding currently causing much debate and controversy (Gan et al., 2010; Stassen and Van den Ackerveken 2011). Kale et al (2010) reported that the RXLR motif, and numerous variants in both oomycete and fungal effectors, is responsible for binding to phosphatidylinositol monophosphates, an activity required for their uptake into a range of animal and plant host cells. However, a recent publication by Yaeno et al. (2011), revealing the nuclear magnetic resonance (NMR) structure of AVR3a4 from *P. capsici*, a homolog of AVR3a from *P. infestans*, and Avr1b from *P. sojae*, demonstrated that phosphatidylinositol monophosphate binding resides in key exposed C-terminal amino acid residues. These authors also showed that the RXLR motif does not bind to phosphatidylinositol monophosphates.

Several reports have indicated that RXLR effectors can suppress host defenses, including PTI. Alleles of *ATR1* and *ATR13* from *H. arabidopsidis* confer enhanced virulence to *Pseudomonas syringae* pathovar *tomato* (*Pst*) DC3000 on susceptible *Arabidopsis* accessions, suggesting that ATR1 and ATR13 positively contribute to pathogen virulence inside plant cells. *ATR13* alleles and *HpRXLR29*, an isolate-specific *H. arabidopsidis* effector, suppress bacterial MAMP-triggered callose deposition in susceptible *Arabidopsis*. Furthermore, expression of a further allele of *ATR13* in plant cells suppressed MAMP-triggered reactive oxygen species production in addition to callose deposition (Cabral et al., 2011; Sohn et al., 2007).

AVR3a from *P. infestans* is able to suppress PCD induced by the *P. infestans* elicitin, INF1, in *N. benthamiana* (Bos et al., 2006, 2009). Moreover, Avr1b from *P. sojae* and the *H. arabidopsidis* Hp*RxL96* can suppress PCD triggered by the mouse Bcl2 associated X (BAX) protein in yeast, soybean, and *N. benthamiana* cells (Dou et al., 2008b). Cell death suppression activity is a common feature of bacterial effectors, and the observation that RXLRs have this ability is thus consistent with expected models of effector function. Recently, the mode of action of AVR3a cell death suppression has been shown to involve direct interaction with, and stabilization of, the host ubiquitin E3 ligase CMPG1 (Bos et al., 2010). CMPG1 is required for PCD triggered by recognition, at the host plasma membrane, of a range of elicitors from fungal, bacterial, and oomycete plant pathogens (Gilroy et al., 2011a; González-Lamothe et al., 2006). AVR3a is able to suppress all of these PCD events (Gilroy et al., 2011a). Interestingly, although

P. infestans suppresses CMPG1-dependent PCD during the biotrophic phase of infection, the pathogen relaxes this suppression and, indeed, provokes CMPG1-dependent PCD during the necrotrophic phase. Recently, Yaeno et al. (2011) have shown that key residues in the C-terminal effector domain of AVR3a are responsible for binding to phosphatidylinositol monophosphates (PIPs). Mutation of these amino acids abolishes PIP binding and reduces the cell death suppression activity of AVR3a and its ability to stabilize CMPG1.

In addition to RXLRs, a further class of oomycete cytoplasmic effectors is one containing the crinkling and necrosis (CRN) proteins (Fig. 20.1). Like RXLR effectors, CRNs possess a modular structure with an N-terminal signal peptide for secretion, followed by a highly conserved domain containing an LXLFLAK motif. Beyond these two domains are variable combinations of domains of unknown function (Haas et al., 2009). The LXLFLAK-containing N-terminal domain mediates translocation of these proteins into host plant cells (Schornack et al., 2010). CRN proteins accumulate in the host nucleus, implicating this as their main site of activity. However, any role of CRN effectors in suppressing host defenses has yet to be determined.

Both *RXLR* and *CRN* genes are numerous and they reside in gene-sparse, repeat-rich, potentially fast-evolving regions of the *P. infestans* genome. It has been speculated that this location, promoting gene mutation, duplication, and rearrangement, facilitates the rapid evasion of detection by host immune receptors (Haas et al., 2009). However, to date, only RXLR effectors have been shown also to be AVR proteins, targeted by cytoplasmic NB-LRR proteins (see below). Compelling evidence was recently provided that the repeat-rich locations of effector genes are associated with higher rates of structural polymorphisms and positive selection, facilitating host jumps, followed by specialization on these new hosts (Raffaele et al., 2010). These authors observed that the repeat-rich regions of the genomes of *P. infestans* and four of its close relatives, each of which infects different host plants, were enriched in genes that were transcriptionally upregulated during infection. This finding implies a role for host adaptation in genome evolution in the oomycetes.

The genome of the obligate biotroph *A. laibachii* possesses both *RXLR* and *CRN* effector genes, although in fewer numbers than in the *Phytophthora* genomes (Kemen et al., 2011). Instead, a further class of potential effectors, containing the conserved motif Cys-His-X-Cys (CHXC), has been revealed (Fig. 20.1). The CHXC motif is sufficient and required for translocation of these effector proteins inside host plant cells. Thus, oomycetes possess a vast arsenal of cytoplasmic effectors. The question is: Why so many and what roles do they play?

Protein–Protein Interactions Reveal a Complex Interlinked Plant Response Network

All work to date on the role of effectors has been based on the activities of single proteins and the way they interact with host cell components. However, this does not take into account the complexity or subtlety of the true biological interaction. In the case of *Ps. syringae*, he bacteria encode approximately 30 potential T3SS effector proteins, and oomycetes potentially encode several hundred cytoplasmic effectors. One would imagine that these are introduced at low levels into host cells and a consequent complex interaction with plant molecular machines results. These interactions in some way suppress the host basal immune response networks and enable a pathogen to complete its reproductive cycle. Therefore, understanding how the effector protein complement interacts with host proteins is key to unraveling this potentially complex system.

The Arabidopsis Interactome Mapping Consortium (2011) have recently published the first comprehensive plant protein interaction network based on yeast two-hybrid analyses. These authors report multiple interactions between more than 8000 *Arabidopsis* proteins. In an accompanying paper, Mukhtar et al. (2011) used the same system to identify a plant–pathogen immune network (PPIN). They screened 99 *H. arabidopsidis* RXLR and 58 *Ps. syringae* T3SS effectors in a yeast two-hybrid screen against 8400 *Arabidopsis* proteins. Additionally, the non-leucine-rich repeat (LRR) domains of nucleotide-binding (NB) site resistance proteins and cytoplasmic domains of RLKs were included in the screen. The *H. arabdiopsidis* effectors interacted with 121 *Arabidopsis* proteins and the *Ps. syringae* effectors interacted with 44 target proteins (Fig. 20.2). Fascinatingly, 19 of these were common host proteins, demonstrating that effector proteins from organisms of highly divergent evolutionary lineages (oomycete and bacterium) have evolved effectors capable of interacting with the same host proteins. This may explain the huge diversity seen in some effector proteins as they generate variation to interact with host target proteins. Knockout mutations in genes encoding these common host interacting proteins either increased susceptibility to *H. arabidopsidis*, demonstrating a role in resistance, or increased resistance, demonstrating a role in susceptibility. The host interacting proteins are enriched in gene ontology (GO) annotations associated with gene transcription and secretion. This is consistent with the role of effectors to alter gene transcription initiated by the host as part of a resistance response and in manipulating the host secretory network targeted toward the invading pathogen.

A remarkable observation reported in Mukhtar et al. (2011) is the level of interconnectedness of the host proteins that are targeted by the effectors. These authors demonstrated that the PPIN contained all the most highly connected regulatory hubs in the *Arabidopsis* interaction network (Arabidopsis Interactome Mapping Consortium, 2011) and that many of the targets of the effectors interact with each other. This demonstrates that the pathogen effectors are interacting with a complex network of proteins rather than targeting specific output processes of immunity (Fig. 20.2). This connectivity may explain the complexity of the effector complements in pathogens as it may be necessary to distort the interactions within this network at many points to prevent signal amplification that results in an immune response. Individual effectors may also target more than one host protein, suggesting a diversity of interaction capability. Again this vision of complexity may hold the key to the diversity of effectors where the generation of great variability across a protein could generate new interactions that alter the response of the immune network.

The PPIN also revealed that the connection between the immune receptors of the host and the effectors is often indirect. Very few effectors interact directly with the NB or RLK cytoplasmic domains. However, there are numerous interactions between effectors and receptors via an intermediate host protein. This is consistent with the guard hypothesis where the receptors monitor a host protein and the interaction with that protein by an effector is detected, triggering a resistance response (Jones and Dangl, 2006).

At every level, this work reveals intrinsic interconnectedness in the immune response protein network. This is consistent with a buffered system allowing many potential inputs but preventing an over-response to any individual input. The effector complements of different pathogens have evolved to interact at many levels with this protein network to alter its ability to respond to pathogen ingress.

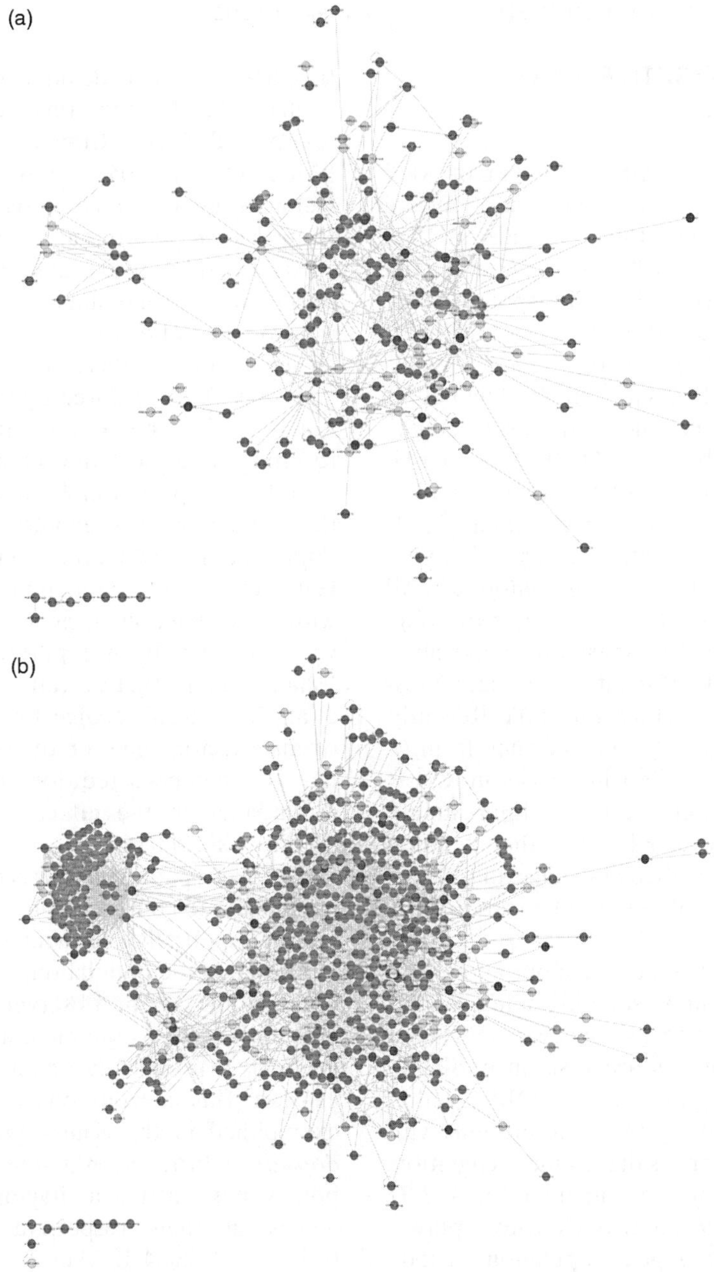

Figure 20.2 Protein–protein interactions between *Arabidopsis* host proteins and pathogen effector protein domains determined by yeast two-hybrid experiments and literature curation (Mukhtar et al., 2011). Pink: *Pseudomonas syringiae* effector; orange: *Hyaloperanospera arabidopsidis* effector; green: *Arabidopsis* immune system protein (NB-LRR, receptor-like kinase, or protein annotated as defense related); gray: other *Arabidopsis* protein. (a) Direct interactions between pathogen effector and *Arabidopsis* host proteins: Host proteins interacting with effectors form a densely connected subnetwork within the host protein–protein interactome. In a number of cases, independently evolved effectors from the two pathogens interact with the same host protein. (b) Direct pathogen effector–host protein interactions in the context of other protein interactions in which the host proteins participate. A number of the host proteins interacting with effectors are major interactome hubs. Interactome hubs as a whole are significantly overrepresented among immune interactors. Many host proteins annotated as immune related interact with proteins interacting with effectors. Host proteins with orthologs among land plants, but not in more distant taxa, are overrepresented in immune interactors. See color insert.

ETI TO OOMYCETE PLANT PATHOGENS

Over the past 7 years, a number of *AVR* genes have been identified from oomycete plant pathogens. These include *AVR3a* (Armstrong et al., 2005), *AVR4* (van Poppel et al., 2008), *AVR-blb1* (Champouret et al., 2009; Vleeshouwers et al., 2008), *AVR-blb2* (Oh et al., 2009) and *AVR2* (Gilroy et al., 2011b) from *P. infestans*; *Avr1b* (Shan et al., 2004), *Avr3c*, (Dong et al., 2009), *Avr3a*, *Avr1a* (Qutob et al., 2009), and *Avr4/6* (Dou et al., 2010) from the soybean pathogen *P. sojae*; and *ATR13* (Allen et al., 2004) and *ATR1* (Rehmany et al., 2005) from the *Arabidopsis* pathogen *H. arabidopsidis*. All are members of the RXLR class of effectors, suggesting that these are the primary effector targets of plant cytoplasmic NBS-LRR proteins that activate ETI. Recently, a third *AVR* gene has been cloned from *H. arabidopsidis*, *ATR5*, which lacks an RXLR motif (Bailey et al., 2011). Interestingly, ATR5 has the dEER motif that is found in many RXLR effectors, suggesting ATR5 is a variant of the canonical RXLR effector.

Each of the AVR effectors above has enhanced our understanding of how oomycete plant pathogens can evade ETI and thus overcome disease resistance. Single-nucleotide polymorphisms (SNPs) within allelic forms may give rise to proteins with amino acid changes that evade recognition. This has been well documented for *AVR3a* from *P. infestans*; only two alleles reported within the pathogen population encode proteins differing in two amino acids (K80E and I103M), which dictate recognition by R3a (Armstrong et al., 2005; Bos et al., 2006). As AVR3a is an essential pathogenicity determinant (Bos et al., 2010), deployment of an *R* gene that targets the virulent form, AVR3a^{EM}, in combination with R3a, which targets the avirulent form, AVR3a^{KI}, would potentially impose strong selection pressure on the pathogen population. SNPs that encode alternative, virulent alleles have also been reported for *ATR1*, *ATR5*, and *ATR13* from *H. arabidopsidis* (Allen et al., 2004; Bailey et al., 2011; Rehmany et al., 2005) and *Avr1b* and *Avr3c* from *P. sojae* (Dong et al., 2009; Shan et al., 2004). In addition to amino acid polymorphisms, which can retain the virulence function of the effector (Bos et al., 2010), virulence on plants containing some *R* genes has been achieved by loss of a functional *AVR* gene. Frameshift mutations, resulting in truncated versions of *AVR4*, have been reported in *P. infestans* isolates that infect potato expressing *R4* (van Poppel et al., 2008). Moreover, differential gene expression, sometimes associated with gene deletion or gene copy number variation, has been reported for *Avr1b* (Shan et al., 2004), *Avr1a*, and *Avr3a* (Qutob et al., 2002) from *P. sojae*. Presumably, loss of an effector gene, or of its expression, may be compensated for by functional redundancy in the effector complement (Birch et al., 2008).

For plant–pathogen interactions that are not subject to constraints imposed by agriculture, naturally occurring populations exist. Within the population, extreme levels of variation can be observed for both *R* and *AVR* genes as the plant and pathogen are locked in a battle of recognition and evasion. This coevolutionary arms race is exemplified in the *Arabidopsis–H. arabidopsidis RPP13–ATR13* interaction, with both genes existing as hyperpolymorphic alleles in their respective populations (Allen et al., 2004; Bakker et al., 2006; Ding et al., 2007; Rose et al., 2004). Initially, it was thought evolutionary forces imposed by a direct RPP13–ATR13 interaction caused the hypervariation, but it has become apparent that the variation may be due to many combinations of R–AVR pairs interacting indirectly. Hall et al. (2009) reported that additional *R* genes are able to recognize *ATR13* alleles and that *RPP13* is able to recognize *ATR* genes other than *ATR13*.

In addition to suppression of PTI, effectors must also be deployed to suppress the effects of ETI. Although the function of the majority of effectors identified to date remains unclear, two from *P. infestans*, IPI04 and SNE1, have recently been shown to do exactly this as they are able to suppress ETI triggered by a fellow effector (Halterman et al., 2010; Kelley et al., 2010). Furthermore, studies of the *Arabidopsis–H. arabidopsidis* interaction have revealed ETI suppressors. However, their identity remains elusive (Rehmany et al., 2005; Sohn et al., 2007).

BIOTROPHY VERSUS NECROTROPHY: GENOME STUDIES REVEAL THE DIFFERENCES

It is clear from the previous sections that to suppress the plant immune system and to manipulate metabolism, which is a key objective of a biotrophic mode of infection, oomycetes possess a diverse array of cytoplasmic effectors. In particular, the RXLR effectors are central to obligate biotrophy in *H. arabidopsidis* and to the biotrophic phases of *Phytophthora* infections. A further class of cytoplasmic effectors in the obligate biotroph *A. laibachii*, CHXC, awaits detailed investigation, although preliminary studies reveal that these also contribute to virulence (Kemen et al., 2011). The necrotroph *Py. ultimum* largely lacks these classes of cytoplasmic effectors suggesting that if it manipulates host signaling/regulatory networks, it does so in different ways. Nevertheless, it is noteworthy that the number of genes in *Py. ultimum* that encode secreted proteases is expanded in comparison to other oomycete genomes (Kemen et al., 2011; Levesque et al., 2010). It will be interesting to see whether these proteases act as apoplastic effectors, targeting secreted defense proteins for degradation.

The hemibiotrophic infection cycle of *P. infestans* is characterized by major transcriptional changes that accompany the transition from biotrophy to necrotrophy. The *RXLR* genes, particularly, are expressed only during the biotrophic phase, the first three days of infection. This phase is distinguished also by the downregulation of many genes encoding MAMPs, such as elicitins and CBELs, presumably to reduce the elicitation of PTI (Haas et al., 2009). The genomes of both of the obligate biotrophic pathogens, *H. arabidopsidis* and *A. laibachii*, have reduced numbers of genes encoding elicitins and CBELs, and also the NPP1-like protein (NLP) family, perhaps as an alternative strategy to achieve stealth during infection. Numbers of genes encoding plant cell wall degrading enzymes are also reduced in both genomes, again decreasing the potential to activate defenses that are elicited by cell damage (Baxter et al., 2010; Kemen et al., 2011).

A further obvious feature of the obligate biotrophic oomycetes is the loss of a number of biosynthetic pathways, presumably because their products are easily obtained from host cells via the haustorial interface (Fig. 20.1). Genes for nitrate and nitrite reductase and sulfite reductase are missing from *H. arabidopsidis* and *A. laibachii* (Baxter et al., 2010; Kemen et al., 2011). The progressive loss of biosynthetic pathways is a well-documented feature of irreversible biotrophy.

THE FUTURE: FROM SYSTEMS BIOLOGY TO TRANSLATIONAL RESEARCH

The evolution of virulence will depend on the lifestyle of the pathogen under study, but for the biotrophs, the interaction with the host is complex and is driven through the effector complement of the pathogen interacting with a web of host proteins defining the immune response network. These interactions will drive alterations in host gene transcription, changes in the distribution of secreted proteins, probably alterations in chloroplast degradation, and, possibly, the release of nutrients to the pathogen (although proof of this remains elusive). Each of these areas in themselves will be rich in novel data, but the consequence of the interactions of many effectors and their host targets and the combined output will be significant in pathology (Pritchard and Birch, 2011). Therefore, new ways of thinking need to be brought to bear on this increasingly data-rich field. Future pathologists will need to be trained in theoretical approaches to analyze data in order to design the relevant experiments that enable mathematical and bioinformatic tools to reveal the important components of the response networks (Pritchard and Birch, 2011). This will enable us to understand how the pathogens have evolved to suppress this network and how we can design more robust components in our crop plants that resist the attack of pathogen effectors.

Research into the interactions between oomycetes and their hosts will allow us a better understanding of the plant immune system and enable the development of new disease control strategies in crops. In particular, it is important to identify which proteins the plants are desperately trying to hide from pathogen effectors and to understand how the effectors are able to manipulate the plant defense system and acquire the nutrients necessary for their growth.

REFERENCES

Allen RL, Bittner-Eddy PD, Grenville-Briggs LJ, Meitz JC, Rehmany AP, Rose LE, Beynon JL. 2004. Host-parasite coevolutionary conflict between *Arabidopsis* and downy mildew. Science 306: 1957–1960.

Arabidopsis Interactome Mapping Consortium. 2011. Evidence for network evolution in an *Arabidopsis* interactome map. Science 333: 601–607.

Armstrong MR, Whisson SC, Pritchard L, Bos JI, Venter E, et al. 2005. An ancestral oomycete locus contains late blight avirulence gene Avr3a, encoding a protein that is recognized in the host cytoplasm. Proc Natl Acad Sci U S A 102: 7766–7771.

Bailey K, Cevik V, Holton N, Byrne-Richardson J, Sohn KH, Coates M, Woods-Tör A, Aksoy HM, Hughes L, Baxter L, Jones JDG, Beynon J, Holub EB, Tör M. 2011. Molecular cloning of ATR5^{Emoy2} from *Hyaloperonospora arabidopsidis*, an avirulence determinant that triggers RPP5-mediated defense in *Arabidopsis*. Mol Plant Microbe Interact 24: 827–838.

Bakker EG, Toomajian C, Kreitman M, Bergelson J. 2006. A genome-wide survey of R gene polymorphisms in Arabidopsis. Plant Cell 18: 1803–1818.

Baxter L, Tripathy S, Ishaque N, Cabral A, Kemen E, et al. 2010. Signatures of adaptation to obligate biotrophy in the *Hyaloperonospora arabidopsidis* genome. Science 330: 1549–1551.

Bhattacharjee S, Hiller NL, Liolios K, Win J, Kanneganti TD, Young C, Kamoun S, Haldar K. 2006. The malarial host-targeting signal is conserved in the Irish potato famine pathogen. PLoS Pathog 2: 453–465.

Birch PRJ, Whisson SC. 2001. *Phytophthora infestans* enters the genomics era. Mol Plant Pathol 2: 257–263.

Birch PRJ, Rehmany AP, Pritchard L, Kamoun S, Beynon JL. 2006. Trafficking arms: Oomycete effectors enter host plant cells. Trends Microbiol 14: 8–11.

Birch PRJ, Boevink PC, Gilroy EM, Hein I, Pritchard L, Whisson SC. 2008. Oomycete RXLR effectors: Delivery, functional redundancy and durable disease resistance. Curr Opin Plant Biol 11: 373–379.

Birch PRJ, Armstrong M, Bos J, Boevink P, Gilroy EM, Taylor RM, Wawra S, Pritchard L, Conti L, Ewan R, Whisson SC, van West P, Sadanandom A, Kamoun S. 2009. Towards understanding the virulence functions of RXLR effectors of the oomycete plant pathogen Phytophthora infestans. J Exp Bot 60: 1133–1140.

Blair JE, Coffey MD, Park S-Y, Geiser DM, Kang S. 2008. A multi-locus phylogeny for *Phytophthora* utilizing markers from complete genome sequences. Fungal Genet Biol 45: 266–277.

Block A, Li G, Fu ZQ, Alfano JR. 2008. Phytopathogen type III effector weaponry and their plant targets. Curr Opin Plant Biol 11: 396–403.

Bos JI, Kanneganti TD, Young C, Cakir C, Huitema E, Win J, Armstrong MR, Birch PR, Kamoun S. 2006. The C-terminal half of *Phytophthora infestans* RXLR effector AVR3a is sufficient to trigger R3a-mediated hypersensitivity and suppress INF1-induced cell death in *Nicotiana benthamiana*. Plant J 48: 165–176.

Bos JIB, Chaparro-Garcia A, Quesada-Ocampo LM, Gardener BBM, Kamoun S. 2009. Distinct amino acids of the *Phytophthora infestans* effector AVR3a condition activation of R3a hypersensitivity and suppression of cell death. Mol Plant Microbe Interact 22: 269–281.

Bos JI, Armstrong MR, Gilroy EM, Boevink PC, Hein I, Taylor RM, Zhendong T, Engelhardt S, Vetukuri RR, Harrower B, Dixelius C, Bryan G, Sadanandom A, Whisson SC, Kamoun S, Birch PRJ. 2010. *Phytophthora infestans* effector AVR3a is essential for virulence and manipulates plant immunity by stabilizing host E3 ligase CMPG1. Proc Natl Acad Sci U S A 107: 9909–9914.

Brunner F, Rosahl S, Lee J, Rudd JJ, Geiler C, Kauppinen S, Rasmussen G, Scheel D, Nürnberger T. 2002. Pep-13, a plant defense-inducing pathogen-associated pattern from *Phytophthora* transglutaminases. EMBO J 21: 6681–6688.

Cabral A, Stassen JHM, Seidl MF, Bautor J, Parker JE, Van Den Ackerveken G. 2011. Identification of *Hyaloperonospora arabidopsidis* transcript sequences expressed during infection reveals isolate-specific effectors. PLoS One 6(5): e19328.

Champouret N, Bouwmeester K, Rietman H, van der Lee T, Maliepaard C, Heupink A, van de Vondervoort PJ, Jacobsen E, Visser RG, van der Vossen EA, Govers F, Vleeshouwers VG. 2009. *Phytophthora infestans* isolates lacking class I *ipiO* variants are virulent on *Rpi-blb1* potato. Mol Plant Microbe Interact 22: 1535–1545.

Chisholm ST, Coaker G, Day B, Staskawicz BJ. 2006. Host-microbe interactions: Shaping the evolution of the plant immune response. Cell 124: 803–814.

Coates ME, Beynon JL. 2010. *Hyaloperonospora arabidopsidis* as a pathogen model. Annu Rev Phytopathol 48: 329–345.

Damasceno CMB, Bishop JG, Ripoll DR, Win J, Kamoun S, Rose JKC. 2008. Structure of the glucanase inhibitor protein (GIP) family from *Phytophthora* species suggests coevolution with plant endo-β-1,3-glucanases. Mol Plant Microbe Interact 21: 820–830.

Ding J, Cheng HL, Jin XQ, Araki H, Yang YH, Tian DC. 2007. Contrasting patterns of evolution between allelic groups at a single locus in Arabidopsis. Genetica 129: 235–242.

Dong S, Qutob D, Tedman-Jones J, Kuflu K, Wang Y, Tyler BM, Gijzen M. 2009. The *Phytophthora sojae* avirulence locus *Avr3c* encodes a multi-copy RXLR effector with sequence polymorphisms among pathogen strains. PLoS One 4(5): e5556.

Dou D, Kale SD, Wang X, Jiang RHY, Bruce NA, Arredondo FD, Zhang X, Tyler BM. 2008a. RXLR-mediated entry of *Phytophthora sojae* effector *Avr1b* into soybean cells does not require pathogen-encoded machinery. Plant Cell 20: 1930–1947.

Dou D, Kale SD, Wang X, Chen Y, Wang Q, Wang X, Jiang RXY, Arredondo FD, Anderson RG, Thakur PB, McDowell JM, Wang Y, Tyler BM. 2008b. Conserved C-terminal motifs required for avirulence and suppression of cell death by *Phytophthora sojae* effector Avr1b. Plant Cell 20: 1118–1133.

Dou D, Kale SD, Liu T, Tang Q, Wang X, Arredondo FD, Basnayake S, Whisson S, Drenth A, Maclean D, Tyler BM. 2010. Different domains of *Phytophthora sojae* effector *Avr4/6* are recognized by soybean resistance genes *Rps4* and *Rps6*. Mol Plant Microbe Interact 23: 425–435.

Fellbrich G, Romanski A, Varet A, Blume B, Brunner F, Engelhardt S, Felix G, Kemmerling B, Krzymowska M, Nürnberger T. 2002. NPP1, a *Phytophthora*-associated trigger of plant defense in parsley and *Arabidopsis*. Plant J 32: 375–390.

Gan PHP, Rafiqi M, Ellis JG, Jones DA, Hardham AR, Dodds PN. 2010. Lipid binding activities of flax rust AvrM and AvrL567 effectors. Plant Signal Behav 5: 1272–1275.

Gaulin E, Jauneau A, Villalba F, Rickauer M, Esquerré-Tugayé MT, Bottin A. 2002. The CBEL glycoprotein of *Phytophthora parasitica* var-*nicotianae* is involved in cell wall deposition and adhesion to cellulosic substrates. J Cell Sci 115: 4565–4575.

Gaulin E, Dramé N, Lafitte C, Torto-Alalibo T, Martinez Y, Ameline-Torregrosa C, Khatib M, Mazarguil H, Villalba-Mateos F, Kamoun S, Mazars C, Dumas B, Bottin A, Esquerré-Tugayé MT, Rickauer M. 2006. Cellulose binding domains of a *Phytophthora* cell wall protein are novel pathogen-associated molecular patterns. Plant Cell 18: 1766–1777.

Gilroy EM, Taylor RM, Hein I, Sadanandom A, Birch PRJ. 2011a. CMPG1-dependent cell death follows perception of pathogen elicitors at the host plasma membrane and is suppressed by *Phytophthora infestans* RXLR effector AVR3a. New Phytol 190: 653–666.

Gilroy EM, Breen S, Whisson S, Squire J, Hein I, Lokossou A, Boevink P, Pritchard L, Avrova AO, Turnbull D, Kaczmarek M, Cano L, Randall E, Lees A, Govers F, van West P, Kamoun S, Vleeshouwers V, Cooke D, Birch PRJ. 2011b. Presence/absence, differential expression and sequence polymorphisms between *PiAVR2* and *PiAVR2-like* in *Phytophthora infestans* determine virulence on *R2* plants. New Phytol 191: 763–766.

Goker M, Vogelmayr H, Riethmuller A, Oberwinkler F. 2007. How do obligate parasites evolve? A multi-layer phyogenetic analysis of downy mildews. Fungal Genet Biol 44: 1543–1546.

González-Lamothe R, Tsitsigiannis DI, Ludwig AA, Panicot M, Shirasu K, Jones JD. 2006. The U-box protein CMPG1 is required for efficient activation of defense mechanisms triggered by multiple resistance genes in tobacco and tomato. Plant Cell 18: 1067–1083.

Haas BJ, Kamoun S, Zody MC, Jiang RH, Handsaker RE, et al. 2009. Genome sequence and analysis of the Irish potato famine pathogen *Phytophthora infestans*. Nature 461: 393–398.

Hall SA, Allen RL, Baumber RE, Baxter LA, Fisher K, Bittner-Eddy PD, Rose LE, Holub EB, Beynon JL. 2009. Maintenance of genetic variation in plants and pathogens involves complex networks of gene-for-gene interactions. Mol Plant Pathol 10: 449–457.

Halterman DA, Chen Y, Sopee J, Berduo-Sandoval J, Sánchez-Pérez A. 2010. Competition between *Phytophthora infestans* effectors leads to increased aggressiveness on plants containing broad-spectrum late blight resistance. PLoS One 5(5): e10536.

Heese A, Hann DR, Gimenez-Ibanez S, Jones AME, He K, Li J, Schroeder JI, Peck SC, Rathjen JP. 2007. The receptor-like kinase SERK3/BAK1 is a central regulator of innate immunity in plants. PNAS 104: 12217–12222.

Hein I, Gilroy EM, Armstrong MR, Birch PRJ. 2009. The zig-zag-zig in oomycete-plant interactions. Mol Plant Pathol 10: 547–562.

Hiller NL, Bhattacharjee S, van Ooij C, Liolios K, Harrison T, Lopez-Estrano C, Haldar K. 2004. A host-targeting signal in virulence proteins reveals a secretome in malarial infection. Science 306: 1934–1937.

Jones JD, Dangl JL. 2006. The plant immune system. Nature 444: 323–329.

Kale SD, Gu B, Capelluto DGS, Dou D, Feldman E, Rumore A, Arredondo FD, Hanlon R, Fudal I, Rouxel T, Lawrence CB, Shan W, Tyler BM. 2010. External lipid PI3P mediates entry of eukaryotic pathogen effectors into plant and animal host cells. Cell 142: 284–295.

Kamoun S. 2003. Molecular genetics of phytopathogenic oomycetes. Eukaryot Cell 2: 191–199.

Kamoun S. 2006. A catalogue of the effector secretome of plant pathogenic oomycetes. Annu Rev Phytopathol 44: 41–60.

Kanzaki H, Saitoh H, Ito A, Fujisawa S, Kamoun S, Katou S, Yoshioka H, Terauchi R. 2003. Cytosolic HSP90 and HSP70 are essential components of INF1-mediated hypersensitive response and non-host resistance to Pseudomonas cichorii in *Nicotiana benthamiana*. Mol Plant Pathol 4: 383–391.

Kanzaki H, Saitoh H, Takahashi Y, Berberich T, Ito A, Kamoun S, Terauchi R. 2008. NbLRK1, a lectin-like receptor kinase protein of *Nicotiana benthamiana*, interacts with *Phytophthora infestans* INF1 elicitin and mediates INF1-induced cell death. Planta 228: 977–987.

Kelley BS, Lee S-J, Damasceno CMB, Chakravarthy S, Kim B-D, Martin GB, Rose JKC. 2010. A secreted effector protein (SNE1) from *Phytophthora infestans* is a broadly acting suppressor of programmed cell death. Plant J 62: 357–366.

Kemen E, Gardiner A, Scultz-Larsen T, Kemen AC, Balmuth AL, Robert-Seilaniantz A, Bailey K, Studholme DJ, MacLean D, Jones JDG. 2011. Gene gain and loss during evolution of obligate parasitism in the white rust pathogen of *Arabidopsis thaliana*. PLoS Biol 9: e1001094.

Lamour KH, Win J, Kamoun S. 2007. Oomycete genomics: New insights and future directions. FEMS Microbiol Lett 274: 1–8.

Levesque CA, De Cock WAM. 2004. Molecular phylogeny and taxonomy of the genus *Pythium*. Mycol Res 108: 1363–1383.

Levesque CA, Brouwer H, Cano L, Hamilton JP, Holt C, et al. 2010. Genome sequence of the necrotrophic plant pathogen *Pythium ultimum* reveals original pathogenicity mechanisms and effector repertoire. Genome Biol 11: R73.

Liu Z, Bos JIB, Armstrong M, Whisson SC, da Cunha L, Torto T, Win J, Avrova AO, Wright F, Birch PRJ, Kamoun S. 2005. Patterns of diversifying selection in the phytotoxin-like *scr74* gene family of *Phytophthora infestans*. Mol Biol Evol 22: 659–672.

Livaja M, Zeidler D, von Rad U, Durner J. 2008. Trancriptional responses of *Arabidopsis thaliana* to the bacterial derived PAMPs harpin and lipopolysaccharide. Immunobiology 213: 161–171.

Marti M, Good RT, Rug M, Knuepfer E, Cowman AF. 2004. Targeting malaria virulence and remodeling proteins to the host erythrocyte. Science 306: 1930–1933.

Mukhtar MS, Carvunis A-R, Dreze M, Epple P, Steinbrenner J, Moore J, Tasan M, Galli M, Hao T, Nishimura MT, Pevzner SJ, Donovan SE, Ghamsari L, Santhanam B, Romero V, Poulin MM, Gebreab F, Gutierrez BJ, Tam S, Monachello D, Boxem M, Harbort CJ, McDonald N, Gai L, Chen H, He Y, European Union Effectoromics Consortium, Vandenhaute J, Roth FP, Hill DE, Ecker JR, Vidal M, Beynon J, Braun P, Dangl JL. 2011. Independently evolved virulence effectors converge onto hubs in a plant immune system network. Science 333: 596–601.

Oh S-K, Young C, Lee M, Oliva R, Bozkurt TO, Cano LM, Win J, Bos JIB, Liu H-Y, van Damme M, Morgan W, Choi D, van der Vossen E, Vleeshouwers V, Kamoun S. 2009. *In planta* expression screens of *Phytophthora infestans* RXLR effectors reveal diverse phenotypes, including activation of the *Solanum bulbocastanum* disease resistance protein Rpi-blb2. Plant Cell 21: 2928–2947.

Orsomando G, Lorenzi M, Raffaelli N, Dalla Rizza M, Mezzetti B, Ruggieri S. 2001. Phytotoxic protein PcF, purification, characterization, and cDNA sequencing of a novel hydroxyproline-containing factor secreted by the strawberry pathogen Phytophthora cactorum. J Biol Chem 276: 21578–21584.

Peart JR, Lu R, Sadanandom A, Malcuit I, Moffett P, Brice DC, Schauser L, Jaggard DA, Xiao S, Coleman MJ, Dow M, Jones JD, Shirasu K, Baulcombe DC. 2002. Ubiquitin ligase-associated protein SGT1 is required for host and nonhost disease resistance in plants. Proc Natl Acad Sci U S A 99: 10865–10869.

Pritchard L, Birch PRJ. 2011. A systems biology perspective on plant-microbe interactions; biochemical and structural targets of pathogen effectors. Plant Sci 180: 584–603.

Qutob D, Kamoun S, Gijzen M. 2002. Expression of a *Phytophthora sojae* necrosis-inducing protein occurs during transition from biotrophy to necrotrophy. Plant J 32: 361–373.

Qutob D, Kemmerling B, Brunner F, Küfner I, Engelhardt S, et al. 2006. Phytotoxicity and innate immune responses induced by Nep1-like proteins. Plant Cell 18: 3721–3744.

Qutob D, Tedman-Jones J, Dong S, Kuflu K, Pham H, Wang Y, Dou D, Kale SD, Arredondo FD, Tyler BM, Gijzen M. 2009. Copy number variation and transcriptional polymorphisms of *Phytophthora sojae* RXLR effector genes *Avr1a* and *Avr3a*. PLoS ONE 4: e5066.

Raffaele S, Farrer RA, Cano L, Studholme DJ, MacLean D, Thines M, Jiang RHY, Zody MC, Kunjeti SG, Donofrio NM, Meyers BC, Nusbaum C, Kamoun S. 2010. Genome evolution following host jumps in the Irish potato famine pathogen lineage. Science 330: 1540–1543.

Rehmany AP, Gordon A, Rose LE, Allen RL, Armstrong MR, Whisson SC, Kamoun S, Tyler BM, Birch PR, Beynon JL. 2005. Differential recognition of highly divergent downy mildew avirulence gene alleles by *RPP1* resistance genes from two *Arabidopsis* lines. Plant Cell 17: 1839–1850.

Rose JK, Ham KS, Darvill AG, Albersheim P. 2002. Molecular cloning and characterization of glucanase inhibitor proteins: Co-evolution of a counter-defense mechanism by plant pathogens. Plant Cell 14: 1329–1345.

Rose LE, Bittner-Eddy PD, Langley CH, Holub EB, Michelmore RW, Beynon JL. 2004. The maintenance of extreme amino acid diversity at the disease resistance gene, *RPP13*, in *Arabidopsis thaliana*. Genetics 166: 1517–1527.

Schornack S, Huitema E, Cano LM, Bozkurt TO, Oliva R, Van Damme M, Schwizer S, Raffaele S, Chaparro-Garcia A, Farrer R, Segretin ME, Bos J, Haas BJ, Zody MC, Nusbaum C, Win J, Thines M, Kamoun S. 2009. Ten things to know about oomycete effectors. Mol Plant Pathol 10: 795–803.

Schornack S, van Damme M, Bozkurt TO, Cano LM, Smoker M, Thines M, Gaulin E, Kamoun S, Huitema E. 2010. Ancient class of translocated oomycete effectors targets the host nucleus. Proc Natl Acad Sci 107: 17421–17426.

Shabab M, Shindo T, Gu C, Kaschani F, Pansuriya T, Chintha R, Harzen A, Colby T, Kamoun S, van der Hoorn RAL. 2008. Fungal effector protein AVR2 targets diversifying defense-related cys proteases of tomato. Plant Cell 20: 1169–1183.

Shan W, Cao M, Leung D, Tyler BM. 2004. The *Avr1b* locus of *Phytophthora sojae* encodes an elicitor and a regulator required for avirulence on soybean plants carrying resistance gene *Rps1b*. Mol Plant Microbe Interact 17: 394–403.

Sohn KH, Lei R, Nemri A, Jones DG. 2007. The downy mildew effector proteins ATR1 and ATR13 promote disease susceptibility in *Arabidopsis thaliana*. Plant Cell 19: 4077–4090.

Song J, Win J, Tian M, Schornack S, Kaschani F, Ilyas M, van der Hoorn R, Kamoun S. 2009. Apoplastic effectors secreted by two unrelated eukaryotic plant pathogens target the tomato defence protease Rcr3. Proc Natl Acad Sci U S A 106: 1654–1659.

Stassen JHM, Van Den Ackerveken G. 2011. How do oomycete effectors interfere with plant life? Curr Opin Plant Biol 14: 1–8.

Takahashi Y, Nasir KH, Ito A, Kanzaki H, Matsumura H, Saitoh H, Fujisawa S, Kamoun S, Terauchi R. 2007. A high-throughput screen of cell-death-inducing factors in *Nicotiana benthamiana* identifies a novel MAPKK that mediates INF1-induced cell death signaling and non-host resistance to *Pseudomonas cichorii*. Plant J 49: 1030–1040.

Thines M, Kamoun S. 2010. Oomycete–plant coevolution: Recent advances and future prospects. Curr Opin Plant Biol 13(4): 427–433.

Tian M, Huitema E, da Cunha L, Torto-Alalibo T, Kamoun S. 2004. A Kazal-like extracellular serine protease inhibitor from *Phytophthora infestans* targets the tomato pathogenesis-related protease P69B. J Biol Chem 279: 26370–26377.

Tian M, Benedetti B, Kamoun S. 2005. A second Kazal-like protease inhibitor from *Phytophthora infestans* inhibits and interacts with the apoplastic pathogenesis-related protease P69B of tomato. Plant Physiol 138: 1785–1793.

Tian M, Win J, Song J, van der Hoorn R, van der Knaap E, Kamoun S. 2007. A *Phytophthora infestans* cystatin-like protein targets a novel tomato papain-like apoplastic protease. Plant Physiol 143: 364–377.

Tyler BM, Tripathy S, Zhang X, Dehal P, Jiang RH, et al. 2006. *Phytophthora* genome sequences uncover evolutionary origins and mechanisms of pathogenesis. Science 313: 1261–1266.

Umemoto N, Kakitani M, Iwamatsu A, Yoshikawa M, Yamaoka M, Ishida I. 1997. The structure and function of a soybean beta glucan-elicitor binding protein. Proc Natl Acad Sci U S A 94: 1029–1034.

Van Loon LC, Rep M, Pieterse CMJ. 2006. Significance of inducible defense-related proteins in infected plants. Annu Rev Phytopathol 44: 135–162.

van Poppel PMJA, Guo J, de Vondervoort PJIV, Jung MWM, Birch PRJ, Whisson SC, Govers F. 2008. The *Phytophthora infestans* avirulence gene *Avr4* encodes an RXLR- dEER effector. Mol Plant Microbe Interact 21: 1460–1470.

Vleeshouwers VG, Rietman H, Krenek P, Champouret N, Young C, et al. 2008. Effector genomics accelerates discovery and functional profiling of potato disease resistance and *Phytophthora Infestans* avirulence genes. PLoS One 3: e2875.

Whisson SC, Boevink PC, Moleleki L, Avrova AO, Morales JG, et al. 2007. A translocation signal for delivery of oomycete effector proteins into host plant cells. Nature 450: 115–119.

Whitaker JW, McConkey GA, Westhead DR. 2009. The transferome of metabolic genes explored: Analysis of the horizontal transfer of enzyme encoding genes in unicellular eukaryotes. Genome Biol 10: R36.

Yaeno T, Li H, Chaparro-Garcia A, Schornack S, Koshiba S, Watanabe S, Kigawa T, Kamoun S, Shirasu K. 2011. Phosphatidylinositol monophosphate-binding interface in the oomycete RXLR effector AVR3a is required for its stability in host cells to modulate plant immunity. Proc Natl Acad Sci 108: 14682–14687.

Yoshioka H, Numata N, Nakajima K, Katou S, Kawakita K, Rowland O, Jones JDG, Doke N. 2003. Nicotiana benthamiana gp91phox homologs NbrbohA and NbrbohB participate in H_2O_2 accumulation and resistance to *Phytophthora infestans*. Plant Cell Online 15: 706–718.

Zipfel C. 2008. Pattern-recognition receptors in plant innate immunity. Curr Opin Immunol 20: 10–16.

CHAPTER 21

EVOLUTION AND GENOMICS OF THE PATHOGENIC *CANDIDA* SPECIES COMPLEX

GERALDINE BUTLER, MICHAEL LORENZ, and NEIL A. R. GOW

INTRODUCTION: THE ORGANISMS AND THEIR PATHOECOLOGY

A limited group of about eight species account for the vast majority of diseases caused by *Candida*. In evolutionary terms, the members of this group are relatively highly diverged, with common ancestry estimated to be at between 100 and 900 million years ago—equivalent to the evolutionary distance between primates and bony fish (Johnson, 2003; Johnston, 2006). These organisms are commonly isolated from the oral and vaginal cavity and less frequently from the gut of healthy individuals. It is estimated that approximately 30–70% of healthy individuals are culture positive for at least one *Candida* species. *Candida* species are all opportunistic pathogens indicating that most infections are associated with disturbance of normal health and immunity. In immunocompetent individuals, they can establish mucosal infections called "thrush" or "yeast infections" and they can be occasional agents of skin and nail disease. In immunocompromised patients with underlying immune suppression due to malignancy, soft or hard organ transplantation, or surgical trauma, they cause life-threatening invasive systemic and bloodstream infections for which crude mortality rates are commonly around 40%, even with antifungal chemotherapy. Low neutrophil levels (neutropenia) and low CD4 counts are risk indicators of invasive and mucosal disease, respectively. Predisposing treatments and conditions that can lead to candidiasis include the administration of long-term courses of antibacterial antibiotics, which result in the suppression of the competitive bacterial microflora of the body, cancer chemotherapy, infancy, and old age. Systemic hematological disease occurs mainly as a secondary consequence of medical treatment for some other conditions or as a result of severe injury and trauma. Systemic disease is most common in regions of the world where advanced medical procedures are common. And while oropharygeal candidiasis in HIV/AIDS is extremely common, the incidence is reducing in the developed world due to the availability of highly active antiviral therapy (HAART). However, *Candida* infections due to HIV/AIDS remain a very significant problem in

Evolution of Virulence in Eukaryotic Microbes, First Edition. Edited by L. David Sibley, Barbara J. Howlett, and Joseph Heitman.

other regions of the world (Pfaller and Diekema, 2007, 2010; Pfaller et al., 2010).

Vulvovaginitis of women remains the most common of the diseases caused by *Candida* species with a majority of women of child-bearing years suffering one or more incidence of this disease. Effective anti-*Candida* antifungals are readily available, safe, and effective, but they do not prevent episodic relapses in many women. Recurrent vaginitis remains a problematic issue causing significant physiological and psychological distress.

There are approximately 314 species categorized as belonging to the *Candida* genus (Lachance et al., 2011). However, the definition of *Candida* is problematic as it refers to imperfect species lacking an observed sexual cycle. The *Candida* genus is therefore polyphyletic, including species that are apparently asexual or parasexual (such as *Candida albicans*) and excluding closely related but fully sexual species (such as *Debaryomyces hansenii*). In fact, as sexual (teleomorphic) forms were discovered, several *Candida* species were renamed. For example, *Candida famata*, *Candida lusitaniae*, and *Candida guilliermondii* are now known as *D. hansenii*, *Clavispora lusitaniae*, and *Meyerozyma guilliermondii*. Some have undergone several name changes (e.g., *C. guilliermondii* to *Pichia guilliermondii* and then to *M. guilliermondii*) as their phylogenetic position was clarified (Kurtzman et al., 2011). However, many of the asexual and sexual relatives of *C. albicans* belong to a single clade and share the evolutionary property that CTG encodes serine rather than leucine (Ohama et al., 1993; Santos and Tuite, 1995). This clade (commonly referred to as the CTG clade) includes most of the species that are pathogenic to man. Notable exclusions are *Candida glabrata* and relatives such as *Candida bracarensis* and *Candida nivariensis*, which are much more closely related to *Saccharomyces cerevisiae* than to other species (Correia et al., 2006; Lockhart et al., 2009; Wong et al., 2003), and *Candida krusei* (also known as *Pichia kudriavzevii* or *Issatchenkia orientalis*), whose exact phylogenetic position remains unclear but which lies within the Saccharomycetaceae and outside the CTG clade (Fig. 21.1).

Four species cause the majority of *Candida* infections in humans and are responsible for 8–10% of the total nosocomial bloodstream infections isolated in U.S. hospitals (Pfaller and Diekema, 2007). Three species (*C. albicans*, *Candida tropicalis*, and *Candida parapsilosis*) are members of the CTG clade and, together with *C. glabrata*, are responsible for about 90% of *Candida* infections (Pfaller and Diekema, 2007). *C. albicans* is the most common cause of infection, but other species are rapidly increasing. In North America, *C. glabrata* is the second most common cause of infection (23.5%), whereas in Latin America, *C. parapsilosis* is more frequently isolated (25.6%) and *C. glabrata* is relatively rare (5.2%) (Pfaller et al., 2010). In Europe, *C. glabrata* and *C. parapsilosis* are isolated at similar frequencies (15.7% and 13.7%, respectively; Pfaller et al., 2010). The increasing incidence of *C. glabrata* infection may be related to an inherent resistance of this species to both azole and echinocandin drugs (Chapeland-Leclerc et al., 2010; Cleary et al., 2008; Panackal et al., 2006; Pfaller and Diekema, 2010; Pfaller et al., 2010). Resistance to echinocandins is usually caused by mutations within the *FKS1* gene (which encodes β-glucan synthase and is the presumed target of the drugs) (Chapeland-Leclerc et al., 2010; Cleary et al., 2008; Thompson et al., 2008). However, the mechanism underlying similar breakthrough resistance observed in *C. parapsilosis* is currently unknown (Pfeiffer et al., 2010).

Relative incidence of specific species in patients is, however, a combination of both inherent virulence (the ability to cause disease in an exposed animal) and

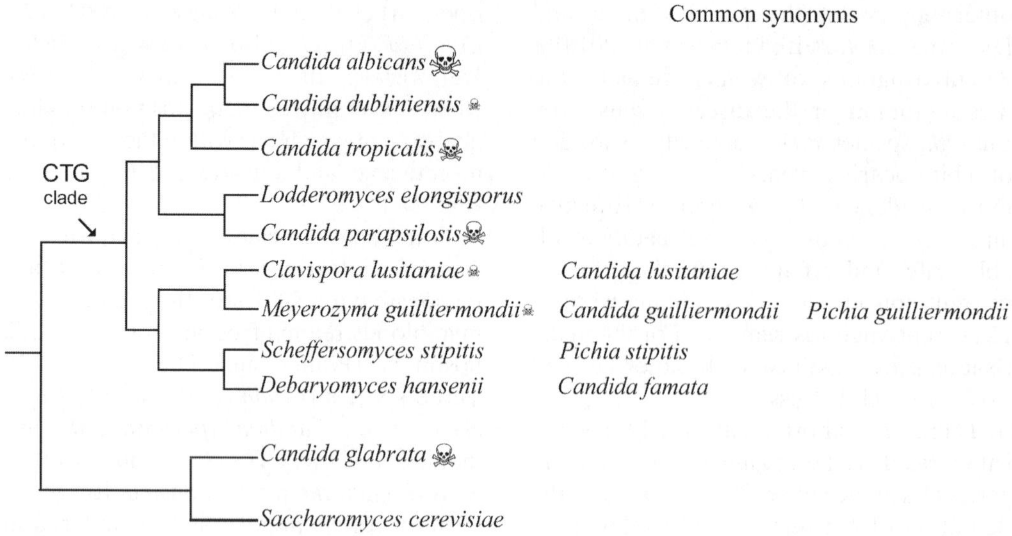

Figure 21.1 Phylogenetic relationship of sequenced *Candida* species. The relationship is based on a supertree constructed as described in Fitzpatrick et al. (2006). The most commonly used synonyms for the species are shown. The most pathogenic species are indicated with a skull and crossbones; the size of the image correlates with their relative importance as pathogens.

the ecological factors that contribute to the likelihood that an animal will be exposed to that species, potentially complicating the evolutionary analysis of virulence traits across species. Careful virulence studies in rodents have elucidated a hierarchy of pathogenicity that broadly follows the patient data. Eight species were tested using a disseminated mouse model, falling into three groups, highly virulent (*C. albicans* and *C. tropicalis*), less virulent (*C. glabrata*, *Cl. lusitaniae*, and *Candida kefyr*), and essentially avirulent (*C. parapsilosis*, *M. guilliermondii, and C. krusei*), with this last group unable to kill even immunosuppressed mice (Arendrup et al., 2002; Bistoni et al., 1984). A similar pattern emerged from an *ex vivo* rat tongue model and in gastrointestinal colonization, though not all species were tested (de Repentigny et al., 1992; Howlett, 1976; Mellado et al., 2000). Thus, there is a reasonable concordance between epidemiological and animal data, though *C. parapsilosis*, in particular, might be less virulent (in mice, at least) than its clinical incidence would suggest.

GENOME ANALYSIS: *C. gLABRATA*

The genome sequence of *C. glabrata* was reported in 2004, enabling comparison with *S. cerevisiae* and other nonpathogenic yeast species (Dujon et al., 2004). The *C. glabrata* genome appears to have undergone a striking genome-wide reduction, with many examples of gene loss. These include the loss of the galactose, phosphate and nicotinic acid metabolism pathways, and loss of several genes involved in nitrogen and sulfur metabolism (Domergue et al., 2005; Dujon et al., 2004). In contrast, there has been a species-specific expansion of the EPA genes in the subtelomeric regions (De Las Penas et al., 2003). EPA genes encode a family of glycolipid proteins that are required for virulence and for adhesion. Similarly, *C. glabrata* contains a large tandem array of *YPS* genes encoding extracellular glycosylphosphatidylinositol (GPI)-linked aspartyl proteases similar to the yapsin proteases of *S. cerevisiae*; this cluster is unique to *C. glabrata* and is also required for virulence (Kaur et al., 2007).

GENOME ANALYSIS: CTG CLADE

The genome sequences of nine *Candida* species from the CTG clade have now been published (Fig. 21.1). The species fall into two major groups: those with diploid genomes (*C. albicans*, *Candida dubliniensis*, *C. tropicalis*, *C. parapsilosis*, *Lodderomyces elongisporus*) and a group that is predominantly haploid and sexual (*M. guilliermondii*, *Cl. lusitaniae*, *D. hansenii*, and *Scheffersomyces stipitis* [previously *Pichia stipitis*]) (Fig. 21.1).

The genome sequence of *C. albicans* was first elucidated to 10.9 × coverage in 2004 (Jones et al., 2004). The diploid genome complicated the assembly, and the original annotation, although useful, contained several errors. Some regions (with little heterozygosity) were presented as single contigs, and in others, a diploid assembly was generated. Many of the annotation errors were corrected by a manual curation published in 2005 (Braun et al., 2005), which described a haploid assembly. In 2007, a relatively complete diploid assembly was reported using optical mapping in conjunction with comparative genome analysis (van het Hoog et al., 2007). Comparison with the *S. cerevisiae* genome led to the identification of a number of gene families that have undergone expansion in *C. albicans* and are associated with virulence (Braun et al., 2005; Jones et al., 2004; van het Hoog et al., 2007). These include secreted aspartyl proteinases (SAPs), secreted lipases (LIPs), iron transporters, oligopeptide transporters, amino acid permeases, and agglutinin-like sequence (ALS) adhesins. The ALS family is particularly well studied. There are eight members in the *C. albicans* genome, each containing a GPI anchor (Hoyer, 2001; Hoyer et al., 2008). The proteins are localized at the cell surface and are required for adhesion. Als1 and Als3 are associated with adherence to endothelial and epithelial cells, as well as to plastic surfaces (Nobile et al., 2006; Zhao et al., 2004). Als3 also has other important roles that contribute to virulence. It is required for endocytosis of *C. albicans* cells via a clathrin-dependent mechanism (Moreno-Ruiz et al., 2009; Phan et al., 2007). Phan et al. (2007) showed that Als3 binds to N-cadherin on endothelial cells and E-cadherin on epithelial cells. In addition, Als3 is required for the uptake of iron from ferritin by hyphal cells and is therefore important for growth during infection of the host (Almeida et al., 2008).

The power of comparative genomic approaches relies on the availability of information from closely related species. In-depth analysis of *C. albicans* was greatly facilitated by the sequence of a second isolate (WO-1), together with the genomes of *C. tropicalis*, *C. parapsilosis*, *L. elongisporus*, *M. guilliermondii*, and *Cl. lusitaniae* (Butler et al., 2009). The comparative analysis included the genome of *D. hansenii*, sequenced in 2004 (Dujon et al., 2004), and genomes of nine members of the *Saccharomyces* clade, including *C. glabrata* (Butler et al., 2009). Comparison of the pathogenic species with the nonpathogenic species revealed that 21 families are enriched in the pathogens (defined as *C. albicans*, *C. tropicalis*, *C. parapsilosis*, *Cl. lusitaniae*, *M. guilliermondii*, and *C. glabrata*). Five families are enriched in the common pathogens (*C. albicans*, *C. tropicalis*, and *C. parapsilosis*) relative to the others.

The cell surface is obviously the primary interface between a pathogen and its host, and the importance of this is obvious in the remarkable concentration of cell wall, plasma membrane, and secreted proteins among the overrepresented families: At least 14 of the 21 pathogen-enriched families fall into this category (Butler et al., 2009). These include several families of adhesins, including the Als proteins, and cell wall remodeling enzymes, but also large families involved in nutrient acquisition, such as oligopeptide and amino acid permeases, ferric reductases, oxidoreductases, and secreted LIPs. Thus, the evolution of pathogenesis has emphasized the expansion of cell surface capabilities.

The genome-wide comparison between *C. albicans* and *C. dubliniensis* is informative regarding the mechanisms that underlie these changes in gene content. In addition to sequence divergence through random mutational drift, genes can be gained or lost by chromosomal translocations or inversions, gene duplication, gene decay (forming pseudogenes) specific deletions, or tandem duplications. There is evidence for all of these mechanisms affecting pathogenesis-related genes. For example, there are 115 pseudogenes in *C. dubliniensis*, many of which are orthologs of intact genes associated with filamentous growth in *C. albicans*. Some of these are found in the IFA family, which has weak similarity to a group of viral leucine repeat proteins; in *C. dubliniensis*, 14 of the 21 predicted IFA sequences are pseudogenes, while only 6 of the 31 *C. albicans* loci are, indicating an ongoing process of gene loss that is occurring more rapidly in the less pathogenic species (Jackson et al., 2009). The cell surface hyphal regulated (*HYR1*) gene, part of the larger IFF gene family, has also been specifically deleted in *C. dubliniensis* (Jackson et al., 2009; Moran et al., 2004). While deletion of the *C. albicans HYR1* does not have a virulence phenotype (Bailey et al., 1996), two other IFF family members, *IFF4* and *IFF11*, do attenuate virulence in the disseminated mouse model when mutated (Bates et al., 2007; Fu et al., 2008; Kempf et al., 2009).

Gene acquisition by duplication is also apparent. *C. albicans* has 15 related, but functionally uncharacterized, genes located primarily in subtelomeric regions and named *TLO* for this reason (van het Hoog et al., 2007), while *C. dubliniensis* has only two such genes, one of which is subtelomeric. Phylogenetic analysis makes clear that this family has expanded from a single precursor (Jackson et al., 2009); the subtelomeric regions are frequently the sites where new genes are gained, as is the case with the *C. glabrata EPA* adhesin family (Kaur et al., 2007). Usefully, the smaller *TLO* family in *C. dubliniensis* allowed functional characterization, which is difficult in *C. albicans* due to the sheer size of the family. Deletion of *CdTLO1* reduced hyphal growth, and this could be complemented by at least two *C. albicans* homologs, providing a plausible link between this family and virulence properties (Jackson et al., 2009).

Chromosome-scale events also drive gene gain and loss. *C. albicans* and *C. dubliniensis* both have multiple *ALS* adhesin genes, but the less pathogenic species lacks *ALS3* (Jackson et al., 2009), encoding a protein involved in biofilm formation, invasion of epithelial and endothelial cells, and iron acquisition in the host; it also mediates a specific interaction with the bacterium *Streptococcus gordonii* (Almeida et al., 2008; Nobile et al., 2006; Phan et al., 2007; Silverman et al., 2010; see Liu and Filler, 2011) for review). *ALS3* apparently arose specifically in *C. albicans* via a translocation/duplication event from the ancestral *ALS1* locus. *ALS1* also gave rise to *CaALS5*, also missing from *C. dubliniensis*, while the latter species has two genes similar to *CaALS2* (Jackson et al., 2009). Similarly, the secreted aspartyl protease family has been expanded via chromosomal inversions, such that *C. albicans* has three genes, *SAP4*, *SAP5*, and *SAP6*, while *C. dubliniensis* has a single homolog, *SAP456* (Jackson et al., 2009).

It is worth noting that differences in gene structure can be seen within species as well. The Als proteins contain a central tandem repeat region of variable length, flanked by an immunoglobulin-like amino terminal domain and a serine/threonine-rich carboxy terminus with a GPI anchoring sequence (Frank et al., 2010; Hoyer, 2001; Loza et al., 2004), similar in structure, if not primary sequence, to the Flo proteins of *S. cerevisiae* and Epa proteins of *C. glabrata* (Cormack et al., 1999; Lo and Dranginis, 1998). The variable repeat region provides allelic diversity: The number of repeats can differ between alleles in the same strain (Oh et al., 2005; Zhao et al., 2003, 2007) and between strains; for instance, over 60 alleles of *ALS7* have been identified (Zhang et al., 2003). These differences have functional consequences in substrate binding properties in the Als proteins (Frank et al., 2010; Oh et al., 2005), as was also shown for *S. cerevisiae FLO11* (Verstrepen et al., 2005).

It has long been assumed that at least most of these expanded gene families are important adaptations that are required for growth and disease progression in the host, and there is some experimental support for this, such as the avirulent phenotypes of the *iff4*Δ and *iff11*Δ mutants, described above (Bates et al., 2007; Fu et al., 2008; Kempf et al., 2009). As suggested above for the *TLO* family, however, the sheer number of these genes often precludes a simple conclusion via the molecular Koch's postulates, despite an often extensive literature (especially for the *CaALS* and *CaSAP* families) attempting to do so. An *als3*Δ mutant, for instance, is fully virulent in the disseminated mouse model (Cleary et al., 2011) despite the variety of seemingly important *in vitro* phenotypes (see Liu and Filler, 2011). Genetic redundancy is presumed to account for the limited *in vivo* effects, and there is some support for this notion from other species; indeed, a *C. glabrata* strain lacking three *EPA* genes (*EPA1*, *6*, *7*) is attenuated in a urinary tract infection model (Domergue et al., 2005). The effect of deleting *C. dubliniensis TLO* genes was described above (Jackson et al., 2009); similarly, deletion of the two adjacent LIP genes (*LIP1*, *2*) in *C. parapsilosis* attenuates virulence (Gacser et al., 2007), indirectly implicating the much larger *C. albicans* family in virulence as well. Nevertheless, the direct evidence for a role for these expanded families remains limited.

In summary, the genomes of the related species *C. albicans* and *C. dubliniensis* are quite plastic, partly explained by the abundance of the major repeat sequences (MRSs), which are found only in these two species and in the next closest relative, *C. tropicalis*, and by evidence of rampant (though seemingly now extinct) retrotransposon activity (Butler et al., 2009; Jackson et al., 2009). Combined with broader phenomena, such as chromosomal rearrangements and subtelomeric "gene factories," this has driven gene gain and loss in critical functions at the host–pathogen interface.

GENE EXPRESSION DIFFERENCES

Clearly, changes in gene content are not the only evolutionary adaptation among pathogenic fungi. Gene expression and transcriptional regulatory networks have also adapted significantly. This was readily apparent in the transcriptional profile analysis of *C. albicans* and *S. cerevisiae* following phagocytosis by macrophages: *C. albicans* differentially regulates more than 10 times the number of genes than the nonpathogenic *S. cerevisiae* (545 vs. 53) (Lorenz et al., 2004). A similar, though not as extreme, pattern was seen in the response of these two species to neutrophils (Rubin-Bejerano et al., 2003). This disparity in the magnitude of transcriptional response was also observed between *C. albicans* and *C. dubliniensis* grown on reconstituted human epithelial (RHE) layers, with about twice as many genes upregulated in the strong pathogen (Spiering et al., 2010). The hypothesis that the robust response of *C. albicans* to relevant host cells is a key contributor to virulence is certainly appealing but, as yet, has not been completely proven.

This is apparent at a single-gene level as well. *NRG1*, a transcriptional repressor of hyphal morphogenesis, is much more tightly controlled following hyphal induction in *C. albicans* than in *C. dubliniensis*, and deletion of *CdNRG1* enhances filamentous growth in this less pathogenic species (Moran et al., 2007). Among the genes differentially regulated between *C. albicans* and *C. dubliniensis* in the RHE profiling experiment described above was *SFL2*, a moderately divergent transcription factor found in both species; deletion of *SFL2* impairs hyphal growth in several inducing conditions and reduces damage to the RHE layer. Expression of *CaSFL2* also enhances filamentation in *C. dubliniensis* (Spiering et al., 2010).

It is tempting to speculate that the expression differences of *SFL2* and *NRG1* between *C. albicans* and *C. dubliniensis* underlies the phenotypic adaptations associated with these two proteins, but the last decade has seen numerous examples of network "rewiring" during evolution whereby highly conserved transcriptional regulators have acquired sometimes radically different sets of target genes. Perhaps the most ironic example is that of Gal4p, the classic regulator of galactose utilization in *S. cerevisiae* (and the heart of the two-hybrid system) that, in *C. albicans*, regulates glycolysis despite recognizing the same DNA sequence in both species (Askew et al., 2009). This is not an isolated example as numerous highly conserved transcription factors have been retasked between these species, including Adr1p, Cat8p, Cbf1p, Hmo1p, Mcm1p, Rap1p, and Tbf1p (Lavoie et al., 2009; Ramirez and Lorenz, 2009; Tuch et al., 2008). In other cases, such as for mating-type specification, the overall logic of the regulatory circuit has been maintained, though the sets of genes under regulation have diverged markedly (Booth et al., 2010). These evolving transcriptional networks thus provide another source of species specificity relevant to pathogenesis.

MORPHOLOGY AND VIRULENCE

All *Candida* species grow predominantly as budding yeasts, although *C. albicans* and *C. dubliniensis* can form true hyphae or germ tubes. *C. krusei*, *C. parapsilosis*, and *C. tropicalis* can grow as "pseudohyphae," which are extended buds that remain attached after cell division. In culture, *C. albicans* and *C. dubliniensis* also can form chlamydospores—the clinical significance of this resting structure is obscure. In *C. albicans*, mutants that are locked in the yeast form or in the filamentous hyphal form are attenuated in virulence (Lo et al., 1997; Sudbery, 2011; Zheng and Wang, 2004), yet other dimorphic fungal pathogens form yeasts exclusively in human tissues. *C. dubliniensis*, which is the nearest evolutionary relative to *C. albicans*, can form true hyphae but is significantly less virulent than *C. albicans* and *C. glabrata*, which is unable to form hyphae or pseudohyphae *in vivo*. In addition, a recent systematic screen of a homozygous mutant library showed that approximately half of the virulence attenuated strains had normal capacity for yeast–hypha morphogenesis (Noble et al., 2010). Therefore, there is no simple relationship between the capacity for filamentous growth of *Candida*, or other fungal pathogens, and the evolution of virulence in fungal species.

ANALYSIS OF THE MATING PATHWAY

Candida species are by definition asexual; they have not been observed to form sexual structures, such as ascospores. *C. albicans* is the best characterized and, until recently, was believed to be completely asexual. However, Hull and Johnson (1999) identified two idiomorphs of a mating type-like locus (*MTL***a** and *MTL*α) from the ongoing genome sequencing project and suggested that mating might occur. Subsequent experiments demonstrated mating between diploid cells that were hemizygous or homozygous for a single *MTL* idiomorph (Hull et al., 2000; Magee and Magee, 2000). Mating efficiencies are greatly increased in cells with an opaque phenotype (Miller and Johnson, 2002). The ability to switch between white and opaque phenotypes was first described in 1985 (Slutsky et al., 1985). The cells have distinct and very different appearances; opaque cells are elongated with a pimply surface, whereas white cells are round. The phenotypes vary in their ability to absorb the dye phloxine B, which is often used a method of distinguishing opaque (pink) from white cells (Anderson et al., 1990).

The white–opaque switch is controlled by the *MTL* locus. Heterozygous *MTL***a**/*MTL*α cells do not switch because the **a**1/α2 regulator produced by these cells represses expression of the *WOR1* transcription factor. In homozygous *MTL***a** or *MTL*α white cells, a stochastic increase in the amount of Wor1 causes a switch to the opaque cell morphology (Huang et al., 2006; Srikantha et al., 2006; Zordan et al., 2006). Wor1 controls its own expression and acts in a feedback loop together with at least three other transcription factors (*WOR2*, *EFG1*, and *CZF1*) to control switching (Zordan et al., 2007).

The ability to mate, and the white–opaque switch, is conserved in *C. dubliniensis*, the closest sequenced relative of *C. albicans* (Pujol et al., 2004). However, mating has not been observed in the other diploid species. The sequenced isolate of *C. tropicalis* is heterozygous at the *MTL* locus (Butler et al., 2009), but mating has not yet been characterized. All tested isolates of *C. parapsilosis* contain only *MTL***a** idiomorphs, and the *MTL***a**1 region is a pseudogene (Butler et al., 2009; Logue et al., 2005; Sai et al., 2011). In *Candida orthopsilosis*, however, which is a very close relative of *C. parapsilosis*, homozygous *MTL***a** and *MTL*α isolates are found in almost equal proportions, suggestive of a mating population (Sai et al., 2011). Despite this, mating has not been detected under various laboratory conditions (Sai et al., 2011). *L. elongisporus*, also a diploid member of the CTG clade, has been described as a homothallic sexual species (Recca and Mrak, 1952; van der Walt, 1966). Surprisingly, however, this species has no *MTL* locus, and the genes coding the **a**-factor pheromone and the pheromone receptor have been lost (Butler et al., 2009). The molecular basis of mating in this species therefore remains to be characterized.

Despite the apparent lack of a sexual cycle in most of the diploid *Candida* species, many of the genes associated with mating and meiosis remain intact (Butler et al., 2009). One possibility is that the pathways are used for cross-species communication. For example, Alby and Bennett (2011) have shown that *C. albicans* responds to mating pheromones from many of the *Candida* species. This response is at the level of transcription, cell morphology, and biofilm formation. Pheromone treatment of white cells of *C. albicans* generally increases biofilm production, irrespective of the source of the pheromone (Alby and Bennett, 2011; Daniels et al., 2006). Pheromone signaling also induces same-sex mating in *C. albicans*, raising the possibility that mating is controlled by environmental factors (Alby and Bennett, 2011; Alby et al., 2009). However, the cell–cell communication appears to be only one way; none of the other species have been shown to respond to pheromone, not even to species-specific pheromone (Alby and Bennett, 2011; Sai et al., 2011).

Although diploid cells of opposite mating types of *C. albicans* do mate, the resulting tetraploid does not undergo meiosis but instead reverts to diploidy through random chromosome losses (Bennett and Johnson, 2003). The *C. albicans* genome is missing some of the key regulators of meiosis described in *S. cerevisiae* (Tzung et al., 2001). However, these are also missing from the fully sexual (and haploid) members of the CTG clade, *M. guilliermondii*, *Cl. lusitaniae*, *Sch. stipitis* and *D. hansenii* (Butler, 2007, 2010; Butler et al., 2009; Reedy et al., 2009). The molecular basis for the lack of a meiotic pathway therefore remains unknown.

Based on the rarity of sex in the human pathogens *Aspergillus fumigatus* and *Cryptococcus neoformans*, and the presence of at best a parasexual cycle in *C. albicans*, Nielsen and Heitman (2007) hypothesized that limiting sexual reproduction allows proliferation of human fungal pathogens in specific environmental niches. Rare sexual reproduction would enable response to stressful conditions. It is difficult to evaluate how important the lack of sex is to virulence. The asexual or parasexual diploid *Candida* species with sequenced genomes are generally more pathogenic than their haploid sexual relatives. The haploid species *C. glabrata* is also a major pathogen, particularly in comparison to other members of the *Saccharomyces* clade. However, this observation is biased to some extent by the fact that fungal pathogens are more likely to have fully sequenced genomes. There are >300 *Candida* species, probably asexual, that are not major pathogens. It is also likely that many species within the *Aspergillus* group mate in a similar manner, whether pathogenic or nonpathogenic (reviewed in Butler, 2010).

In *C. albicans*, however, there are clear associations between virulence and mating, at least via white–opaque switching. White cells are more virulent in a tail vein model of infection (Kvaal et al., 1999). They also secrete a chemoattractant for human polymorphonuclear leukocytes (PMNs) (Geiger et al., 2004; Kvaal et al., 1999). It is therefore possible that opaque cells have developed mechanisms for evading the host immune system. This is supported by the observation that opaque cells are less efficiently phagocytosed than white cells (Lohse and Johnson, 2008). On the other hand, white cells adhere better to human buccal epithelial cells, whereas opaque cells adhere better to skin cells (Kennedy et al., 1988; Kvaal et al., 1997; Lachke et al., 2003). Anaerobic conditions and high CO_2 levels stabilize opaque cells (Dumitru et al., 2007; Huang et al., 2009). The white–opaque switch may therefore have evolved as a mechanism of adaptation to different host niches (Morschhauser, 2010). Opaque cells are also more stable at lower temperatures and mass convert to white cells at 37°C. Mating has been observed when competent cells were inoculated on the skin of mice, which is cooler compared with core body temperature (Lachke et al., 2003). Homothallic (same-sex) mating has been observed in mutants lacking the secreted Bar1 protease, which inactivates the α-type mating pheromone (Alby et al., 2009). White cells are also more adhesive than opaque cells and form larger biofilms (Daniels et al., 2006). Although white cells are mating incompetent, exposure to mating pheromone induces changes in expression of genes that are directly involved in biofilm formation (Sahni et al., 2009). The signal transduction pathway is generally shared with the mating pathway except for the final step, which requires the transcription factor Tec1 (Sahni et al., 2010). However, this biofilm-specific pathway is restricted to *C. albicans* and *C. dubliniensis* and is unlikely to occur in the other pathogenic *Candida* species (Yi et al., 2009). It is therefore unlikely to represent a general virulence mechanism.

TYPING, POPULATION ECOLOGY, AND EVOLUTION

Although *C. albicans* and most pathogenic *Candida* species do not apparently exhibit meiosis, they do not establish clonal populations. Even without the genetic reshuffling associated with meiosis, *C. albicans* can generate genomic diversity by generating and losing heterozygosity, via chromosomal rearrangements, the creation of aneuploid states, and the formation of isochromosomes. Molecular typing methods and karyotype analyses illuminated this genomic plasticity, and much work is now focused on understanding the phenotypic consequences of this plasticity and its implications in the epidemiology and pathology of disease.

Recent molecular typing methods have included DNA fingerprinting, microsatellite analyses, single-nucleotide polymorphism (SNP) arrays, and, most recently, multilocus sequence typing (MLST) (Odds, 2010; Odds and Jacobsen, 2008; Soll, 2000). MLST is particular well suited to studies of molecular ecology and evolution since the information generated can be archived on servers and made available worldwide. The methodology involves sequencing 400- to 600-bp regions of six to eight housekeeping genes and defining strain types by unique SNP patterns on the relevant alleles (Odds and Jacobsen, 2008). In diploid organisms, each strain can be assigned a diploid sequence type (DST), which is unique to a specific fingerprint of SNPs. MLST protocols have been developed for *C. albicans* (Bougnoux et al., 2002, 2003, 2008), *C. dubliniensis* (McManus et al., 2008), *C. glabrata* (Dodgson et al., 2003), *C. tropicalis* (Tavanti et al., 2005b), and *C. krusei* (Jacobsen et al., 2007). MLST failed to generate usable phylogenies in *C. parapsilosis* but revealed that three distinguishable cryptic species now called *C. parapsilosis*, *C. orthopsilosis*, and *Candida metapsilosis* (Tavanti et al., 2005a). *C. parapsilosis* has an extremely low level of genetic variability and consequently was untypable by MLST. *C. orthopsilosis* isolates are more genetically variable, suggesting that *C. parapsilosis* may have evolved recently from the former and that it has been very rapidly disseminated worldwide (Tavanti et al., 2005a). However, it has recently been recognized that some isolates originally categorized as *C. orthopsilosis* are likely to belong to a different, though closely related, species, which may explain the sequence variability observed (Sai et al., 2011).

C. albicans MLST analyses have shown that the global population of strains is defined mainly by clonal reproduction but with high frequency background recombination. Hence, there are many examples of singleton DSTs, indicative of background recombination in the genomes of *C. albicans*, *C. tropicalis*, and *C. krusei*, but also groups of radiating clonal clusters that are highly related to one another. Large genomic tracts have been observed in three diploid *Candida* species that are homozygous, which suggest possible recent recombination events (Butler et al., 2009). High background recombination in *C. albicans* is also indicated by the fact that phylogenetic trees constructed from SNPs on individual chromosomes give differing lineages, indicating that the genes on chromosomes were evolving independently (Odds et al., 2007). However, polymorphic nucleotide positions were shown to be in Hardy–Weinberg disequilibrium with an excess of heterozygosities, suggesting that mating within clades occurs infrequently in *C. albicans* (Bougnoux et al., 2008). MLST analysis of the near relative, *C. dubliniensis*, showed a much reduced frequency of SNPs, indicating a much more clonal distribution of strains than in *C. albicans* (McManus et al., 2008).

Over 2000 strains of *C. albicans* have now been typed by MLST. The most

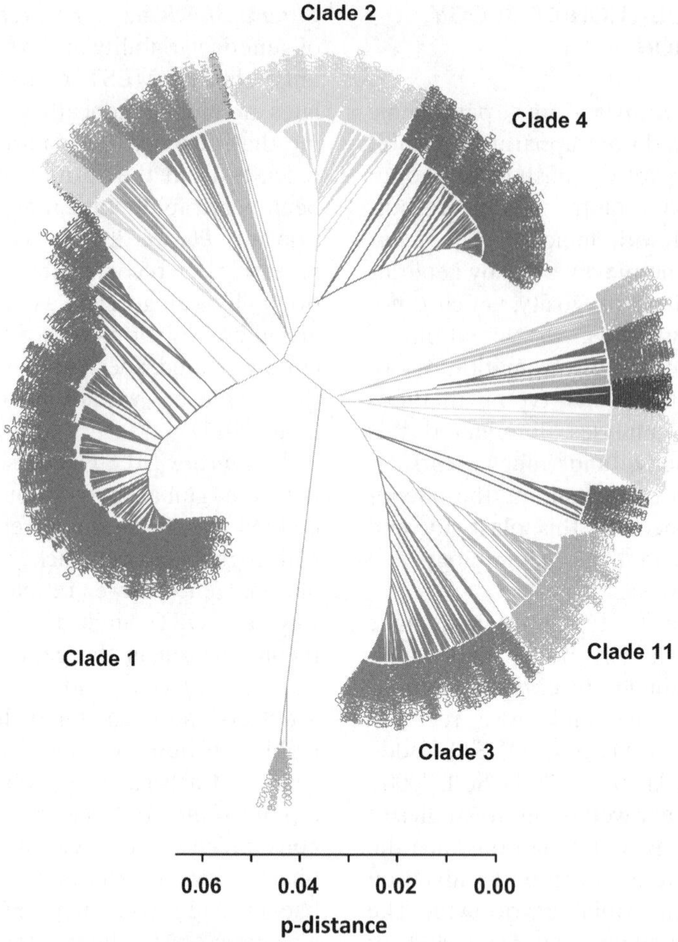

Figure 21.2 UPGMA dendrogram based on 1410 MLST diploid sequence types of *C. albicans* showing the five major clades (1, 2, 3, 4, and 11) and the more distantly related clade 13 (in brown at 6 o'clock). See color insert.

common DSTs have been found universally in all continents, perhaps indicative of strain mixing due to the explosion of global travel in the last 30 years. MSLT data displayed as UPGMA dendrograms show that most of these strains are accommodated by five major clades and several minor clades (Fig. 21.2). The fully sequenced strains SC5314 and WO-1 reside in clade 1 and clade 6, respectively. Clade 1 contains about a third of all global isolates and is somewhat enriched with strains recovered from superficial infections and strains exhibiting higher levels of resistance to 5-flurocytosine due to a single mutation in the *FUR1* allele and to terbinafine (Odds, 2010). Clade 13 defines a small group of closely related isolates that are the most distantly related to the other clades. These organisms were once called *Candida africana* strains, but they are clearly not sufficiently diverged to represent a separate species. This clade is the only one with a significant geographic bias, with a majority of isolates coming from genital isolations from people in Africa and Europe. There are no major phenotypic traits, including virulence in mice models, that segregate with clades, and transcript profiles also show extremely similar patterns between representatives strains from different clades (Odds et al., 2007).

MLST and other forms of DNA-based typing systems have revealed some clear aspects of *Candida* disease epidemiology. For example, most patients carry a single strain at multiple sites and in serial samples taken from the same site. However, it is not uncommon to observe mixed *Candida* species being carried by a single individual. Unlike in many examples from medical bacteriology, nosocomial transmission in hospitals is uncommon for most *Candida* species, although outbreaks of *C. parapsilosis* in neonatal units have been reported (Pfaller and Diekema, 2007, 2010).

CONCLUSION

The genus *Candida* is paraphyletic and contains a large number of species, of which a small group of four (*C. albicans*, *C. glabrata*, *C. parapsilosis*, and *C. tropicalis*) is responsible for >95% of all *Candida* infections and four others (*C. krusei*, *Cl. lusitaniae*, *M. guilliermondii*, and *C. dubliniensis*) account for most of the remaining 5%. The group contains haploid and diploid species, and genomic analysis has revealed significant differences in the mating locus, degree of genetic diversity, and levels of underlying recombination between isolates. Comparative genome analysis shows that virulence is associated with the expansion of a number of key gene families, particularly in the species in the CTG clade. However, no universal genome features that explain the differences in relative pathogenicity observed between all *Candida* species have yet been identified.

ACKNOWLEDGMENTS

G.B. is supported by Science Foundation Ireland; M.L. by the National Institutes of Health Award R01AI075091; and N.G. by the Wellcome Trust, BBSRC (CRISP/SABR), EC (ALLFUN, Ariadne). We are grateful to Prof. Frank Odds for providing Figure 21.2.

REFERENCES

Alby K, Bennett RJ. 2011. Interspecies pheromone signaling promotes biofilm formation and same-sex mating in *Candida albicans*. Proc Natl Acad Sci U S A 108: 2510–2515.

Alby K, Schaefer D, Bennett RJ. 2009. Homothallic and heterothallic mating in the opportunistic pathogen *Candida albicans*. Nature 460: 890–893.

Almeida RS, Brunke S, Albrecht A, Thewes S, Laue M, Edwards JE, Filler SG, Hube B. 2008. The hyphal-associated adhesin and invasin Als3 of *Candida albicans* mediates iron acquisition from host ferritin. PLoS Pathog 4: e1000217.

Anderson J, Mihalik R, Soll DR. 1990. Ultrastructure and antigenicity of the unique cell wall pimple of the *Candida* opaque phenotype. J Bacteriol 172: 224–235.

Arendrup M, Horn T, Frimodt-Moller N. 2002. *In vivo* pathogenicity of eight medically relevant *Candida* species in an animal model. Infection 30: 286–291.

Askew C, Sellam A, Epp E, Hogues H, Mullick A, Nantel A, Whiteway M. 2009. Transcriptional regulation of carbohydrate metabolism in the human pathogen *Candida albicans*. PLoS Pathog 5: e1000612.

Bailey DA, Feldmann PJ, Bovey M, Gow NA, Brown AJ. 1996. The *Candida albicans HYR1* gene, which is activated in response to hyphal development, belongs to a gene family encoding yeast cell wall proteins. J Bacteriol 178: 5353–5360.

Bates S, de la Rosa JM, MacCallum DM, Brown AJ, Gow NA, Odds FC. 2007. *Candida albicans* Iff11, a secreted protein required for cell wall structure and virulence. Infect Immun 75: 2922–2928.

Bennett RJ, Johnson AD. 2003. Completion of a parasexual cycle in *Candida albicans* by induced chromosome loss in tetraploid strains. EMBO J 22: 2505–2515.

Bistoni F, Vecchiarelli A, Cenci E, Sbaraglia G, Perito S, Cassone A. 1984. A comparison of experimental pathogenicity of *Candida* species in cyclophosphamide-immunodepressed mice. Sabouraudia 22: 409–418.

Booth LN, Tuch BB, Johnson AD. 2010. Intercalation of a new tier of transcription regulation into an ancient circuit. Nature 468: 959–963.

Bougnoux ME, Morand S, Enfert C. 2002. Usefulness of multilocus sequence typing for characterization of clinical isolates of *Candida albicans*. J Clin Microbiol 40: 1290–1297.

Bougnoux ME, Tavanti A, Bouchier C, Gow NA, Magnier A, Davidson AD, Maiden MC, D'Enfert C, Odds FC. 2003. Collaborative consensus for optimized multilocus sequence typing of *Candida albicans*. J Clin Microbiol 41: 5265–5266.

Bougnoux ME, Pujol C, Diogo D, Bouchier C, Soll DR, d'Enfert C. 2008. Mating is rare within as well as between clades of the human pathogen *Candida albicans*. Fungal Genet Biol 45: 221–231.

Braun BR, van Het Hoog M, d'Enfert C, Martchenko M, Dungan J, Kuo A, Inglis DO, Uhl MA, Hogues H, Berriman M, et al. 2005. A human-curated annotation of the *Candida albicans* genome. PLoS Genet 1: e1.

Butler G. 2007. The evolution of MAT: The ascomycetes. In: Heitman J, Kronstad JW, Taylor JW, Casselton LA, eds. *Sex in Fungi*. ASM Press, Washington, DC, pp. 3–18.

Butler G. 2010. Fungal sex and pathogenesis. Clin Microbiol Rev 23: 140–159.

Butler G, Rasmussen MD, Lin MF, Santos MA, Sakthikumar S, Munro CA, Rheinbay E, Grabherr M, Forche A, Reedy JL, et al. 2009. Evolution of pathogenicity and sexual reproduction in eight *Candida* genomes. Nature 459: 657–662.

Chapeland-Leclerc F, Hennequin C, Papon N, Noel T, Girard A, Socie G, Ribaud P, Lacroix C. 2010. Acquisition of flucytosine, azole, and caspofungin resistance in *Candida glabrata* bloodstream isolates serially obtained from a hematopoietic stem cell transplant recipient. Antimicrob Agents Chemother 54: 1360–1362.

Cleary IA, Reinhard SM, Miller CL, Murdoch C, Thornhill MH, Lazzell AL, Monteagudo C, Thomas DP, Saville SP. 2011. *Candida albicans* adhesin Als3p is dispensable for virulence in the mouse model of disseminated candidiasis. Microbiology 157: 1806–1815.

Cleary JD, Garcia-Effron G, Chapman SW, Perlin DS. 2008. Reduced *Candida glabrata* susceptibility secondary to an *FKS1* mutation developed during candidemia treatment. Antimicrob Agents Chemother 52: 2263–2265.

Cormack BP, Ghori N, Falkow S. 1999. An adhesin of the yeast pathogen *Candida glabrata* mediating adherence to human epithelial cells. Science 285: 578–582.

Correia A, Sampaio P, James S, Pais C. 2006. *Candida bracarensis* sp. nov., a novel anamorphic yeast species phenotypically similar to *Candida glabrata*. Int J Syst Evol Microbiol 56: 313–317.

Daniels KJ, Srikantha T, Lockhart SR, Pujol C, Soll DR. 2006. Opaque cells signal white cells to form biofilms in *Candida albicans*. EMBO J 25: 2240–2252.

De Las Penas A, Pan SJ, Castano I, Alder J, Cregg R, Cormack BP. 2003. Virulence-related surface glycoproteins in the yeast pathogen *Candida glabrata* are encoded in subtelomeric clusters and subject to RAP1- and SIR-dependent transcriptional silencing. Genes Dev 17: 2245–2258.

de Repentigny L, Phaneuf M, Mathieu LG. 1992. Gastrointestinal colonization and systemic dissemination by *Candida albicans* and *Candida tropicalis* in intact and immunocompromised mice. Infect Immun 60: 4907–4914.

Dodgson AR, Pujol C, Denning DW, Soll DR, Fox AJ. 2003. Multilocus sequence typing of *Candida glabrata* reveals geographically enriched clades. J Clin Microbiol 41: 5709–5717.

Domergue R, Castano I, De Las Penas A, Zupancic M, Lockatell V, Hebel RJ, Johnson D, Cormack BP. 2005. Nicotinic acid limitation regulates silencing of *Candida* adhesins during UTI. Science 388: 866–870.

Dujon B, Sherman D, Fischer G, Durrens P, Casaregola S, Lafontaine I, De Montigny J, Marck C, Neuveglise C, Talla E, et al. 2004. Genome evolution in yeasts. Nature 430: 35–44.

Dumitru R, Navarathna DH, Semighini CP, Elowsky CG, Dumitru RV, Dignard D, Whiteway M, Atkin AL, Nickerson KW. 2007. *In vivo* and *in vitro* anaerobic mating in *Candida albicans*. Eukaryot Cell 6: 465–472.

Fitzpatrick DA, Logue ME, Stajich JE, Butler G. 2006. A fungal phylogeny based on 42 complete genomes derived from supertree and combined gene analysis. BMC Evol Biol 6: 99.

Frank AT, Ramsook CB, Otoo HN, Tan C, Soybelman G, Rauceo JM, Gaur NK, Klotz SA, Lipke PN. 2010. Structure and function of glycosylated tandem repeats from *Candida albicans* Als adhesins. Eukaryot Cell 9: 405–414.

Fu Y, Luo G, Spellberg BJ, Edwards JE, Jr, Ibrahim AS. 2008. Gene overexpression/suppression analysis of candidate virulence factors of *Candida albicans*. Eukaryot Cell 7: 483–492.

Gacser A, Trofa D, Schafer W, Nosanchuk JD. 2007. Targeted gene deletion in *Candida parapsilosis* demonstrates the role of secreted lipase in virulence. J Clin Invest 117: 3049–3058.

Geiger J, Wessels D, Lockhart SR, Soll DR. 2004. Release of a potent polymorphonuclear leukocyte chemoattractant is regulated by white-opaque switching in *Candida albicans*. Infect Immun 72: 667–677.

Howlett JA. 1976. The infection of rat tongue mucosa in vitro with five species of *Candida*. J Med Microbiol 9: 309–316.

Hoyer LL. 2001. The ALS gene family of *Candida albicans*. Trends Microbiol 9: 176–180.

Hoyer LL, Green CB, Oh SH, Zhao X. 2008. Discovering the secrets of the *Candida albicans* agglutinin-like sequence (*ALS*) gene family—A sticky pursuit. Med Mycol 46: 1–15.

Huang G, Wang H, Chou S, Nie X, Chen J, Liu H. 2006. Bistable expression of *WOR1*, a master regulator of white-opaque switching in *Candida albicans*. Proc Natl Acad Sci U S A 103: 12813–12818.

Huang G, Srikantha T, Sahni N, Yi S, Soll DR. 2009. CO(2) regulates white-to-opaque switching in *Candida albicans*. Curr Biol 19: 330–334.

Hull CM, Johnson AD. 1999. Identification of a mating type-like locus in the asexual pathogenic yeast *Candida albicans*. Science 285: 1271–1275.

Hull CM, Raisner RM, Johnson AD. 2000. Evidence for mating of the "asexual" yeast *Candida albicans* in a mammalian host. Science 289: 307–310.

Jackson AP, Gamble JA, Yeomans T, Moran GP, Saunders D, Harris D, Aslett M, Barrell JF, Butler G, Citiulo F, et al. 2009. Comparative genomics of the fungal pathogens *Candida dubliniensis* and *C. albicans*. Genome Res 19: 2231–2244.

Jacobsen MD, Gow NA, Maiden MC, Shaw DJ, Odds FC. 2007. Strain typing and determination of population structure of *Candida krusei* by multilocus sequence typing. J Clin Microbiol 45: 317–323.

Johnson A. 2003. The biology of mating in *Candida albicans*. Nat Rev Microbiol 1: 106–116.

Johnston N. 2006. So much diversity, such littel cells. Scientist 20: 65–66.

Jones T, Federspiel NA, Chibana H, Dungan J, Kalman S, Magee BB, Newport G, Thorstenson YR, Agabian N, Magee PT, et al. 2004. The diploid genome sequence of *Candida albicans*. Proc Natl Acad Sci U S A 11: 7329–7334.

Kaur R, Ma B, Cormack BP. 2007. A family of glycosylphosphatidylinositol-linked aspartyl proteases is required for virulence of *Candida glabrata*. Proc Natl Acad Sci U S A 104: 7628–7633.

Kempf M, Cottin J, Licznar P, Lefrancois C, Robert R, Apaire-Marchais V. 2009. Disruption of the GPI protein-encoding gene IFF4 of *Candida albicans* results in decreased adherence and virulence. Mycopathologia 168: 73–77.

Kennedy MJ, Rogers AL, Hanselmen LR, Soll DR, Yancey RJ, Jr. 1988. Variation in adhesion and cell surface hydrophobicity in

Candida albicans white and opaque phenotypes. Mycopathologia 102: 149–156.

Kurtzman CP, Fell JW, Boekhout T. 2011. *The Yeast, a Taxonomic Study*, 5th edition. Elsevier, Amsterdam.

Kvaal C, Lachke SA, Srikantha T, Daniels K, McCoy J, Soll DR. 1999. Misexpression of the opaque-phase-specific gene *PEP1* (*SAP1*) in the white phase of *Candida albicans* confers increased virulence in a mouse model of cutaneous infection. Infect Immun 67: 6652–6662.

Kvaal CA, Srikantha T, Soll DR. 1997. Misexpression of the white-phase-specific gene *WH11* in the opaque phase of *Candida albicans* affects switching and virulence. Infect Immun 65: 4468–4475.

Lachance M-A, Boekhout T, Scorzetti G, Fell JW, Kurtzmann CP. 2011. *Candida* Berkhout (1923). In: Kurtzman CP, Fell JW, Boekhout T, eds. *The Yeasts, a Taxonomic Study*. Elsevier, Amsterdam, pp. 987–1278.

Lachke SA, Lockhart SR, Daniels KJ, Soll DR. 2003. Skin facilitates *Candida albicans* mating. Infect Immun 71: 4970–4976.

Lavoie H, Hogues H, Whiteway M. 2009. Rearrangements of the transcriptional regulatory networks of metabolic pathways in fungi. Curr Opin Microbiol 12: 655–663.

Liu Y, Filler SG. 2011. *Candida albicans* Als3: A multifunctional adhesin and invasin. Eukaryot Cell 10: 168–173.

Lo HJ, Kohler JR, DiDomenico B, Loebenberg D, Cacciapuoti A, Fink GR. 1997. Nonfilamentous *C. albicans* mutants are avirulent. Cell 90: 939–949.

Lo WS, Dranginis AM. 1998. The cell surface flocculin *Flo11* is required for pseudohyphae formation and invasion by *Saccharomyces cerevisiae*. Mol Biol Cell 9: 161–171.

Lockhart SR, Messer SA, Gherna M, Bishop JA, Merz WG, Pfaller MA, Diekema DJ. 2009. Identification of *Candida nivariensis* and *Candida bracarensis* in a large global collection of *Candida glabrata* isolates: Comparison to the literature. J Clin Microbiol 47: 1216–1217.

Logue ME, Wong S, Wolfe KH, Butler G. 2005. A genome sequence survey shows that the pathogenic yeast *Candida parapsilosis* has a defective *MTLa1* allele at its mating type locus. Eukaryot Cell 4: 1009–1017.

Lohse MB, Johnson AD. 2008. Differential phagocytosis of white versus opaque *Candida albicans* by *Drosophila* and mouse phagocytes. PLoS One 3: e1473.

Lorenz MC, Bender JA, Fink GR. 2004. Transcriptional response of *Candida albicans* upon internalization by macrophages. Eukaryot Cell 3: 1076–1087.

Loza L, Fu Y, Ibrahim AS, Sheppard DC, Filler SG, Edwards JE, Jr. 2004. Functional analysis of the *Candida albicans ALS1* gene product. Yeast 21: 473–482.

Magee BB, Magee PT. 2000. Induction of mating in *Candida albicans* by construction of *MTLa* and *MTLalpha* strains. Science 289: 310–313.

McManus BA, Coleman DC, Moran G, Pinjon E, Diogo D, Bougnoux ME, Borecka-Melkusova S, Bujdakova H, Murphy P, d'Enfert C, et al. 2008. Multilocus sequence typing reveals that the population structure of *Candida dubliniensis* is significantly less divergent than that of *Candida albicans*. J Clin Microbiol 46: 652–664.

Mellado E, Cuenca-Estrella M, Regadera J, Gonzalez M, Diaz-Guerra TM, Rodriguez-Tudela JL. 2000. Sustained gastrointestinal colonization and systemic dissemination by *Candida albicans*, *Candida tropicalis* and *Candida parapsilosis* in adult mice. Diagn Microbiol Infect Dis 38: 21–28.

Miller MG, Johnson AD. 2002. White-opaque switching in *Candida albicans* is controlled by mating-type locus homeodomain proteins and allows efficient mating. Cell 110: 293–302.

Moran G, Stokes C, Thewes S, Hube B, Coleman DC, Sullivan D. 2004. Comparative genomics using *Candida albicans* DNA microarrays reveals absence and divergence of virulence-associated genes in *Candida dubliniensis*. Microbiology 150: 3363–3382.

Moran GP, MacCallum DM, Spiering MJ, Coleman DC, Sullivan DJ. 2007. Differential regulation of the transcriptional repressor *NRG1* accounts for altered host-cell interactions in *Candida albicans* and *Candida dubliniensis*. Mol Microbiol 66: 915–929.

Moreno-Ruiz E, Galan-Diez M, Zhu W, Fernandez-Ruiz E, d'Enfert C, Filler SG, Cossart P, Veiga E. 2009. *Candida albicans* internalization by host cells is mediated by a clathrin-dependent mechanism. Cell Microbiol 11: 1179–1189.

Morschhauser J. 2010. Regulation of white-opaque switching in *Candida albicans*. Med Microbiol Immunol 199: 165–172.

Nielsen K, Heitman J. 2007. Sex and virulence of human pathogenic fungi. Adv Genet 57: 143–173.

Nobile CJ, Andes DR, Nett JE, Smith FJ, Yue F, Phan QT, Edwards JE, Filler SG, Mitchell AP. 2006. Critical role of Bcr1-dependent adhesins in *C. albicans* biofilm formation *in vitro* and *in vivo*. PLoS Pathog 2: e63.

Noble SM, French S, Kohn LA, Chen V, Johnson AD. 2010. Systematic screens of a *Candida albicans* homozygous deletion library decouple morphogenetic switching and pathogenicity. Nat Genet 42: 590–598.

Odds FC. 2010. Molecular phylogenetics and epidemiology of *Candida albicans*. Future Microbiol 5: 67–79.

Odds FC, Jacobsen MD. 2008. Multilocus sequence typing of pathogenic *Candida* species. Eukaryot Cell 7: 1075–1084.

Odds FC, Bougnoux ME, Shaw DJ, Bain JM, Davidson AD, Diogo D, Jacobsen MD, Lecomte M, Li SY, Tavanti A, et al. 2007. Molecular phylogenetics of *Candida albicans*. Eukaryot Cell 6: 1041–1052.

Oh SH, Cheng G, Nuessen JA, Jajko R, Yeater KM, Zhao X, Pujol C, Soll DR, Hoyer LL. 2005. Functional specificity of *Candida albicans* Als3p proteins and clade specificity of *ALS3* alleles discriminated by the number of copies of the tandem repeat sequence in the central domain. Microbiology 151: 673–681.

Ohama T, Suzuki T, Mori M, Osawa S, Ueda T, Watanabe K, Nakase T. 1993. Non-universal decoding of the leucine codon CUG in several *Candida* species. Nucleic Acids Res 21: 4039–4045.

Panackal AA, Gribskov JL, Staab JF, Kirby KA, Rinaldi M, Marr KA. 2006. Clinical significance of azole antifungal drug cross-resistance in *Candida glabrata*. J Clin Microbiol 44: 1740–1743.

Pfaller MA, Diekema DJ. 2007. Epidemiology of invasive candidiasis: A persistent public health problem. Clin Microbiol Rev 20: 133–163.

Pfaller MA, Diekema DJ. 2010. Epidemiology of invasive mycoses in North America. Crit Rev Microbiol 36: 1–53.

Pfaller MA, Castanheira M, Messer SA, Moet GJ, Jones RN. 2010. Variation in *Candida* spp. distribution and antifungal resistance rates among bloodstream infection isolates by patient age: Report from the SENTRY Antimicrobial Surveillance Program (2008–2009). Diagn Microbiol Infect Dis 68: 278–283.

Pfeiffer CD, Garcia-Effron G, Zaas AK, Perfect JR, Perlin DS, Alexander BD. 2010. Breakthrough invasive candidiasis in patients on micafungin. J Clin Microbiol 48: 2373–2380.

Phan QT, Myers CL, Fu Y, Sheppard DC, Yeaman MR, Welch WH, Ibrahim AS, Edwards JE, Jr, Filler SG. 2007. Als3 is a *Candida albicans* invasin that binds to cadherins and induces endocytosis by host cells. PLoS Biol 5: e64.

Pujol C, Daniels KJ, Lockhart SR, Srikantha T, Radke JB, Geiger J, Soll DR. 2004. The closely related species *Candida albicans* and *Candida dubliniensis* can mate. Eukaryot Cell 3: 1015–1027.

Ramirez MA, Lorenz MC. 2009. The transcription factor homolog *CTF1* regulates {beta}-oxidation in *Candida albicans*. Eukaryot Cell 8: 1604–1614.

Recca J, Mrak E. 1952. Yeasts occurring in citrus products. Food Technol 6: 450–454.

Reedy JL, Floyd AM, Heitman J. 2009. Mechanistic plasticity of sexual reproduction and meiosis in the *Candida* pathogenic species complex. Curr Biol 19: 891–899.

Rubin-Bejerano I, Fraser I, Grisafi P, Fink GR. 2003. Phagocytosis by neutrophils induces an amino acid deprivation response in *Saccharomyces cerevisiae* and *Candida albicans*. Proc Natl Acad Sci U S A 100: 11007–11012.

Sahni N, Yi S, Daniels KJ, Srikantha T, Pujol C, Soll DR. 2009. Genes selectively up-regulated by pheromone in white cells are involved in biofilm formation in *Candida albicans*. PLoS Pathog 5: e1000601.

Sahni N, Yi S, Daniels KJ, Huang G, Srikantha T, Soll DR. 2010. Tec1 mediates the pheromone response of the white phenotype of *Candida albicans*: Insights into the evolution of new signal transduction pathways. PLoS Biol 8: e1000363.

Sai S, Holland L, McGee CF, Lynch DB, Butler G. 2011. Evolution of mating within the *Candida parapsilosis* species group. Eukaryot Cell 10: 578–587.

Santos MA, Tuite MF. 1995. The CUG codon is decoded in vivo as serine and not leucine in *Candida albicans*. Nucleic Acids Res 23: 1481–1486.

Silverman RJ, Nobbs AH, Vickerman MM, Barbour ME, Jenkinson HF. 2010. Interaction of *Candida albicans* cell wall Als3 protein with *Streptococcus gordonii* SspB adhesin promotes development of mixed-species communities. Infect Immun 78: 4644–4652.

Slutsky B, Buffo J, Soll DR. 1985. High-frequency switching of colony morphology in *Candida albicans*. Science 230: 666–669.

Soll DR. 2000. The ins and outs of DNA fingerprinting the infectious fungi. Clin Microbiol Rev 13: 332–370.

Spiering MJ, Moran GP, Chauvel M, Maccallum DM, Higgins J, Hokamp K, Yeomans T, d'Enfert C, Coleman DC, Sullivan DJ. 2010. Comparative transcript profiling of *Candida albicans* and *Candida dubliniensis* identifies *SFL2*, a *C. albicans* gene required for virulence in a reconstituted epithelial infection model. Eukaryot Cell 9: 251–265.

Srikantha T, Borneman AR, Daniels KJ, Pujol C, Wu W, Seringhaus MR, Gerstein M, Yi S, Snyder M, Soll DR. 2006. *TOS9* regulates white-opaque switching in *Candida albicans*. Eukaryot Cell 5: 1674–1687.

Sudbery PE. 2011. Growth of *Candida albicans* hyphae. Nat Rev Microbiol 9: 737–748.

Tavanti A, Davidson AD, Gow NA, Maiden MC, Odds FC. 2005a. *Candida orthopsilosis* and *Candida metapsilosis* spp. nov. to replace *Candida parapsilosis* groups II and III. J Clin Microbiol 43: 284–292.

Tavanti A, Davidson AD, Johnson EM, Maiden MC, Shaw DJ, Gow NA, Odds FC. 2005b. Multilocus sequence typing for differentiation of strains of *Candida tropicalis*. J Clin Microbiol 43: 5593–5600.

Thompson GR, 3rd, Wiederhold NP, Vallor AC, Villareal NC, Lewis JS, 2nd, Patterson TF. 2008. Development of caspofungin resistance following prolonged therapy for invasive candidiasis secondary to *Candida glabrata* infection. Antimicrob Agents Chemother 52: 3783–3785.

Tuch BB, Galgoczy DJ, Hernday AD, Li H, Johnson AD. 2008. The evolution of combinatorial gene regulation in fungi. PLoS Biol 6: e38.

Tzung KW, Williams RM, Scherer S, Federspiel N, Jones T, Hansen N, Bivolarevic V, Huizar L, Komp C, Surzycki R, et al. 2001. Genomic evidence for a complete sexual cycle in *Candida albicans*. Proc Natl Acad Sci U S A 98: 3249–3253.

van der Walt JP. 1966. *Lodderomyces*, a new genus of the Saccharomycetacea. Antonie Van Leeuwenhoek 32: 1–5.

van het Hoog M, Rast TJ, Martchenko M, Grindle S, Dignard D, Hogues H, Cuomo C, Berriman M, Scherer S, Magee BB, et al. 2007. Assembly of the *Candida albicans* genome into sixteen supercontigs aligned on the eight chromosomes. Genome Biol 8: R52.

Verstrepen KJ, Jansen A, Lewitter F, Fink GR. 2005. Intragenic tandem repeats generate functional variability. Nat Genet 37: 986–990.

Wong S, Fares MA, Zimmermann W, Butler G, Wolfe KH. 2003. Evidence from comparative genomics for a complete sexual cycle in the "asexual" pathogenic yeast *Candida glabrata*. Genome Biol 4: R10.

Yi S, Sahni N, Pujol C, Daniels KJ, Srikantha T, Ma N, Soll DR. 2009. A *Candida albicans*-specific region of the alpha-pheromone receptor plays a selective role in the white cell pheromone response. Mol Microbiol 71: 925–947.

Zhang N, Harrex AL, Holland BR, Fenton LE, Cannon RD, Schmid J. 2003. Sixty alleles of the *ALS7* open reading frame in *Candida albicans*: *ALS7* is a hypermutable contingency locus. Genome Res 13: 2005–2017.

Zhao X, Pujol C, Soll DR, Hoyer LL. 2003. Allelic variation in the contiguous loci encoding *Candida albicans ALS5*, *ALS1* and *ALS9*. Microbiology 149: 2947–2960.

Zhao X, Oh SH, Cheng G, Green CB, Nuessen JA, Yeater K, Leng RP, Brown AJ, Hoyer LL. 2004. *ALS3* and *ALS8* represent a single locus that encodes a *Candida albicans* adhesin; functional comparisons between Als3p and Als1p. Microbiology 150: 2415–2428.

Zhao X, Oh SH, Hoyer LL. 2007. Deletion of *ALS5*, *ALS6* or *ALS7* increases adhesion of *Candida albicans* to human vascular endothelial and buccal epithelial cells. Med Mycol 45: 429–434.

Zheng X, Wang Y. 2004. Hgc1, a novel hypha-specific G1 cyclin-related protein regulates *Candida albicans* hyphal morphogenesis. EMBO J 23: 1845–1856.

Zordan RE, Galgoczy DJ, Johnson AD. 2006. Epigenetic properties of white-opaque switching in *Candida albicans* are based on a self-sustaining transcriptional feedback loop. Proc Natl Acad Sci U S A 103: 12807–12812.

Zordan RE, Miller MG, Galgoczy DJ, Tuch BB, Johnson AD. 2007. Interlocking transcriptional feedback loops control white-opaque switching in *Candida albicans*. PLoS Biol 5: e256.

CHAPTER 22

EVOLUTION OF *ENTAMOEBA HISTOLYTICA* VIRULENCE

UPINDER SINGH and CHRISTOPHER D. HUSTON

INTRODUCTION

Entamoeba histolytica, the enteric protozoan responsible for invasive amebiasis, causes millions of symptomatic infections each year (WHO, 1997). Current estimates of the overall burden of disease are limited by crude epidemiologic data that were collected prior to distinction of *E. histolytica* from the morphologically identical nonpathogenic species *Entamoeba dispar* in 1993 (Diamond and Clark, 1993). Epidemiologic studies that accurately distinguished these species have been conducted in endemic regions and emphasize the true magnitude of the problem. For example, in Mirpur, Bangladesh, where diarrheal diseases are the leading causes of childhood mortality, nearly 50% of children have serologic evidence of *E. histolytica* infection by age 5 years, and *E. histolytica* infection has been identified in 8.7% of cases of dysentery in young children (Haque et al., 1999, 2003b). The presence of *E. histolytica* and closely related nonpathogenic (or in the case of *Entamoeba moshkovskii*, less pathogenic) species provides an opportunity for comparative studies aimed at understanding both the basis for *E. histolytica* virulence and the evolution of virulence among *Entamoebae*.

Most genomic diversity between *E. histolytica* and *E. dispar* is due to sequence variation within genes that contain repetitive elements, such as the genes encoding the serine-rich *E. histolytica*/*E. dispar* protein and the tRNA-short tandem repeat (STR) loci (Ayeh-Kumi et al., 2001; Weedall and Hall, 2011; Zaki and Clark, 2001). In contrast to these polymorphic repetitive genes, genome-wide hybridization studies and comparative sequence analyses suggest that genome-wide sequence diversity between *E. histolytica* and *E. dispar* is modest and indicate that the two species likely diverged from a common ancestor relatively recently (Ghosh et al., 2000; Shah et al., 2005; Weedall and Hall, 2011). Both organisms infect only humans and some higher nonhuman primates. Thus, divergence of these species has apparently proceeded within the human host, and the virulence of *E. histolytica* relative to *E. dispar* has thus far been maintained. This comparison offers circumstantial evidence counter to the traditional notion that patho-

Evolution of Virulence in Eukaryotic Microbes, First Edition. Edited by L. David Sibley, Barbara J. Howlett, and Joseph Heitman.

gens necessarily evolve toward commensalism or symbiosis and suggests positive selection of virulence attributes in *E. histolytica*. In this chapter, we review current knowledge of genetic diversity among *Entamoebae* and mechanisms of amebic virulence, and the abilities and limitations of currently available methods to study the evolution of *E. histolytica* virulence.

GENETIC DIVERSITY

Epidemiological studies of *E. histolytica* indicate that significant genetic diversity exists among clinical isolates. The tools utilized for these studies include sequence analysis of polymorphic loci, microarray-based comparative genomic hybridizations, and more recently, genomic sequencing. Using randomly amplified polymorphic DNA (RAPD) to estimate genetic variability among 14 *E. histolytica* (EH) strains from different geographic locations, a high degree of divergence was identified (Gomes et al., 2000). In a study of eight *E. histolytica* isolates using RNA arbitrarily primed (RAP)-PCR analysis, similar results were obtained (Valle et al., 2000). PCR, restriction fragment length polymorphism (RFLP), and sequence analyses of the serine-rich *E. histolytica* protein (SREHP) and chitinase genes (two protein-coding polymorphic loci) have also revealed significant genetic diversity among strains (Clark and Diamond, 1993; Ghosh et al., 2000; Zaki and Clark, 2001; Zaki et al., 2002). In a study of clinical EH isolates from Bangladesh using nested PCR of SREHP and small subunit of rRNA genes, strains with different clinical manifestations (asymptomatic intestinal colonization, diarrhea, and amebic liver abscesses) were analyzed. Twenty-five distinct DNA banding patterns were observed among 42 stool isolates and 9 distinct patterns among 12 liver abscess isolates (Ayeh-Kumi et al., 2001). Analysis of genetic polymorphisms of two noncoding loci as well as genes coding for chitinase and SREHP among isolates obtained from geographically diverse regions of Japan, Thailand, and Bangladesh has also been performed (Haghighi et al., 2003). The extreme genetic variability of these strains was evidenced in the fact that 63 isolates were delineated into 53 genotypes. However, studies evaluating clustering methods to group these variable genotypes have not been performed. Genetic loci associated with tRNA genes have also been utilized for molecular typing. The tRNA genes constitute 13% of the sequences obtained by the *E. histolytica* genome project and are present in clusters of one to five distinct types, interspersed with A/T-rich STRs, repeated to form long arrays (Zaki and Clark, 2001). The sequence divergence in tRNA genes has allowed species differentiation and strain typing (Ali et al., 2005; Zaki et al., 2002). Using this approach, one study compared *E. histolytica* samples from patients in Bangladesh with asymptomatic colonization, diarrhea and/or dysentery, and amebic liver abscess. The parasite genotype prevalence in each of the three groups was significantly different from each other, leading to the suggestion that the parasite genotype plays a role in the outcome of infection (Ali et al., 2007).

Although these studies are important first steps in studying *E. histolytica* epidemiology, there were some significant limitations to these approaches. The main concerns with these studies were (i) that they focused almost exclusively on analyzing a small number of highly polymorphic genetic loci; (ii) most had very little sequence data; and (iii) that for a number of these studies, the clinical strains had been cultured *in vitro*, introducing issues about parasite diversity bottlenecks during culture adaptation. Highly polymorphic regions are obvious targets for strain typing; however, there are some inherent biases in relying exclusively on these loci. Some genes may be subject to strong selection pressures or can have patterns of mutations that are atypical of the genome as a whole. For example, the SREHP is highly immunogenic, may have a role in conferring immunity to amebic liver disease (Zhang et al., 1994), and may thus be under different selection pressures than other genetic loci. Additionally, both SREHP and chitinase genes have repeat regions, which can be highly prone to incorporating polymorphisms due to DNA slippage.

A few studies have looked at sequence analysis of genetic loci other than the highly polymorphic regions discussed above. Not surprisingly, the data from these studies, although very limited, indicate more limited genetic diversity. The Gal/GalNAc lectin gene was sequenced from three clinical isolates from Bangladesh and one isolate from Tbilisi, Georgia. All four clinical isolates had unique SREHP RFLP patterns and appeared to be genetically distinct; however, only a slight genetic diversity was observed in the lectin gene sequences (Beck et al., 2002). Whether this reflects that the lectin gene is under strong functional selection pressure or that the high degree of polymorphism identified earlier is not a genome-wide phenomenon remains to be determined. The sequence of the intergenic region between the superoxide dismutase and actin3 genes from a number of clinical isolates also revealed no polymorphisms (Ghosh et al., 2000). Another study detected 0.007% single-nucleotide polymorphisms (SNPs) in coding regions (of genes likely to be conserved as well as those under positive selection pressure) among the *E. histolytica* strains, with noncoding region SNPs being fivefold more frequent (Bhattacharya et al., 2005). Thus, although limited, the sequence data indicate that the high degree of polymorphism identified in the SREHP, chitinase, and tRNA/STR loci may represent a biased data set. Overall, the data indicate that the choice of the genetic loci used in sequence analysis is important since different chromosomal regions are under variable genetic pressures. In particular, in highly AT-rich organisms, codon bias restricts the extent of genetic variability that can occur in coding regions. With recent advances in high throughput sequencing technology, whole-genome sequences will soon be available for several additional strains of *E. histolytica*. This will rapidly lead to the evaluation and implementation of SNP-based typing methodologies for molecular epidemiological studies.

Another approach to study diversity is comparative genomic hybridizations. This method was applied to the study of *Entamoeba* using a genomic DNA microarray and comparative genomic hybridizations of *E. histolytica* virulent and nonvirulent strains as well as *E. dispar* (Shah et al., 2005). The study identified significant genomic differences (copy number variations as well as sequence divergence) between the virulent and nonvirulent species *E. histolytica* and *E. dispar*. Additionally, analysis of different *E. histolytica* strains indicated genomic separation of *E. histolytica* virulent strains (HM-1:IMSS, 200:NIH, and HK-9) from a nonvirulent strain (Rahman). Further investigations of genomic differences are now under way with genome sequencing of *E. histolytica* clinical isolates by different groups. These data should add greatly to our understanding of amebic genetic diversity and will shed light on both the organism's population structure as well as genetic fingerprints associated with virulence phenotypes.

Our current understanding of the *E. histolytica* population structure is that significant polymorphism (at least in sections of the genome) exists among clinical isolates. Using strain typing techniques, it has been shown that "outbreaks" are usually caused by a single strain (with person-to-person transmission) and that long-term carriage is usually with the same strain. Geographic locations with increased incidence of liver disease and colitis have been identified, although the molecular/genetic basis for this is unknown.

LINK BETWEEN GENOTYPE AND VIRULENCE

There are limited data on phenotype–genotype associations for *E. histolytica*. It has long been observed that most people who become colonized with *E. histolytica* do not progress to invasive disease. Additionally, geographically variable disease predilection has also been seen in studies of amebic epidemiology. Invasive disease in Egypt is predominantly amebic colitis (Abd-Alla and Ravdin, 2002), whereas in South Africa, disease is mostly amebic liver abscess (Stauffer and Ravdin, 2003). In Hue City, Vietnam, the estimated frequency of amebic liver abscess is as high as 21 cases per 100,000 inhabitants (Blessmann et al., 2002). Whether these observations are indicative of differences in host or parasite genetics is not currently clear. To date, most studies have not found a correlation between genotypes of *E. histolytica* strains and their clinical manifestations. However, two studies (one using RAPD analysis of amebic isolates and one using SHREP PCR and RFLP) did give some preliminary indication that it may be feasible to find a genotypic pattern predictive of a phenotypic outcome (Ayeh-Kumi et al., 2001; Valle et al., 2000). Using comparative genomic hybridzations, one study reported that a nonvirulent *E. histolytica* strain was divergent compared to virulent *E. histolytica* strains (Shah et al., 2005). Another study compared *E. histolytica* samples from patients in Bangladesh with asymptomatic colonization, diarrhea and/or dysentery, and amebic liver abscess. The parasite genotype prevalence in each of the three groups was significantly different from each other, leading to the suggestion that the parasite genotype plays a role in the outcome of infection (Ali et al., 2007). To assess tissue-specific tropism, one study used size and sequencing of tRNA-linked STRs from paired stool and liver abscess pus samples from 18 patients with amebic liver abscess. Unexpectedly, the data revealed that for any given patient, the parasite genotypes were different from the intestinal sample compared to the liver abscess pus samples (Ali et al., 2008). The implications of this finding are not yet clear but could indicate that either the intestinal parasite population contained multiple genotypes, with only a subpopulation migrating (or capable of migrating) to the liver. Alternatively, the authors suggested that it is possible that DNA reorganization or recombination events take place during migration of the amebas from the intestine to the liver, or that DNA reorganization or recombination gives rise to certain strains that are more likely to cause invasive hepatic disease. Utilization of next-generation sequencing approaches will be crucial to shed further light on this topic.

VIRULENCE MECHANISMS

Infection with *E. histolytica* results from ingestion of infectious cysts in fecally contaminated food or water (reviewed in Haque et al., 2003a). Excystation occurs in the lumen of the small intestine, with each quadrinucleate cyst giving rise to eight motile trophozoites. The majority of infections remain asymptomatic, with trophozoites dividing asexually within the colonic lumen and, in response to as yet unknown cues, re-encysting to complete the life cycle. In the subset of symptomatic infections, trophozoites invade the colonic epithelium, causing dysentery and, more rarely, hematogenous spread and abscesses at distant sites such as the liver.

What determines the outcome of a given infection is not known, but it is believed from histopathologic studies of natural and experimental infections that *E. histolytica* trophozoites invade tissue and cause symptomatic disease through sequential steps of adherence, cell killing, protease-dependent tissue destruction, and migration into tissues (Beaver et al., 1988; Chadee and Meerovitch, 1985; Huston and Petri, 1998; Ravdin, 1989; Takeuchi and Phillips, 1975). Phagocytosis of host nucleated cells and erythrocytes are also prominent during tissue invasion, and erythrophagocytosis can be used clinically to distinguish *E. histolytica* from *E. dispar* in fecal samples (Gonzalez-Ruiz et al., 1994; Griffin, 1972). *In vitro* assays for each of these characteristic histophathogic features of invasive amebiasis exist and have been used in combination with plasmid-based methods for exogenous gene expression and gene silencing to characterize a number of *E. histolytica* genes required for each step (see Table 22.1) (Bracha et al., 1999, 2003; Hamann et al., 1995, 1997; Huston et al., 2003; Ramakrishnan et al., 1997; Ravdin and Guerrant, 1981; Vines et al., 1995). Similarly, rodent models of amebic liver abscess and colitis following direct inoculation of trophozoites enable tests of the requirement for a given gene product for *in vivo* virulence (Chadee and Meerovitch, 1984; Cieslak et al., 1992; Houpt et al., 2002; Seydel and Stanley, 1998). There is currently no method to produce infectious *E. histolytica* cysts *in vitro* and no animal model that enables encystation and studies of disease transmission. Thus, it is not possible to test the effect of virulence on transmission efficiency in the laboratory, which is the necessary experiment to test the traditional view that commensalism should be favored during evolution (i.e., that maintenance of a viable host will necessarily favor increased disease transmission and be selected for over time) (Levin, 1996).

Although it is not possible using current *in vitro* and rodent models to assess effects on transmission of *E. histolytica*, several *in vitro* studies have addressed the possibility that *E. histolytica* virulence is the result of coincidental evolution. This general notion is supported by similar lifestyle necessities for pathogenic and nonpathogenic *Entamoebae*, suggesting that virulence of *E. histolytica* may have evolved due to factors other than increasing transmissibility. For example, phagocytosis of colonic bacteria is believed to be the major source of nutrients for both *E. histolytica* and *E. dispar* (Jacobs, 1947; Nakamura, 1953; Rees et al., 1941). One study was specifically intended to address the possibility that amebic virulence is coincidental, and demonstrated that similar cellular machinery is necessary for phagocytosis of both bacteria and host cells (Ghosh and Samuelson, 1997). Specifically, $p21^{racA}$, phosphoinositide 3-kinase, and vacuoloar ATPase are all required for efficient phagocytosis of both *Escherichia coli* and erythrocytes. The presence of orthologs of each of these genes in *E. dispar* further indicates that, though required for the virulence-associated phenotype of phagocytosis, the presence of these genes was selected for functions not specific to virulence. Similarly, at least some of the ligands that stimulate uptake of bacteria also stimulate uptake of host cells, suggesting the presence of amebic receptors capable of mediating both. Collectins (i.e., C-type lectins of the innate immune system, such as mannose-binding lectin) are present in intestinal secretions, bind to both apoptotic cells and bacteria, and stimulate *E. histolytica* phagocytosis (Teixeira et al., 2008). These studies collectively suggest that phagocytosis of bacteria is a requirement for nutrient acquisition and that phagocytosis of host cells during tissue invasion may merely be coincidental.

High sequence conservation of accepted virulence factors among pathogenic and nonpathogenic *Entamoebae* further supports the concept of coincidental evolution. One example is the heterodimeric Gal/GalNAc-specific surface lectin, which is the most well-characterized *E. histolytica* cell surface adhesin and is widely accepted to

TABLE 22.1 *E. histolytica* Proteins Implicated in Virulence

Protein Name	Function/Virulence-Associated Phenotype(s)	Present in *E. dispar*?	Evidence *in vitro*	Evidence *in vivo*	References
Amoebapore A	Cytotoxicity	No	Yes	Yes, liver abscess	Bracha et al. (1999) and Zhang et al. (2004)
C2-domain protein kinase (EhC2PK)	Phagocytosis	Yes	Yes	No	Somlata et al. (2011)
Calcium binding protein 1 (EhCaBP1)	Cytotoxicity, phagocytosis	Yes	Yes	No	Jain et al. (2008)
Cysteine proteinase 1 (EhCp-A1)	Cytotoxicity, phagocytosis	No	Yes	Yes, cecal infection	Hirata et al. (2007) and Melendez-Lopez et al. (2007)
Cysteine proteinase 2 (EhCp-A2)	Cytoxicity, phagocytosis	Yes	Yes	No	Hirata et al. (2007)
Cysteine proteinase 4 (EhCp-A4)	Unknown	Yes	No	Yes, cecal infection	He et al. (2010)
Cysteine proteinase 5 (EhCp-A5)	Cytotoxicity, phagocytosis, mucus degradation, induction of inflammation	No	Yes	Yes, cecal infection	Ankri et al., (1999b), Moncada et al. (2006), and Zhang et al. (2000)
Cysteine proteinase 7 (EhCp-A7)		No	Yes	No	Irmer et al. (2009)
EhRab11B	Cysteine proteinase secretion	Yes	Yes	No	Mitra et al. (2007)
EhRab5	Phagocytosis	Yes	Yes	No	Saito-Nakano et al. (2004)
EhRabA	Adherence, cytotoxicity, cysteine proteinase release	Yes	Yes	No	Welter et al. (2005)
FYVE-finger domain protein 4 (EhFP4)	Phagocytosis	Yes	Yes	No	Nakada-Tsukui et al. (2009)
Gal/GalNAc lectin heavy subunit (Hgl)	Adherence, cytotoxicity, phagocytosis, complement resistance	Yes	Yes	Yes, liver abscess	Braga et al. (1992), Petri et al. (1987), and Vines et al. (1998)
Gal/GalNAc lectin light subunit (Lgl)	Adherence, cytotoxicity, phagocytosis	Yes	Yes	Yes, liver abscess	Ankri et al. (1999a) and Katz et al. (2002)
GlcNAc-phosphatidylinositol deacetylase (PIG-L)	Glycophosphatidyl inositol biosynthesis, adherence, endocytosis	Yes	Yes	No	Vats et al. (2005)

(*Continued*)

TABLE 22.1 (*Continued*)

Protein Name	Function/Virulence-Associated Phenotype(s)	Present in *E. dispar*?	Evidence *in vitro*	Evidence *in vivo*	References
Guanine nucleotide exhange factor 1 (EhGEF1)	Motility, cytotoxicity	Yes	Yes	No	Aguilar-Rojas et al. (2005)
Lysine- and glutamic-acid rich protein 1 (KERP1)	Adherence	No	Yes	Yes, liver abscess	Santi-Rocca et al. (2008) and Seigneur et al. (2005)
Mannosyltransferase	Glycophosphatidyl inositol biosynthesis, complement resistance	Yes	Yes	Yes, liver abscess	Weber et al. (2008)
Myosin IB	Phagocytosis	Yes	Yes	No	Marion et al. (2004)
Myosin II	Motility	Yes	Yes	No	Coudrier et al. (2005)
p21 racA	Motility, phagocytosis	Yes	Yes	No	Ghosh and Samuelson (1997) and Labruyere et al. (2003)
Peroxiredoxin	Oxygen-free radical scavenger	Yes	No	Yes, cecal infection	Davis et al. (2006)
Phagosome-associated transmembrane kinase (PATMK)	Phagocytosis	Yes	Yes	Yes, cecal infection	Boettner et al. (2008)
Phosphatidyl-inositol-3-kinase (PI3K)	Phagocytosis	Yes	Yes	No	Byekova et al. (2010) and Ghosh and Samuelson (1997)
Rhomboid protease 1 (EhROM1)	Adherence, phagocytosis	Yes	Yes	No	Baxt et al. (2010)
Serine-rich *E. histolytica* protein (SREHP)	Adherence, phagocytosis	Yes	Yes	No	Teixeira and Huston (2008)
Serine-, threonine-, and isoleucine-rich protein (EhSTIRP)	Adherence, cytotoxicity	Yes	Yes	No	MacFarlane and Singh, (2007)
Transmembrane kinase 39 (TMK39)	Phagocytosis	Yes	Yes	No	Buss et al. (2010)
Upstream regulatory element 3-binding protein (URE3-BP)	Regulation of virulence-associated gene expression	Yes	Yes	Yes, cecal infection and liver abscess	Gilchrist et al. (2010)
Vacuolar ATPase	Phagocytosis	Yes	Yes	No	Ghosh and Samuelson (1997)

be a critical virulence factor (reviewed in Petri et al., 2002). The GalNAc-specific lectin is highly conserved in *E. histolytica* strains isolated from geographically diverse locations, and its heavy and light subunits are 86% and 79% identical in *E. dispar*, respectively (Beck et al., 2002; Dodson et al., 1997). These data suggest that the lectin evolved to play a critical role in *Entamoeba* biology (e.g., adherence by trophozoites to colonic mucus) and that the protein's role in virulence is coincidental.

Despite the many similarities and the apparent maintenance of numerous factors required for virulence by virtue of a central role in essential processes, it is clear that coincidental evolution cannot explain the difference in virulence of *E. histolytica* and *E. dispar* since significant phenotypic and genetic differences between pathogenic and nonpathogenic *Entamoebae* have been recognized. Though *E. dispar* expresses a highly conserved GalNAc-specific surface lectin, it is expressed on the cell surface at lower levels, and *E. dispar* adheres to and kills host cells less efficiently *in vitro* than *E. histolytica* (Dodson et al., 1997; Pillai et al., 2001). This difference in cell killing ability appears to be functionally important. The ability to induce host cell apoptosis or loss of erythrocyte membrane asymmetry facilitates phagocytosis by inducing exposure of phosphatidylserine, which stimulates *E. histolytica* particle uptake (Bailey et al., 1987; Boettner et al., 2005; Huston et al., 2003). *E. dispar* is less efficient at inducing erythrocyte phosphatidylserine exposure than *E. histolytica* and, thus, phagocytoses fresh erythrocytes less efficiently (Boettner et al., 2005). However, it phagocytoses Ca^{2+} ionophore-treated erythrocytes (Ca^{2+} ionophore treatment induces phosphatidylserine exposure) with equal efficiency. This capability suggests that differences in cell killing ability may at least partially explain *E. histolytica* virulence and that, though an essential activity, phagocytic ability is not the basis.

Virulence for many gram-negative bacterial pathogens can be established by the presence of a single piece of DNA, such as the gene encoding shiga toxin in strains of *E. coli* that cause hemolytic uremic syndrome or a pathogenicity island in *Salmonella enterica* (Groisman and Ochman, 1997). The *E. dispar* genome is currently estimated to have 415 more genes than are present in *E. histolytica*, and only a handful of genes have been identified in *E. histolytica* that are not found in *E. dispar* (Weedall and Hall, 2011). Unlike the case for gram-negative bacterial pathogens, it therefore seems unlikely that a single genetic determinant of *E. histolytica* virulence exists. It is possible that *E. histolytica* has lost a determinant present in *E. dispar* that prevents virulence or, perhaps more likely, that the major difference is in gene regulation. Nevertheless, axenic culture of *E. dispar* and stable transfection systems have been developed, which enable testing the effect on phenotype of transferring *E. histolytica* genes to *E. dispar* (Clark, 1995; Moshitch-Moshkovitch et al., 1996).

This reverse genetics approach has been used in an attempt to study several *E. histolytica* cysteine proteinases (Hellberg et al., 2001). The family of *E. histolytica* cysteine proteinases has been shown to disrupt polymerization of MUC2 (the major protein component of human colonic mucus), and, consistent with an important protective role for colonic mucus, *E. histolytica* infection of *Muc2*$^{-/-}$ mice results in significantly higher intestinal inflammation compared with wild-type mice (Hou et al., 2010; Moncada et al., 2003). In addition to MUC2, the cysteine proteinases degrade extracellular matrix proteins and immunoglobulins and contribute to immune-mediated tissue damage through cleavage and activation of preinterleukin 1-β (Keene et al., 1986; Li et al., 1995; Thran et al., 1998; Zhang et al., 2000). Differences in the complement of cysteine proteinase genes and their expression patterns in *E. histolytica* and *E. dispar* have been well described, including degenerate genes for two *E. histolytica* cysteine proteinases (EhCP1 and EhCP5) in *E. dispar* (Bruchhaus et al., 2003); furthermore, clinical *E. histolytica* isolates release 10- to 1000-fold more cysteine proteinase activity into culture supernatants than *E. dispar* isolates (Reed et al., 1989). Attempts to express EhCP1 and EhCP5 in *E. dispar* were unsuccessful (Hellberg et al., 2001). EhCp2, which is normally expressed at low levels in *E. dispar*, was successfully overexpressed, leading to an overall sevenfold increase in *E. dispar* cysteine proteinase activity (Hellberg et al., 2001). This conferred an increased ability to destroy tissue culture monolayers *in vitro*; however, there was no effect on liver abscess formation in gerbils, suggesting that qualitative differences in cysteine proteinase activity may be more important than quantitative differences *in vivo*. To date, similar transgenic experiments examining other virulence-associated genes have not been conducted.

SUMMARY, LIMITATIONS, AND OPPORTUNITIES FOR FUTURE WORK

The primary evolutionary pressure on *E. histolytica* trophozoites is the ability to survive within the intestinal lumen, and this logically requires an ability to adhere and to compete successfully with other microbes for nutrients. Since it is hard to imagine that tissue invasion per se confers an evolutionary advantage (i.e., trophozoites present in a liver abscess will never reach a new host), the idea that many amebic abilities associated with virulence have been maintained for killing and phagocytosis of bacteria is appealing (Ghosh and Samuelson, 1997). Conservation of the mechanisms for many processes (e.g., secretion of antimicrobial peptides that are both bactericidal and cytolytic, maintenance of adhesins, and the general mechanisms of phagocytosis) is consistent with the notion that *E. histolytica* virulence is the result of coincidental evolution. However, important phenotypic differences between virulent and avirulent *Entamoeba* strains/species indicate that *E. histolytica* virulence cannot be fully explained by coincidence nor does it appear from comparative genomic studies that *E. histolytica* virulence can be explained by a single genetic determinant. Rather, it seems most likely that *E. histolytica* virulence relative to *E. dispar* is predominantly due to differences in gene regulation. Although it is probable that host factors and the local microbial community are important, the fact that only a minority of infections due to any specific *E. histolytica* strain result in disease is consistent with this possibility.

The major limitation to formally testing models of evolution beyond the idea of coincidental evolution is the inability to determine if transmissibility and virulence are positively correlated using currently available animal models. It may be possible to address this question with molecular epidemiologic studies, but no such study has been conducted to date. For example, molecular typing methods have enabled association of different *E. histolytica* clinical strains with the clinical outcomes of asymptomatic infection, diarrhea, and hepatic abscess (Ali et al., 2007). It should be possible to use similar methods to intensively measure infectious cyst numbers present in feces for individuals infected with *E. histolytica* and to correlate this with strain type and clinical symptoms. Although confounders may limit conclusions, it may also be possible to assess the number of infected household members for symptomatic versus asymptomatic *E. histolytica* infections and to infer transmission rates associated with each. Similarly, for individuals coinfected with *E. dispar* and *E. histolytica*, quantitative reverse transcriptase (qRT)-PCR using cyst-specific transcripts might enable direct comparison of cyst production. Finally, at least one theory of evolution suggests that host crowding, by virtue of eliminating the need for host mobility and prolonged survival to maximize transmission to new hosts, should favor selection of parasites with increased virulence (Levin, 1996). Molecular epidemiologic studies could be used to determine if host crowding favors parasites of higher virulence.

REFERENCES

Abd-Alla MD, Ravdin JI. 2002. Diagnosis of amoebic colitis by antigen capture ELISA in patients presenting with acute diarrhoea in Cairo, Egypt. Tropical Medicine and International Health 7: 365–370.

Aguilar-Rojas A, Almaraz-Barrera J, Krzeminski M, Robles-Flores M, Hernandez-Rivas R, Guillen N, Maroun RC, Vargas M. 2005. *Entamoeba histolytica*: Inhibition of cellular functions by overexpression of EhGEF1, a novel Rho/Rac guanine nucleotide exchange factor. Experimental Parasitology 109: 150–162.

Ali IK, Zaki M, Clark CG. 2005. Use of PCR amplification of tRNA gene-linked short tandem repeats for genotyping *Entamoeba histolytica*. Journal of Clinical Microbiology 43: 5842–5847.

Ali IK, Solaymani-Mohammadi S, Akhter J, Roy S, Gorrini C, Calderaro A, Parker SK, Haque R, Petri WA, Clark CG. 2008. Tissue invasion by *Entamoeba histolytica*: Evidence of genetic selection and/or DNA reorganization events in organ tropism. PLoS Neglected Tropical Diseases 2: e219.

Ali IKM, Mondal U, Roy S, Haque R, Petri WA, Clark CG. 2007. Evidence for a link between parasite genotype and outcome of infection with *Entamoeba histolytica*. Journal of Clinical Microbiology 45: 285–289.

Ankri S, Padilla-Vaca F, Stolarsky T, Koole L, Katz U, Mirelman D. 1999a. Antisense inhibition of expression of the light subunit (35 kDa) of the Gal/GalNac lectin complex inhibits *Entamoeba histolytica* virulence. Molecular Microbiology 33: 327–337.

Ankri S, Stolarsky T, Bracha R, Padilla-Vaca F, Mirelman D. 1999b. Antisense inhibition of expression of cysteine proteinases affects *Entamoeba histolytica*-induced formation of liver abscess in hamsters. Infection and Immunity 67: 421–422.

Ayeh-Kumi PF, Ali IM, Lockhart LA, Gilchrist CA, Petri WA, Haque R. 2001. *Entamoeba histolytica*: Genetic diversity of clinical isolates from Bangladesh as demonstrated by polymorphisms in the serine-rich gene. Experimental Parasitology 99: 80–88.

Bailey GB, Day DB, Nokkaew C, Harper CC. 1987. Stimulation by target cell membrane lipid of actin polymerization and phagocytosis by *Entamoeba histolytica*. Infection and Immunity 55: 1848–1853.

Baxt LA, Rastew E, Bracha R, Mirelman D, Singh U. 2010. Downregulation of an *Entamoeba histolytica* rhomboid protease reveals roles in regulating parasite adhesion and phagocytosis. Eukaryotic Cell 9: 1283–1293.

Beaver PC, Blanchard JL, Seibold HR. 1988. Invasive amebiasis in naturally infected New World and Old World monkeys with and without clinical disease. American Journal of Tropical Medicine and Hygiene 39: 343–352.

Beck DL, Tanyuksel M, Mackey AJ, Haque R, Trapaidze N, Pearson WR, Loftus B, Petri WA. 2002. *Entamoeba histolytica*: Sequence conservation of the Gal/GalNAc lectin from clinical isolates. Experimental Parasitology 101: 157–163.

Bhattacharya D, Haque R, Singh U. 2005. Coding and noncoding genomic regions of *Entamoeba histolytica* have significantly different rates of sequence polymorphisms: Implications for epidemiological studies. Journal of Clinical Microbiology 43: 4815–4829.

Blessmann J, Van Linh P, Nu PA, Thi HD, Muller-Myhsok B, Buss H, Tannich E. 2002. Epidemiology of amebiasis in a region of high incidence of amebic liver abscess in central Vietnam. American Journal of Tropical Medicine and Hygiene 66: 578–583.

Boettner DR, Huston CD, Sullivan JA, Petri WA. 2005. *Entamoeba histolytica* and *Entamoeba dispar* utilize externalized phosphatidylserine for recognition and phagocytosis of erythrocytes. Infection and Immunity 73: 3422–3430.

Boettner DR, Huston CD, Linford AS, Buss SN, Houpt E, Sherman NE, Petri WA. 2008. *Entamoeba histolytica* phagocytosis of human erythrocytes involves PATMK, a member of the transmembrane kinase family. PLoS Pathogens 4: 122–133.

Bracha R, Nuchamowitz Y, Leippe M, Mirelman D. 1999. Antisense inhibition of amoebapore expression in *Entamoeba histolytica* causes a decrease in amoebic virulence. Molecular Microbiology 34: 463–472.

Bracha R, Nuchamowitz Y, Mirelman D. 2003. Transcriptional silencing of an amoebapore gene in *Entamoeba histolytica*: Molecular analysis and effect on pathogenicity. Eukaryotic Cell 2: 295–305.

Braga LL, Ninomiya H, Mccoy JJ, Eacker S, Wiedmer T, Pham C, Wood S, Sims PJ, Petri WA. 1992. Inhibition of the complement membrane attack complex by the galactose-specific adhesin of *Entamoeba histolytica*. Journal of Clinical Investigation 90: 1131–1137.

Bruchhaus I, Loftus BJ, Hall N, Tannich E. 2003. The intestinal protozoan parasite *Entamoeba histolytica* contains 20 cysteine protease genes, of which only a small subset is expressed during in vitro cultivation. Eukaryotic Cell 2: 501–509.

Buss SN, Hamano S, Vidrich A, Evans C, Zhang Y, Crasta OR, Sobral BW, Gilchrist CA, Petri WA. 2010. Members of the *Entamoeba histolytica* transmembrane kinase family play non-redundant roles in growth and phagocytosis. International Journal for Parasitology 40: 833–843.

Byekova YA, Powell RR, Welter BH, Temesvari LA. 2010. Localization of phosphatidylinositol (3,4,5,)-trisphosphate to phagosomes in *Entamoeba histolytica* achieved using glutathione S-transferase- and green fluorescent protein-tagged lipid biosensors. Infection and Immunity 78: 125–137.

Chadee K, Meerovitch E. 1984. The pathogenesis of experimentally induced amebic liver abscess in the gerbil. American Journal of Pathology 117: 71–79.

Chadee K, Meerovitch E. 1985. Entamoeba histolytica: Early progressive pathology in the cecum of the gerbil (Meriones unguiculatus). American Journal of Tropical Medicine and Hygiene 34: 283–291.

Cieslak PR, Virgin HWT, Stanley SL, Jr. 1992. A severe combined immunodeficient (SCID) mouse model for infection with *Entamoeba histolytica*. Journal of Experimental Medicine 176: 1605–1609.

Clark CG. 1995. Axenic cultivation of *Entamoeba dispar* Brumpt 1925, *Entamoeba insolita* Geiman and Wichterman 1937 and *Entamoeba ranarum* Grassi 1879. Journal of Eukaryotic Microbiology 42: 590–593.

Clark CG, Diamond LS. 1993. *Entamoeba histolytica*: A method for isolate identification. Experimental Parasitology 77: 450–455.

Coudrier E, Amblard F, Zimmer C, Roux P, Olivo-Marin JC, Rigothier MC, Guillen N. 2005. Myosin II and the Gal-GalNAc lectin play a crucial role in tissue invasion by *Entamoeba histolytica*. Cellular Microbiology 7: 19–27.

Davis PH, Zhang X, Guo J, Townsend RR, Stanley SL. 2006. Comparative proteomic analysis of two *Entamoeba histolytica* strains with different virulence phenotypes identifies peroxiredoxin as an important component of amoebic virulence. Molecular Microbiology 61: 1523–1532.

Diamond LS, Clark CG. 1993. A redescription of *Entamoeba histolytica* Schaudinn, 1903 (Emended Walker, 1911) separating it from *Entamoeba dispar* Brumpt, 1925. Journal of Eukaryotic Microbiology 40: 340–344.

Dodson JM, Clark CG, Lockhart LA, Leo BM, Schroeder JW, Mann BJ. 1997. Comparison of adherence, cytotoxicity, and Gal/GalNAc lectin gene structure in *Entamoeba histolytica* nad *Entamoeba dispar*. Parasitology International 28: 168–169.

Ghosh S, Frisardi M, Ramirez-Avila L, Descoteaux S, Sturm-Ramirez K, Newton-Sanchez OA, Santos-Preciado JI, Ganguly C, Lohia A, Reed S, Samuelson J. 2000. Molecular epidemiology of *Entamoeba* spp.: Evidence of a bottleneck (Demographic sweep) and transcontinental spread of diploid parasites. Journal of Clinical Microbiology 38: 3815–3821.

Ghosh SK, Samuelson J. 1997. Involvement of p21racA, phosphoinositide 3-kinase, and vacuolar ATPase in phagocytosis of bacteria and erythrocytes by *Entamoeba histolytica*: Suggestive evidence for coincidental evolution of amebic invasiveness. Infection and Immunity 65: 4243–4249.

Gilchrist CA, Moore ES, Zhang Y, Bousquet CB, Lannigan JA, Mann BJ, Petri WA. 2010. Regulation of virulence of *Entamoeba histolytica* by the URE3-BP transcription factor. MBio 1: e00057–e00010.

Gomes MA, Melo MN, Macedo AM, Furst C, Silva EF. 2000. RAPD in the analysis of isolates of *Entamoeba histolytica*. Acta Tropica 75: 71–77.

Gonzalez-Ruiz A, Haque R, Aguirre A, Castanon G, Hall A, Guhl F, Ruiz-Palacios G, Miles MA, Warhurst DC. 1994. Value of microscopy in the diagnosis of dysentery associated with invasive *Entamoeba histolytica*. Journal of Clinical Pathology 47: 236–239.

Griffin JL. 1972. Human amebic dysentery: Electron microscopy of *Entamoeba histolytica* contacting, ingesting, and digesting inflammatory cells. American Journal of Tropical Medicine and Hygiene 21: 895–906.

Groisman EA, Ochman H. 1997. How *Salmonella* became a pathogen. Trends in Microbiology 5: 343–349.

Haghighi A, Kobayashi S, Takeuchi T, Thammapalerd N, Nozaki T. 2003. Geographic diversity among genotypes of *Entamoeba histolytica* field isolates. Journal of Clinical Microbiology 41: 3748–3756.

Hamann L, Nickel R, Tannich E. 1995. Transfection and continuous expression of heterologous genes in the protozoan parasite *Entamoeba histolytica*. Proceedings of the National Academy of Sciences of the United States of America 92: 8975–8979.

Hamann L, Buss H, Tannich E. 1997. Tetracycline-controlled gene expression in *Entamoeba histolytica*. Molecular and Biochemical Parasitology 84: 83–91.

Haque R, Ali IM, Petri WA, Jr. 1999. Prevalence and immune response to *Entamoeba histolytica* infection in preschool children in Bangladesh. American Journal of Tropical Medicine and Hygiene 60: 1031–1034.

Haque R, Huston CD, Hughes M, Houpt E, Petri WA. 2003a. Amebiasis. New England Journal of Medicine 348: 1565–1573.

Haque R, Mondal D, Kirkpatrick BD, Akther S, Farr BM, Sack RB, Petri WA. 2003b. Epidemiologic and clinical characterisitics of acute diarrhea with emphasis on *Entamoeba histolytica* infections in preschool children in an urban slum of Dhaka, Bangladesh. American Journal of Tropical Medicine and Hygiene 69: 398–405.

He C, Nora GP, Schneider EL, Kerr ID, Hansell E, Hirata K, Gonzalez D, Sajid M, Boyd SE, Hruz P, Cobo ER, Le C, Liu WT, Eckmann L, Dorrestein PC, Houpt ER, Brinen LS, Craik CS, Roush WR, Mckerrow J, Reed SL. 2010. A novel *Entamoeba histolytica* cysteine proteinase, EhCP4, is key for invasive amebiasis and a therapeutic target. Journal of Biological Chemistry 285: 18516–18527.

Hellberg A, Nickel R, Lotter H, Tannich E, Bruchhaus I. 2001. Overexpression of cysteine proteinase 2 in *Entamoeba histolytica* or *Entamoeba dispar* increases amoeba-induced monolayer destruction *in vitro* but does not augment amoebic liver abscess formation in gerbils. Cellular Microbiology 3: 13–20.

Hirata KK, Que X, Melendez-Lopez SG, Debnath A, Myers S, Herdman DS, Orozco E, Bhattacharya A, Mckerrow JH, Reed SL. 2007. A phagocytosis mutant of *Entamoeba histolytica* is less virulent due to deficient proteinase expression and release. Experimental Parasitology 115: 192–199.

Hou Y, Mortimer L, Chadee K. 2010. *Entamoeba histolytica* cysteine proteinase 5 binds integrin on colonic cells and stimulates NF-kB-mediated pro-inflammatory responses. Journal of Biological Chemistry 285: 35497–35504.

Houpt ER, Glembocki DJ, Obrig TG, Moskaluk CA, Lockhart LA, Wright RL, Seaner RM, Keepers TR, Wilkins TD, Petri WA. 2002. The mouse model of amebic colitis reveals mouse strain susceptibility to infection and exacerbation of disease by CD4+ T cells. Journal of Immunology 169: 4496–4503.

Huston CD, Petri WA, Jr. 1998. Host-pathogen interaction in amebiasis and progress in vaccine development. European Journal of

Clinical Microbiology and Infectious Diseases 17: 601–614.

Huston CD, Boettner DR, Miller-Sims V, Petri WA. 2003. Apoptotic killing and phagocytosis of host cells by the parasite *Entamoeba histolytica*. Infection and Immunity 71: 964–972.

Irmer H, Tillack M, Biller L, Handal G, Leippe M, Roeder T, Tannich E, Bruchhaus I. 2009. Major cysteine peptidases of *Entamoeba histolytica* are required for aggregation and digestion of erythrocytes but are dispensable for phagocytosis and cytopathogenicity. Molecular Microbiology 72: 658–667.

Jacobs L. 1947. The elimination of viable bacteria from cultures of *Endamoeba histolytica* and the subsequent maintenance of such cultures. American Journal of Hygiene 46: 172–176.

Jain R, Santi-Rocca J, Padhan N, Bhattacharya S, Guillen N, Bhattacharya A. 2008. Calcium-binding protein 1 of *Entamoeba histolytica* transiently associates with phagocytic cups in a calcium-dependent manner. Cellular Microbiology 10: 1373–1389.

Katz U, Ankri S, Stolarsky T, Nuchamowitz Y, Mirelman D. 2002. *Entamoeba histolytica* expressing a dominant negative N-truncated light subunit of its gal-lectin are less virulent. Molecular Biology of the Cell 13: 4256–4265.

Keene WE, Petitt MG, Allen S, Mckerrow JH. 1986. The major neutral proteinase of *Entamoeba histolytica*. Journal of Experimental Medicine 163: 536–549.

Labruyere E, Zimmer C, Galy V, Olivo-Marin JC, Guillen N. 2003. EhPAK, a member of the p21-activated kinase family, is involved in the control of *Entamoeba histolytica* migration and phagocytosis. Journal of Cell Science 116: 61–71.

Levin BR. 1996. The evolution and maintenance of virulence in microparasites. Emerging Infectious Diseases 2: 93–102.

Li E, Yang WG, Zhang T, Stanley SL, Jr. 1995. Interaction of laminin with *Entamoeba histolytica* cysteine proteinases and its effect on amebic pathogenesis. Infection and Immunity 63: 4150–4153.

Macfarlane RC, Singh U. 2007. Identification of an *Entamoeba histolytica* serine-, threonine, and isoleucine-rich protein with roles in adhesion and cytotoxicity. Eukaryotic Cell 6: 2139–2146.

Marion S, Wilhelm C, Voigt H, Bacri JC, Guillen N. 2004. Overexpression of myosin IB in living *Entamoeba histolytica* enhances cytoplasm viscosity and reduces phagocytosis. Journal of Cell Science 117: 3271–3279.

Melendez-Lopez SG, Herdman S, Hirata K, Choi MH, Choe Y, Craik C, Caffrey CR, Hansell E, Chavez-Munguia B, Chen YT, Roush WR, Mckerrow J, Eckmann L, Guo J, Stanley SL, Reed SL. 2007. Use of recombinant *Entamoeba histolytica* cysteine proteinase 1 to identify a potent inhibitor of amebic invasion in a colonic model. Eukaryotic Cell 6: 1130–1136.

Mitra BN, Saito-Nakano Y, Nakada-Tsukui K, Sato D, Nozaki T. 2007. Rab11B small GTPase regulates secretion of cysteine proteases in the enteric protozoan parasite *Entamoeba histolytica*. Cellular Microbiology 9: 2112–2125.

Moncada D, Keller K, Chadee K. 2003. *Entamoeba histolytica* cysteine proteinases disrupt the polymeric structure of colonic mucin and alter its protective function. Infection and Immunity 71: 838–844.

Moncada D, Keller K, Ankri S, Mirelman D, Chadee K. 2006. Antisense inhibition of *Entamoeba histolytica* cysteine proteases inhibits colonic mucus degradation. Gastroenterology 130: 721–730.

Moshitch-Moshkovitch S, Stolarsky T, Mirelman D, Alon RN. 1996. Stable episomal transfection and gene expression in Entamoeba dispar. Molecular and Biochemical Parasitology 83: 257–261.

Nakada-Tsukui K, Okada H, Mitra BN, Nozaki T. 2009. Phosphatidylinositol-phosphates mediate cytoskeletal reorganization during phagocytosis via a unique modular protein consisting of RhoGEF/DH and FYVE domains in the parasitic protozoan *Entamoeba histolytica*. Cellular Microbiology 11: 1471–1491.

Nakamura M. 1953. Nutrition and physiology of *Endamoeba histolytica*. Bacteriology Reviews 17: 189–212.

Petri WA, Haque R, Mann BJ. 2002. The bittersweet interface of parasite and host: Lectin-carbohydrate interactions during human invasion by the parasite *Entamoeba histolytica*. Annual Review of Microbiology 56: 39–64.

Petri WA, Jr, Smith RD, Schlesinger PH, Murphy CF, Ravdin JI. 1987. Isolation of the galactose binding lectin that mediates the in vitro adherence of *Entamoeba histolytica*. Journal of Clinical Investigation 80: 1238–1244.

Pillai DR, Kobayashi S, Kain KC. 2001. *Entamoeba dispar*: Molecular characterization of the galactose/N-acetyl-D-galactosamine lectin. Experimental Parasitology 99: 226–234.

Ramakrishnan G, Vines RR, Mann BJ, Petri WAJ. 1997. A tetracycline-inducible gene expression system in *Entamoeba histolytica*. Molecular and Biochemical Parasitology 84: 93–100.

Ravdin JI. 1989. Amebiasis now. American Journal of Tropical Medicine and Hygiene 41: 40–48.

Ravdin JI, Guerrant RL. 1981. Role of adherence in cytopathic mechanisms of *Entamoeba histolytica*. Study with mammalian tissue culture cells and human erythrocytes. Journal of Clinical Investigation 68: 1305–1313.

Reed SL, Keene WE, Mckerrow JH. 1989. Thiol proteinase expression correlates with pathogenicity of *Entamoeba histolytica*. Journal of Clinical Microbiology 27: 2772–2777.

Rees CW, Reardon LV, Jacobs L. 1941. The cultivation of the parasitic protozoa without bacteria. American Journal of Tropical Medicine 21: 695–716.

Saito-Nakano Y, Yasuda T, Nakada-Tsukui K, Leippe M, Nozaki T. 2004. Rab5-associated vacuoles play a unique role in phagocytosis of the enteric protozoan parasite *Entamoeba histolytica*. Journal of Biological Chemistry 279: 49497–49507.

Santi-Rocca J, Weber C, Guigon G, Sismeiro O, Coppee JY, Guillen N. 2008. The lysine- and glutamic acid-rich protein KERP1 plays a role in *Entamoeba histolytica* liver abscess pathogenesis. Cellular Microbiology 10: 202–217.

Seigneur M, Mounier J, Prevost MC, Guillen N. 2005. A lysine- and glutamic acid-rich protein, KERP1, from *Entamoeba histolytica* binds to human enterocytes. Cellular Microbiology 7: 569–579.

Seydel KB, Stanley SL, Jr. 1998. *Entamoeba histolytica* induces host cell death in amebic liver abscess by a non-Fas-dependent, non-tumor necrosis factor alpha-dependent pathway of apoptosis. Infection and Immunity 66: 2980–2983.

Shah PH, Macfarlane RC, Bhattacharya D, Matese JC, Demeter J, Stroup SE, Singh U. 2005. Comparative genomic hybridizations of Entamoeba strains reveal unique genetic fingerprints that correlate with virulence. Eukaryotic Cell 4: 504–515.

Somlata V, Bhattacharya S, Bhattacharya A. 2011. A C2 domain protein kinase initiates phagocytosis in the protozoan parasite *Entamoeba histolytica*. Nature Communications 2: 230.

Stauffer W, Ravdin JI. 2003. *Entamoeba histolytica*: An update. Current Opinion in Infectious Diseases 16: 479–485.

Takeuchi A, Phillips BP. 1975. Electron microscope studies of experimental *Entamoeba histolytica* infection in the guinea pig. I. Penetration of the intestinal epithelium by trophozoites. American Journal of Tropical Medicine and Hygiene 24: 34–48.

Teixeira JE, Huston CD. 2008. Participation of the serine-rich *Entamoeba histolytica* protein in amebic phagocytosis of apoptotic host cells. Infection and Immunity 76: 959–966.

Teixeira JE, Heron BT, Huston CD. 2008. C1q- and collectin-dependent phagocytosis of apoptotic host cells by the intestinal protozoan *Entamoeba histolytica*. Journal of Infectious Diseases 198: 1062–1070.

Thran VQ, Herdman DS, Torian BE, Reed SL. 1998. The neutral cysteine proteinase of *Entamoeba histolytica* degrades IgG and prevents its binding. Journal of Infectious Diseases 177: 508–511.

Valle PR, Souza MB, Pires EM, Silva EF, Gomes MA. 2000. Arbitrarily primed PCR fingerprinting of RNA and DNA in Entamoeba histolytica. Revista Do Instituto de Medicina Tropical de Sao Paulo 42: 249–253.

Vats D, Vishwakarma RA, Bhattacharya S, Bhattacharya A. 2005. Reduction of cell surface glycosylphosphatidylinositol conjugates in *Entamoeba histolytica* by antisense blocking of *E. histolytica* GlcNAc-phosphatidylinositol deacetylase expression: Effect on cell proliferation, endocytosis, and adhesion to target cells. Infection and Immunity 73: 8381–8392.

Vines RR, Purdy JE, Ragland BD, Samuelson J, Mann BJ, Petri WA, Jr. 1995. Stable episomal transfection of *Entamoeba histolytica*. Molecular and Biochemical Parasitology 71: 265–267.

Vines RR, Ramakrishnan G, Rogers JB, Lockhart LA, Mann BJ, Petri WA, Jr. 1998. Regulation of adherence and virulence by the *Entamoeba histolytica* lectin cytoplasmic domain, which contains a beta2 integrin motif. Molecular Biology of the Cell 9: 2069–2079.

Weber C, Blazquez S, Marion S, Ausseur C, Vats D, Krzeminski M, Rigothier MC, Maroun RC, Bhattacharya A, Guillen N. 2008. Bioinformatics and functional analysis of an *Entamoeba histolytica* mannosyltransferase necessary for parasite complement resistance and hepatical infection. PLoS Neglected Tropical Diseases 2: e165.

Weedall GD, Hall N. 2011. Evolutionary genomics of *Entamoeba*. Research in Microbiology 162: 637–645.

Welter BH, Powell RR, Leo M, Smith CM, Temesvari LA. 2005. A unique Rab GTPase, EhRabA, is involved in motility and polarization of *Entamoeba histolytica* cells. Molecular and Biochemical Parasitology 140: 161–173.

WHO. 1997. WHO/PAHO/UNESCO report. A consultation with experts on amoebiasis. Mexico City, Mexico 28–29 January, 1997. WHO Epidemiological Bulletin 18: 13–14.

Zaki M, Clark CG. 2001. Isolation and characterization of polymorphic DNA from *Entamoeba histolytica*. Journal of Clinical Microbiology 39: 897–905.

Zaki M, Meelu P, Sun W, Clark CG. 2002. Simultaneous differentiation and typing of *Entamoeba histolytica* and *Entamoeba dispar*. Journal of Clinical Microbiology 40: 1271–1276.

Zhang T, Cieslak PR, Foster L, Kunz-Jenkins C, Stanley SL, Jr. 1994. Antibodies to the serine rich *Entamoeba histolytica* protein (SREHP) prevent amoebic liver abscess in severe combined immunodeficient (SCID) mice. Parasite Immunology 16: 225–230.

Zhang X, Zhang Z, Alexander D, Bracha R, Mirelman D, Stanley SL. 2004. Expression of amoebapores is required for full expression of *Entamoeba histolytica* virulence in amebic liver abscess but is not necessary for the induction of inflammation or tissue damage in amebic colitis. Infection and Immunity 72: 678–683.

Zhang Z, Wang L, Seydel KB, Li E, Ankri S, Mirelman D, Stanley SL, Jr. 2000. *Entamoeba histolytica* cysteine proteinases with interleukin-1 beta converting enzyme (ICE) activity cause intestinal inflammation and tissue damage in amoebiasis. Molecular Microbiology 37: 542–548.

CHAPTER 23

SEX AND VIRULENCE IN BASIDIOMYCETE PATHOGENS

GUUS BAKKEREN, EMILIA K. KRUZEL, and CHRISTINA M. HULL

INTRODUCTION TO THE BASIDIOMYCOTA

Among the fungi, the largest and most explored groups are the Ascomycota and the Basidiomycota, which comprise the Dikarya. There are over 40,000 described species of Ascomycota, including the model yeast *Saccharomyces cerevisiae*, many filamentous fungi, and most of the human fungal pathogens (Hawksworth, 2001; Mitchell, 2005; Morrow and Fraser, 2009). The Basidiomycota contains approximately 30,000 described species, including mushrooms, yeasts, rust, and smut fungi. These fungi are known for their ubiquitous nature in the environment; they are found in nearly all terrestrial ecosystems and are the major wood-degrading organisms on earth, essential for maintaining the earth's carbon cycle (Hibbett, 2006; Kirk et al., 2001; Mitchell, 2005). Some, such as certain mycorrhizal symbionts, are universal in these ecosystems where they are essential for forest health and may have been fundamental to land colonization by plants (Martin et al., 2008). While no single morphological feature or property defines all basidiomycetes, the production of several specialized cell types occurs in the majority of the basidiomycetes described thus far. Stable dikaryons (the occurrence of two haploid nuclei in the same cell), clamp cells, and basidia all characterize these fungi and play important roles in sexual development (Casselton and Olesnicky, 1998; Raper, 1983). In turn, sexual development is closely linked with the success of the basidiomycetes, particularly those that cause disease in plants and animals. Approximately 8000 rust fungi and smut species are plant pathogens, and at least 40 basidiomycetes cause disease in mammals (Blackwell, 2011; Kirk et al., 2001; Mitchell, 2005). Two genera represent the models most utilized for studying the pathogenesis of the basidiomycetes. Smut fungi of the genus *Ustilago* have been studied extensively for their plant pathogenic properties, and members of the *Cryptococcus* species complex are among the leading causes of fungal meningitis in humans (Banuett, 1995; Park et al., 2009). In this chapter, we discuss the roles of sexual development in virulence of these two representative pathogen-containing groups. We also

Evolution of Virulence in Eukaryotic Microbes, First Edition. Edited by L. David Sibley, Barbara J. Howlett, and Joseph Heitman.

present information emerging from recent studies of the cereal rust fungi, which are among the most devastating crop pathogens (Bolton et al., 2008; Brown and Hovmøller, 2002; Goellner et al., 2010; Singh et al., 2011).

Mating-Type Loci

Sexual development of fungi is controlled by a specialized region of the genome known as a mating-type (*MAT*) locus (Fraser and Heitman, 2003). Akin to the sex chromosomes of larger eukaryotes, this region of the genome houses the genes required to establish differences between sex cells (aka mating types in fungi). Thus, mating-type determination by *MAT* is critical for establishing which cells will engage in productive fusion events. There is striking variation among fungal *MAT* loci; however, ascomycetes generally have a single *MAT* locus that houses transcription factors that govern cell type. This results in two mating types (a bipolar mating system) (Herskowitz, 1988; Kronstad and Staben, 1997). Basidiomycetes are more flexible in their *MAT* configurations. For these fungi, it is common to have two *MAT* loci: one encoding homeodomain transcription factors and one encoding pheromones and pheromone receptors (Casselton and Olesnicky, 1998; Raper, 1966). In principle, this allows for four mating types (a tetrapolar mating system). However, many basidiomycetes have multiple alleles of the genes housed in these loci, increasing the potential production of many more mating types in a population—up to tens of thousands of potential mating types for higher mushrooms. To undergo sexual reproduction, it is critical that distinct cell types be established, and the *MAT* locus is the lynchpin in this process. Although the specific nature of contributions of *MAT* loci to sexual development varies among species, in all cases described, developmental cell types are controlled by the expression of different pheromones and pheromone receptors and the function of a heterodimeric transcription factor complex composed of a pair of *MAT*-encoded homeodomain proteins with one member contributed by each parent (Casselton and Olesnicky, 1998; Hsueh and Heitman, 2008; Kronstad and Staben, 1997). This paradigm governs *MAT* control of sexual development and provides a common platform from which to explore the diversity of basidiomycete *MAT* loci in greater detail.

Sexual Development

Consistent with the diversity of *MAT* loci, sexual development among basidiomycetes varies in its details. A relative constant, however, is that sexual development is initiated when cells of compatible mating types fuse with one another, and this fusion eventually results in the formation of a fruiting body known as a basidium (after which this group is named) on which sexual spores are produced (Casselton and Olesnicky, 1998). A second characteristic feature of basidiomycetes is the production of a prolonged binucleate cell known as a dikaryon, which results from the fusion of compatible cells containing genetically distinct nuclei. Although cell fusion and cytoplasmic mixing occur, the nuclei do not undergo immediate karyogamy and remain distinct. The nuclei are replicated relatively synchronously but are maintained as single entities throughout this growth phase (Iwasa et al., 1998; Raper, 1983). In some basidiomycetes, the dikaryon can grow indefinitely, but in others, the dikaryotic state is transient. In both cases, the process of dikaryon replication and maintenance is coordinated to ensure that each cell contains two nuclei, one from each parent. A cell type critical for this process is the clamp cell, a specialized protrusion that must fuse with the filament to maintain two nuclei in each filament cell aligned at the cell midline (Inada et al., 2001; Scherer et al., 2006). The selective pressures leading to the maintenance of dikaryons are essentially unknown. The complex nuclear migration events that occur require significant input of energy, and the outcome is eventually the same: To complete sexual development, the two nuclei in the terminal filament cell eventually fuse with one another and undergo meiosis. Studies of mushrooms indicate that dikaryons diverge phenotypically to a much greater extent than monokaryons do, suggesting that dikaryotic growth could provide an increased potential for genetic and phenotypic variation, resulting in a fitness advantage (Clark and Anderson, 2004).

Pathogens of Plants and Animals

Most human fungal pathogens are ascomycetes; however, basidiomycete pathogens of humans are a growing group, in large part because of increasing numbers of immunocompromised individuals due to other diseases, malnutrition, and complex medical procedures and drug regimes (Böhme and Karthaus, 1999; Brown, 2004; Mitchell and Perfect, 1995; Park et al., 2009). Some of these species may have adapted to life in human/animal hosts via the acquisition of tolerance mechanisms that provide resistance to various hostile environmental conditions (higher temperatures, pH, iron deficiency, etc.) (Casadevall et al., 2003; Roetzer et al., 2011a,b). The best-studied of these "new" human basidiomycete pathogens is the *Cryptococcus* species complex consisting of *Cryptococcus neoformans*, *Cryptococcus gattii*, *Cryptococcus albidus*, and *Cryptococcus laurentus* (Casadevall and Perfect, 1998). *C. neoformans* is the model for this group as well as for basidiomycete human pathogens overall. Very abundant, but not immediately life threatening, is a basidiomycete causing skin disease, *Malassezia globosa*, implicated in human dandruff and atopic eczema. The genome of *M. globosa* has recently been sequenced, and although the sexual cycle has not yet been observed in nature or under laboratory conditions, the genome sequence revealed conserved genes involved in mating and meiosis in other fungi, and what may represent a bipolar *MAT* locus, suggesting that this pathogen may be capable of sex (Xu et al., 2007). Other basidiomycete species implicated in opportunistic invasive diseases belong to the *Trichosporon*, *Rhodotorula*, and *Sporobolomyces* genera.

Among fungal species adapted to life in plants or trees, basidiomycetes account for several pathogen groups that cause major destruction worldwide. These include many rust fungi (Pucciniales) on many plant and tree hosts, and smut fungi (Ustilaginaceae) and bunts (Tilletiaceae) on grasses, including cereal crops (Bolton et al., 2008; Brown and Hovmøller, 2002; Goellner et al., 2010; Mitchell, 2005). Many of these fungi have come to prominence because they attack important agriculture and forestry crops; for instance, the wheat stem rust fungus has been observed since biblical times (Chester, 1946). Surveys of cereal rust fungi and targeted breeding of plants for resistance have contributed greatly over the last 60 years to efforts to maintain production levels in the presence of this unabating disease pressure (Li and Wang, 2009; Singh et al., 2011). However, targeted breeding has lead to rounds of introductions of disease resistance genes, monoculture production, and, consequently, the evolution of new strains of fungi overcoming introduced resistance properties. The emergence of novel, very virulent wheat stem rust isolates in Uganda in East Africa in the late 1990s (Ug99 and derivatives), likely through sexual recombination, has proven to be a severe threat to wheat production worldwide (Singh et al., 2011). In addition, large amounts of fungicides are applied annually to control smuts, bunts, and rusts, a costly necessity often not available in many parts of the world and often devastating to the environment.

SMUT FUNGI

In smut fungi, pathogenicity is tightly associated with sex because in these plant pathogenic fungi, mating between haploid partners of different mating types is necessary to form the pathogenic dikaryotic cell type. Diploid teliospores that are the dispersal and survival propagules of these plant pathogens germinate to produce the basidium in which meiosis occurs and four haploid basidiospores are produced. This cell type is nonpathogenic and can be easily germinated and maintained as budding cells and also manipulated in the laboratory. Mating type has segregated in these progenies, and only basidiospores having different mating types initiate the formation of a conjugation tube in response to each other, which can fuse. This then brings two parental haploid nuclei together in the same cell, forming the dikaryotic state, which is maintained throughout the rest of the life cycle up to spore formation and which is characteristic for many basidiomycetes. Combining nuclei with mating-type genes of different allelic specificities causes a major change in the regulation of many genes, resulting in a switch to filamentous growth now capable of infecting plant host tissue. This cell type represents the obligate parasitic stage of the life cycle since it cannot easily be maintained *in vitro* and needs the host for completion of its life cycle to produce the teliospores.

Mating-Type Loci

The complete genomes of three smut species have been sequenced and their mating-type loci analyzed. The corn smut fungus *Ustilago maydis* remains the model organism for this group and for basidiomycete plant pathogens in general (Brefort et al., 2009; Kamper et al., 2006). It is closely related to the genome of the corn and sorghum head smut fungus *Sporisorium reilianum* (Schirawski et al., 2010). The third sequenced smut genome is that of *Ustilago hordei*, which causes covered smut on barley and oats (Bakkeren et al., 2006; J. Laurie, S. Ali, G. Bakkeren, J. Schirawski, R. Kahmann, unpublished data). These fungi had been known for a long time to represent two different groups with respect to mating types: ones that possess a bipolar mating system and ones that are governed by a tetrapolar mating system. These designations were based on assessment of mating types of numerous progenies generated from many different natural smut fungal isolates that had been paired in various combinations. In species such as *U. hordei*, only one genetic locus determined mating type, and two allelic specificities seemed to segregate, designating it as a bipolar mating system. In other species such as *U. maydis*, mating type was governed by two unlinked genetic loci, designating it as tetrapolar (Fisher and Holton, 1957; Raper, 1966). In the latter mating system, one locus, termed the *a* locus, seemed to have two specificities (Bolker et al., 1992), although three specificities have been described in *S. reilianum* (Schirawski et al., 2005). For the other locus, termed the *b* locus, many allelic specificities have been found in nature.

Molecular analysis of *U. maydis* has elucidated functions underlying these genetic findings. The *a* locus spans no more than 10 kb and encodes two main components: a pheromone mating factor a (Mfa) and pheromone receptor A (PRA) (Fig. 23.1). The *a1* locus encodes Mfa1 and PRA1, and the *a2* locus encodes Mfa2 and PRA2. Mfa1 stimulates cells having PRA2 (and the *a2* specificity) and vice versa. Limited allelic variation has been found, although more than two alleles appear to be present in *S. reilianum* (Schirawski et al., 2005). The *b* locus has two divergently transcribed genes spanning no more than 4 kb and codes for heterodimeric homeodomain-containing transcription regulators, bE and bW. Component bE1 from mating

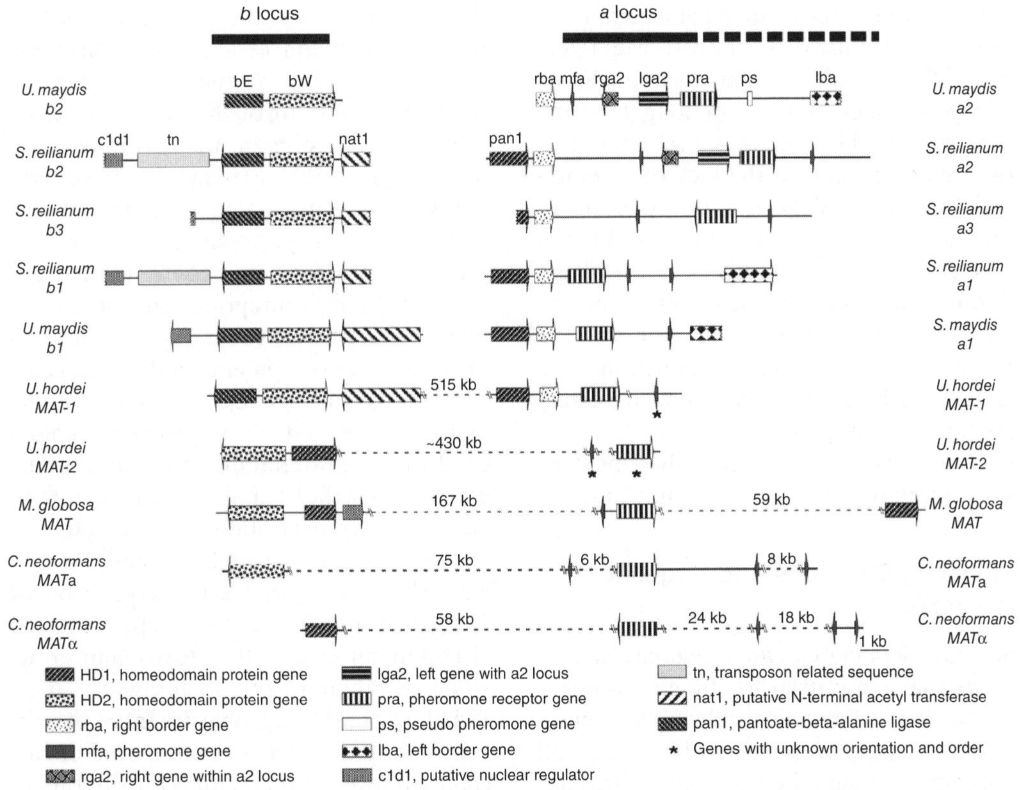

Figure 23.1 Genetic organization of the mating-type loci of selected basidiomycetes. Genes are indicated by arrows with the arrow denoting direction of transcription. Related genes are denoted by the same shading and respective gene functions are explained in the lower part of the figure. * indicates that the relative order and orientation of these genes have not been determined. In the tetrapolar species *Ustilago maydis* and *Sporisorium reilianum*, the *a*- and *b*-specific sequences reside on different chromosomes while they are linked by spacer regions (which are not drawn to scale and whose length is indicated) in the bipolar species *Ustilago hordei* and *Cryptococcus neoformans*, as well as in *Malassezia globosa*. The black bars on top of the figure indicate the regions of the *b* locus, which covers the two homeodomain protein genes *bE* and *bW*, and the *a* locus (which expands to different lengths in the different fungi, indicated by a broken line) from the *lba* gene to the *rba* gene. Sequence information was obtained from the following GenBank accessions: AF043940, AM118080, AACP01000083, AACP01000013, AJ884588, AJ884583, AJ884590, AJ884585, AJ884589, AJ884584, U37796, M84182, AF184070, AF184069, Z18531, AAYY01000003, AF542530, and AF542531. Reprinted from Bakkeren et al. (2008), with permission from Elsevier.

specificity 1 interacts, after cell fusion, with bWx (not 1) from cells having a different allelic specificity. Many different allelic variants have been found, and amino acid changes are responsible for altered interaction and subsequent downstream gene activation (Yee and Kronstad, 1998; reviewed in Brefort et al., 2009).

In the bipolar smut fungus *U. hordei*, the same *a* and *b* mating-type components are present, but they are physically linked or reside on the genetically described single *MAT* locus (Bakkeren and Kronstad, 1994). This linkage seems to be the defining difference between bipolar and tetrapolar mating systems. Astoundingly, the *a* and *b* mating-type complexes are separated by 527 kb at the *MAT-1* locus (Fig. 23.1). This region is suppressed for recombination with the *MAT-2* region in cells from the opposite mating type and has inversions (at least at the *b* gene complex), deletions (the region between the *a* and *b* gene complexes at the *MAT-2* locus is 70 kb shorter), and over 50% of coding capacity is taken up by transposable elements and long terminal repeat (LTR)-like sequences. All these features may account for the lack of recombination and possibly for the genesis of specialized sex chromosomes (Bakkeren et al., 2006; Lee et al., 1999). A translocation of a chromosome segment harboring the *a* mating-type complex to link it to the *b* mating-type complex to constitute the large *MAT* locus appears to have occurred, possibly in a progenitor, and thus resulting in a new evolutionary clade that includes small grain-infecting bipolar smut fungi.

Signaling Cascades and Regulatory Networks

Several signaling cascades regulate sets of genes that are crucial for fungal pathogenicity or play a major role in virulence. Many components seem to be conserved also among nonpathogens, but their deletion does not always affect an *in vitro*/saprobic lifestyle. It is therefore likely that as species evolved to become pathogenic, they found novel ways of relaying and responding to or taking advantage of (danger) signals from host environments during infections.

As mentioned, in the smut fungi, the binding of mating pheromone Mfa to the correct PRA initiates conjugation tube formation and cell fusion (mating) between partners of different allelic specificities, *a1* and *a2*. Many components of two major signal transduction pathways involved, a G-protein-ras-adenylate cyclase cascade acting through cyclic AMP (cAMP), and a mitogen-activated protein kinase (MAPK) module of hierarchical kinases, have been identified (reviewed in Brefort et al., 2009; García-Pedrajas et al., 2008; Lee et al., 2003). Mating and pathogenic filamentation are initiated upon signaling by both the cAMP and pheromone MAPK cascades that converge on the Prf1 transcription factor (Hartmann et al., 1999). The PRA receptor, active in response to pheromone binding, signals through sequential MAP kinases, and results in the activating phosphorylation of Prf1. Signaling via the cAMP pathway activates a second kinase whose additional phosphorylation of Prf1 is required for complete activity (Hartmann et al., 1999). Prf1 integrates nutritional and mating signals, thus governing the initiation of sexual development and pathogenic growth. When active, Prf1 induces the expression of genes that promote haploid cell fusion and simultaneously activates the expression of the bE and bW homeodomain-containing transcription factors that heterodimerize upon cellular fusion. At this stage, bE/bW regulates the expression of 347 genes, inducing 212 and repressing 135 (Brachmann et al., 2001; Hartmann et al., 1996; Kahmann and Schirawski, 2007; Urban et al., 1996). Downstream of bE/bW lies a regulatory network composed of cascading transcription factors contributing to differential aspects of pathogenesis and biotrophic growth. Of the direct targets of bE/bW, Rbf1, a zinc finger transcription factor, is responsible for most *b*-dependent transcriptional and morphological effects. Rbf1 controls the expression of secondary transcription factors Hdp1 and Biz1, which contribute to propagation *in planta* and cell cycle regulation of dikaryotic growth (Flor-Parra et al., 2006; Heimel et al., 2010b). Among other genes regulated by bE/bW is MAPK *kpp6* involved in invasive growth of the newly formed dikaryotic filament (Brachmann et al., 2003).

One important aspect of the life cycle is cell cycle control in relation to morphogenesis and, hence, infection and pathogenicity. Growth by budding of the basidiospores, switching to the formation of conjugation tubes, mating, filamentous growth, and the production of teliospores are momentous changes. Initiation of mating requires cells to be arrested in G2 (Garcia-Muse et al.,

2003), which rearranges the cytoskeleton and prepares the cell for (fast) polar growth; such cycling between cell proliferation and G2 arrest to allow filamentous growth might be a hallmark of hyphae penetrating host cells during infection (Brefort et al., 2009; Flor-Parra et al., 2006, 2007; Perez-Martin et al., 2006). The dimorphic switch to hyphal growth is essential for pathogenicity in the smut fungi, and the disruption of many genes regulating or involved in cell polarity, such as cyclin-dependent kinases and microtubule kinesin, myosin, and dynein motor proteins, affects pathogenicity or virulence (Castillo-Lluva et al., 2007; Steinberg, 2007). Interestingly, DNA damage checkpoint kinases Chk1 and Atr1, involved in a widely conserved signaling cascade, may have additional roles regulating cell type by controlling cell cycle arrest and maintenance of the dikaryotic state during infection (de Sena-Tomás et al., 2011).

Another signaling pathway responding to environmental signals such as sensing pH (H^+) and other ions, high temperature, Ca^{2+} levels (through Ca^{2+}-binding calmodulin), cell wall stress, and so on, acts through calcineurin, a calcium/calmodulin-regulated serine/threonine phosphatase consisting of a catalytic subunit A (CNA) and regulatory subunit B (CNB). In pathogenic fungi, including *Ustilago* species, disrupting this central regulator affects virulence (Cervantes-Chávez et al., 2011; Egan et al., 2009; Fox et al., 2001). A comprehensive comparative genomic study of MAPK and calcium–calcineurin signaling components in plant and human pathogenic fungi was published recently (Rispail et al., 2009).

Pathogenesis and Virulence

To identify pathogenicity and virulence genes, functional assays and pathogenicity tests, based on mating tests and host interaction and disease ratings, have been employed to screen mutants generated by chemical mutagens or radiation, random insertional mutagenesis, or by reverse genetic approaches through gene deletion of candidate genes. Many genes have been found to impact the ability of the smut fungi to cause disease (pathogenicity factors) or the degree of disease and symptom formation (virulence factors). Studies, mainly in *U. maydis*, are too numerous to list here and the reader is referred to the reviews listed below. As mentioned, the haploid basidiospores can be easily manipulated experimentally, but upon mating, which in nature usually occurs on the leaf or coleoptile (emerging seedling) surface, the resulting dikaryotic hyphae in the smut fungi represent biotrophic cell types that cannot be cultured easily and need the host for completion of the life cycle, that is, the sporulation process. Therefore, in mutational screens, extensive use has also been made of pathogenic haploid strains to circumvent the need for mating (see reviews Brefort et al., 2009; Kahmann and Kamper, 2004). Distinguishing between virulence factors *sensu stricto*, that is, contributing to the disease process such as penetration and suppression of the host defense responses, and genes merely involved in fitness and/or general metabolism, is extremely difficult in the biotrophs and is often a matter of definition and interpretation. However, the acquisition of such factors led saprobic fungi to evolve pathogenic lifestyles. In general, the environment in which fungi can thrive and reproduce is likely to be unimportant to them, but the emergence of pathogenicity during evolution of the smut fungi needing host infection to produce progeny, or the rust fungi that have become strict biotrophs (see further), must have provided selective advantages such as a guaranteed source of nutrients, a superior method of producing progeny, and a dispersal mechanism.

Virulence factors include secreted effectors, which are part of the so-called secretome, and proteins predicted to be secreted, often having an N-terminal signal peptide. Effectors in oomycetes and fungi are relatively small (<350 amino acids and can be smaller than 80 aa), and most do not match sequences with annotated functions in databases. Effectors can be conserved among related fungi, but subsets are more genus or even species specific. Some have multiple cysteine residues in a conserved spacing pattern among homologs, possibly allowing for disulfide bridge formation and characteristic folding for function and/or protection against proteases encountered in the host environment. Their highly variable nature and the fact that many often reside in series of paralogous family members, often in physical clusters, suggest that at least a subset is under diversifying selection, likely under selection pressure from interactions with host components.

Large sets of small secreted proteins (SSPs) have been identified in basidiomycete pathogen genomes, and numbers range from 350 to 400 in *U. maydis* (Mueller et al., 2008), *S. reilianum* (Schirawski et al., 2010), and *U. hordei* (J. Laurie, S. Ali, R. Kahmann, J. Schirawski, and G. Bakkeren, unpublished data), and as many as 1184 in the poplar leaf rust fungus *Melampsora larici-populina* (*Mlp*) and 1386 in the wheat stem rust fungus *Puccinia graminis* f. sp. *tritici* (*Pgt*) (Duplessis et al., 2011; Joly et al., 2010). In smut fungi, several clusters of SSP genes are highly upregulated once the fungus is inside the host and are needed for full virulence (Kamper et al., 2006; Schirawski et al., 2010). Individual effectors, such as PEP1, PIT2, and UhAVR1, suppress host defense responses (Doehlemann et al., 2009, 2011; X. Song, S. Ali, J. Laurie, and G. Bakkeren, unpublished data), and PEP1 is also essential for penetration. In some *U. hordei* species, UhAVR1 triggers resistance in barley host cultivars having the resistance gene allele that recognizes this effector.

As stated, smut fungi have various mating-type architectures (reviewed in Bakkeren et al., 2008), and these probably have affected the evolution of virulence. For example, unlinked *a* and *b* loci might have led to diversification of allelic specificities of the *b* genes, resulting in the many mating types observed for *U. maydis*. This is thought to favor outcrossing and hence increases possible recombination and diversification of virulence factors such as predicted secreted effectors, among mates. As mentioned, in smut fungi, many effectors are in clusters of related paralogous family members. In contrast, in the bipolar smut fungus *U. hordei*, the single *MAT* locus has been hypothesized to favor inbreeding, possibly stimulating other means of diversifying virulence components. Indeed, early analysis of an effector cluster in this fungus revealed fewer family members (S. Ali, J. Laurie, and G. Bakkeren, unpublished data).

CRYPTOCOCCUS SPECIES

C. neoformans is a budding yeast saprophyte that grows clonally via budding or undergoes a filamentous phase during sexual development (Casadevall and Perfect, 1998; Kwon-Chung, 1976). Unlike the plant pathogenic basidiomycetes, pathogenesis is not tightly associated with sex because full sexual development can take place outside the host, the yeast form is infectious in animal models, and there is no evidence that sexual development occurs in the host environment under any conditions. The link between sexual development and pathogenesis among *Cryptococcus* species is through the role that sex plays in the development of spores and in the generation of genetic diversity (Giles et al., 2009; Velagapudi et al., 2009).

C. neoformans is a bipolar fungus with two stable mating types (**a** and α) that can be grown and manipulated in the laboratory (Hull and Heitman, 2002; Kwon-Chung, 1976). When *C. neoformans* cells of opposite mating type encounter one another under appropriate nutrient conditions, they can signal via pheromones and pheromone receptors and fuse with one another (Davidson et al., 2000; McClelland et al., 2004; Shen et al., 2002; Stanton et al., 2010). After cell fusion, filamentous dikaryotic growth ensues and continues until the formation of a basidium in which nuclear fusion and meiosis occur. The meiotic products are then repeatedly replicated mitotically, and the resulting products are packaged into spores. These basidiospores are then budded onto the surface of the basidium in four long chains (Alspaugh et al., 2000; Idnurm, 2010).

Another form of sexual development in *C. neoformans* is known as haploid fruiting (Lin et al., 2005, 2006). During this process, α strains, in response to severe desiccation and nutrient limitation, undergo monokaryotic filament formation, basidium formation, and sporulation. The resultant spores are recombinant, arising from an early endoduplication or same-sex fusion event (Bui et al., 2008; Lin et al., 2005). The capacity to carry out a sexual cycle in the absence of a mating partner may confer a fitness advantage because strains could respond to undesirable environmental conditions by forming spores for dispersal in the absence of a suitable mating partner.

Mating-Type Loci

To date, five *Cryptococcus* genomes have been sequenced to completion and analyzed, including those of *C. neoformans* var. *neoformans*, *C. neoformans* var. *grubii*, and *C. gattii* (D'Souza et al., 2011; Loftus et al., 2005). In each case, the genome contains a single, unusually large *MAT* locus over 100 kb in size that harbors upward of 20 genes (Lengeler et al., 2002). One end of the locus encodes pheromones and pheromone receptors responsible for specifying haploid cell identity. These factors mediate mate recognition and cell fusion and show no allelic variation; that is, **a** cells encode Ste3**a** (the functional homolog of *Ustilago* PRA) and MF**a**, and α cells encode Ste3α and MFα, and the specificities of the pheromones and receptors do not vary (Fig. 23.1) (Chung et al., 2002; Davidson et al., 2000; McClelland et al., 2002; Moore and Edman, 1993). This limited allelic repertoire is also seen at the opposite end of the locus that encodes the homeodomain transcription factors Sxi1α and Sxi2**a** (the functional homologs of *Ustilago* bE and bW proteins, respectively; Fig. 23.1) (Hull et al., 2004, 2005). These transcriptional regulators are responsible for specifying the dikaryotic and diploid cell types, and unlike in many other basidiomycetes, they do not vary (Casselton and Olesnicky, 1998). As a result, *C. neoformans* is a bipolar fungus with only two mating types. The midsection of the *C. neoformans MAT* locus contains ~15 genes of varying types, including genes of unknown function and highly conserved housekeeping genes. The roles of these genes in sexual development (if any) are unknown, and most appear to have been captured via a series of transposon-mediated rearrangements (Fraser et al., 2004; Lengeler et al., 2002; Loftus et al., 2005). These rearrangements have led to a region suppressed in recombination, facilitating additional divergence between the *MAT* alleles. The resulting locus thus appears to represent an evolutionary intermediate in a tetrapolar to bipolar evolutionary transition, akin to the locus of *U. hordei* in which a tetrapolar mating system has been reduced to a bipolar arrangement via the linking of two *MAT* loci (Hsueh et al., 2008; Lee et al., 1999).

Signaling Cascades and Regulatory Networks

Many genes and signaling networks that contribute to development have been identified in *C. neoformans*. Interestingly, numerous signaling pathways contribute to both the ability to undergo sexual development and survival within the host (Kozubowski et al., 2009). The evolutionary linkage between development and pathogenesis is not understood as *C. neoformans* exhibits a saprobic lifestyle in the environment and is not communicable host to host. Virulence traits are therefore unlikely to be under selection of the host environment. Despite the distinct natures of these two processes (sexual development and virulence) in the lifestyle of *C. neoformans*, a number of important signaling pathways and regulatory networks are in common, with many components contributing to both sexual morphogenesis and pathogenesis.

In *C. neoformans*, the protein kinase A (PKA)/cAMP signaling cascade contributes to mating, sexual differentiation, and the production of the virulence factors melanin and capsule. The cAMP pathway operates via a network of signaling proteins that regulate the activity of Cac1, adenylyl cyclase, which converts ATP to the signaling molecule cAMP. Mutants unable to produce cAMP (*gpa1*, *cac1*) show defects in melanin production, sexual filamentation, and capsule formation (Alspaugh et al., 1997, 2002). Specifically, cAMP levels regulate the expression of the laccase genes *LAC1* and *LAC2*, both critical for melanin production (Pukkila-Worley et al., 2005). cAMP levels influence these downstream events presumably via cAMP-dependent kinases such as Pka1. Interestingly, *pka1* mutant phenotypes are less severe than those of *gpa1* or *cac1*, suggesting a more redundant signaling network downstream of cAMP levels (D'Souza et al., 2001; Hu et al., 2007b). The only known Pka1 phosphorylation target is the transcription factor Nrg1, which is required for sexual development and influences the expression of *UGD1*, a gene involved in capsule production (Cramer et al., 2006). Additional targets of Pka1 phosphorylation are unknown, and much remains to be described regarding the mechanisms of cAMP signaling influencing downstream effectors that are important for morphogenesis and pathogenesis.

The *C. neoformans* MAPK cascades are also critical for pathogenesis and sexual morphogenesis. The high osmolarity glycerol (HOG) MAPK pathway is central to the resistance of *C. neoformans* to host stress conditions but contributes minimally to sexual development (Bahn et al., 2005, 2007). The pheromone-activated MAPK cascade influences phenotypes relating to mating, but certain components affect both morphogenic and pathogenic phenotypes, suggesting an interesting link between these two growth stages. During sexual development, pheromone binding to its complementary receptor activates its associated G protein; this activation then triggers an intracellular MAPK cascade (Davidson et al., 2003). The pheromone signal transduction pathway requires the sequential activity of the kinases Ste20**a**/α (PAK kinase), Ste11**a**/α (MAPK kinase kinase), Ste7 (MAPK kinase), and Cpk1 (MAPK) (Clarke et al., 2001; Davidson et al., 2003; Nichols et al., 2004; Wang et al., 2002). A number of these components are required for sexual development, and deletion mutants of *gpb1*, *ste11*, *ste7*, and *cpk1* exhibit dominant sterile phenotypes during development and cannot undergo cellular fusion with a wild-type mating partner. When assessed for their capacity to cause disease in murine models, the same mutations have little to no effect on virulence (Clarke et al., 2001; Davidson et al., 2003; Wang et al., 2000). However, Ste20 is required for full virulence in *C. neoformans* var. *grubii* (Nichols et al., 2004; Wang et al., 2002).

Downstream of the pheromone MAPK signaling cascade, several transcription factors mediate the cellular response to a mating partner. The recently identified Mat2 regulator appears to play a central role during morphogenesis and is a potential direct target of Cpk1 phosphorylation, but when Mat2 is deleted, virulence is unaffected (Lin et al., 2010). Downstream of Mat2 regulation are the mating-type specific regulators Ste12**a** and Ste12α, named for their sequence similarity to the pheromone response and pseudohyphal growth regulator Ste12 of *S. cerevisiae* (Chang et al., 2001; Fields and Herskowitz, 1985; Wickes et al., 1997; Yue et al., 1999). In *C. neoformans*, they contribute to cellular fusion and hyphal growth but are not required, and comprise a branch of the MAPK regulatory network that functions in parallel to currently unknown components (Davidson et al., 2003). Interestingly, Ste12**a** and Ste12α are important for pathogenesis and have been linked to the full expression of certain virulence factors (melanin, capsule). They influence expression of the virulence genes *LAC1* and *CAP59* (Chang et al., 2000; Clarke et al., 2001; Wickes et al., 1997).

The Mat2 transcription factor also activates the expression of the downstream master regulatory genes *SXI1*α and *SXI2***a** (Lin et al., 2010). Despite their induction during early development, they are dispensable for mating and cellular fusion (Hull et al., 2005). Current models suggest that the *SXI* genes are induced prefusion so that their gene products can function immediately postfusion, when they heterodimerize to form an active complex. The Sxi1α–Sxi2**a** heterodimer then acts to initiate the fusant to filament transition via the transcriptional regulation of target effector genes (Stanton et al., 2009). The downstream effectors of Sxi1α-Sxi2**a** activity are unknown, aside from the likely direct target *CLP1* (Ekena et al., 2008). Microarray studies suggest that the Sxi1α–Sxi2**a** complex influences the expression of hundreds of genes during sexual development. Studies suggest that many of these changes may be indirect, and that a number of transcription factors may be among the direct Sxi-regulon (M.E. Mead and C.M. Hull, unpublished data). Further studies should elucidate whether Sxi1α–Sxi2**a** fits the model established in *U. maydis*, where the bE/bW heterodimer acts to initiate a tiered transcription factor cascade controlling dikaryotic growth and the full sexual/pathogenic cycle leading to spore production (Heimel et al., 2010a,b).

Pathogenesis and Virulence

Humans and other animals acquire *Cryptococcus* from the environment via a respiratory route of infection. *C. neoformans* cells, either desiccated yeast or spores, are inhaled into the lung alveoli where they generally cause little or no respiratory disease but can disseminate to other tissues (Casadevall and Perfect, 1998; Garcia-Hermoso et al., 1999; Goldman et al., 2001). Dissemination to the brain in humans results in the development of meningoencephalitis, which is fatal without treatment (Casadevall and Perfect, 1998; Hull and Heitman, 2002). Most cryptococcal diseases occur in immunocompromised people; approximately 1 million cases and over 600,000 deaths are estimated to occur worldwide each year (Park et al., 2009). Numerous virulence properties have been identified in *C. neoformans*, including growth at high temperature, the α mating type, the ability to produce a polysaccharide capsule, and the ability to produce melanin (Chang and Kwon-Chung, 1994; Kwon-Chung et al., 1992a,b). Furthermore, specific genes associated with these pathways have also been identified and tested in a mouse model of cryptococcosis.

Interestingly, over 95% of isolates of *Cryptococcus* isolated from both patients and the environment are of the α mating type (Erke, 1976; Lin et al., 2005, 2006; Wickes et al., 1996). This bias suggests that sexual development itself or genes associated with sexual identity or development somehow influence fitness, persistence, and/or pathogenesis. In fact, some α strains of *C. neoformans* var. *neoformans* have been shown to be more virulent than the **a** counterparts in a mouse model of infection

(Kwon-Chung et al., 1992a). One hypothesis is that haploid fruiting (which occurs preferentially in α cells) could lead to better dispersal of α strains, thus leading to more opportunities to persist in the environment and to infect hosts (Lin et al., 2005, 2006). A caveat to this proposal is that α fruiting has only been detected in *C. neoformans* var. *neoformans*, and the vast majority of isolates in the world are *C. neoformans* var. *grubii*, in which α fruiting has not been detected (Hull and Heitman, 2002; Wickes et al., 1996). Thus, it remains to be determined what parameters of *C. neoformans* growth lead to differences in virulence and how sexual reproduction contributes to this. Among the human fungal pathogens, *Cryptococcus* is the only one with a sexual cycle that can be readily manipulated in the laboratory. The ability to induce sexual development and link genotype to phenotype through classical genetic studies is a hallmark of the system, resulting in a relatively large number of known virulence genes.

Sexual development of *Cryptococcus* is additionally important because spores are suspected infectious particles in human disease. Recently, spores were purified for the first time in numbers sufficient for biochemical, immunological, and virulence studies (Botts et al., 2009). (All previous studies of *Cryptococcus* had been carried out with the yeast form of the organism.) Spores have now been shown to be resistant to many environmental stresses and to cause disease in the mouse inhalation model of cryptococcosis (Botts et al., 2009; Giles et al., 2009; Velagapudi et al., 2009). Furthermore, spores interact with the mammalian immune system in a fundamentally different manner than the genetically identical yeast form does (Giles et al., 2009). These intriguing findings imply that *Cryptococcus* could infect mammals as spores, as yeast, or as a mixture, ultimately leading to differences in disease profile and outcome. As more fertile strains are recovered from patients and the environment, the full spectrum of spore production and virulence can be determined (Litvintseva et al., 2003). One area of great interest is in the role that sexual development and resulting spore production are playing in human infections in sub-Saharan Africa. The largest concentration of people with AIDS is in this area, in which there is strong evidence for *Cryptococcus* sexual development and resulting strain diversity (Litvintseva et al., 2003; Park et al., 2009). Because sexual development creates genetic diversity, the opportunity for more virulent strains to arise in susceptible hosts may be imminent. Clearly, sexual development and pathogenesis are linked in *Cryptococcus* in ways that have not been fully recognized previously. Future studies promise to elucidate these links and to provide insights into how the relationship evolved in this and other basidiomycetes.

RUST FUNGI

Sex in Relation to Virulence

In contrast with smut fungi, the cereal rust fungi have a complex life cycle including five different spore types and two hosts (macrocyclic). For example, the wheat leaf rust fungus *Puccinia triticina* (*Pt*), formerly called *Puccinia recondita* f. sp. *tritici*, produces brown-colored urediniospores from which the rust obtained its name, on wheat (*Triticum aestivum* L.). They are the asexual dikaryotic, infectious propagules, which are easily carried long distances by prevailing winds and spread on wheat through reinfection, which can lead to epidemics. On senescing wheat plants, teliospores can be produced, which are primarily two celled, with each cell containing two haploid nuclei that have paired, if not fused, to form the diploid state (Mendgen, 1984). Similar to smuts, in rust fungi, teliospores are the survival structures and produce haploid basidiospores. A third mitotic division can result in basidiospores having two nuclei while being genetically monokaryons, containing two nuclei of the same type (Anikster, 1983). These spore types are ephemeral and infect meadow rue (*Thalictrum speciosissimum* L.), the so-called alternate host on which the fungus completes its sexual stage. Upon infection through direct penetration of the leaf surface, monokaryotic hyphae produce specialized pycnia, which generate pycniospores embedded in nectar. Because they originated from haploid meiotic products, the pycniospores in a specific pycnium represent one mating type and need to cross-fertilize (heterothallism), often through the action of insects attracted by the nectar. During fertilization, one pycniospore fuses with a receptive hyphum in the pycnium of a different mating type. This cellular fusion is followed by nuclear transfer, and the newly formed dikaryon undergoes developmental reprogramming. The resultant mycelium traverses the leaf and forms aecia on the underside, in which dikaryotic aeciospores develop. Aeciospores are dispersal propagules that will infect the primary wheat host (Bolton et al., 2008; Horton et al., 2005; Samborski, 1985). Very similar life cycles can be described for wheat stem rust, *Pgt*, and wheat stripe rust, *Puccinia striiformis* f. sp *tritici* (*Pst*), although these have their sexual cycles on *Berberis* species; this was very recently discovered for *Pst* (Jin et al., 2010).

Unlike the smuts, the rust fungi do not require sex for infection (of wheat) and the asexual cycle can produce such enormous populations that changes in virulence are thought to occur through mutation or parasexual interactions. The latter hypothesis was recently substantiated when vegetative or parasexual recombination was shown to occur between hyphae of germinated *Pt* urediniospores (Wang and McCallum, 2009). Large reductions in *Pt* populations due to the planting of resistant wheat cultivars and control of the disease through the use of large amounts of fungicides have correlated with reductions in the variability of virulence. The *Thalictrum* species that serve as good hosts for the sexual reproduction of *Pt* are mainly found in the Mediterranean basin and in the Middle East, the regions where wheat and progenitors coevolved with their rust pathogens. *Thalictrum* species indigenous in North America are widespread and can support sexual reproduction but to limited degrees (Saari et al., 1968). A striking correlation exists between the availability of the alternate host for *Pgt* and *Pst*, *Berberis* species that can support sexual reproduction, and the occurrence of isolates (races) with a high genetic variability and diverse virulence spectrum (Jin et al., 2010). Indeed, barberry bush eradication campaigns in the Great Plains in the United States since the 1930s coincided with a steady decline of race variation, probably because the populations became asexual. It seems therefore that virulence increases through sexual recombination in the grass-infecting rust fungi and as such is dependent on the action of the mating-type genes, but not in the same sense as in the smuts where mating is a prerequisite for pathogenicity. In these rusts, reshuffling genes likely leads to new combinations of effector repertoires, which can overcome defenses put up by the grass hosts.

Whether these correlations and findings extend to the rusts infecting dicots has not been researched to the same extent as for the cereal rusts, where large collections of genetically typed pathogens and hosts are available. In the poplar leaf rust *Melampsora medusae*, more genetic variabilities seem to be present in populations where hosts for both the telial (*Populus* spp.) and the aecial (*Larix* spp.) are present, although this does not correlate with virulence profiles (Bourassa et al., 2007). In *M. larici-populina*, increased variability in virulence is indicated in populations where the sexual host is also present (Gerard et al., 2006). Various populations of *Melampsora lini*, which causes rust of flax (*Linum* species), have different levels of genetic variability and virulence, but it is unclear whether this is solely correlated with the presence of sexual reproduction. In addition, like the cereal rust fungi, this fungus is macrocyclic, but in contrast to the cereal rust fungi, all stages occur on the same host, flax (autoecious rust). It is currently unknown whether this particular property affects possible correlations of sex and the evolution of virulence compared to the cereal rust fungi (reviewed in Lawrence et al., 2007). In the bean rust fungus *Uromyces appendiculatus*, there is no clear indication of fewer genetic polymorphisms in asexual populations, based on a limited set of virulence and isoenzyme markers (Groth et al., 1995). Economically important legume-infecting *Phakopsora* species, such as the soybean rust pathogen *Phakopsora pachyrhizi*, have not been observed to produce teliospores (and therefore to go through a sexual cycle) in North America. A teliospore stage has been found in Asia, but germination has not been reported in nature and it is not known whether this fungus is autoecious or heteroecious (reviewed in Goellner et al., 2010). Analyses of Nigerian asexual populations of *P. pachyrhizi* revealed low genetic differentiation and a limited variability in pathotypes (Twizeyimana et al., 2011).

The Search for Functional Homologs Related to Sex

Molecular genetic analysis of the rust fungi has only recently received a boost mainly due to several large-scale genome sequencing projects. Although molecular genetic research has been performed for the last 20 years by a few tenacious researchers, the fact that the rust fungi are obligate biotrophic pathogens has made analyses of gene function very difficult. The lack of transformation systems made using reverse genetic or mutation complementation techniques impossible. Studies were limited to the isolation of (conserved) candidate genes and expression analyses, and *in situ* localization of gene products and comparative analyses to genes of other pathogens.

Several studies have attempted to shed light on the mating-type system in rust fungi. Conclusions and speculations vary from rust fungi having a simple bipolar system in several *Puccinia* and *Uromyces* species (Anikster et al., 1999) to a more complicated tetrapolar system with multiple allelic specificities in *M. lini* (Lawrence, 1980) and related oat crown rust pathogen, *P. coronata* (Narisawa et al., 1994). Different rust fungi may have different mating systems and harbor similar gene complexes in various arrangements as in the smuts and mushrooms. Indeed, searches of the public *Pgt* genome identified smut *a*-like pheromone (*mfa*) and pheromone receptor (*Pra* or *STE3*) genes and one set of divergently transcribed, *b*-like homeodomain-containing genes. Examination of the partial *Pt* genome reveals homologs of PRA1 and PRA2, and of two *bE* and *bW* pairs likely representing the two allelic specificities in the two haploid nuclei, bE1/bW1 and bE2/bW2 (G. Bakkeren, L. Szabo, J. Fellers, C. Cuomo, et al., unpublished data).

The Search for Functional Homologs Related to Pathogenicity

The study of effectors with avirulence functions, that is, effectors that trigger resistance in hosts carrying resistance gene alleles that recognize them, has been a focus in plant pathology for the last 30 years. Despite genetic evidence of these in the rust fungi since 1942 (Flor, 1942), molecular work aimed at identifying and cloning them in the rust fungi is rather recent (Zambino et al., 2000). Candidate gene approaches, using specific cDNA libraries from haustoria (specialized feeding structures established after plant infection), has accelerated research into such effectors from bean (Hahn and Mendgen, 1997; Link and Voegele, 2008; Puthoff et al., 2008) and flax rust fungi (Catanzariti et al., 2006; Dodds et al., 2004) and from *Mlp* (Joly et al., 2010), *Pst* (Yin et al., 2009), and *Pt* (Xu et al., 2011). Candidate effectors have also been revealed in haustoria using a proteomic approach (Song et al., 2011). As mentioned, many effectors are predicted in the *Mlp* and *Pgt* genomes (Duplessis et al., 2011).

It is still challenging to investigate the functionality of identified genes in rust fungi, although progress has been made in developing transformation systems (Lawrence et al., 2010; Webb et al., 2006). With regard to effectors, an RNA silencing approach was used in *M. lini* to demonstrate an avirulence function that at the same time served as selection for genetic transformation (Lawrence et al., 2010). Also in populations of this fungus, allelic variation at effector gene loci correlated with pathogenicity. This is a direct demonstration of how evolution of effector loci can change virulence profiles in natural populations that are likely under selection pressure of coevolving hosts (Barrett et al., 2009).

Having several complete rust fungal genome sequences and expressed sequence tag (EST) collections from several more species, many candidate pathogenicity and virulence genes other than effectors have been identified. These are often based on proven functionality in other pathosystems that are more amenable to molecular genetic manipulation. However, in a promising approach using two somewhat related basidiomycete pathogens, Hu et al. expressed a *Pt* MAP kinase in *U. maydis* strains in which the homologous MAPK genes, *Ubc3/Kpp2* and *Kpp6*, were deleted, and showed that heterologous expression can restore mating and pathogenicity defects (Hu et al., 2007a). This approach opens up avenues along which to functionally analyze genes from biotrophs. *U. maydis* seems particularly suited for exploitation because it is a well-researched model system for which numerous deletion mutants exist. In another approach, RNA silencing, when induced in the host plant but targeted to fungal genes, can suppress the expression of such genes, thus allowing their functional analysis. Initially demonstrated in the powdery mildew fungus–barley pathosystem (Nowara et al., 2010), this technique seems also functional in the *Pst*– and *Pgt*–wheat pathosystem, presumably by uptake of siRNA molecules via haustoria (Yin et al., 2011). More importantly, this approach can also be used to target *Pt*, *Pst*, and *Pgt* fungal genes such as the mentioned MAPK involved in pathogenicity in other systems, and in suppression of disease development in wheat (V. Panwar and G. Bakkeren, unpublished data).

SUMMARY

Basidiomycetes comprise an eclectic mixture of saprobic organisms and facultative pathogens able to live saprobic lifestyles but able to engage in opportunistic invasive growth on or within hosts. A subset has evolved to become highly specialized, true biotrophic pathogens that establish intercellular hyphae and haustoria while inflicting minimal damage to host cells and maintaining the integrity of the host tissue (Brefort et al., 2009; Hibbett, 2006; Kirk et al., 2001). The adaptation of saprobic fungi to pathogenic lifestyles, utilizing host resources and exhibiting varying degrees of virulence, requires coevolution with hosts on a population level and also requires the ability to subdue or subvert host defense responses. It is an open question of whether such adaptations have occurred or are occurring more quickly in organisms displaying sex (Heitman, 2011; Morrow and Fraser, 2009). From detailed studies in the individual model organisms and other systems, it is becoming apparent that molecular determinants involved in sex are a part of gene networks impinging on and interacting with other networks involved in pathogenicity (Sahni et al., 2009, 2010). We are only beginning to understand the implications emanating from these discoveries, but, apart from satisfying academic interests and shedding light on evolutionary questions, knowledge about these gene networks and the molecular basis underpinning pathogenicity and virulence will allow for the design of novel strategies for disease control in humans, animals, and crops.

REFERENCES

Alspaugh J, Davidson R, Heitman J. 2000. Morphogenesis of *Cryptococcus neoformans*. Contrib Microbiol 5: 217–238.

Alspaugh JA, Perfect JR, Heitman J. 1997. *Cryptococcus neoformans* mating and virulence are regulated by the G-protein alpha subunit *GPA1* and cAMP. Genes Dev 11: 3206–3217.

Alspaugh JA, Pukkila-Worley R, Harashima T, Cavallo LM, Funnell D, Cox GM, Perfect JR, Kronstad JW, Heitman J. 2002. Adenylyl cyclase functions downstream of the Gα protein Gpa1 and controls mating and pathogenicity of *Cryptococcus neoformans*. Eukaryot Cell 1: 75–84.

Anikster Y. 1983. Binucleate basidiospores; a general rule in rust fungi. Trans Br Mycol Soc 81: 624–626.

Anikster Y, Eilam T, Mittelman L, Szabo LJ, Bushnell WR. 1999. Pycnial nectar of rust fungi induces cap formation on pycniospores of opposite mating type. Mycologia 91: 858–870.

Bahn YS, Kojima K, Cox GM, Heitman J. 2005. Specialization of the HOG pathway and its impact on differentiation and virulence of *Cryptococcus neoformans*. Mol Biol Cell 16: 2285–2300.

Bahn YS, Geunes-Boyer S, Heitman J. 2007. Ssk2 mitogen-activated protein kinase kinase kinase governs divergent patterns of the stress-activated Hogl signaling pathway in *Cryptococcus neoformans*. Eukaryot Cell 6: 2278–2289.

Bakkeren G, Kronstad JW. 1994. Linkage of mating-type loci distinguishes bipolar from tetrapolar mating in basidiomycetous smut fungi. Proc Natl Acad Sci U S A 91: 7085–7089.

Bakkeren G, Jiang G, Warren RL, Butterfield Y, Shin H, Chiu R, Linning R, Schein J, Lee N, Hu G, Kupfer DM, Tang Y, Roe BA, Jones S, Marra M, Kronstad JW. 2006. Mating factor linkage and genome evolution in basidiomycetous pathogens of cereals. Fungal Genet Biol 43: 655–666.

Bakkeren G, Kämper J, Schirawski J. 2008. Sex in smut fungi: Structure, function and evolution of mating-type complexes. Fungal Genet Biol 45: S15–S21.

Banuett F. 1995. Genetics of *Ustilago maydis*, a fungal pathogen that induces tumors in maize. Annu Rev Genet 29: 179–208.

Barrett LG, Thrall PH, Dodds PN, van der Merwe M, Linde CC, Lawrence GJ, Burdon JJ. 2009. Diversity and evolution of effector loci in natural populations of the plant pathogen *Melampsora lini*. Mol Biol Evol 26: 2499–2513.

Blackwell M. 2011. The fungi: 1, 2, 3. 5.1 million species? Am J Bot 98: 426–438.

Böhme A, Karthaus M. 1999. Systemic fungal infections in patients with hematologic malignancies: Indications and limitations of the antifungal armamentarium. Chemotherapy 45: 315–324.

Bolker M, Urban M, Kahmann R. 1992. The *a* mating type locus of *Ustilago maydis* specifies cell signaling components. Cell 68: 441–450.

Bolton MD, Kolmer JA, Garvin DF. 2008. Wheat leaf rust caused by *Puccinia triticina*. Mol Plant Pathol 9: 563–575.

Botts M, Giles S, Gates M, Kozel T, Hull C. 2009. Isolation and characterization of *Cryptococcus neoformans* spores reveal a critical role for capsule biosynthesis genes in spore biogenesis. Eukaryot Cell 8: 595–605.

Bourassa M, Bernier L, Hamelin RC. 2007. Genetic diversity in poplar leaf rust. *Melampsora medusae* f. sp. *deltoidae*. in the zones of host sympatry and allopatry. Phytopathology 97: 603–610.

Brachmann A, Weinzierl G, Kaemper J, Kahmann R. 2001. Identification of genes in the bW/bE regulatory cascade in *Ustilago maydis*. Mol Microbiol 42: 1047–1063.

Brachmann A, Schirawski J, Mueller P, Kahmann R. 2003. An unusual MAP kinase is required for efficient penetration of the plant surface by *Ustilago maydis*. EMBO J 22: 2199–2210.

Brefort T, Doehlemann G, Mendoza-Mendoza A, Reissmann S, Djamei A, Kahmann R. 2009. *Ustilago maydis* as a pathogen. Annu Rev Phytopathol 47: 423–445.

Brown JK, Hovmøller MS. 2002. Aerial dispersal of pathogens on the global and continental scales and its impact on plant disease. Science 297: 537–541.

Brown JM. 2004. Fungal infections in bone marrow transplant patients. Curr Opin Infect Dis 17: 347–352.

Bui T, Lin X, Malik R, Heitman J, Carter D. 2008. Isolates of *Cryptococcus neoformans* from infected animals reveal genetic exchange in unisexual, α mating type populations. Eukaryot Cell 7: 1771–1780.

Casadevall A, Perfect JR. 1998. *Cryptococcus neoformans*. ASM Press, Washington, DC, p. 541.

Casadevall A, Steenbergen JN, Nosanchuk JD. 2003. "Ready made" virulence and "dual use" virulence factors in pathogenic environmental fungi—The *Cryptococcus neoformans* paradigm. Curr Opin Microbiol 6: 332–337.

Casselton L, Olesnicky N. 1998. Molecular genetics of mating recognition in basidiomycete fungi. Microbiol Mol Biol Rev 62: 55–70.

Castillo-Lluva S, Alvarez-Tabares I, Weber I, Steinberg G, Perez-Martin J. 2007. Sustained cell polarity and virulence in the phytopathogenic fungus *Ustilago maydis* depends on an essential cyclin-dependent kinase from the Cdk5/Pho85 family. J Cell Sci 120: 1584–1595.

Catanzariti AM, Dodds PN, Lawrence GJ, Ayliffe MA, Ellis JG. 2006. Haustorially expressed secreted proteins from flax rust are highly enriched for avirulence elicitors. Plant Cell 18: 243–256.

Cervantes-Chávez JA, Ali S, Bakkeren G. 2011. Response to environmental stresses, cell-wall integrity, and virulence are orchestrated through the calcineurin pathway in *Ustilago hordei*. Mol Plant Microbe Interact 24: 219–232.

Chang YC, Kwon-Chung KJ. 1994. Complementation of a capsule-deficient mutation of *Cryptococcus neoformans* restores its virulence. Mol Cell Biol 14: 4912–4919.

Chang YC, Wickes BL, Miller GF, Penoyer LA, Kwon-Chung KJ. 2000. *Cryptococcus neoformans STE12*α regulates virulence but is not essential for mating. J Exp Med 191: 871–882.

Chang YC, Penoyer LA, Kwon-Chung KJ. 2001. The second *STE12* homologue of *Cryptococcus neoformans* is *MATa*-specific and plays an important role in virulence. Proc Natl Acad Sci U S A 98: 3258–3263.

Chester K. 1946. *The Nature and Prevention of the Cereal Rusts as Exemplified in the Leaf Rust of Wheat*. Chronica Botanica, Waltham, MA.

Chung S, Karos M, Chang YC, Lukszo J, Wickes BL, Kwon-Chung KJ. 2002. Molecular analysis of *CPR*α, a *MAT*α-specific pheromone receptor gene of *Cryptococcus neoformans*. Eukaryot Cell 1: 432–439.

Clark T, Anderson J. 2004. Dikaryons of the basidiomycete fungus *Schizophyllum commune*: Evolution in long-term culture. Genetics 167: 1663–1675.

Clarke DL, Woodlee GL, McClelland CM, Seymour TS, Wickes BL. 2001. The *Cryptococcus neoformans STE11*α gene is similar to other fungal mitogen-activated protein kinase kinase kinase (MAPKKK) genes but is mating type specific. Mol Microbiol 40: 200–213.

Cramer KL, Gerrald QD, Nichols CB, Price MS, Alspaugh JA. 2006. Transcription factor Nrg1 mediates capsule formation, stress response, and pathogenesis in *Cryptococcus neoformans*. Eukaryot Cell 5: 1147–1156.

D'Souza CA, Alspaugh JA, Yue C, Harashima T, Cox GM, Perfect JR, Heitman J. 2001. Cyclic AMP-dependent protein kinase controls virulence of the fungal pathogen *Cryptococcus neoformans*. Mol Cell Biol 21: 3179–3191.

D'Souza CA, Kronstad JW, Taylor G, Warren R, Yuen M, Hu G, Jung WH, Sham A, Kidd SE, Tangen K, Lee N, Zeilmaker T, Sawkins J, McVicker G, Shah S, Gnerre S, Griggs A, Zeng Q, Bartlett K, Li W, Wang X, Heitman J, Stajich JE, Fraser JA, Meyer W, Carter D, Schein J, Krzywinski M, Kwon-Chung KJ, Varma A, Wang J, Brunham R, Fyfe M, Ouellette BF, Siddiqui A, Marra M, Jones S, Holt R, Birren BW, Galagan JE, Cuomo CA. 2011. Genome variation in *Cryptococcus gattii*, an emerging pathogen of immunocompetent hosts. MBio 2: e00342–e00310.

Davidson R, Nichols C, Cox G, Perfect J, Heitman J. 2003. A MAP kinase cascade composed of cell type specific and non-specific elements controls mating and differentiation of the fungal pathogen *Cryptococcus neoformans*. Mol Microbiol 49: 469–485.

Davidson RC, Moore TD, Odom AR, Heitman J. 2000. Characterization of the MFα pheromone of the human fungal pathogen *Cryptococcus neoformans*. Mol Microbiol 38: 1017–1026.

de Sena-Tomás C, Fernández-Álvarez A, Holloman WK, Pérez-Martín J. 2011. The DNA damage response signaling cascade regulates proliferation of the phytopathogenic fungus *Ustilago maydis in planta*. Plant Cell 23: 1654–1665.

Dodds PN, Lawrence GJ, Catanzariti AM, Ayliffe MA, Ellis JG. 2004. The *Melampsora lini* AvrL567 avirulence genes are expressed in haustoria and their products are recognized inside plant cells. Plant Cell 16: 755–768.

Doehlemann G, van der Linde K, Assmann D, Schwammbach D, Hof A, Mohanty A, Jackson D, Kahmann R. 2009. Pep1, a secreted effector protein of *Ustilago maydis*, is required for successful invasion of plant cells. PLoS Pathog 5: e1000290.

Doehlemann G, Reissmann S, Assmann D, Fleckenstein M, Kahmann R. 2011. Two linked genes encoding a secreted effector and a membrane protein are essential for *Ustilago maydis*-induced tumour formation. Mol Microbiol 81: 751–766.

Duplessis S, Cuomo CA, Lin Y-C, Aerts A, Tisserant E, Veneault-Fourrey C, Joly DL, Hacquard S, Amselem J, Cantarel BL, Chiu R, Coutinho PM, Feau N, Field M, Frey P, Gelhaye E, Goldberg J, Grabherr MG, Kodira CD, Kohler A, Kües U, Lindquist EA, Lucas SM, Mago R, Mauceli E, Morin E, Murat C, Pangilinan JL, Park R, Pearson M, Quesneville H, Rouhier N, Sakthikumar S, Salamov AA, Schmutz J, Selles B, Shapiro H, Tanguay P, Tuskan GA, Henrissat B, Van De Peer Y, Rouzé P, Ellis JG, Dodds PN, Schein JE, Zhong S, Hamelin RC, Grigoriev IV, Szabo LJ, Martin F. 2011. Obligate biotrophy features unraveled by the genomic analysis of rust fungi. Proc Natl Acad Sci U S A 108: 9166–9171.

Egan JD, Garcia-Pedrajas MD, Andrews DL, Gold SE. 2009. Calcineurin is an antagonist to PKA protein phosphorylation required for postmating filamentation and virulence, while PP2A is required for viability in *Ustilago maydis*. Mol Plant Microbe Interact 22: 1293–1301.

Ekena J, Stanton B, Schiebe-Owens J, Hull C. 2008. Sexual development in *Cryptococcus neoformans* requires *CLP1*, a target of the homeodomain transcription factors Sxi1α and Sxi2a. Eukaryot Cell 7: 49–57.

Erke KH. 1976. Light microscopy of basidia, basidiospores, and nuclei in spores and hyphae of *Filobasidiella neoformans* (*Cryptococcus neoformans*). J Bacteriol 128: 445–455.

Fields S, Herskowitz I. 1985. The yeast *STE12* product is required for expression of two sets of cell-type specific genes. Cell 42: 923–930.

Fisher GW, Holton CS. 1957. *Biology and Control of the Smut Fungi*. Ronald Press, New York.

Flor HH. 1942. Inheritance of pathogenicity in *Melampsora lini*. Phytopathology 32: 653–669.

Flor-Parra I, Vranes M, Kamper J, Perez-Martin J. 2006. Biz1, a Zinc Finger Protein required for plant invasion by *Ustilago maydis*, regulates the levels of a mitotic cyclin. Plant Cell 18: 2369–2387.

Flor-Parra I, Castillo-Lluva S, Perez-Martin J. 2007. Polar Growth in the Infectious hyphae of the phytopathogen *Ustilago maydis* depends on a virulence-specific cyclin. Plant Cell 19: 3280–3296.

Fox DS, Cruz MC, Sia RA, Ke H, Cox GM, Cardenas ME, Heitman J. 2001. Calcineurin regulatory subunit is essential for virulence and mediates interactions with FKBP12-FK506 in *Cryptococcus neoformans*. Mol Microbiol 39: 835–849.

Fraser JA, Heitman J. 2003. Fungal mating-type loci. Curr Biol 13: R792–R795.

Fraser JA, Diezmann S, Subaran RL, Allen A, Lengeler KB, Dietrich FS, Heitman J. 2004. Convergent evolution of chromosomal sex-determining regions in the animal and fungal kingdoms. PLoS Biol 2: e384.

Garcia-Hermoso D, Janbon G, Dromer F. 1999. Epidemiological evidence for dormant *Cryptococcus neoformans* infection. J Clin Microbiol 37: 3204–3209.

Garcia-Muse T, Steinberg G, Perez-Martin J. 2003. Pheromone-induced G2. Arrest in the phytopathogenic fungus *Ustilago maydis*. Eukaryot Cell 2: 494–500.

García-Pedrajas MD, Nadal M, Bölker M, Gold SE, Perlin MH. 2008. Sending mixed signals: Redundancy vs. uniqueness of signaling components in the plant pathogen, *Ustilago maydis*. Fungal Genet Biol 45: S22–S30.

Gerard PR, Husson C, Pinon J, Frey P. 2006. Comparison of genetic and virulence diversity of *Melampsora larici-populina* populations on wild and cultivated poplar and influence of the alternate host. Phytopathology 96: 1027–1036.

Giles S, Dagenais T, Botts M, Keller N, Hull C. 2009. Elucidating the pathogenesis of spores from the human fungal pathogen *Cryptococcus neoformans*. Infect Immun 77: 3491–3500.

Goellner K, Loehrer M, Langenbach C, Conrath U, Koch E, Schaffrath U. 2010. *Phakopsora pachyrhizi*, the causal agent of Asian soybean rust. Mol Plant Pathol 11: 169–177.

Goldman DL, Khine H, Abadi J, Lindenberg DJ, Pirofski LA, Niang R, Casadevall A. 2001. Serologic evidence for *Cryptococcus neoformans* infection in early childhood. Pediatrics 107: e66.

Groth JV, McCain JW, Roelfs AP. 1995. Virulence and isozyme diversity of sexual versus asexual collections of *Uromyces appendiculatus* (bean rust fungus). Heredity 75: 234–242.

Hahn M, Mendgen K. 1997. Characterization of in planta-induced rust genes isolated from a haustorium -specific cDNA library. Mol Plant Microbe Interact 10: 427–437.

Hartmann HA, Kahmann R, Bolker M. 1996. The pheromone response factor coordinates filamentous growth and pathogenicity in *Ustilago maydis*. EMBO J 15: 1632–1641.

Hartmann HA, Kruger J, Lottspeich F, Kahmann R. 1999. Environmental signals controlling sexual development of the corn Smut fungus *Ustilago maydis* through the transcriptional regulator Prf1. Plant Cell 11: 1293–1306.

Hawksworth D. 2001. The magnitude of fungal diversity: The 1.5 million species estimate revisited. Mycol Res 105: 1422–1432.

Heimel K, Scherer M, Schuler D, Kämper J. 2010a. The *Ustilago maydis* Clp1 protein orchestrates pheromone and *b*-dependent signaling pathways to coordinate the cell cycle and pathogenic development. Plant Cell 22: 2908–2922.

Heimel K, Scherer M, Vranes M, Wahl R, Pothiratana C, Schuler D, Vincon V, Finkernagel F, Flor-Parra I, Kämper J. 2010b. The transcription factor Rbf1 is the master regulator for *b*-mating type controlled pathogenic development in *Ustilago maydis*. PLoS Pathog 6: e1001035.

Heitman J. 2011. Microbial pathogens in the fungal kingdom. Fungal Biol Rev 25: 48–60.

Herskowitz I. 1988. Life cycle of the budding yeast *Saccharomyces cerevisiae*. Microbiol Rev 52: 536–553.

Hibbett D. 2006. A phylogenetic overview of the Agaricomycotina. Mycologia 98: 917–925.

Horton JS, Bakkeren G, Klosterman SJ, Garcia-Pedrajas M, Gold SE. 2005. Genetics of morphogenesis in Basidiomycetes. In: Arora DK, Berka R, eds. *Applied Mycology and Biotechnology Genes and Genomics*. Elsevier, Dordrecht, pp. 353–422.

Hsueh YP, Heitman J. 2008. Orchestration of sexual reproduction and virulence by the fungal mating-type locus. Curr Opin Microbiol 11: 517–524.

Hsueh YP, Fraser JA, Heitman J. 2008. Transitions in sexuality: Recapitulation of an ancestral tri- and tetrapolar mating system in

Cryptococcus neoformans. Eukaryot Cell 7: 1847–1855.

Hu G, Kamp A, Linning R, Naik S, Bakkeren G. 2007a. Complementation of *Ustilago maydis* MAPK mutants by a wheat leaf rust, *Puccinia triticina* homolog: Potential for functional analyses of rust genes. Mol Plant Microbe Interact 20: 637–647.

Hu G, Steen BR, Lian T, Sham AP, Tam N, Tangen KL, Kronstad JW. 2007b. Transcriptional regulation by protein kinase A in *Cryptococcus neoformans*. PLoS Pathog 3: e42.

Hull C, Heitman J. 2002. Genetics of *Cryptococcus neoformans*. Annu Rev Genet 36: 557–615.

Hull C, Cox G, Heitman J. 2004. The α-specific cell identity factor Sxi1α is not required for virulence of *Cryptococcus neoformans*. Infect Immun 72: 3643–3645.

Hull C, Boily M, Heitman J. 2005. Sex-specific homeodomain proteins Sxi1α and Sxi2a coordinately regulate sexual development in *Cryptococcus neoformans*. Eukaryot Cell 4: 526–535.

Idnurm A. 2010. A tetrad analysis of the basidiomycete fungus *Cryptococcus neoformans*. Genetics 185: 153–163.

Inada K, Morimoto Y, Arima T, Murata Y, Kamada T. 2001. The *clp1* gene of the mushroom *Coprinus cinereus* is essential for A-regulated sexual development. Genetics 157: 133–140.

Iwasa M, Tanabe S, Kamada T. 1998. The two nuclei in the dikaryon of the homobasidiomycete *Coprinus cinereus* change position after each conjugate division. Fungal Genet Biol 23: 110–116.

Jin Y, Szabo LJ, Carson M. 2010. Century-old mystery of *Puccinia striiformis* life history solved with the Identification of *Berberis* as an alternate host. Phytopathology 100: 432–435.

Joly D, Feau N, Tanguay P, Hamelin R. 2010. Comparative analysis of secreted protein evolution using expressed sequence tags from four poplar leaf rusts. *Melampsora* spp. BMC Genomics 11: 422.

Kahmann R, Kamper J. 2004. *Ustilago maydis*: How its biology relates to pathogenic development. New Phytol 164: 31–42.

Kahmann R, Schirawski J. 2007. Mating in the smut fungi: From a to b to the downstream cascades. In: Heitman J, Kronstad JW, Taylor AJ, Casselton LA, eds. *Sex in Fungi; Molecular Determination and Evolutionary Principles*. ASM Press, Washington, DC, pp. 377–388.

Kamper J, Kahmann R, Bolker M, Ma L-J, Brefort T, Saville BJ, Banuett F, Kronstad JW, Gold SE, Muller O, Perlin MH, Wosten HAB, de Vries R, Ruiz-Herrera J, Reynaga-Pena CG, Snetselaar K, McCann M, Perez-Martin J, Feldbrugge M, Basse CW, Steinberg G, Ibeas JI, Holloman W, Guzman P, Farman M, Stajich JE, Sentandreu R, Gonzalez-Prieto JM, Kennell JC, Molina L, Schirawski J, Mendoza-Mendoza A, Greilinger D, Munch K, Rossel N, Scherer M, Vranes M, Ladendorf O, Vincon V, Fuchs U, Sandrock B, Meng S, Ho ECH, Cahill MJ, Boyce KJ, Klose J, Klosterman SJ, Deelstra HJ, Ortiz-Castellanos L, Li W, Sanchez-Alonso P, Schreier PH, Hauser-Hahn I, Vaupel M, Koopmann E, Friedrich G, Voss H, Schluter T, Margolis J, Platt D, Swimmer C, Gnirke A, Chen F, Vysotskaia V, Mannhaupt G, Guldener U, Munsterkotter M, Haase D, Oesterheld M, Mewes H-W, Mauceli EW, DeCaprio D, Wade CM, Butler J, Young S, Jaffe DB, Calvo S, Nusbaum C, Galagan J, Birren BW. 2006. Insights from the genome of the biotrophic fungal plant pathogen *Ustilago maydis*. Nature 444: 97–101.

Kirk P, Cannon P, David J, Stalpers J. 2001. *Ainsworth & Bisby's Dictionary of the Fungi*. CABI Publishing, Wallingford.

Kozubowski L, Lee SC, Heitman J. 2009. Signalling pathways in the pathogenesis of *Cryptococcus*. Cell Microbiol 11: 370–380.

Kronstad J, Staben C. 1997. Mating type in filamentous fungi. Annu Rev Genet 31: 245–276.

Kwon-Chung KJ. 1976. Morphogenesis of *Filobasidiella neoformans*, the sexual state of *Cryptococcus neoformans*. Mycologia 68: 821–833.

Kwon-Chung KJ, Edman JC, Wickes BL. 1992a. Genetic association of mating types and virulence in *Cryptococcus neoformans*. Infect Immun 60: 602–605.

Kwon-Chung KJ, Wickes BL, Stockman L, Roberts GD, Ellis D, Howard DH. 1992b.

Virulence, serotype, and molecular characteristics of environmental strains of *Cryptococcus neoformans var. gattii*. Infect Immun 60: 1869–1874.

Lawrence GJ, Dodds PN, Ellis JG. 2007. Rust of flax and linseed caused by *Melampsora lini*. Mol Plant Pathol 8: 349–364.

Lawrence GJ, Dodds PN, Ellis JG. 2010. Transformation of the flax rust fungus, *Melampsora lini*: Selection via silencing of an avirulence gene. Plant J 61: 364–369.

Lawrence GL. 1980. Multiple mating-type specificities in the flax rust *Melampsora lini*. Science 209: 501–503.

Lee N, Bakkeren G, Wong K, Sherwood JE, Kronstad JW. 1999. The mating-type and pathogenicity locus of the fungus *Ustilago hordei* spans a 500-kb region. Proc Natl Acad Sci U S A 96: 15026–15031.

Lee N, D'Souza CA, Kronstad JW. 2003. Of smuts, blasts, mildews, and blights: cAMP signaling in phytopathogenic fungi. Annu Rev Phytopathol 41: 399–427.

Lengeler KB, Fox DS, Fraser JA, Allen A, Forrester K, Dietrich FS, Heitman J. 2002. Mating-type locus of *Cryptococcus neoformans*: A step in the evolution of sex chromosomes. Eukaryot Cell 1: 704–718.

Li H, Wang X. 2009. *Thinopyrum ponticum* and *Th. intermedium*: The promising source of resistance to fungal and viral diseases of wheat. J Genet Genomics 36: 557–565.

Lin X, Hull C, Heitman J. 2005. Sexual reproduction between partners of the same mating type in *Cryptococcus neoformans*. Nature 434: 1017–1021.

Lin X, Huang JC, Mitchell TG, Heitman J. 2006. Virulence attributes and hyphal growth of *C. neoformans* are quantitative traits and the *MAT*α allele enhances filamentation. PLoS Genet 2: e187.

Lin X, Jackson JC, Feretzaki M, Xue C, Heitman J. 2010. Transcription factors Mat2 and Znf2 operate cellular circuits orchestrating opposite- and same-sex mating in *Cryptococcus neoformans*. PLoS Genet 6: e1000953.

Link TI, Voegele RT. 2008. Secreted proteins of *Uromyces fabae*: Similarities and stage specificity. Mol Plant Pathol 9: 59–66.

Litvintseva AP, Marra RE, Nielsen K, Heitman J, Vilgalys R, Mitchell TG. 2003. Evidence of sexual recombination among *Cryptococcus neoformans* serotype A isolates in sub-Saharan Africa. Eukaryot Cell 2: 1162–1168.

Loftus BJ, Fung E, Roncaglia P, Rowley D, Amedeo P, Bruno D, Vamathevan J, Miranda M, Anderson IJ, Fraser JA, et al. 2005. The genome of the basidiomycetous yeast and human pathogen *Cryptococcus neoformans*. Science 307: 1321–1324.

Martin F, Aerts A, Ahrén D, Brun A, Danchin EG, Duchaussoy F, Gibon J, Kohler A, Lindquist E, Pereda V, et al. 2008. The genome of *Laccaria bicolor* provides insights into mycorrhizal symbiosis. Nature 452: 88–92.

McClelland CM, Fu J, Woodlee GL, Seymour TS, Wickes BL. 2002. Isolation and characterization of the *Cryptococcus neoformans MATa* pheromone gene. Genetics 160: 935–947.

McClelland CM, Chang YC, Varma A, Kwon-Chung KJ. 2004. Uniqueness of the mating system in *Cryptococcus neoformans*. Trends Microbiol 12: 208–212.

Mendgen K. 1984. Development and physiology of teliospores. In: Bushnell WR, Roelfs AP, eds. *The Cereal Rusts*. Academic Press, Inc., New York, pp. 375–398.

Mitchell T. 2005. Kingdom Fungi: Fungal phylogeny and systematics. In: Merz W, Hay R, eds. *Topley & Wilson's Microbiology and Microbial Infections*. Hoddler Arnold, London, pp. 43–68.

Mitchell T, Perfect J. 1995. Cryptococcosis in the era of AIDS—100 years after the discovery of *Cryptococcus neoformans*. Clin Microbiol Rev 8: 515–548.

Moore TD, Edman JC. 1993. The α-mating type locus of *Cryptococcus neoformans* contains a peptide pheromone gene. Mol Cell Biol 13: 1962–1970.

Morrow CA, Fraser JA. 2009. Sexual reproduction and dimorphism in the pathogenic basidiomycetes. FEMS Yeast Res 9: 161–177.

Mueller O, Kahmann R, Aguilar G, Trejo-Aguilar B, Wu A, de Vries RP. 2008. The secretome of the maize pathogen *Ustilago maydis*. Fungal Genet Biol 45: S63–S70.

Narisawa K, Yamaoka Y, Katsuya K. 1994. Mating type of isolates derived from the spermogonial state of *Puccinia coronata* var. *coronata*. Mycoscience 35: 131–135.

Nichols C, Fraser J, Heitman J. 2004. PAK kinases Ste20 and Pak1 govern cell polarity at different stages of mating in *Cryptococcus neoformans*. Mol Biol Cell 15: 4476–4489.

Nowara D, Gay A, Lacomme C, Shaw J, Ridout C, Douchkov D, Hensel G, Kumlehn J, Schweizer P. 2010. HIGS: Host-induced gene silencing in the obligate biotrophic fungal pathogen *Blumeria graminis*. Plant Cell 22: 3130–3141.

Park BJ, Wannemuehler KA, Marston BJ, Govender N, Pappas PG, Chiller TM. 2009. Estimation of the current global burden of cryptococcal meningitis among persons living with HIV/AIDS. AIDS 23: 525–530.

Perez-Martin J, Castillo-Lluva S, Sgarlata C, Flor-Parra I, Mielnichuk N, Torreblanca J, Carbo N. 2006. Pathocycles: *Ustilago maydis* as a model to study the relationships between cell cycle and virulence in pathogenic fungi. Mol Genet Genomics 276: 211–229.

Pukkila-Worley R, Gerrald QD, Kraus PR, Boily MJ, Davis MJ, Giles SS, Cox GM, Heitman J, Alspaugh JA. 2005. Transcriptional network of multiple capsule and melanin genes governed by the *Cryptococcus neoformans* cyclic AMP cascade. Eukaryot Cell 4: 190–201.

Puthoff DP, Neelam A, Ehrenfried ML, Scheffler BE, Ballard L, Song Q, Campbell KB, Cooper B, Tucker ML. 2008. Analysis of Expressed Sequence Tags from *Uromyces appendiculatus* hyphae and haustoria and their comparison to sequences from other rust fungi. Phytopathology 98: 1126–1135.

Raper CA. 1983. Controls for development and differentiation of the dikaryon of basidiomycetes. In: Bennett J and Ciegler A, eds. *Secondary Metabolism and Differentiation in Fungi*. Dekker, New York, pp. 195–238.

Raper JR. 1966. *Genetics of Sexuality in Higher Fungi*. Ronald Press, New York.

Rispail N, Soanes DM, Ant C, Czajkowski R, Grünler A, Huguet R, Perez-Nadales E, Poli A, Sartorel E, Valiante V, Yang M, Beffa R, Brakhage AA, Gow NAR, Kahmann R, Lebrun M-H, Lenasi H, Perez-Martin J, Talbot NJ, Wendland J, Di Pietro A. 2009. Comparative genomics of MAP kinase and calcium-calcineurin signalling components in plant and human pathogenic fungi. Fungal Genet Biol 46: 287–298.

Roetzer A, Gabaldón T, Schüller C. 2011a. From *Saccharomyces cerevisiae* to *Candida glabrata* in a few easy steps: Important adaptations for an opportunistic pathogen. FEMS Microbiol Lett 314: 1–9.

Roetzer A, Klopf E, Gratz N, Marcet-Houben M, Hiller E, Rupp S, Gabaldón T, Kovarik P, Schüller C. 2011b. Regulation of *Candida glabrata* oxidative stress resistance is adapted to host environment. FEBS Lett 585: 319–327.

Saari EE, Young HC, Kernkamp MF. 1968. Infection of north American *Thalictrum* spp. with *Puccinia recondita* f.sp. *tritici*. Phytopathology 58: 939–943.

Sahni N, Yi S, Daniels KJ, Srikantha T, Pujol C, Soll DR. 2009. Genes selectively up-regulated by pheromone in white cells are involved in biofilm formation in *Candida albicans*. PLoS Pathog 5: e1000601.

Sahni N, Yi S, Daniels KJ, Huang G, Srikantha T, Soll DR. 2010. Tec1 mediates the pheromone response of the white phenotype of *Candida albicans*: Insights into the evolution of new signal transduction pathways. PLoS Biol 8: e1000363.

Samborski DJ. 1985. Wheat leaf rust. In: Roelfs AP, Bushnell WR, eds. *The Cereal Rusts. Vol II: Diseases, Distribution, Epidemiology and Control*. Academic Press, Orlando, FL, pp. 39–59.

Scherer M, Heimel K, Starke V, Kämper J. 2006. The Clp1 protein is required for clamp formation and pathogenic development of *Ustilago maydis*. Plant Cell 18: 2388–2401.

Schirawski J, Heinze B, Wagenknecht M, Kahmann R. 2005. Mating type loci of *Sporisorium reilianum*: Novel pattern with three *a* and multiple *b* specificities. Eukaryot Cell 4: 1317–1327.

Schirawski J, Mannhaupt G, Münch K, Brefort T, Schipper K, Doehlemann G, Di Stasio M,

Rössel N, Mendoza-Mendoza A, Pester D, Müller O, Winterberg B, Meyer E, Ghareeb H, Wollenberg T, Münsterkötter M, Wong P, Walter M, Stukenbrock E, Güldener U, Kahmann R. 2010. Pathogenicity determinants in smut fungi revealed by genome comparison. Science 330: 1546–1548.

Shen WC, Davidson RC, Cox GM, Heitman J. 2002. Pheromones stimulate mating and differentiation via paracrine and autocrine signaling in *Cryptococcus neoformans*. Eukaryot Cell 1: 366–377.

Singh RP, Hodson DP, Huerta-Espino J, Jin Y, Bhavani S, Njau P, Herrera-Foessel S, Singh PK, Singh S, Govindan V. 2011. The emergence of Ug99 races of the stem rust fungus is a threat to world wheat production. Annu Rev Phytopathol 49: 465–481.

Song X, Rampitsch C, Soltani B, Mauthe W, Linning R, Banks T, McCallum B, Bakkeren G. 2011. Proteome analysis of wheat leaf rust fungus, *Puccinia triticina*, infection structures enriched for haustoria. Proteomics 11: 944–963.

Stanton B, Giles S, Kruzel E, Warren C, Ansari A, Hull C. 2009. Cognate site identifier analysis reveals novel binding properties of the sex inducer homeodomain proteins of *Cryptococcus neoformans*. Mol Microbiol 72: 1334–1347.

Stanton BC, Giles SS, Staudt MW, Kruzel EK, Hull CM. 2010. Allelic exchange of pheromones and their receptors reprograms sexual identity in *Cryptococcus neoformans*. PLoS Genet 6: e1000860.

Steinberg G. 2007. On the move: Endosomes in fungal growth and pathogenicity. Nat Rev Microbiol 5: 309–316.

Twizeyimana M, Ojiambo PS, Haudenshield JS, Caetano-Anollés G, Pedley KF, Bandyopadhyay R, Hartman GL. 2011. Genetic structure and diversity of *Phakopsora pachyrhizi* isolates from soyabean. Plant Pathol 70: 719–729.

Urban M, Kahmann R, Bolker M. 1996. Identification of the pheromone response element in *Ustilago maydis*. Mol Gen Genet 251: 31–37.

Velagapudi R, Hsueh YP, Geunes-Boyer S, Wright JR, Heitman J. 2009. Spores as infectious propagules of *Cryptococcus neoformans*. Infect Immun 77: 4345–4355.

Wang P, Perfect JR, Heitman J. 2000. The G-protein beta subunit *GPB1* is required for mating and haploid fruiting in *Cryptococcus neoformans*. Mol Cell Biol 20: 352–362.

Wang P, Nichols CB, Lengeler KB, Cardenas ME, Cox GM, Perfect JR, Heitman J. 2002. Mating-type-specific and nonspecific PAK kinases play shared and divergent roles in *Cryptococcus neoformans*. Eukaryot Cell 1: 257–272.

Wang X, McCallum B. 2009. Fusion body formation, germ tube anastomosis, and nuclear migration during the germination of urediniospores of the wheat leaf rust fungus, *Puccinia triticina*. Phytopathology 99: 1355–1364.

Webb CA, Szabo LJ, Bakkeren G, Garry C, Staples RC, Eversmeyer M, Fellers JP. 2006. Transient expression and insertional mutagenesis of *Puccinia triticina* using biolistics. Funct Integr Genomics 6: 250–260.

Wickes BL, Mayorga ME, Edman U, Edman JC. 1996. Dimorphism and haploid fruiting in *Cryptococcus neoformans*: Association with the α-mating type. Proc Natl Acad Sci U S A 93: 7327–7331.

Wickes BL, Edman U, Edman JC. 1997. The *Cryptococcus neoformans STE12α* gene: A putative *Saccharomyces cerevisiae STE12* homologue that is mating type specific. Mol Microbiol 26: 951–960.

Xu J, Saunders CW, Hu P, Grant RA, Boekhout T, Kuramae EE, Kronstad JW, DeAngelis YM, Reeder NL, Johnstone KR, Leland M, Fieno AM, Begley WM, Sun Y, Lacey MP, Chaudhary T, Keough T, Chu L, Sears R, Yuan B, Dawson TL, Jr. 2007. Dandruff-associated *Malassezia* genomes reveal convergent and divergent virulence traits shared with plant and human fungal pathogens. Proc Natl Acad Sci U S A 104: 18730–18735.

Xu J, Linning R, Fellers J, Dickinson M, Zhu W, Antonov I, Joly DL, Donaldson ME, Eilam T, Anikster Y, Banks T, Munro S, Mayo M, Wynhoven B, Ali J, Moore R, McCallum B, Borodovsky M, Saville B, Bakkeren G. 2011.

Gene discovery in EST sequences from the wheat leaf rust fungus *Puccinia triticina* sexual spores, asexual spores and haustoria, compared to other rust and corn smut fungi. BMC Genomics 12: 161.

Yee AR, Kronstad JW. 1998. Dual sets of chimeric alleles identify specificity sequences for the *bE* and *bW* mating and pathogenicity genes of *Ustilago maydis*. Mol Cell Biol 18: 221–232.

Yin C, Chen X, Wang X, Han Q, Kang Z, Hulbert S. 2009. Generation and analysis of expression sequence tags from haustoria of the wheat stripe rust fungus *Puccinia striiformis* f. sp. *tritici*. BMC Genomics 10: 626.

Yin C, Jurgenson JE, Hulbert SH. 2011. Development of a host-induced RNAi system in the wheat stripe rust fungus *Puccinia striiformis* f. sp. *tritici*. Mol Plant Microbe Interact 24: 554–561.

Yue C, Cavallo LM, Alspaugh JA, Wang P, Cox GM, Perfect JR, Heitman J. 1999. The *STE12*α homolog is required for haploid filamentation but largely dispensable for mating and virulence in *Cryptococcus neoformans*. Genetics 153: 1601–1615.

Zambino PJ, Kubelik AR, Szabo LJ. 2000. Gene action and linkage of avirulence genes to DNA markers in the rust fungus *Puccinia graminis*. Phytopathology 90: 819–826.

CHAPTER 24

EMERGENCE OF THE CHYTRID FUNGUS *BATRACHOCHYTRIUM DENDROBATIDIS* AND GLOBAL AMPHIBIAN DECLINES

MATTHEW C. FISHER, JASON E. STAJICH, and RHYS A. FARRER

INTRODUCTION

Emerging pathogenic fungi are causing an increasing threat to wild animal populations and species (Fisher et al., 2012). For example, little brown bats, *Myotis lucifugus*, across the United States are increasingly affected by a previously unknown ascomycete fungus, recently named *Geomyces destructans* (Frick et al., 2010; Gargas et al., 2009), with implications for the long-term survival of this bat species (Frick et al., 2010). However, the greatest impact that any pathogen has had on its host species owes to the emergence of a pathogen belonging to a basal lineage of fungi, the Chytridiomycota. *Batrachochytrium dendrobatidis* (*Bd*) was first discovered in 1997 (Berger et al., 1998) and was defined as a species in 1999 (Longcore et al., 1999). The infection causes a rapidly progressing and sometimes fatal cutaneous disease, chytridiomycosis, in anuran (frog-like) and caudate (salamander-like) amphibians. Prior to the discovery of *Bd*, it was recognized that amphibians were facing an extinction crisis that threatened around a third of all species (Stuart et al., 2004). Many amphibian species declines were found occurring in pristine protected environments where known threats (such as habitat loss or species overharvesting) did not occur; these mysterious declines were recorded as "enigmatic declines" by the IUCN Red List. *Bd* was subsequently discovered as the proximate driver of these multiple-species enigmatic declines (Lotters et al., 2009; Skerratt et al., 2007) following the observation of simultaneous waves of population and species loss across North and Central America, Australia, and Europe (Berger et al., 1998; Fisher et al., 2009b; Lips et al., 2006; Walker et al., 2010). The fungus has now been found infecting over 442 species of amphibians in 49 countries and on all continents except the Antarctic (Fisher et al., 2009b). This chapter considers the evolutionary underpinnings that have led to the remarkable emergence of this fungus as a primary driver of biodiversity loss.

Evolution of Virulence in Eukaryotic Microbes, First Edition. Edited by L. David Sibley, Barbara J. Howlett, and Joseph Heitman.

HOST–PATHOGEN INTERACTIONS AND PATHOGENICITY IN *Bd*

The Chytridiomycota are widely distributed across soil, aquatic, and high altitude ecosystems (Freeman et al., 2009). Many species of chytrids are aquatic saprobes, where they catabolize macromolecules such as chitin and cellulose. While *Bd* is the only species of chytrid that parasitizes vertebrates, many species of chytrid parasitize phyto- and zooplankton, fungi, invertebrates, and plants (Gleason et al., 2008). During its reproductive, parasitic phase, *Bd* zoosporangia release flagellate zoospores into water where they exhibit chemotaxis toward a number of substrates including sugars, proteins, and amino acids (Moss et al., 2008). Following attachment to amphibian skin cells, likely facilitated by carbohydrate-binding proteins, cell penetration by zoospores occurs and the formation of new sporangia, completing the infectious life cycle.

While *Bd* is not a dermatophyte *sensu stricto*, it has close similarities to other cutaneous fungi of vertebrates in its utilization of keratin as a primary metabolic substrate. This feature limits the distribution of infection to the *stratum corneum* and the *stratum granulosum* of adult amphibian skin, where sporangia develop within keratinized tissues. In larval amphibians (tadpoles), the infection is localized to the keratinized mouthparts, and as a consequence, they do not suffer the dramatic pathologies that are associated with adult infection (Garner et al., 2009b). While the mechanism of infection has not been fully determined, comparative analysis of the *Bd* genome compared to other nonpathogenic fungi has shown that fungalysin metallopeptidase (also known as peptidase M36) and serine protease gene families have undergone extensive expansions in the *Bd* genome (Rosenblum et al., 2010). Metallopeptidases have similarly undergone expansion in the human-infecting dermatophytes *Trichophyton* and *Microsporum*, where they are highly upregulated in infection, accounting for up to 36% of total secreted protein extracts (Burmester et al., 2011; Jousson et al., 2004). Dermatophyte fungi, like *Bd*, are keratinophilic, and comparative genomics of *Arthroderma benhamiae* and *Trichophyton verrucosum* show over 235 predicted protease-encoding genes, many of which are shared with other closely related Onygenales such as *Coccidioides immitis*, which also has been shown to have an expansion of metalloproteases (Burmester et al., 2011; Sharpton et al., 2009). RNA-seq of *A. benhamiae* after coinoculation with and without keratinocytes demonstrated the differential expression of over 40 genes encoding putatively secreted proteins showing the capacity of skin-infecting fungi to modify gene expression according to their metabolic substrate (Burmester et al., 2011); such secreted proteins are therefore prime candidates as virulence factors in *Bd* as well as other skin-infecting fungi. Proteinases have also been detected in analyses of the *Bd* proteome for a number of different isolates, showing that these open reading frames are translated at high enough levels to be detected by 2-D gel approaches (Fisher et al., 2009a). The fungalysin metallopeptidase gene family was shown to be differentially expressed between two different life history stages of *Bd*, the zoospore and the sporangia, lending support to their putatively key role in the infection process (Rosenblum et al., 2008).

In common with vertebrate dermatophyte fungi, many of the mechanisms underlying the interactions between *Bd* and its hosts are currently unclear. For example, there appears to be a minimal host reaction to infection other than hyperplasia and hyperkeratosis of the stratum corneum, and noticeable lesions are usually not observed. Death of the host appears to result from pathophysiological changes related to electrolyte imbalances; intact skin function is essential for amphibians, owing to their need to actively maintain hyperosmotic internal environments as a consequence of having highly permeable skins. In diseased individuals, pronounced imbalance in electrolytes occurs as a consequence of epidermal sodium and chloride channels becoming inhibited, leading to hypokalemia (low plasma potassium) and hyponatremia (low plasma sodium). The ultimate cause of death is cardiac arrest resulting from the electrolytic imbalances (Voyles et al., 2009). However, whether the release of a fungal toxin or direct damage to infected host cells results in the disruption of osmoregulatory function is not yet known. Further, no genes involved in pathogenesis have yet been described that interact directly or indirectly with any host response to infection.

COMPARATIVE EVOLUTIONARY GENOMICS OF *Bd*

The application of whole-genome sequencing to delineate the genetic components of *Bd* has allowed for a comparative inventory of genes that the species shares with other fungi in order to craft hypotheses about possible virulence factors and their evolutionary history. Gene family size changes and lineage-specific expansions are important modes of evolution of pathogenesis factors in fungi (Moran et al., 2011). Some comparisons of the *Bd* genome have revealed expansions of gene copies in the previously described metalloprotease and peptidase M36 and S8 families (Rosenblum et al., 2010), but also in gene families such as the chitin-binding family CBM18 (Abramyan and Stajich, 2011) and the presence of an expanded group of genes that share some homology with the C-terminal of the Crinkler (Crn) family of effector proteins found in the *Phytophthora* genus (Joneson et al., 2011; Sun et al., 2011). However, the identification of meaningful differences is limited to the evolutionary proximity of the compared genomes and in the contrast between a pathogen and avirulent species. To achieve this, a recent study used the sequence of a genome of the closest known relative to *Bd*, *Homolaphylyctis polyrhiza* (*Hp*), to identify genes or gene families that appear to be unique to the pathogenic chytrid (Joneson et al., 2011). The comparisons, summarized in Table 24.1, indicate that the expansions of M36 and S41 families, CBM18, and Crn domain are all recent changes that occur on the evolutionary branch to *Bd* as they are absent in *Hp* and other more distant chytrid species (Joneson et al., 2011).

Bd grows intracellularly in close contact with its host, and it seems likely that it would secrete proteins that might affect the host. These effectors may target host defense mechanisms to enable the microbe to gain access to the host cell or to avoid

TABLE 24.1 Gene Copy Number of Those Containing Domains That Appear Expanded in *Bd* Relative to Other Related Fungi

Family	*Bd*	*Hp*	*S. punctatus*	*A. macrogynus*
Aspartyl protease	87	16	9	2
Metalloprotease M36	29	5	3	31
Serine-type protease S41	33	3	3	0
Chitin-binding module CBM18	19	4	3	4
Crinkler	62	0	0	0

These copy number expansions are confirmed by phylogenetic analyses; data from Abramyan and Stajich (2011) and Joneson et al. (2011).

detection by the host response. *Bd* has a collection of Crn-like genes, a unique property currently not found in any other fungi (Joneson et al., 2011; Sun et al., 2011), which may play a role in host–pathogen interface as Crns containing genes are able to translocate plant cells and target the nuclei (Schornack et al., 2010). However, the absence of a recognizable secretion signal suggests that they remain intracellular. Because they are unique to *Bd* and are not found in other fungi, it has been suggested that the Crn genes were gained recently, perhaps through horizontal transfer (Sun et al., 2011). However, experimental confirmation is needed to establish the link these Crns may play in the virulence of *Bd* as their functionality has yet to be proven.

Microbes may interact with host cells through a variety of other secreted proteins, and computational tools suggest that out of the roughly 8700 genes in the genome, about 1500 (17%) are predicted to be secreted. This is in comparison to the 12% predicted in *Saccharomyces cerevisiae* and on par with the 17% in the plant pathogen *Fusarium graminearum*, but more than the human pathogenic fungi *C. immitis* (6%) and *Histoplasma capsulatum* (5%). Which proteins in the secretome of *Bd* interact with host amphibian host cells and play a role in pathogenesis remains an open area of research, but the potentially large fraction dedicated to secreted proteins may be an important component in determining how the pathogen modulates host reponses.

In many fungi, melanin is an important virulence factor providing protection by absorbing free radicals produced by the host immune cells or UV radiation (Casadevall et al., 2000). Genome analysis of *Bd* reveals a large copy number for tyrosinase genes (E. M. Medina, J. E. Stajich, S. Restrepo, unpublished data), a critical enzyme in the melanin biosynthesis pathway. While not all 11 copies of the family may be involved in this synthesis, the potential redundancy in the enzyme repertoire may suggest an important role for the family. Copy number alone is not sufficient to determine if the expansion is recent, and comparisons with other fungal genomes, such as another basal fungus *Allomyces macrogynus* (Blastocladiales) and the basidiomycete mushroom *Coprinopsis cinerea*, revealed many copies in these genomes as well. A phylogenetic tree of the gene copies indicated that the tyrosinase expansion is recent to *Bd* (as all members cluster together in the tree) but that frequent expansion of this family is seen in other species and is perhaps involved in pigmentation in *Allomyces* or formation of black spores in *Coprinopsis*. As there is currently no documented role for melanin in *Bd* fungi, more work is still needed to understand if these copies provide a defensive or offensive weapon.

Genome comparisons also revealed that the cell walls of *Bd* have a different composition from Dikarya fungi (J.E. Stajich, unpublished data). An analysis of the gene families related to cell wall biosynthesis components revealed expansions of chitin synthase genes in three subfamilies, but that *Bd* was also lacking a key enzyme for synthesis of 1,3-beta-glucan, the homolog of Fks1 in *S. cerevisiae*. Fks1 is a highly conserved protein found in all examined Dikarya fungi and Zygomycetes but missing in three species of the Chytridiomycota: *Spizellomyces punctatus*, *Bd*, and *Hp*. This surprising result suggests that cell walls in this group of fungi and perhaps in the ancestral fungus look very different from those seen in extant Dikarya molds, yeasts, and mushrooms. Additional analysis of cell wall genes revealed an expansion of a subfamily of chitin-binding module genes called CBM18. This expansion appears to be unique to *Bd* as *Hp* has only a few copies, and a gene tree analysis shows a recent and rapid expansion of copies of this gene. In addition, the CBM18 domain is found amplified within these genes to as many 11 copies within one locus, while the domain is only rarely found to occur in more than two copies per locus in most other fungi. The function of CBM18 is to bind chitin so it may be that the proteins bind chitin in the *Bd* cell wall and thus cloak the chitin from recognition by a host immunity system. Another interpretation is that the CBM18 genes bind another carbohydrate and provide an adhesive role for *Bd* cell attachment. The repeating domain architecture could be a simple mechanism that has increased the binding efficiency of the proteins. Taken together, it appears that the cell walls of *Bd* may exhibit differences, some of which could play a role in virulence through adhesion, protection, or avoidance of detection.

GLOBAL VECTORS OF *Bd*

The contemporary global spread of *Bd* is very likely driven by the global trade in infected amphibians. A meta-analysis by Fisher and Garner showed that at least 28 species of amphibians are known to be carriers of *Bd* and to have invaded novel ecosystems as introduced alien species (Fisher and Garner, 2007). Several of these species are known to support asymptomatic infections of *Bd*, and to have been widely introduced outside of their native ranges; examples are the African clawed frog *Xenopus laevis*, the North American bullfrog *Lithobates catesbeianus* (formerly *Rana catesbeiana*), and the South American cane toad *Rhinella marina* (formerly *Bufo marinus*). These species have established nonnative alien populations in the Americas, Europe, Australia, Asia as well as in many oceanic (Japan) and coastal islands (Caribbean; Monstserrat and Dominica, Mediterranean; Mallorca; and Sardinia), and are associated with high prevalences of infection of *Bd* both across their native ranges and in regions where they have been introduced (Fisher and Garner, 2007; Garner et al., 2006). North American bullfrogs are important vectors of *Bd* as they act as "superspreaders" of infection. This is because the species is widely infected by *Bd* across its native range in the United States (Longcore et al., 2007) and tolerates high burdens of infection; therefore, it is asymptomatic when moved by trade. Outside of North America, the species is farmed in huge quantities where infection by *Bd* is amplified (Mazzoni et al., 2003) and live infected individuals are globally exported (Garner et al., 2006). Following translocations, the bullfrogs have become widely invasive outside of its native range following uncontrolled introduction (Garner et al., 2009a). Analyses of polymorphisms in the ribosomal ITS have shown that specific haplotypes of *Bd* are associated with bullfrogs and have been spread into native populations of amphibians following introduction (Fisher, 2009; Goka et al., 2009), demonstrating the status of this species as a vector of the pathogen into Southeast Asia, as well as Europe (Cunningham et al., 2005). Similar superspreader status has been awarded to *X. laevis*, which, like bullfrogs, are widely infected and farmed, globally transported for research purposes, and tolerate high burdens of infection. Direct "spillover" of *Bd* from wild-caught African *Xenopus* species into a naive *Bd*-susceptible Mallorcan midwife toad *Alytes muletensis* has been demonstrated within the environs of zoos, followed by the introduction of *Bd*-infected *A. muletensis* onto the balearic island of Mallorca where infection has now become established (Walker et al., 2008).

Attempts to trace the origins of *Bd* by detecting the pathogen in preserved museum specimens have shown infections of African pipid frogs occurring as early as 1933 (*Xenopus fraseri* Cameroon 1933) (Soto-Azat et al., 2010) and 1938 (*X. laevis* South Africa 1938) (Weldon et al., 2004). Studies from other continents show no early (pre-1970s) evidence of infection, and consequentially an "out of Africa" origin has been proposed for the emergence of *Bd*, vectored by the widespread trade of pipid frogs in the mid 20th century (Weldon et al., 2004).

POPULATION GENOMICS OF *Bd* AND THE ORIGINS OF THE *Bd* PANZOOTIC

Newly emerging pathogenic fungi are characterized by having low levels of genetic variation relative to their point of origin; this owes to the multiple population bottlenecks that are associated with rapid pathogen spread. Therefore, attempts to identify the origin of the *Bd* panzootic have focused on developing genetic tools to analyze phylogenetic patterns of relatedness between isolates of *Bd*. The first marker-based study on the population genetics of *Bd* used 17 sequence-based markers typed in 59 *Bd* isolates from five continents and included 30 host amphibian species (James et al., 2009). This comprehensive study showed that the entire global diversity of *Bd* could be explained by the dispersal of a single diploid individual, and levels of genetic diversity were among the lowest recorded for a eukaryotic pathogen (James et al., 2009). While these findings are consistent with a recent globalization of *Bd*, they did not support Weldon et al.'s out of Africa hypothesis as levels of genetic diversity in isolates from North America were found to be as high as they were in Africa. The paucity of polymorphisms across these loci coupled with a sampling strategy that was targeted at New World populations of amphibians suggested that a greater depth of both sampling and genotyping would be necessary to more effectively address the question of *Bd*'s origin (James et al., 2009).

Subsequent studies now focus on investigating the evolutionary epidemiology of *Bd* using next-generation sequencing (NGS) to enable whole-genome sequence typing of the pathogen. Such approaches have already found use in molecular epidemiological applications for other pathogenic fungi, such as by determining identity between cases of coccidioidomycosis disseminated by organ transplants (Engelthaler et al., 2011). Platforms such as Illumina HiSeq, ABI SOLiD, or Roche 454 can provide the entire genomic sequences for numerous isolates of *Bd*, which can then be assembled or aligned to either of the two publically available genomes of *Bd* (JEL423, http://www.broadinstitute.org/, or JAM81, http://www.jgi.doe.gov/), revealing polymorphic sites among those samples across the whole genome. This whole-genome approach vastly increases the analytical power of population genetic analyses as tens of thousands of single-nucleotide polymorphisms (SNPs) are scored, as opposed to the dozens that typify earlier approaches.

Recent NGS approaches used ABI SOLiD sequencing of a global panel of *Bd* isolates. Twenty isolates were sequenced from five continents (Europe, North and Central America, South Africa, and Australia) and from 11 amphibian host species and 8 epizootics (Farrer et al., 2011). This sequencing was successful in characterizing 51,915 homozygous SNPs and 87,121 heterozygous positions, of which 21% of the homozygous SNP positions and 19% of the heterozygous positions were covered in ≥4 reads in all 20 samples and were used for phylogenetic analysis. Sixteen of the twenty samples, including the reference strain JEL423, were >99.9% genetically identical and fell within a single highly supported clade (Fig. 24.1). This *Bd* "global panzootic lineage" (*Bd*GPL) includes all previously genotyped isolates of *Bd* and all the isolates in the panel that were associated with regional epizootics, recovered from five continents; this lineage therefore contains the set of isolates described by James et al. (2009). The remaining four sequenced isolates formed two novel, deeply divergent lineages. The "*Bd* Cape lineage" (*Bd*CAPE)

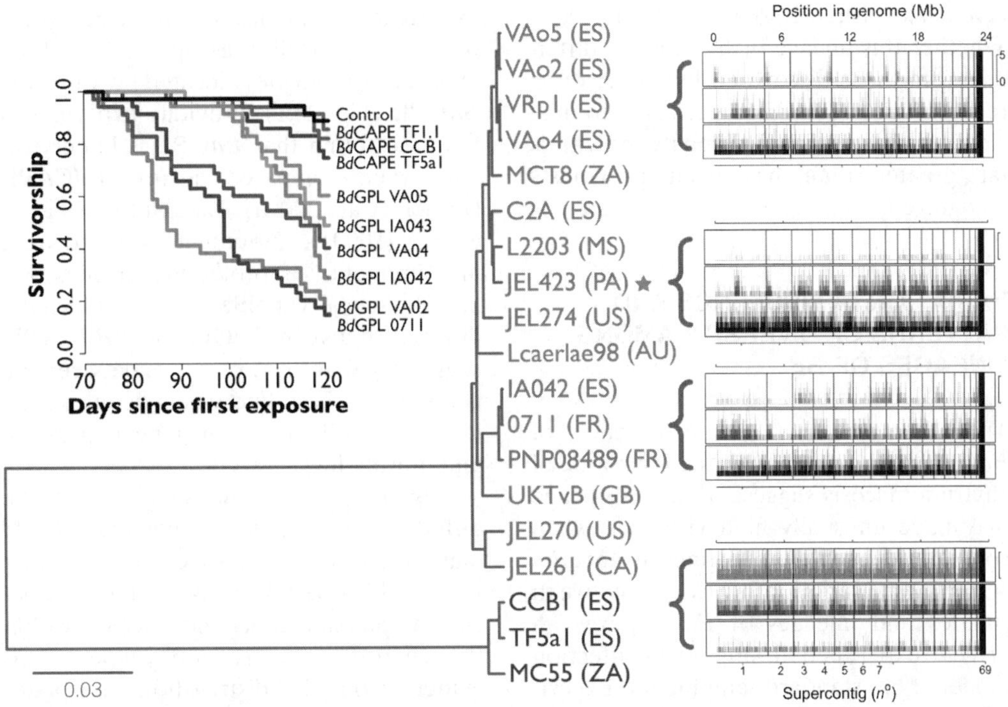

Figure 24.1 A tree of 19 *Bd* nuclear genomes made using the UPGMA algorithm in PAUP with Kaplan–Meier survival curves and representations of the polymorphic sites within the genome. The figure demonstrates two divergent lineages (*Bd*GPL shown in red and *Bd*CAPE shown in blue) and the significant difference between postmetamorphic survival of toadlets (*Bufo bufo*) exposed to isolates from those two lineages. To the right of the tree shows a tally of homozygous SNPs (red) and heterozygous SNPs (blue) identified from alignments to the *Bd*GPL genome isolate JEL423, and homozygous SNPs from alignments to a *Bd*CAPE consensus sequence (gray) using a nonoverlapping 1.4-kb sliding window across the genome. SNPs are nonrandomly distributed (in clusters) across isolates belonging to *Bd*GPL, indicative of recombination. This genomic feature does not appear in the divergent *Bd*CAPE lineage. * signifies the JEL423 genome isolate. See color insert.

included two isolates from the island of Mallorca and one from the Cape Province, South Africa. A third novel lineage ("*Bd* Swiss lineage," *Bd*CH) was composed of a single isolate derived from a common midwife toad (*Alytes obstetricans*), sampled from a pond in Switzerland.

These data illustrate three important points about the origins of *Bd*. The first is that isolates that are associated with epizootics in Australia, America, and Europe all fall within the *Bd*GPL, suggesting that the emergence lineage is responsible for these *Bd*-driven collapses in biodiversity. Second, there appear to be multiple, genetically highly diverged, varieties of *Bd* that have gone previously undetected (*Bd*CAPE, *Bd*CH). Third, based on this limited sampling of global *Bd* diversity, Africa and Europe are both centers of diversity for *Bd* as they contain not only *Bd*GPL but the other newly diagnosed lineages too. However, further insights into whether these regions really represent centers of *Bd* diversity await much more in-depth sampling of the chytrids infecting their amphibian fauna. Indeed, recent reports from Southeast Asia suggest that there are divergent lineages of *Bd* that are associated with Japanese giant salamanders *Andrias* *japonicus* (Fisher, 2009; Goka et al., 2009). Whether this lineage is the same or different from those discovered by Farrer et al. remains to be seen; however, it is becoming clear that the global diversity of *Bd* is far greater than has been previously recognized.

VARIATION IN VIRULENCE AND PATTERNS OF DIVERSITY AMONG LINEAGES OF *Bd*

The discovery that the emergence of *Bd*GPL is associated with the panzootic of chytridiomycosis suggests that this lineage may have unusually high virulence when compared against other lineages of *Bd*. Ascertaining variation in virulence between isolates and lineages of *Bd* requires an *in vivo* experimental amphibian infection model. The standard amphibian experimental genetic models, *X. laevis* and *Silurana tropicalis*, are not particularly amenable to infection by *Bd* as both species are tolerant of high infectious burdens and do not readily manifest disease in captivity unless artificially stressed, for instance, by reducing temperature to below the animals' optima (Ribas et al., 2009; Rosenblum et al., 2008). Progress in assessing virulence has relied on the development of new animal models, and the European common toad *Bufo bufo* is increasingly being used owing to the species high susceptibility to infection, small size, and relatively simple husbandry, thus enabling the use of large numbers of animals (Garner et al., 2009b, 2011). To ascertain if *Bd* virulence recapitulated the NGS phylogeny, common toad tadpoles were exposed to repeated, high concentration doses of *Bd* using multiple isolates from the two lineages (*Bd*GPL and *Bd*CAPE) and from two clusters within the *Bd*GPL lineage (Valencia and the Pyrenees). These experiments showed that postmetamorphic survival of animals exposed to isolates derived from *Bd*CAPE experienced significantly reduced infection and mortality compared to those exposed to isolates located within the *Bd*GPL (Fig. 24.1). Previous experiments had also found that *Bd*GPL is hypervirulent when compared against *Bd*CAPE (Fisher et al., 2009a). The study by Fisher et al. (2009a), moreover, showed that measurements of proteomic profiles and morphological variables were significantly different between *Bd*GPL and *Bd*CAPE, suggesting that the divergence between the genomes of these lineages is deep enough to realize differences in gene expression and morphology.

Together, these data suggest that the differentiation between lineages of *Bd* may be suggestive of speciation and that what NGS is describing is a set of conspecific amphibian-associated species within the chytrid order Rhizophydiales. It is evident from the distribution of isolates that at least two lineages of *Bd* occur widely in different continents (*Bd*GPL and *Bd*CAPE), and both of these lineages are associated with intercontinental transmission via the amphibian trade; *Bd*GPL via North American bullfrog introductions (Cunningham et al., 2005; Goka et al., 2009) and *Bd*CAPE via *Xenopus* translocations (Walker et al., 2008). However, virulence data suggests that *Bd*GPL is hypervirulent and has been causing epizootics worldwide, while other lineages may be acting more as commensal organisms rather as than pathogens. This last feature perhaps explains why these hypovirulent lineages have remained undetected until recently.

Given these data, we asked the question: Are there any genomic features that are unique to *Bd*GPL that may perhaps explain the enhanced virulence of this lineage? One striking difference is in the distribution of SNPs that exist in the heterozygous versus the homozygous state. Within *Bd*CAPE and *Bd*CH, heterozygous and homozygous polymorphisms are distributed randomly throughout the genome

(Fig. 24.1; Farrer et al., 2011). However, for *Bd*GPL, a highly uneven distribution of homozygous SNPs and heterozygous positions was identified across all isolates. This loss of heterozygosity (LOH) had been previously reported by James et al. (2009) and is hypothesized to be caused by mitotic recombination resulting in gene-conversion events leading to LOH. Strikingly, shared LOH events are seen spanning all of the sequenced *Bd*GPL isolates to date; for example, Figure 24.1 shows an LOH event that spans the entirety of chromosome 2. If LOH is unique to *Bd*GPL, then this may be associated with the hypervirulence of this lineage. This idea is not without precedence as phenotypic changes in other fungi are known to be associated with LOH, such as changes in phenotype for the human pathogen *Candida albicans* (Magee et al., 2005).

The remarkably similar pattern of LOH events across *Bd*GPL suggests that there must have been a single origin of this lineage followed by global transmission of the clonally reproducing progeny. Given the lack of SNPs that distinguish between *Bd*GPL isolates (less than 1 per 1000 bases), it appears the lineage arose via the mating between two reasonably closely related parental genotypes. The increased virulence of the lineage may then owe to a "heterosis" effect, also known as hybrid vigor, leading to its extraordinarily rapid, and destructive, emergence worldwide. The paucity of polymorphisms in *Bd*GPL is also consistent with a recent emergence of the lineage. This idea was tested by dating the emergence of *Bd*GPL using Bayesian approaches, with a range of 35–257 years ago being recovered depending on the region of the genome for which the dating was undertaken (Farrer et al., 2011). These dates are compatible with the hypothesis that the globalization of *Bd*GPL occurred sometime within the 20th century and that this lineage is associated with increases in the global trade in amphibians. While it is interesting to speculate on the geographic origins of *Bd*GPL, the data do not support any specific hypothesis as yet. However, it is clear that the combination of NGS, global mapping and comparative genomics has yielded a powerful set of tools that will likely yield an answer to this question in the near future.

SUMMARY POINTS

The emergence of *Bd* during the mid-20th century comprises one of the most devastating pan-global outbreaks of infection yet ever witnessed, with untold consequences for amphibian biodiversity and their associated role in supporting healthy ecosystems. Genome, spatial mapping, and virulence studies have shown that the emergence of panzootic chytridiomycosis owes to the 20th century emergence of a single, hypervirulent, lineage, currently known as *Bd*GPL. However, other less aggressive, closely related lineages of *Bd*, potentially comprising cryptic species, have been detected, suggesting that there exists an undescribed diversity of amphibian-associated chytrids held within the genus *Batrachochytrium*. Uniquely, *Bd*GPL appears to be characterized by frequent LOH, but whether this occurs via by meitotic or mitotic recombination has yet to be described. *In vivo* virulence models show wide variation in virulence among lineages of *Batrachochytrium*, suggesting that LOH may determine virulence to some extent. Comparative genomics approaches are identifying key genes and expanded gene families that are unique to *Batrachochytrium*, suggesting potential virulence mechanisms and modes of host–pathogen interaction. While much remains to be understood about virulence and pathogenicity in *Bd* and the origin of the panzootic, next-generation tools have accelerated the pace of research and recent developments have shed much new light on this previously highly enigmatic novel pathogen.

ACKNOWLEDGMENTS

M.C.F. and R.A.F. are funded by the Biodiversa project *RACE*: Risk Assessment of Chytridiomycosis to European Amphibian Biodiversity (http://www.bd-maps.eu)

REFERENCES

Abramyan J, Stajich JE. 2011. Species-specific chitin binding module 18 (CBM18) expansion in the amphibian pathogen *Batrachochytrium dendrobatidis*, submitted.

Berger L, Speare R, Daszak P, Green DE, Cunningham AA, Goggin CL, Slocombe R, Ragan MA, Hyatt AH, Mcdonald KR, Hines HB, Lips KR, Marantelli G, Parkes H. 1998. Chytridiomycosis causes amphibian mortality associated with population declines in the rain forests of Australia and Central America. Proceedings of the National Academy of Science of the United States of America 95: 9031–9036.

Burmester A, Shelest E, Glockner G, Heddergott C, Schindler S, Staib P, Heidel A, Felder M, Petzold A, Szafranski K, Feuermann M, Pedruzzi I, Priebe S, Groth M, Winkler R, Li W, Kniemeyer O, Schroeckh V, Hertweck C, Hube B, White TC, Platzer M, Guthke R, Heitman J, Wostemeyer J, Zipfel PF, Monod M, Brakhage AA. 2011. Comparative and functional genomics provide insights into the pathogenicity of dermatophytic fungi. Genome Biology 12: R7.

Casadevall A, Rosas AL, Nosanchuk JD. 2000. Melanin and virulence in *Cryptococcus neoformans*. Current Opinion in Microbiology 3: 354–358.

Cunningham AA, Garner TWJ, Aguilar-Sanchez V, Banks B, Foster J, Sainsbury AW, Perkins M, Walker SF, Hyatt AD, Fisher MC. 2005. Emergence of amphibian chytridiomycosis in Britain. Veterinary Record 157: 386–387.

Engelthaler DM, Chiller T, Schupp JA, Colvin J, Beckstrom-Sternberg SM, Driebe EM, Moses T, Tembe W, Sinari S, Beckstrom-Sternberg JS, Christoforides A, Pearson JV, Carpten J, Keim P, Peterson A, Terashita D, Balajee SA. 2011. Next-generation sequencing of *Coccidioides immitis* isolated during cluster investigation. Emerging Infectious Diseases 17: 227–232.

Farrer RA, Weinert LA, Bielby J, Garner TWJ, Balloux F, Clare F, Bosch J, Cunningham AA, Weldon C, du Preez LH, Anderson L, Kosakovsky Pond SL, Shahar-Golan R, Henk DA, Fisher MC. 2011. Multiple emergences of genetically diverse amphibian-infecting chytrids include a globalised hypervirulent recombinant lineage. Proc Natl Acad Sci USA 108(46): 18732–18736.

Fisher MC. 2009. Endemic and introduced haplotypes of *Batrachochytrium dendrobatidis* in Japanese amphibians: Sink or source? Molecular Ecology 18: 4731–4733.

Fisher MC, Garner TWJ. 2007. The relationship between the introduction of *Batrachochytrium dendrobatidis*, the international trade in amphibians and introduced amphibian species. Fungal Biology Reviews 21: 2–9.

Fisher MC, Bosch J, Yin Z, Stead DA, Walker J, Selway L, Brown AJP, Walker LA, Gow NAR, Stajich JE, Garner TWJ. 2009a. Proteomic and phenotypic profiling of the amphibian pathogen *Batrachochytrium dendrobatidis* shows that genotype is linked to virulence. Molecular Ecology 18: 415–429.

Fisher MC, Garner TWJ, Walker SF. 2009b. Global emergence of *Batrachochytrium dendrobatidis* and amphibian *Chytridiomycosis* in space, time, and host. Annual Review of Microbiology 63: 291–310.

Fisher MC, Henk DA, Briggs C, Brownstein JS, Madoff L, McCraw SL, Gurr S. 2012. Emerging fungal threats to animal, plant and ecosystem health. Nature 484: 186–194.

Freeman KR, Martin AP, Karki D, Lynch RC, Mitter MS, Meyer AF, Longcore JE, Simmons DR, Schmidt SK. 2009. Evidence that chytrids dominate fungal communities in high-elevation soils. Proceedings of the National Academy of Sciences of the United States of America 106: 18315–18320.

Frick WF, Pollock JF, Hicks AC, Langwig KE, Reynolds DS, Turner GG, Butchkoski CM, Kunz TH. 2010. An emerging disease causes regional population collapse of a common North American bat species. Science 329: 679–682.

Gargas A, Trest MT, Christensen M, Volk TJ, Blehert DS. 2009. *Geomyces destructans* sp. nov. associated with bat white-nose syndrome. Mycotaxon 108: 147–154.

Garner TWJ, Perkins M, Govindarajulu P, Seglie D, Walker S, Cunningham AA, Fisher MC. 2006. The emerging amphibian pathogen *Batrachochytrium dendrobatidis* globally infects introduced populations of the North American bullfrog, *Rana catesbeiana*. Biology Letters 2: 455–459.

Garner TWJ, Stephen I, Wombwell E, Fisher MC. 2009a. The amphibian trade: Bans or best practice? Ecohealth 6: 148–151.

Garner TWJ, Walker S, Bosch J, Leech S, Rowcliffe JM, Cunningham AA, Fisher MC. 2009b. Life history trade-offs influence mortality associated with the amphibian pathogen *Batrachochytrium dendrobatidis*. Oikos 118: 783–791.

Garner TWJ, Rowcliffe JM, Fisher MC. 2011. Climate change, chytridiomycosis or condition: An experimental test of amphibian survival. Global Change Biology 17: 667–675.

Gleason FH, Kagami M, Lefevre E, Sime-Ngando T. 2008. The ecology of chytrids in aquatic ecosystems: Roles in food web dynamics. Fungal Biology Reviews 22(1): 17–25.

Goka K, Yokoyama J, Une Y, Kuroki T, Suzuki K, Nakahara M, Kobayashi A, Inaba S, Mizutani T, Hyatt A. 2009. Amphibian chytridiomycosis in Japan: Distribution, haplotypes and possible route of entry into Japan. Molecular Ecology 18(23): 4757–4774.

James TY, Litvintseva A, Vilgalys R, Morgan JA, Taylor JW, Fisher MC, Berger L, Weldon C, Du Preez LH, Longcore J. 2009. Rapid expansion of an emerging fungal disease into declining and healthy amphibian populations. PLoS Pathogens 5: e1000458.

Joneson S, Stajich JE, Shiu S-H, Rosenblum EB. 2011. Genomic transition to pathogenicity in chytrid fungi. PLoS Pathogens 7(11): e1002338.

Jousson O, Lechenne B, Bontems O, Capoccia S, Mignon B, Barblan J, Quadroni M, Monod M. 2004. Multiplication of an ancestral gene encoding secreted fungalysin preceded species differentiation in the dermatophytes *Trichophyton* and *Microsporum*. Microbiology-Sgm 150: 301–310.

Lips KR, Brem F, Brenes R, Reeve JD, Alford RA, Voyles J, Carey C, Livo L, Pessier AP, Collins JP. 2006. Emerging infectious disease and the loss of biodiversity in a neotropical amphibian community. Proceedings of the National Academy of Sciences of the United States of America 103: 3165–3170.

Longcore JE, Pessier AP, Nichols DK. 1999. *Batrachochytrium dendrobatidis* gen et sp. nov., a chytrid pathogenic to amphibians. Mycologia 91: 219–227.

Longcore JR, Longcore JE, Pessier AP, Halteman WA. 2007. Chytridiomycosis widespread in anurans of northeastern United States. Journal of Wildlife Management 71: 435–444.

Lotters S, Kielgast J, Bielby J, Schmidtlein S, Bosch J, Veith M, Walker SF, Fisher MC, Rodder D. 2009. The link between rapid enigmatic amphibian decline and the globally emerging chytrid fungus. Ecohealth 6: 358–372.

Magee PT, Forche A, May G. 2005. Demonstration of loss of heterozygosity by single-nucleotide polymorphism microarray analysis and alterations in strain morphology in *Candida albicans* strains during infection. Eukaryotic Cell 4: 156–165.

Mazzoni R, Cunningham AA, Daszak P, Apolo A, Perdomo E, Speranza G. 2003. Emerging pathogen of wild amphibians in frogs (*Rana catesbeiana*) farmed for international trade. Emerging Infectious Diseases 9: 995–998.

Moran GP, Coleman DC, Sullivan DJ. 2011. Comparative genomics and the evolution of pathogenicity in human pathogenic fungi. Eukaryotic Cell 10: 34–42.

Moss AS, Reddy NS, Dortaj IM, Francisco MJS. 2008. Chemotaxis of the amphibian pathogen *Batrachochytrium dendrobatidis* and its response to a variety of attractants. Mycologia 100: 1–5.

Ribas L, Li M-S, Doddington BJ, Robert J, Seidel JA, Kroll JS, Zimmerman LB, Grassly NC, Garner TWJ, Fisher MC. 2009. Expression profiling the temperature-dependent amphibian response to infection by *Batrachochytrium denrobatidis*. PLoS ONE 4(12): e8408.

Rosenblum EB, Stajich JE, Maddox N, Eisen MB. 2008. Global gene-expression profiles for life stages of the deadly amphibian pathogen *Batrachochytrium dendrobatidis*. Proceedings of the National Academy of Sciences of the United States of America 105: 17034–17039.

Rosenblum EB, Fisher MC, James TY, Stajich JE, Longcore J. 2010. A molecular perspective on the biology of the emerging pathogen *Batrachochytrium dendrobatidis*. Diseases of Aquatic Organisms 92(2): 131–147.

Schornack S, Van Damme M, Bozkurt TO, Cano LM, Smoker M, Thines M, Gaulin E, Kamoun S, Huitema E. 2010. Ancient class of translocated oomycete effectors targets the host nucleus. Proceedings of the National Academy of Sciences of the United States of America 107: 17421–17426.

Sharpton TJ, Stajich JE, Rounsley SD, Gardner MJ, Wortman JR, Jordar VS, Maiti R, Kodira CD, Neafsey DE, Zeng Q, Hung CY, Mcmahan C, Muszewska A, Grynberg M, Mandel MA, Kellner EM, Barker BM, Galgiani JN, Orbach MJ, Kirkland TN, Cole GT, Henn MR, Birren BW, Taylor JW. 2009. Comparative genomic analyses of the human fungal pathogens *Coccidioides* and their relatives. Genome Research 19: 1722–1731.

Skerratt LF, Berger L, Speare R, Cashins S, Mcdonald KR, Phillott AD, Hines HB, Kenyon N. 2007. Spread of chytridiomycosis has caused the rapid global decline and extinction of frogs. Ecohealth 4: 125–134.

Soto-Azat C, Clarke BT, Poynton JC, Cunningham AA. 2010. Widespread historical presence of *Batrachochytrium dendrobatidis* in African pipid frogs. Diversity and Distributions 16: 126–131.

Stuart SN, Chanson JS, Cox NA, Young BE, Rodrigues AS, Fischman DL, Waller RW. 2004. Status and trends of amphibian declines and extinctions worldwide. Science 306: 1783–1786.

Sun G, Yang Z, Kosch T, Summers K, Huang J. 2011. Evidence for acquisition of virulence effectors in pathogenic chytrids. BMC Evolutionary Biology 11: 195.

Voyles J, Young S, Berger L, Campbell C, Voyles WF, Dinudom A, Cook D, Webb R, Alford RA, Skerratt LF, Speare R. 2009. Pathogenesis of chytridiomycosis, a cause of catastrophic amphibian declines. Science 326: 582–585.

Walker S, Bosch J, James TY, Litvintseva A, Valls JAO, Pina S, Garcia G, Abadie-Rosa G, Cunningham AA, Hole S, Griffiths R, Fisher MC. 2008. Invasive pathogens threaten species recovery programs. Current Biology 18: R853–R854.

Walker SF, Bosch J, Gomez V, Garner TWJ, Cunningham AA, Schmeller DS, Ninyerola M, Henk DA, Ginestet C, Arthur CP, Fisher MC. 2010. Factors driving pathogenicity vs. prevalence of amphibian panzootic chytridiomycosis in Iberia. Ecology Letters 13: 372–382.

Weldon C, Du Preez LH, Hyatt AD, Muller R, Speare R. 2004. Origin of the amphibian chytrid fungus. Emerging Infectious Diseases 10: 2100–2105.

INTERNET RESOURCES

Global *Bd*-Mapping Project: http://www.bd-maps.net/

Risk Assessment of Chytridiomycosis to European Amphibian Biodiversity: http://www.bd-maps.eu

CHAPTER 25

IMPACT OF HORIZONTAL GENE TRANSFER ON VIRULENCE OF FUNGAL PATHOGENS OF PLANTS

BARBARA J. HOWLETT and RICHARD P. OLIVER

INTRODUCTION

Fungi, being heterotrophic, are ultimately dependent on carbon fixed by plants for their nutrition. The most common form of nutrition is saprophytic growth on dead and decaying plant material, and the evolution of symbiosis and pathogenesis on plants gave fungi access to richer nutrient sources. Evolution followed several paths as fungal lineages adopted different feeding strategies, including, for example, necrotrophy, in which host tissue is killed; biotrophy, in which the host is living; and hemibiotrophy, which consists of phases of biotrophy and necrotrophy. Other adaptations included penetration site (roots, crowns, stomatal apertures, or epidermis), feeding structure (e.g., haustoria), and symptom (wilts, rots, and blotches, among others) variation. Another important property that evolved is host specificity. Plant pathogenic fungi can have broad or narrow host ranges; some are even host genotype (cultivar) specific. In contrast, mammalian fungal pathogens seem to be more homogeneous with their requirements and lifestyles, being linked by a common ability to survive at temperatures of 37°C and to evade host immune responses (Sexton and Howlett, 2006).

A range of mechanisms is responsible for genomic adaptations that allow or enhance virulence and/or the host range of a fungus. One of these is horizontal gene transfer (HGT) between organisms. This term is often used interchangeably with lateral or interspecific gene transfer to describe the transmission of genetic material between species by any mechanism other than the routine processes of sexual (meiotic) or asexual (mitotic) reproduction. Horizontal transfer can involve single genes, gene clusters, or even whole chromosomes; such chromosomes are also described as "dispensable" or supernumerary or B type.

The mere transfer of DNA into another cell may have trivial consequences, but stable integration into a recipient chromosome enables its inheritance. If the transfer is to have positive consequences, the recipient strain must acquire a new property that enables it to outcompete both its

Evolution of Virulence in Eukaryotic Microbes, First Edition. Edited by L. David Sibley, Barbara J. Howlett, and Joseph Heitman.

unaffected siblings and other organisms in its environment. In other words, the transferred gene must confer a selective advantage to the recipient.

Reports of HGT where fungi are donors and/or recipients have increased significantly in recent years, especially as genome sequencing has become routine (Richards et al., 2006, 2011). Additionally, instances of HGT between plant pathogenic ascomycetes that affect host range have been recently reviewed (Mehrabi et al., 2011). There are many more reports of HGT in ascomycetes than in basidiomycetes, but it is not yet clear whether this is a real difference or if it reflects the number of genomes sequenced from each phylum. In this chapter, we describe horizontal transfer of genes involved in virulence between fungi and highlight a situation where the transfer of a toxin-encoding gene has led to the recipient fungus becoming a major pathogen of wheat.

EVIDENCE USED TO SUPPORT THE OCCURRENCE OF HGT

The major requirement for HGT to be proposed is that phylogenetic trees of the gene purported to have been transferred and of other genes of the host organism will be incongruent. Indeed, Richards et al. (2011) regard phylogenetic evidence as the "gold standard" for HGT. Let us consider a phylogenetic tree with five equidistant taxa: A to E. A particular gene in taxon A is similar to one in taxon E but absent in taxa B through D. The gene is inferred to have transferred horizontally from A to E (or vice versa). The strength of this evidence and hence the validity of this hypothesis depends on a number of factors. The similarity of the genes in A and E and the degree of certainty that the gene is missing in B through D should be considered. Similarity between genes in different organisms can be quantified by the BLAST and similar algorithms. When genes are uniformly present, the bit scores of genes to their "best hits" in related species will be closely related to the evolutionary distance between the species. In our example, the bit score of the gene in A matched with E will be high, whereas the scores between the genes in A or E versus B, C, or D will be low. The ratio of these scores generates a list of HGT candidates. After identifying the suspect genes, sophisticated tree building and congruence tests can be applied (Marcet-Houben and Gabaldon, 2009; Richards et al., 2006, 2011; Slot and Rokas, 2011).

The alternative hypothesis to gene gain is that taxa A through E possessed the gene when they diverged from their common ancestor, but the gene was subsequently lost in taxa B, C, and D. Quickly evolving genes may still be present but would have diverged to the point of becoming unrecognizable. This scenario can be tested by estimating the divergence time of the gene in taxa A and E (Schoch et al., 2009). If the divergence time of the gene is not significantly different from the time that the species differentiated, the HGT hypothesis is not supported. Complete genome sequences are essential for these analyses, not only to identify genes that are present but to define those that are absent as well. Furthermore, genome sequences provide ample numbers of orthologous but neutrally selected genes that can be used to estimate divergence times.

Convergent evolution may also confound evidence supporting the presence of HGT. We can envisage a situation in which two unrelated pathogens both evolve a similar molecule, such as an effector, to overcome a particular host defense mechanism. Such genes might have close enough blast matches to warrant consideration as HGT candidates. Unlike HGT candidates, genes that are convergently evolving would not have matching intron positions but would have GC contents and codon usage indices characteristic of the rest of their genome. Furthermore, we would not expect flanking genes to be conserved.

In contrast, horizontally transferred genes are likely to share intron positions, GC content, and codon usage indices. These features can suggest the direction of transfer because horizontally transferred genes will be more similar to other genes of the donor than to genes of the recipient genome. The proximity of transposable elements to the gene adds weight to the evidence for HGT of single genes or gene clusters. Active transposons could allow excision or integration of the gene(s) from or into the genome. Additionally, in some organisms, gene clusters are located near telomeres, unstable locations within chromosomes. Such genomic locations are likely to facilitate the transfer of DNA into and from these regions by HGT.

FREQUENCY OF HGT AND PROPOSED MECHANISMS OF TRANSFER

Long lists of HGT candidates are usually generated following the release and bioinformatics analysis of filamentous fungal pathogen genomes. Such lists should be treated with caution because when compared to intensively studied groups, such as flowering plants or vertebrates, filamentous fungi display very high genome sequence diversity, even between individuals of the same species (e.g., between isolates of the rice blast fungus *Magnaporthe oryzae*) (Yoshida et al., 2009). Thus, a high level of genomic diversity coupled with limited taxonomic sampling means that a priori evidence of HGT between species arises frequently.

Nonetheless, the weight of evidence for HGT in fungi is becoming abundant; therefore, we are confident that HGT has had a profound impact on the evolution of fungi in general and pathogens in particular (Richards et al., 2011). HGT may occur very frequently and if so, many combinations of genes could have been tested in nature during evolution. Hence, the critical factor in the acquisition of horizontally transferred genes is not likely to be mechanics of DNA transfer or even integration into genome but rather the dynamic selection of recombinants as ecological niches change over time. Agricultural plant pathogens have certainly experienced profound and rapid niche changes, particularly since the domestication of crops in the last 10,000 years. The last century has seen the widespread sowing of monocultural crops, the widespread usage of fungicides (particularly in Europe), and the adoption of practices such as limited or minimum tillage in which the soil is not cultivated before sowing the crop, and stubble (trash) from previous crops is not removed. These scenarios all contribute to new niches that could have selected for HGT-adapted pathogens.

There are many examples of HGT between prokaryotes (Doolittle, 1999) and the first definitive evidence for HGT from bacteria to fungi was provided by Hall et al. (2005), who reported transfer of 1 bacterial gene into *Ashbya gossypii* and 10 into *Saccharomyces cerevisiae*. These proposed transfers would have enabled the fungi to broaden their metabolic range. Since then, many cases of HGT from prokaryotes or oomycetes to fungi have been described. For instance, Marcet-Houben and Gabaldon (2009) analyzed 60 fully sequenced fungal genomes and reported phylogenetic evidence that >700 prokaryotic genes had been transferred by HGT to ascomycetes, with particularly high rates in the Pezizomycotina lineage (the filamentous ascomycetes). Transferred genes included those that facilitate the adaptation of the fungi to different nutritional substrates. Highlighting the widespread occurrence of HGT, Richards et al. (2011) reanalyzed 340 proposed examples of HGT into and between fungi using updated fungal taxon sampling and recalculating phylogenetic trees. In 323 cases, tree topologies were consistent with the original data and the proposed HGT event, with most of these cases involving members of the Pezizomycotina.

No cases of HGT have been experimentally verified such that the mechanism of transfer is unambiguously elucidated. However, bacterial conjugative plasmids have been show to mobilize DNA transfer between bacteria and *S. cerevisiae* (Heinemann and Sprague, 1989). It is extremely unlikely that an HGT event occurring in nature would ever be observed. As such, we can only speculate on the basis of very incomplete data. As noted above, transfer between prokaryotes is very common and this is ascribed to the presence of conjugation tubes or pili and the lack of barrier for DNA to be acquired from the medium. Fungi, unlike some other eukaryotes, generally are not thought to engulf organisms such as bacteria as a means of acquiring DNA. An exception to this is the zygomycete *Rhizopus microsporus*, which harbors an endosymbiont bacterium, *Burkholderia* spp. (Partida-Martinez and Hertweck, 2005). Eukaryotes such as plants and fungi have complex cell walls through which DNA cannot easily penetrate, and all eukaryotes have a nuclear membrane. Furthermore, sex cells are separated from autosomal cells in many eukaryotes, and sex cells need to include the gene in order for an HGT event to be heritable.

A likely site for gene transfer between fungi is anastomosis tubes that form between hyphae and especially between conidia, for example, in the bread mold *Neurospora crassa* (Roca et al., 2005). Furthermore, hyphae from different species often contact each other, although there is often a robust vegetative incompatibility system governed by several heterokaryon incompatibility (*het*) loci (Glass and Kaneko, 2003). However, the incompatibility reaction and cell death may not ensue quickly enough to arrest all anastomoses. Only one transfer event in millions of contacts between species is needed for DNA exchange to occur and to be inherited via conventional means (mitotically or meiotically) thereafter.

Fungi that exchange DNA must occupy similar environmental niches. Many pathogens, particularly those that can infect animals, normally subsist as soil saprophytes on decaying vegetable or animal matter. Many necrotrophic pathogens cause premature senescence of the leaf that they infect, and this leaf can be subsequently colonized by secondary saprophytes. Thus, gene transfers occurring on dead and dying leaves or stubble could account for the reason why HGT has been noted between and within saprophytic and necrotrophic fungi. In contrast, there are few, if any, reports of HGT into or between biotrophs. Infection by biotrophs requires a healthy plant and usually involves the development of complex infection structures, such as appressoria, and intricate feeding structures, such as haustoria; therefore, intimate communication between such organisms is less likely than between necrotrophs.

HGT OF SINGLE GENES

As noted above, genes that are proposed to have undergone HGT include those that allow an organism to adapt to a changed environment. Thus, it is not surprising that some fungal effectors, a class of molecules that plays a key role in plant–pathogen interactions, appear to have been transferred between fungi via HGT. Avirulence effector genes are complementary to resistance genes in host plants, and interaction between key pathogen and plant molecules determines whether disease will occur (Jones and Dangl, 2006). Avirulence effector genes are often highly polymorphic within species and generally do not have obvious homologs in other species. However, a homolog of an avirulence gene, *Avr4*, of *Cladosporium fulvum*, which causes tomato wilt, is present in another dothideomycete, *Mycosphaerella fijiensis*, a pathogen of banana. The *M. fijiensis* gene is able to complement *Avr4* activity in *C. fulvum* (Stergiopoulos et al., 2010). The sequencing of additional fungal genomes will undoubtedly uncover homologs of other avirulence effector genes in different species and also indicate whether such avirulence effectors have arisen by HGT.

A molecule within the effector class of proteinaceous host-specific toxins (now called necrotrophic effectors) provides the best example of HGT of a single gene. *ToxA*, which encodes a small, secreted proteinaceous toxin. Many of the techniques described above used to support arguments for HGT have been validated in the case of *ToxA*. This gene was first discovered in the tan spot pathogen *Pyrenophora tritici-repentis* of wheat (Ciuffetti et al., 1997), where it is a major determinant of virulence and is present in about 80% of worldwide isolates. ToxA interacts with molecules in wheat cultivars that carry the complementary susceptibility *Tsn1* gene to produce a necrotic cell death response that promotes asexual sporulation of the pathogen. *Tsn1* encodes a protein with similar domains to disease resistance genes even though it acts as a susceptibility factor in the host (Faris et al., 2010).

Tan spot is now the most economically important disease of wheat in Australia (Murray and Brennan, 2009). The rise in disease incidence coincided with the adoption of limited tillage farming practices to preserve top soil in the 1970s. Such practices, where stubble is not removed or cultivated into the soil, promote the persistence of diseases such as tan spot because sexual reproduction occurs on stubble and the resulting airborne inocula of sexual spores are released. Tan spot was first noted in Australia only in the 1950s and in other countries in 1941. The pathogen had been described in the 1920s but only as the cause of a minor and occasional disease of wheat and related grasses (Friesen et al., 2006a).

ToxA was identified in a related member of the dothideomycete family Pleosporales, *Stagonospora* (*Septoria*, *Phaeosphaeria*) *nodorum*, when the genome of this fungus was sequenced (Friesen et al., 2006b; Hane et al., 2007). This fungus is a wheat pathogen that causes necrotic leaf, stem, and glume blotches. It was comprehensively described in the 19th century and ascribed a significant role in wheat disease losses (Shaw et al., 2008). Although both fungi are Pleosporales, the genera are quite distinct and are believed to be separated by 200 million years (Hane et al., 2007). In the sequenced isolate of *S. nodorum*, the ToxA homolog had only five conservative changes in amino acids (Friesen et al., 2006b; Hane et al., 2007). However, homologs could not be sought in other Pleosporales species initially because genomes had not been sequenced. However, PCR experiments with conserved primer sequences failed to detect *ToxA*. The absence of *ToxA* was confirmed when the genomes of other species became available, for example, *Leptosphaeria maculans* (Rouxel et al., 2011) and *Mycosphaerella graminicola* (Goodwin et al., 2011). These results fulfill a requirement for the hypothesis HGT described above—that closely related species lack the gene.

Convergent evolution of *ToxA* between *P. tritici-repentis* and *S. nodorum* was considered but discounted. Initially, an 11-kb region surrounding *ToxA* was analyzed in the two species and this revealed 80–100% sequence similarity despite the presence of only one other gene and about 7 kb of noncoding DNA (Friesen et al., 2006b). That the other gene was a putative transposase added to the weight of evidence for HGT. The direction of transfer was suggested by the finding that *ToxA* in *S. nodorum* was highly polymorphic between isolates, while the gene was monomorphic in *P. tritici–repentis*.

The previously mentioned data indicated that a DNA fragment of at least 11 kb had been transferred from *S. nodorum* to *P. tritici-repentis*. The recent emergence of tan spot as an important disease of wheat around 1941 suggests a very recent timescale for this event. Subsequent analyses are consistent with this hypothesis and have indeed suggested that the gene was acquired by *S. nodorum* within the last 300 years (Stukenbrock and McDonald, 2007). *ToxA* sequences are absent from about 60% of *S. nodorum* isolates surveyed in a worldwide collection. In isolates with *ToxA*, the sequence variability is inconsistent with the biogeography of the species as defined by neutrally evolving genes. Among sampled populations, the center of origin appears to be the Fertile Crescent, whereas maximum sequence diversity of *ToxA* is found in South African isolates. It is tempting to speculate that *ToxA* was acquired when wheat was introduced to South Africa by Dutch colonizers in the 17th century (Stukenbrock and McDonald, 2009).

The primary source of *ToxA* has not been determined. The timing of the transfer does not appear to be related to evolution of the resistance gene complement of wheat cultivars, as cultivars carrying *Tsn1* have been sown at least since the 19th century (Oliver et al., 2008). Coinfection of wheat leaves is common (Bhathal et al., 2003), which would provide appropriate conditions for transfer. We favor the idea that limited tillage farming practices greatly increased the acreage of coinfected wheat stubble and led in subsequent seasons to selection of an aggressive strain of *P. tritici-repentis* with the acquired gene. The mechanism of transfer has not been elucidated, but conidial anastomosis tubes have been observed between *S. nodorum* isolates in the laboratory (K. Rybak and R. Oliver, unpublished data).

The *S. nodorum*–wheat interaction is controlled by several host-specific toxins. The finding that *ToxA* has undergone HGT has focused attention on other effector toxins in these two Pleosporales species (*S. nodorum* and *P. tritici-repentis*), including two toxin effectors (*Tox3* and *Tox1*) that have been cloned (Liu et al., 2009) (T. Friesen, pers. comm.) and are unique to *S. nodorum*. Both genes are highly polymorphic and are missing in many isolates; however, a species with a homolog showing some degree of monomorphism has not been described, so HGT cannot yet be inferred. ToxB, another toxin that plays a major role in the virulence of *P. tritici-repentis*, has been characterized (Andrie et al., 2007). The gene encoding ToxB is present in some isolates of *P. tritici-repentis*, sometimes in several copies, and the degree of virulence is proportional to *ToxB* copy number. Interestingly, a slightly modified version of this gene is found in *Pyrenophora bromi*, a pathogen of brome grass (Andrie and Ciuffetti, 2011). Explanations for its presence in *P. bromi* are HGT or, alternatively, a situation in which the gene was amplified in some isolates of *P. tritici-repentis* and lost in others during or after speciation from *P. bromi*; however, there is not yet enough evidence to support one hypothesis over the other.

An association between the recent emergence of fungal plant diseases (within the last 100 years or less) that use necrotrophic effectors and HGT has been proposed (Oliver and Solomon, 2010). Necrotrophic fungal pathogens are mainly in the Pleosporales and include genera such as *Leptosphaeria*, *Cochliobolus*, *Alternaria*, *Pyrenophora*, and *Stagonospora*. Sequencing of the genomes of these species has been completed and when the genomes are annotated and available for analyses, it will be interesting to see how many other examples similar to the *ToxA* transfer are indicated.

Unlike the dothideomycete pathogens of plants, there are as yet few reports of HGT of single genes between fungal pathogens of animals. Infections of animals are often opportunistic and unlike plant pathogens, there is generally little host specificity or coadaptation with their hosts. A review of comparative genomic studies of human pathogenic fungi has concluded that HGT has not played an important role in evolution of fungal virulence on humans (Moran et al., 2011). Intriguingly, the presence of polymorphic effectors in isolates of any fungal pathogen of animals has not been reported, although there are few analyses of multiple isolates. There is evidence for acquisition of two virulence effectors from oomycetes via HGT by the chytrid fungus *Batrachochytrium dendrobatidis*, which is responsible for amphibian declines worldwide (Sun et al., 2011). These are a crinkler protein, which is involved in virulence in oomycetes, and a serine peptidase, which has a potential role in degrading host antimicrobial peptides.

HGT OF CLUSTERS OF GENES ENCODING ENZYMES FOR BIOSYNTHESIS OF SECONDARY METABOLITE TOXINS

Secondary metabolite toxins have been often proposed to play a key role in virulence in fungi; therefore, one should consider the evolutionary origin of their biosynthetic genes. Such genes are generally clustered, which allows coregulation and rapid pathway flux. Clustering also facilitates the gain or loss of metabolic pathways, as with the galactose utilization pathway (Slot and Rokas, 2010).

There are many reports of transfer of secondary metabolite gene clusters among bacteria (Boucher et al., 2003). Until recently, such gene clusters in fungi were thought to have arisen only by transfer from bacteria. However, with the proliferation of genome sequences, there are now many examples of gene clusters with extremely similar sequences and microsynteny that are discontinuously distributed in phylogenetically diverse fungi, with no obvious bacterial homologs. These gene clusters include many that encode non-host-specific toxins, such as sterigmatocystin. This molecule has a biosynthetic gene cluster of 23 genes and is present in a large number of ascomycetes (Rank et al., 2011). A detailed study of this gene cluster in *Aspergillus* (a eurotiomycete) and *Podospora* (a sordariomycete) led to a proposal that HGT accounts for the distribution of this gene cluster in the ascomycetes (Slot and Rokas, 2011).

Another example of HGT of secondary metabolite gene clusters in fungi is that of epipolythiodioxopiperazine (ETP) gene clusters, which synthesize sirodesmin in *L. maculans* (a dothideomycete) and gliotoxin in both *Aspergillus fumigatus* (a eurotiomycete) and *Trichoderma virens* (a sordariomycete) (Patron et al., 2007). Putative ETP clusters were identified in 14 ascomycete taxa but were absent in genomes of numerous other ascomycetes examined. Phylogenetic analyses of six of the ETP cluster genes in these taxa suggested that the clusters have a single origin and have been inherited relatively intact rather than assembling independently in the different ascomycete lineages. The discontinuous distribution in ascomycete lineages of these clusters is suggestive of multiple instances of independent cluster loss and HGT of gene clusters between lineages. A similar situation has been inferred for the *Ace1* gene cluster, which confers avirulence in the rice blast fungus *M. oryzae*, but whose product is as yet unknown (Khaldi et al., 2008). Like the ETP cluster, the *Ace1* cluster is discontinuously distributed in a range of ascomycetes. A significant number of these secondary metabolite gene clusters are close to telomeres and/or transposable elements, chromosomal locations that may facilitate their transfer by HGT. In the cases described above (sterigmatocystin, ETP, and *Ace1* clusters), the hypothesis that HGT is responsible for the distribution of the gene clusters has been supported by more detailed phylogenetic analyses (Richards et al., 2011).

Until recently, fungal secondary metabolite toxins such as the ones described above were thought to be very important in fungal virulence in plant or animal hosts. Now that pathways and gene clusters have been unequivocally identified, biosynthetic genes can be mutated to see if the resultant mutant causes a similar degree of disease to that caused by wild type. In many cases, toxins appear to be involved in disease but are not crucial for virulence on plant or animal hosts. For instance, the ETP sirodesmin contributes only partially to the virulence of *L. maculans* on stems of canola (Elliott et al., 2007). In addition, the related ETP gliotoxin is a virulence factor for *A. fumigatus* in mice that have retained neutrophil function after immunosuppression by corticosteroids alone but is not a virulence factor in neutropenic mice (Kwon-Chung and Sugui, 2009). Thus, the role of gliotoxin in virulence can depend on the immune status of the mouse.

An increasingly popular hypothesis among fungal biologists is that these secondary metabolite toxins enable the fungi that produce them to survive in their environmental niches. Fungi often live in soil or dead and decaying plant and animal matter. Therefore, they need to protect themselves and their food sources. Secondary metabolite toxins may be important for virulence toward a diverse range of organisms such as insects, amoebas, nematodes, mites, and earthworms in their environmental niches but not necessarily crucial for virulence toward plant and animal hosts that the fungi infect, often opportunistically (Mylonakis et al., 2007).

In some cases, the secondary metabolite toxin is not made by the fungus but rather by a symbiont. For instance, *R. microsporus* hosts the bacterium *Burkholderia rhizoxinia* in its cytoplasm (Partida-Martinez and Hertweck, 2005). *B. rhizoxinia* synthesizes and secretes the mitosis-inhibiting macrocyclic polyketide rhizoxin, which is a virulence factor for *R. microsporus* during infection of wheat. Thus, the presence of a gene cluster in the symbiont achieves the same outcome as a situation in which the cluster is within the fungus. This situation could represent a prelude to HGT of the gene cluster from bacteria to fungus.

Unlike the examples described above, some gene clusters encode enzymes involved in the biosynthesis of secondary metabolites that are host specific. Several of these are located on dispensable chromosomes that are inherited in a non-Mendelian fashion. Transfers of these chromosomes and implications for virulence and host specificity are discussed in the next section.

HORIZONTAL CHROMOSOME TRANSFER

Dispensable (supernumerary) chromosomes can be transferred within and between organisms. A recent example of this is the transfer of lineage-specific chromosomes between *Fusarium oxysporum* (Ma et al., 2010). This fungus has several *formae speciales*, which are taxonomic groupings of organisms that are extremely similar morphologically but are adapted to a specific host. The lineage-specific chromosomes appear to have a different evolutionary origin than those inherited in a Mendelian fashion, and their presence is associated with virulence on different plants. Furthermore, cocultivation of pathogenic and nonpathogenic *formae speciales* has been shown to result in new pathotypes. This example is discussed in detail in another chapter of this book (Ma et al., 2012). Another example of a change in host range is the imperfect fungus *Alternaria alternata*, which contains secondary metabolite gene clusters in dispensable chromosomes. This fungus has several variant strains that produce host-specific toxins of diverse chemical structure; that is, the production of particular toxins leads to disease on different hosts (Hatta et al., 2002). For instance, the tomato pathotype has a 1.0-Mb chromosome with genes encoding enzymes for the biosynthesis of the sphinganine-analogue toxin AAL, while the strawberry pathotype has a 1.05-Mb chromosome with genes encoding enzymes for the biosynthesis of AF toxin. Cocultivation of protoplasts of these two pathotypes gave rise to a strain that contained both chromosomes, produced both toxins, and was pathogenic on both strawberries and tomatoes (Akagi et al., 2009). Although these two examples of chromosome transfer described above were artificially generated in the laboratory, one could envision that this process might occur in nature to extend host ranges of pathogens.

CONCLUSIONS

HGT appears to be a common occurrence into and between fungi, particularly members of the Pezizomycotina. Organisms colocated in similar niches may have the opportunity to exchange genetic material, but mechanisms for the transfers are yet to be discovered. Fungi that live on dead or decaying material on stubble or in soil are well represented in taxa reported to undergo HGT. As genome sequences of more fungi are analyzed and compared, many more such instances will be uncovered, some of which will affect virulence on a particular host. It will be particularly interesting to see if such events are common in obligate biotrophic pathogenic fungi whose lifestyle generally precludes the presence of other microorganisms.

ACKNOWLEDGMENT

We thank the Australian Grains Research and Development Corporation for funds that supported our research.

REFERENCES

Akagi Y, Akamatsu H, Otani H, Kodama M. 2009. Horizontal chromosome transfer, a mechanism for the evolution and differentiation of a plant-pathogenic fungus. Eukaryotic Cell 8: 1732–1738.

Andrie RM, Ciuffetti LM. 2011. *Pyrenophora bromi*, causal agent of brownspot of bromegrass, expresses a gene encoding a protein with homology and similar activity to Ptr ToxB, a host-selective toxin of wheat. Molecular Plant-Microbe Interactions 24: 359–367.

Andrie RM, Pandelova I, Ciuffetti LM. 2007. A combination of phenotypic and genotypic characterization strengthens *Pyrenophora tritici-repentis* race identification. Phytopathology 97: 694–701.

Bhathal JS, Loughman R, Speijers J. 2003. Yield reduction in wheat in relation to leaf disease from yellow (tan) spot and *Septoria nodorum* blotch. European Journal of Plant Pathology 109: 435–443.

Boucher Y, Douady CJ, Papke RT, Walsh DA, Boudreau ME, Nesbo CL, Case RJ, Doolittle WF. 2003. Lateral gene transfer and the origins of prokaryotic groups. Annual Review of Genetics 37: 283–328.

Ciuffetti LM, Tuori RP, Gaventa JM. 1997. A single gene encodes a selective toxin causal to the development of tan spot of wheat. The Plant Cell 9: 135–144.

Doolittle WF. 1999. Lateral genomics. Trends in Cell Biology 9: M5–M8.

Elliott CE, Gardiner DM, Thomas G, Cozijnsen A, Van de Wouw A, Howlett BJ. 2007. Production of the toxin sirodesmin PL by *Leptosphaeria maculans* during infection of

Brassica napus. Molecular Plant Pathology 8: 791–802.

Faris JD, Zhang Z, Lu H, Lu S, Reddy L, Cloutier S, Fellers JP, Meinhardt SW, Rasmussen JB, Xu SS, Oliver RP, Simons KJ, Friesen TL. 2010. A unique wheat disease resistance-like gene governs effector-triggered susceptibility to necrotrophic pathogens. Proceedings of the National Academy of Sciences of the United States of America 107: 13544–13549.

Friesen TL, Faris JD, Lai Z, Steffenson BJ. 2006a. Identification and chromosomal location of major genes for resistance to *Pyrenophora teres* in a doubled-haploid barley population. Genome 49: 855–859.

Friesen TL, Stukenbrock EH, Liu Z, Meinhardt S, Ling H, Faris JD, Rasmussen JB, Solomon PS, Mcdonald BA, Oliver RP. 2006b. Emergence of a new disease as a result of interspecific virulence gene transfer. Nature Genetics 38: 953–956.

Glass NL, Kaneko I. 2003. Fatal attraction: Nonself recognition and heterokaryon incompatibility in filamentous fungi. Eukaryotic Cell 2: 1–8.

Goodwin SB, Ben M'Barek S, Dhillon B, Wittenberg AH, Crane CF, Hane JK, Foster AJ, Van Der Lee TA, Grimwood J, Aerts A, Antoniw J, Bailey A, Bluhm B, Bowler J, Bristow J, Van Der Burgt A, Canto-Canche B, Churchill AC, Conde-Ferraez L, Cools HJ, Coutinho PM, Csukai M, Dehal P, De Wit P, Donzelli B, Van De Geest HC, Van Ham RC, Hammond-Kosack KE, Henrissat B, Kilian A, Kobayashi AK, Koopmann E, Kourmpetis Y, Kuzniar A, Lindquist E, Lombard V, Maliepaard C, Martins N, Mehrabi R, Nap JP, Ponomarenko A, Rudd JJ, Salamov A, Schmutz J, Schouten HJ, Shapiro H, Stergiopoulos I, Torriani SF, Tu H, De Vries RP, Waalwijk C, Ware SB, Wiebenga A, Zwiers LH, Oliver RP, Grigoriev IV, Kema GH. 2011. Finished genome of the fungal wheat pathogen *Mycosphaerella graminicola* reveals dispensome structure, chromosome plasticity, and stealth pathogenesis. PLoS Genetics 7: e1002070.

Hall C, Brachat S, Dietrich FS. 2005. Contribution of horizontal gene transfer to the evolution of *Saccharomyces cerevisiae*. Eukaryotic Cell 4: 1102–1115.

Hane JK, Lowe RG, Solomon PS, Tan KC, Schoch CL, Spatafora JW, Crous PW, Kodira C, Birren BW, Galagan JE, Torriani SF, Mcdonald BA, Oliver RP. 2007. Dothideomycete plant interactions illuminated by genome sequencing and EST analysis of the wheat pathogen *Stagonospora nodorum*. The Plant Cell 19: 3347–3368.

Hatta R, Ito K, Hosaki Y, Tanaka T, Tanaka A, Yamamoto M, Akimitsu K, Tsuge T. 2002. A conditionally dispensable chromosome controls host-specific pathogenicity in the fungal plant pathogen *Alternaria alternata*. Genetics 161: 59–70.

Heinemann JA, Sprague GF, JR. 1989. Bacterial conjugative plasmids mobilize DNA transfer between bacteria and yeast. Nature 340: 205–209.

Jones JD, Dangl JL. 2006. The plant immune system. Nature 444: 323–329.

Khaldi N, Collemare J, Lebrun MH, Wolfe KH. 2008. Evidence for horizontal transfer of a secondary metabolite gene cluster between fungi. Genome Biology 9: R18.

Kwon-Chung KJ, Sugui JA. 2009. What do we know about the role of gliotoxin in the pathobiology of *Aspergillus fumigatus*? Medical Mycology 47(Suppl. 1): S97–S103.

Liu Z, Faris JD, Oliver RP, Tan KC, Solomon PS, Mcdonald MC, Mcdonald BA, Nunez A, Lu S, Rasmussen JB, Friesen TL. 2009. *SnTox3* acts in effector triggered susceptibility to induce disease on wheat carrying the *Snn3* gene. PLoS Pathogens 5: e1000581.

Ma L, Kistler HC, Rep M. 2012. Evolution of plant pathogenicity in *Fusarium* species.

Ma LJ, Van Der Does HC, Borkovich KA, Coleman JJ, Daboussi MJ, Di Pietro A, Dufresne M, Freitag M, Grabherr M, Henrissat B, Houterman PM, Kang S, Shim WB, Woloshuk C, Xie X, Xu JR, Antoniw J, Baker SE, Bluhm BH, Breakspear A, Brown DW, Butchko RA, Chapman S, Coulson R, Coutinho PM, Danchin EG, Diener A, Gale LR, Gardiner DM, Goff S, Hammond-Kosack KE, Hilburn K, Hua-Van A, Jonkers W, Kazan K, Kodira CD, Koehrsen M, Kumar L, Lee YH, Li L, manners JM, Miranda-Saavedra D, Mukherjee M, Park G, Park J, Park SY, Proctor RH, Regev A, Ruiz-Roldan MC, Sain D, Sakthikumar S, Sykes S, Schwartz

DC, Turgeon BG, Wapinski I, Yoder O, Young S, Zeng Q, Zhou S, Galagan J, Cuomo CA, Kistler HC, Rep M. 2010. Comparative genomics reveals mobile pathogenicity chromosomes in *Fusarium*. Nature 464: 367–373.

Marcet-Houben M, Gabaldon T. 2009. Acquisition of prokaryotic genes by fungal genomes. Trends in Genetics 26: 5–8.

Mehrabi R, Bahkali AH, Abd-Elsalam KA, Moslem M, Ben M'Barek S, Gohari AM, Jashni MK, Stergiopoulos I, Kema GH, De Wit PJ. 2011. Horizontal gene and chromosome transfer in plant pathogenic fungi affecting host range. FEMS Microbiology Reviews 35: 542–554.

Moran GP, Coleman DC, Sullivan DJ. 2011. Comparative genomics and the evolution of pathogenicity in human pathogenic fungi. Eukaryotic Cell 10: 34–42.

Murray GM, Brennan JP. 2009. Estimating disease losses to the Australian wheat industry. Australasian Plant Pathology 38: 558–570.

Mylonakis E, Casadevall A, Ausubel FM. 2007. Exploiting amoeboid and non-vertebrate animal model systems to study the virulence of human pathogenic fungi. PLoS Pathogens 3: e101.

Oliver RP, Solomon PS. 2010. New developments in pathogenicity and virulence of necrotrophs. Current Opinion in Plant Biology 13: 415–419.

Oliver RP, Lord A, Rybak K, Faris JD, Solomon PS. 2008. Emergence of tan spot disease caused by toxigenic *Pyrenophora tritici-repentis* in Australia is not associated with increased deployment of toxin-sensitive cultivars. Phytopathology 98: 488–491.

Partida-Martinez LP, Hertweck C. 2005. Pathogenic fungus harbours endosymbiotic bacteria for toxin production. Nature 437: 884–888.

Patron NJ, Waller RF, Cozijnsen AJ, Straney DC, Gardiner DM, Nierman WC, Howlett BJ. 2007. Origin and distribution of epipolythiodioxopiperazine (ETP) gene clusters in filamentous ascomycetes. BMC Evolutionary Biology 7: 174.

Rank C, Nielsen KF, Larsen TO, Varga J, Samson RA, Frisvad JC. 2011. Distribution of sterigmatocystin in filamentous fungi. Fungal Biology Reviews 115: 406–420.

Richards TA, Dacks JB, Jenkinson JM, Thornton CR, Talbot NJ. 2006. Evolution of filamentous plant pathogens: Gene exchange across eukaryotic kingdoms. Current Biology 16: 1857–1864.

Richards TA, Leonard G, Soanes DM, TALBOT NJ. 2011. Gene transfer into the fungi. Fungal Biology Reviews 25: 98–110.

Roca MG, Arlt J, Jeffree CE, Read ND. 2005. Cell biology of conidial anastomosis tubes in *Neurospora crassa*. Eukaryotic Cell 4: 911–919.

Rouxel T, Grandaubert J, Hane JK, Hoede C, Van De Wouw AP, Couloux A, Dominguez V, Anthouard V, Bally P, Bourras S, Cozijnsen AJ, Ciuffetti LM, Degrave A, Dilmaghani A, Duret L, Fudal I, Goodwin SB, Gout L, Glaser N, Linglin J, Kema GH, Lapalu N, Lawrence CB, May K, Meyer M, Ollivier B, Poulain J, Schoch CL, Simon A, Spatafora JW, Stachowiak A, Turgeon BG, Tyler BM, Vincent D, Weissenbach J, Amselem J, Quesneville H, Oliver RP, Wincker P, Balesdent MH, Howlett BJ. 2011. Effector diversification within compartments of the *Leptosphaeria maculans* genome affected by repeat-induced point mutations. Nature Communications 2: n202.

Schoch CL, Sung GH, Lopez-Giraldez F, Townsend JP, Miadlikowska J, Hofstetter V, Robbertse B, Matheny PB, Kauff F, Wang Z, Gueidan C, Andrie RM, Trippe K, Ciufetti LM, Wynns A, Fraker E, Hodkinson BP, Bonito G, Groenewald JZ, Arzanlou M, De Hoog GS, Crous PW, Hewitt D, Pfister DH, Peterson K, Gryzenhout M, Wingfield MJ, Aptroot A, Suh SO, Blackwell M, Hillis DM, Griffith GW, Castlebury LA, Rossman AY, Lumbsch HT, Lucking R, Budel B, Rauhut A, Diederich P, Ertz D, Geiser DM, Hosaka K, Inderbitzin P, Kohlmeyer J, Volkmann-Kohlmeyer B, Mostert L, O'Donnell K, Sipman H, Rogers JD, Shoemaker RA, Sugiyama J, Summerbell RC, Untereiner W, Johnston PR, Stenroos S, Zuccaro A, Dyer PS, Crittenden PD, Cole MS, Hansen K, Trappe JM, Yahr R, Lutzoni F, Spatafora JW. 2009. The Ascomycota tree of life: A phylum-wide phylogeny clarifies the origin and evolution of fundamental reproductive and ecological traits. Systematic Biology 58: 224–239.

Sexton AC, Howlett BJ. 2006. Parallels in fungal pathogenesis on plant and animal hosts. Eukaryotic Cell 5: 1941–1949.

Shaw MW, Bearchell SJ, Fitt BD, Fraaije BA. 2008. Long-term relationships between environment and abundance in wheat of *Phaeosphaeria nodorum* and *Mycosphaerella graminicola*. The New Phytologist 177: 229–238.

Slot JC, Rokas A. 2010. Multiple GAL pathway gene clusters evolved independently and by different mechanisms in fungi. Proceedings of the National Academy of Sciences of the United States of America 107: 10136–10141.

Slot JC, Rokas A. 2011. Horizontal transfer of a large and highly toxic secondary metabolic gene cluster between fungi. Current Biology 21: 134–139.

Stergiopoulos I, Van Den Burg HA, Okmen B, Beenen HG, Van Liere S, Kema GH, De Wit PJ. 2010. Tomato *Cf* resistance proteins mediate recognition of cognate homologous effectors from fungi pathogenic on dicots and monocots. Proceedings of the National Academy of Sciences of the United States of America 107: 7610–7615.

Stukenbrock EH, Mcdonald BA. 2007. Geographical variation and positive diversifying selection in the host-specific toxin *SnToxA*. Molecular Plant Pathology 8: 321–332.

Stukenbrock EH, Mcdonald BA. 2009. Population genetics of fungal and oomycete effectors involved in gene-for-gene interactions. Molecular Plant-Microbe Interactions 22: 371–380.

Sun G, Yang Z, Kosch T, Summers K, Huang J. 2011. Evidence for acquisition of virulence effectors in pathogenic chytrids. BMC Evolutionary Biology 11: 195.

Yoshida K, Saitoh H, Fujisawa S, Kanzaki H, Matsumura H, Tosa Y, Chuma I, Takano Y, Win J, Kamoun S, Terauchi R. 2009. Association genetics reveals three novel avirulence genes from the rice blast fungal pathogen *Magnaporthe oryzae*. The Plant Cell 21: 1573–1591.

CHAPTER 26

EVOLUTION OF PLANT PATHOGENICITY IN *FUSARIUM* SPECIES

LI-JUN MA, H. CORBY KISTLER, and MARTIJN REP

INTRODUCTION

Fungal pathogenicity may simply be defined as the capacity of a fungus to produce disease through interaction with a host. But the process itself is not simple. Infection requires the pathogen to sense the presence of the host and to develop a specialized infection structure at the host–pathogen interface that facilitates penetration through physical barriers. Once the interaction starts, a plant can usually quickly recognize microbe-derived nonself components called pathogen- (or microbe-) associated molecule patterns (PAMPs/MAMPs) and activate its immune response to prevent further intrusion. To prevent or overcome this host response, fungi, like other pathogens, secrete proteinaceous and/or small molecule effectors that alter the plant immune response, allowing growth within the host. Selection on resistance/susceptibility genes in the host and effector genes in pathogens has led to antagonistic host–pathogen coevolution.

Interestingly, fungus–plant interactions (not necessarily parasitic interactions) probably arose as early as the first land plant appeared on earth, around 450 million years ago (Kenrick and Crane, 1997), by which time multiple fungal lineages had already emerged (Berbee et al., 2000; Taylor and Berbee, 2006). Apparently, the establishment of such interactions must have successfully occurred multiple times from diverse saprobe ancestors since in clades with saprobes, endophytes, and pathogenic strains, the saprophytic state appears to be ancestral (Spatafora et al., 2007).

The genus *Fusarium* is mostly known for causing plant diseases in crops, trees, and ornamentals. Among the most well-known pathogens in the genus are *Fusarium graminearum* (*Fg*), a toxin-producing pathogen of cereals (Goswami and Kistler, 2004), and *Fusarium oxysporum* (*Fo*), a species complex that harbors vascular wilt and root rot pathogens for many vegetable crops and flowers (Michielse and Rep, 2009; Recorbet et al., 2003). Despite their negative impact on human food and health, it is important to acknowledge that *Fusarium* species are also often found as apparently harmless saprobes and endophytes, sometimes even as suppressors of plant diseases (Fravel et al., 2003).

Much of the molecular research on *Fusarium* species over the last decade has

Evolution of Virulence in Eukaryotic Microbes, First Edition. Edited by L. David Sibley, Barbara J. Howlett, and Joseph Heitman.

been dedicated to uncovering the molecular basis of pathogenicity toward different plant hosts. In practice, this has meant identifying genes required for pathogenicity and gaining insight into the molecular mechanisms through which the encoded proteins act to promote virulence. Roughly, these genes can be divided into two broad categories: (i) genes widely conserved in fungi (or a large portion of fungi such as the ascomycetes or pezizomycetes) and (ii) genes that seem to be associated with specific host–pathogen interactions.

Investigation of the molecular functions of the genes in the first class has revealed basic cellular processes, such as signal transduction routes, cell wall modification, general stress adaptation, and metabolic capabilities, which are required for the ability to invade (living) plants. Although required for pathogenicity, these functions are generally also present in nonpathogens and do not, therefore, define pathogenicity as such. In contrast, genes in the second category may be called "determinants of pathogenicity," or simply virulence factors, and are often required for virulence toward particular host plants. It is this class of genes and their associated functions in *Fusarium* species that are the primary focus of this chapter.

The dichotomy above may appear straightforward, but it is not since (i) pathogenicity is the result of a complex interplay between general cellular functions and a combination of many different virulence factors; (ii) many present-day virulence factors may have been derived from genes that previously had another function and are therefore still related to genes in nonpathogens; (iii) many present-day "nonpathogens" may have relatively recent pathogenic ancestors and can therefore still harbor virulence factors; and (iv), virulence factors may be shared with nonpathogenic endophytes because the latter also invade living plants, requiring functions related to virulence such as suppression of plant immunity and detoxification of plant antifungal compounds. These complications help explain why a general "pathogenicity footprint" cannot be simply derived from genome sequences.

Clearly, then, the line is blurred between, on the one hand, basic conserved functions required for pathogenicity and, on the other, specific determinants of pathogenicity or virulence factors. Nevertheless, in the spectrum defined by these two extremes, it is genes closer to the latter that are of primary interest for this review because their study most particularly enhances the understanding of (i) host specificity, (ii) direct interactions between molecules of plants and fungi, and (iii) the evolution of pathogenicity. We will first review the virulence factors that have been identified in diverse *Fusarium* species and the methods leading to their identification. Then we will see how their phylogenetic distribution and genomic context provide clues for the evolution of pathogenicity within the genus *Fusarium*.

DETERMINANTS OF PATHOGENICITY WITHIN THE GENUS *FUSARIUM*

Toxin-Driven Virulence: *F. graminearum* and the *Gibberella fujikuroi* Species Complex

Gibberellins. A rice disease caused by *Fusarium fujikuroi* (*G. fujikuroi*) is one of the classic examples of involvement of fungal secondary metabolites in plant pathogenesis. Bakanae or "foolish seedling" disease of rice is characterized by hypertrophic seedling development resulting in overelongation of tillers, leaf chlorosis, and poor grain fill in panicles. Infected plants grow well above the canopy of healthy plants and are prone to lodging due to their greater height. Studies in Japan in the early 20th century (summarized in Boemke and Tudzynski, 2009) were the first to describe the "toxic" substances produced by the fungus. These diterpenoid substances were called gibberellins, after the fungal genus *Gibberella*. Years later, these same gibberellic acid compounds (GAs) were discovered in plants (Kato et al., 1962; Radley, 1956) and were shown to be endogenous phytohormones regulating growth and plant development. The biosynthetic enzymes, pathways, and regulatory machinery for the production of GAs in plants and *Fusarium* are considerably different, suggesting independent evolution of GA biosynthesis in these different kingdoms of life (Bomke and Tudzynski, 2009). In *F. fujikuroi*, unlike in plants, GA biosynthetic genes are clustered and encode seven coregulated transcripts specific for GA biosynthesis (Tudzynski and Holter, 1998).

Based on DNA sequence similarity and gene proximity, additional GA biosynthetic gene clusters have been identified in other members of the *G. fujikuroi* species complex, even in species not previously known to produce GAs such as *Fusarium sacchari* and *Fusarium subglutinans*, where gibberellin accumulation in culture has been confirmed (Malonek et al., 2005; Troncoso et al., 2010). The entire gene cluster also is present, although nonfunctional due to several mutations, in strains of *Fusarium proliferatum* that are pathogenic to maize and sorghum (Malonek et al., 2005). Interestingly, a nonpathogenic endophytic strain of *F. proliferatum* has a functional GA biosynthetic gene cluster and can produce gibberellins in culture, which has led to speculation that this phytohormone may play different roles in pathogenesis or symbiosis, depending on species or the spectrum of gibberellins produced (Tsavkelova et al., 2008).

Considerable variation exists within species for the gibberellin biosynthetic gene cluster and the spectrum of GAs produced. While over 100 GA derivatives are known to exist, only a few are thought to behave as phytohormones (Bomke and Tudzynski, 2009). Undoubtedly, there is much more to learn about the biological activity of the spectrum of GAs and their role in *Fusarium*–plant interactions.

Fumonisins. The fumonisins are a group of related polyketide-derived mycotoxins produced by *Fusarium verticillioides* (*Fv*) and related *Fusarium* species (Proctor et al., 2004, 2008; Rheeder et al., 2002). Fumonisins, especially fumonisin B1 (FB1), have been studied largely because of their deleterious effects on animals, which may ingest the toxins through their feed. These mycotoxins induce leukoencephalomalacia in horses and rabbits, pulmonary edema in swine, and can cause liver and kidney damage and cancer in rats (Bennett and Klich, 2003). Fumonisins are structurally similar to sphingosine and act by inhibiting ceramide synthase, resulting in reduced sphingolipid biosynthesis (Wang et al., 1991). Sphingolipids are essential components of cell membranes and participate in a variety of cell signaling pathways. Fumonisins trigger apoptosis in both plant and animal cells apparently by altering membrane composition and interfering with conserved signaling events (Gilchrist, 1997; Wang et al., 1996).

In *Fv*, genes for fumonisin biosynthesis are arranged as a large cluster (Proctor et al., 2003) consisting of 17 genes (Alexander et al., 2009). While most *Fv* strains synthesize a predictable spectrum of fumonisin derivatives, variation in fumonisin profiles exists for some strains (Desjardins et al., 1996), and there are also strains that naturally produce little or no fumonisins (Glenn et al., 2008). Comparison of DNA polymorphisms in the fumonisin gene cluster has revealed the genetic basis for these metabolite polymorphisms (Glenn et al., 2008; Proctor et al., 2006).

Fumonisin biosynthesis does not appear to determine the outcome of maize ear rot disease in the field (Desjardins et al., 2002), but it does have a significant impact on foliar necrosis and maize seedling blight (Glenn et al., 2008). It should be noted that *Fv* is a pervasive nonsymptomatic colonizer of maize (Bacon et al., 2008) and in its endophytic stage may have complex interactions with other fungi, including other pathogens (Lee et al., 2009), or even harmful insects (Schulthess et al., 2002). Future studies on *Fv* likely will focus on the endophytic phase of the fungus and the potential role fumonisins and other secondary metabolites may play in the composition of endophytic microbial communities (Pan and May, 2009).

Trichothecenes. Trichothecenes are sesquiterpenoid compounds produced by at least seven distantly related fungal genera; over 150 trichothecene derivatives have been identified (Rocha et al., 2005). Perhaps most notable are trichothecenes produced by members of the *Fg* species complex because of their economic impact. *Fg sensu stricto* and related species cause *Fusarium* head blight disease on wheat and barley as well as maize ear rot. They also contaminate grain with the mycotoxins deoxynivalenol (DON) or nivalenol (NIV) and their esterified derivatives (Goswami and Kistler, 2004). Trichothecenes cause numerous injurious acute symptoms in animals including contact dermatitis, emesis, diarrhea, hemorrhage, neural disorders, and spontaneous abortion. While DON is not considered a mutagen or carcinogen, it is a potent inhibitor of protein synthesis and alters membrane structure in both plants and animals. Chronic symptoms of DON toxicosis include decreased body mass and immunomodulation (Pestka, 2010).

In *Fg*, genes for trichothecene biosynthesis are arranged as a single large cluster of approximately 25–30 kb as well as two additional genes each on separate chromosomes (Rep and Kistler, 2010). The trichothecene gene cluster consists of 10 or 12 genes, depending on type of trichothecene produced, that are coregulated (Seong et al., 2009). Comparison of the trichothecene clusters from strains that produce different spectra of trichothecenes (different "chemotypes") revealed that polymorphisms track more readily with chemotype than with species of *Fusarium* producing the toxin (Ward et al., 2002). Indeed, gene polymorphisms within the trichothecene cluster have been confirmed as the biochemical basis for chemotype (Alexander et al., 2011; Lee et al., 2002).

Trichothecenes produced by *Fg* affect disease development in some plants but not in others. Mutant strains with deletions in the gene for trichodiene synthase (Tri5), the enzyme catalyzing the first step specific to trichothecene biosynthesis, do not make trichothecenes. Pathogenicity tests of *tri5* mutants unable to make DON show that they are still able to cause disease on wheat, barley, and maize but are greatly reduced in aggressiveness on wheat. Similar *tri5* mutants of NIV strains, unable to make NIV, are equally able to cause disease on wheat and barley but are reduced in their ability to cause maize ear rot (Jansen et al., 2005; Maier et al., 2006). In addition to being a crop pathogen, *Fg* also grows endophytically in maize and noncultivated grass species without causing symptoms (Goswami and Kistler, 2004; Pan and May, 2009). The role trichothecenes may play in endophytic associations (if any) or their potential effect on plant-associated microbial communities remains to be determined.

Overcoming Toxic Plant Barriers: *F. solani*, *F. sambucinum*, and *F. oxysporum*

Plants produce a variety of antifungal compounds, either constitutively or in response to microbial challenge (Morrissey and Osbourn, 1999). These compounds, often called phytoalexins or phytoanticipins, may represent a barrier to plant infection. Fungal pathogens tend to be more resistant to these compounds than are nonpathogens, and mechanisms that confer tolerance to plant antifungal compounds have been shown to be essential for maximum virulence of some *Fusarium* species to particular plant hosts.

Pea-infecting strains of *F. solani* (*Nectria haematococca* MPVI) (*Fs*) are highly tolerant to the pea phytoalexin pisatin. Early studies indicated that tolerance resulted, at least in part, from the ability to metabolically detoxify pisatin by a cytochrome P450 monooxygenase-catalyzed demethylation (Kistler and VanEtten, 1984; Weltring et al., 1988). Disruption of *PDA1*, the gene for this monooxygenase, resulted in quantitative reduction in both tolerance to pisatin and virulence to pea (Ciuffetti and VanEtten, 1996; Wasmann and VanEtten, 1996). *PDA1* was found to be part of a larger cluster of coexpressed genes on a supernumerary chromosome, each incrementally enhancing virulence toward pea (Han et al., 2001).

A second gene conferring both pisatin tolerance and virulence toward pea has recently been described for *Fs* f. sp. *pisi* on conserved chromosome 8 (Coleman et al., 2011). *NhABC1* is a gene for an ABC transporter and its deletion only slightly reduces pisatin tolerance and virulence. However, a Δ*NhABC1*, Δ*PDA1* double-mutant strain was more sensitive to pisatin and less pathogenic to pea than strains with mutations in either gene alone, suggesting that maximum virulence requires multiple mechanisms of antimicrobial resistance (Coleman et al., 2011). Interestingly, this gene is part of a *Fusarium*-specific lineage of ABC transporters that includes *GpABC1* from *Fusarium sambucinum* (*Gibberella pulicaris*), the pathogen causing dry rot of potato tubers (Fleissner et al., 2002). Deletion of *GpABC1* from this wound pathogen resulted in reduced virulence and greatly increased sensitivity to the potato phytoalexin rishitin. Unlike the wild-type *F. sambucinum*, mutant strains could not colonize potato tubers beyond the point of inoculation (Fleissner et al., 2002). It will be interesting to determine the potential role of *ABC1* orthologs in other *Fusarium* pathosystems such as in *Fo* (FOXG_13653.2), *Fg* (FGSG_04580.3), and *Fv* (FVEG_11089.3).

Fs also harbors strains that cause root rot disease of chickpea (*Cicer arietinum*). These strains have the ability to detoxify the chickpea phytoalexin maackiain, by oxidative mechanisms catalyzed by cytochrome P450 monooxygenases, and this oxidative detoxification is important for full virulence toward the host (Enkerli et al., 1998; Miao and VanEtten, 1992). Maackiain detoxification (MAK) genes also are located on supernumerary chromosomes (Covert et al., 1996).

Tomatoes constitutively produce a toxic steroidal glycoalkaloid, called α-tomatine, in several plant tissues. Many pathogens of tomato have been found to detoxify α-tomatine by the activity of a variety of glycosyl hydrolase enzymes. An extracellular β-glycosyl hydrolase called tomatinase from *F. oxysporum* f. sp. *lycopersici* (*Fol*) hydrolyzes the entire oligosaccharide portion of tomatine, rendering it less toxic (Roldan-Arjona et al., 1999). Placing the gene for this tomatinase (*TOM1*) under the control of a strong constitutive promoter increases the aggressiveness of *Fol* strains toward tomato, whereas deletion of *TOM1* delays the onset of disease symptoms for these mutants (Pareja-Jaime et al., 2008). Although *TOM1*-deletion mutants are more sensitive to inhibition by α-tomatine than wild type or strains constitutively expressing *TOM1*, several other genes for tomatine detoxification are known to exist in the same genetic background. Redundancy in the mechanisms for overcoming toxicity of plant metabolites may be common in fungal plant pathogens.

Small Secreted Proteins: *F. oxysporum*

Fo is the only *Fusarium* species known to exhibit gene-for-gene interactions with some of its hosts. The first indication of this was the discovery of monogenic resistance in host plants for 9 of more than 100 known *formae speciales* (pathogenic forms) of *Fo* (Michielse and Rep, 2009). Until now, definitive proof of gene-for-gene resistance has been obtained only for *Fol*, the tomato-infecting pathogenic form. Three resistance genes against *Fol*, *I*, *I-2*, and *I-3*, have been introgressed into cultivated tomato from wild relatives, and for each of these, the corresponding avirulence factor has been identified as a small, *in xylem* secreted protein (Takken and Rep, 2010). These small proteins, along with other small secreted proteins (called Six proteins for "secreted in xylem"), as well as several enzymes, have been identified using mass spectrometry of proteins extracted from xylem sap of *Fol*-infected tomato plants (Houterman et al., 2007).

Apart from being targets for surveillance (R) proteins of the tomato immune system, small secreted proteins of *Fol* are also virulence factors. Deletion of *AVR2* (*SIX3*), *AVR3* (*SIX1*), *SIX2*, and *SIX6* reduces, but does not altogether eliminate, virulence toward a generally susceptible tomato line (Houterman et al., 2009; Rep et al., 2004) (P.M. Houterman and M. Rep, unpublished data). A special case is *AVR1* (*SIX4*): deletion of *AVR1* in a race 1 background (only race 1 contains *AVR1*) does not affect virulence toward generally a susceptible tomato line, but surprisingly, virulence toward *I-2* and *I-3*-lines is gained. This appears to be because Avr1 normally suppresses *I-2*- and *I-3*-mediated resistance (Houterman et al., 2008). Avr1 is therefore a virulence factor only for tomato plants containing resistance genes *I-2* and/or *I-3* but lacking *I* (since the latter causes resistance in the presence of Avr1).

Avr2 can activate the intracellular I-2 protein inside plant cells (Houterman et al., 2009) and appears to be taken up by plant cells after secretion by *Fol* (L. Ma and F. Takken, pers. comm.). The observation that intracellularly produced Avr1 can suppress the hypersensitive response (HR) in *Nicotiana benthamiana* leaves induced by combined expression of *AVR2* and *I-2* also suggests uptake of Avr1 into plant cells (F. Gawehns and F. Takken, pers. comm.). This is not exceptional; uptake by plant cells of small proteins secreted by pathogenic fungi has by now been demonstrated or strongly suggested for several plant–fungus interactions (Catanzariti et al., 2007; Houterman et al., 2009; Kale et al., 2010; Manning and Ciuffetti, 2005). How the various effectors of *Fol* promote virulence is unknown.

The genes for effector proteins of *Fol* are largely unique to this *forma specialis* and are rarely found in other pathogenic forms, although homologs of *SIX1*, *SIX4*, *SIX6*, *SIX7*, and *SIX8* have been found in a few other *formae speciales* (Chakrabarti et al., 2011; Lievens et al., 2009) (our unpublished data). This (almost) exclusive presence of *Fol* effector genes in *Fol* underscores their involvement in host-specific virulence. Presumably, other *formae speciales* employ their own sets of effectors, some of which may be homologous to *Fol* effectors.

Such diversity of effectors within a single species complex is stunning considering that more than a hundred *formae speciales* are known (Michielse and Rep, 2009). Genome sequencing has now revealed that individual isolates have unique extra, lineage-specific (LS) chromosomes and chromosome extensions harboring hundreds of genes. In the case of the sequenced strain of *Fol*, one such chromosome harbors all known effector genes (Ma et al., 2010). Even within a *forma specialis* such as f. sp. *lycopersici*, this extra material as a whole can vary greatly between different clonal lines even though the chromosome that carries the effector genes is (almost) identical (our unpublished data). This remarkable richness of extra genetic material combined with identical effector genes within a *forma specialis* appears to be the result of a combination of long-term evolution of extra chromosomes and horizontal chromosome transfer events (see next section for a more detailed discussion of the evolutionary aspects of this).

In *Fol*, at least *AVR2* and *AVR3* are specifically expressed upon entry of living root tissue or close proximity to living plant cells (van der Does et al., 2008a) (L. Ma, pers. comm.). In an insertional mutagenesis screen aimed at finding genes in *Fol* required for pathogenicity (Michielse et al., 2009a), a transcriptional regulator called Sge1 ("Six gene expression") was identified, which is required for expression of all effector genes investigated (*AVR2*, *AVR3*, *SIX2*, and *SIX5*) (Michielse et al., 2009b). Interestingly, Sge1 is conserved in ascomycetes and is related to transcriptional regulators in human pathogenic fungi that are required for morphological switches implied in virulence (Michielse et al., 2009b). Except for reduced production of microconidia, *sge1* mutants of *Fol* have no apparent metabolic or morphological defects, but they are completely nonpathogenic and fail to proliferate inside roots (Michielse et al., 2009b). Apart from being required for expression of effector genes, Sge1 is also required for the production of certain secondary metabolites (Michielse et al., 2009b) and other genes (W. Jonkers and C. Kistler, unpublished data). Taken together, it appears likely that the nonpathogenic phenotype of the *sge1* mutant is due to impaired production of effectors and perhaps other proteins and/or compounds upon entry of plant roots. The basal immune system of the host would then suffice to prevent extensive colonization, just as it does when a naturally nonpathogenic strain attempts to penetrate roots (Olivain et al., 2006).

EVOLUTION OF PATHOGENICITY REVEALED THROUGH *FUSARIUM* COMPARATIVE GENOMICS

The genomes of several *Fusarium* species have been sequenced and comparative genomics has shown features that have contributed to the evolution of pathogenicity. *Fg* has four chromosomes and syntenic alignments against genome assemblies of *Fg*, *Fv*, and *Fo* are presented in Figure 26.1. The alignments display end-to-end synteny in large blocks with the exception of chromosome ends. *Fg* has multiple chromosome fusions and the fusion sites correspond to the previously described highly polymorphic regions of *Fg* (Cuomo et al., 2007), where many secondary metabolite gene clusters have been identified.

Comparative studies of these genomes and that of *Fs* have highlighted two different mechanisms that contribute to the evolution of pathogenicity, namely, the horizontal transfer of pathogenicity chromosomes that carry host specificity effectors and the diversification of secondary metabolite biosynthesis gene clusters (Coleman et al., 2009; Ma et al., 2010).

Large-scale chromosomal polymorphisms within the species complexes of *Fo* and *Fs* were well documented before the genomics era (Boehm et al., 1994; Miao et al., 1991b; Migheli et al., 1995; Temporini and VanEtten, 2002). In association with chromosomal polymorphism, supernumerary chromosomes were first reported as the site of pea pathogenicity (PEP) determinants (Han et al., 2001; Miao et al., 1991a) and for rhizosphere niche adaptation in *Fs* (Rodriguez-Carres et al., 2008).

Comparative studies not only confirmed the presence of supernumerary chromosomes in *Fo* (Ma et al., 2010) and *Fs* (Coleman et al., 2009) genomes but also provided strong evidence for the horizontal acquisition of these chromosomes and their role in determining host-specific virulence. The sequenced *Fol* genome contains four

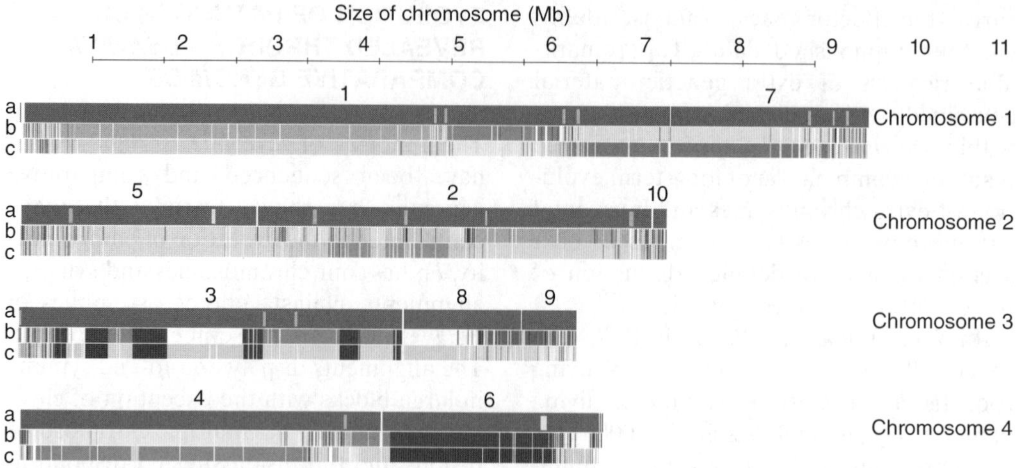

Figure 26.1 Global view of syntenic alignments of *F. graminearum* (*Fg*) to *F. verticillioides* (*Fv*) and *F. oxysporum* (*Fo*) using *Fg* chromosomes as reference. For each chromosome, row a represents the *Fg* genomic scaffolds positioned on the chromosomes separated by scaffold breaks. The vertical bars on the reference (*Fg*) chromosomes represent the locations of potential secondary metabolite biosynthesis gene clusters (yellow: orthologous clusters present in all three species, red: not conserved). Rows b and c display the syntenic mapping of *Fv* and *Fo* chromosomes. The alignments display end-to-end synteny in large blocks with the exception of chromosome ends and reveal multiple chromosome fusions in *Fg*. These fusion sites also correspond to the previously described highly polymorphic regions of *Fg* (Cuomo et al., 2007), where many secondary metabolite gene clusters have been identified. See color insert.

LS chromosomes when compared to the closely related and otherwise largely syntenic genome of *Fv*, as well as LS extensions to the two largest chromosomes. In agreement with the concept of dispensable chromosomes, the supernumerary chromosomes in both *Fs* and *Fo* are deficient in genes for housekeeping functions but enriched for genes whose predicted function can be associated with host-specific virulence or niche adaptation in general (Coleman et al., 2009; Ma et al., 2010). Among the over 3000 proteins predicted to be encoded in *Fol* LS regions, about 20% could be functionally classified on the basis of homology to known proteins. These proteins are significantly enriched for the functional categories "secreted effectors and virulence factors," "transcription factors," and "proteins involved in signal transduction." As described above, most *Fol* effector ("Six") proteins are encoded on chromosome 14. LS region-encoded proteins with a predicted function related to pathogenicity also include necrosis and ethylene-inducing peptides as well as a variety of secreted enzymes predicted to degrade or modify plant or fungal cell walls that are expressed during early stages of tomato root infection (Ma et al., 2010).

The 1.6-Mb chromosome 14 of *Fs* contains the PEP locus that encodes several genes that influence lesion size on epicotyls of the host *Pisum sativum* (Han et al., 2001). This chromosome also carries genes for utilization of homoserine, an amino acid particularly enriched in the pea rhizosphere. A strain of *Fs* containing the homoserine utilization (HUT) locus of chromosome 14 was more competitive in the pea rhizosphere than a HUT mutant (Rodriguez-Carres et al., 2008). Further comparison with *Fo* strains that are pathogenic to other plant hosts reveals the existence of unique sets of LS chromosomes associated with different hosts, which are very likely to contribute to virulence and host specificity as observed in the *Fol* strain (L.-J. Ma, unpublished data). Remarkably, supernumerary chromosomes 14 of *Fo* and *Fs* exhibit mesosynteny with each other but not with chromosomes of other Sordariomycetes genomes (Hane et al., 2011), suggesting that these supernumerary chromosomes, although not syntenic, share related sequences and may have evolved from a common, ancestral "pathogenicity" chromosome.

Horizontal transfer of the *Fol* supernumerary chromosomes was experimentally confirmed through simple coincubation between two otherwise genetically isolated strains of *F. oxyporum* (Ma et al., 2010). This shows that, in a single event, an entire suite of genes required for host compatibility can be transferred into a new genetic lineage, explaining the observed multiplicity of clonal lines in many host-specific *formae speciales* of *Fo* (O'Donnell et al., 1998) that can harbor the same virulence genes (van der Does et al., 2008b). Such a capacity potentially equips the fungus to rapidly adapt to a novel host. If a strain that has received a pathogenicity chromosome had an environmental adaptation different from the chromosome donor strain, transfer could increase the overall incidence of disease in a plant species by introducing pathogenicity into a genetic background preadapted to a local environment. Alternatively, (re)combination with already present LS chromosomes may lead to emergence of strains with new characteristics, such as pathogenicity toward new hosts.

EVOLUTION OF SECONDARY METABOLITE GENE CLUSTERS ASSOCIATED WITH PATHOGENICITY

Fungal genes for biosynthesis of secondary metabolites, including toxic compounds produced in plants, often are clustered at a single locus and are coexpressed (Keller et al., 2005). Using this principle, whole-genome sequencing has revealed a wealth of potential gene clusters among three sequenced *Fusarium* genomes (18 in *Fg*, 16 in *Fv*, and 12 in *Fol*) that have the potential to produce secondary metabolites, referred to as secondary metabolite biosynthetic gene clusters (SMB), summarized in Table 26.1. These predicted gene clusters include some characterized toxin gene clusters, such as those for fumonisin, trichothecenes, zearalenone, and fusarin C, as well as many novel ones. As described above, some of these compounds enhance virulence toward some plant hosts. On average, each cluster contains eight genes and spans over 24 kb. Despite its much larger genome size and larger set of predicted protein-encoding genes, *Fol* has the lowest number of SMB gene clusters identified through this approach. In contrast, the Fg genome is the smallest among the three investigated (36 Mb, about 60% the size of the *Fol* genome and 86% the size of the Fv genome), but contains the highest number of SMB gene clusters. This may indicate that secondary metabolites play a more significant role in the life of Fg.

As a way to evaluate the potential functional importance of the SMB gene clusters encoded in *Fg*, expression of these clusters was examined under multiple conditions, including during sexual reproduction, infection of multiple plant hosts, and growth in complete medium. This exercise revealed coregulation of the genes within 14 out of 18 predicted SMB gene clusters in Fg, which is strong evidence for their functional unity. Three are specifically expressed *in planta*, including the previously described trichothecene (FG3_21) and fusarin C (FG3_19) clusters, as well as a novel cluster of unknown function (FG3_26). In addition, the novel terpene synthase (TS)-associated SMB cluster FG3_29 is expressed in wheat but not in barley and is also expressed in most culture

TABLE 26.1 Secondary Metabolite Biosynthetic Gene Clusters in Three *Fusarium* Species[a]

Class	PKSs	Cluster-ID	Size (bp)	Genes	Expression[b]	Known Cluster
	FGSG_04588	**FG3_26**	**38179**	**17**	*In planta*	
	FGSG_03964	FG3_25	29205	9	Poor	
	FGSG_02395/ FGSG_12126	FG3_34	29521	8	Poor	Zearalenone
Nonreducing PKS	FGSG_09182	FG3_15	21303	8	Sexual	PGL1
	FVEG_03695	FV3_16	19281	7		
	FOXG_05816	FO2_8	19694	7		
	FGSG_02324	FG3_33	25247	10	In culture	Aurofusarin
	FVEG_03379	FV3_12	18788	7		
	FOXG_04757	FO2_19	18624	8		
	FVEG_05537	FV3_32	24127	8		
	FVEG_08425	FV3_41	33102	13		
	FOXG_02884	FO2_31	19644	7		
	FVEG_01736	FV3_10	19430	7		
	FVEG_00316	FV3_21	49188	16		
	FGSG_01790	FG3_45	17914	5	Sexual	
	FVEG_13715	FV3_37	9568	5		
	FGSG_10464	FG3_35	27100	11	Sexual	
	FVEG_12610	FV3_27	22802	5		
	FOXG_15296	FO2_23	22800	5		
	FVEG_11932	FV3_5	35392	11		
	FOXG_03945	FO2_7	42649	16		
Reducing PKS	FOXG_14587	FO2_13	32469	10		
	FVEG_10497	FV3_36	28933	9		
	FOXG_01189	FO2_14	26645	8		
	FVEG_11086	FV3_17	27587	9		Fusarin C
	FGSG_07798	**FG3_19**	**26974**	**9**	*In planta*	Fusarin C
	FOXG_02741	FO2_30	22204	9		
	FVEG_09961	FV3_19	12742	5		
	FVEG_00079	FV3_24	14557	4		
	FGSG_04694	FG3_1	14340	4		
	FVEG_12523	FV3_28	23139	7		
	FOXG_15248	FO2_22	25367	8		
	FGSG_08208	FG3_20	31544	7	Sexual	
	FGSG_08795	FG3_28	22447	6	Sexual	
	FVEG_01914	FV3_8	21577	5		
	FOXG_03051	FO2_33	21645	5		
	FGSG_10548	FG3_38	17838	4	Poor	
	FOXG_10805	FO2_21	28119	9		
	FGSG_05794	FG3_12	19827	6	Constitutively expressed	
		FG3_13	19701	12	Constitutively expressed	
		FG3_21	**27016**	**12**	*In planta*	**Trichothecene**
		FG3_29	21271	10	*In wheat but not barley*	
TS		FV3_33	21271	9		
		FO2_17	18689	8		
		FG3_40	29834	11	Sexual	
		FG3_44	26907	8	poor	

[a] Shaded SMB clusters share orthologous PKS genes as well as the flanking genes identified in the cluster.

[b] Expression of the Fg clusters under various experimental conditions using an *Fg* microarray (Ma et al., 2010). The clusters in italics are specifically expressed *in planta*. The clusters labeled "Sexual" are specifically expressed during sexual development.

conditions examined. Trichothecenes are phytotoxic and play an important role in the aggressiveness of *Fusarium* species toward certain plant hosts. Whether the fusarin C and the novel clusters contribute to pathogenicity of Fg is still unclear.

Among the identified polyketide synthase (PKS) SMB gene clusters, only 2 (FG3_15 and FG3_28) out of 18 appear to be orthologous between all three *Fusarium* genomes examined (Table 26.1; yellow bars in Fig. 26.1, row a). In contrast to the overall sequence conservation among these three genomes (80% sequence identity between *Fg* and *Fv*/*Fol* and 90% between *Fol* and *Fv*), such low level of conservation reflects a highly variable nature of secondary metabolite gene clusters, as observed for many fungal species (Rosewich and Kistler, 2000). Consistent with the high variability, the SMB gene clusters are also localized in the most variable genomic regions of the *Fg* genome, where multiple chromosome fusions were proposed in the *Fg* lineage and a higher rate of recombination was observed (Cuomo et al., 2007). In *Aspergillus fumigatus*, most of the SMB gene clusters are reported to be localized near telomeric regions (Nierman et al., 2005).

Because of the discontinuous presentation of SMB gene clusters within the fungal kingdom, as well as disagreement of the phylogeny of the genes encoded in clusters with the overall species phylogeny, horizontal gene transfer was proposed as the most parsimonious explanation for some gene clusters (Patron et al., 2007). Still, most cluster evolution in fungi seems to occur by way of cluster duplication and diversification combined with frequent cluster loss (Khaldi et al., 2008). Strikingly, both of the orthologous gene clusters present in all three *Fusarium* species (FG3_15 and FG3_28) are preferentially expressed during perithecial formation in *Fg*, which is remarkable since *Fo* has never been observed to produce sexual structures and is presumed to be asexual.

CONCLUSION

As well-characterized plant pathogens, many studies of species within the genus *Fusarium* have focused on understanding the diversity, origins, and function of virulence factors. In addition to gaining a deeper understanding of virulence factors that play a significant role in plant–fungus interactions, such studies have also shed light on horizontal transfer as an evolutionary mechanism, enabling the potentially rapid adaptation of some species, such as *Fo* and *Fs*. The compelling illustration of the scale and ease of transfer of whole chromosomes conferring pathogenicity suggests that such transfer may happen frequently in nature. However, how genomic subregions that are inherited vertically versus horizontally interact and what the evolutionary consequences are of such processes remain intriguing questions for continued investigation. *Fusarium* genomes will serve as outstanding model systems for such studies.

REFERENCES

Alexander NJ, Proctor RH, McCormick SP. 2009. Genes, gene clusters, and biosynthesis of trichothecenes and fumonisins in *Fusarium*. Toxin Reviews 28: 198–215.

Alexander NJ, McCormick SP, Waalwijk C, Van Der Lee T, Proctor RH. 2011. The genetic basis for 3-ADON and 15-ADON trichothecene chemotypes in *Fusarium*. Fungal Genetics and Biology 48: 485–495.

Bacon CW, Glenn AE, Yates IE. 2008. *Fusarium verticillioides*: Managing the endophytic association with maize for reduced fumonisins accumulation. Toxin Reviews 27: 411–446.

Bennett JW, Klich M. 2003. Mycotoxins. Clinical Microbiology Reviews 16: 497.

Berbee ML, Carmean DA, Winka K. 2000. Ribosomal DNA and resolution of branching order among the ascomycota: How many nucleotides are enough? Molecular Phylogenetics and Evolution 17: 337–344.

Boehm EWA, Ploetz RC, Kistler HC. 1994. Statistical-analysis of electrophoretic karyotype variation among vegetative compatibility groups of *Fusarium oxysporum* f. sp. *cubense*. Molecular Plant-Microbe Interactions 7: 196–207.

Boemke C, Tudzynski B. 2009. Diversity, regulation, and evolution of the gibberellin biosynthetic pathway in fungi compared to plants and bacteria. Phytochemistry 70: 1876–1893.

Catanzariti AM, Dodds PN, Ellis JG. 2007. Avirulence proteins from haustoria-forming pathogens. FEMS Microbiology Letters 269: 181–188.

Chakrabarti A, Rep M, Wang B, Ashton A, Dodds P, Ellis J. 2011. Variation in potential effector genes distinguishing Australian and non-Australian isolates of the cotton wilt pathogen *Fusarium oxysporum* f. sp. *vasinfectum*. Plant Pathology 60: 232–243.

Ciuffetti LM, VanEtten HD. 1996. Virulence of a pisatin demethylase-deficient *Nectria haematococca* MPVI isolate is increased by transformation with a pisatin demethylase gene. Molecular Plant-Microbe Interactions 9: 787–792.

Coleman JJ, Rounsley SD, Rodriguez-Carres M, Kuo A, Wasmann CC, Grimwood J, et al. 2009. The genome of *Nectria haematococca*: Contribution of supernumerary chromosomes to gene expansion. PLoS Genetics 5: e1000618.

Coleman JJ, White GJ, Rodriguez-Carres M, VanEtten HD. 2011. An ABC transporter and a cytochrome P450 of *Nectria haematococca* MPVI are virulence factors on pea and are the major tolerance mechanisms to the phytoalexin pisatin. Molecular Plant-Microbe Interactions 24: 368–376.

Covert SF, Enkerli J, Miao VPW, VanEtten HD. 1996. A gene for maackiain detoxification from a dispensable chromosome of *Nectria haematococca*. Molecular & General Genetics 251: 397–406.

Cuomo CA, Güldener U, Xu JR, Trail F, Turgeon BG, Di Pietro A, et al. 2007. The *Fusarium graminearum* genome reveals a link between localized polymorphism and pathogen specialization. Science 317: 1400–1402.

Desjardins AE, Plattner RD, Proctor RH. 1996. Linkage among genes responsible for fumonisin biosynthesis in *Gibberella fujikuroi* mating population A. Applied and Environmental Microbiology 62: 2571–2576.

Desjardins AE, Munkvold GP, Plattner RD, Proctor RH. 2002. FUM1—A gene required for fumonisin biosynthesis but not for maize ear rot and ear infection by *Gibberella moniliformis* in field tests. Molecular Plant-Microbe Interactions 15: 1157–1164.

Enkerli J, Bhatt G, Covert SF. 1998. Maackiain detoxification contributes to the virulence of *Nectria haematococca* MP VI on chickpea. Molecular Plant-Microbe Interactions 11: 317–326.

Fleissner A, Sopalla C, Weltring KM. 2002. An ATP-binding cassette multidrug-resistance transporter is necessary for tolerance of *Gibberella pulicaris* to phytoalexins and virulence on potato tubers. Molecular Plant-Microbe Interactions 15: 102–108.

Fravel D, Olivain C, Alabouvette C. 2003. *Fusarium oxysporum* and its biocontrol. The New Phytologist 157: 493–502.

Gilchrist DG. 1997. Mycotoxins reveal connections between plants and animals in apoptosis and ceramide signaling. Cell death and differentiation 4: 689–698.

Glenn AE, Zitomer NC, Zimeri AM, Williams LD, Riley RT, Proctor RH. 2008. Transformation-mediated complementation of a FUM gene cluster deletion in *Fusarium verticillioides* restores both fumonisin production and pathogenicity on maize seedlings. Molecular Plant-Microbe Interactions 21: 87–97.

Goswami RS, Kistler HC. 2004. Heading for disaster: *Fusarium graminearum* on cereal crops. Molecular Plant Pathology 5: 515–525.

Han Y, Liu XG, Benny U, Kistler HC, VanEtten HD. 2001. Genes determining pathogenicity to pea are clustered on a supernumerary chromosome in the fungal plant pathogen *Nectria haematococca*. The Plant Journal 25: 305–314.

Hane JK, Rouxel T, Howlett BJ, Kema GH, Goodwin SB, Oliver RP. 2011. A novel mode of chromosomal evolution peculiar to filamentous Ascomycete fungi. Genome Biology 12: R45.

Houterman PM, Speijer D, Dekker HL, de Koster CG, Cornelissen BJC, Rep M. 2007. The mixed proteome of *Fusarium oxysporum*-infected tomato xylem vessels. Mol Plant Pathol 8: 215–221.

Houterman PM, Cornelissen BJC, Rep M. 2008. Suppression of plant resistance gene-based immunity by a fungal effector. PLoS Pathogens 4: e1000061.

Houterman PM, Ma L, van Ooijen G, de Vroomen MJ, Cornelissen BJC, Takken FLW, Rep M. 2009. The effector protein Avr2 of the xylem-colonizing fungus *Fusarium oxysporum* activates the tomato resistance protein I-2 intracellularly. The Plant Journal 58: 970–978.

Jansen C, Von Wettstein D, Schafer W, Kogel KH, Felk A, Maier FJ. 2005. Infection patterns in barley and wheat spikes inoculated with wild-type and trichodiene synthase gene disrupted *Fusarium graminearum*. Proceedings of the National Academy of Sciences of the United States of America 102: 16892–16897.

Kale SD, Gu B, Capelluto DG, Dou D, Feldman E, Rumore A, et al. 2010. External lipid PI3P mediates entry of eukaryotic pathogen effectors into plant and animal host cells. Cell 142: 284–295.

Kato J, Purves WK, Phinney BO. 1962. Gibberellin-like substances in plants. Nature 196: 687–688.

Keller NP, Turner G, Bennett JW. 2005. Fungal secondary metabolism—From biochemistry to genomics. Nature Reviews. Microbiology 3: 937–947.

Kenrick P, Crane PR. 1997. The origin and early evolution of plants on land. Nature 389: 33–39.

Khaldi N, Collemare J, Lebrun MH, Wolfe KH. 2008. Evidence for horizontal transfer of a secondary metabolite gene cluster between fungi. Genome Biology 9: R18.

Kistler HC, VanEtten HD. 1984. Three nonallelic genes for pisatin demethylation in the fungus *Nectria haematococca*. Journal of General Microbiology 130: 2595–2603.

Lee K, Pan JJ, May G. 2009. Endophytic *Fusarium verticillioides* reduces disease severity caused by *Ustilago maydis* on maize. FEMS Microbiology Letters 299: 31–37.

Lee T, Han YK, Kim KH, Yun SH, Lee YW. 2002. *Tri13* and *Tri7* determine deoxynivalenol-and nivalenol-producing chemotypes of *Gibberella zeae*. Applied and Environmental Microbiology 68: 2148–2154.

Lievens B, Houterman PM, Rep M. 2009. Effector gene screening allows unambiguous identification of *Fusarium oxysporum* f. sp. *lycopersici* races and discriminationfrom other *formae speciales*. FEMS Microbiology Letters 300: 201–215.

Ma LJ, van der Does HC, Borkovich KA, Coleman JJ, Daboussi MJ, Di Pietro A, et al. 2010. Comparative genomics reveals mobile pathogenicity chromosomes in *Fusarium*. Nature 464: 367–373.

Maier FJ, Miedaner T, Hadeler B, Felk A, Salomon S, Lemmens M, et al. 2006. Involvement of trichothecenes in fusarioses of wheat, barley and maize evaluated by gene disruption of the trichodiene synthase (Tri5) gene in three field isolates of different chemotype and virulence. Molecular Plant Pathology 7: 449–461.

Malonek S, Bomke C, Bornberg-Bauer E, Rojas MC, Hedden P, Hopkins P, Tudzynski B. 2005. Distribution of gibberellin biosynthetic genes and gibberellin production in the *Gibberella fujikuroi* species complex. Phytochemistry 66: 1296–1311.

Manning VA, Ciuffetti LM. 2005. Localization of Ptr ToxA produced by *Pyrenophora tritici-repentis* reveals protein import into wheat mesophyll cells. The Plant Cell 17: 3203–3212.

Miao VPW, VanEtten HD. 1992. Genetic analysis of the role of phytoalexin detoxification in virulence of the fungus *Nectria haematococca* on chickpea (*Cicer arietinum*). Applied and Environmental Microbiology 58: 809–814.

Miao VP, Covert SF, VanEtten HD. 1991a. A fungal gene for antibiotic-resistance on a dispensable (B) chromosome. Science 254: 1773–1776.

Miao VPW, Matthews DE, VanEtten HD. 1991b. Identification and chromosomal locations of a family of cytochrome P-450 genes for pisatin detoxification in the fungus *Nectria haematococca*. Molecular & General Genetics 226: 214–223.

Michielse CB, Rep M. 2009. Pathogen profile update: *Fusarium oxysporum*. Molecular Plant Pathology 10: 311–324.

Michielse CB, van Wijk R, Reijnen L, Cornelissen BJC, Rep M. 2009a. Insight into the molecular requirements for pathogenicity of *Fusarium oxysporum* f. sp. *lycopersici* through large-scale insertional mutagenesis. Genome Biology 10: R4.

Michielse CB, van Wijk R, Reijnen L, Manders EMM, Boas S, Olivain C, et al. 2009b. The nuclear protein *Sge1* of *Fusarium oxysporum* is required for parasitic growth. PLoS Pathogens 5: e1000637.

Migheli Q, Berio T, Gullino ML, Garibaldi A. 1995. Electrophoretic karyotype variation among pathotypes of *Fusarium oxysporum* f. sp. *dianthi*. Plant Pathology 44: 308–315.

Morrissey JP, Osbourn AE. 1999. Fungal resistance to plant antibiotics as a mechanism of pathogenesis. Microbiology and Molecular Biology Reviews 63: 708–724.

Nierman WC, Pain A, Anderson MJ, Wortman JR, Kim HS, Arroyo J, et al. 2005. Genomic sequence of the pathogenic and allergenic filamentous fungus *Aspergillus fumigatus*. Nature 438: 1151–1156.

O'Donnell K, Kistler HC, Cigelnik E, Ploetz RC. 1998. Multiple evolutionary origins of the fungus causing Panama disease of banana: Concordant evidence from nuclear and mitochondrial gene genealogies. Proceedings of the National Academy of Sciences of the United States of America 95: 2044–2049.

Olivain C, Humbert C, Nahalkova J, Fatehi J, L'Haridon F, Alabouvette C. 2006. Colonization of tomato root by pathogenic and nonpathogenic *Fusarium oxysporum* strains inoculated together and separately into the soil. Applied and Environmental Microbiology 72: 1523–1531.

Pan JJ, May G. 2009. Fungal-fungal associations affect the assembly of endophyte communities in maize (*Zea mays*). Microbial Ecology 58: 668–678.

Pareja-Jaime Y, Roncero MIG, Ruiz-Roldan MC. 2008. Tomatinase from *Fusarium oxysporum* f. sp. *lycopersici* is required for full virulence on tomato plants. Molecular Plant-Microbe Interactions 21: 728–736.

Patron NJ, Waller RF, Cozijnsen AJ, Straney DC, Gardiner DM, Nierman WC, Howlett BJ. 2007. Origin and distribution of epipolythiodioxopiperazine (ETP) gene clusters in filamentous ascomycetes. BMC Evolutionary Biology 7: 174.

Pestka JJ. 2010. Toxicological mechanisms and potential health effects of deoxynivalenol and nivalenol. World Mycotoxin Journal 3: 323–347.

Proctor RH, Brown DW, Plattner RD, Desjardins AE. 2003. Co-expression of 15 contiguous genes delineates a fumonisin biosynthetic gene cluster in *Gibberella moniliformis*. Fungal Genetics and Biology 38: 237–249.

Proctor RH, Plattner RD, Brown DW, Seo JA, Lee YW. 2004. Discontinuous distribution of fumonisin biosynthetic genes in the *Gibberella fujikuroi* species complex. Mycological Research 108: 815–822.

Proctor RH, Plattner RD, Desjardins AE, Busman M, Butchko RAE. 2006. Fumonisin production in the maize pathogen *Fusarium verticillioides*: Genetic basis of naturally occurring chemical variation. Journal of Agricultural and Food Chemistry 54: 2424–2430.

Proctor RH, Busman M, Seo JA, Lee YW, Plattner RD. 2008. A fumonisin blosynthetic gene cluster in *Fusarium oxysporum* strain O-1890 and the genetic basis for B versus C fumonisin production. Fungal Genetics and Biology 45: 1016–1026.

Radley M. 1956. Occurrence of substances similar to gibberellic acid in higher plants. Nature 178: 1070–1071.

Recorbet G, Steinberg C, Olivain C, Edel V, Trouvelot S, Dumas-Gaudot E, et al. 2003. Wanted: Pathogenesis-related marker molecules for *Fusarium oxysporum*. The New Phytologist 159: 73–92.

Rep M, Kistler HC. 2010. The genomic organization of plant pathogenicity in *Fusarium* species. Current Opinion in Plant Biology 13: 420–426.

Rep M, van der Does HC, Meijer M, van Wijk R, Houterman PM, Dekker HL, de Koster CG, Cornelissen BJC. 2004. A small, cysteine-rich protein secreted by Fusarium oxysporum during colonization of xylem vessels is required for I-3-mediated resistance in tomato. Mol Microbiol 53: 1373–1383.

Rheeder JP, Marasas WF, Vismer HF. 2002. Production of fumonisin analogs by *Fusarium* species. Applied and Environmental Microbiology 68: 2101–2105.

Rocha O, Ansari K, Doohan FM. 2005. Effects of trichothecene mycotoxins on eukaryotic cells: A review. Food Additives and Contaminants 22: 369–378.

Rodriguez-Carres A, White G, Tsuchiya D, Taga M, VanEtten HD. 2008. The supernumerary chromosome of *Nectria haematococca* that carries pea-pathogenicity-related genes also carries a trait for pea rhizosphere competitiveness. Applied and Environmental Microbiology 74: 3849–3856.

Roldan-Arjona T, Perez-Espinosa A, Ruiz-Rubio M. 1999. Tomatinase from *Fusarium oxysporum* f. sp. *lycopersici* defines a new class of saponinases. Molecular Plant-Microbe Interactions 12: 852–861.

Rosewich UL, Kistler HC. 2000. Role of horizontal gene transfer in the evolution of fungi. Annual Review of Phytopathology 38: 325–363.

Schulthess F, Cardwell KF, Gounou S. 2002. The effect of endophytic *Fusarium verticillioides* on infestation of two maize varieties by lepidopterous stemborers and coleopteran grain feeders. Phytopathology 92: 120–128.

Seong K, Pasquali M, Song J, Hilburn K, McCormick S, Dong Y, et al. 2009. Global gene regulation by *Fusarium* transcription factors *Tri6* and *Tri10* reveals adaptations for toxin biosynthesis. Molecular Microbiology 72: 354–367.

Spatafora JW, Sung GH, Sung JM, Hywel-Jones NL, White JF. 2007. Phylogenetic evidence for an animal pathogen origin of ergot and the grass endophytes. Molecular Ecology 16: 1701–1711.

Takken FLW, Rep M. 2010. The arms race between tomato and *Fusarium oxysporum*. Mol Plant Pathol 11: 309–314.

Taylor JW, Berbee ML. 2006. Dating divergences in the fungal tree of life: Review and new analyses. Mycologia 98: 838–849.

Temporini ED, VanEtten HD. 2002. Distribution of the pea pathogenicity (*PEP*) genes in the fungus *Nectria haematococca* mating population VI. Current Genetics 41: 107–114.

Troncoso C, Gonzalez X, Bomke C, Tudzynski B, Gong F, Hedden P, Rojas MC. 2010. Gibberellin biosynthesis and gibberellin oxidase activities in *Fusarium sacchari*, *Fusarium konzum* and *Fusarium subglutinans* strains. Phytochemistry 71: 1322–1331.

Tsavkelova EA, Bomke C, Netrusov AI, Weiner J, Tudzynski B. 2008. Production of gibberellic acids by an orchid-associated *Fusarium proliferatum* strain. Fungal Genetics and Biology 45: 1393–1403.

Tudzynski B, Holter K. 1998. Gibberellin biosynthetic pathway in *Gibberella fujikuroi*: Evidence for a gene cluster. Fungal Genetics and Biology 25: 157–170.

van der Does HC, Duyvesteijn RGE, Goltsteijn PM, van Schie CCN, Manders EMM, Cornelissen BJC, Rep M. 2008a. Expression of effector gene *SIX1* of *Fusarium oxysporum* requires living plant cells. Fungal Genet Biol 45: 1257–1264.

van der Does HC, Lievens B, Claes L, Houterman PM, Cornelissen BJC, Rep M. 2008b. The presence of a virulence locus discriminates *Fusarium oxysporum* isolates causing tomato wilt from other isolates. Environmental Microbiology 10: 1475–1485.

Wang E, Norred WP, Bacon CW, Riley RT, Merrill AH, Jr. 1991. Inhibition of sphingolipid biosynthesis by fumonisins. Implications for diseases associated with *Fusarium moniliforme*. The Journal of Biological Chemistry 266: 14486–14490.

Wang W, Jones C, Ciacci-Zanella J, Holt T, Gilchrist DG, Dickman MB. 1996. Fumonisins and *Alternaria alternata lycopersici* toxins: Sphinganine analog mycotoxins induce apoptosis in monkey kidney cells. Proceedings of the National Academy of Sciences of the United States of America 93: 3461–3465.

Ward TJ, Bielawski JP, Kistler HC, Sullivan E, O'Donnell K. 2002. Ancestral polymorphism and adaptive evolution in the trichothecene mycotoxin gene cluster of phytopathogenic *Fusarium*. Proceedings of the National Academy of Sciences of the United States of America 99: 9278–9283.

Wasmann CC, VanEtten HD. 1996. Transformation-mediated chromosome loss and disruption of a gene for pisatin demethylase decrease the virulence of *Nectria haematococca* on pea. Molecular Plant-Microbe Interactions 9: 793–803.

Weltring KM, Turgeon BG, Yoder OC, VanEtten HD. 1988. Isolation of a phytoalexin-detoxification gene from the plant pathogenic fungus *Nectria haematococca* by detecting its expression in *Aspergillus nidulans*. Gene 68: 335–344.

CHAPTER 27

GENETIC, GENOMIC, AND MOLECULAR APPROACHES TO DEFINE VIRULENCE OF *ASPERGILLUS FUMIGATUS*

LAETITIA MUSZKIETA, WILLIAM J. STEINBACH, and JEAN-PAUL LATGE

INTRODUCTION

Aspergillus fumigatus is a ubiquitous saprophytic fungus that lives on decaying vegetables and is essential for carbon and nitrogen recycling (Fig. 27.1). Dissemination is achieved through the release into the air of haploid conidia, which are produced by the fungus after starvation of the mycelium, the vegetative growth form of *A. fumigatus*. Conidia dispersed by wind or water are able to germinate and grow when they land on a new nutritive substratum. An estimated 1–100 *A. fumigatus* conidia per cubic meter have been reported in normal environments (Latgé and Steinbach, 2009). Although analysis of the *A. fumigatus* genome has revealed the presence of >200 genes associated with the sexual reproduction of a heterothallic species, including a *MAT1-2* high mobility group domain and a *MAT1-1* alpha domain located at the mating-type locus, as well as genes for pheromone production and sensing (Paoletti et al., 2005). The presence of a sexual stage has only been recently observed in laboratory conditions (O'Gorman et al., 2009). Cleistothecia and ascospores (teleomorph *Neosartorya fumigata*) form and are maintained for several months following contact between two compatible strains with complementary mating type (*MAT*) genes (*MAT1-1* and *MAT1-2*) (O'Gorman et al., 2009). Although it has never been observed in nature, this sexual stage may be responsible for the genotypic recombination observed in this species (Pringle et al., 2005). However, this recombination seems to be limited and could instead be due to a parasexual cycle between two compatible strains that can then lead to genetic exchanges through the formation of a heterokaryon (Fig. 27.1).

The inhalation of conidia can result in two major different types of *Aspergillus* disease: allergy (in the absence of mycelial growth) or invasive aspergillosis infection with the extent of invasion depending on the level of immunosuppression of the host (Ben-Ami et al., 2010). In immunocompetent individuals, alveolar macrophages are primarily responsible for the phagocytosis and killing of inhaled *A. fumigatus* conidia

Evolution of Virulence in Eukaryotic Microbes, First Edition. Edited by L. David Sibley, Barbara J. Howlett, and Joseph Heitman.

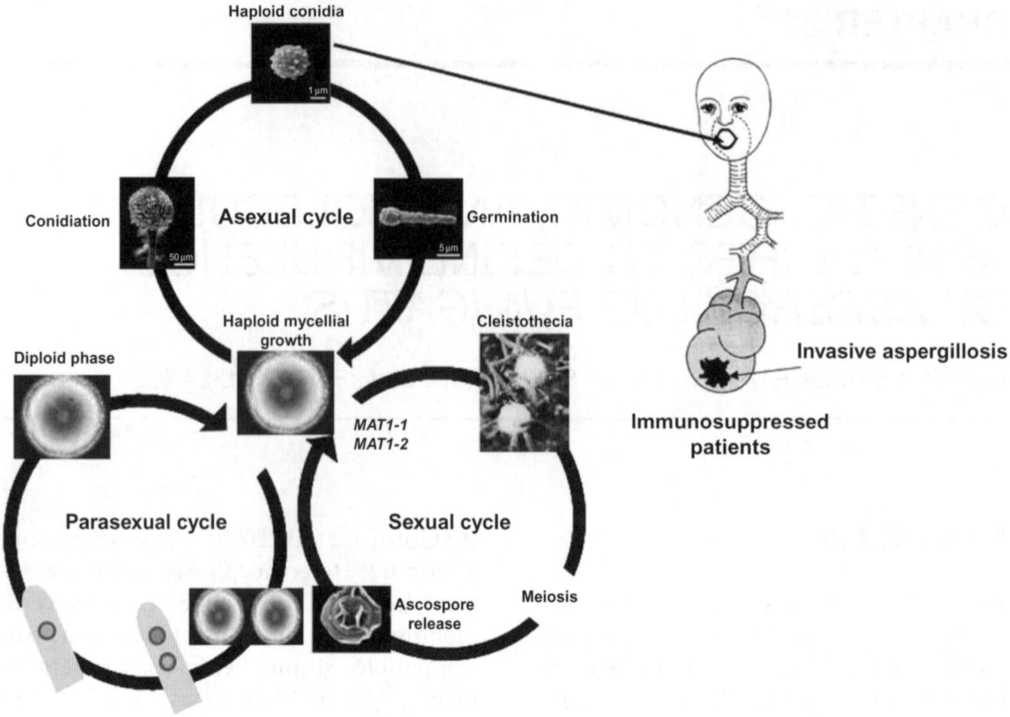

Figure 27.1 Life cycles of *Aspergillus fumigatus*. The asexual cycle is the common form of *A. fumigatus* growth. Fungi grow on decaying vegetal material until nutrient deprivation, which induces conidiation. Haploid conidia are dispersed to the air and can then colonize a new substrate. *A. fumigatus* possesses all genes required to complete a sexual cycle, and the formation of cleistothecia and ascospores are observed in laboratory conditions even though sexual reproduction has never been observed in natural conditions. *A. fumigatus* can be manipulated in a diploid form to undergo a parasexual cycle also not observed in nature. The return to haploid form can be induced with chemicals. Humans are continuously exposed to small-sized haploid conidia that can infiltrate pulmonary alveoli. In immunosuppressed patients, the immune system fails to eliminate conidia, which can then germinate, colonize lung tissue, and cause invasive aspergillosis. See color insert.

as well as the initiation of a proinflammatory response that recruits neutrophils, complement, and cytokines to the site of infection (Ben-Ami et al., 2010). Conidia that evade macrophage killing and germinate then become the target of infiltrating neutrophils, which are able to destroy germinated hyphae by releasing reactive oxygen species (ROS) and other toxic molecules. The risk of developing invasive aspergillosis results primarily from a dysfunction in these innate host defenses. As the number of immunocompromised patients has increased in recent decades, mainly due to immunosuppressive therapies for malignancy or organ and stem cell transplantation, the incidence of invasive aspergillosis has been continuously increasing (Latgé and Steinbach, 2009). In spite of a high mortality rate, the understanding of the pathobiology of this disease remains very poor. This chapter will review the progress recently made in the genetics of *A. fumigatus* and the development of "omic" approaches as they have both improved our understanding of the key factors in the infectious process and identified new molecular mechanisms explaining the survival and growth of this opportunistic fungal pathogen.

Molecular Genetic Analysis of *A. fumigatus*

Progress in the molecular analysis of genes that contribute to the pathogenicity of *A. fumigatus* has resulted from recent technical developments. The sequencing of the *A. fumigatus* genome has accelerated gene structure and function studies (Nierman et al., 2005) and offers the opportunity to make comparative genomic and proteomic analyses with other sequenced *Aspergillus* species. Additionally, many experimental tools derived from those used in the genetic model *Aspergillus nidulans* have emerged and allowed us to quite efficiently manipulate the *A. fumigatus* genome.

Comparative Genomics. The filamentous fungal genus *Aspergillus* consists of over 250 saprophytic species (Geiser et al., 2007). Although some species, such as *Aspergillus niger*, *Aspergillus terreus*, and *Aspergillus oryzae*, are exploited commercially for the production of enzymes, pharmaceuticals, and traditional Asian foods and beverages (Hendrickson et al., 1999; Machida et al., 2005; Pel et al., 2007), others such as *A. fumigatus* are capable of colonizing and infecting humans (Latgé and Steinbach, 2009). Many of these individual species have been sequenced and comparative genomics has been undertaken to determine whether genes unique to human pathogenic species, such as *A. fumigatus*, represent virulence factors.

The three most closely related *Aspergilli* are *A. fumigatus*, *Neosartorya fischeri*, and *Aspergillus clavatus*, but they are only as close evolutionary as the relationship between humans, mice, and birds (Fedorova et al., 2008; Galagan et al., 2005; Rokas et al., 2007). The *A. fumigatus* genome is 29.4 Mb in size and consists of eight chromosomes containing 9926 predicted genes. The *A. clavatus* genome (27.9 Mb) is the smallest to date among the sequenced *Aspergilli*. The *N. fischeri* genome (32.6 Mb, 10,407 protein-coding genes) is 10–15% larger than the *A. clavatus* and *A. fumigatus* genomes. Species-specific transposable elements (Girardin et al., 1994) may have contributed to this genome size expansion. Fedorova et al. have identified 7514 orthologous core and 818, 1402, and 1151 species-specific genes in the *A. fumigatus*, *N. fischeri*, and *A. clavatus* genomes, respectively. These species-specific genes are significantly smaller in size, contain fewer exons, and exhibit a subtelomeric bias compared with the core genes. At least 20% of *A. fumigatus*-specific genes appear to be functional and could constitute virulence determinants. However, it is overly simplistic to assume that these unique genes could be virulence determinants as experimental invasive aspergillosis can result from infection with organisms generally regarded as less pathogenic, such as *A. oryzae*, *A. niger*, or *N. fisheri* (J.P. Latgé, unpublished data), showing that these saprophytic species have the necessary attributes to infect a mammal. To date, comparisons of *Aspergillus* genomes highlight the potential usefulness of genomics for the accurate identification of species boundaries in the genus *Aspergillus* but not for the identification of genes essential for fungal virulence.

Gene Targeting Strategies to Identify Virulence-Determining Factors. The complete inactivation of its encoding gene has become a classical procedure to assess the cellular function of an uncharacterized protein. This candidate approach is a prerequisite to testing defined strains in experimental models of aspergillosis, specifically to determine genes that encode putative virulence factors. The most utilized approach is the complete deletion of the target locus by replacement. However, a limited number of resistance markers are available to select transformants. Common markers used for *A. fumigatus* are the *Escherichia coli* phosphostranferase-encoding *hph* gene, which confers hygromycin resistance, the *Streptoalloteichus hindustanus ble* gene, which confers phleomycin resistance (Punt and van den Hondel, 1992), or the *A. oryzae ptrA* gene, which confers pyrithiamine resistance. In contrast to *A. nidulans*, *BAR* from *Streptomyces hygroscopicus* cannot be used to select glufosinate-resistant transformants because *A. fumigatus* is naturally resistant to this compound, as well as to nourseothricin (Alcazar-Fuoli et al., 2011), eliminating two selectable markers commonly used in other fungi. Since the availability of dominant selectable markers is limited, other techniques must be used to produce multiple deletion strains because *A. fumigatus* is very rich in multigenic families (Hartmann et al., 2010). To avoid a counterselection, the "pyrG–blaster" and the "self-excising marker" methods have been used. These methods are based on a recyclable marker that allows repetitive targeting in filamentous fungi (Hartmann et al., 2010). The self-excising marker method uses a β-recombinase under the control of the endoxylanase promoter that is placed in front of the selectable marker; this construction is flanked by six direct recognition sequences and borders the target gene. Transformants bearing resistance to the selectable marker are then cultivated in the presence of xylose to induce the expression of the recombinase, allowing the excision of the selection marker.

DNA constructs can be introduced into *A. fumigatus* wild-type strains by transforming protoplasts or by electroporation (d'Enfert et al., 1999). However, gene targeting in filamentous fungi results in a low frequency of homologous recombination (especially in electroporation, which makes this technique unsuitable for wild-type mutants) and results in ectopic integration of the transforming DNA fragment. Tremendous progress resulted from a unique molecular approach that improved the frequency of homologous recombination based on the deletion of genes encoding a subunit of the Ku heterodimer (*akuB*^{KU80}), which is responsible for binding the Ku complex to broken DNA ends and for stimulating DNA repair in wild-type strains (da Silva Ferreira et al., 2006).

The use of a regulatable promoter that drives the transcription of target genes through the replacement of the endogenous promoter may provide a means to determine gene essentiality. However, the total loss of transcription in repression conditions has never been observed in *A. fumigatus*. Less "leaky" conditional promoters that allow growth in inducible conditions include pNiia and the pXyl (Lambou et al., 2010; Zadra et al., 2000). Attempts have also been undertaken to show the essentiality of target genes by RNA interference (RNAi), allowing the downregulation of a target gene without full inhibition of expression. However, RNAi transformants are frequently genetically unstable, and partial deletion of RNAi constructs from the genome has been observed (Henry et al., 2007). Gene essentiality can now be verified by heterokaryon rescue (Monteiro and de Lucas, 2010); a stable diploid strain disrupted for one allele of the target gene can be used, and a diploid strain can be induced to become a haploid strain. If strains are recovered with the target locus deleted, the deletion is presumed to be nonessential; however, if a strain without the deleted locus cannot be recovered, the deletion is presumed to be lethal.

***An Experimental Model System to Determine the Virulence of* A. fumigatus.** The second essential tool indispensable for the identification of virulence factors is appropriate experimental model systems to test the virulence of a deletion mutant. The most popular models of invasive aspergillosis in mice involve the induction of neutropenia by cyclophosphamide or/and corticosteroid treatment, which reduces the efficacy of the innate immune system. The most commonly used routes of infection in mice are intranasal or aerosol inoculations to mimic the natural route of infection, but a comparative analysis with an intravenous route has not been thoroughly investigated. Arthropod models, such as *Drosophilia melanogaster* (Lionakis and Kontoyiannis, 2005, 2010) and *Galleria mellonella* (Renwick et al., 2006), have potential for use as high throughput screening methods, but clearly mimic the infected patient less appropriately than a mammalian model. Due to the variability often seen in the animal response, the most efficient protocol comparing the virulence of a mutant and its parental strain is through the joint inoculation of both strains in the same animal, either through inhalation or subcutaneous injection in a mildly immunocompromised model (Ben-Ami et al., 2010; Maubon et al., 2005).

"Omics". Global transcriptional analyses provide insights into fungal gene expression during host colonization. Analysis of the *A. fumigatus* genome sequence has allowed the design of DNA microarrays and subsequent genome transcriptional profiling during saprophytic or pathogenic growth (Lamarre et al., 2008; McDonagh et al., 2008; Sugui et al., 2008). However, the number of transcribed genes is very limited (<1000), and this system will be replaced in the near future by RNA sequencing, in which a maximum of transcribed genes and a quantitative estimation of their level of transcription can be determined. This is crucial to truly understand the meaning of results, as recent published data showed that the transcription level of 9000 genes varied between two culture conditions (Gibbons et al., 2011). Proteins produced under a particular condition can also be identified by 2-D gel electrophoresis or liquid chromatography coupled to MS–MS analysis (Vödisch et al., 2009). Proteome analysis identified less than 500 proteins, most soluble. Although the coupling of these genomic and proteomic methods has led to a large amount of data, they have not yet led to the specific identification of proteins or metabolic pathways controlling *A. fumigatus* virulence.

Molecular Bases of Virulence of *A. fumigatus*

Heat Tolerance and Hypoxia. *A. fumigatus* is one of the most thermophilic fungi on earth and is able to grow actively even at 55°C and to survive temperatures as high as 75°C (Tekaia and Latgé, 2005). This thermotolerance is necessary to occupy its ecological niche and explains why it is the only fungus found in compost, where high temperatures occur during the fermentation process. Although DNA microarray analyses have highlighted a distinct pattern of gene expression at 37°C, very few genes have been shown to be required for thermotolerance (Nierman et al., 2005). Among them, a deletion of the gene encoding the ribosome biogenesis protein CgrA is the only one that impairs the growth at 37°C, suggesting that the defect in ribosome biogenesis caused by the loss of CgrA is compatible with the limited physiological demands at room temperature but not with the metabolic requirements at 37°C (Bhabhra et al., 2004). The impact of deletion of other genes, such as those coding for the mannosyltransferases, Pmt4 and Mnt1, on thermotolerance is seen at much higher temperatures such as 48–50°C, suggesting that a tight correlation exists between permissive temperature and the deleted gene (Mouyna et al., 2010; Wagener et al., 2008). Understanding thermotolerance remains a major research objective because it is postulated that this focus on such a unique *A. fumigatus* biological trait will result in the identification of new drug targets.

Another essential environmental parameter is the unique hypoxic condition that the fungus encounters not only in the soil but also at the site of infection where inflammation, generated by the immune response, is responsible for a local decrease in oxygen level (Willger et al., 2008). Although *A. fumigatus* is considered to be an obligate aerobic organism, it was shown to be able to grow at low oxygen concentrations between 0.1% (v/v) and 0.5% (v/v) on agar plates. A recent comparison of the proteome analysis of *A. fumigatus* mycelia grown under normoxic (21% oxygen) or hypoxic (0.2% oxygen) conditions led to the identification of 117 proteins with an altered abundance under hypoxic conditions (Vödisch et al., 2011). Although the changes in metabolism following growth in hypoxic conditions have not been fully elucidated, the transcription factor SrbA was shown to mediate oxygen-dependent sterol biosynthesis and growth under hypoxic conditions (Willger et al., 2008). Deletion of *srbA* produces a mutant strain that grows normally in atmospheric oxygen (21%) but is incapable of growth under hypoxic conditions and is also avirulent (Willger et al., 2008). More work in the area of carbonic anhydrase metabolism could lead to essential discoveries in the understanding of the environmental physiology of *A. fumigatus*. These results show that these two environmental factors (heat and lack of oxygen) are essential to control pathogenic growth of *A. fumigatus*.

Counteracting Host Defense Mechanisms: The Example of ROS. ROS produced by cells of the innate immune system are essential for the killing of *A. fumigatus* in the immunocompetent host (Philippe et al., 2003). Accordingly, it was expected that the enzymes produced by *A. fumigatus* that are able to scavenge ROS would be essential for fungal virulence, as is the case in bacterial pathogens. Several ROS scavenging enzymes have been identified and characterized, including three catalases (one conidial and two mycelial enzymes) that are able to break down H_2O_2 (Paris et al., 2003). Based on BLAST analysis, four superoxide dismutases (SODs) catalyzing the breakdown of superoxide radicals (a cytosplamic Cu/ZnSOD, a mitochondrial MnSOD, a cytoplasmic MnSOD, and a gene encoding a MnSOD C-terminal domain) were described in *A. fumigatus* (Lambou et al., 2010). The transcription factor Yap1, which controls most of the antioxidant reactive species, has also been studied (Lessing et al., 2007). Other scavengers, such as those of the glutathione pathway, have not been analyzed to date, though single and multiple mutants have been constructed. Despite the increased sensitivity to H_2O_2 of catalase mutants, menadione of Δ*sod* strains, both oxidants of the Δ*yap1*strain, and killing by the phagocyte of an immunocompetent host for all mutants, none of these mutants displays attenuated virulence in severely immunocompromised murine models of experimental aspergillosis using heavy inocula (Lambou et al., 2010; Paris et al., 2003). A recent cutaneous model of aspergillosis using only moderately immunocompromised mice was able to show a difference in the pathogenicity of these reactive oxidant mutants (Ben-Ami et al., 2010). This result indicated that the amount of ROS produced by the cyclophosphamide/glucocorticoid-treated animals is so low that it does not affect the fungal growth, regardless of the level of scavenging enzymes.

Cell Wall and Immunity. The fungal cell wall is the first barrier against the hostile environment, providing integrity and physical protection to the cell (Latgé and Steinbach, 2009). The cell wall is also the first fungal organelle to be in contact with the host, as well as the primary target of the host immune system during infection. The cell wall of *A. fumigatus* is principally composed of α(1,3)- and β-(1,3)-glucans, chitin, and galactomanan organized in a three-dimensional network (Gastebois et al., 2009). However, this composition varies according to the morphotype; for example, the conidial cell wall is covered by an outer layer of hydrophobin and melanin, which is absent from the mycelial cell wall (Latgé, 2010). The rodlet layer, composed of the conidial hydrophobin RodA, has an organized structure that facilitates conidial dispersion in the atmosphere but also plays an essential role in the host immune response (Thau et al., 1994). The hydrophobin layer protects the conidia to an inflammatory reaction that could be detrimental for both the host and fungus. Below this rodlet layer, the melanin pigment can also counteract host defense reactions (Langfelder et al., 2003). The melanin responsible for the green color of the conidia is a secondary metabolite synthesized from acetate by the enzymatic products of six genes (Tsai et al., 1999). Mutants lacking polyketide synthase PksP, which facilitates the first step in the melanin biosynthesis pathway, have white conidia with attenuated virulence (Thywißen et al., 2002). Several explanations for the observed reduced virulence have been proposed: (i) conidia of the Δ*pksP* strain induce neutrophils to release greater quantities of ROS than wild-type conidia; (ii) white conidia are more sensitive to ROS *in vitro*; and (iii) Δ*pksP* conidia are more rapidly trafficked to the acidified phagolysosomal compartment (Langfelder et al., 2003).

The constitutive polysaccharides of the cell wall also trigger the immune response. In a vaccination model, α(1,3)-glucan and β(1,3)-glucan protect the host from infection by stimulating Th1 Treg or Th17 responses (Ben-Ami et al., 2010). In contrast, galactomannan promotes fungal growth through the activation of a Th2/Th17 response. The phagocyte must sense the cell wall; however, the surface receptors of the immune cells are poorly studied, and the only receptor identified to date involved in this response is Dectin-1, which recognizes β(1,3)-glucan. This area of research is in its infancy, though essential, as it could lead to future vaccination protocols, and would benefit from the use of mutants and knockout mice.

Divalent Cation Acquisition. Divalent cations (iron, magnesium, zinc, etc.) are essential for fungal growth because they are necessary cofactors for many biosynthetic pathways. In addition, iron plays an important role in resistance against oxidative stress. Very low levels free divalent cations are available in the host, and *A. fumigatus*, like other pathogens, has developed mechanisms to extract the required amounts from host proteins.

The instability of free iron and iron sequestration *in vivo* by host defense mechanisms severely limits iron availability in the mammalian lung (Schrettl et al., 2004). The capacity to acquire iron from the host is a necessary requisite for *A. fumigatus* growth. Two high affinity iron uptake systems have been identified in *A. fumigatus*: reductive iron assimilation and siderophore (small ferric iron-specific chelators)-assisted iron uptake (Schrettl et al., 2004). The growth of *A. fumigatus* in restrictive iron concentrations is fully dependent on siderophore-assisted iron uptake and intracellular storage (Schrettl et al., 2004). Among the four known siderophores, fusarinine C and triacetylfusarinine are able to capture extracellular iron, whereas ferricrocin and hydroxyferricrocin are intracellular siderophores used to store iron in hyphae and conidia (Schrettl et al., 2007). Deletion of genes from the pathway controlling siderophore biosynthesis from L-ornithine has shown that both intracellular and extracellular siderophores are essential for *in vivo* growth under low iron concentrations and germination in the lung environment. This growth defect leads to a partial or full reduction in virulence, which is in agreement with an earlier transcriptional analysis of *A. fumigatus* growing in the lungs of infected mice in a neutropenic model that showed the upregulation of genes encoding siderophores (McDonagh et al., 2008).

At least six *A. fumigatus* transporters (ZrfA-H) that control zinc influx have been identified *in silico*, and the impact of each of these transporters is currently being studied (Amich et al., 2009, 2010). *A. fumigatus* zinc regulation is mediated by the transcription factor ZafA that is expressed only under zinc-limiting conditions, where it becomes essential for growth (Moreno et al., 2007). Accordingly, the Δ*zafA* mutant is avirulent because it cannot germinate *in vivo* in the lung space.

Magnesium is the most abundant divalent cation in cells. *A. fumigatus* requires at least 10 times more magnesium than iron or zinc to grow (Tekaia and Latgé, 2005); however, magnesium influx has been poorly investigated in *A. fumigatus*. An ortholog of the bacterial gene *MGTC*, which regulates magnesium transport and is required for growth at low magnesium concentrations, has been acquired in *A. fumigatus* by horizontal transfer but does not share the same function as bacterial MgtC proteins (Gastebois et al., 2011). Other putative magnesium transporters, such as orthologs of the Alr1 family in yeast, have not yet been investigated.

Production of Secondary Metabolites. *A. fumigatus* can synthesize several secondary metabolites, among which the best known are fumitremorgin, fumagillin, and gliotoxin (Bok et al., 2006; Maiya et al., 2006; Spikes et al., 2008). Most of these secondary metabolites have toxic properties that have been implicated in *A. fumigatus* virulence (Müllbacher et al., 1988). A feature of these secondary metabolites is that their production is under the control of gene clusters (Keller and Hohn, 1997). The *A. fumigatus* genome contains 22 clusters that include 14 clusters containing polyketide synthase genes, 14 clusters containing nonribosomal peptide synthase genes, 1 cluster containing fatty acid synthase gene, and 7 clusters containing dimethylallyl tryptophan synthase genes (Nierman et al., 2005). Among all *A. fumigatus*-specific mycotoxins, gliotoxin has attracted the most interest because it is the only toxin that has been isolated from patients or animals with invasive aspergillosis (Lewis et al., 2005; Pardo et al., 2006). Gliotoxin is primarily an antioxidant but also exerts immunosuppressive properties on host leukocytes by blocking phagocytosis and transcription of inflammatory mediators and inducing apoptosis of neutrophils and monocytes (Eichner et al., 1986; Stanzani et al., 2005). Studies of the gliotoxin nonproducing mutants produced by deleting the genes encoding nonribosomal peptide synthetase *GliP* or the transcription factor gene *GliZ* revealed a limited impact of gliotoxin on fungal virulence, seen only in mice immunosuppressed with corticosteroids alone (Bok et al., 2006; Spikes et al., 2008). These results suggested that gliotoxin increases the virulence of *A. fumigatus* only in the presence of neutrophils, raising the possibility that neutrophils are a target of this toxin (Spikes et al., 2008). This hypothesis is supported by the observation of reduction of neutrophil apoptosis in a fungal lesion in mice infected with a Δ*gliP* strain. The gliotoxin biosynthethic cluster comprises 12 genes (Gardiner and Howlett, 2005), and interestingly, the deletion of a single gene, *gliT*, renders the organism highly sensitive to exogenous gliotoxin and completely disrupts gliotoxin secretion (Schrettl et al., 2010). The expression of *gliT* appears to be independently regulated compared to all other cluster components, which suggests that expression of all members of the clusters identified *in silico* may not be expressed together, a scenario suggested by many authors.

The specific roles of other secondary metabolites, such as clavine ergot alkaloid, fumagilin, helvolic acid or fumitremorgin, on *Aspergillus* virulence have not been tested (Lodeiro et al., 2009; Maiya et al.,

2006; Mitchell et al., 1997). However, a global repression of secondary metabolite production following deletion of the *laeA* gene, which encodes a methyltransferase that regulates multiple secondary metabolite clusters, resulted in impaired virulence in a neutropenic model of invasive aspergillosis (Bok et al., 2005). The Δ*laeA* strain did not display an altered growth or germination phenotype, suggesting that the role of secondary metabolites is in addition to its role in virulence control. These data are in agreement with the transcriptional profiling of mRNA extracted from murine lung following infection, which showed that a significant proportion of secondary metabolite genes under the control of LaeA were upregulated *in vivo* (Perrin et al., 2007).

Secreted Hydrolytic Enzymes and Nutrient Uptake. *A. fumigatus*, like other *Aspergillus* species, can secrete an extensive arsenal of extracellular degradative enzymes, including cellulases, pectinases, and hemicellulases that all are related to the saprophytic growth of the fungus on decaying vegetables. Similarly, *A. fumigatus* produces many proteases that contribute to growth in the lung parenchyma, which serves largely as a protein sponge. A serine protease and a metalloproteinase are the two major proteases of *A. fumigatus* involved in the degradation of lung collagen and elastin (Jaton-Ogay et al., 1994). Despite conflicting results, it seems that these two proteases are dispensable for growth in the lung, which may be explained by the following reasons: (i) >100 proteases, some of them with relatively broad specificity, have been identified in the *A. fumigatus* genome that are potent antigens detected in the lung and can compensate for the deletion of one or two protease genes; (ii) while the deletion of proteases reduces the growth of *A. fumigatus* on a protein substratum *in vitro*, it may not affect the growth rate *in vivo*, which is a key factor in tissue invasion.

Although it is clear that the functional redundancy of proteolytic enzymes does not allow easy identification of the individual role of each protease in pathogenicity, it is obvious that the secretion of active hydrolases that is controlled by the unfolded protein response (UPR) is essential for growth (Richie et al., 2009). The disruption of the UPR by deletion of the *hacA* gene results in a growth defect in mammalian lung tissue. This mutant was significantly impaired in experimental models of invasive aspergillosis, reflecting the importance of the UPR to regulate protein secretion needed for tissue degradation and nutrient acquisition *in vivo*. Other hydrolases, such as lipases, that are essential among bacterial colonizers of the lung have not been investigated in *A. fumigatus*.

Primary carbon and nitrogen metabolism directly influences the pathogenicity of *A. fumigatus*, as seen in the example of the growth defect of several auxotrophic strains with folate (*pabA*), uridine/uracil (*pyrG*), or lysine (*lysF*) deletions (Brown et al., 2000; d'Enfert, 1996; Liebmann et al., 2004a,b). Although *A. fumigatus* can utilize many different carbon sources from hexoses to fatty acids or ethanol and amino acids, the role of central transcription factors such as CreA in controlling carbon catabolite repression has not been investigated relative to fungal pathogenicity. Similarly, the progressive destruction of host tissues by degrading enzymes is likely to release amino acids that will be used by primary metabolism to support growth. When the gene encoding the GATA transcription factor AreA, which controls the assimilation of primary and secondary nitrogen sources, is deleted (Hensel et al., 1998), the deletion strain is as virulent as the wild-type strain in a neutropenic model of invasive aspergillosis and only displays a growth delay in lung tissues. Under nitrogen starvation conditions while growing on epithelial cells, expression of the RheB protein RhbA, which is required for virulence, is induced (Panepinto et al., 2002). In contrast, the transcription factor CpcA is also highly expressed under nitrogen starvation but is not associated with fungal virulence (Krappmann et al., 2004). The role of the TOR pathway, which also contributes to nitrogen assimilation, has not been investigated. Adequate nutrient supply for the fungus is required to produce aspergillosis, and many basic aspects of carbon and nitrogen metabolism must be investigated in depth to identify novel transport or metabolic pathways that are important for growth *in vivo*. This task may be very difficult due to the overlapping catabolic pathways responsible for the nutritional versatility of this fungus that are required for its saprotrophic lifestyle.

Signaling Pathways and Morphogenesis. The ability of *A. fumigatus* to colonize different ecological niches, including mammalian lung tissues, requires both sensing of and adaptative responses to environmental changes. This perception requires an effective communication strategy mediated by signal transduction pathways that promote adaptive responses to environmental conditions. The majority of these signaling pathways are also involved in morphogenesis programs, such as conidial germination. Therefore, the inactivation of these signaling pathways primarly affects growth and, consequently, virulence.

The calcium-dependent signaling pathway plays an important role under stress conditions (Cramer et al., 2008; Soriani et al., 2008). Calcineurin is a highly conserved calcium/calmodulin-regulated protein phosphatase that is an important mediator of calcium signaling. Deletion of *cnaA*, which encodes the catalytic A subunit of calcineurin, is not lethal in *A. fumigatus* but is required for polarized growth, normal septation, tissue invasion, and virulence (Juvvadi et al., 2008; Steinbach et al., 2006). Similarly, a mutant lacking the calcineurin-dependent transcription factor CrzA is attenuated for virulence, although it grows normally *in vitro* (Cramer et al., 2008; Soriani et al., 2008). These observations argue that the calcineurin pathway is essential for both hyphal growth and virulence.

A. fumigatus must be able to adapt to neutral pH of the lung and the acidic pH of macrophages (Bignell et al., 2005). The metabolic cascade that controls pH adaptation in *A. fumigatus* is similar to the one extensively analyzed in *A. nidulans*. Specifically, deletion mutants of *pacC* and *palB* are significantly attenuated in virulence in murine models due to their inability to grow at low pH (Bignell et al., 2005).

The cyclic adenosine monophosphate (cAMP)-dependent signaling pathway regulates the growth, development, and morphogenesis of several fungal pathogens. This pathway also participates in the sensing of environmental stress (Liebmann et al., 2003). Adenylyl cyclase (AcyA), under the regulation of a G-protein alpha subunit (GpaB), generates cAMP, which acts as a second messenger to the regulatory protein kinase A subunit (PkaR) (Liebmann et al., 2003; Liebman et al., 2004; Zhao et al., 2006). This interaction releases the two PKA catalytic subunits, PkaC1 and PkaC2, which then phosphorylate downstream targets and trigger appropriate adaptive responses (Grosse et al., 2008). All mutants of this signaling pathway (Δ*gpaB*, Δ*pkaR*, Δ*pkaC1*, and Δ*pkaC2*) are affected in morphogenesis and are less virulent than the wild-type strain in murine models of invasive aspergillosis. In the Δ*gpaB*, Δ*acyA*, and Δ*pkaR* strains, an increase in sensitivity to ROS was demonstrated *in vitro*, even though all mutants contained melanin.

Mitogen-activated protein kinase (MAPK) pathways are found in all eukaryotes and regulate cellular physiology in developmental programs and environmental changes. The existence of at least three MAPK-mediated signal transduction pathways was predicted from the *A. fumigatus* genome based on *Saccharomyces cerevisiae* homologs for osmoadaption (SakA and MpkC), mating/hyphal growth (MpkB), and the regulation of cell integrity (May, 2008). Deletion of MAP-kinase MpkA results in heightened sensitivity to cell wall stress and a considerable growth defect on standard medium; however, this mutant was as virulent as the wild-type strain in murine models of invasive aspergillosis (Valiante et al., 2007). Like Δ*mpkA* deletion, the deletion of an upstream regulator of the high osmolarity glycerol (HOG)–MAPK pathway, Δ*sho1*, also has impaired the *in vitro* growth rate without affecting virulence (Ma et al., 2008). Infection with Δ*sakA* resulted in no decrease in virulence (Du et al., 2006).

CONCLUSION

Molecular data are continuously accumulating for *A. fumigatus*. Proteome, transcriptome, comparative genomics, and RNA sequencing analysis tools are now available and increasingly used to improve the understanding of the fungal growth in the lung parenchyma. However, the sophistication of molecular biology methods has shed insight as to why *A. fumigatus* is pathogenic. Is it only because this fungal species is present in the atmosphere in high concentration and patients at risk are heavily immunocompromised? The inability to identify specific virulence factors may be simply due to the fact that this fungus does not have true virulence factors. If we consider the classical definition of virulence factor as a factor that promotes disease (in other words, a protein for which the deletion of the encoding gene only induces a reduction of growth *in vivo* and not *in vitro*), it would be difficult to identify any *A. fumigatus* gene as a true virulence factor. Pathogenicity of *A. fumigatus* is complex and multifactorial and involves mechanisms that allow adaptation to its ecological niche (Fig. 27.2). The formidable

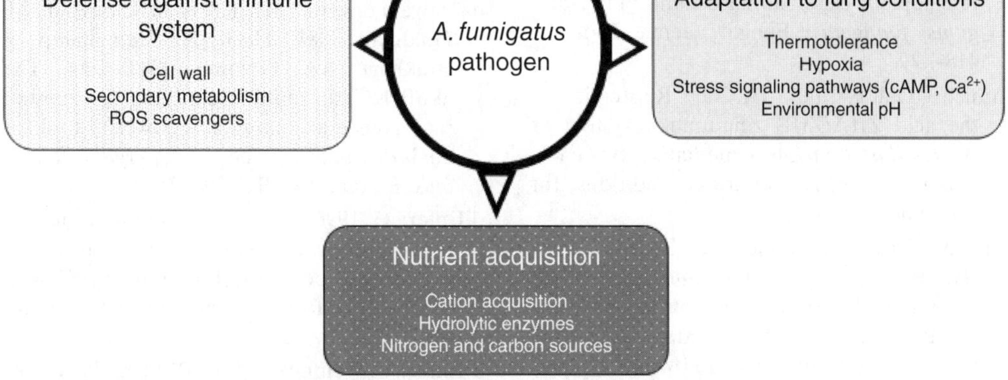

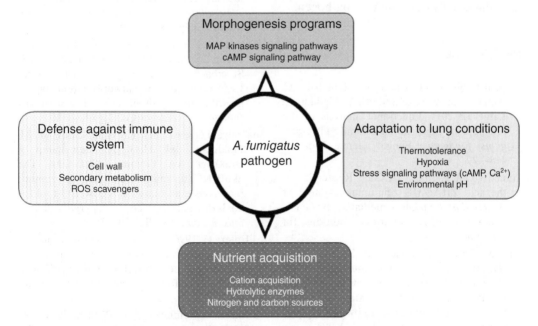

Figure 27.2 Pathogenicity of *A. fumigatus* is multifactorial. Successful *A. fumigatus* infection is dependent on multiple parameters involving several biological processes adapted to a saprophytic lifestyle. These processes permit the fungus to adapt to lung conditions, defend against the immune system, and use available nutrients for morphogenesis programs.

ability of *A. fumigatus* to adapt to many ecological niches facilitates its colonization in lung tissues. Global transcriptional analyses that are recently emerging will also analyze the adaptation mechanisms required for the infectious stage as compared to the saprophytic stage. The development of newer molecular biological techniques will allow us to realize all the multiple deletions required to determine the function of classes of genes with the same product activity, as the presence of many redundant activities considerably complicates the implication of potential virulence factors by single-gene mutagenesis. If the accumulation of this newer molecular data is not useful to identify virulence determinants, it will be essential to pinpoint specific and essential metabolic pathways. A better molecular understanding of the infectious cycle of *A. fumigatus* will also lead to new diagnostic tools and the identification of new drug targets for the subsequent development of new antifungal agents that are urgently needed to fight this deadly opportunistic pathogen.

REFERENCES

Alcazar-Fuoli L, Clavaud C, Lamarre C, Aimanianda V, Seidl-Seiboth V, Mellado E, Latgé JP. 2011. Functional analysis of the fungal/plant class chitinase family in *Aspergillus fumigatus*. Fungal Genet Biol 48: 418–429.

Amich J, Leal F, Calera JA. 2009. Repression of the acid ZrfA/ZrfB zinc-uptake system of *Aspergillus fumigatus* mediated by PacC under neutral, zinc-limiting conditions. Int Microbiol 12: 39–47.

Amich J, Vicente Franqueira R, Leal F, Calera JA. 2010. *Aspergillus fumigatus* survival in alkaline and extreme zinc-limiting environments relies on the induction of a zinc homeostasis system encoded by the *zrfC* and *aspf2* genes. Eukaryot Cell 9: 424–437.

Ben-Ami R, Lewis RE, Leventakos K, Latgé JP, Kontoyiannis DP. 2010. Cutaneous model of invasive aspergillosis. Antimicrob Agents Chemother 54: 1848–1854.

Bhabhra R, Miley MD, Mylonakis E, Boettner D, Fortwendel J, Panepinto JC, Postow M, Rhodes JC, Askew DS. 2004. Disruption of the *Aspergillus fumigatus* gene encoding nucleolar protein CgrA impairs thermotolerant growth and reduces virulence. Infect Immun 72: 4731–4740.

Bignell E, Negrete-Urtasun S, Calcagno AM, Haynes K, Arst HN Jr, Rogers T. 2005. The *Aspergillus* pH-responsive transcription factor PacC regulates virulence. Mol Microbiol 55: 1072–1084.

Bok JW, Balajee SA, Marr KA, Andes D, Nielsen KF, Frisvad JC. 2005. LaeA, a regulator of morphogenetic fungal virulence factors. Eukaryot Cell 4: 1574–1582.

Bok JW, Chung D, Balajee SA, Marr KA, Andes D, Nielsen KF. 2006. GliZ, a transcriptional regulator of gliotoxin biosynthesis, contributes to *Aspergillus fumigatus* virulence. Infect Immun 74: 6761–6768.

Brown JS, Aufauvre-Brown A, Brown J, Jennings JM, Arst H Jr, Holden DW. 2000. Signature-tagged and directed mutagenesis identify PABA synthetase as essential for *Aspergillus fumigatus* pathogenicity. Mol Microbiol 36: 1371–1380.

Cramer RA Jr, Perfect BZ, Pinchai N, Park S, Perlin DS, Asfaw YG, Heitman J, Perfect JR, Steinbach WJ. 2008. Calcineurin target CrzA regulates conidial germination, hyphal growth, and pathogenesis of *Aspergillus fumigatus*. Eukaryot Cell 7: 1085–1097.

da Silva Ferreira ME, Kress MR, Savoldi M, Goldman MH, Härtl A, Heinekamp T, Brakhage AA, Goldman GH. 2006. The akuB(KU80) mutant deficient for nonhomologous end joining is a powerful tool for analyzing pathogenicity in *Aspergillus fumigatus*. Eukaryot Cell 5: 207–211.

d'Enfert C. 1996. Selection of multiple disruption events in *Aspergillus fumigatus* using the orotidine-5'-decarboxylase gene, *pyrG*, as a unique transformation marker. Curr Genet 30: 76–82.

d'Enfert C, Weidner G, Mol PC, Brakhage AA. 1999. Transformation systems of *Aspergillus fumigatus*: New tools to investigate fungal virulence. Contrib Microbiol 2: 149–166.

Du C, Sarfati J, Latge JP, Calderone R. 2006. The role of the *sakA* (Hog1) and *tcsB* (sln1) genes in the oxidant adaptation of *Aspergillus fumigatus*. Med Mycol 44: 211–218.

Eichner RD, Al Salami M, Wood PR, Mullbacher A. 1986. The effect of gliotoxin upon macrophage function. Int J Immunopharmacol 8: 789–797.

Fedorova ND, Khaldi N, Joardar VS, Maiti R, Amedeo P, Anderson MJ, Crabtree J, Silva JC, Badger JH, Albarraq A, Angiuoli S, Bussey H, Bowyer P, Cotty PJ, Dyer PS, Egan A, Galens K, Fraser-Liggett CM, Haas BJ, Inman JM, Kent R, Lemieux S, Malavazi I, Orvis J, Roemer T, Ronning CM, Sundaram JP, Sutton G, Turner G, Venter JC, White OR, Whitty BR, Youngman P, Wolfe KH, Goldman GH, Wortman JR, Jiang B, Denning DW, Nierman WC. 2008. Genomic islands in the pathogenic filamentous fungus *Aspergillus fumigatus*. PLoS Genet 4: e1000046.

Galagan JE, Calvo SE, Cuomo C. 2005. Sequencing of *Aspergillus nidulans* and comparative analysis with *A. fumigatus* and *A. oryzae*. Nature 438: 1105–1115.

Gardiner DM, Howlett BJ. 2005. Bioinformatic and expression analysis of the putative gliotoxin biosynthetic gene cluster of *Aspergillus fumigatus*. FEMS Microbiol Lett 248: 241–248.

Gastebois A, Clavaud C, Aimanianda V, Latgé JP. 2009. *Aspergillus fumigatus*: Cell wall polysaccharides, their biosynthesis and organization. Future Microbiol 4: 583–595.

Gastebois A, Blanc Potard AB, Gribaldo S, Beau R, Latgé JP, Mouyna I. 2011. Phylogenetic and functional analysis of *Aspergillus fumigatus* MGTC, a fungal protein homologous to a bacterial virulence factor. Appl Environ Microbiol 77: 4700–4703.

Geiser DM, Klich MA, Frisvad JC, Peterson SW, Varga J, Samson RA. 2007. The current status of species recognition and identification in *Aspergillus*. Stud Mycol 59: 1–10.

Gibbons JG, Beauvais A, Beau R, McGary KL, Latge JP, Rokas A. 2011. Global transcriptome changes underlying colony growth in the opportunistic human pathogen *Aspergillus fumigatus*. Eukaryot Cell 11: 68–78.

Girardin H, Sarfati J, Kobayashi H, Bouchara JP, Latgé JP. 1994. Use of DNA moderately repetitive sequence to type *Aspergillus fumigatus* isolates from aspergilloma patients. J Infect Dis 169: 683–685.

Grosse C, Heinekamp T, Kniemeyer O, Gehrke A, Brakhage AA. 2008. Protein kinase A regulates growth, sporulation, and pigment formation in *Aspergillus fumigatus*. Appl Environ Microbiol 74: 4923–4933.

Hartmann T, Dümig M, Jaber BM, Szewczyk E, Olbermann P, Morschhäuser J, Krappmann S. 2010. Validation of a self-excising marker in the human pathogen *Aspergillus fumigatus* by employing the beta-rec/six site-specific recombination system. Appl Environ Microbiol 18: 6313–6317.

Hendrickson L, Davis CR, Roach C, Nguyen DK, Aldrich T, McAda PC, Reeves CD. 1999. Lovastatin biosynthesis in *Aspergillus terreus*: Characterization of blocked mutants, enzyme activities and a multifunctional polyketide synthase gene. Chem Biol 6: 429–439.

Henry C, Mouyna I, Latgé JP. 2007. Testing the efficacy of RNA interference constructs in *Aspergillus fumigatus*. Curr Genet 51: 277–284.

Hensel M, Arst HN Jr, Aufauvre-Brown A, Holden DW. 1998. The role of the *Aspergillus fumigatus* AreA gene in invasive pulmonary aspergillosis. Mol Gen Genet 258: 553–557.

Jaton-Ogay K, Paris S, Huerre M, Quadroni M, Falchetto R, Togni G, Latgé JP, Monod M. 1994. Cloning and disruption of the gene encoding an extracellular metalloprotease of *Aspergillus fumigatus*. Mol Microbiol 14: 917–928.

Juvvadi PR, Fortwendel JR, Pinchai N, Perfect BZ, Heitman J, Steinbach WJ. 2008. Calcineurin localizes to the hyphal septum in *Aspergillus fumigatus*: Implications for septum formation and conidiophore development. Eukaryot Cell 7: 1606–1610.

Keller NP, Hohn TM. 1997. Metabolic pathway gene clusters in filamentous fungi. Fungal Genet Biol 21: 17–29.

Krappmann S, Bignell EM, Reichard U, Rogers T, Haynes K, Braus GH. 2004. The *Aspergillus fumigatus* transcriptional activator CpcA contributes significantly to the virulence of this fungal pathogen. Mol Microbiol 52: 785–799.

Krappmann S, Sasse C, Braus GH. 2006. Gene targeting in *Aspergillus fumigatus* by homologous recombination is facilitated in a nonhomologous end-joining-deficient genetic background. Eukaryot Cell 1: 212–215.

Lamarre C, Sokol S, Debeaupuis JP, Henry C, Lacroix C, Glaser P, Coppée JY, François JM, Latgé JP. 2008. Transcriptomic analysis of the exit from dormancy of *Aspergillus fumigatus* conidia. BMC Genomics 9: 417.

Lambou K, Lamarre C, Beau R, Dufour N, Latge JP. 2010. Functional analysis of the superoxide dismutase family in *Aspergillus fumigatus*. Mol Microbiol 75: 910–923.

Langfelder K, Streibel M, Jahn B, Haase G, Brakhage AA. 2003. Biosynthesis of fungal melanins and their importance for human pathogenic fungi. Fungal Genet Biol 38: 143–158.

Latgé JP. 2010. Tasting the fungal cell wall. Cell Microbiol 12: 863–872.

Latgé JP, Steinbach WJ. 2009. Aspergillus *and Aspergillosis*. ASM Press, Washington, DC.

Lessing F, Kniemeyer O, Wozniok I, Loeffler J, Kurzai O, Haertl A, Brakhage AA. 2007. The *Aspergillus fumigatus* transcriptional regulator AfYap1 represents the major regulator for defense against reactive oxygen intermediates but is dispensable for pathogenicity in an intranasal mouse infection model. Eukaryot Cell 6: 2290–2302.

Lewis RE, Wiederhold NP, Chi J, Han XY, Komanduri KV, Kontoyiannis DP, Prince RA. 2005. Detection of gliotoxin in experimental and human aspergillosis. Infect Immun 73: 635–637.

Liebmann B, Gattung S, Jahn B, Brakhage AA. 2003. cAMP signaling in *Aspergillus fumigatusis* involved in the regulation of the virulence gene pksP and in defense against killing by macrophages. Mol Genet Genomics 269: 420–435.

Liebmann B, Müller M, Braun A, Brakhage AA. 2004a. The cyclic AMP-dependent protein kinase a network regulates development and virulence in *Aspergillus fumigatus*. Infect Immun 72: 5193–5203.

Liebmann B, Muhleisen TW, Muller M, Hecht M, Weidner G, Braun A. 2004b. Deletion of the *Aspergillus fumigatus* lysine biosynthesis gene lysF encoding homoaconitase leads to attenuated virulence in a low-dose mouse infection model of invasive aspergillosis. Arch Microbiol 181: 378–383.

Lionakis MS, Kontoyiannis DP. 2005. Fruit flies as a minihost model for studying drug activity and virulence in *Aspergillus*. Med Mycol 43: S111–S114.

Lionakis MS, Kontoyiannis DP. 2010. The growing promise of Toll-deficient *Drosophila melanogaster* as a model for studying *Aspergillus* pathogenesis and treatment. Virulence 1: 488–499.

Lodeiro S, Xiong Q, Wilson WK, Ivanova Y, Smith ML, May GS, Matsuda SP. 2009. Protostadienol biosynthesis and metabolism in the pathogenic fungus *Aspergillus fumigatus*. Org Lett 11: 1241–1244.

Ma Y, Qiao J, Liu W, Wan Z, Wang X, Calderone R, Li R. 2008. The sho1 sensor regulates growth, morphology, and oxidant adaptation in *Aspergillus fumigatus* but is not essential for development of invasive pulmonary aspergillosis. Infect Immun 76: 1695–1701.

Machida M, Asai K, Sano M, Tanaka T, Kumagai T, Terai G, Kusumoto K, Arima T, Akita O, Kashiwagi Y, Abe K, Gomi K, Horiuchi H, Kitamoto K, Kobayashi T, Takeuchi M, Denning DW, Galagan JE, Nierman WC, Yu J, Archer DB, Bennett JW, Bhatnagar D, Cleveland TE, Fedorova ND, Gotoh O, Horikawa H, Hosoyama A, Ichinomiya M, Igarashi R, Iwashita K, Juvvadi PR, Kato M, Kato Y, Kin T, Kokubun A, Maeda H, Maeyama N, Maruyama J, Nagasaki H, Nakajima T, Oda K, Okada K, Paulsen I, Sakamoto K, Sawano T, Takahashi M, Takase K, Terabayashi Y, Wortman JR, Yamada O, Yamagata Y, Anazawa H, Hata Y, Koide Y, Komori T, Koyama Y, Minetoki T, Suharnan S, Tanaka A, Isono K, Kuhara S, Ogasawara N, Kikuchi H. 2005. Genome sequencing and analysis of *Aspergillus oryzae*. Nature 438: 1157–1161.

Maiya S, Grundmann A, Li SM, Turner G. 2006. The fumitremorgin gene cluster of *Aspergillus fumigatus*: Identification of a gene encoding brevianamide F synthetase. Chembiochem 7: 1062–1069.

Maubon D, Park S, Tanguy M, Huerre M, Schmitt C, Prévost MC, Perlin DS, Latgé JP, Beauvais

A. 2005. AGS3, an alpha(1-3)glucan synthase gene family member of *Aspergillus fumigatus*, modulates mycelium growth in the lung of experimentally infected mice. Fungal Genet Biol 43: 366–375.

May GS. 2008. Mitogen-activated protein kinase pathways in *Aspergilli*. In: Goldman GH, Osmani SA, eds. *The* Aspergilli. *Genomics, Medical Aspects, Biotechnology, And Research Methods*. CRC Press, Boca Raton, FL, p. 127.

McDonagh A, Fedorova ND, Crabtree J, Yu Y, Kim S, Chen D, Loss O, Cairns T, Goldman G, Armstrong-James D, Haynes K, Haas H, Schrettl M, May G, Nierman WC, Bignell E. 2008. Sub-telomere directed gene expression during initiation of invasive aspergillosis. PLoS Pathog 4: e1000154.

Mitchell CG, Slight J, Donaldson K. 1997. Diffusible component from the spore surface of the fungus *Aspergillus fumigatus* which inhibits the macrophage oxidative burst is distinct from gliotoxin and other hyphal toxins. Thorax 52: 796–801.

Monteiro MC, De Lucas JR. 2010. Study of the essentiality of the *Aspergillus fumigatus triA* gene, encoding RNA triphosphatase, using the heterokaryon rescue technique and the conditional gene expression driven by the *alcA* and *niiA* promoters. Fungal Genet Biol 47: 66–79.

Moreno MA, Ibrahim-Granet O, Vicentefranqueira R, Amich J, Ave P, Leal F. 2007. The regulation of zinc homeostasis by the ZafA transcriptional activator is essential for *Aspergillus fumigatus* virulence. Mol Microbiol 64: 1182–1197.

Mouyna I, Kniemeyer O, Jank T, Loussert C, Mellado E, Aimanianda V, Beauvais A, Wartenberg D, Sarfati J, Bayry J, Prévost MC, Brakhage AA, Strahl S, Huerre M, Latgé JP. 2010. Members of protein O-mannosyltransferase family in *Aspergillus fumigatus* differentially affect growth, morphogenesis and viability. Mol Microbiol 76: 1205–1221.

Müllbacher A, Moreland AF, Waring P, Sjaarda A, Eichner RD. 1988. Prevention of graft-versus-host disease by treatment of bone marrow with gliotoxin in fully allogeneic chimeras and their cytotoxic T cell repertoire. Transplantation 46: 120–125.

Nierman WC, Pain A, Anderson MJ, Wortman JR, Kim HS, Arroyo J, Berriman M, Abe K, Archer DB, Bermejo C, Bennett J, Bowyer P, Chen D, Collins M, Coulsen R, Davies R, Dyer PS, Farman M, Fedorova N, Fedorova N, Feldblyum TV, Fischer R, Fosker N, Fraser A, García JL, García MJ, Goble A, Goldman GH, Gomi K, Griffith-Jones S, Gwilliam R, Haas B, Haas H, Harris D, Horiuchi H, Huang J, Humphray S, Jiménez J, Keller N, Khouri H, Kitamoto K, Kobayashi T, Konzack S, Kulkarni R, Kumagai T, Lafon A, Latgé JP, Li W, Lord A, Lu C, Majoros WH, May GS, Miller BL, Mohamoud Y, Molina M, Monod M, Mouyna I, Mulligan S, Murphy L, O'Neil S, Paulsen I, Peñalva MA, Pertea M, Price C, Pritchard BL, Quail MA, Rabbinowitsch E, Rawlins N, Rajandream MA, Reichard U, Renauld H, Robson GD, Rodriguez de Córdoba S, Rodríguez-Peña JM, Ronning CM, Rutter S, Salzberg SL, Sanchez M, Sánchez-Ferrero JC, Saunders D, Seeger K, Squares R, Squares S, Takeuchi M, Tekaia F, Turner G, Vazquez De Aldana CR, Weidman J, White O, Woodward J, Yu JH, Fraser C, Galagan JE, Asai K, Machida M, Hall N, Barrell B, Denning DW. 2005. Genomic sequence of the pathogenic and allergenic filamentous fungus *Aspergillus fumigatus*. Nature 438: 1151–1156.

O'Gorman CM, Fuller HT, Dyer PS. 2009. Discovery of a sexual cycle in the opportunistic fungal pathogen *Aspergillus fumigatus*. Nature 457: 471–474.

Panepinto JC, Oliver BG, Amlung TW, Askew DS, Rhodes JC. 2002. Expression of the *Aspergillus fumigatus* rheb homologue, *rhbA*, is induced by nitrogen starvation. Fungal Genet Biol 36: 207–214.

Paoletti M, Rydholm C, Schwier EU, Anderson MJ, Szakacs G, Lutzoni F, Debeaupuis JP, Latgé JP, Denning DW, Dyer PS. 2005. Evidence for sexuality in the opportunistic fungal pathogen *Aspergillus fumigatus*. Curr Biol 15: 1242–1248.

Pardo J, Urban C, Galvez EM, Ekert PG, Muller U, Kwon-Chung J, Lobigs M, Mullbacher A, Wallich R, Borner C. 2006. The mitochondrial protein Bak is pivotal for gliotoxin-induced apoptosis and a critical host factor of *Aspergillus fumigatus* virulence in mice. J Cell Biol 174: 509–519.

Paris S, Wysong D, Debeaupuis JP, Shibuya K, Philippe B, Diamond RD, Latgé JP. 2003. Catalases of *Aspergillus fumigatus*. Infect Immun 71: 3551–3562.

Pel HJ, de Winde JH, Archer DB, Dyer PS, Hofmann G, Schaap PJ, Turner G, de Vries RP, Albang R, Albermann K, Andersen MR, Bendtsen JD, Benen JA, van den Berg M, Breestraat S, Caddick MX, Contreras R, Cornell M, Coutinho PM, Danchin EG, Debets AJ, Dekker P, van Dijck PW, van Dijk A, Dijkhuizen L, Driessen AJ, Enfert C, Geysens S, Goosen C, Groot GS, de Groot PW, Guillemette T, Henrissat B, Herweijer M, van den Hombergh JP, van den Hondel CA, van der Heijden RT, van der Kaaij RM, Klis FM, Kools HJ, Kubicek CP, van Kuyk PA, Lauber J, Lu X, van der Maarel MJ, Meulenberg R, Menke H, Mortimer MA, Nielsen J, Oliver SG, Olsthoorn M, Pal K, van Peij NN, Ram AF, Rinas U, Roubos JA, Sagt CM, Schmoll M, Sun J, Ussery D, Varga J, Vervecken W, van de Vondervoort PJ, Wedler H, Wösten HA, Zeng AP, van Ooyen AJ, Visser J, Stam H. 2007. Genome sequencing and analysis of the versatile cell factory *Aspergillus niger* CBS 513.88. Nat Biotechnol 25: 221–231.

Perrin RM, Fedorova ND, Bok JW, Cramer RA, Wortman JR, Kim HS, Nierman WC, Keller NP. 2007. Transcriptional regulation of chemical diversity in *Aspergillus fumigatus* by LaeA. PLoS Pathogen 3: e50.

Philippe B, Ibrahim-Granet O, Prévost MC, Gougerot-Pocidalo MA, Sanchez Perez M, Van Der Meeren A, Latgé JP. 2003. Killing of *Aspergillus fumigatus* by alveolar macrophages is mediated by reactive oxidant intermediates. Infect Immun 71: 3034–3042.

Pringle A, Baker DM, Platt JL, Wares JP, Latgé JP, Taylor JW. 2005. Cryptic speciation in the cosmopolitan and clonal human pathogenic fungus *Aspergillus fumigatus*. Evolution 59: 1886–1899.

Punt PJ, Van Den Hondel CA. 1992. Transformation of filamentous fungi based on hygromycin B and phleomycin resistance markers. Methods Enzymol 216: 447–457.

Renwick J, Daly P, Reeves EP, Kavanagh K. 2006. Susceptibility of larvae of *Galleria mellonella* to infection by *Aspergillus fumigatus* is dependent upon stage of conidial germination. Mycopathologia 161: 377–384.

Richie DL, Hartl L, Aimanianda V, Winters MS, Fuller KK, Miley MD, White S, McCarthy JW, Latgé JP, Feldmesser M, Rhodes JC, Askew DS. 2009. A role for the unfolded protein response (UPR) in virulence and antifungal susceptibility in *Aspergillus fumigatus*. PLoS Pathog 5: e1000258.

Rokas A, Payne G, Fedorova ND, Baker SE, Machida M, Yu J, Georgianna DR, Dean RA, Bhatnagar D, Cleveland TE, Wortman JR, Maiti R, Joardar V, Amedeo P, Denning DW, Nierman WC. 2007. What can comparative genomics tell us about species concepts in the genus *Aspergillus*? Stud Mycol 59: 11–17.

Schrettl M, Bignell E, Krag LC, Joech LC, Rogers T, Arst HN Jr. 2004. Siderophore biosynthesis but not reductive iron assimilation is essential for *Aspergillus fumigatus* virulence. J Exp Med 200: 1213–1219.

Schrettl M, Bignell E, Kragl C, Sabiha Y, Loss O, Eisendle M. 2007. Distinct roles for intra- and extracellular siderophores during *Aspergillus fumigatus* infection. PLoS Pathog 3: 1195–1207.

Schrettl M, Carberry S, Kavanagh K, Haas H, Jones GW, O'Brien J, Nolan A, Stephens J, Fenelon O, Doyle S. 2010. Self-protection against gliotoxin a component of the gliotoxin biosynthetic cluster, GliT, completely protects *Aspergillus fumigatus* against exogenous gliotoxin. PLoS Pathog 6: e1000952.

Soriani FM, Malavazi I, Da Silva Ferreira ME, Savoldi M, Von Zeska Kress MR, De Souza Goldman MH, Loss O, Bignell E, Goldman GH. 2008. Functional characterization of the *Aspergillus fumigatus* Crz1 homologue CrzA. Mol Microbiol 67: 1274–1291.

Spikes S, Xu R, Nguyen CK, Chamilos G, Kontoyiannis DP, Jacobson RH. 2008. Gliotoxin production in *Aspergillus fumigatus* contributes to host-specific differences in virulence. J Infect Dis 197: 479–486.

Stanzani M, Orciuolo E, Lewis R, Kontoyiannis DP, Martins SL, StJohn LS. 2005. *Aspergillus fumigatus* suppresses the human cellular immune response via gliotoxin-mediated apoptosis of monocytes. Blood 105: 2258–2265.

Steinbach WJ, Cramer RA Jr, Perfect BZ, Asfaw YG, Sauer TC, Najvar LK, Kirkpatrick WR, Patterson TF, Benjamin DK Jr, Heitman J. 2006. Calcineurin controls growth, morphology, and pathogenicity in *Aspergillus fumigatus*. Eukaryot Cell 5: 1091–1103.

Sugui JA, Kim HS, Zarember KA, Chang YC, Gallin JI, Nierman WC, Kwon-Chung KJ. 2008. Genes differentially expressed in conidia and hyphae of *Aspergillus fumigatus* upon exposure to human neutrophils. PLoS ONE 3: e2655.

Tekaia F, Latgé JP. 2005. *Aspergillus fumigatus*: Saprophyte or pathogen? Curr Opin Microbiol 8: 385–392.

Thau N, Monod M, Crestani B, Rolland C, Tronchin G, Latgé JP, Paris S. 1994. *rodletless* mutants of *Aspergillus fumigatus*. Infect Immun 62: 4380–4388.

Thywißen A, Heinekamp T, Dahse HM, Schmaler-Ripcke J, Nietzsche S, Zipfel PF, Brakhage AA. 2002. Conidial dihydroxynaphtalene melanin of the human pathogenic fungus *Aspergillus fumigatus* interferes with the host endocytosis pathway. Front Microbiol 2: 96.

Tsai HF, Wheeler MH, Chang YC, Kwon-Chung KJ. 1999. A developmentally regulated gene cluster involved in conidial pigment biosynthesis in *Aspergillus fumigatus*. J Bacteriol 181: 6469–6477.

Valiante V, Heinekamp T, Jain R, Hartl A, Brakhage AA. 2007. The mitogen-activated protein kinase MpkA of *Aspergillus fumigatus* regulates cell wall signaling and oxidative stress response. Fungal Genet Biol 45: 618–627.

Vödisch M, Scherlach K, Winkler R, Hertweck C, Braun HP, Roth M, Haas H, Werner ER, Brakhage AA, Kniemeyer O. 2011. *Analysis of the Aspergillus fumigatus* proteome reveals metabolic changes and the activation of the pseurotin A biosynthesis gene cluster in response to hypoxia. J Proteome Res 10: 2508–2524.

Vödisch M, Albrecht D, Lessing F, Schmidt AD, Winkler R, Guthke R, Brakhage AA, Kniemeyer O. 2009. Two-dimensional proteome reference maps for the human pathogenic filamentous fungus *Aspergillus fumigatus*. Proteomics 9: 1407–1415.

Wagener J, Echtenacher B, Rohde M, Kotz A, Krappmann S, Heesemann J, Ebel F. 2008. The putative alpha-1,2-mannosyltransferase AfMnt1 of the opportunistic fungal pathogen *Aspergillus fumigatus* is required for cell wall stability and full virulence. Eukaryot Cell 7: 1661–1673.

Willger SD, Puttikamonkul S, Kim KH, Burritt JB, Grahl N, Metzler LJ, Barbuch R, Bard M, Lawrence CB, Cramer RA Jr. 2008. A sterol-regulatory element binding protein is required for cell polarity, hypoxia adaptation, azole drug resistance, and virulence in *Aspergillus fumigatus*. PLoS Pathog 4: e1000200.

Zadra I, Abt B, Parson W, Haas H. 2000. xylP promoter-based expression system and its use for antisense down regulation of the *Penicillium chrysogenum* nitrogen regulator NRE. Appl Environ Microbiol 66: 4810–4816.

Zhao W, Panepinto JC, Fortwendel JR, Fox L, Oliver BG, Askew DS, Rhodes JC. 2006. Deletion of the regulatory subunit of protein kinase A in *Aspergillus fumigatus* alters morphology, sensitivity to oxidative damage, and virulence. Infect Immun 74: 4865–4874.

CHAPTER 28

CRYPTOSPORIDIUM: COMPARATIVE GENOMICS AND PATHOGENESIS

SATOMI KATO and JESSICA C. KISSINGER

INTRODUCTION

Cryptosporidium parasites are unicellular eukaryotes that belong to the phylum Apicomplexa. *Cryptosporidium* was first described in mice in the early 1900s (Tyzzer, 1907), but the first cases in humans were not reported until 1976 (Nime et al., 1976) including a case associated with immunosuppression (Meisel et al., 1976). Interest in human *Cryptosporidium* infections increased significantly as a result of its noted association with immunocompromised individuals (Anonymous, 1982). *Cryptosporidium* parasites cause an enteric disease, cryptosporidiosis, in both humans and animals.

Transmission of *Cryptosporidium* is via the oral–fecal route. Infection with *Cryptosporidium* causes self-limiting diarrhea, nausea, headache, and vomiting. Infection begins with ingestion of an oocyst. Sporozoites, which are released from the oocysts, infect intestinal epithelial cells. Following several rounds of replication and development to produce merozoite and gametocyte stages, oocysts are formed. Two types of oocysts are produced; one form is autoinfective and can perpetuate the infection *in situ*. The second form is released into the environment to await ingestion and completion of the life cycle (Borowski et al., 2010; Fayer, 2008).

Cryptosporidium is a significant AIDS-related pathogen. For immunocompetent individuals, the symptoms last only for approximately 10–14 days; however, for immunocompromised individuals, the symptoms are chronic and often life threatening (Chalmers and Davies, 2010). Cryptosporidiosis is a particular problem in children in developing countries where malnutrition, complicated by diarrhea, results in stunted growth. The World Health Organization lists *Cryptosporidium* as an important protozoan pathogen of concern to public health in its guidelines for drinking water quality (WHO, 2006).

The transmissible stage of this parasite, the oocyst, is quite resistant to environmental stress including chlorination of water; therefore, disseminated oocysts can survive for a long period of time in the environment, especially in water. The long survival of oocysts facilitates infection through contaminated water, food, or fomites, and by person-to-person (anthroponotic transmission) or animal-to-person (zoonotic trans-

Evolution of Virulence in Eukaryotic Microbes, First Edition. Edited by L. David Sibley, Barbara J. Howlett, and Joseph Heitman.

TABLE 28.1 Internet Resources for *Cryptosporidium* Research

Resource	Internet URL	Data Available
CryptoDB	http://CryptoDB.org	Genomic, transcriptomic, proteomic, SNP, isolates
J. Craig Venter Institute	http://gsc.jcvi.org/projects/msc/cryptosporidium_muris/	*Cryptosporidium muris* genome and transcriptome
Cryptosporidium outbreak data	http://www.cdc.gov/foodborneoutbreaks/Default.aspx	Searchable Centers for Disease Control and Prevention outbreak database
Full-Length cDNA	http://fullmal.hgc.jp/index_cp_ajax.html	*Cryptosporidium* full-length cDNAs
World Health Organization	http://www.who.int/water_sanitation_health/publications/cryptoRA/en/index.html	2009 water risk assessment report
Centers for Disease Control and Prevention	http://www.cdc.gov/parasites/crypto/	Current information on risks and treatment
National Institutes of Health Category A–C Priority Pathogens	http://www.niaid.nih.gov/topics/biodefenserelated/biodefense	Complete list of pathogens prioritized as A–C and the National Institute of Allergy and Infectious Diseases biodefense overview

mission) contact. Cryptosporidiosis is endemic in developing countries, whereas epidemic waterborne and foodborne outbreaks have occurred in developed countries. Because of the risk of contaminated drinking water sources, *Cryptosporidium* spp. are listed as Category B Priority Pathogens for Bioterrorism (Table 28.1).

Species and Genotypes

Currently, 21 species and greater than 40 genotypes of *Cryptosporidium* have been recognized. They infect not only humans but also a wide variety of domestic and wild animals, including ruminants, primates, felines, and canines (Fayer et al., 2009) (Table 28.2). Historically, new species and genotypes/subtypes have been defined when *Cryptosporidium* oocysts are isolated from new host species. However, this approach does not address potential cryptic species or species that may be able to infect multiple hosts. Therefore, it is argued that species should be determined on the basis of genetic markers. It is therefore imperative to identify markers appropriate for the identification of *Cryptosporidium* spp.

Cryptosporidium parasites have been organized into 21 recognized species and numerous recognized genotypes based on molecular sequence tags (Fayer et al., 2009; Xiao and Fayer, 2008; Plutzer and Karanis, 2009). Of these, human infections have been observed with eight species: *Cryptosporidium hominis*, *Cryptosporidium parvum*, *Cryptosporidium meleagridis*, *Cryptosporidium felis*, *Cryptosporidium canis*, *Cryptosporidium suis*, *Cryptosporidium muris*, and *Cryptosporidium andersoni*, and seven of the recognized genotypes: *C. hominis* monkey genotype, *C. parvum* mouse genotype, and skunk, horse, rabbit, cervine, and chipmunk I genotypes (Ajjampur et al., 2007; Chalmers and Davies, 2010; Robinson et al., 2008; Xiao and Ryan, 2008). Two species, *C. hominis* and *C. parvum*, are responsible for the majority of human infections. *C. hominis* primarily infects humans, whereas *C. parvum* is mainly a parasite of young ruminants but can be transmitted by both zoonotic and anthroponotic routes.

TABLE 28.2 ***Cryptosporidium*** **Species, Hosts, and Observed Geographic Location**

Species	Subtype (*gp60*)	Host Range	Predominant Areas of Isolation	Representative References
C. hominis	Ia	Human, environment	United States, Kenya, Peru, India, Bangladesh	Cama et al., (2008), Gatei et al. (2007), Hira et al. (2011), Peng et al. (2001), Strong et al. (2000)
	Ib	Human, lamb, environment	Australia, France, United States, China, United Kingdom, South Africa	Cohen et al. (2006), Waldron et al. (2009a,b) and Chalmers et al. (2009), Feng et al. (2009), Leav et al. (2002), Strong et al. (2000)
	Id	Human, calf, environment	Australia, Kenya, China, South Africa, Peru, India	Gatei et al. (2007), Leav et al. (2002), Sturbaum et al. (2003), Waldron et al. (2009a,b)
	Ie	Human, environment	United States, Kenya, Bangladesh, Peru, China, Japan	Hira et al. (2011), Peng et al. (2001)
	If	Human, environment	South Africa, Bangladesh, Australia, China	Feng et al. (2009), Hira et al. (2011), Leav et al. (2002)
	Ig	Human	United Kingdom, Australia	Chalmers et al. (2008), Glaberman et al. (2002), Jex et al. (2008b)
	Ih	Child	Nigeria	Molloy et al. (2010)
	Ii	Human	Bangladesh	Hira et al. (2011)
C. parvum	IIa	Human, animal, environment	United States, Australia, United Kingdom, The Netherlands, Slovenia, Canada, New Zealand, Saudi Arabia	Al-Brikan et al. (2008), Grinberg et al. (2008), Soba and Logar (2008), Strong et al. (2000), Trotz-Williams et al. (2006), Waldron et al. (2009a,b), Wielinga et al., (2008)
	IIb	Human	Portugal	Alves et al. (2003), Peng et al. (2001)
	IIc Further divided into subgroups a—i	Human, environment	South Africa, Australia, Germany, Portugal, Peru, United States	Cama et al. (2008); Dyachenko et al. (2010); Leav et al. (2002)
	IId	Human, animal	Spain, Kuwait, Portugal, Sweden, United Kingdom, The Netherlands, Hungary	Alves et al. (2006), Leoni et al. (2007b), Plutzer and Karanis (2007), Quilez et al. (2008a,b), Silverlås et al. (2010), Sulaiman et al. (2005), Wielinga et al. (2008)
	IIe	Human	Bangladesh, Malawi, Uganda	Hira et al. (2011), Jex et al. (2008b), Peng et al. (2003a)
	IIf	Child	Kuwait	Sulaiman et al. (2005)

	IIg	Child	Uganda	Akiyoshi et al. (2006)
	IIh	Child	Uganda	Akiyoshi et al. (2006)
	Iii	Child	Uganda	Akiyoshi et al. (2006)
	IIk	Raccoon dog	Japan	Abe et al. (2006)
	III	Human, animal	Serbia, Slovenia, Lithuania	Misc and Abe (2007), Soba and Logar (2008), Wielinga et al. (2008)
C. meleagridis	IIIa	Human, turkey	Kenya, United States	Glaberman et al. (2001)
	IIIb	Human, turkey	Peru, Kenya, United States	Glaberman et al. (2001)
	IIIc	Child	Peru	Glaberman et al. (2001)
	IIId	Human	India	Glaberman et al., (2001)
	IIIe	Child	India	Ajjampur et al. (2007)
	IIIf	Human	Australia	Jex et al. (2007b)
C. fayeri	IVa	Kangaroo	Australia	Power (2010)
	IVb	Koala, bandicoot	Australia	Power (2010)
	IVc	Kangaroo	Australia	Power (2010)
	IVd	Kangaroo	Australia	Power (2010)
	IVe	Wallaby	Australia	Power (2010)
	IVf	Kangaroo	Australia	Power (2010)
C. cuniculus	Va	Human, rabbit, environment	United Kingdom	Chalmers et al. (2009b), Robinson and Chalmers (2010)
	Vb	Rabbit	China, Czech Republic	Chalmers et al. (2009b)
Horse genotype	Via	Calf, horse	United Kingdom, Czech Republic	Xiao et al. (2009), Thompson et al. (2007)
	VIb	Human	United States	Xiao et al. (2009)
Hedgehog genotype	VIIa	European hedgehog	Germany	Dyachenko et al. (2010)
C. ubiquitum		Human, ruminant, rodent, carnivore	United States	Chalmers et al. (2011b), Fayer et al. (2010), Shen et al. (2011), Sweeny et al. (2011)
C. bovis		Bovine	United States	Fayer et al. (2005)
C. felis		Feline	Peru, United Kingdom	Cama et al. (2008), Elwin et al. (2012)
C. canis		Canine	Peru, United Kingdom	Cama et al. (2008), Elwin et al. (2012)
C. baileyi		Chicken	United States, Japan, China	Abe and Makino (2010); Amer (2010); Current et al. (1986); Nakamura et al. (2009)

Infection with *Cryptosporidium*, Diagnosis, and Treatment

Experimental human infections with *Cryptosporidium* have been studied at the University of Texas Health Science Center in Houston (Okhuysen et al., 1999). *C. parvum* oocysts were isolated from cattle in Maine, Iowa, and Texas, and ID_{50} levels were determined using human volunteers. The studies revealed that the ID_{50} for each isolate, Maine, Iowa, and Texas, were 1042, 87, and 9 oocysts, respectively. The ID_{50} for the IOWA strain was previously shown to be 132 oocysts (Chappell et al., 1996). In contrast, the ID_{50} for *C. hominis* has been shown to be 10 oocysts for healthy adults (Chappell et al., 2006). Taken together, these studies indicate that the infectious dose is not determined solely by the species of *Cryptosporidium* and involves, as of yet, undetermined, virulence factors.

Experimental infection of gnotobiotic pigs with *C. hominis* and *C. parvum* revealed that the infection with *C. parvum* exhibits moderate to severe villus/mucosal attenuation with lymphoid hyperplasia of the entire small intestines, whereas the infection with *C. hominis* exhibited mild to moderate attenuation and lesions that were localized to the ileum and colon (Pereira et al., 2002).

The most commonly used and cost-effective diagnostic method for cryptosporidiosis is identification of excreted *Cryptosporidium* oocysts by microscopic observation with acid-fast staining. Immunoflorescence antibody staining can also be used. Contaminated water can easily be screened for *Cryptosporidium* oocysts using immunomagnetic bead separation techniques (USEPA, 2005a,b). Since it is impossible to distinguish the species of *Cryptosporidium* microscopically, molecular diagnostic methods such as a polymerase chain reaction combined with restriction fragment length polymorphisms (PCR-RFLP) of the small subunit ribosomal RNA gene are used extensively. Acquired immunological responses are also identified using enzyme-linked immunosorbent assay (ELISA) and Western blotting; these methods are used to identify *Cryptosporidium* exposure history (Egorov et al., 2010; Petry et al., 2010).

There is no effective treatment thus far for treating all cases of cryptosporidiosis. Nitazoxanide was approved relatively recently for the treatment of cryptosporidiosis in immunocompetent children and adults. However, it is not approved for treatment in infants and in immunocompromised adults (Rossignol, 2010; Rossignol et al., 2001), a very large segment of the infected population. Prophylaxis and chemotherapy options for both humans and animals are reviewed in Rossignol (2010) and in Stockdale et al. (2008). Because of the difficulty of working with *Cryptosporidium* (see below), the development and testing of new therapeutic compounds is greatly inhibited but not impossible (Sharling et al., 2010). There are a few leads for potential new targets (Macpherson et al., 2010; Sullivan et al., 2005; Umejiego et al., 2004) and therapeutic compounds (Gargala et al., 2010). The development of an effective vaccine for the treatment of cryptosporidiosis will require the development of a more detailed understanding of the immune response to *Cryptosporidium* infection (Mead, 2010).

Barriers to the Investigation of *Cryptosporidium*

A major impediment to the study of *Cryptosporidium* host–parasite interactions and pathogenesis is the lack of an *in vitro* propagation system. Although completion of the life cycle of both *C. parvum* and *C. hominis* in cell-free culture has been reported (Hijjawi, 2010; Hijjawi et al., 2004, 2010), researchers have had problems reproducing the results (Girouard et al.,

2006). Excysted sporozoites can infect mammalian epithelial cell lines such as HCT-8 cells; however, the life cycle of *Cryptosporidium* cannot be completed and researchers cannot propagate oocysts for further research (Wanyiri and Ward, 2006). Thus, *in vitro* genetic manipulation of *Cryptosporidium* has not yet been achieved. However, genetic crosses *in vivo* are possible and will prove useful once significant markers for the relevant strains are developed. The community is also lacking a transient DNA transfection system effectively limiting approaches for the study of protein functions and protein–protein interactions in this parasite. Finally, many *Cryptosporidium* surface proteins are glycosylated and improperly produced in recombinant systems (Wanyiri and Ward, 2006).

In Vivo Experimental Systems

In order to propagate *Cryptosporidium* parasites, a bovine system (Arrowood and Sterling, 1987) and a murine system (Petry et al., 1995; Tzipori et al., 1994) have been used. In the bovine system, neonatal calves are used. In the mouse model, neonatal and several immunodeficient rodent models have been used (Tzipori and Widmer, 2008). For *C. hominis*, gnotobiotic piglets have been used in order to avoid contamination with other bovine *Cryptosporidium* species (Akiyoshi et al., 2002).

PATHOGENESIS AND VIRULENCE

Little is known about the pathogenesis of *Cryptosporidium* and no relationship has been observed between the histological intensity of *Cryptosporidium* infection and the intensity of clinical symptoms (Manabe et al., 1998; Thompson, 2009). There are many steps and many interactions that occur along the path from ingestion of *Cryptosporidium* oocysts to infection and clinical symptoms.

Excystation

Excystation is the natural process that results in the release of infectious sporozoites from oocysts following ingestion. Ingested oocysts will excyst in human/animal intestines and produce motile infective sporozoites. Only sporozoites can infect host epithelial cells. *In vitro* excystation can be triggered by physical and chemical stimulations, many of which mimic ingestion by a host, such as exposure to increased temperatures, changes in pH, and bile salts (Borowski et al., 2008; Upton et al., 1994). Following excystation, upregulation of heat shock proteins has been reported in sporozoites (Cohn et al., 2010). There is preferential transcription of genes encoding structural proteins; genes involved in signaling, proliferation, and metabolism, and many proteins of unknown function (Jenkins et al., 2011).

Sporozoite Cell Invasion

The primary infection site in most organisms is the small intestine; however, the infection can be spread in the entire gastrointestinal tract and in extraintestinal locations such as the lungs, biliary tract, and pancreas (Chalmers and Davies, 2010; Fayer, 2008). Sporozoites will recognize intestinal epithelial cells with their anterior (apical) end and invade the epithelial cells using their apical complex (Wetzel et al., 2005). Components of the apical complex, the conoid and a variety of secretory organelles, play an important role in host cell invasion. Proteins are secreted or released from rhoptories, micronemes, and electron-dense granules at the apical end of the parasite (Fayer and Xiao, 2008). Several surface and apical complex proteins have been identified that are involved in host–parasite interactions including important gliding roles during sporozoite cell invasion (Table 28.3).

TABLE 28.3 ***Cryptosporidium* Surface Complex Receptor Proteins Related to Host–Parasite Interactions and Their Characteristics**

Name	Localization	Size	Possible Function(s) and Characteristics	References
CPA-135	Apical region of zoites, micronemes	135 kDa (566 aa)	Gliding motility, homology with other apicomplexan proteins	Dessens et al. (2004), Tosini et al. (2004)
CP2	Surface of sporozoites, parasitophorous vacuole membrane (PVM)	82 kDa (711 aa)	Host cell invasion	Chen et al. (2004b), O'Hara et al. (2004)
CP12	Oocyst surface, apical region of sporozoites	12 kDa (104 aa)	Host cell adhesion	Yao et al. (2007)
CP47	Apical region of zoites	47-kDa gene not identified	Host cell invasion	Nesterenko et al. (1998)
Circumsporozoite-like glycoprotein (CSL)	Apical region of zoites	1300 kDa, gene not identified	Zoite infectivity (virulens), host cell adhesion	Langer and Riggs (1999), Langer et al. (2001), Riggs et al. (1997)
Gal/GalNAc lectin (P30)	Surface of sporozoites	30 kDa	Host cell adhesion, Binding to intestinal mutin	Bhat et al. (2007), Hashim et al. (2006)
GP15/17 (C17)	Surface of zoites	14–16 kDa (326-aa precursor)	Gliding motility, host cell adhesion	Cevallos et al. (2000), O'connor et al. (2007), Wanyiri et al. (2007)
GP25-200	Surface pellicle and apical region of zoites	25–200 kDa	Zoite infectivity (virulence)	Arrowood et al. (1989), Riggs et al. (2002)
GP40	Surface and apical region of zoites	40 kDa (326 aa percursor)	Host cell adhesion, gliding motility	Cevallos et al. (2000), O'connor et al. (2007), Wanyiri et al. (2007)
GP900	Apical region of zoites, micronemes	900 kDa (1832 aa)	Gliding motility, host cell adhesion, binding to intestinal mucin	Barnes et al. (1998), Bonnin et al. (2001), Cevallos et al. (2000b)
P23/27	Surface of zoites	23/27 kDa (213 aa)	Gliding motility	Perryman et al. (1996), Shirafuji et al. (2005), Takashima et al. (2003)
TRAP-C1/P786	Apical region of sporozoites	190 kDa/76 kDa	Gliding motility, host cell adhesion, homology with other apicomplexan proteins	Deng et al. (2002), Okhuysen et al. (2004), Spano et al. (1998)

Data are summarized from Borowski et al. (2008), Boulter-Bitzer et al. (2007), and Wanyiri and Ward (2006).

Attachment. Although the mechanism of sporozoite attachment to host epithelial cells is not fully understood, there are several surface and apical complex proteins that are believed to mediate sporozoite attachment to the mucus barrier on the ileum (Borowski et al., 2008). *Cryptosporidium* surface proteins that play a role in host–parasite interactions have been summarized in Wanyiri and Ward (2006) and in Fayer et al. (2009). Proteins and protein complexes that mediate attachment to mucus include GP900, a high molecular weight glycoprotein localized to micronemes; circumsporozoite-like glycoprotein (CSL); and GP15 and GP40 (Table 28.3).

Sporozoites contain P30, a 30-kDa Gal/GalNAc-specific lectin on their surface. It is associated with GP900 and GP40 and mediates sporozoite attachment to host mucus (Bhat et al., 2007). Other surface protein complexes that are recognized in the process of the sporozoite attachment to host cells are CPA135, thrombospondin (TSP)-related adhesive protein of *Cryptosporidium*-1 (TRAP-C1), CP47, and CP2 (Table 28.3). A recent study suggested that mucin-like sporozoite surface proteins such as GP900, GP40, and P30 tether sporozoites to the inner surface of the oocyst wall (Chatterjee et al., 2010). It is suggested that proteases *C. parvum* subtilase-1 (CpSUB1) and *C. hominis* subtilase-1 (ChSUB1) play an important role in the parasite infection to host cells by processing GP40/15 (Table 28.3).

Gliding Motility. Gliding motility relies on a conserved actin–myosin motor for motility and its widely conserved feature of the Apicomplexa, which requires receptor–host cell ligand interactions (Sibley, 2004; Wetzel et al., 2005; Yao et al., 2007). In addition, to mediate gliding motility, protein receptors need to be exhibited on the zoite surface by an apical organelle discharge (Chen et al., 2004b). One of the sporozoite micronemal proteins that shed by gliding sporozoites is the TRAP-C1 that promotes gliding of infective zoites (Deng et al., 2002; Putignani et al., 2008; Spano et al., 1998). Other receptor proteins that are believed to work for gliding motility are GP900, CPA135, and GP40 (Table 28.3).

Invasion. There are several parasite-derived enzymes that may be involved in the sporozoite cell invasion process: serine and cysteine proteases (Forney et al., 1996; Snelling et al., 2007), arginine aminopeptidase (Okhuysen et al., 1994), and secretory phospholipase A_2 ($_{S}PLA_2$) (Pollok et al., 2003). Secretory phospholipase A_2 is a virulence factor for infectious microorganisms and also an important enzyme for sporozoite epithelial cell invasion by triggering the release of rhoptry proteins in the apical region of sporozoites (Borowski et al., 2008; Wanyiri and Ward, 2006).

Actin Remodeling and the Rho-GTPase, CDC42. The role of actin polymerization in both the parasite and host cell have been studied (Chen et al., 2004a,c, 2005; Elliott and Clark, 2000; Hashim et al., 2006; Perkins et al., 1999; Pollok et al., 2003; Wetzel et al., 2005). As mentioned above, only parasite actin polymerization appears to be important in invasion (Wetzel et al., 2005). However, following invasion, host actin polymerization appears to have a role in forming the membranous protruding structures that eventually envelop the parasite at the attachment site (Chen et al., 2004a,c; O'Hara et al., 2008). Also, *in vitro* studies have suggested that *Cryptosporidium* invasion of epithelial cells can trigger a host cell signaling pathway for phosphatidylinositol 3-kinase (PI3K) and activates CDC42 and downstream neural Wiskott–Aldrich syndrome protein (N-WASP) and P34-ARC that induce actin remodeling to produce a plaque formation on the host cell apical membrane (Borowski et al., 2008; Chen et al., 2004a,c; O'Hara et al., 2008)

Parasitophorous Vacuole. The parasite shows a unique characteristic of localized in an extracytoplasmic or epicellular location within host cells, connected via a myzocytosis-like feeding/attachment structure. This structure is one of the characteristics that *Cryptosporidium* shares with gregarines. Within the parasitophorous vacuole on the surface of intestinal epithelial cells, *Cryptosporidium* undergoes replication via asexual and sexual cycles (Borowski et al., 2008). Thick-walled and thin-walled oocysts are produced after the sexual cycle; thick-walled oocysts will be excreted in feces and thin-walled oocysts will excyst in the lumen to self-infect epithelial cells. The infection causes villus flattening that may result in malabsorption and diarrhea (Chalmers and Davies, 2010).

Feeder Organelle. Feeding organelle may exhibit the ancestral myzocytotic morphological modification (Borowski et al., 2008, 2010; Huang et al., 2004). The feeder organelle is considered to be a highly membranous "tunnel-like" structure that is used to obtain necessary nutrients from infected host cells (Borowski et al., 2008; Thompson et al., 2005). The *C. parvum* ATP-binding cassette protein (CpABC), a membrane protein, has been shown to be located at the host–parasite interface. It has been suggested that the feeder organelle is formed by the CpABC for transporting substrates (Borowski et al., 2008; Perkins et al., 1999).

Host Response

Human immune responses to *Cryptosporidium* infection are not completely understood because of the complex innate and adaptive immune responses involved. There are limited studies of mucosal immune responses in human *Cryptosporidium* infections. Mucosal epithelial immunity is the first defense mechanism that may play an important role in host–parasite interactions (Gong et al., 2010). Several *in vitro* studies and an *in vivo* mouse study revealed that upon recognition of *Cryptosporidium* cell invasion, toll-like receptors (TLRs) activate a set of adaptor myeloid differentiation protein 88 (MyD88) proteins that regulate the expression of cytokine and chemokines, which then regulate the nuclear translocation of nuclear factor κB (NF-κB), the Janus kinase (JAK) signaling pathway, c-Jun activation, and the expression of inflammatory cytokine genes (Chen et al., 2005, 2007; Hu et al., 2010; Rogers et al., 2006). In addition, *in vitro* studies have revealed the immune defense regulatory mechanisms of microRNAs (miRNAs), endogenous small regulatory RNAs, upon *Cryptosporidium* infection of epithelial cells (Gong et al., 2010; Hu et al., 2010; Zhou et al., 2009).

Cell-mediated immune responses are a major component of the human immune response to *Cryptosporidium* (Board and Ward, 2010). The AIDS patient studies done by Pozio et al. (1997) and Schmidt et al. (2001) suggested that T-cell immune response pathways that are mediated by CD4+ T-cell immune responses play an important role for suppressing cryptosporidiosis, although the mechanisms of this immune response is not clearly understood (Petry et al., 2010).

In human volunteer studies, mucosal interferon-gamma (IFN-γ) production from CD4+ and CD8+ cells was increased in volunteers with *Cryptosporidium* infection compared to volunteers with no previous infection and *ex vivo* stimulation, indicating that in the case of reinfection with *Cryptosporidium*, infection can be greatly suppressed by IFN-γ-mediated memory responses (Gomez Morales et al., 1999; Preidis et al., 2007; White et al., 2000).

There are several studies of the humoral immune response. Following infection with *Cryptosporidium*, antibody responses have been observed. Several *Cryptosporidium* proteins have been shown to mediate host–parasite immune reactions: CP21 (Yao et al., 2007; Yu et al., 2010), CpP2 (Priest et al., 2010), CP41 (Kjos et al., 2005), S16 (Boulter-Bitzer et al., 2010), CP12, CPS-500, Gal/GalNac lectin (p30) (Borowski et al., 2008), CP2, CP47, Cpa135 (Borowski et al., 2008; Wanyiri and Ward, 2006), GP15/17, GP40, GP900, P23/27/CP23, GP25-200, CP15, CSL, and thrombospondin-related adhesive protein (TRAP)-C1/P786 (Boulter-Bitzer et al., 2007). The characteristics of these antigens are nicely summarized in Borowski et al. (2008), Boulter-Bitzer et al. (2007), and Wanyiri and Ward (2006).

EVOLUTION, GENOMICS, AND POPULATION STRUCTURE

The exact taxonomic affiliation of *Cryptosporidium* has changed over time as methods for morphological, developmental, and molecular assessment have evolved. The parasite itself contains very few morphological features that facilitate classification at the species level. Cryptosporidia contain an apical complex (conoid, and associated secretory organelles) clearly identifying it as an apicomplexan. Within the Apicomplexa, it has historically been grouped with other spore-producing, gut-infecting parasites in the class Coccidea. However, there are morphological and developmental features of *Cryptosporidium* that are similar to gregarines. The feeder organelle of *Cryptosporidium* is very similar to the epimerite of gregarines (Tyzzer, 1907; Valigurova et al., 2007). There are also features of *Cryptosporidium* that are quite different from the coccidia even though all of these parasites inhabit the same environment (Barta and Thompson, 2006). For example, even though *Cryptosporidium* is intracellular, it is extracytoplasmic (Thompson et al., 2005). When combined, the morphological and developmental observations are prompting the community to seriously reevaluate their views and assumptions about *Cryptosporidium* (Thompson, 2009).

Phylogenetic Analyses

Molecular analyses have also challenged the taxonomic classification of *Cryptosporidium*. The first molecular evidence came from analyses of the small subunit of ribosomal RNA (Carreno et al., 1999), which suggested that *Cryptosporidium* was more closely related to gregarines than coccidia. Subsequent analyses included additional genes, including protein-encoding genes and, importantly, additional apicomplexan species, including diverse gregarines (Leander et al., 2003a,b). These studies confirmed that *Cryptosporidium* did not group with the coccidia and instead placed them as the sister group to the gregarines. In 2010, a genome survey study of *Ascogregarina taiwanensis* supported the sister group relationship of *Cryptosporidium* with gregarines but also revealed significant metabolic differences, often a lack of particular pathways in *Cryptosporidium* associated with mitochondria, nucleotide, and amino acid synthesis (Templeton et al., 2010). A recent survey of phylogenetic relationship within the larger alveolate taxonomic super group identified *Cryptosporidium* as the most difficult to place taxon, and the author specifically tested hypotheses that would place *Cryptosporidium* within the coccidia (Bachvaroff et al., 2011). It is possible that the inconclusive results stem from the lack of gregarine sequences contained in the analysis.

Comparative Genomic Insights

Genome sequences are available for three species of *Cryptosporidium*: *C. parvum* (Abrahamsen et al., 2004; Bankier et al., 2003), *C. hominis* (Xu et al., 2004); and *C. muris* (Carneiro et al., in preparation). The *Cryptosporidium* genome is small, ~9.1 Mb contained in eight chromosomes. The *C. parvum* sequence, IOWA strain, is the most complete and has been reassembled into eight scaffolds ranging is size from 0.875 to 1.3 Mb each. The *C. hominis* genome sequence exists in 1422 contigs and *C. muris* is in 45 contigs. The data are available at CryptoDB (Heiges et al., 2006) and at JCVI (Table 28.1). The genomes are highly reduced in gene number with only ~3900 protein-encoding genes per genome, 1500–2300 of which are hypothetical and of unknown function depending upon the species. *C. parvum* and *C. hominis* are distinct species with host preferences, yet at the genomic level, they are 95–98% identical, depending upon the region of the genome examined. Their gene contents (as best as they can be determined with incomplete genome sequences) are nearly identical (Xu et al., 2004). *C. mur*is is 70–85% identical to *C. parvum*, depending upon the region.

The genome sequences confirmed the highly reduced metabolic repertoire of each species and the heavy dependence of the parasite upon uptake of nutrients including nucleobases and nucleosides as well as numerous amino acids from their host via an expanded number of transporters. *Cryptosporidium* has twice as many transporters as *Plasmodium falciparum*, nearly 70 (Abrahamsen et al., 2004).

No large families of subtelomeric genes involved in antigenic variation were discovered. A few families of proteins rich in serine and threonine and containing signal peptides were discovered, but they are of unknown function (Abrahamsen et al., 2004). The genome sequences confirmed the lack of both a mitochondrial and apicoplast genome. With the loss of these genomes and in the case of the apicoplast, the organelle itself (a degenerate mitochondrial organelle is present), many potential and known drug targets in other Apicomplexa are absent.

Gene Expression

Expressed sequence tag data were available prior to generation of the genome sequences (Strong and Nelson, 2000). The EST sequences provided several insights into possible therapeutic targets including S-adenosylhomocysteine hydrolase, histone deacetylase, polyketide/fatty acid synthases, various cyclophilins, TSP-related cysteine-rich protein and ATP-binding cassette transporters. Several sequences containing signal peptides were also identified.

In concert with the *C. muris* genome project and following the *C. parvum* genome project, several cDNA libraries were generated and sequenced. Because all life cycle stages of the parasite are intracellular except for the oocyst and the sporozoite stages, most cDNA libraries are generated from these stages. There is one 24–48 postinfection library that was sequenced and submitted to the GenBank. Currently, there are a total of nine EST sequence data sets available for searching on CryptoDB that were generated by several different research groups from several different strains (Strong and Nelson, 2000) (Table 28.1). These data are a rich source of data for single-nucleotide polymorphism (SNP) detection. In addition to transcript expression data sets, there are 14 proteomic mass spectrometry data sets available for different protein fractions, each of which focuses on a different aspect of *Cryptosporidium* biology and/or structure.

Whole-genome sequences and some expression data from *Cryptosporidium* species have provided the necessary data for researchers to develop microsatellite and minisatellite markers for advanced population genetic studies to elucidate population genetic structures, temporal and geographic spread, transmission dynamics, and potentially, the pathobiology of *Cryptosporidium* species (see Fig. 28.1).

Figure 28.1 Screenshot of the *Cryptosporidium* Database CryptoDB. Most molecular data (genomic, transcriptomic, proteomic, etc.) are housed and available for comparative analysis and complex querying at CryptoDB. In this figure, the second panel, "Identify Other Data Types," is expanded to reveal a rich set of tools that can be used to query isolate data from epidemiological studies. Particular isolates can be identified by geographic region, species, molecular marker used for typing, host species, and environmental source, for example, drinking water, feces, and so on. As a service to the community, common restriction fragment length polymorphism (RFLP) profiles are also available to facilitate isolate identification. The database maintains a Basic Local Alignment Search Tool (BLAST) server of reference type sequences for each species to facilitate quick identification of and polymorphisms in common molecular sequence tags used by the community. See color insert.

Host–Parasite Interactions

The available genome data were used to examine orthologous *C. parvum* and *C. hominis* protein-encoding genes for signs of positive selection. Twenty-seven genes (1.1% of the orthologs) were found to be under positive selection. Most of these genes are hypothetical genes of unknown function that contain signal peptides. A few are transporters (Ge et al., 2008). Since *C. parvum* and *C. hominis* have different but overlapping host ranges (Table 28.2), these genes warrant further investigation.

Host susceptibility to *Cryptosporidium* and other enteric pathogens has been studied by Flores and Okhuyssen (2009). They discovered that host genetic diversity with respect to major histocompatibility complex (MHC) class II antigen recognition can result in different short-term and long-term complications of diarrhea caused by parasite infections. Their data indicate that identification of polymorphisms in particular proteins used for recognition and attachment of parasites can provide a better understanding of outcomes and facilitate treatment and vaccine strategies, thus improving protective strategies against parasite infection. *Cryptosporidium*, primarily, but not solely, infects intestinal epithelial cells. Research on the many proteins listed in Table 28.3 provides evidence for proteins involved in recognition processes, especially GP60. Once the parasite begins attachment, a very special relationship with the host is established. The host cell membrane on each side of the parasite envelops the parasite in a parasitophorous vacuole, but it remains extracytoplasmic and thus protected from many host cell defenses. The highly membranous feeder organelle forms the interface through which molecular interaction with the host cell cytoplasm must occur (Fayer, 2008).

An *in vitro* microarray study of host cell response revealed that *C. parvum* infection of HCT-8 cells caused significant changes in host biochemical processes, including gene transcription, signal transduction, and cell metabolism (Deng et al., 2004). Currently, little is known about how, or with which molecules, *Cryptosporidium* interacts with its host and affects different pathogenic outcomes.

Population Structure

Multilocus Sequence Typing (MLST). Only a few studies have examined the geographic distribution of *Cryptosporidium*. The global population structures of *C. hominis* and *C. parvum* isolated in seven countries were studied using a nine-locus DNA subtyping scheme. The results of cluster analysis among recently introduced species and between the various countries (Uganda, the United States, the United Kingdom, New Zealand, Turkey, Israel, and Serbia) indicated a quasi-complete phylogenetic segregation, suggesting that gene flow is not sufficient to compensate for the genetic divergence among geographically segregated populations. Also, their study revealed that overall *Cryptosporidium* has both clonal and panmictic population structures with panmictic structures observed more often in areas where infection is endemic. *C. parvum* and *C. hominis* are reproductively separate and often, genotypes found within a country are more closely related to each other than to populations from other countries. Nine markers were sufficient to detect recently introduced isolates (Tanriverdi et al., 2008; Widmer and Sullivan, 2012).

Multilocus analyses targeting microsatellite and minisatellite elements in *Cryptosporidium* genomes have been used to identify *C. hominis* and *C. parvum* species subtypes. MLST using several microsatellite and minisatellite markers has been used to analyze the population structure and geographic segregation of *Cryptosporidium* isolates (reviewed in Widmer and Sullivan, 2012). Markers include the serine repeat antigen gene (MSC6-7) minisatellite, hydroxyproline-rich glycoprotein gene (DZHRGP) minisatellite and microsatellite, 47-kDa protein gene (CP47) microsatellite, and 70-kDa heat shock protein (HSP70) locus. The results of MLST analysis suggested that there is a limited recombination between species indicative of a clonal population structure (Gatei et al., 2007; Mallon et al., 2003a,b; Ngouanesavanh et al., 2006).

However, recombination events between *C. hominis* and *C. parvum* species are suspected. Experimental studies on recombination between *C. parvum* strains in INF-gamma knockout mice revealed recombination (Feng et al., 2002). In a more recent study, 40 loci spread across all eight chromosomes revealed high levels of recombination. Sixteen recombinant lines were obtained. Recombination was seen across the genome except for a 476-kb region of chromosome V (Feng et al., 2002; Tanriverdi et al., 2007).

***gp60* Subtyping.** Sequence analysis of the *Cryptosporidium 60-kDa* glycoprotein gene (also known as *s60*, *gp60*, or *cpgp40/15*) has been increasingly used to identify *Cryptosporidium* species and subtypes in human infections (Robinson et al., 2010; Sulaiman et al., 2005; Waldron et al., 2009a,b; Xiao et al., 2009) (Table 28.3). The *gp60* locus is the most polymorphic sequence identified thus far, and it has been used widely in the identification of intraspecies variation (Widmer, 2009). Cama et al. reported differences in the clinical manifestation of *C. parvum* (chronic diarrhea and vomiting) and *C. hominis* (diarrhea) infections in children and HIV-seropositive persons in Peru, indicating that the *gp60* gene may serve as a marker for virulence (Cama et al., 2007, 2008).

Glycoprotein 60 (GP60) encodes a precursor protein that is cleaved to produce a mature glycoprotein, GP45, and a fragment, GP15. Both are surface glycoproteins that play important roles in the invasion of epithelial cells (Cevallos et al., 2000; Winter et al., 2000). GP60 is a target protein in the development of *Cryptosporidium* vaccines because of its surface location in the apical region of sporozoites, the invasive life cycle stage of *Cryptosporidium*. As a mucin, the GP60 protein has long stretch serine residues in the N-terminus of the GP45 fragment, which is encoded by tandem repeats of a trinucleotide microsatellite sequence, TCA/TCG/TCT. It has been postulated that extensive polymorphisms observed in *gp60* might be caused by selective pressure exerted by the host immune response.

Although the *gp60*-subtyping system has increasingly been used for molecular epidemiological studies in order to identify transmission routes and tracking of *Cryptosporidium* outbreaks, the *gp60*-based subtyping system may not be an ideal system for elucidating population structure and genetic diversity of this globally distributed pathogen. New subtypes have been constantly deposited in the GenBank. Currently, there is no complete reference sequence database to identify *gp60*-based subtypes. The lack of standardized subtype terminology and subtyping protocols reduces the power of molecular epidemiologic tools in characterizing *Cryptosporidium* transmission.

Recombination has been observed experimentally in *Cryptosporidium* (Feng et al., 2002). A meta-analysis done by Widmer (2009) suggested that the *gp60*-subtyping approach is incompatible since there could be a recombination event between species/genotypes. It has been suggested that an MLST or multilocus genotyping (MLG) approach would be a better approach to elucidate the genetic diversity of *Cryptosporidium* rather than the use of a single-locus subtyping system such as the *gp60* subtyping (Widmer and Lee, 2010).

Meta-analyses of the global distribution of *Cryptosporidium* species and subtypes have been performed using the *gp60* gene (Jex and Gasser, 2010; Plutzer and Karanis, 2009; Widmer, 2009). However, the locus was not sufficient to observe geographic diversity. Although the subtyping analysis of *gp60* has limitations for studying population structures and evolution, it may still be advantageous because of its ease of use. Additionally, the sequence from the *gp60* locus may then be combined with other data in an MLST analysis in order to reveal the population structure and evolution of *Cryptosporidium*.

SNP. In an attempt to identify new genetic markers, Bouzid et al. (2010) analyzed incomplete regions and missing genes in the genomes of *C. hominis* and *C. parvum*. They discovered new SNPs. The majority of them are synonymous (64.2%), species specific, and stable within species and subtypes. They indicated that these SNPs can be used to further analyze the phylogenetic distances and evolutionary relationships between and within species. Their paired SNP analysis can be the basis for future analyses of genetic diversity.

Repetitive Sequences. A few studies have analyzed the characteristics of *gp60* serine repeats, and have shown that the human-infectious *C. parvum* IIc subtype family sequences contain shorter serine repeats than the IIc subtype family that is isolated from animals (Widmer, 2009), indicating the selective pressure of the *gp60* serine repeats in order to be able to infect different species (Fig. 28.3).

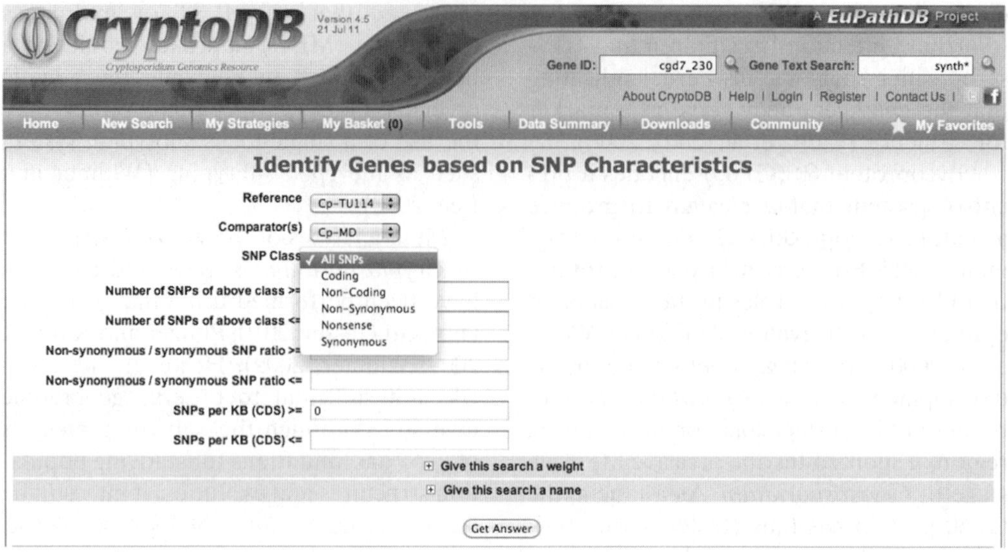

Figure 28.2 Screenshot of a CryptoDB single-nucleotide polymorphism (SNP) search between *Cryptosporidium* strains. SNP data are available for three *C. parvum* strains. Users of the database can identify SNPs between the strains based upon a number of criteria including the location of the SNP in a coding or noncoding region. If an SNP is located in a protein-encoding region, the SNP's potential effect (synonymous or nonsynonymous substitution) on the resulting amino acid can be queried. Searches by SNP density and dN/dS ratio are possible. See color insert.

Epidemiology and Markers for Virulence

The long survival of oocysts in the environment makes it easier for this pathogen to infect humans by the fecal–oral routes through contaminated water, food, or fomites, and by person-to-person (anthroponotic transmission) or animal-to-person (zoonotic transmission) contact. Zoonotic transmissions are well documented (Chalmers and Davies, 2010; Xiao and Ryan, 2008). Infections with *C. meleagridis*, *C. canis*, and *C. felis* were reported in India, Peru, Jamaica, and the United Kingdom (Ajjampur et al., 2007; Cama et al., 2007; Chalmers et al., 2009a; Gatei et al., 2007), and *C. cuniculus* infections were observed in the United Kingdom (Chalmers et al., 2011a,b; Elwin et al., 2012; Robinson et al., 2008, 2010). The infections in India, Peru, and Jamaica were identified in children and in HIV-positive patients, while the infection of *C. cuniculus* was reported in immunocompetent humans.

The development of molecular markers permits tests for association with virulence. The results have been mixed. Some researchers have reported an association between a single marker, the *gp60* genotype (Cama et al., 2007, 2008; Iqbal et al., 2011), and others could not find an association (Ajjampur et al., 2010). Much more work is needed in this area.

Outbreaks

Waterborne and foodborne outbreaks of cryptosporidiosis have been recognized in many countries, and endemic cryptosporidiosis is common in the world (Chalmers and Davies, 2010). Waterborne outbreaks of cryptosporidiosis occur with increased frequency in Ireland, the United Kingdom, and Belgium (Semenza and Nichols, 2007).

Recent foodborne outbreaks were reported in Denmark and Finland (Ethelberg et al., 2009; Pnka et al., 2009). In the United States, the largest waterborne outbreak was reported in 1993 in Milwaukee, Wisconsin, where more than 2000 people were affected and several died (Mackenzie et al., 1995). Waterborne and foodborne outbreaks of cryptosporidiosis have been reported in many different U.S. states, and the incidence of cryptosporidiosis has increased significantly in recent years (Valderrama et al., 2009; Yoder and Beach, 2010). A seasonal increase in reported cases is observed from late summer through early fall, caused mainly by recreational water exposure (Yoder and Beach, 2010).

CHALLENGES AND UNANSWERED QUESTIONS

There are several significant challenges facing the *Cryptosporidium* community. The lack of continuous *in vitro* culture hampers most genetic and molecular genetic approaches used in other apicomplexan systems. Currently, it is not possible to transform *Cryptosporidium* with a selectable marker or green fluorescent protein (GFP). Some localization studies are possible in short-term *in vitro* culture, but the life cycle cannot be completed and the parasites die around 72 hours. Protein localization and gene regulation can be studied prior to 72 hours, but it is not known how similar the *in vitro* patterns are to *in vivo*, especially as the parasites begin to die. However, in order to address the function of the 1700 hypothetical proteins in *C. parvum*, alternative approaches or a significant technological and biological breakthrough is needed.

The lack of a continuous culture system also greatly impacts the ease with which potential therapeutic compounds can be screened and the ease with which *Cryptosporidium* parasites can be obtained in significant numbers. Oocysts for *C. parvum* remain infective for up to 90 days; after this window, fresh oocysts must be obtained from an animal model system. Cryopreservation of *Cryptosporidium* does not exist.

Since the parasites must be propagated in order to be studied, we do not know the extent to which clinical and environmental isolates have become reduced in their genetic diversity as the result of passage in what may be a nonnatural host.

There is a significant need for additional molecular markers for *Cryptosporidium*. Currently, epidemiology studies utilize only a handful of molecular markers that cover, at most, nine loci in the genome. Also, the community has not settled on a definitive set of markers to use for genotyping studies. The consequences of this situation for the field are numerous; first, the results of different studies cannot be combined and compared to each other because different markers were utilized. Second, the small number of markers is insufficient to adequately type an isolate if recombination is occurring. Finally, the small number of markers is insufficient to adequately address association of a marker with virulence. Given the lack of an *in vitro* experimental genetic system in *Cryptosporidium*, association of genetic markers with phenotypes can only be made if the genome is saturated with markers.

The best way to develop the much-needed markers is to address the remaining challenge, namely, a lack of knowledge regarding the global distribution and population structure of *Cryptosporidium* species. The technological breakthroughs associated with next-generation sequencing offer the prospect of significant, cost-effective sequence generation from numerous isolates without passage through an animal model system. The sequence data generated will permit the development of numerous markers that can be used for species typing, epidemiology studies, detection of zoonoses, and association of markers with various phenotypes such as host specificity.

REFERENCES

Abe N, Makino I. 2010. Multilocus genotypic analysis of *Cryptosporidium* isolates from cockatiels, Japan. Parasitology Research 106: 1491–1497.

Abe N, Matsubayashi M, Kimata I, Iseki M. 2006. Subgenotype analysis of *Cryptosporidium parvum* isolates from humans and animals in Japan using the 60-kDa glycoprotein gene sequences. Parasitology Research 99: 303–305.

Abrahamsen M, Templeton T, Enomoto S, Abrahante J, Zhu G, Lancto C, Deng M, Liu C, Widmer G, Tzipori S, Buck G, Xu P, Bankier A, Dear P, Konfortov B, Spriggs H, Iyer L, Anantharaman V, Aravind L, Kapur V. 2004. Complete genome sequence of the apicomplexan, *Cryptosporidium parvum*. Science 304: 441–445.

Ajjampur SSR, Gladstone BP, Selvapandian D, Muliyil JP, Ward H, Kang G. 2007. Molecular and spatial epidemiology of cryptosporidiosis in children in a semiurban community in South India. Journal of Clinical Microbiology 45: 915–920.

Ajjampur SSR, Sarkar R, Sankaran P, Kannan A, Menon VK, Muliyil J, Ward H, Kang G. 2010. Symptomatic and asymptomatic *Cryptosporidium* infections in children in a semiurban slum community in Southern India. The American Journal of Tropical Medicine and Hygiene 83: 1110–1115.

Akiyoshi DE, Feng X, Buckholt MA, Widmer G, Tzipori S. 2002. Genetic analysis of a *Cryptosporidium parvum* human genotype 1 isolate passaged through different host species. Infection and Immunity 70: 5670–5675.

Akiyoshi DE, Tumwine JK, Bakeera-Kitaka S, Tzipori S. 2006. Subtype analysis of *Cryptosporidium* isolates from children in Uganda. The Journal of Parasitology 92: 1097–1100.

Al-Brikan FA, Salem HS, Beeching N, Hilal N. 2008. Multilocus genetic analysis of *Cryptosporidium* isolates from Saudi Arabia. Journal of the Egyptian Society of Parasitology 38: 645–658.

Alves M, Xiao L, Sulaiman I, Lal AA, Matos O, Antunes F. 2003. Subgenotype analysis of *Cryptosporidium* isolates from humans, cattle, and zoo ruminants in Portugal. Journal of Clinical Microbiology 41: 2744–2747.

Alves M, Xiao L, Antunes F, Matos O. 2006. Distribution of *Cryptosporidium* subtypes in humans and domestic and wild ruminants in Portugal. Parasitology Research 99: 287–292.

Amer S. 2010. First detection of *Cryptosporidium baileyi* in Ruddy Shelduck (*Tadorna ferruginea*) in China. The Journal of Veterinary Medical Science 72: 935–938.

Anonymous. 1982. Cryptosporidiosis, assessment of chemotherapy of males with acquired immune deficiency syndrome (AIDS). Morbidity and Mortality Weekly Report 31: 89–102.

Arrowood MJ, Sterling CR. 1987. Isolation of *Cryptosporidium* oocysts and sporozoites using discontinuous sucrose and isopycnic percoll gradients. The Journal of Parasitology 73: 314–319.

Arrowood MJ, Mead JR, Mahrt JL, Sterling CR. 1989. Effects of immune colostrum and orally administered antisporozoite monoclonal antibodies on the outcome of *Cryptosporidium parvum* infections in neonatal mice. Infection and Immunity 57: 2283–2288.

Bachvaroff TR, Handy SM, Place AR, Delwiche CF. 2011. Alveolate phylogeny inferred using concatenated ribosomal proteins. The Journal of Eukaryotic Microbiology 58: 223–233.

Bankier AT, Spriggs HF, Fartmann B, Konfortov BA, Madera M, Vogel C, Teichmann SA, Ivens A, Dear PH. 2003. Integrated mapping, chromosomal sequencing and sequence analysis of *Cryptosporidium parvum*. Genome Research 13: 1787–1799.

Barnes DA, Bonnin A, Huang J-X, Gousset L, Wu J, Gut J, Doyle P, Dubremetz J-F, Ward H, Petersen C. 1998. A novel multi-domain mucin-like glycoprotein of *Cryptosporidium parvum* mediates invasion. Molecular and Biochemical Parasitology 96: 93–110.

Barta JR, Thompson RC. 2006. What is *Cryptosporidium*? Reappraising its biology and phylogenetic affinities. Trends in Parasitology 22: 463–468.

Bhat N, Joe A, Pereiraperrin M, Ward HD. 2007. *Cryptosporidium* p30, a galactose/n-acetylgalactosamine-specific lectin, mediates infection *in vitro*. The Journal of Biological Chemistry 282: 34877–34887.

Board A, Ward H. 2010. Human immune responses in cryptosporidiosis. Future Microbiology 5: 507–519.

Bonnin A, Ojcius D, Souque P, Barnes D, Doyle P, Gut J, Nelson R, Petersen C, Dubremetz J-F. 2001. Characterization of a monoclonal antibody reacting with antigen-4 domain of gp900 in *Cryptosporidium parvum* invasive stages. Parasitology Research 87: 589–592.

Borowski H, Clode PL, Thompson RCA. 2008. Active invasion and/or encapsulation? A reappraisal of host-cell parasitism by *Cryptosporidium*. Trends in Parasitology 24: 509–516.

Borowski H, Thompson RCA, Armstrong T, Clode PL. 2010. Morphological characterization of *Cryptosporidium parvum* life-cycle stages in an *in vitro* model system. Parasitology 137: 13–26.

Boulter-Bitzer JI, Lee H, Trevors JT. 2007. Molecular targets for detection and immunotherapy in *Cryptosporidium parvum*. Biotechnology Advances 25: 13–44.

Boulter-Bitzer JI, Lee H, Trevors JT. 2010. Single chain variable fragment antibodies selected by phage display against the sporozoite surface antigen S16 of *Cryptosporidium parvum*. Experimental Parasitology 125: 124–129.

Bouzid M, Tyler K, Christen R, Chalmers R, Elwin K, Hunter P. 2010. Multi-locus analysis of human infective *Cryptosporidium* species and subtypes using ten novel genetic loci. BMC Microbiology 10: 213.

Cama VA, Ross JM, Crawford S, Kawai V, Chavez-Valdez R, Vargas D, Vivar A, Ticona E, Navincopa M, Williamson J, Ortega Y, Gilman RH, Bern C, Xiao L. 2007. Differences in clinical manifestations among *Cryptosporidium* species and subtypes in HIV-infected persons. The Journal of Infectious Diseases 196: 684–691.

Cama VA, Bern C, Roberts J, Cabrera L, Sterling CR, Ortega Y, Gilman RH, Xiao L. 2008. *Cryptosporidium* species and subtypes and clinical manifestations in children, Peru. Emerging Infectious Diseases 14: 1567–1574.

Carreno RA, Martin DS, Barta JR. 1999. *Cryptosporidium* is more closely related to the gregarines than to Coccidia as shown by phylogenetic analysis of apicomplexan parasites inferred using small-subunit ribosomal RNA gene sequences. Parasitol Research 85: 899–904.

Cevallos AM, Zhang X, Waldor MK, Jaison S, Zhou X, Tzipori S, Neutra MR, Ward HD. 2000. Molecular cloning and expression of a gene encoding *Cryptosporidium parvum* glycoproteins gp40 and gp15. Infection and Immunity 68: 4108–4116.

Cevallos AM, Bhat N, Verdon R, Hamer DH, Stein B, Tzipori S, Pereira MEA, Keusch GT, Ward HD. 2000b. Mediation of *Cryptosporidium parvum* infection *in vitro* by mucin-like glycoproteins defined by a neutralizing monoclonal antibody. Infection and Immunity 68: 5167–5175.

Chalmers RM, Davies AP. 2010. Minireview: Clinical cryptosporidiosis. Experimental Parasitology 124: 138–146.

Chalmers R, Elwin K, Thomas A, Guy E, Mason B. 2009a. Long-term *Cryptosporidium* typing reveals the aetiology and species-specific epidemiology of human cryptosporidiosis in England and Wales, 2000 to 2003. Euro Surveillance 14(2): pii=19086.

Chalmers RM, Robinson G, Elwin K, Hadfield SJ, Xiao L, Ryan U, Modha D, Mallaghan C. 2009b. *Cryptosporidium* sp. Rabbit genotype, a newly identified human pathogen. Emerging Infectious Diseases 15: 829–830.

Chalmers RM, Hadfield SJ, Jackson CJ, Elwin K, Xiao L, Hunter P. 2008. Geographic linkage and variation in *Cryptosporidium hominis*. Emerging Infectious Diseases 14: 496–498.

Chalmers RM, Elwin K, Hadfield SJ, Robinson G. 2011a. Sporadic human cryptosporidiosis caused by *Cryptosporidium cuniculus*, United Kingdom, 2007–2008. Emerging Infectious Diseases 17: 536–538.

Chalmers RM, Smith R, Elwin K, Clifton-Hadley FA, Giles M. 2011b. Epidemiology of anthroponotic and zoonotic human cryptosporidiosis in England and Wales, 2004–2006. Epidemiology and Infection 139: 700–712.

Chappell CL, Okhuysen PC, Sterling CR, Dupont HL. 1996. *Cryptosporidium parvum*: Intensity of infection and oocyst excretion patterns in healthy volunteers. The Journal of Infectious Diseases 173: 232–236.

Chappell CL, Okhuysen PC, Langer-Curry R, Widmer G, Akiyoshi DE, Tanriverdi S, Tzipori S. 2006. *Cryptosporidium hominis*: Experimental challenge of healthy adults. The American Journal of Tropical Medicine and Hygiene 75: 851–857.

Chatterjee A, Banerjee S, Steffen M, O'Connor RM, Ward HD, Robbins PW, Samuelson J. 2010. Evidence for mucin-like glycoproteins that tether sporozoites of *Cryptosporidium parvum* to the inner surface of the oocyst wall. Eukaryotic Cell 9: 84–96.

Chen X-M, Huang BQ, Splinter PL, Orth JD, Billadeau DD, McNiven MA, Larusso NF. 2004a. Cdc42 and the actin-related protein/neural Wiskott-Aldrich syndrome protein network mediate cellular invasion by *Cryptosporidium parvum*. Infection and Immunity 72: 3011–3021.

Chen X-M, O'Hara SP, Huang BQ, Nelson JB, Lin JJ-C, Zhu G, Ward HD, Larusso NF. 2004b. Apical organelle discharge by *Cryptosporidium parvum* is temperature, cytoskeleton, and intracellular calcium dependent and required for host cell invasion. Infection and Immunity 72: 6806–6816.

Chen X-M, Splinter PL, Tietz PS, Huang BQ, Billadeau DD, Larusso NF. 2004c. Phosphatidylinositol 3-kinase and frabin mediate *Cryptosporidium parvum* cellular invasion via activation of Cdc42. The Journal of Biological Chemistry 279: 31671–31678.

Chen X-M, O'Hara SP, Huang BQ, Splinter PL, Nelson JB, Larusso NF. 2005. Localized glucose and water influx facilitates *Cryptosporidium parvum* cellular invasion by means of modulation of host-cell membrane protrusion. Proceedings of the National Academy of Sciences of the United States of America 102: 6338–6343.

Chen X-M, Splinter PL, O'Hara SP, Larusso NF. 2007. A cellular micro-RNA, let-7i, regulates Toll-like receptor 4 expression and contributes to cholangiocyte immune responses against *Cryptosporidium parvum* infection. The Journal of Biological Chemistry 282: 28929–28938.

Cohen S, Dalle F, Gallay A, Di Palma M, Bonnin A, Ward HD. 2006. Identification of Cpgp40/15 Type Ib as the predominant allele in isolates of *Cryptosporidium* spp. from a waterborne outbreak of gastroenteritis in South Burgundy, France. Journal of Clinical Microbiology 44: 589–591.

Cohn B, Manque P, Lara A, Serrano M, Sheth N, Buck G. 2010. Putative *cis*-regulatory elements associated with heat shock genes activated during excystation of *Cryptosporidium parvum*. PLoS ONE 5: e9512–e9512.

Current WL, Upton SJ, Haynes TB. 1986. The life cycle of *Cryptosporidium baileyi* n. sp. (Apicomplexa, Cryptosporidiidae) infecting chickens. The Journal of Protozoology 33: 289–296.

Deng M, Templeton TJ, London NR, Bauer C, Schroeder AA, Abrahamsen MS. 2002. *Cryptosporidium parvum* genes containing thrombospondin type 1 domains. Infection and Immunity 70: 6987–6995.

Deng M, Rutherford MS, Abrahamsen MS. 2004. Host intestinal epithelial response to *Cryptosporidium parvum*. Advanced Drug Delivery Reviews 56: 869–884.

Dessens JT, Sinden RE, Claudianos C. 2004. LCCL proteins of apicomplexan parasites. Trends in Parasitology 20: 102–108.

Dyachenko V, Kuhnert Y, Schmaeschke R, Etzold M, Pantchev N, Daugschies A. 2010. Occurrence and molecular characterization of *Cryptosporidium* spp. genotypes in European hedgehogs (*Erinaceus europaeus l.*) in Germany. Parasitology 137: 205–216.

Egorov A, Montuori Trimble L, Ascolillo L, Ward H, Levy D, Morris R, Naumova E, Griffiths J. 2010. Recent diarrhea is associated with elevated salivary IgG responses to *Cryptosporidium* in residents of an eastern Massachusetts community. Infection 38: 117–123.

Elliott DA, Clark DP. 2000. *Cryptosporidium parvum* induces host cell actin accumulation at the host-parasite interface. Infection and Immunity 68: 2315–2322.

Elwin K, Hadfield SJ, Robinson G, Chalmers RM. 2012. The epidemiology of sporadic human infections with unusual cryptosporidia detected during routine typing in England and Wales, 2000–2008. Epidemiology and Infection 140: 673–683.

Ethelberg S, Lisby M, Vestergaard LS, Enemark HL, Olsen KEP, Stensvold CR, Nielsen HV, Porsbo LJ, Plesner AM, Mølbak K. 2009. A foodborne outbreak of *Cryptosporidium hominis* infection. Epidemiology and Infection 137: 348–356.

Fayer R. 2008. General biology. In: Fayer R, Xiao L, eds. *Cryptosporidium and Cryptosporidiosis*. CRC Press, Boca Raton, FL, pp. 1–42.

Fayer R, Xiao L, eds. 2008. *Cryptosporidium and Cryptosporidiosis*, 2nd edition. CRC Press, Boca Raton, FL.

Fayer R, Santin M, Xiao L. 2005. *Cryptosporidium bovis* n. sp. (Apicomplexa: Cryptosporidiidae) in cattle (*Bos taurus*). The Journal of Parasitology 91: 624–629.

Fayer R, Orlandi P, Perdue ML. 2009. Virulence factor activity relationships for hepatitis E and *Cryptosporidium*. Journal of Water and Health 7: S55–S63.

Fayer R, Santin M, Macarisin D. 2010. *Cryptosporidium ubiquitum* n. sp. in animals and humans. Veterinary Parasitology 172: 23–32.

Feng X, Rich SM, Tzipori S, Widmer G. 2002. Experimental evidence for genetic recombination in the opportunistic pathogen *Cryptosporidium parvum*. Molecular and Biochemical Parasitology 119: 55–62.

Feng Y, Li N, Duan L, Xiao L. 2009. *Cryptosporidium* genotype and subtype distribution in raw wastewater in Shanghai, China: Evidence for possible unique *Cryptosporidium hominis* transmission. Emerging Infectious Diseases 47: 153–157.

Flores J, Okhuysen PC. 2009. Genetics of susceptibility to infection with enteric pathogens. Current Opinion in Infectious Diseases 22: 471–476.

Forney JR, Yang S, Healey MC. 1996. Protease activity associated with excystation of *Cryptosporidium parvum* oocysts. The Journal of Parasitology 82: 889–892.

Gargala G, Le Goff L, Ballet J-J, Favennec L, Stachulski AV, Rossignol J-F. 2010. Evaluation of new thiazolide/thiadiazolide derivatives reveals nitro group-independent efficacy against *in vitro* development of *Cryptosporidium parvum*. Antimicrobial Agents and Chemotherapy 54: 1315–1318.

Gatei W, Das P, Dutta P, Sen A, Cama V, Lal AA, Xiao L. 2007. Multilocus sequence typing and genetic structure of *Cryptosporidium hominis* from children in Kolkata, India. Infection, Genetics and Evolution 7: 197–205.

Ge G, Cowen L, Feng X, Widmer G. 2008. Protein coding gene nucleotide substitution pattern in the apicomplexan protozoa *Cryptosporidium parvum* and *Cryptosporidium hominis*. Comparative and Functional Genomics 2008: 879023.

Girouard D, Gallant J, Akiyoshi DE, Nunnari J, Tzipori S. 2006. Failure to propagate *Cryptosporidium* spp. in cell-free culture. The Journal of Parasitology 92: 399–400.

Glaberman S, Sulaiman IM, Bern C, Limor J, Peng MM, Morgan U, Gilman R, Lal AA, Xiao L. 2001. A multilocus genotypic analysis of *Cryptosporidium meleagridis*. The Journal of Eukaryotic Microbiology 48(Suppl.): 19s–22s.

Glaberman S, Moore JE, Lowery CJ, Chalmers RM, Sulaiman I, Elwin K, Rooney PJ, Millar BC, Dooley JS, Lal AA, Xiao L. 2002. Three drinking-water-associated cryptosporidiosis outbreaks, Northern Ireland. Emerging Infectious Diseases 8: 631–633.

Gomez Morales MA, La Rosa G, Ludovisi A, Onori AM, Pozio E. 1999. Cytokine profile induced by *Cryptosporidium* antigen in peripheral blood mononuclear cells from immunocompetent and immunosuppressed persons with cryptosporidiosis. The Journal of Infectious Diseases 179: 967–973.

Gong A-Y, Zhou R, Hu G, Liu J, Sosnowska D, Drescher KM, Dong H, Chen X-M. 2010. *Cryptosporidium parvum* induces B7-H1 expression in cholangiocytes by downregulating microRNA-513. The Journal of Infectious Diseases 201: 160–169.

Grinberg A, Learmonth J, Kwan E, Pomroy W, Villalobos NL, Gibson I, Widmer G. 2008. Genetic diversity and zoonotic potential of *Cryptosporidium parvum* causing foal diarrhea. Journal of Clinical Microbiology 46: 2396–2398.

Hashim A, Mulcahy G, Bourke B, Clyne M. 2006. Interaction of *Cryptosporidium hominis* and *Cryptosporidium parvum* with primary human and bovine intestinal cells. Infection and Immunity 74: 99–107.

Heiges M, Wang H, Robinson E, Aurrecoechea C, Gao X, Kaluskar N, Rhodes P, Wang S, He C-Z, Su Y, Miller J, Kraemer E, Kissinger JC. 2006. CryptoDB: A *Cryptosporidium* bioinformatics resource update. Nucleic Acids Research 34: D419–D422.

Hijjawi N. 2010. *Cryptosporidium*: New developments in cell culture. Experimental Parasitology 124: 54–60.

Hijjawi N, Estcourt A, Yang R, Monis P, Ryan U. 2010. Complete development and multiplication of *Cryptosporidium hominis* in cell-free culture. Veterinary Parasitology 169: 29–36.

Hijjawi NS, Meloni BP, Ng'anzo M, Ryan UM, Olson ME, Cox PT, Monis PT, Thompson RC. 2004. Complete development of *Cryptosporidium parvum* in host cell-free culture. International Journal for Parasitology 34: 769–777.

Hira KG, Mackay MR, Hempstead AD, Ahmed S, Karim MM, O'Connor RM, Hibberd PL, Calderwood SB, Ryan ET, Khan WA, Ward HD. 2011. Genetic diversity of *Cryptosporidium* spp. from Bangladeshi children. Journal of Clinical Microbiology 49: 2307–2310.

Hu G, Zhou R, Liu J, Gong A-Y, Chen X-M. 2010. MicroRNA-98 and let-7 regulate expression of suppressor of cytokine signaling 4 in biliary epithelial cells in response to *Cryptosporidium parvum* infection. The Journal of Infectious Diseases 202: 125–135.

Huang BQ, Chen XM, Larusso NF. 2004. *Cryptosporidium parvum* attachment to and internalization by human biliary epithelia *in vitro*: A morphologic study. The Journal of Parasitology 90: 212–221.

Iqbal J, Khalid N, Hira PR. 2011. Cryptosporidiosis in Kuwaiti children: Association of clinical characteristics with *Cryptosporidium* species and subtypes. Journal of Medical Microbiology 60: 647–652.

Jenkins M, O'Brien C, Miska K, Schwarz R, Karns J, Santin M, Fayer R. 2011. Gene expression during excystation of *Cryptosporidium parvum* oocysts. Parasitology Research 109: 509–513.

Jex A, Pangasa A, Campbell B, Whipp M, Hogg G, Sinclair M, Stevens M, Gasser R. 2008b. Classification of *Cryptosporidium* species from patients with sporadic cryptosporidiosis by use of sequence-based multilocus analysis following mutation scanning. Journal of Clinical Microbiology 46: 2252–2262.

Jex AR, Gasser RB. 2010. Genetic richness and diversity in *Cryptosporidium hominis* and *C. parvum* reveals major knowledge gaps and a need for the application of "next generation" technologies—Research review. Biotechnology Advances 28: 17–26.

Jex AR, Whipp M, Campbell BE, Caccìò SM, Stevens M, Hogg G, Gasser RB. 2007b. A practical and cost-effective mutation scanning-based approach for investigating genetic variation in *Cryptosporidium*. Electrophoresis 28: 3875–3883.

Kjos SA, Jenkins M, Okhuysen PC, Chappell CL. 2005. Evaluation of recombinant oocyst protein CP41 for detection of *Cryptosporidium*-specific antibodies. Clinical and Diagnostic Laboratory Immunology 12: 268–272.

Langer RC, Riggs MW. 1999. *Cryptosporidium parvum* apical complex glycoprotein CSL contains a sporozoite ligand for intestinal epithelial cells. Infection and Immunity 67: 5282–5291.

Langer RC, Schaefer DA, Riggs MW. 2001. Characterization of an intestinal epithelial cell receptor recognized by the *Cryptosporidium parvum* sporozoite ligand CSL. Infection and Immunity 69: 1661–1670.

Leander BS, Clopton RE, Keeling PJ. 2003a. Phylogeny of gregarines (Apicomplexa) as inferred from small-subunit rDNA and beta-tubulin. International Journal of Systematic and Evolutionary Microbiology 53: 345–354.

Leander BS, Harper JT, Keeling PJ. 2003b. Molecular phylogeny and surface morphology of marine aseptate gregarines (Apicomplexa): *Selenidium* spp. and *Lecudina* spp. The Journal of Parasitology 89: 1191–1205.

Leav BA, Mackay MR, Anyanwu A, O' Connor RM, Cevallos AM, Kindra G, Rollins NC, Bennish ML, Nelson RG, Ward HD. 2002. Analysis of sequence diversity at the highly polymorphic Cpgp40/15 locus among *Cryptosporidium* isolates from human immunodeficiency virus-infected children in South Africa. Infection and Immunity 70: 3881–3890.

Leoni F, Mallon ME, Smith HV, Tait A, Mclauchlin J. 2007b. Multilocus analysis of *Cryptosporidium hominis* and *Cryptosporidium parvum* isolates from sporadic and outbreak-related human cases and *C. parvum* isolates from sporadic livestock cases in the United Kingdom. Journal of Clinical Microbiology 45: 3286–3294.

MacKenzie WR, Schell WL, Blair KA, Addiss DG, Peterson DE, Hoxie NJ, Kazmierczak JJ, Davis JP. 1995. Massive outbreak of waterborne *Cryptosporidium* infection in Milwaukee, Wisconsin: Recurrence of illness and risk of secondary transmission. Clinical Infectious Diseases 21: 57–62.

MacPherson IS, Kirubakaran S, Gorla SK, Riera TV, D'Aquino JA, Zhang M, Cuny GD, Hedstrom L. 2010. The structural basis of *Cryptosporidium*-specific IMP dehydrogenase inhibitor selectivity. Journal of the American Chemical Society 132: 1230–1231.

Mallon M, MacLeod A, Wastling J, Smith H, Reilly B, Tait A. 2003a. Population structures and the role of genetic exchange in the zoonotic pathogen *Cryptosporidium parvum*. Journal of Molecular Evolution 56: 407–417.

Mallon ME, MacLeod A, Wastling JM, Smith H, Tait A. 2003b. Multilocus genotyping of *Cryptosporidium parvum* Type 2: Population genetics and sub-structuring. Infection, Genetics and Evolution 3: 207–218.

Manabe YC, Clark DP, Moore RD, Lumadue JA, Dahlman HR, Belitsos PC, Chaisson RE, Sears CL. 1998. Cryptosporidiosis in patients with AIDS: Correlates of disease and survival. Clinical Infectious Diseases 27: 536–542.

Mead JR. 2010. Challenges and prospects for a *Cryptosporidium* vaccine. Future Microbiology 5: 335–337.

Meisel JL, Perera DR, Meligro C, Rubin CE. 1976. Overwhelming watery diarrhea associated with a *Cryptosporidium* in an immunosuppressed patient. Gastroenterology 70: 1156–1160.

Misc Z, Abe N. 2007. Subtype analysis of *Cryptosporidium parvum* isolates from calves on farms around Belgrade, Serbia and Montenegro, using the 60 kDa glycoprotein gene sequences. Parasitology 134: 351–358.

Molloy SF, Smith HV, Kirwan P, Nichols RAB, Asaolu SO, Connelly L, Holland CV. 2010. Identification of a high diversity of *Cryptosporidium* species genotypes and subtypes in a pediatric population in Nigeria. The American Journal of Tropical Medicine and Hygiene 82: 608–613.

Nakamura AA, Simoes DC, Antunes RG, Da Silva DC, Meireles MV. 2009. Molecular characterization of *Cryptosporidium* spp. from fecal samples of birds kept in captivity in Brazil. Veterinary Parasitology 166: 47–51.

Nesterenko MV, Woods K, Upton SJ. 1998. Receptor/ligand interactions between *Cryptosporidium parvum* and the surface of the host cell. Biochimica et Biophysica Acta (BBA)—Molecular Basis of Disease 1454: 165–173.

Ngouanesavanh T, Guyot K, Certad G, Le Fichoux Y, Chartier C, Verdier R-I, Cailliez J-C, Camus D, Dei-Cas E, Bañuls A-L. 2006. *Cryptosporidium* population genetics: Evidence of clonality in isolates from France and Haiti. The Journal of Eukaryotic Microbiology 53(Suppl. 1): S33–S36.

Nime FA, Burek JD, Page DL, Holscher MA, Yardley JH. 1976. Acute enterocolitis in a human being infected with the protozoan *Cryptosporidium*. Gastroenterology 70: 592–598.

O'Hara S, Yu J-R, Lin J-C. 2004. A novel *Cryptosporidium parvum* antigen, CP2, preferentially associates with membranous structures. Parasitology Research 92: 317–327.

O'Connor RM, Wanyiri JW, Cevallos AM, Priest JW, Ward HD. 2007. *Cryptosporidium parvum* glycoprotein gp40 localizes to the sporozoite surface by association with gp15. Molecular and Biochemical Parasitology 156: 80–83.

O'Hara SP, Small AJ, Chen XM, Larusso NF. 2008. Host cell actin remodeling in response to *Cryptosporidium*. Sub-cellular Biochemistry 47: 92–100.

Okhuysen PC, Dupont HL, Sterling CR, Chappell CL. 1994. Arginine aminopeptidase, an integral membrane protein of the *Cryptosporidium parvum* sporozoite. Infection and Immunity 62: 4667–4670.

Okhuysen PC, Chappell CL, Crabb JH, Sterling CR, Dupont HL. 1999. Virulence of three distinct *Cryptosporidium parvum* isolates for healthy adults. The Journal of Infectious Diseases 180: 1275–1281.

Okhuysen PC, Rogers GA, Crisanti A, Spano F, Huang DB, Chappell CL, Tzipori S. 2004. Antibody response of healthy adults to recombinant thrombospondin-related adhesive protein of *Cryptosporidium* 1 after experimental exposure to *Cryptosporidium* oocysts. Clinical and Diagnostic Laboratory Immunology 11: 235–238.

Peng M, Matos O, Gatei W, Das P, Stantic-Pavlinic M, Bern C, Sulaiman IM, Glaberman S, Lal A, Xiao L. 2001. A comparison of *Cryptosporidium* subgenotypes from several geographic regions. The Journal of Eukaryotic Microbiology 48(Suppl.): 28s–31s.

Peng MM, Meshnick SR, Cunliffe NA, Thindwa BD, Hart A, Broadhead RL, Xiao L. 2003. Molecular epidemiology of cryptosporidiosis in children in Malawi. The Journal of Eukaryotic Microbiology 50(Suppl.): 557–559.

Pereira SJ, Ramirez NE, Xiao L, Ward LA. 2002. Pathogenesis of human and bovine *Cryptosporidium parvum* in gnotobiotic pigs. The Journal of Infectious Diseases 186: 715–718.

Perkins ME, Riojas YA, Wu TW, Le BSM. 1999. CpABC, a *Cryptosporidium parvum* ATP-binding cassette protein at the host-parasite boundary in intracellular stages. Proceedings of the National Academy of Sciences of the United States of America 96: 5734–5739.

Perryman LE, Jasmer DP, Riggs MW, Bohnet SG, McGuire TC, Arrowood MJ. 1996. A cloned gene of *Cryptosporidium parvum* encodes neutralization-sensitive epitopes. Molecular and Biochemical Parasitology 80: 137–147.

Petry F, Robinson HA, Mcdonald V. 1995. Murine infection model for maintenance and amplification of *Cryptosporidium parvum* oocysts. Journal of Clinical Microbiology 33: 1922–1924.

Petry F, Jakobi V, Tessema TS. 2010. Host immune response to *Cryptosporidium parvum* infection. Experimental Parasitology 126: 304–309.

Plutzer J, Karanis P. 2007. Genotype and subtype analyses of *Cryptosporidium* isolates from cattle in Hungary. Veterinary Parasitology 146: 357–362.

Plutzer J, Karanis P. 2009. Genetic polymorphism in *Cryptosporidium* species: An update. Veterinary Parasitology 165: 187–199.

Pollok RCG, McDonald V, Kelly P, Farthing MJG. 2003. The role of *Cryptosporidium parvum*-derived phospholipase in intestinal epithelial cell invasion. Parasitology Research 90: 181–186.

Pönka A, Kotilainen P, Rimhanen-Finne R, Hokkanen P, Hänninen ML, Kaarna A, Meri T, Kuusi M. 2009. A foodborne outbreak due to *Cryptosporidium parvum* in Helsinki, November 2008. Euro Surveillance 14(28): pii=19269.

Power ML. 2010. Biology of *Cryptosporidium* from marsupial hosts. Experimental Parasitology 124: 40–44.

Pozio E, Rezza G, Boschini A, Pezzotti P, Tamburrini A, Rossi P, Di Fine M, Smacchia C, Schiesari A, Gattei E, Zucconi R, Ballarini P. 1997. Clinical cryptosporidiosis and human immunodeficiency virus (HIV)-induced immunosuppression: Findings from a longitudinal study of HIV-positive and HIV-negative former injection drug users. The Journal of Infectious Diseases 176: 969–975.

Preidis GA, Wang H-C, Lewis DE, Castellanos-Gonzalez A, Rogers KA, Graviss EA, Ward HD, White AC, Jr. 2007. Seropositive human subjects produce interferon gamma after stimulation with recombinant *Cryptosporidium hominis* gp15. The American Journal of Tropical Medicine and Hygiene 77: 583–585.

Priest JW, Kwon JP, Montgomery JM, Bern C, Moss DM, Freeman AR, Jones CC, Arrowood MJ, Won KY, Lammie PJ, Gilman RH, Mead JR. 2010. Cloning and characterization of the acidic ribosomal protein P2 of *Cryptosporidium parvum*: A new 17-kDa antigen. Clinical and Vaccine Immunology 17: 954–965.

Putignani L, Possenti A, Cherchi S, Pozio E, Crisanti A, Spano F. 2008. The thrombospondin-related protein CpMIC1 (CpTSP8) belongs to the repertoire of micronemal proteins of *Cryptosporidium parvum*. Molecular and Biochemical Parasitology 157: 98–101.

Quilez J, Torres E, Chalmers RM, Hadfield SJ, Del Cacho E, Sanchez-Acedo C. 2008a.

Cryptosporidium genotypes and subtypes in lambs and goat kids in Spain. Applied and Environmental Microbiology 74: 6026–6031.

Quilez J, Torres E, Chalmers RM, Robinson G, Del Cacho E, Sanchez-Acedo C. 2008b. *Cryptosporidium* species and subtype analysis from dairy calves in Spain. Parasitology 135: 1613–1620.

Riggs M, Stone A, Yount P, Langer R, Arrowood M, Bentley D. 1997. Protective monoclonal antibody defines a circumsporozoite-like glycoprotein exoantigen of *Cryptosporidium parvum* sporozoites and merozoites. Journal of Immunology 158: 1787–1795.

Riggs MW, Schaefer DA, Kapil SJ, Barley-Maloney L, Perryman LE. 2002. Efficacy of monoclonal antibodies against defined antigens for passive immunotherapy of chronic gastrointestinal cryptosporidiosis. Antimicrobial Agents and Chemotherapy 46: 275–282.

Robinson G, Chalmers RM. 2010. The European rabbit *Oryctolagus cuniculus*, a source of zoonotic cryptosporidiosis. Zoonoses and Public Health 57: e1–e13.

Robinson G, Elwin K, Chalmers RM. 2008. Unusual *Cryptosporidium* genotypes in human cases of diarrhea. Emerging Infectious Diseases 14: 1800–1802.

Robinson G, Wright S, Elwin K, Hadfield SJ, Katzer F, Bartley PM, Hunter PR, Nath M, Innes EA, Chalmers RM. 2010. Redescription of *Cryptosporidium cuniculus* Inman and Takeuchi, 1979 (Apicomplexa: Cryptosporidiidae): Morphology, biology and phylogeny. International Journal for Parasitology 40: 1539–1548.

Rogers KA, Rogers AB, Leav BA, Sanchez A, Vannier E, Uematsu S, Akira S, Golenbock D, Ward HD. 2006. MyD88-dependent pathways mediate resistance to *Cryptosporidium parvum* infection in mice. Infection and Immunity 74: 549–556.

Rossignol J-F. 2010. *Cryptosporidium* and *Giardia*: Treatment options and prospects for new drugs. Experimental Parasitology 124: 45–53.

Rossignol JF, Ayoub A, Ayers MS. 2001. Treatment of diarrhea caused by *Cryptosporidium parvum*: A prospective randomized, double-blind, placebo-controlled study of nitazoxanide. The Journal of Infectious Diseases 184: 103–106.

Schmidt W, Wahnschaffe U, Schafer M, Zippel T, Arvand M, Meyerhans A, Riecken EO, Ullrich R. 2001. Rapid increase of mucosal CD4 T cells followed by clearance of intestinal cryptosporidiosis in an AIDS patient receiving highly active antiretroviral therapy. Gastroenterology 120: 984–987.

Semenza JC, Nichols G. 2007. Cryptosporidiosis surveillance and water-borne outbreaks in Europe. Euro Surveillance 12: E13–E14.

Sharling L, Liu X, Gollapalli DR, Maurya SK, Hedstrom L, Striepen B. 2010. A screening pipeline for antiparasitic agents targeting *Cryptosporidium* inosine monophosphate dehydrogenase. PLoS Neglected Tropical Diseases 4: e794.

Shen Y, Yin J, Yuan Z, Lu W, Xu Y, Xiao L, Cao J. 2011. The identification of the *Cryptosporidium ubiquitum* in pre-weaned ovines from Aba Tibetan and Qiang Autonomous Prefecture in China. Biomedical and Environmental Sciences 24: 315–320.

Shirafuji H, Xuan X, Kimata I, Takashima Y, Fukumoto S, Otsuka H, Nagasawa H, Suzuki H. 2005. Expression of P23 of *Cryptosporidium parvum* in *Toxoplasma gondii* and evaluation of its protective effects. The Journal of Parasitology 91: 476–479.

Sibley LD. 2004. Intracellular parasite invasion strategies. Science 304: 248–253.

Silverlås C, Näslund K, Björkman C, Mattsson JG. 2010. Molecular characterisation of *Cryptosporidium* isolates from Swedish dairy cattle in relation to age, diarrhoea and region. Veterinary Parasitology 169: 289–295.

Snelling WJ, Lin Q, Moore JE, Millar BC, Tosini F, Pozio E, Dooley JSG, Lowery CJ. 2007. Proteomics analysis and protein expression during sporozoite excystation of *Cryptosporidium parvum* (Coccidia, Apicomplexa). Molecular & Cellular Proteomics 6: 346–355.

Soba B, Logar J. 2008. Genetic classification of *Cryptosporidium* isolates from humans and calves in Slovenia. Parasitology 135: 1263–1270.

Spano F, Putignani L, Naitza S, Puri C, Wright S, Crisanti A. 1998. Molecular cloning and expression analysis of a *Cryptosporidium parvum* gene encoding a new member of the thrombospondin family. Molecular and Biochemical Parasitology 92: 147–162.

Stockdale HD, Spencer JA, Blagburn BL. 2008. Prophylaxis and chemotherapy. In: Fayer R, Xiao L, eds. *Cryptosporidium and Cryptosporidiosis*. CRC Press, Boca Raton, FL, pp. 255–287.

Strong WB, Nelson RG. 2000. Preliminary profile of the *Cryptosporidium parvum* genome: An expressed sequence tag and genome survey sequence analysis. Molecular and Biochemical Parasitology 107: 1–32.

Strong WB, Gut J, Nelson RG. 2000. Cloning and sequence analysis of a highly polymorphic *Cryptosporidium parvum* gene encoding a 60-kilodalton glycoprotein and characterization of its 15- and 45-kilodalton zoite surface antigen products. Infection and Immunity 68: 4117–4134.

Sturbaum GD, Jost BH, Sterling CR. 2003. Nucleotide changes within three *Cryptosporidium parvum* surface protein encoding genes differentiate genotype I from genotype II isolates. Molecular and Biochemical Parasitology 128: 87–90.

Sulaiman IM, Hira PR, Zhou L, Al-Ali FM, Al-Shelahi FA, Shweiki HM, Iqbal J, Khalid N, Xiao L. 2005. Unique endemicity of cryptosporidiosis in children in Kuwait. Journal of Clinical Microbiology 43: 2805–2809.

Sullivan WJ, Jr, Dixon SE, Li C, Striepen B, Queener SF. 2005. IMP dehydrogenase from the protozoan parasite *Toxoplasma gondii*. Antimicrobial Agents and Chemotherapy 49: 2172–2179.

Sweeny JP, Ryan UM, Robertson ID, Jacobson C. 2011. *Cryptosporidium* and *Giardia* associated with reduced lamb carcase productivity. Veterinary Parasitology 182: 127–139.

Takashima Y, Xuan X, Kimata I, Iseki M, Kodama Y, Nagane N, Nagasawa H, Matsumoto Y, Mikami T, Otsuka H. 2003. Recombinant bovine herpesvirus-1 expressing p23 protein of *Cryptosporidium parvum* induces neutralizing antibodies in rabbits. The Journal of Parasitology 89: 276–282.

Tanriverdi S, Blain JC, Deng B, Ferdig MT, Widmer G. 2007. Genetic crosses in the apicomplexan parasite *Cryptosporidium parvum* define recombination parameters. Molecular Microbiology 63: 1432–1439.

Tanriverdi S, Grinberg A, Chalmers RM, Hunter PR, Petrovic Z, Akiyoshi DE, London E, Zhang L, Tzipori S, Tumwine JK, Widmer G. 2008. Inferences about the global population structure of *Cryptosporidium parvum* and *Cryptosporidium hominis*. Applied and Environmental Microbiology 74: 7227–7234.

Templeton TJ, Enomoto S, Chen W-J, Huang C-G, Lancto CA, Abrahamsen MS, Zhu G. 2010. A genome-sequence survey for *Ascogregarina taiwanensis* supports evolutionary affiliation but metabolic diversity between a gregarine and *Cryptosporidium*. Molecular Biology and Evolution 27: 235–248.

Thompson HP, Dooley JSG, Kenny J, McCoy M, Lowery CJ, Moore JE, Xiao L. 2007. Genotypes and subtypes of *Cryptosporidium* spp. in neonatal calves in Northern Ireland. Parasitology Research 100: 619–624.

Thompson RC, Olson ME, Zhu G, Enomoto S, Abrahamsen MS, Hijjawi NS. 2005. *Cryptosporidium* and cryptosporidiosis. Advances in Parasitology 59: 77–158.

Thompson RCA. 2009. *Echinococcus*, *Giardia* and *Cryptosporidium*: Observational studies challenging accepted dogma. Parasitology 136: 1529–1535.

Tosini F, Agnoli A, Mele R, Gomez Morales MA, Pozio E. 2004. A new modular protein of *Cryptosporidium parvum*, with ricin B and LCCL domains, expressed in the sporozoite invasive stage. Molecular and Biochemical Parasitology 134: 137–147.

Trotz-Williams LA, Martin DS, Gatei W, Cama V, Peregrine AS, Martin SW, Nydam DV, Jamieson F, Xiao L. 2006. Genotype and subtype analyses of *Cryptosporidium* isolates from dairy calves and humans in Ontario. Parasitology Research 99: 346–352.

Tyzzer EE. 1907. A sporozoan found in the peptic glands of the common mouse. Proceedings of the Society for Experimental Biology and Medicine 5: 12–13.

Tzipori S, Widmer G. 2008. A hundred-year retrospective on cryptosporidiosis. Trends in Parasitology 24: 184–189.

Tzipori S, Rand W, Griffiths J, Widmer G, Crabb J. 1994. Evaluation of an animal model system for cryptosporidiosis: Therapeutic efficacy of paromomycin and hyperimmune bovine colostrum-immunoglobulin. Clinical and Diagnostic Laboratory Immunology 1: 450–463.

Umejiego NN, Li C, Riera T, Hedstrom L, Striepen B. 2004. *Cryptosporidium parvum* IMP dehydrogenase: Identification of functional, structural, and dynamic properties that can be exploited for drug design. The Journal of Biological Chemistry 279: 40320–40327.

Upton SJ, Tilley M, Nesterenko MV, Brillhart DB. 1994. A simple and reliable method of producing in vitro infections of *Cryptosporidium parvum* (Apicomplexa). FEMS Microbiology Letters 118: 45–49.

USEPA. 2005a. USEPA method 1622: *Cryptosporidium* in water by filtration/IMS/FA.

USEPA. 2005b. USEPA method 1623: *Cryptosporidium* and *Giardia* in water by filtration/IMS/FA.

Valderrama AL, Hlavsa MC, Cronquist A, Cosgrove S, Johnston SP, Roberts JM, Stock ML, Xiao L, Xavier K, Beach MJ. 2009. Multiple risk factors associated with a large statewide increase in cryptosporidiosis. Epidemiology and Infection 137: 1781–1788.

Valigurova A, Hofmannova L, Koudela B, Vavra J. 2007. An ultrastructural comparison of the attachment sites between *Gregarina steini* and *Cryptosporidium muris*. The Journal of Eukaryotic Microbiology 54: 495–510.

Waldron LS, Ferrari BC, Gillings MR, Power ML. 2009a. Terminal restriction length polymorphism for identification of *Cryptosporidium* species in human feces. Applied and Environmental Microbiology 75: 108–112.

Waldron LS, Ferrari BC, Power ML. 2009b. Glycoprotein 60 diversity in *C. hominis* and *C. parvum* causing human cryptosporidiosis in NSW, Australia. Experimental Parasitology 122: 124–127.

Wanyiri J, Ward H. 2006. Molecular basis of *Cryptosporidium*-host cell interactions: Recent advances and future prospects. Future Microbiology 1: 201–208.

Wanyiri JW, O'Connor R, Allison G, Kim K, Kane A, Qiu J, Plaut AG, Ward HD. 2007. Proteolytic processing of the *Cryptosporidium* glycoprotein gp40/15 by human furin and by a parasite-derived furin-like protease activity. Infection and Immunity 75: 184–192.

Wetzel DM, Schmidt J, Kuhlenschmidt MS, Dubey JP, Sibley LD. 2005. Gliding motility leads to active cellular invasion by *Cryptosporidium parvum* sporozoites. Infection and Immunity 73: 5379–5387.

White AC, Robinson P, Okhuysen PC, Lewis DE, Shahab I, Lahoti S, Dupont HL, Chappell CL. 2000. Interferon expression in jejunal biopsies in experimental human cryptosporidiosis correlates with prior sensitization and control of oocyst excretion. The Journal of Infectious Diseases 181: 701–709.

WHO. 2006. WHO guidelines for drinking water quality *Cryptosporidium*.

Widmer G. 2009. Meta-analysis of a polymorphic surface glycoprotein of the parasitic protozoa *Cryptosporidium parvum* and *Cryptosporidium hominis*. Epidemiology and Infection 137: 1800–1808.

Widmer G, Lee Y. 2010. Comparison of single- and multilocus genetic diversity in the protozoan parasites *Cryptosporidium parvum* and *C. hominis*. Applied and Environmental Microbiology 76: 6639–6644.

Widmer G, Sullivan S. 2012. Genomics and population biology of *Cryptosporidium* species. Parasite Immunology 34: 61–71.

Wielinga PR, De Vries A, Van Der Goot TH, Mank T, Mars MH, Kortbeek LM, Van Der Giessen JWB. 2008. Molecular epidemiology of *Cryptosporidium* in humans and cattle in The Netherlands. International Journal for Parasitology 38: 809–817.

Winter G, Gooley AA, Williams KL, Slade MB. 2000. Characterization of a major sporozoite surface glycoprotein of *Cryptosporidum parvum*. Functional & Integrative Genomics 1: 207–217.

Xiao L, Fayer R. 2008. Molecular characterisation of species and genotypes of *Cryptosporidium* and *Giardia* and assessment of zoonotic transmission. International Journal for Parasitology 38: 1239–1255.

Xiao L, Ryan UM. 2008. Molecular epidemiology. In: Fayer R, Xiao L, eds. *Cryptosporidium and Cryptosporidiosis*. CRC Press, Boca Raton, FL, pp. 119–163.

Xiao L, Hlavsa MC, Yoder J, Ewers C, Dearen T, Yang W, Nett R, Harris S, Brend SM, Harris M, Onischuk L, Valderrama AL, Cosgrove S, Xavier K, Hall N, Romero S, Young S, Johnston SP, Arrowood M, Roy S, Beach MJ. 2009. Subtype analysis of *Cryptosporidium* specimens from sporadic cases in four U.S. states in 2007: The wide occurrence of one *Cryptosporidium hominis* subtype and a case report of an infection with the *Cryptosporidium* horse genotype. Journal of Clinical Microbiology 47: 3017–3020.

Xu P, Widmer G, Wang Y, Ozaki LS, Alves JM, Serrano MG, Puiu D, Manque P, Akiyoshi D, Mackey AJ, Pearson WR, Dear PH, Bankier AT, Peterson DL, Abrahamsen MS, Kapur V, Tzipori S, Buck GA. 2004. The genome of *Cryptosporidium hominis*. Nature 431: 1107–1112.

Yao L, Yin J, Zhang X, Liu Q, Li J, Chen L, Zhao Y, Gong P, Liu C. 2007. *Cryptosporidium parvum*: Identification of a new surface adhesion protein on sporozoite and oocyst by screening of a phage-display cDNA library. Experimental Parasitology 115: 333–338.

Yoder JS, Beach MJ. 2010. *Cryptosporidium* surveillance and risk factors in the United States. Experimental Parasitology 124: 31–39.

Yu Q, Li J, Zhang X, Gong P, Zhang G, Li S, Wang H. 2010. Induction of immune responses in mice by a DNA vaccine encoding *Cryptosporidium parvum* Cp12 and Cp21 and its effect against homologous oocyst challenge. Veterinary Parasitology 172: 1–7.

Zhou R, Hu G, Liu J, Gong A-Y, Drescher KM, Chen X-M. 2009. NF-kappaB p65-dependent transactivation of miRNA genes following *Cryptosporidium parvum* infection stimulates epithelial cell immune responses. PLoS Pathogens 5: e1000681.

INDEX

Evolution of Virulence in Eukaryotic Microbes, First Edition. Edited by L. David Sibley, Barbara J. Howlett, and Joseph Heitman.